AF545090

V. Hessel, H. Löwe,
A. Müller, G. Kolb
Chemical Micro Process Engineering

Further Titles of Interest

Hessel, V., Hardt, S., Löwe, H.

Chemical Micro Process Engineering

Fundamentals, Modelling and Reactions

2004

ISBN 3-527-30741-9

Ehrfeld, W., Hessel, V., Löwe, H.

Microreactors

New Technology for Modern Chemistry

2000

ISBN 3-527-29590-0

Menz, W., Mohr, J., Paul, O.

Microsystem Technology

2001

ISBN 3-527-29634-4

Sanchez Marcano, J. G., Tsotsis, Th. T.

Catalytic Membranes and Membrane Reactors

2002

ISBN 3-527-30277-8

Dobre, T. Gh., Sanchez Marcano, J. G.

Chemical Engineering

Modelling, Simulation and Similitude

2005

ISBN 3-527-30607-2

Sundmacher, K., Kienle, A. (Eds.)

Reactive Distillation

Status and Future Directions

2003

ISBN 3-527-30579-3

Nunes, S. P., Peinemann, K.-V. (Eds.)

Membrane Technology in the Chemical Industry

2001

ISBN 3-527-28485-0

Volker Hessel, Holger Löwe, Andreas Müller, Gunther Kolb

Chemical Micro Process Engineering

Processing and Plants

WILEY-VCH Verlag GmbH & Co. KGaA

Authors:

Dr. Volker Hessel
Dr. Holger Löwe
Dr. Andreas Müller
Dr. Gunther Kolb
IMM – Institut für Mikrotechnik Mainz GmbH
Carl-Zeiss-Str. 18–20
55129 Mainz
Germany

Library of Congress Card No.: Applied for

British Library Cataloging-in-Publication Data:
A catalogue record for this book is available from the British Library.

Bibliographic information published by Die Deutsche Bibliothek
Die Deutsche Bibliothek lists this publication in the Deutsche Nationalbibliografie; detailed bibliographic data is available in the internet at http://dnb.ddb.de.

Printed in the Federal Republic of Germany
Printed on acid-free paper

Typesetting Manuela Treindl, Laaber
Printing betz-druck GmbH, Darmstadt
Bookbinding J. Schäffer GmbH i. G., Grünstadt

ISBN-13 978-3-527-30998-6
ISBN-10 3-527-30998-5

Contents

Chemical Micro Process Engineering. V. Hessel, H. Löwe, A. Müller, G. Kolb.

ISBN: 3-527-30998-5

Preface

About one year ago the book "Chemical Micro Process Engineering – Fundamentals, Modelling, and Reactions" was released by the author team Hessel, Hardt and Löwe. It described the fundamentals of the new technology and presented in detail applications concerning organic and inorganic reactions as well as gas-phase reactions. Thus, it provided insights for the readers how mature are today's micro reactors for real-world applications. What was missing is the processing before and after such experiments. "Before" are the unit operations which are part of every reaction, e.g. mixing and heat transfer. "After" is the combination of several (unit) operations including product purification to a complete process and the erection of a plant for this purpose, as e.g. given for the field of fuel processing and respective auxiliary power units. From a commercial perspective, there is a need to have a description about "micro-reactor process design" for finally gathering cost calculations, now that the first part of the book series about "micro-reactor design" has been released – thus, the new book part is subtitled "Processing and Plants".

Facing the (unit) operations of chemical engineering, microstructured devices offer improvements of existing processing and even completely novel possibilities, not covered by today's apparatus. At best, one would have given directly a compendium on all unit operations including their several sub-versions. For mixing, for example, this would have to encompass blending, emulsification, foaming, gas absorption, suspension, and more. Having a glance on the current literature in this field, it is evident that there are more than enough scientific papers to fill a book on the topic mixing in all its particulars. The same holds for the "after"-reaction processing. Thus, a decision had to be made and topics of major interest had to be selected.

"Mixing" is the key to many improvements concerning the performance of reactions, in particular in the field of organic synthesis. Here, the focus was drawn on mixing of miscible fluids, since this alone serves to fill an extended chapter. Heat transfer would have been the second most important subject; but books on micro-channel heat transfer are already on the market. "Fuel processing" is the current main application of heterogeneous gas-phase processing using micro-structured reactors. The corresponding target application, which is energy generation, is regarded to be of crucial relevance for our society and industry, as e.g. the funding policies of the EU or of the U.S. demonstrate. The development of catalysts is central for having highly active reformers, water-gas shift reactors, or gas-

Chemical Micro Process Engineering. V. Hessel, H. Löwe, A. Müller, G. Kolb.

ISBN: 3-527-30998-5

purification reactors; thus, to have a further chapter on catalyst screening was quite logical. As a last chapter, plant concepts are being discussed. Albeit this development is in its infancy, it is advised to have it as a separate topic here. Microstructured reactors pose their individual needs on interface development, process control, tubing and other peripherals, and all other aspects of plant built-up. The provision of robust plants is the final step needed for a commercial acceptance of the novel technology. The series "Chemical Micro Process Engineering" is, accordingly, not complete now; further descriptions of unit operations and their combinations have to be added in future. But the subjects discussed now are given in a comprehensive manner and on the latest state of the art. Other topics of relevance such as heat transfer or dispersion mixing may be considered later.

In the meantime, the topic chemical micro processing engineering and micro-reactor technology have gained even more attention in the scientific world. There is hardly any conference in chemical engineering which has not micro-reactor technology as key topic and gives an own session to it. Scientific publishing houses know that micro-reactor articles are of major interest for their audience. Special issues of their journals are prepared on the subject. Industry is not only testing micro-reactor technology, but is actually starting using it. In particular, micro-structured reactor plant concepts are on the way to be realized. The interest is world-wide and indeed countries such as Japan, Korea and China have become active besides France, U.K., Germany, U.S., and The Netherlands. The micro-reactor technology has become a true part of globalization. Still, there are believers and those who are more reluctant to this new technology. But the developments are not on a national level anymore. It is everyone's own fortune how to handle the new possibilities. Chemical micro process engineering is deeply interdisciplinary and a knowledge-based technology; there is considerable capital investment on the side of the traditional chemical engineering. One cannot simply "buy" the new technology; it is more a strategic and long-lasting decision to do so. On the other side, it is sometimes very simple to have profits from micro reactors, as e.g. for organometallic reactions or processes with explosive media. There is simply needed the willingness to change old habits; management decision and action should be in favor of lengthy discussions about the "pros" and "cons" which will never come to a unanimous decision.

We would like to thank Lea Widarto for administrative work, Tobias Hang for administrative and graphical works, Friedhelm Schönfeld for scientific advice. We also acknowledge the help of the publishing house Wiley-VCH, the STM-books team and especially here Karin Sora and Rainer Münz. Last but not least, we appreciate the patience of our wives during 'ad inifinitum' lasting works on the book during the last year.

Mainz, January 2005

Volker Hessel, Holger Löwe,
Andreas Müller, Gunther Kolb

Abbreviations and Symbols

a	Channel dimension
A/D	Analog~/digital~
AC	Alternate current
AISP	Aluminium triisopropylate
AMG	Algebraic multi-grid
AMOS	Automated multiplex oligonucleotide synthesizer
AMR	Autothermal methanol reforming
ANN	Artificial neural net
ANSI	American National Standards Institute
APS	Average particle size
APU	Auxiliary power units
ASCII	American Standard Code for Information Interchange
ASD	Anodic spark deposition
ASE	Advanced silicon etching
ASIC	Application specific integrated circuit
ASME	American Society of Mechanical Engineers
ATP	Adenosine triphosphate
ATR	Autothermal reforming
AuMμRes	Automated micro reaction system
BET	Brunauer, Emmett, Teller (surface area)
BMBF	German Ministry of Education and Science
BOE	Buffered oxide etching
c	Concentration
CAC	Catalytic combustion of alcohol fuels
CAD	Computer aided design
CAE	Computer aided engineering
CAPD	Computer aided plant design
CCD	Charge-coupled device
CE	European certification
CEC	Capillary electrochromatography column
CER	Coupled electrorotation
CFD	Computational fluid dynamics

CHC	Catalytic hydrogen combustion
CHCC	Catalytic hydrocarbon combustion
CNC	Computerized Numeric Control
COC	Cyclo olefin copolymer
CPAC	Center for Process Analytical Chemistry
CPR	Catalytic plate reactor
CPU	Central processing unit
CVD	Chemical vapor deposition
CW	Continous wave
D	Diffusion coefficient
d	Internal diameter
DC	Direct current
DEMiS	Demonstration and Evaluation of Microreaction Technology in Industrial Systems
DEP	Dielectrophoresis
d_h	Hydraulic diameter
DIN	Deutsche Industrienorm
DMFC	Direct methanol fuel cell
DNA	Desoxyribonucleic acid
DoE	Design of experiments
DRIE	Deep reactive ion etching
E	Activation energy
$E(t)$	Exit-age distribution function
EDM	Electro-discharge machining
EDTA	Ethylene-diamine-tetraacetic acid
EDX	Energy dispersive X-ray
EHD	Electrohydrodynamic
EKI	Electrokinetic instability
EO	Electroosmotic~
EOF	Electroosmotic flow
ESI-MS	Electrospray ionization mass spectrometry
E_z	Electric field
f	Pulsing frequency
FAMOS	Fraunhofer-Allianz Modulares Mikroreaktionssystem
FIA	Flow injection analysis
FID	Flame ionization detector
FMG	Fluorescein mono-β-D-galactopyranoside
FTIR	Fourier transform infrared
FTOL	Fluorescence-turn-off length
FVM	Finite volume method
GA	Generic algorithm

GC	Gas chromatography
GHSV	Gas hourly space velocity
h	Channel height
h	Viscosity
HCC	Homogeneous combustion in micro-channels
HCR	Hydrocarbon reforming
h_{disp}	Averaged signal height
HEPES	(4-(2-Hydroxyethyl)-piperazine-1-ethane-sulfonic acid
HEX	Heat-exchanger
HPLC	High performance liquid chromatography
I/O	in/out
ICS	Integrated chemical synthesizer
IFP	Integrated fuel processor
IR	Infrared
ISBL	Inside battery limit
ISM	Integrated systems fuelled by methane
ISMol	Integrated systems fuelled by methanol
ISV	Systems running on various fuels
ITO	Indium tin oxide
k	Rate constant
K	Dean number
k_0	Pre-exponential term
KBS	Knowledge-based system
KISS	Keep it simple and stupid
k_v	Dispersion coefficient
l	Length
l	Layer thickness
L	Hydraulic diameter
L_0 and $L(t)$	Characteristic dimensions of the interfacial area
LAN	Local area network
LCD	Liquid crystal display
LHV	Lower heating value
LIGA	German acronym for lithography, electroforming, moulding (Lithograpie, Galvanik, Abformung)
LPCVD	Low pressure chemical vapour deposition
MCFC	Molten carbonate fuel cells
MEMS	Micro-electro-mechanical systems
MiRTH-e	Microreactor Technology for Hydrogen and Electricity
MMem	Micro-structured membranes for CO Clean-up
MR	Membrane reactor

MRP	Microstructured reactor plant
MRT	Micro reaction technology
MS	Mass spectrometry
MSR	Methanol steam reforming
n	Numbers of SAR steps
ND	Numerical diffusion
NEDC	New European Drive Cycle
NeSSI	New Sampling Sensor Initiative
NIR	Near infrared
OSBL	Outside battery limit
P	Power output
p	Pressure
p_0	Pressure drop of one SAR step
PCHE	Printed circuit heat-exchanger
PCR	Printed circuit reactor
PCS	Process control system
PDF	Probability density function
PDMS	Poly-dimethylsiloxane
Pe	Peclet-number
PEM	Proton exchange membrane
PEMFC	Proton exchange membrane fuel cell
PET	Poly-ethylene terephthalate
pH	Potentia Hydrogenii (measure for acid and base strength)
PMMA	Poly-methylmethacrylate
PNGV	Partnership of New Generation Vehicles
PO	Propylene oxide
POx	Partial oxidation
PPMA	Poly-methyl methacrylate
PrOx	Preferential carbon monoxide oxidation
PTFE	Poly-tetrafluorethylene
PVD	Physical vapor deposition
PZT	Lead-zirconate-titanate
Q	Total flow rate
R	Gas constant
R	Radius of curvature
Re	Reynolds number
r_{eff}	Measured reaction rate
RIE	Reactive ion etching
RNG	ReNormalization Group

ROMP	Ring-opening metathesis polymerization
RTD	Residence time distribution
RWGS	Reverse water-gas shift reaction
S/C	Steam to carbon ratio
SAM	Sensor analytical manager
SAR	Split-and-recombine
SDS	Sodium dodecyl sulphate
SEM	Scanning electron microscopy
SGS	Simultaneous gradient-sputtering
SHM	Staggered herringbone mixer
SME	Static mixing elements
SNMS	Secondary neutral particles mass spectrometry
SOI	Silicon on insulator
SPS	Spark plasma sintering
SR	Steam reforming
St	Strouhal number
STP	Standard pressure
t	Finite time
T	Temperature
T	Periodicity
TAP	Temporal analysis of products
t_{disp}	Critical temperature
TEM	Transmission electron microscopy
TEOS	Tetraethyl-ortho-silane
TG	Thermo gravimetry
THF	Tetrahydrofurane
TOF	Turnover frequency
TOS	Time on stream
TPR	Temperature programmed reduction
U^+, U^-, and $U^\pm$	Dimensionless slip velocities
u	Mean velocity
UV	Ultraviolet
V	Average velocity
VDMA	Verband Deutscher Maschinen- und Anlagenbauer
V_{i}	Fluid velocity
Vis	Visible
w	Fluid volume
w	Channel width
WGS	Water-gas shift reaction
WHSV	Weight hourly space velocity

X	Axial coordinate
x_1	Axial coordinate
XPS	X-ray photoelectron spectrum
XRD	X-ray diffraction
Y	Axial coordinate
α	Heat transfer coefficient
α	Angle
ΔH_r	Reaction enthalpy
Δp	Pressure drop
ε	Extinction coefficient
ϕ	Thiele modulus
Φ	Photometric brightness
λ	Interfacial stretching
μEDM	Micro electro discharge machining
μ	Dynamic fluid viscosity
μPVT	Industrial platform for modular micro process engineering
μTAS	Micro total analysis system
ρ	Density
σ	Lyapunov exponent
Ψ	Stream function
ζ	Zeta potential

1
Mixing of Miscible Fluids

1.1
Mixing in Micro Spaces – Drivers, Principles, Designs and Uses

1.1.1
'Mixing Fields', a Demand Towards a more Knowledge-based Approach – Room for Micro Mixers?

Many unsolved challenges remain in the field of mixing [1]. The diversity of mixing tasks is large and so is their industrial importance. Mixing is a good example how equipment dominates the type of processing solution chosen (see Figure 1.1) [1]. Mixing has been carried out in stirred tanks over decades and all mixing problems were solved using this assumption as a starting condition.

Meanwhile, there is a slight paradigm shift apart from such equipment design dominance to a more knowledge-based approach (see Figure 1.2), with the mixing objective in the focus, i.e. the so-called 'mixing fields', related to the rate and scales of segregation destruction (for a detailed definition see [1]) [1]. This demands a

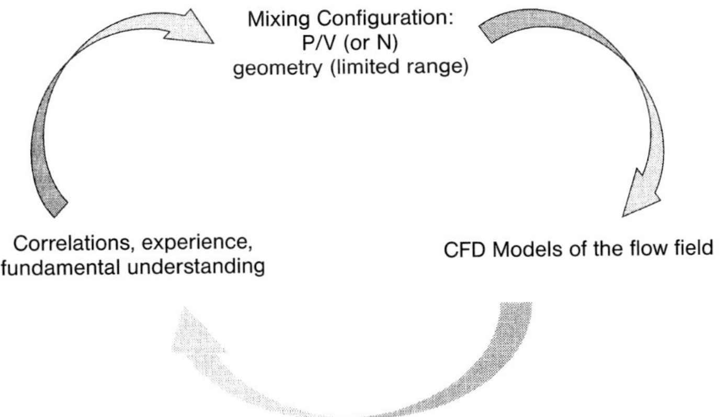

Figure 1.1 Design and development cycle for equipment-based design: using pre-decided equipment, a mixing configuration is chosen by correlations and experience. For stirred tanks this configuration is given, e.g., by the power-to-volume ratio P/V and the impeller diameter N. Then, CFD models are made to describe the flow field [1].

Chemical Micro Process Engineering. V. Hessel, H. Löwe, A. Müller, G. Kolb.

ISBN: 3-527-30998-5

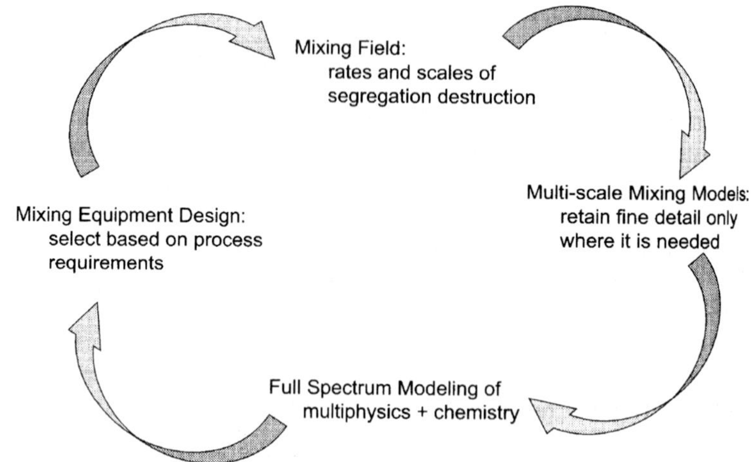

Figure 1.2 Design and development cycle for mixing-field based design: there is a selection of the mixing equipment based on the process requirements. This leads to the specification of a mixing field. CFD simulations give the flow field reduced to a multi-scale mixing model. The mixing field is integrated with other models of the key process mechanisms, aiming at giving an entire picture of the process [1].

greater variety of equipment solutions and in particular specially equipment, tailor-made for one specific mixing task. Recently, this has led industry to use rotor stators, static mixers, multi-shaft mixers, extruders and pulping machines [1]. It stands to reasons that such development may also pave the way for using microstructured mixers for industrial applications.

Experts predict a trend from stirred-vessel mixing to the use of continuous mixing, e.g. by in-line mixers [1]. This again provides a chance for many microstructured mixers.

1.1.2
Drivers for Mixing in Micro Spaces

Many passive microstructured mixers (see e.g. [2, 3]) follow design principles used at the macro-scale for static mixers with internal packings [4]. It stands to reason that some of the advantages in processing claimed for conventional static mixer also apply or may be even more pronounced when using static mixers [4]:

- compactness and low capital cost
- low energy consumption and other operating expenses
- negligible wear and no moving parts, which minimizes maintenance
- lack of penetrating shafts and seals, which provides closed-system operation
- short mixing time and well-defined mixing behavior
- narrow residence-time distribution
- performance independent of pressure and temperature.

In addition, the following specific chemical engineering drivers may govern the decision to use a micro or microstructured mixer:

- enabling technology for niche mixing, where conventional mixers fail
- enabling technology in particular for mixing under laminar-flow conditions in minute spaces
- fast mixing for even faster reactions in chemical synthesis (see e.g. [5])
- analytical processing of fast reactions, e.g. for quench-flow analysis (< 1 ms) to study rapid biological transformations [6]
- laminar mixing of viscous media [7], as most micro-flow processing is anyway in that regime
- mixing at only small overall internal device volumes, e.g. for
  - handling of rare, precious samples in analysis or synthesis
  - handling and screening of numerous samples on a small format in chemical and biological analysis [8–15])
- mixing below threshold dimensions and at small partial internal volumes to ensure safety [16–19], for both mechanistic and thermal reasons, respectively [16]
- mixing of a flow of high structural regularity [20], e.g. to enhance predictability of modeling and to improve scaling-/numbering-up.

1.1.3 Mixing Principles

Mixing in minute spaces can basically rely only on two principles which are diffusion and convection. Diffusion between short distances, establishing high concentration gradients (see e.g. [21]), was initially the most frequently applied principle by simply making the channels themselves smaller and smaller. Soon, the limits of that strategy, also in terms of robustness (fouling) and costs (complex microfabrication), became obvious. In recent years, various methods were developed to overcome the limits by diffusion mixing, all of them based on the induction of secondary-flow (convective) patterns which are superposed on the main flow, often in the vertical direction to the flow axis. This includes recirculation patterns, chaotic advection and swirling flows, just to name a few. Convection is effective for mixing, since it serves to enlarge mixing interfaces. Convections of 'gross' mass portions can be used at a much larger scale to 'stir' complete chamber volumes, e.g. by ultrasound, by elektrokinetic instability or acoustic means. At high Reynolds numbers, turbulent mixing can be utilized; however, this is often not practicable, as this implies achievement of unrealistic large flow velocities. The few specially equipment known to use turbulence rely either on free-guided flows or guide through meso-scale channels.

1.1.4
Means for Mixing of Micro Spaces

The means of mixing can be classified as either *active* or *passive*. *Passive micro mixers* use part of the flow energy for feeding and thereby generate special flow schemes with ultra-thin flow compartments such as lamellae for diffusion mixing or utilize chaotic advection by secondary flows to enlarge the interfaces. *Active micro mixers* rely on moving parts or externally applied forcing functions such as pressure or electric field.

External energy sources for *active mixing* are, for example, ultrasound [22], acoustic, bubble-induced vibrations [23, 24], electrokinetic instabilities [25], periodic variation of flow rate [26–28], electrowetting induced merging of droplets [29], piezoelectric vibrating membranes [30], magneto-hydrodynamic action [31], small impellers [32], integrated micro valves/pumps [33] and many others, which are listed in detail in Section 1.2.

Devices relying on *passive mixing* utilize the flow energy, e.g. due to pumping action or hydrostatic potential, to restructure a flow in a way which results in faster mixing. For example, thin multi-lamellae can be created in one step in special feed arrangements, termed interdigital [20, 34–42]. A serial way of creating multi-lamellae can be achieved by split-and-recombine (SAR) flow guidance [7, 43, 140]. Chaotic mixing results from superimposed recirculation flow patterns (such as helical flows), with an exponential increase in specific interfaces [27, 28, 44–50]. The injection of many sub-streams, e.g. via an array of nozzles, into one main stream can create micro-plumes with large interfaces [51]. Turbulent mixing can be achieved by collision of jets [52–54]. A number of specially flow guidances are known as well. For example, re-directed flows create eddies which are exploited in Coanda effect mixers [55] and in other recycle-flow mixers [56]. These and more passive principles are described in Section 1.3.

There are more reports about and more different types for passive than for active micro mixers. This is understandable, since for many applications flow energy is given. Also, active micro mixers may be more difficult to fabricate, as they require special additional elements besides the normal fluid pathway as in the case of passive devices. This also demands control of these functions, i.e. further external equipment may be needed. All this implies greater complexity for active devices. On the other hand, these tools are specially designed for mixing tasks which passive mixers cannot accomplish, i.e. mixing at very low flow velocities and/or of large fluid chambers. Sometimes, active mixing devices may consume a much smaller footprint area than passive ones with all their fluid feed channel architecture and large inlet and outlet ports. The complexity of active micro mixers may not be a problem any more for future devices, when microsystem integration is brought to a more advanced level.

1.1.5 Generic Microstructured Elements for Micro-mixer Devices

The above-mentioned micro-mixer means have to be 'transformed' into physical objects, i.e. the microstructures which then perform the mixing (see Figure 1.3). Bi-lamination can be achieved in T- and Y-flow structures [6, 57], which are miniaturized analogues of conventional mixing tees. Multi-lamination is done via structures with alternate feeds. The latter are realized either by interdigital [20, 34–41] or bifurcation [42] structures. To speed up mixing, thinning of the multi-lamellae flow via geometric focusing zones can be utilized [20, 34–39]. Several types of flow dividing and recombining structures were developed for SAR-type mixing, including fork-like, stack-like, Möbius-type and 3-D curved caterpillar designs [7, 43, 125, 126, 140, 141]. Chaotic mixing was first achieved by alternately arranged slanted grooves, so-called herringbone structures, in a micro channel [44, 45]. Barrier-embedded structures may be added and will further improve the mixing efficiency [3, 58]. Later, other structures such as simple curved channels and zig-zag channels were used as well [27, 28, 46–50, 59]. Micro-plume injection is done by multi-hole plates adjacent to a mixing chamber, as simple through-holes with straight injection [51] or complex oblique arrays with tilted injection [54].

These are just a few among other examples of microstructured designs which are discussed in detail in the next two chapters. More information about these micro-mixer designs can be obtained from reviews, e.g. [60–66].

In addition to grouping the mixers according to their mixing principles and their generic microstructure designs, a practically oriented classification refers to the complexity of the fluid network [25]. So-called in-plane mixers rely on streams which are divided and mixed in a fluid network confined to one level (i.e. a pattern that can be projected on to a single plane) [25]. In turn, out-of-plane mixers rely on a

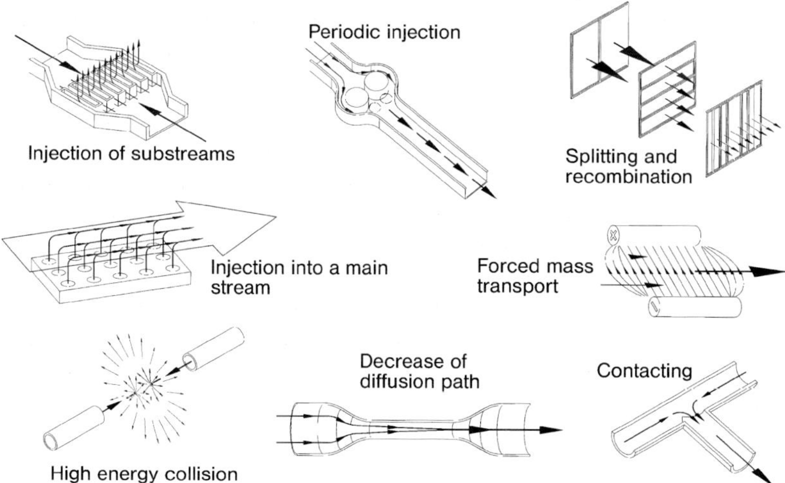

Figure 1.3 Schematic diagrams of selected passive and active micro mixing principles [66] (source IMM).

more complex, three-dimensional fluid network. Examples of in-plane mixers are bi-laminating, T- and cross-channel mixers as described [57, 67–71]. SAR and most multi-lamination mixers act out-of-plane; examples of these device types are given in [7, 43] and [20, 36, 37, 39], respectively. Another type of out-of-plane mixer is based on micro-plume array injection [51].

It is customary, mainly owing to fabrication needs, that for biological applications chip-like systems with two-layer construction are used, thus being in-plane. Mixers used for the same purposes have to adjust to this fact. Since multi-lamination typically needs several layers to achieve the proper feeding pathways, other mixers with simpler designs such as the electrokinetic instability mixer need to be applied [25]. In contrast, chemical applications, where the mixer is only associated with part of the plant and not integrated in a small, flat device, do not pose such preferences; indeed, multi-layer microfabrication architectures have been used.

1.1.6
Experimental Characterization of Mixing in Microstructured Devices

For simulation characterization, the reader should refer to Chapter 2, *Modeling and Simulation of Micro Reactors,* in the first volume of this series [72].

For experimental characterization, flow visualization by colored or fluorescent streams is the most facile method. *Dilution-type experiments* contact dyed and pure water streams (passive mixing) or standing volume portions (active mixing) in a type of photometric experiment. This is usually monitored with the aid of microscopic, photo, video or high-speed camera techniques (see e.g. [20]).

Reaction-type experiments underlie mixing with a very fast reaction so that mixed regions spontaneously indicate the result of the reaction (see e.g. [20]). Besides using 'normal' fast organic reactions with color formation, change or quenching [20, 73], the usage of acid–base reactions with a pH-sensitive dye or a pH indicator is common. More detailed information is given by *competitive reactions,* i.e. two parallel reactions [74–78]. These reactions develop differently with varying pH, solvent, etc., which is influenced via mixing. Such reactions were first applied for determining mixing efficiency in stirred batch reactors and later adapted to the needs of micro mixer devices [36]. Still later, optimized protocols were developed for micro-mixer testing giving more accurate and more reproducible results [79].

Concentration profiling uses on- or in-line measurements of optical properties, typically not done for the whole volume, but along lines such as the channel cross-section (see e.g. [20]). Concentrations are accessible by photometric, electric or fluorescence measurements. Furthermore, vibrational analysis such as IR and Raman spectroscopy can be used for the same task [80, 81]. Concentration profiling can also be achieved simply by gray-scale or comparable image analysis for quantitative data extraction from microscopy images of colored flows [20, 37, 68].

These techniques are the most often used and simplest ways to characterize mixing in microstructured mixers. Certainly, many more were used in the past. Information on such specially techniques given in the next two chapters where the respective mixer is discussed.

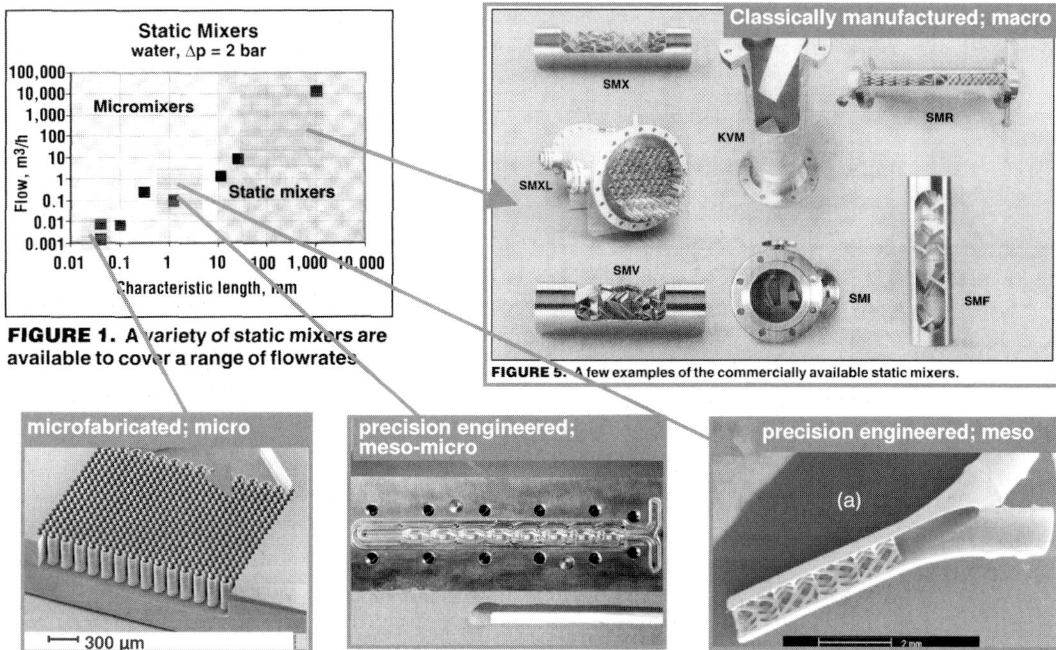

Figure 1.4 Micro mixers (laboratory scale) and micro structured mixers (pilot scale) close the gap with static mixers, yielding apparatus for a multi-scale concept. Today's microstructured devices achieve mixing at up to about 1 $m^3\,h^{-1}$ liquid throughput [2, 64] (by courtesy of RSC and Chemical Engineering).

1.1.7 Application Fields and Types of Micro Channel Mixers

Generally, application fields of micro channel based mixers encompass both modern, specialised issues such as sample preparation for analysis and traditional, widespread usable mixing tasks such as reaction, gas absorption, emulsification, foaming and blending [63, 64, 66, 72, 82] (see also [83–90]). For novel and modern chemical and biochemical analysis, typically *micro mixer elements* serve as mixing units within credit card-sized fluidic chips, often being complex integrated systems. Chip-like micro mixer components *(micro mixers)* are employed for the more conventional chemical and chemical engineering applications at the laboratory scale. At pilot or even production scales, much bigger components are applied for the same mixing tasks, typically comprising microstructures in a large housing, therefore being more correctly termed *micro structured mixers.*

Micro mixer elements, micro mixers and micro structured mixers typically have flows in the ml h^{-1}, 1 l h^{-1} and 1000 l h^{-1} ranges, respectively, thus covering the whole flow range up to the conventional static mixers and being amenable to analysis and chemical production as well (see Figure 1.4). When used at the upper flow limit, microstructured mixers can act as process-intensification (PI) equipment.

1.2
Active Mixing

In the following the state of the art of microstructured mixing devices is presented. Only the mixing of miscible liquids (and gases) is considered; the same micro mixers, however, can usually be used for making liquid/liquid and gas/liquid dispersions, which is outside the scope of this chapter, but certainly is worth consideration in the future. If not otherwise mentioned, liquid mixing is involved. The few examples on gas mixing are explicitly dealt with.

This chapter is on mixing principles, their respective devices and their characterization and is intended to give the reader an idea of how well they already function. This chapter is not really on mixing theory and physics of micro mixers.

The contents are presented in a structured, hierarchical order similar to an encyclopedia style. Hence not one description or work simply follows another, but are intertwined for better comparison. It is aimed at giving a comprehensive picture of the field.

First, the mixing principle is explained in a generic fashion. Then, a device section follows, describing all the different versions of microstructured devices actually realized. Details on the way in which the generic principle is applied are given as well as details on microfabrication and design specifications.

In the *Mixing Characterization Protocols/Simulation* sections, details on the methods of mixing characterization and simulation are provided.

In the *Typical Results* sections, it is pointed out how ready the use of such devices for industry is at present. The reactor and the protocol/simulation are not explained any further, but there is a reference to the respective section. The results are divided according to the special topics they refer to, as indicated by a heading. It is aimed to achieve a logical order of the sequence in which the topics are presented.

1.2.1
Electrohydrodynamic Translational Mixing

Most relevant citations

Peer-reviewed journals: [25, 91, 92]; proceedings contributions: [48, 93, 94].

Mixing can be accomplished by electric forces, when fluids with different electric properties such as conductivity and/or permittivity are exposed [91]. In MEMS devices, electric fields of relatively large amplitude can be generated by means of low voltages so that respective mixing effects should be as pronounced as in theory. Further, MEMS fabrication techniques are fairly advanced concerning the patterning and integration of electrode structures. No moving parts are required for micro mixers of this type, which generally is not easily accomplished at the micro scale.

Electric-force mixing is adequate for mixing flows at very low Reynolds number (~1). By lamination of two fluids with different electric conductivity and/or permittivity in a micro mixer, a steep cross-sectional gradient of the respective properties can be established [91]. The electric field may be parallel or perpendicular to the fluid interface, which is also a boundary where electric properties abruptly change. In the presence of a sufficiently large electric field, a transversal flow secondary flow can be stirred. Direct (DC) and alternating (AC) currents can be applied for this purpose. The latter have oscillating frequency, e.g. of square or sinusoidal type.

Translational motion, i.e. secondary flow lateral to the flow direction, can be achieved by spatially homogeneous or inhomogeneous fields, respectively termed electrophoresis and dielectrophoresis. Rotational motion can be achieved as well, when a dipole is induced and a torque is exerted on that dipole. Then, the rotating entity may be a solid object, e.g. a microsphere, which actively mixes and is not the fluid to be mixed. Such rotating objects generates secondary flow in their vicinity and are described in Section 1.2.2.

Electric fields may interact with flows fed by hydrostatic or pumping action [91]. Flows driven by electroosmotic means may be mixed as well by the action of fluctuating electric fields, which creates oscillating electroosmotic flows and has been termed electrokinetic instability (EKI) [25, 93]. In this way, rapid stretching and folding of material lines are induced, not unlike the effect of stirring. In one realized example, comparatively low frequencies, below ~100 Hz, and electric field strengths in excess of 100 V mm^{-1} are applied for channels with dimensions of about 50 μm [25].

1.2.1.1 Mixer 1 [M 1]: Electrohydrodynamic Micro Mixer (I)

This micro mixer, named electrohydrodynamic (EHD) microfluidic mixer, comprises a simple T-channel structure (see Figure 1.5) [91]. After passing the T-junction, a bi-laminated stream is realized. Following a downstream zone for such flow establishment, a channel zone with several electrode wires on both sides of the channel is located. In this way, an electric field perpendicular to the fluid interface is generated. Thereafter, an electrode-free zone of the channel is situated for completion of the mixing initiated.

The electrodes face each other with a differential potential between the two sets, one set grounded and the other energized [91].

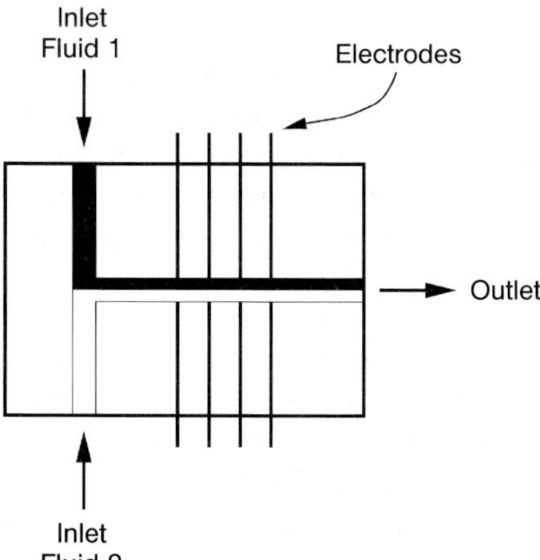

Figure 1.5 Schematic of the electrohydrodynamic mixer with T-channel and pairs of electrodes adjacent to the channel and perpendicular to the fluid interface generated [91] (by courtesy of RSC).

Device manufacture was made by conventional precision engineering machining [91]. A thin plate was glued on a thick substrate to make it immobile. The slots for the wires were made by saw blading. The electrodes were then press fitted into these slots. A layer of epoxy glue was used for sealing of the electrodes within the slots. The flow channel was milled using a 0.25 mm carbide endmill. A microscope glass cover slip was glued on top of the device to seal the micro channel, but to still allow visual inspection.

Mixer type	Electrohydro-dynamic mixer	Wire electrode diameter	250 μm
Mixer material	Lexan	Spacing between electrodes	500 μm
Plate thickness	3.175 mm	Spacing between electrodes	500 μm
Mixing channel width, depth, length	250 μm, 250 μm, 30 mm	Electrode material	Titanium
Slots for wire electrodes	250 μm × 250 μm	Plate dimensions	12 mm × 15 mm

1.2.1.2 Mixer 2 [M 2]: Electrohydrodynamic Micro Mixer (II)

This electrohydrodynamic (EHD) mixer (Figure 1.6) device provides a simple flow-through chamber which has an upper and lower electrode for generating a electromagnetic field. The chamber channel is given by a sandwich of two plates, one being microstructured [94]. The bottom plate contains a trapezoid channel. Two electrode layers are deposited on parts of the channel bottom and channel top and on the top part of this plate so that they reach the outside for external electrical contact.

By electromagnetic means, surface charges can be induced and accumulated on the boundary of a dielectric material [94]. Liquid samples can be treated the same way. For a non-uniform external electric field, interfacial shear stress in liquids is generated, inducing flow motion which tends to eliminate this stress. In this way, new interfaces are formed and mixing can be achieved.

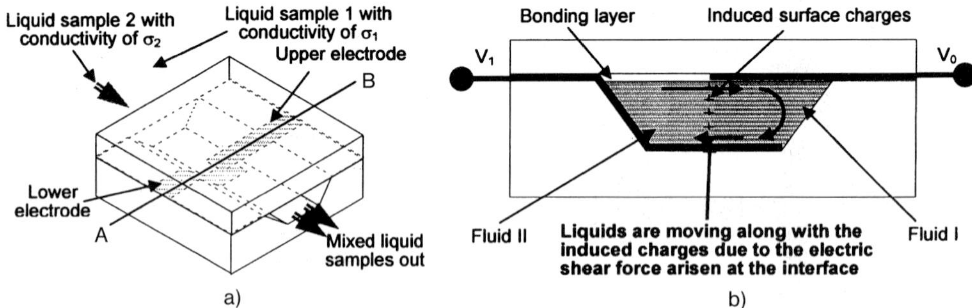

Figure 1.6 Schematic design of the *electrohydrodynamic (EHD) mixer* (left) and cross-sectional view giving the electrode arrangement and sketching the flow motion induced by interfacial shear (right) [94] (by courtesy of Kluwer Academic Publishers).

So far, only a design study of the EHD mixer has been provided [94], hence no device specifications can be given here.

1.2.1.3 Mixer 3 [M 3]: Electrokinetic Instability Electroosmotic Flow Micro Mixer, First-generation Device

This device, named electrokinetic instability (EKI) mixer, is driven by electroosmotic flow (EOF) and thus termed an EKI-EOF micro mixer in the following [25, 93]. It contains two channels which are arranged cross-wise, each having two reservoirs at the ends (see Figure 1.7) The two fluids to be mixed come from the two ends of the shorter channel and merge when entering the longer channel, moving along it. A simple version was cast in polydimethylsiloxane (PDMS) which was considered to be the first-generation device to [M 4] (see below). A high-voltage amplifier coupled with a function generator is connected to one reservoir of the longer channel, generating the electroosmotic flow and causing the whole flow direction to be towards the fourth reservoir at the other end of the longer channel.

The mixer was made by casting, molding and curing [25, 93]. Then, the channels were covered by glass slides. Two ports were placed at the inlet reservoir feeds. The liquids were pumped by hydrostatic pressure. Platinum electrodes were deposited in the upstream and downstream reservoirs by AC excitation. The EKI then occurs along the entire channel.

The frequency and the applied voltage of the AC field were 10 Hz and 1 kV, respectively [25].

Mixer type	Electrokinetic instability electro osmotic flow mixer, 1st-generation device	EKI micro channel: width, depth, length	1000 μm, 300 μm, 7 mm
Mixer material	PDMS	Plate dimensions	12 mm × 15 mm
Electrode spacing	9 mm		

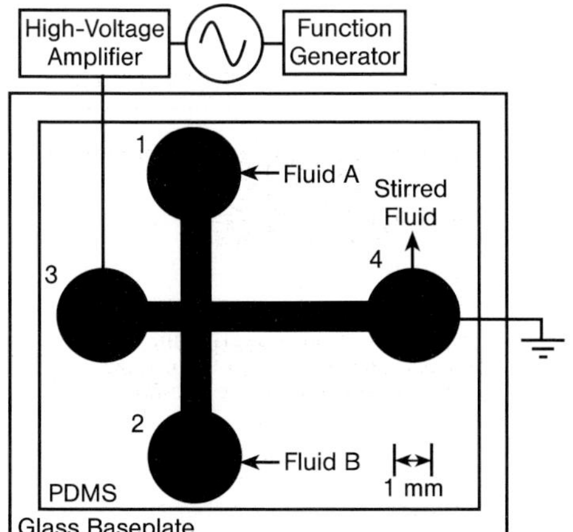

Figure 1.7 Design of an electrokinetic instability EOF micro mixer, first-generation device [25] (by courtesy of ACS).

1.2.1.4 Mixer 4 [M 4]: Electrokinetic Instability Electroosmotic Flow Micro Mixer, Second-generation Device

This micro mixer was built after first tests with the [M 3] device as a more robust and practical successor device (see Figure 1.8) [25, 93]. Two channels were arranged cross-wise, each having two reservoirs at the ends. Now, the two liquids are pumped into two feed channels which merge in a Y-type configuration. From there, they flow concurrently in one channel and are collected in one reservoir. At the beginning of this channel a quadratic chamber intersects in which the flow is mixed by EKI action. For this purpose, the chamber is connected to two channels with two reservoirs at their ends. These reservoirs are each in contact with electrodes, which equal to [M 3], are connected to a high-voltage amplifier coupled with a function generator.

The mixer is made by standard photolithography and wet etching in glass [25]. The fluidic network was sealed by thermal bonding with another glass plate. Holes were drilled through the cover plate.

The AC field was generated with a sine wave fed into a high-voltage amplifier (0 ± 10 kV) [25]. The frequency and the applied voltage were 5 Hz and 4 kV, respectively.

Mixer type	Electrokinetic instability (EKI) mixer	Mixing chamber: width, length, depth	1000 μm, 1000 μm; 100 μm
Mixer material	Borofloat glass	Mixing chamber volume	0.1 μl
EKI micro channel width, depth	300 μm, 100 μm	Plate dimensions	15 mm × 22 mm

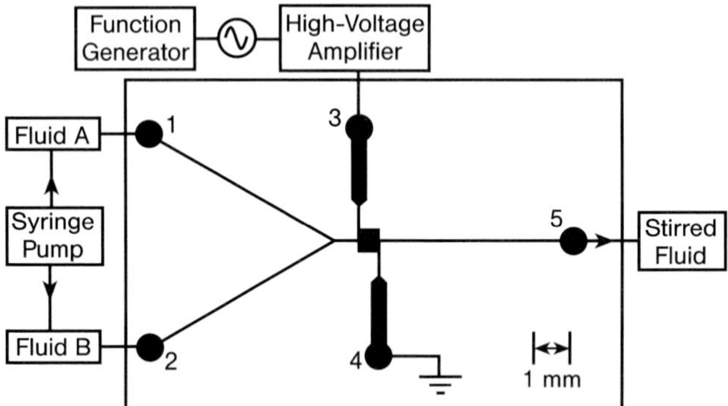

Figure 1.8 Design of an electrokinetic instability micro mixer, second-generation device, based on the results obtained with the first design given in Figure 1.7. The electrokinetic instability is confined to the square mixing chamber shown in the center of the schematic and, to a small extent, to fluid channel regions attached to it [25] (by courtesy of ACS).

The major improvement of design [M 4] over [M 3] is that now mixing is largely confined to a small mixing space, i.e. the mixing chamber, while before a more extended region was addressed [25]. Actually, the region of EKI instability is confined to the mixing chamber and a small part of the adjacent channels, being no longer than two micro channel widths. Input flow streams can now be either pressure or electric-field driven. Another major improvement was the use of porous, dielectric frits with 0.5 μm pores which serve to separate the external fluid reservoirs from the internal micro channel flows.

1.2.1.5 Mixer 5 [M 5]: Electrokinetic Instability Micro Mixer by Zeta-potential Variation

This electrokinetically driven micro mixer uses localized capacitance effects to induce zeta potential variations along the surface of silica-based micro channels [92]. The zeta potential variations are given near the electrical double layer region of the electroosmotic flow utilized for species transport. Shielded ('buried') electrodes are placed underneath the channel structures for the fluid flow in separate channels, i.e. they are not exposed to the liquid. The potential variations induce flow velocity changes in the fluid and thus promote mixing [92].

The microstructure consists of a cross-shaped feed structure with one longer mixing channel, connecting to the waste [92]. Two of the other three channels serve for feed of the two solutions to be mixed, the third connects to waste. The long channel is placed over nine buried electrodes, having a rectangular shape.

The mixer was made by wet-chemical etching of glass substrates, following a photolithographic step, yielding channels for the fluid flow and for the electrodes. The electrodes (100 Ω resistors) were made by an electron-beam evaporation process generating thin metallic films in the etched channels [92]. Holes were drilled into the cover plate for inlet and outlet connection. Fusion bonding was used to seal the plates.

Mixer type	Electrokinetic instability micro mixer with zeta potential variation	Shielding electrode channel width, depth	250 μm, 25 μm
Mixer material	Glass	Mixing channel width, length	150 μm, 34 mm
Number of electrodes	9	Side channel length of the cross	10 mm
Distance electrodes–mixing channel	130 μm	Top channel length of the cross	4.4 mm
Electrode materials and thickness	Cr: 50 nm; Au: 0.4 μm	Diameter of reservoirs	1.5 mm

1.2.1.6 Mixer 6 [M 6]: Electrokinetic Dielectrophoresis Micro Mixer

This dielectrophoresis (DEP) mixer, specially designed for mixing of dielectric particles was made with a rectangular chamber having one inlet and outlet [48]. Pairs of micromachined electrodes generate the electric field.

Dielectrophoresis is the translational motion of neutral matter owing to polarization effects in a non-uniform electric field. Depending on matter or electric parameters, different particle populations can exhibit different behavior, e.g. following attractive or repulsive forces. DEP can be used for mixing of charged or polarizable particles by electrokinetic forces [48]. In particular, dielectric particles are mixed by dielectrophoretic forces induced by AC electric fields, which are periodically switched on and off.

The mixer was made by using MEMS technology [48]. Inlet and outlet holes were made by anisotropic etching with KOH from the backside (see Figure 1.9) [48]. The electrodes with pads were patterned on the top side and were coated with an insulation layer. The channel walls and the chamber were made by using SU-8 photoresist technology. The cover plate was sealed using a special bonding technique. Electrical connections are placed on the top side, whereas fluidic connections are on the rear.

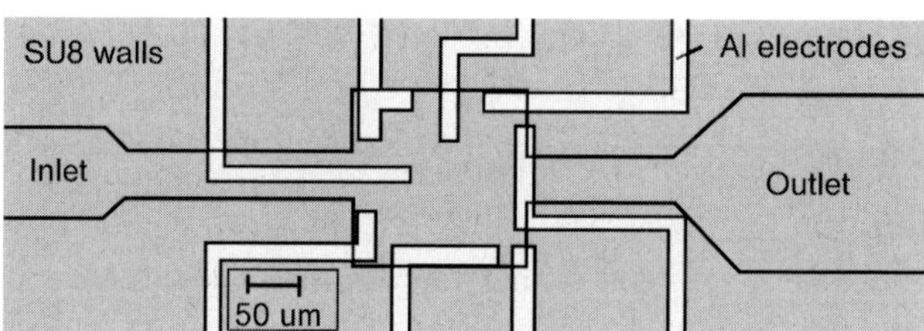

Figure 1.9 Top view of the electrokinetic dielectrophoresis mixer [48] (by courtesy of Springer-Verlag).

Mixer type	Electrokinetic dielectrophoresis mixer	Bonding layer material	SU-8
Mixer material	Silicon	Inlet and outlet channel width	50 μm
Electrode material	Aluminum	Mixing chamber width, depth, length	200 μm, 200 μm, 25 μm
Insulation layer material	SiO_2		

1.2.1.7 Mixing Characterization Protocols/Simulation

Here and throughout, protocols are designated [P 1], [P 2], etc.

[P 1] The electrodes were connected to a signal generator and DC power supply for continuous-voltage operation and an amplifier for alternating-voltage operation [91]. The set-up allowed one to vary the frequency and the potential and to measure their precise values.

The use of pumps was abandoned owing to their pulsation. Instead, hydrostatic pumping was selected and turned out to be sufficient for the flow rates envisaged [91]. The total average velocity was 4.2 mm s^{-1}.

One of the fluids was pure Mazola corn oil and the other was the same oil colored with oil-based Teal dye and doped with oil-miscible antistatic Stadis® 450 to increase the conductivity and permittivity [91]. The latter values were measured with a broadband dielectric spectrometer in a spatially uniform low electric field for frequencies of 0.5–1 kHz.

The flow was monitored by the use of a microscope and a video camera to detect the color changes [91]. A gray-scale level analysis was performed.

[P 2] Deionized water and KCl aqueous solution (1 mM) containing a dye are mixed in a dilution-type experiment [94].

Simulations were performed using CFD-ACE+ [94].

[P 3] Both dilution and chemical reaction techniques were used for flow visualization in [25] (see also [93]), amenable to optical techniques. In a dilution experiment, a dye dissolved in a medium is mixed with a non-dyed solution, typically the medium of the dyed solution itself. In a reacting experiment, a dye is either created or converted ('quenched') to a non-dye state. In a concrete case, fluorescein was used as dye.

For the reaction experiment, a fast acid–base reaction was applied which affects the fluorescence quantum yield of a dye by changing the pH. Fluorescein again was used [25].

A high-resolution charge-coupled device (CCD) camera records the stirring and diffusion of fluorescein from an initially non-mixed state [25]. The fluorescence intensity is integrated over measurement volumes (voxels), thereby yielding a spatial distribution of the mixing degree. The resolution was 2.7×2.7 μm^2. Ensemble-averaged probability density functions (PDFs) and power spectra of the instantaneous spatial intensity profiles were used to quantify the mixing processes. The use of PDFs was regarded as superior as it takes into account the two-dimensional standard deviation [25]; the latter can have statistical fluctuations of fluorescence intensity. Power spectra display the spectral content of the image intensity fields. Energy at high spatial frequencies equals rapid mixing, while low-frequency components of image-power spectra are associated with both unresolved stirring and well-diffused concentration fields.

[P 4] For details on the simulation, see [92].

The mixing experiments were performed using a mercury lamp-induced fluorescence method [92]. A microscope, a photomultiplier and a CCD camera were used for image monitoring. Sodium borate (10 mM) as buffer and Rhodamine B as sample were used. A gray-scale analysis was performed to obtain data on the concentration distribution.

[P 5] The electric field are generated at 1 MHz and 10 V AC voltage [48]. Different DEP domains for polystyrene particles in aqueous suspensions were investigated.

Kinematic simulations were undertaken to describe the folding of material lines and identify parameter settings which give chaotic mixing [48].

1.2.1.8 Typical Results

Flow perturbation upon continuous (DC) current operation

[M 1] [P 1] Microscopic observations of the flow were made to detect the changes induced by switching on an electric field for DC operation [91]. Initially a bi-laminated structure resulted owing to the T-junction contact. When the fluids passed the region of the adjacent electrodes, perturbations of the flow became clearly visible, i.e. the interfaces were deformed. The result was a color, i.e. species, distributed in a cross-sectional direction.

Mixing vs. field strength for DC operation

[M 1] [P 1] The mixing index, defined via coefficients of the electric field, increased as a function of the field strength for DC operation and then approached a plateau at about $7\ 10^5$ V m^{-1} [91]. As threshold value for initiation of mixing a field intensity of $2 \cdot 10^5$ V m^{-1} was detected.

Flow perturbation upon alternating (AC) current operation

[M 1] [P 1] For AC operation, frequencies of 0.5, 10, and 100 Hz were applied [91]. As to be expected from the DC experiments and the derived relationship between the electric field and the mixing, the same holds for a non-constant electric field. The maximum action is achieved for maximum electric fields. This becomes evident when looking at the corresponding microscopy images and the deformations of the fluid interface induced (see Figure 1.10).

The dynamic changes actually lead to the formation of a pulse, which is more or less elongated depending on the frequency [91]. This pulse is followed by a hardly deformed zone so that mixed and unmixed zones are created in an alternating

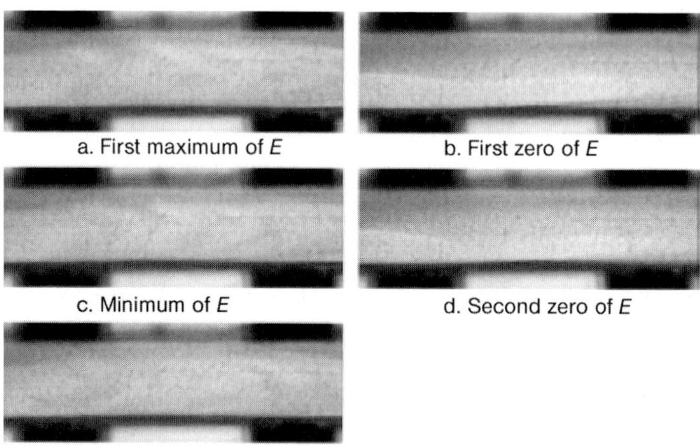

Figure 1.10 Flow visualization by means of a dilution-type experiment after applying an AC electric field of intensity of $4.24 \cdot 10^5$ V m^{-1} at a frequency of 0.5 Hz at various times corresponding to (a) first maximum, (b) first zero, (c) first minimum, (d) second zero and (e) second maximum [91] (by courtesy of RSC).

manner. It is important if the flow can fully develop or not between the pulses to achieve axial mixing or to have a more segregated fluid stream. At low frequency the flow can fully develop, whereas at high frequency ait cannot.

Mixing vs. field strength for AC operation

[M 1] [P 1] The mixing index, defined via coefficients of the electric field, increases as a function of the field strength for AC operation in the range $0.8\text{–}5.7 \cdot 10^5$ V m^{-1} [91].

Mixing vs. frequency for AC operation

[M 1] [P 1] The mixing index, defined via coefficients of the electric field, decreases as a function of the frequency for AC operation in the range 1–5000 Hz (see Figure 1.11) [91].

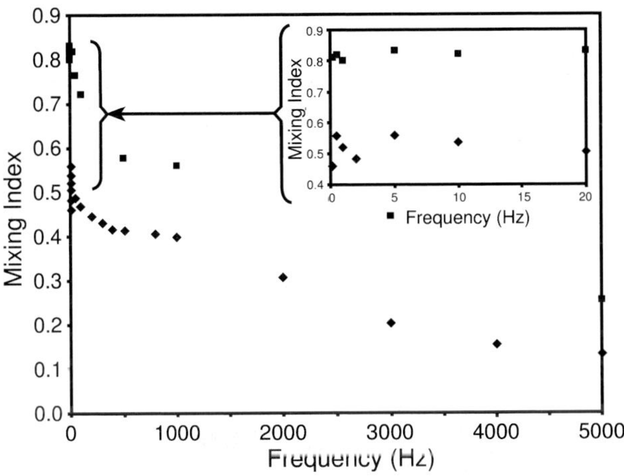

Figure 1.11 Variation of the mixing index with the frequency for an electric field intensity of $4.24 \cdot 10^5$ V m^{-1} for an AC sinusoidal (diamonds) and square (squares) electric fields [91] (by courtesy of RSC).

Square vs. sinusoidal waves for AC operation

[M 1] [P 1] The analysis of the dependence of the mixing index on the field strength for AC operation clearly shows that high field strengths are favorable for mixing (see *Mixing vs. frequency for AC operation* above); therefore, the use of square waves, i.e. a change between positive and negative maximum values, seems to be advisable [91]. Indeed, it could be shown by microscopy and mixing-index analysis that square waves cause superior deformations of the fluid interface as compared with sinusoidal waves (see Figure 1.11).

Multi-electrode operation

[M 1] [P 1] The flow was exposed to three electrode pairs instead of one, as used for all experiments described above [91]. It was shown that a successive progression of the interface deformation and thus of mixing could be achieved in this way. After

passage of the three electrode pairs the mixing seemed to be complete, as indicated by the homogeneous color texture. An advantage of using three instead of one electrode is that the same mixing effect can be achieved at lower field strength. Microscope images of the interface deformation again prove comparable effects for $2.834 \cdot 10^5$ V m^{-1} applied to each of the three electrode pairs as for $4.24 \cdot 10^5$ V m^{-1} in the case of one electrode pair only.

Velocity field change upon electric field turning-on

[M 2] [P 2] A numerical simulation of the velocity field after applying the electric field was made at three times, 1, 10 and 10 ms after the start of the turning-on of the field [94]. The velocity field is considerably changed on this short time scale, which usually is indicative of good mixing.

Electric potential distribution change upon electric field turning-on

[M 2] [P 2] A numerical simulation of the electrical potential distribution after applying the electric field was made at two times, 1 and 10 ms after the start of the turning-on of the field [94]. The electric potential distribution changes considerably, as does the velocity field (see *Velocity field change upon electric field turning-on* above).

Mixing flow visualization

[M 2] [P 2] Injecting a dyed and a colorless stream into the EHD mixer yields a visually homogeneous solution after passage through the mixing chamber channel with the electrodes [94].

Feasibility of EKI mixing under EOF conditions

[M 3] [P 3] Fluorescence images at various times were taken in the main channel, i.e. along the direction of the electric field, of the first-generation micro mixer [25, 93]. After a period of 2 s, the flow becomes unstable and transverse velocities stretch and fold material lines in the flow. The initial seeded/unseeded interface becomes rapidly deformed. Finally after about 13 s, a random distribution of the tracer transverse to the applied AC field is achieved. EKI action is visible throughout the whole channel length of 7 mm. Thus, feasibility of EKI action for micro mixing has been demonstrated.

Improved mixing by confining the EKI region to a small chamber

[M 4] [P 3] Full-field images of the entire mixing chamber of the improved EKI mixer device with parts of the inlet, outlet and side excitation channels were taken (see Figure 1.12). In the mixing chamber, rapid stretching and folding of the fluorescence tracer were observed [25]. Consequently, a homogeneous fluorescence texture in the outlet channel is found. The full-field images also prove that EKI in all the channels attached to the mixing chamber is small and confined to the vicinity to the chamber; hence no uncontrolled and undesired mixing takes place. The mixing time of 2.5 s is superior to the performance of the prototype device [M 3], needing about 13 s for a similar purpose (see *Feasibility of EKI mixing under EOF conditions* above).

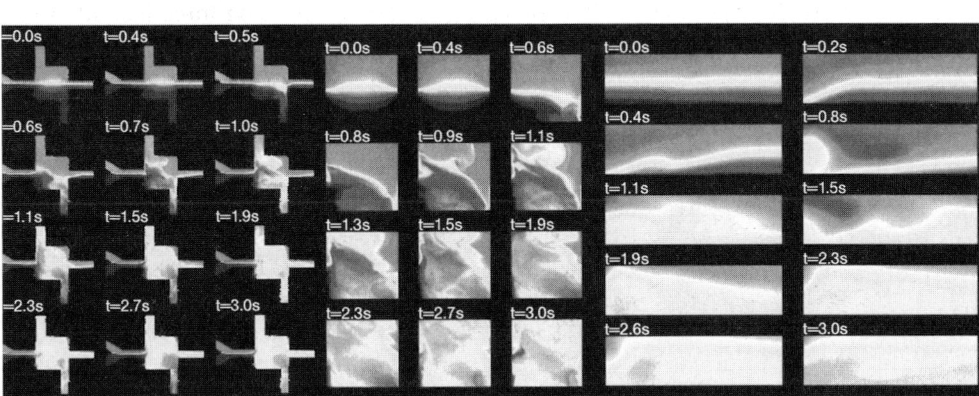

Figure 1.12 Images obtained by a dilution-type fluorescence mixing experiment for the electrokinetic instability micro mixer, second-generation device. (a) 4× objective images of the mixing chamber with inlet, outlet and side channels. The fluid interface is disturbed when starting the AC field; mixing takes place. (b) 10× objective images of the mixing chamber. Complex fluid motions rapidly distribute the dye throughout the majority of the mixing chamber. (c) 10× objective images of the outlet channel, in close vicinity of the mixing chamber. The near-uniform intensity profile evidences the well-stirred fluid exiting the mixing chamber [25] (by courtesy of ACS).

Mixing time for EKI mixing under EOF conditions

[M 4] [P 3] Using the improved EKI mixer device, mixing times of about 2.5 s are obtained [25]. This was deduced from time-resolved images showing the point when a randomly distribution of a fluorescence tracer is achieved (see also *Flow perturbation upon alternating (AC) current operation* above for the lower performance of the first-generation device).

Flow velocities and flow range (*Re*) for EKI mixing under EOF conditions

[M 4] [P 3] Using the improved EKI mixer device, bulk-averaged flow velocities in the outlet channel and in the mixing chamber were 0.5 and 0.16 mm s^{-1}, respectively [25]. The corresponding Reynolds number (*Re*) is 1.5.

Particle tracer experiments for EKI mixing under EOF donditions

[M 4] [P 3] 490 nm particles were used to display the fluid motions in the improved EKI mixer device [25]. Three-dimensional motions were detected. This particle trajectory demonstrates the instability of the electric field and is not expected in a stable electric field.

Ensemble-averaged temporal PDF evolution for EKI mixing under EOF conditions

[M 4] [P 3] The ensemble-averaged temporal evolution of voxel-averaged spatial intensity PDFs was followed in the improved EKI mixer device, both in the mixing

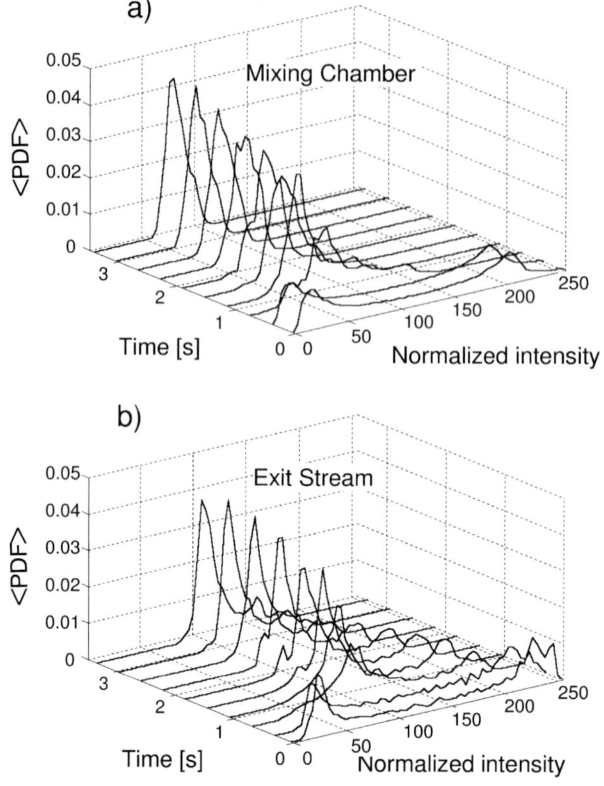

Figure 1.13 Ensemble-averaged temporal evolution of voxel-averaged spatial intensity PDFs for the electrokinetic instability micro mixer, second-generation device. Each ensemble consists of nine realizations. The dimensions of the fluid voxels correspond to 100 μm depth and to about 30 nm width. (a) Mixing chamber: the initial bimodal distribution becomes unimodal upon application of the AC field; (b) exit stream: a similar behavior is observed, leaving, however, a slight bimodality [25] (by courtesy of ACS).

chamber and in the exit stream (see Figure 1.13) [25, 93]. Each ensemble consisted of nine realizations. The spatial intensity fields were further binned to 4×4, to produce superpixels $100.7 \times 100.7\ \mu m^2$ in the image plane. Thus, the dimensions of the fluid voxel correspond to the channel depth of 100 μm and have an approximate diameter of 30 μm.

The initial bimodal distribution in the mixing chamber changes to a unimodal distribution after about 2.5 ms of application of the AC field [25, 93]. A similar behavior is found for the outlet channel, downstream of the mixing chamber. A slight bimodality is, however, still present, caused by an unmixed stagnant layer in the corner of the mixing chamber.

The ensemble-averaged temporal evolution of voxel-averaged spatial intensity PDFs was very reproducible from run to run [25, 93].

Reproducibility of ensemble-averaged temporal PDF evolution, i.e. mixing, for EKI mixing under EOF conditions

[M 4] [P 3] The standard deviation of the voxel-averaged spatial intensity PDFs can be taken as a measure of the level of mixing [25]. The temporal evolution of the above-mentioned figure shows a sharp decrease in the mixing chamber after about 0.5 s of action of the AC field; thus, mixing takes place after a short induction period and being completed in a short time scale. For the flow in the outlet channel downstream of the chamber a more continuous decrease of the standard deviation is found. Error bars of the data reflect the 95% confidence intervals across the realizations and demonstrate the high degree of reproducibility of the PDF development.

Image power spectra for EKI mixing under EOF conditions

[M 4] [P 3] Image, i.e. non-voxel averaged, power spectra complete the information given by direct imaging (see *Improved mixing by confining the EKI region to a small chamber*) or voxel-averaged spatial intensity PDF evolution (see *Reproducibility of ensemble-averaged temporal PDF evolution, i.e. mixing*) [25]. In the initial unmixed state with no AC field, the image power spectrum is characterized by a frequency band slightly elongated in the vertical direction (see Figure 1.14); this is indicative of higher spatial frequencies transverse to the interface and corresponds well to

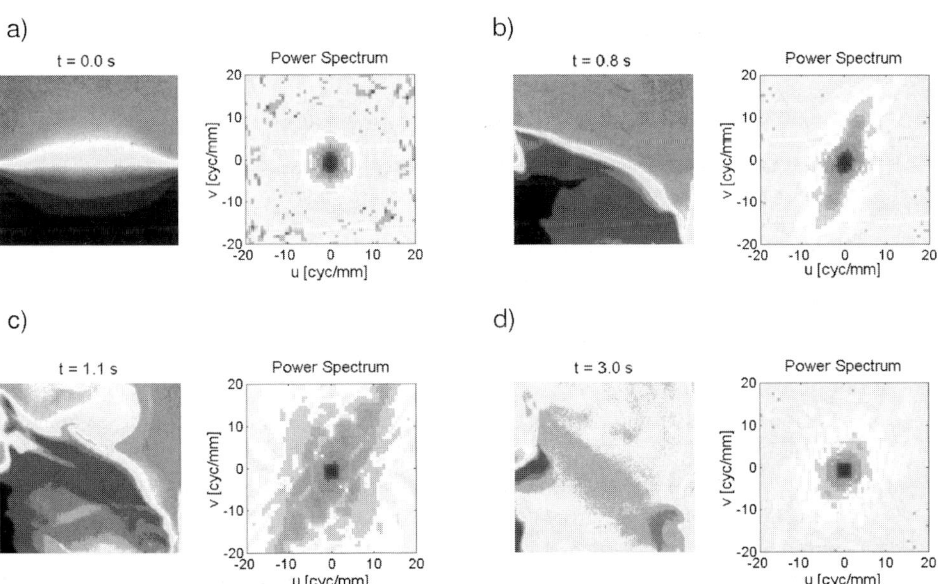

Figure 1.14 Two-dimensional power spectra of various mixing chamber images for the electrokinetic instability micro mixer, second-generation device. (a) Large frequency components along the vertical direction owing to the initial layered distribution of the dye. (b) Larger spatial frequencies are introduced by the EKI stirring within the chamber. (c) The attenuation of large spatial frequencies corresponds to a nearly homogeneous intensity profile [25] (by courtesy of ACS).

the initial horizontal orientation of the interface, being diffused to a certain extent, however. In the state of mixing development, advective flux arises which creates high spatial frequency gradients in the power spectra. New, undiffused fluid interface lengths are generated in the flow. In the final, mixed state, these high-frequency bands are damped. The corresponding well-stirred power spectra are hence isotropic.

Ensemble-averaged power spectra for EKI mixing under EOF conditions

[M 4] [P 3] An analysis of the ensemble-averaged history of the voxel-averaged intensity image bandwidth completes the information given by image power spectra (see Figure 1.15 and *Image power spectra*) [25]. The intensity bandwidth starts at an intermediate value of 3.1 cycles min^{-1} (unmixed state), increases to a maximum at 4.5 cycles min^{-1}, decreases to a minimum (mixing development) and then levels off at the minimum at less than 2.7 cycles min^{-1} (mixed state). The large-magnitude, low-temporal frequency fluctuations are due to the momentary, large-scale displacements of the top and bottom regions of the mixer fluid volume into the side ports. These fluctuations are quickly damped upon increasing stirring in the chamber.

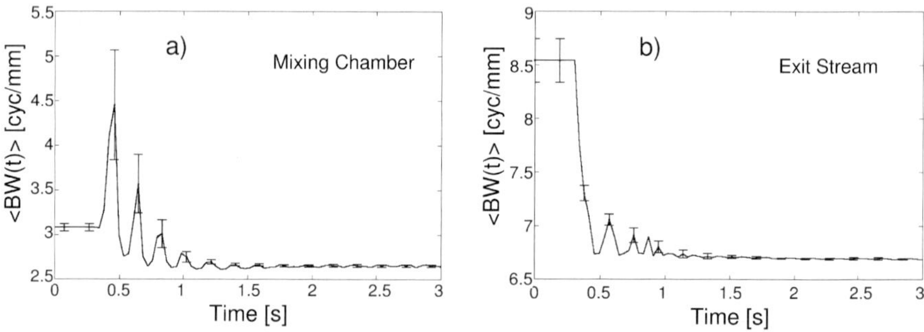

Figure 1.15 Ensemble-averaged –20 dB bandwidth power spectra for the electrokinetic instability micro mixer, second-generation device. (a) Mixing chamber; (b) exit stream. Error bars are shown with 95% confidence intervals. EKI stirring is immediately observed after application of the AC field at $t = 0.4$ s. The initial large oscillation is indicative of large-scale fluid displacements. After 1 s of actuation, the fluid is sufficiently stirred and the bandwidth remains constant [25] (by courtesy of ACS).

Heat generation during EKI for EKI mixing under EOF conditions

[M 4] [P 3] Temperature differences of 5–10 °C result by the action of the AC field owing to Joule heating for the 3 s duration of the pulse [25]. The data were gathered from both thermocouple measurements and capacitance-based heat transfer analysis.

Comparison of electrokinetic (EKI) with electrohydrodynamic (EHD) instability

[M 4] [P 3] The EHD instability is usually observed for fluids with electrical conductivities at least 2–3 orders of magnitude lower than for the EKI instability mixing [25].

Applicability of EKI mixing with regard to construction material and media

[M 4] [P 3] EKI mixing was demonstrated for mixers made of polymer materials such as PDMS or PMMA and of glass [25]. Processing with various electrolytes could be successfully applied, including deionized water and borate and HEPES buffers, with electrical conductivities ranging from 5 to 250 μS cm^{-1}.

Flow patterns and concentration profiles for mixing by zeta potential variation

[M 5] [P 4] Images of the distribution of a fluorescent species were monitored experimentally and calculated numerically within and behind the mixing section (1/6 Hz, 300 V for electroosmotic flow, 800 V to the electrodes) [92]. Starting from an initial bi-laminating pattern, a transverse movement of the species is detectable and a homogeneous fluorescent texture is observed behind the mixing section.

Concentration profiles were derived by gray-scale analysis from the images mentioned above [92]. Whereas without use of zeta potential variation no difference in concentration profiles between the upstream and downstream positions is visible, a much more flattened, i.e. mixed, profile results under electrokinetically driven conditions in the downstream position.

Impact of control potential, EOF field and frequency for the zeta potential variation

[M 5] [P 4] With increasing applied control voltage increased mixing efficiency is achieved (0–700 V) [92]. At about 600 V, a plateau is achieved.

Mixing performance is decreased with higher EOF driven field shows decreasing as a consequence of the shorter residence times [92]. The higher the frequency, the lower is the mixing performance [92]. A threshold value at about 1/8 Hz is found.

Simulation of material line folding

[M 6] [P 5] The folding and stretching of material lines when entering an cavity with electrodes and thus exposure to a non-uniform electric field were investigated [48]. At small amplitude only weak oscillations of the material lines along the channel are found. In contrast, all particles are trapped around the electrodes at large amplitude. At medium amplitude, the desired folding and stretching occur, resulting in enhancement of mixing.

Particle motion owing to the electric field

[M 6] [P 5] The particles are trapped at the edges of the electrodes, align with field lines and form pearl chains at a frequency of 100 kHz [48]. At a frequency of 10 MHz, particles are repelled from the edges of the electrodes towards the field gradient minima.

Imaging of particle motion

[M 6] [P 5] At zero electric field, a bi-laminated system with straight interface can be seen for contacting deionized water and a particle solution (see Figure 1.16) [48]. After switching on of the electric field, folding and stretching takes place.

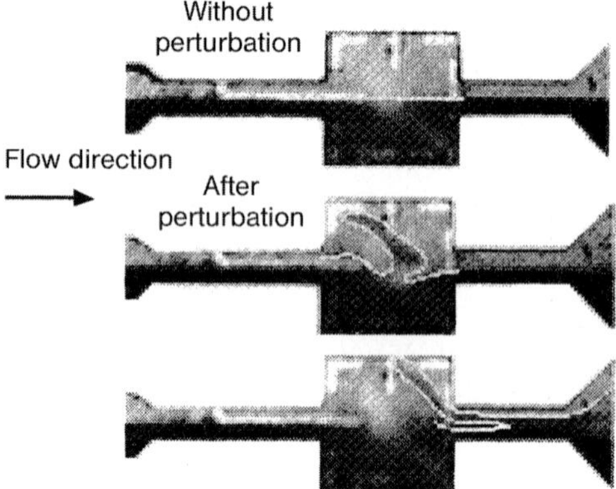

Figure 1.16 Imaging of the folding and stretching of particle lines when an electrokinetic perturbation is applied. The bright line indicates the interface between the particle solution and pure water [48] (by courtesy of Springer-Verlag).

1.2.2 Electro Rotational Mixing

Most Relevant Citations
Peer-reviewed journals: [95].

Objects having a dipole can be set into rotational motion by applying a torque by means of an electric field [95]. Electrorotation is the rotation of particles as a consequence of the induction of dipole moments and torque exertion by a rotating electric field. Coupled electrorotation (CER) uses static external fields which are spatially fixed to induce dipoles in two or more adjacent particles. This creates oscillating components of the electric field, finally resulting in a rotating electric field (for more details, refer to the original literature [95]).

If one of the objects cannot rotate, e.g. because it is fixed in space, CER still can induce rotation of the other object [95]. That is the basis for the mixing principle presented here. A microsphere is positioned nearby a (fixed) microstructure, typically of the same or similar material. By application of an electric field, the above-mentioned interaction takes place and the microsphere rotates.

1.2.2.1 Mixer 7 [M 7]: Coupled Electrorotation Micro Mixer

This coupled electrorotation mixer was actually not built as a complete device, but realized as a prototype version with simple microstructures, which were made by sputtering on to glass cover slips [95]. By photolithography a small gap was etched in the middle of the coated slip to separate it into two equal halves. The latter are used as two electrodes of equal size. A wedge, freshly cut from the corner of a raw material piece, is placed on the gap as a kind of fixed microstructure to perform as the static 'dipole object' so that an adjacent microsphere can be set into rotation.

Mixer type	Coupled electro-rotation micro mixer	Electrode layer thickness	~100 nm
Mixer base material	Glass cover slip	Gap separating the electrodes	10 μm
Cover slip size	24 mm × 60 mm	Wedge material	PDMS
Sputtered electrode material	Gold	Microsphere material	Polybead carboxylate microspheres

1.2.2.2 Mixing Characterization Protocols/Simulation

[P 6] A ~50 μl droplet with 2 μm latex spheres suspended in water was spread over both electrodes on to the chip. The mixing was followed by a microscope with an oil immersion objective [95]. Evaporation of the droplet solution has to be minimized, as this notably affects the electrorotation.

The radiofrequency output from a function generator was coupled directly to the electrodes on the chip [95]. Wires are attached to electrodes using a conductive epoxy. A 500 kHz electric field sine wave (60 kV cm^{-1}) was applied.

Mixing was visualized by additional fluorescent 500 nm tracer particles (yellow–green carboxylate FluoSpheres), marking by their tracer trajectories the flow pattern [95].

1.2.2.3 Typical Results

Tracer trajectories giving the flow field around a rotating microsphere

[M 7] [P 6] By coupled electrorotation of a microsphere local circulation of the surrounding liquid can be induced, as proven by the trajectories of fluorescent particle tracers around the (larger) rotating microsphere [95]. Hence the mixing effect was shown qualitatively. No details on quantitative results such as mixing time were given.

Some thoughts on future applications of rotating microspheres

[M 7] [P 6] The fact that probably efficient mixing in the close vicinity of the sphere can be induced (see *Tracer trajectories giving the flow field around a rotating microsphere* above) suggests its use for mixing in real miniature fluid compartments, e.g. with characteristic diameter below 1 μm, which hardly can be mixed be other means [95]. No physical contact is then required, unlike other types of miniature rotating actors. The mixer itself, i.e. the microsphere, does not need to be microfabricated, but can be purchased commercially. Mixing can be switched on and off, unlike for static fluid mixing. Positioning of the microsphere is easily achieved to have control over the region which is exposed to mixing.

1.2.3 Chaotic Electroosmotic Stirring Mixing

Most Relevant Citations

Peer-reviewed journals: [28].

When electroosmotic flows in micro channels are superimposed by temporal modulation of non-uniform ζ potentials along the conduits' walls, it is possible to induce chaotic advection which can lead to mixing by stirring-analogue material transport [28]. Non-uniform ζ potentials can be obtained by coating the channel's walls with different materials or by using different buffer solutions. Both spatial and temporal control of the ζ potential can be achieved by imposing an electric field perpendicular to the solid–liquid interface. In practice, the generation of such normal electric fields can be accomplished by placing electrodes beneath the solid–liquid interface, provided that these electrodes are insulated from the liquid. Alternatively, photosensitive surfaces can be surface charged with light.

In the following, the use of periodical ζ potentials will be described; the periodicity will be denoted T [28]. Switching between two ore more flow patterns is performed inducing chaotic advection. One flow field is maintained in one time interval and another flow field in a second interval. This is repeated with the period T. The switching of the flow fields is accomplished by controlling the distribution of the ζ potential created by the electrodes. By flow field alternation, particles virtually expose a zig-zag path, thereby distributing material all over the channel's cross-section. Such transport is similar to efficient stirring.

1.2.3.1 Mixer 8 [M 8]: Chaotic Electroosmotic Micro Mixer

This chaotic electroosmotic micro mixer has not been realized so far; only a theoretical study on the principle of the device was made [28]. A schematic was given which can serve as base for the future realization of such a device (see Figure 1.17).

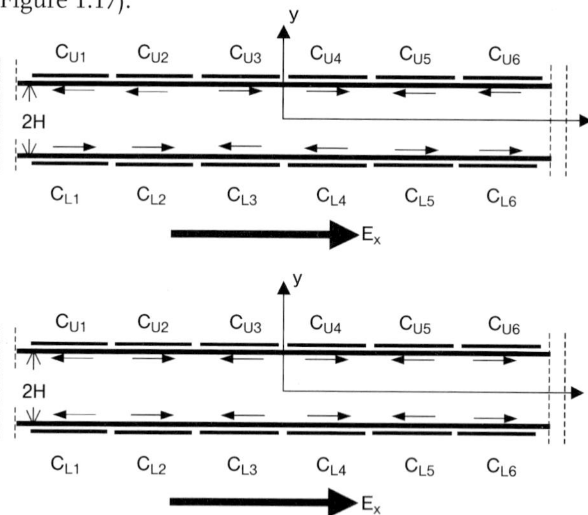

Figure 1.17 Schematic of a micro channel equipped with many electrodes at the upper (U_i) and lower (L_i) walls for control of the ζ potential. The arrows in the channel denote the directions of the electroosmotic velocities creating one type of flow pattern, here a counter-current arrangement of top and bottom flows (top); alternating-flow arrangement, demonstrating another type of control over the ζ potential (bottom) [28] (by courtesy of ACS).

A micro channel of height 2 H is equipped with electrodes at the upper (U_i) and lower (L_i) walls [28]. These electrodes are used to control the ζ potential at the solid–liquid interface. In this way, the direction of the electroosmotic flow near the interface can be changed locally. The external electric field is given as E_x.

1.2.3.2 Mixing Characterization Protocols/Simulation

[P 7] The topic has only been treated theoretically so far [28]. A mathematical model was set up; slip boundary conditions were used and the Navier–Stokes equation was solved to obtain two-dimensional electroosmotic flows for various distributions of the ζ potential. The flow field was determined analytically using a Fourier series to allow one tracking of passive tracer particles for flow visualization. It was chosen to study the asymptotic behavior of the series' components to overcome the limits of Fourier series with regard to slow convergence. In this way, with only a few terms highly accurate solutions are yielded. Then, alternation between two flow fields is used to induce chaotic advection. This is achieved by periodic alteration of the electrodes' potentials.

1.2.3.3 Typical Results

Time-independent electroosmotic flows

[M 8] [P 7] Flow fields were analyzed for time-independent electroosmotic flows at a variety of ζ potential distributions (see Figure 1.18) [28]. The four basic flow structures Ψ_{EO}, Ψ_{EE}, Ψ_{OO} and Ψ_{OE} are given in [28]. Ψ denotes a stream function and the subscripts *EO*, *EE*, *OO*, and *OE* denote, respectively, even in X and odd in Y, even in X and even in Y, odd in X and odd in Y, and odd in X and even in Y. U^+, U^-, and $U^{\pm}$ are dimensionless slip velocities and provide the following boundary conditions:

(a) $U^{\pm}(X) - U^{-}(X) - 1$;
(b) $U^{+}(X) = -U^{-}(X) = 1$;
(c) $U^{\pm}(X) = -U^{\pm}(-X)$, $U^{+}(X) = -U^{-}(X) = 1$;
(d) $U^{+}(X) = -U^{+}(-X) = 1$; $U^{\pm}(X) = -U^{\pm}(-X)$, $U^{+}(X) = U^{-}(-X)$ and $U^{+}(X) = -U^{+}(-X) = 1$.

where X and Y are the longitudinal and transverse dimensions in the channel.

For case (a), the flow field consists of a single convective cell which circulates around an elliptic point (center) located at $X = Y = 0$ [28]. For case (b) the flow field consists of two counter-rotating cells with centers at $X = 0$ and $Y = \pm 0.58$. The cells are separated by the surface $Y = 0$. For case (c), the flow field is similar to (b) in the sense that the flow field consists of two counter-rotating cells separated by the surface at $X = 0$. The centers of rotation are at $X = \pm 1$ and $Y = 0$. For case (d), the flow field consists of four counter-rotating cells separated by two surfaces at $X = 0$ and $Y = 0$.

Thus, complex flow patterns can be generated for non-uniform distributions of the ζ potential in a time-independent manner [28]. However, the flow is highly regular and no transport transverse to the streamlines is given. Particles will only follow the streamlines which does not give as good mixing as the action of stirring.

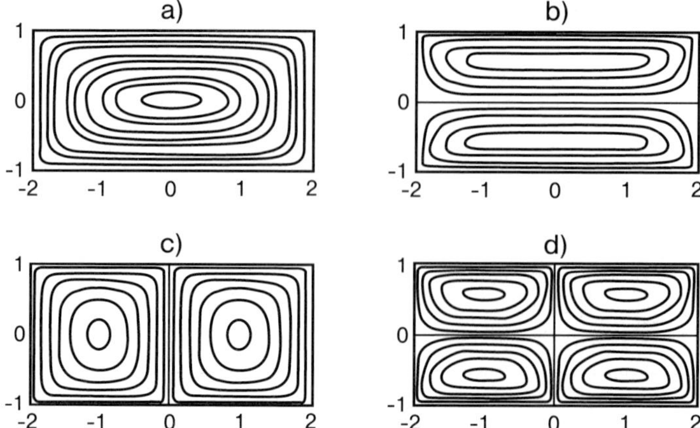

Figure 1.18 Streamline patterns for various electroosmotic flows when $h = 2$ (h being the periodicity of arranging the electrodes). (a) $U^{\pm}(X) = -U^{-}(X) = 1$; (b) $U^{+}(X) = -U^{-}(X) = 1$; (c) $U^{\pm}(X) = -U^{\pm}(-X)$, $U^{+}(X) = -U^{-}(X) = 1$, (d) $U^{+}(X) = -U^{+}(-X) = 1$; $U^{\pm}(X) = -U^{\pm}(-X)$, $U^{+}(X) = U^{-}(-X)$ and $U^{+}(X) = -U^{+}(-X) = 1$ [28] (by courtesy of ACS).

Time-dependent electroosmotic flows – chaotic advection

[M 8] [P 7] The effect of switching between the flow fields (a) and (d) given in Figure 1.18 at various periods T was analyzed (see Figure 1.19) [28]. For very high alterations, a simple superposition of the flow fields is achieved. Elliptic fixed points surrounded by closed orbits (tori) of various periods are found.

For larger T ($T = 1.6$), chaotic behavior arises, the hyperbolic fixed point is disrupted and the tori are perturbed (see Figure 1.20) [28]. A chaotic region appears with homoclinic tangle and formation of new hyperbolic and elliptic points.

On further increasing T to 2, 4 and 6, the complexity of the flow becomes more pronounced (see Figure 1.21) [28]. First, the particles wander around the superimposed image. Then, particles stray further away from the 'regular path' and sample most of the cell's area. Chaotic advection is now present.

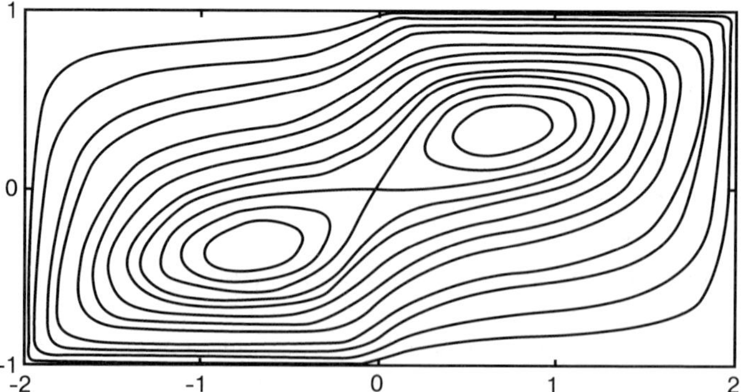

Figure 1.19 Streamline pattern for the superimposed flow structure, by switching between the flows (a) and (d) given in Figure 1.18, for $h = 2$ [28] (by courtesy of ACS).

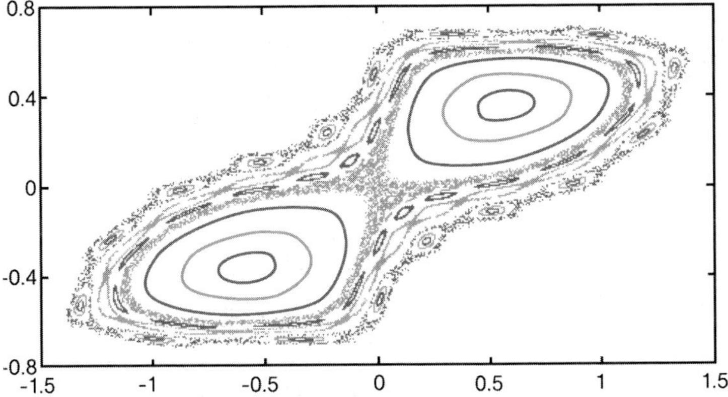

Figure 1.20 Pointcaré section when $T = 1.6$ and $h = 2$. Passive tracer particles were placed at selected places, given in [28] (by courtesy of ACS).

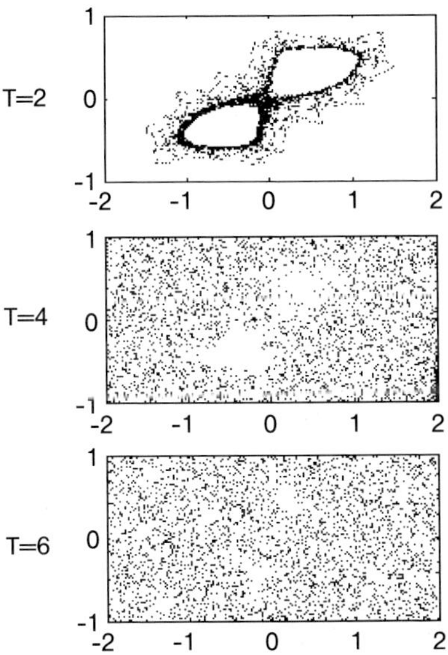

Figure 1.21 Pointcaré sections for various periods $T = 2$, 4 and 6 and $h = 2$. A passive tracer particle was initially inserted at $(x_0, y_0) = (0, 0.01)$, and its motions were followed by 3000 periods [28] (by courtesy of ACS).

Deformation of a blob by chaotic advection – simulation of a stirring process

[M 8] [P 7] In order to simulate a stirring process, a square material blob is placed in the center of the channel (see Figure 1.22) [28]. This is equivalent to inserting a drop of a dye in the channel. Rapid stretching and folding processes, characteristic of chaotic advection, arise on periodically switching between flow fields. Eventually, the blob is spread over the entire cell's area. Hence efficient stirring is given.

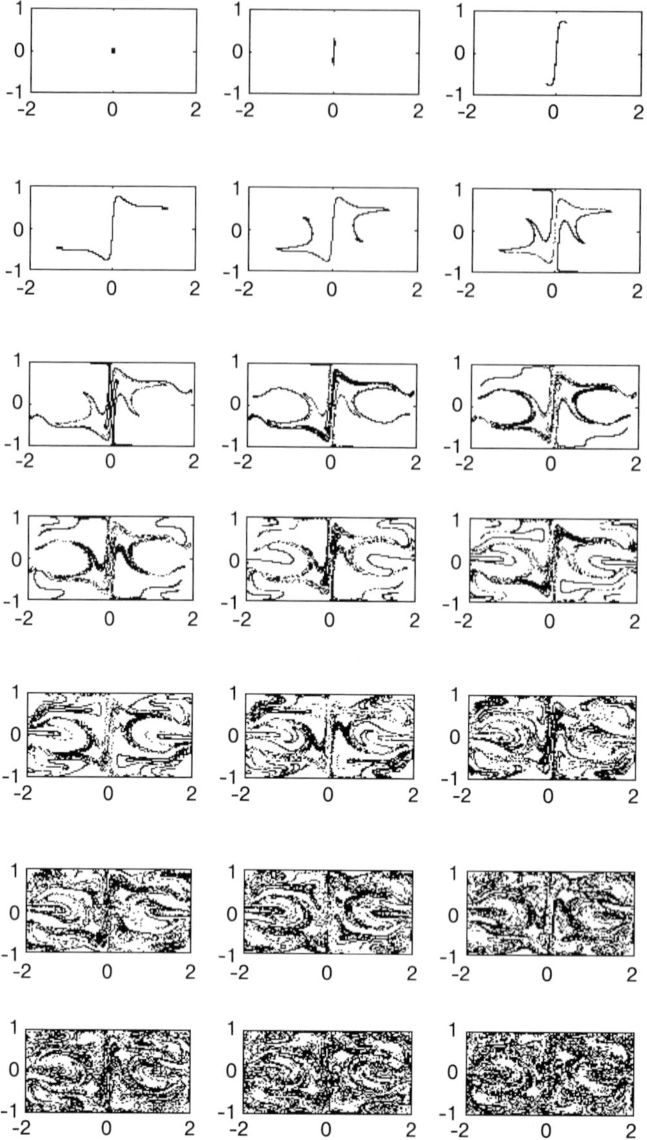

Figure 1.22 Deformation of a blob with an initial edge of 0.1. Initially ($t = 0$) centered at (0,0). $T = 4$. $t/T = 0, 1, \ldots, 17$ [28] (by courtesy of ACS).

Variation of the number of electrodes

[M 8] [P 7] The above-mentioned simulations were made with two electrodes each on the top and bottom of the walls [28]. When using four embedded electrodes, even more complex flow fields are achieved, which are in parts similar to those generated by switching with two electrodes. Superimposed fields are found and also chaotic patterns.

1.2.4 Magnetohydrodynamic Mixing

Most Relevant Citations

Peer-reviewed journals: [96].

The above-mentioned examples have proven the suitability of using electric fields generated by electrode arrays for mixing of liquids. In the following example, a micro channel is also equipped with an array of electrodes which can generate a complex electric field [96]. By alternate potential differences, currents are induced in various directions. This electric field can be coupled to a magnetic field yielding Lorentz body forces for fluid mixing. In this way, cellular motion is achieved which stretches and deforms material lines.

1.2.4.1 Mixer 9 [M 9]: Magnetohydrodynamic Micro Mixer

This device consists of a rectangular channel which contains uniformly spaced electrodes, transversely oriented to the flow direction (see Figure 1.23) [96]. The electrodes are connected to the positive and negative poles of a DC power supply. The direction of the electric current varies from one location to another. A uniform magnetic field normal to the channel bottom is applied. By coupling of the electric and the magnetic forces, a Lorentz body force is yielded which is perpendicular to the electric and magnetic fields and directed towards the side walls. The direction of this force also alternates. As a result, the fluid is moved up- and downwards. The net effect is eddy-type convection.

The device was built from a bottom plate with a five-electrode structure, five plates which altogether form a rectangular mixing chamber, and a cover plate [96]. All plates contain vertical vias for electrical connection. The plates were realized as ceramic green tapes which a fired to a solid stack in the following way.

Thin green tapes are cast and microstructured by milling. Alternatively, laser machining or photolithography may be applied [96]. Rectangles of the desired size were blanked from these tapes. According to the information given above,

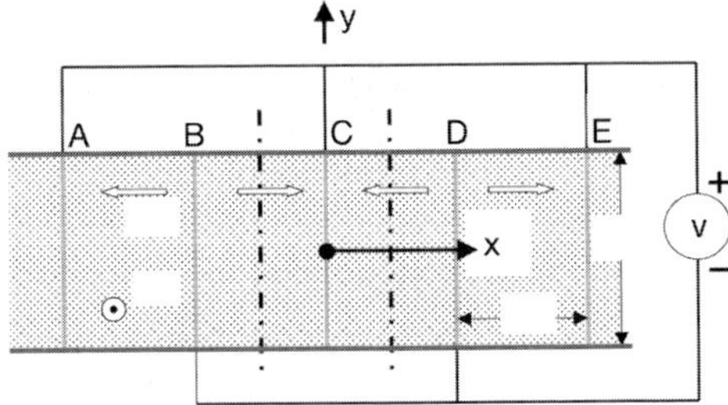

Figure 1.23 Schematic of the top view of the magnetohydrodynamic mixer. The electrodes are denoted A–E. The electrodes are connected to positive and negative poles of a power supply in an alternate manner. The *x*-direction corresponds to the long axis of the mixing chamber. The magnetic field is normal to the figure's plane [96] (by courtesy of Elsevier Ltd.).

seven tapes were fired. By firing, the organic material was removed and sintering took place, leaving a solid body. If needed, many more tapes (> 80) can be stacked together and co-fired to yield even more complex assemblies with channels and electrodes. Transparent windows can also be attached to allow inspection of the flow. Metal paths were hand-printed to form electrodes. Three electrodes were connected to a single conductor and two to a second conductor.

A permanent magnet was placed below the mixer [96]. The size of the magnet was chosen to be much larger than the mixer internals to result in a homogeneous magnetic field.

Mixer type	Magnetohydrodynamic micro mixer	Gold electrode paste material	DuPont 5734
Mixer material	LTCC 951AT	Paste filler for vias	DuPont 6141
Tape thickness	~250 μm	Soldering pad material	DuPont 6146
Mixing chamber width, length, depth	4.7 mm, 22.3 mm, 1 mm	Magnet material	Neodymium
Electrode depth, width, distance	20 μm, 700 μm, 3.3 mm	Magnet load capacity	12–15 lb

1.2.4.2 Mixing Characterization Protocols/Simulation

[P 8] The resistance between the electrodes was measured in situ. The power supply was operated in constant-voltage mode [96]. A low-volume tracer was injected with a syringe and then applied a voltage of 4 V across the channel. The current was of the order of μA and was below the resolution level of the power supply (10 mA).

Details on the simulations based on the Navier–Stokes equation with a body force are given in [96].

1.2.4.3 Typical Results

Simulation of circular motions

[M 9] [P 8] Simulations show that magnetic forces induce circular motion within the liquid [96]. This is shown by several findings, e.g. by a streamline analysis, a plot of the velocity versus dimensions of the channel, constant pressure contour lines and the pressure distribution.

Simulation of interface stretching

[M 9] [P 8] Simulations predict the extent to which and by which pattern the interfaces are stretched, which is a qualitative measure of mixing efficiency (see Figure 1.24) [96]. The evolution of the interface is given as a function of the dimensionless time. An initial bi-layered system forms a vortex-type structure with increasing spiral winding upon time.

A quantitative analysis shows that the interface increases slightly faster than a linear function of time [96]. This is better than for having diffusion only, but is behind the performance of chaotic advection.

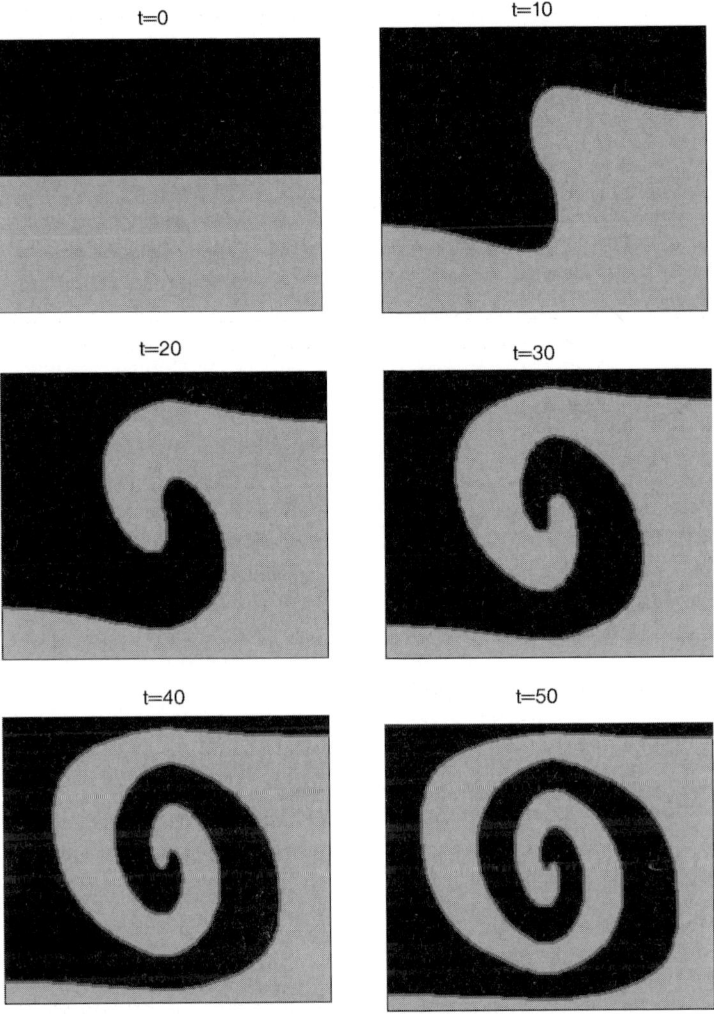

Figure 1.24 Evolution of the flow pattern and corresponding interface stretching as a function of dimensionless time [96] (by courtesy of Elsevier Ltd.).

Experimental flow observation

[M 9] [P 8] Owing to the DC potential difference, visible flow was clearly induced for a water-filled chamber with a central dyed line by magnetohydrodynamic action [96]. Upwards and downwards fluid motion in a rotating fashion was observed.

Reversal of electrode polarity

[M 9] [P 8] After reversal of the polarity of the electrodes, a formerly deformed dyed line remains to the original state, a straight line and finally reaches a mirror-imaged deformed dyed line (see Figure 1.25) [96]. After some time, the dye is fairly homogeneously distributed over the mixing chamber.

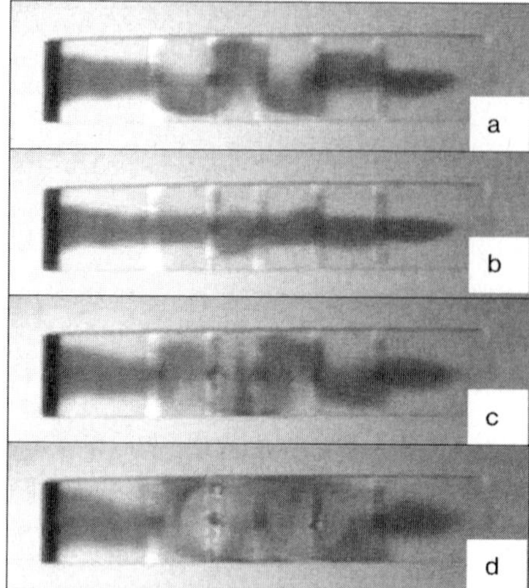

Figure 1.25 Dye mixing visualization experiments starting from a water-filled mixing chamber with a central dyed line. (a) Initial deformation of the dyed line by action of the Lorentz force; (b) change of electrode polarity leads to deformation of the dyed line which returns to the original state, the straight dyed line; (c) deformation mirror-imaged to (a) by continuing action of the Lorentz force after change of electrode polarity; (d) eddy formation becomes apparent on continuing the Lorentz-force action given in (d) [96] (by courtesy of Elsevier Ltd.).

Experimental and computed mixing time

[M 9] [P 8] Experimental and simulated times were compared for the movement of a tracer from one location to another [96]. Only dimensionless times were given. The computed times were smaller than the experimentally found values. This may be due, for example, to incorrect estimation of the magnetic field strength.

1.2.5
Air-bubble Induced Acoustic Mixing

Most Relevant Citations

Peer-reviewed journals: [23, 24].

An air bubble in a liquid medium can act as actuator, i.e. the bubble surface behaves like a vibrating membrane, when it is energized by an acoustic field [23, 24]. This bubble actuation is largely determined by the bubble resonance characteristics (for more details hitherto, see [23, 24]). Bubble vibration due to a sound field induces friction forces at the air/liquid interface which cause a bulk fluid flow around the air bubble, termed *cavitation microstreaming* or *acoustic microstreaming*. Circulatory flows lead to global convection flows with a 'tornado'-type pattern which enhances mixing. At low driving amplitudes, the insonation frequency has to meet the resonance frequency for pulsation. The bubble then has to be fixed at a solid boundary. The frequency of acoustic microstreaming is, as expected, strongly dependent on the bubble radius and vice versa.

Acoustic microstreaming is a method of mixing fluids in micro chambers having a certain, relatively large internal volume [23, 24]. Typically, mixing in such volumes solely by the aid of diffusion would require several hours' mixing time. A special design with micromachined air pockets was made to entrap air bubbles. By use of a commercially available piezoelectric (PZT) disk, fluid motion can be generated which leads to mixing. In this way, the mixing time can be reduced from hours to a few seconds.

1.2.5.1 Mixer 10 [M 10]: Acoustic Microstreaming Micro Mixer, Version 1

This micro device consists of a round or square micro chamber machined in a bottom plate sealed by a cover slip with double-sided adhesive tape [23]. Sound irradiation was achieved by a PZT disk below. The PZT disk was bonded by super glue to the surface of the micro chamber. The PZT disk was driven by an HP functional generator

The micro chamber was fabricated using a milling machine [23].

The device [M 10] was fabricated in three versions, differing in the number of air pockets [23].

- Version (a): one bubble only held in a conduit, situated at the chamber boundary, adjacent to the inlets; round micro chamber.
- Version (b): four bubbles, or air pockets, placed symmetrically at the arc of the circle; round micro chamber.
- Version (c): multiple (4 × 4 array) bubbles, or air pockets, placed as a symmetric matrix all above the micro chamber; square micro chamber.

Mixer type	Acoustic microstreaming micro mixer, version 1	Mixing chamber, version (c): width, length, depth	12 mm,15 mm, 125 μm
Base plate material	Polycarbonate	Air pockets version (b): depth, diameter	2 mm, 300 μm
Cover slip	Polycarbonate	Air pockets version (c): depth, diameter	500 μm, 500 μm
Thickness of adhesive tape	125 μm	PZT disk: diameter	15 mm
Mixing chambers, version (a) + (b): diameter, depth	15 mm, 300 μm		

1.2.5.2 Mixer 11 [M 11]: Acoustic Microstreaming Micro Mixer, Version 2

This micro device consists of a square micro chamber which has as a bottom plate a conventional DNA micro-array chip sealed by a cover slip with double-sided adhesive tape (see Figure 1.26) [24]. The adhesive tape serves as a spacing gasket to define the shape and dimensions of the chamber. The cover slip contains the air pockets with a uniform, pitched distribution. The air pockets trap the air bubbles

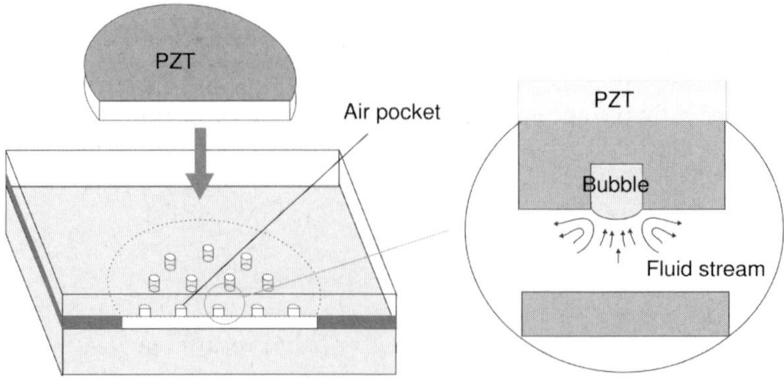

Figure 1.26 Schematic of a cover slip with an array of air pockets sealed to a DNA biochip chamber by adhesive tape. (a) Overview; (b) detail with side view of one air pocket [23] (by courtesy of RSC).

in the reaction solution. A PZT disk was bonded on the cover slip using super glue.

The air pockets were fabricated using a milling machine [24].

The sound was generated by the PZT disk driven by an HP functional generator [24].

Mixer type	Acoustic micro-streaming micro mixer, version 2	Mixing chamber: width, length, depth	16 mm, 16 mm, 200 μm
Microarray material	Glass	Air pockets: depth, diameter	500 μm, 500 μm
Cover slip	Polycarbonate	PZT disk: diameter	15 mm
Thickness of adhesive tape	200 μm		

1.2.5.3 Mixer 12 [M 12]: Design Case Studies for Micro Chambers of Acoustic Microstreaming Micro Mixer, Version 2

Three artificial micro chambers were proposed for the simulations, described under [P 11] [23]. They comprised mixing chambers with one bubble pocket in the center (a), four bubble pockets at the corners (b), and five air pockets with four at the corners and one in the center (c).

1.2.5.3 Mixing Characterization Protocols/Simulation

[P 9] Visual observations were made using a stereoscope [23]. Half of the chamber was filled with deionized water and the other with an aqueous solution composed of the dye phenolphthalein and sodium hydroxide, giving a red color. A sinusoidal wave of 2 kHz frequency with a peak-to-peak amplitude of 5 V was used for device types [M 10] (a) and (b). A sinusoidal wave of 5 kHz frequency with a peak-to-peak amplitude of 5 V was used for device type [M 10] (b).

[P 10] All conditions being the same as for [P 9]; only a square wave with 5 kHz frequency with a peak-to-peak amplitude of 40 V was used [24].

[P 11] Simulations were carried with a simplified chamber and air-bubble pocket geometry. Details on this geometry and the several assumptions taken for describing the fluid dynamics can be found in [23] and are not described further here. Generally, the experimental known fluid dynamic features were taken into account, e.g. the convective motion based on vortices was assumed also in the model.

1.2.5.4 Typical Results

Sonic irradiation without air bubbles

[M 11] [P 9] Sonic irradiation without the use of air bubbles causes only little fluid motion [23, 24].

Induction time

[M 12] [P 11] A certain induction time was found for artificial bubble configurations in a micro chamber [23]. This period was 10 wall motion cycles or 100 time steps.

Flow visualization for acoustic microstreaming

[M 10] [M 11] [P 9] [P 10] Sonic irradiation with the use of air bubbles caused considerable gross fluid motion for a dye solution [23, 24]. Owing to the uniform distribution of the air pockets, mixing is induced in the complete micro chamber and not localized to one part only.

When starting the PZT, single microstreaming fields around the bubbles are generated and interact with each other [23, 24]. At the gas/liquid interface churning motion was observed, inducing a convection streaming with a 'tornado' like pattern. A narrow stream occurs which moves rapidly towards the bubble surface (see Figure 1.27). In the close vicinity of the bubble, the stream velocity is decreased and the stream spreads into the liquid bulk.

[M 10a] [P 9] In a round micro chamber with one bubble only held at a small conduit, the streaming field is composed of orderly patterns with a symmetry about an axis perpendicular to the solid wall through the center of the bubble [23]. Fluidic

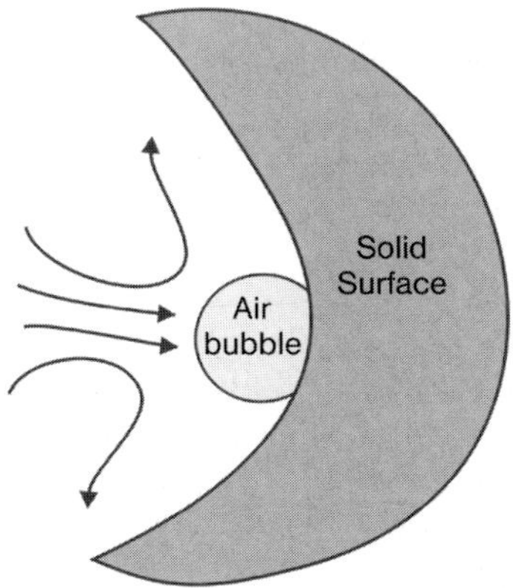

Figure 1.27 Schematic of flow patterns, being part of acoustic microstreaming, induced by an air bubble resting on a solid wall and actuated by a piezoelectric disk [23] (by courtesy of RSC).

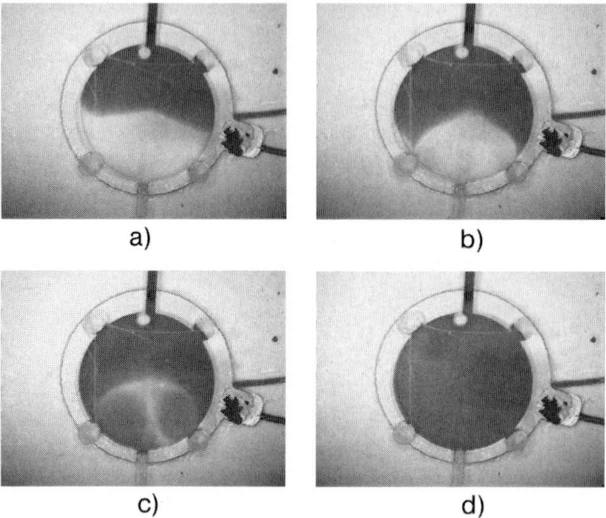

Figure 1.28 Photographs showing acoustic microstreaming in a micro chamber (with 300 μm depth and 15 mm diameter) which has four air pockets (300 μm depth and 2 mm diameter) at time (a) 0, (b) 10, (c) 25 and (d) 45 s [23] (by courtesy of RSC).

elements move towards the center of the bubble towards this axis. Here, also the above-mentioned spreading of the stream is observed, when approaching the bubble. The maximum speed of these streams was estimated to be ~5 mm s^{-1}.

[M 10b] [P 9] In a round micro chamber with four bubbles held in four air pockets, notable gross liquid motion and orderly vortex-like patterns near the bubbles were seen (see Figure 1.28) [23]. Churning motion was also present.

[M 10c] [M 11] [P 9] [P 10] In multiple-bubble micro chambers, even tighter control of fluid motion and mixing is feasible [23, 24]. The microstreaming fields around each bubble interfere with the other. The dye is moved from one side of the chamber to the other.

Mixing time

[M 10a] [P 9] By one-bubble microstreaming mixing, the dye completely fills the micro chamber within about 110 s [23].

[M 10b] [P 9] By four-bubble microstreaming mixing, the dye completely fills the micro chamber within about 45 s [23]. This is an improvement in mixing time of about 40% compared with the *[M 10a]* device with one bubble only.

[M 10c] [P 9] By multiple-bubble microstreaming mixing, the dye completely fills the micro chamber within about 105 s (5 kHz; 5 V) [23]. Mixing based on pure diffusion in the same volume would have taken about 6 h.

[M 11] [P 9] Increasing the peak-to-peak amplitude further enhances mixing. By multiple-bubble microstreaming mixing at 5 kHz and 40 V, the dye completely fills the micro chamber within about 6 s (see Figure 1.29) [24]. Mixing based on pure diffusion in the same volume would have taken about 8 h.

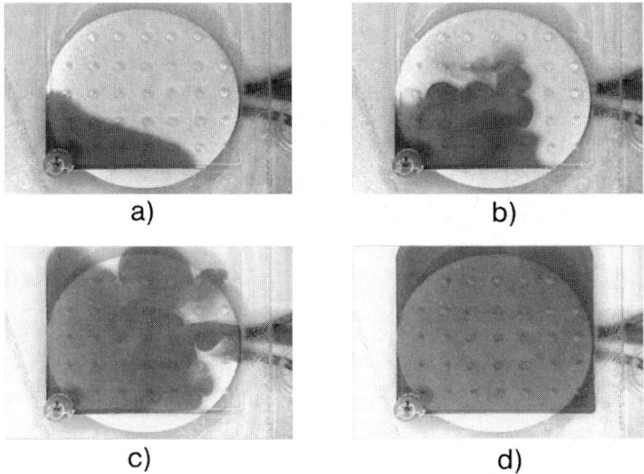

Figure 1.29 Photographs of the mixing of an aqueous dye solution with de-ionized water by acoustic microstreaming (7 × 5 top bubbles; 5 kHz; 40 V) in a 16 × 16 micro chamber (0.2 mm^3 volume). (a) 0, (b) 2, (c) 4 and (d) 6 s [24] (by courtesy of ACS).

Type of wave

[M 10] [M 11] [P 9] [P 10] Square waves give faster mixing than sinusoidal waves [23, 24].

Height of amplitude

[M 10] [M 11] [P 9] [P 10] The higher the amplitude, the more improved is the mixing [23, 24]. At 5 kHz and 5 V, multiple-bubble microstreaming mixing is completed within about 105 s [23]. Increasing the peak-to-peak amplitude further to 40 V enhances mixing. By multiple-bubble microstreaming mixing at 5 kHz and 40 V, mixing is completed within about 6 s (see Figure 1.29) [24].

Critical stirred volume/critical micro chamber dimensions

[M 10] [M 11] [P 9] [P 10] Bubbles with radii of 0.5 mm can stir fluid volumes in a distance with a radius smaller than 2 mm [23, 24]. Thus, any micro chamber is suitable for acoustic microstreaming with a depth < 2 mm and with an air pocket pitch < 4 mm.

Enhancement of immunomagnetic cell capture experiments – shear strain field

[M 10c] [P 9] Immunomagnetic cell capture experiments need mixing of the bacterial cell (*Escherichia coli* K12) matrix suspended in blood with magnetic capture beads, which results in highly effective immunomagnetic cell capture. Bacterial viability assay experiments demonstrated that acoustic microstreaming mixing has a relatively low shear strain field [23]. The capture efficiencies of acoustic mixing (90%, at best) were as high as for conventional vortex mixing (91%, at best). They were much larger than for non-mixed samples (4%, at best). Double staining tests with SYTO 9 green fluorescence and propidium iodide red fluorescence showed that the blood cells and bacteria remained intact after mixing.

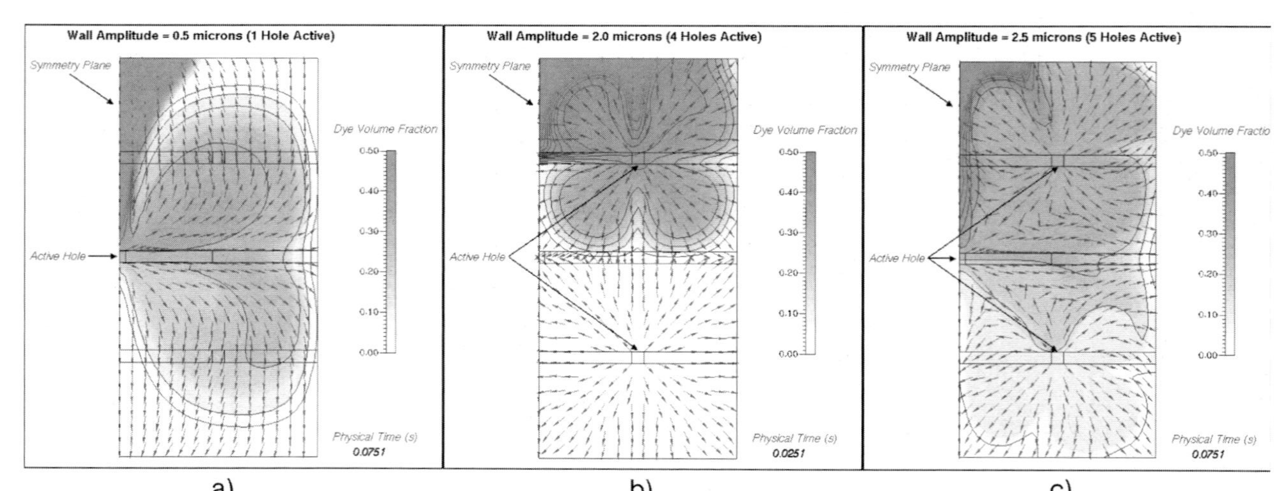

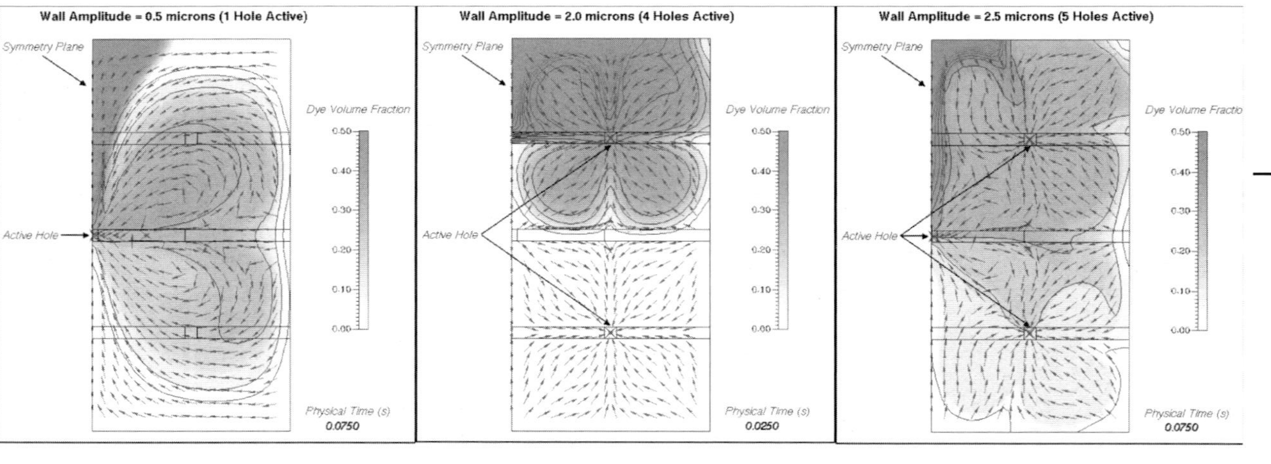

Figure 1.30 Dye penetration results for acoustic mixing by (a) one hole, one bubble, (b) four holes, four bubbles, and (c) five holes, five bubbles. Also shown is the flow-field geometry given by the velocity vectors at the inflow and outflow portions of the wall oscillation cycle [23] (by courtesy of RSC).

Enhancement of hybridization performance – interplay mixing–reaction

[M 11] [P 10] Since the acoustic microstreaming micro mixer was part of a DNA microarray chip, the enhancement of the hybridization performance by the micro mixing was investigated [24]. By acoustic microstreaming, the hybridization signal was increased by a factor of five. The signal uniformity was also improved compared with diffusion-based (2 h) chips. In addition, the kinetics were accelerated by a factor of five when acoustic microstreaming was applied. Further advantages of the use of acoustic microstreaming as the mixing mechanism are the simplicity of the apparatus, the ease of implementation, the low power consumption (~2 mW) and the low cost.

Simulation of the effect of the number of holes and their positions

[M 12] [P 11] To analyze the effect of the number of holes and their positions on the acoustic mixing, three artificial mixing micro chambers were proposed, the first with one bubble pocket in the center (a), the second with four bubble pockets at the corners and the third with five air pockets with four at the corners and one in the center [23]. It was found that acoustic mixing depends largely on the number of holes and their positions (see Figure 1.30).

The holes act as sources and sinks for fluid motion [23]. They draw the dye from in the surrounding region and expel it in a jet-like action into other parts of the liquid. These jet-like structures are mainly responsible for mixing. Where flow symmetries delineated by the formation of internal stagnation zones exist, mixing is retarded. Best mixing results are obtained for the five-bubble arrangement.

An analysis of the flow patterns created on the basis of velocity vectors shows that the dye spread along with velocity vectors twice during the wall motion cycle. with its suction and blowing stages [23].

A certain induction time was found for all three configurations [23]. This period was 10 wall motion cycles or 100 time steps.

1.2.6
Ultrasonic Mixing

Most Relevant Citations

Peer-reviewed journals: [22].

Mixing can be achieved by ultrasound using lead zirconate titanate (PZT), a piezoelectric ceramic, operated in the kHz region [22]. In this way, liquid streams can be moved and even turbulent-like eddies are induced. Favorably, ultrasonic action is coupled into a closed volume, a micro chamber. Here, the creation of standing waves has been reported.

Compared with other means of creating turbulence in closed chambers, such as mixing by valves, ultrasonic mixing is said to have advantages with regard to sensitivity towards the presence of bubbles and can freely choose the flow ratio between the two liquids to be mixed (which is not given for valve mixing) [22].

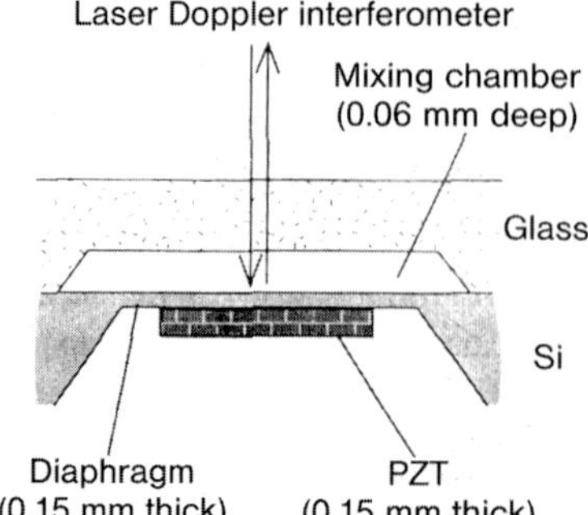

Figure 1.31 Schematic of the cross-section of the mixing chamber. An etched glass substrate is joined to a silicon wafer by anodic bonding. The silicon plate is etched from the backside to yield a diaphragm, and a PZT is attached to the oscillating diaphragm [22] (by courtesy of Elsevier Ltd.).

1.2.6.1 **Mixer 13 [M 13]: Ultrasonic Micro Mixer**

This micro device for ultrasonic mixing was made from two plates, one of which has a shallow open volume to give a micro chamber after assembly (see Figure 1.31) [22]. The area that actually covers the micro chamber is considerably thinned compared with the remaining thickness of the top plate. On this diaphragm, a thin PZT membrane is deposited, which acts as ultrasonic wave generator. The micro chamber is connected to two inlet channels and one outlet channel.

The excitation frequency was up to 60 kHz [22]. A function generator connected to a power amplifier was used to generate a square wave (50 V peak-to-peak at 60 kHz) for the PZT excitation.

The micro chamber was made by standard photolithographic methods and HF isotropic etching in the framework of bulk silicon micromachining [22]. Anodic bonding of silicon to glass was used for sealing the device. For fabrication of the diaphragm anisotropic silicon etching was used. A piece of bulk piezoelectric PZT ceramic was attached directly on the diaphragm by using an epoxy resin.

Mixer type	Ultrasonic micro mixer	PZT ceramic: width, length, thickness	5 mm, 4 mm, 150 μm
Base plate material (with mixing chamber)	Glass	Excitation frequency	60 kHz
Cover plate material (with diaphragm)	Silicon	Wave type	square; 50 V peak-to-peak
Mixing chamber: width, length, depth	6 mm, 6 mm, 60 μm		

1.2.6.2 **Mixing Characterization Protocols/Simulation**

[P 12] The mixing performance was analyzed by a dilution-type experiment. Here, water and the fluorescent dye uranine, sodium fluorescein [22]. The latter is water-soluble. External pressure was applied to the liquids using a fluid dispenser. The flow rates of both water and the aqueous uranine solution were 5 μl min^{-1}. A fluorescent microscope with a digital CCD camera with a 1.25× objective was used for optical mixing analysis. Fluorescent filters at 460–490 and 515–550 nm

for excitation and absorption, respectively, were used. The images were inverted, thereby changing the fluorescent bright zones to dark ones and vice versa with the non-fluorescent zones. The temperature was monitored with a thermo-inspector. Temperature measurements were done in a non-flow mode. A water-filled chamber was exposed for 5 min to ultrasonic action at room temperature. Diaphragm displacements were measured by using a laser Doppler interferometer.

1.2.6.3 Typical Results

Flow patterns by ultrasonic mixing

[M 13] [P 12] In the absence of ultrasonic mixing, two stable fluid regions with the separated water liquid and the uranine solution were found in the mixing chamber [22]. A straight interface was given; only a low degree of mixing was observed, being limited to the interface by slow molecular diffusion.

Upon ultrasonic action, turbulence occurred, moving material throughout the whole mixing chamber [22]. After termination of ultrasonic mixing, the initial pattern rapidly developed, the two zones being separated by a straight interface.

Mixing time

[M 13] [P 12] From video observation, a mixing time of about 2 s was judged [22]. A more exact determination was not possible, as images could be taken only every 7 s; hence it had to rely on direct microscope observation.

Fluorescence intensity monitoring

[M 13] [P 12] Fluorescence intensity measurements were made using data from the CCD camera, although these figures are not proportional to the uranine concentration, as no calibration could be performed with the existing equipment [22]. The intensities were taken along a cross-sectional line in the (left) feed side of the mixing chamber. Accordingly, the intensity plot without ultrasonic action shows two zones, one with maximal and one with zero uranine intensity signal, i.e. unmixed zones (see Figure 1.32). In between, a steep rise of concentration is

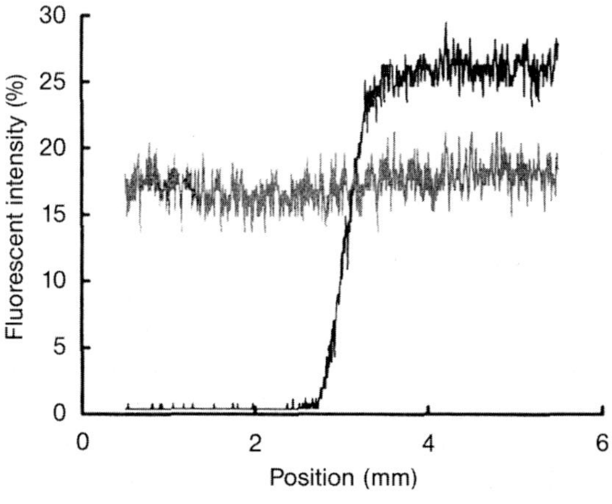

Figure 1.32 Qualitative mixing performance by plotting the non-calibrated fluorescence intensity measured near the outlet of the mixing chamber along a cross-sectional line as a function of the position on this line. The mixing performances are shown before (gray curve) and after (black curve) the ultrasonic action. Position 0 refers to the top of the mixing [22] (by courtesy of Elsevier Ltd.).

observed, giving rise to an interface, diffused over about 0.4 mm. After onset of ultrasonic mixing, a medium intensity at all positions of the measurements, i.e. also in the edges of the mixing chamber, is achieved, thus complete mixing has taken place. The average intensity now, however, is not exactly half of the former maximum intensity signal, which is explainable owing to the missing calibration.

Relationship between PZT excitation frequency and ultrasonic mixing

[M 13] [P 12] A comparison of the frequency plots of the mixing performance and the diaphragm displacement displays no similarities [22]. The relationship between mixing performance, roughly measured by the extent of the mixed area, and frequency was complex. No mixing could be observed until a frequency of 8 kHz. The mixed area became larger until about 15 kHz. From then on, the mixed area remained at the same size until ~90 kHz, then decreased. At ~130 kHz, an increase was observed again.

The mixing performance–frequency plot does not resemble the diaphragm displacement–frequency characteristics of the PZT piezoelectric ceramic [22].

Relationship between mixing performance and input power

[M 13] [P 12] The mixed area increased on increasing the input power, as evidenced by the mixing of red ink in ethanol under 50, 60, 70, and 90 V [22]. However, the mixing speed was not altered by changing the input power.

Role of diaphragm oscillation

[M 13] [P 12] The oscillation of the diaphragm has no obvious effect on the mixing and is not essential for performing ultrasonic mixing [22]. The oscillating diaphragm, however, is useful to prevent the ultrasonic irradiation escaping into other parts of the micro device and to focus it into the mixing chamber. In a control experiment without a diaphragm, no mixing could be achieved when operating the PZT piezoelectric ceramic.

1.2.7
Moving- and Oscillating-droplet Mixing by Electrowetting

Most Relevant Citations

Peer-reviewed journals: [29, 97] (see also [98] for the basic microfluidics).

Some modern microfluidic approaches rely on the movement of discrete droplets rather than handling continuously flowing streams (see e.g. [97, 98]). In this way, flexible chemical protocols can be carried out, not unlike the traditional processing of batch systems. Especially with regard to μTAS applications, the footprint area the sample volume for fluidic handling are notably decreased.

The movement of droplets is based on an electrostatic method which changes the interfacial tension of the droplets by voltage, which is known as electrowetting (see Figure 1.33) [97, 98]. The nature of the liquid to be moved has to be polarizable and/or conductive. Application of an electric field on only one side of the droplet creates an imbalance of interfacial tension which can drive bulk flow of the droplet.

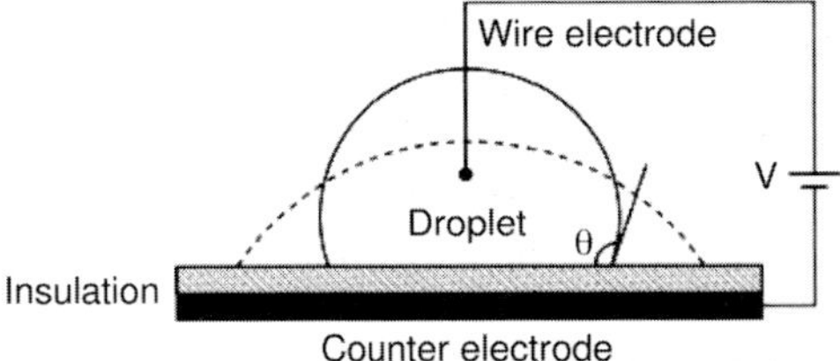

Figure 1.33 The electrowetting effect. A droplet of conducting liquid has a contact angle θ with a solid hydrophobic insulator (solid contour). The solid/liquid interfacial energy is reduced on applying a voltage V between the droplet and a counter-electrode below the insulator. This decreases θ and leads to improved wetting of the solid by the droplet (dashed contour) [98] (by courtesy of RSC).

Passive ('moving') droplet mixing uses simple droplet merging and the merged droplet rests on the electrode, relying on diffusion as the mixing mechanism [97, 98]. Active ('oscillating') bubble mixing relies on the fast movement of the merged droplet between many electrodes. In this way, convections are induced that further promote mixing [97].

1.2.7.1 Mixer 14 [M 14]: Moving- and Oscillating-droplet Micro Mixer

This device consists of two glass plates, held at a fixed distance by a spacer, which sandwich the droplets (see Figures 1.34 and 1.35) [97, 98]. The bottom glass plate carries an array of independently addressable control electrodes patterned in a thin layer of chromium. The electrodes are interdigitated for better contact with the droplets, albeit non-interdigitated electrodes were also applied. The electrodes were coated with an insulating material. The top plate was coated with a conducting, optically transparent layer to form the ground electrode. Both top and bottom glass plate were coated with hydrophobic layers.

The height of the gap between the two glass plates, i.e. the droplet height, has a considerable influence on droplet motion and mixing, as is to be expected [97].

Referring to Figure 1.34, several electrode configurations are possible owing to different arrangements of the ground and control electrodes and their combinations

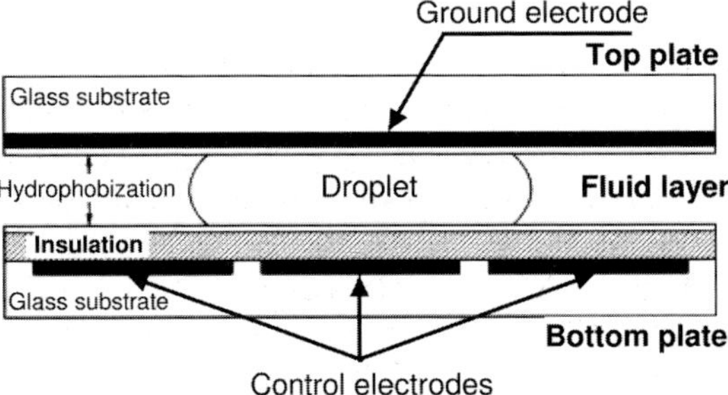

Figure 1.34 Schematic of the cross-section of the electrowetting chip. A droplet is sandwiched between two glass plates carrying electrodes [98] (by courtesy of RSC).

[98]. The ground electrode may be placed in the top glass part and an array of control electrodes in the bottom glass part. The ground electrode may be either insulated or directly in contact with the fluid, which affects its capacitance per area and thus the amount of electrostatic energy at the interface. Even configurations with no ground electrode are possible. Here, the pitch of the control electrodes may be enhanced. The ground and control electrodes may be arranged coplanar, i.e. in the same bottom plate. The ground electrode has a larger pitch than the control electrode and is positioned underneath the latter within the bottom plate, whereas the control electrode is on the surface. Finally, control and ground electrodes may be arranged as areas in the top and bottom plates.

Mixer type	Moving- and oscillating-droplet mixer	Insulator material	Parylene C
Mixer base material	Glass	Insulator layer thickness	800 nm
Electrode material	Chromium	Conducting top material	Indium tin oxide (ITO)
Electrode layer thickness	200 nm	Hydrophobic layer material	Teflon AF
Gap between electrodes (passive mixing)	800 μm	Hydrophobic layer thickness	50 nm
Gap between electrodes (active mixing)	600 μm	Spacer material	Glass

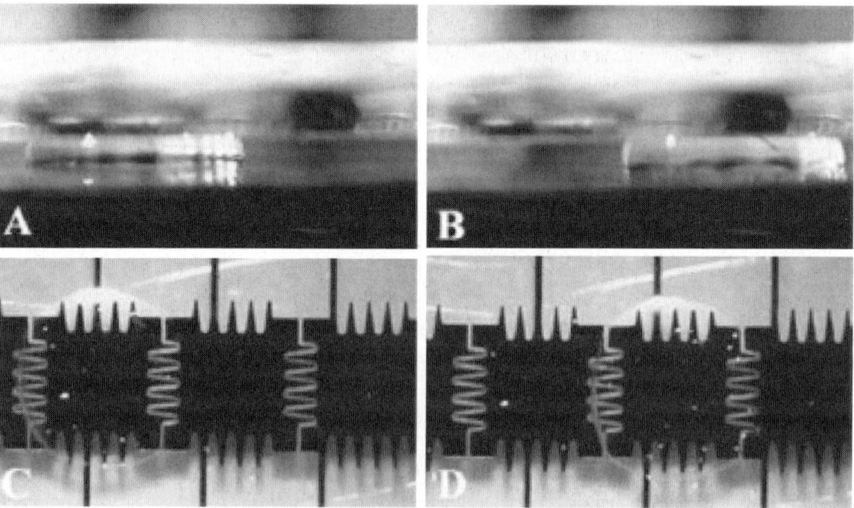

Figure 1.35 Photographs of droplet transfer. Moving droplets from the side (A), (B) and top (C), (D) at 66 ms intervals. Side-view experiments: droplet in air; top-view experiments: droplet in silicone oil [98] (by courtesy of RSC).

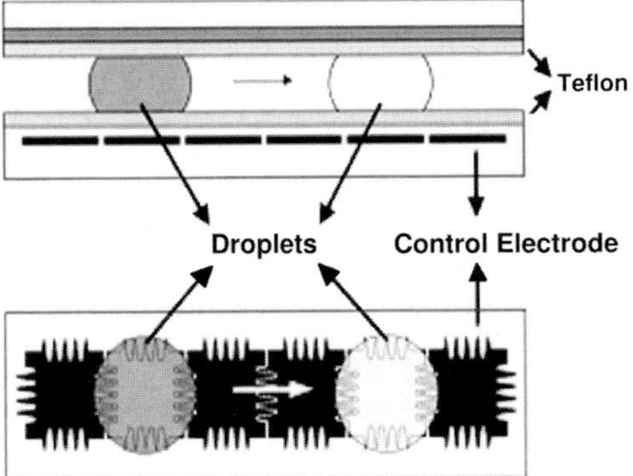

Figure 1.36 Schematic of top and side views of a central part of the electrowetting-based mixer [97] (by courtesy of RSC).

1.2.7.2 **Mixing Characterization Protocols/Simulation**

[P 13] The volume of the droplets is set slightly larger than the pitch of the electrodes (see Figure 1.36) [97]. Thereby, an overlap to adjacent electrodes is achieved. An immiscible filler fluid, 1 cSt silicone oil, surrounds the droplets to prevent evaporation and to reduce the voltage needed. The electrodes were individually switched by using a custom controller.

Flow and mixing visualization were achieved by means of fluorescence dilution imaging [97]. A fluorescent droplet (1 mM fluorescein, 0.125 M KCl, 0.125 M sodium hydroxide) was merged with a non-fluorescent droplet (0.125 M KCl, 0.125 M sodium hydroxide). The volume of the droplet in passive mixing was 1.75 μl, while 1.32 μl droplets were applied for active mixing. The actuation voltage was 30 V. The fluidic properties of the filler and droplet fluids were closely matched. The interfacial tension between the two liquids was 36 and 37 dyn cm^{-1} for the non-fluorescent and the fluorescent droplets, respectively.

Droplet visualization was achieved both from top and side views [97]. The latter is essential, since erroneous conclusions can be drawn when only inspecting from the top view owing to integration of intensity profiles along the optical axis (see Figure 1.37). In this way, non-mixed liquids can have a homogeneous fluorescence texture which could suggest mixing, but indeed only resembles a segregated structure. Details on the fluorescence set-up can be found in [97]. An image-processing toolkit in MATLAB was applied to judge mixing times and a calibration based on the intensity of the mixed state was performed.

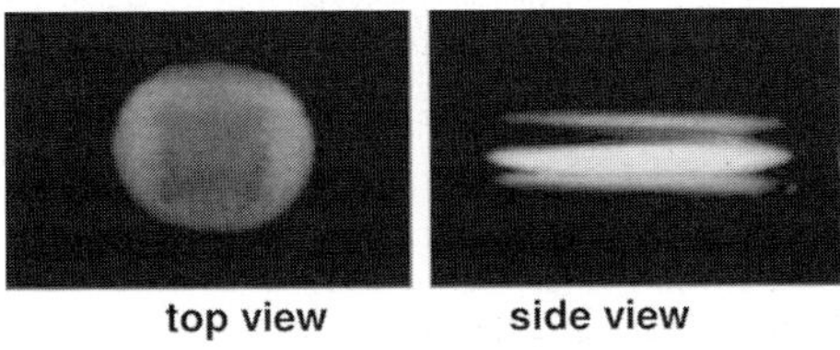

Figure 1.37 Top and side views taken 15 s after merging of two droplets, one being fluorescent and the other transparent. Having only the top view can lead to the erroneous conclusion that mixing is completed, while the side view shows that actually a segregated, layered fluid system exists [97] (by courtesy of RSC).

[P 14] Measurements of droplet displacement were made by video imaging techniques, by which the droplet velocity could also be determined. Further information concerning the experiments is given in [98].

1.2.7.3 Typical Results

Velocity profile upon droplet displacement

[M 14] [P 14] Video-image analysis of the droplet displacement shows that an S-shaped dependence over time is yielded, i.e. the initial and final velocities are low, having the highest velocity after an induction period for the moving droplet [98].

Threshold voltages for droplet displacement in air and silicone oil

[M 14] [P 14] A certain threshold voltage has to be passed before the droplet is moved [98]. This threshold is dependent on the medium surrounding the droplet. The voltage needed is higher in air than in silicon oil (air 48 V, silicone oil 13 V for a 900 nl droplet of 0.1 M aq. KCl (see Figure 1.38); further details can be found in [98]).

Scaling properties of threshold voltage

[M 14] [P 14] For electrode arrays of different pitches with fixed geometric ratios the threshold voltage and the droplet transport, i.e. the velocity-voltage function, did not change [98].

Droplet dispensing

[M 14] [P 14] Dispensing of unit-sized droplets can be achieved from a larger initial droplet or from an external source through a series of binary splitting operations [98]. Asymmetric dispensing is also possible.

Passive mixing of fluidically matched droplets

[M 14] [P 13] The fluorescent droplet is moved towards the non-fluorescent droplet and the coalesced droplet is held in place [97] (see also an initial experiment in [98]). By this action, the fluorescent droplet moves underneath the non-fluorescent one. From a top view, a homogeneous texture is yielded; however, a vertically segregated fluidic system exists (see Figure 1.37). Mixing takes then place by diffusion and needs about 90 s to be completed.

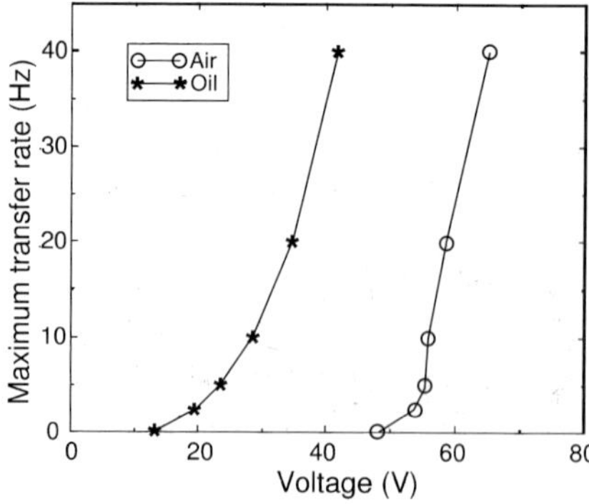

Figure 1.38 Effect of the medium on droplet transport. The threshold value for droplet movement (900 nl; 0.1 M aq. KCl) is given by the *x*-axis intercept. [97] (by courtesy of RSC).

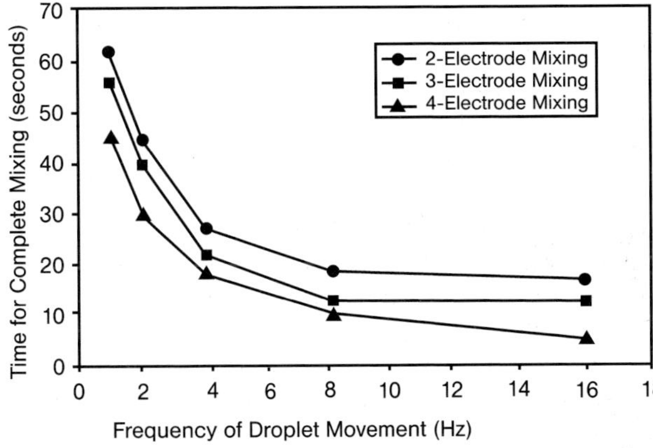

Figure 1.39 Liquid mixing time as a function of the droplet oscillation frequency, given for two-, three- and four-electrode structures [97] (by courtesy of RSC).

Passive mixing of droplets with different fluidic properties
[M 14] [P 13] On changing the fluidic properties by giving the non-fluorescent droplet a different composition (KCl solution only) than the fluorescent droplet, a donut shape is evident from the top view [97]. Now, the side view gives a homogeneous texture. The fluorescent droplet engulfs the non-fluorescent KCl droplet. A mixing time of 90 s was determined, which is equal to the matched-droplet mixing, basically because the diffusion distances are the same in both cases.

Active mixing of droplets – droplet frequency and number of electrodes
[M 14] [P 13] Mixing times were determined as a function of the frequency (1–16 Hz) of droplet movement and the number of electrodes (2, 3, 4) [97]. The mixing times decrease on increasing either the frequency or the electrode number. This is due to improve oscillation ('shaking') of the droplets inducing secondary flows (see Figure 1.39). For example, at 1 Hz and using one electrode a mixing time of about 60 s is found, whereas only 4.6 s are needed for 16 Hz and four-electrode mixing. Frequencies larger than 16 Hz give no further gain.

Active mixing of droplets – two-electrode flow imaging
[M 14] [P 13] Starting from one engulfed coalesced droplet, two-electrode active mixing leads to protrusions of fluorescent and non-fluorescent fluid compartments [97]. In this way, new interfacial areas are created and mixing is promoted. However, owing to the laminar regime, flow reversibility is observed, that is, achievements in protrusion are (partly) undone on moving the droplet back to the prior electrode position.

Active mixing of droplets – three-electrode flow imaging
[M 14] [P 13] Three-electrode active mixing creates in a similar way protrusions (see Figure 1.40) [97]. Despite the flow reversibility also observed, a better mixing performance than for the two-electrode mixing is given. This can be explained by the higher number of configurations of the droplet movement yielding more complex patterns and generating larger interfaces.

a)

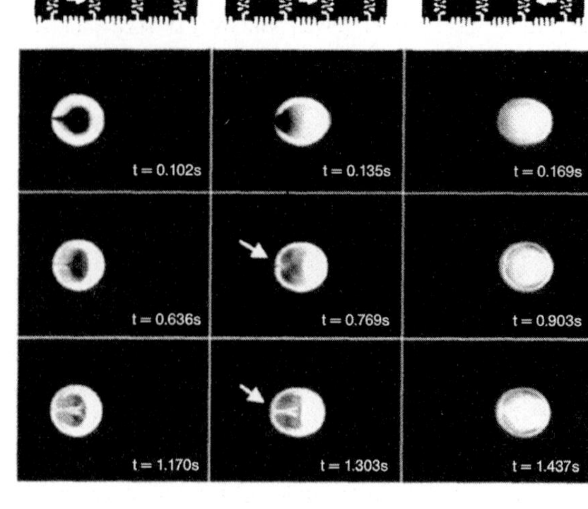

b)

Figure 1.40 Images for the active mixing of fluorescent and non-fluorescent droplets taken at various times using a three-electrode structure (8 Hz) [97] (by courtesy of RSC).

Active mixing of droplets – four-electrode flow imaging

[M 14] [P 13] Four-electrode active mixing gives more droplet configurations and achieves better mixing than two- and three-electrode mixing [97]. Flow reversibility is still present but to a reduced extent.

Effect of aspect ratio

[M 14] [P 14] The mixing in the droplets was investigated as a function of the aspect ratio, i.e. the ratio of the gap height to the electrode pitch size [29]. It was known before that this aspect ratio has a large influence on droplet formation and splitting; hence it could be anticipated that the same holds for the mixing of a merged droplet.

The flow patterns and the mixing times differ remarkably depending on the aspect ratio (see Figure 1.41) [29]. It is found that mixing times decrease from 15 to 6 s on raising the aspect ratio from 0.1 to 0.5. Thereafter, a slight increase in mixing

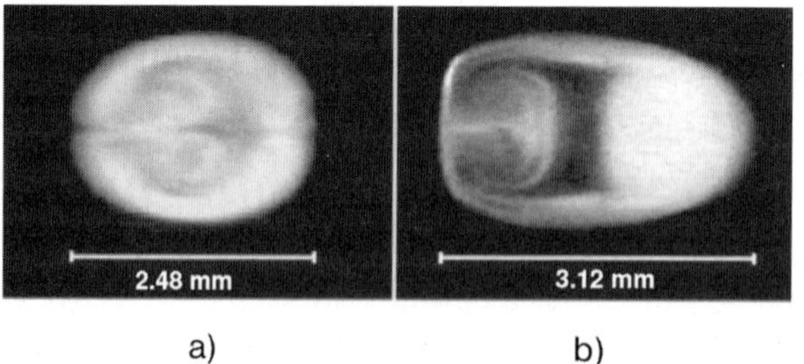

a) b)

Figure 1.41 Top view of droplets mixed by a linear four-electrode array with aspect ratio (gap height to the electrode pitch size) of (a) 0.4 and (b) 0.2 after 1.2 s (16 Hz switching frequency) [29] (by courtesy of RSC).

time is observed again. This is due to the formation of complex flow patterns at aspect ratios of 0.4–0.5, whereas only bi-layered structures are found for small aspect ratios. The latter results in smaller interfacial areas, and hence leads to longer mixing times. The increase in mixing time for aspect ratios > 0.5 is to the larger volume, i.e. the slightly less efficient generation of interfaces under such conditions.

Mixing strategies for droplets formed at low aspect ratio/split and merge mixing
[M 14] [P 14] Splitting of droplets can only occur at aspect ratios < 0.2 (for definition see *Effect of aspect ratio*) [29]. Since this is an important microfluidic action, it was worth finding suitable novel mixing strategies to overcome this limitation.

The simplest solution is to use droplet splitting and then oscillation ('shaking') followed by re-merging and second splitting [29]. This is named split-and-merge mixing. A three-electrode structure is the simplest engineering solution to achieve such multiple splitting and merging actions.

Avoidance of flow reversibility at high aspect ratio – 2 × 2 electrode array
[M 14] [P 14] It was shown above that mixing of droplets is fairly fast at an aspect ratio of about 0.4, in the order of 4–5 s [29]. It was assumed that a further reduction of mixing time is not possible, as the flow reversibility poses limits here, which is the 'relaxation' of a new flow pattern, with increased or fresh interface, to the original state.

For a such case, linear arrays of electrodes may be used; however, this will lead to extended structures when using, e.g., more than four electrodes. Instead of using such unidirectional motion, the droplet may be moved in circular fashion by a square-like 2 × 2 array of electrodes. Indeed, faster mixing times compared with simple droplet merging can be achieved, albeit not faster than for the respective four-electrode linear structure. Actually, a small portion of the droplet remains unmixed. This was explained as due to the droplet pivoting around the array center. For this reason, a non-symmetric array (2 × 3) was developed as mentioned below.

Avoidance of flow reversibility at high aspect ratio – 2 × 3 electrode array
[M 14] [P 14] In order to have the droplet turning about a moving pivot point, a 2 × 3 array of electrodes was developed [29]. An average mixing time of 6.1 s was thus achieved, which is a better performance as compared with the 2 × 2 array with a 9.95 s mixing time. The better mixing was related to additional translation between the pivot points. Indeed, the flow patterns exhibit different forms during one loop passage (see Figure 1.42). An unmixed part, as for the 2 × 2 array, is give no longer. However, the mixing performance is still worse compared with the four-electrode linear array.

Combining the positive aspects of linear and circular arrays – 2 × 4 electrode array
[M 14] [P 14] In order to combine the effect of efficient oscillation of the extended linear arrays and the avoidance of flow reversibility of the circular arrays, an extended circular array (2 × 4) of electrodes was developed [29]. This has three possible pivot points with up to two translations. This enhances the number of possible paths for the droplet which may result in enhanced mixing.

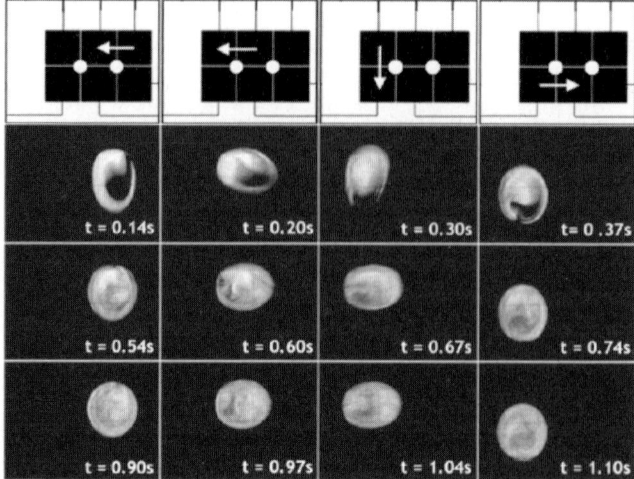

Figure 1.42 Schematic of the pivot point around which the droplet moves in a 2 × 3 mixer array (top row). Images of droplet mixing in a 2 × 3 array taken at various times (16 Hz switching frequency) [29] (by courtesy of RSC).

A mixing time of only 2.9 s was derived [29]. More complex flow patterns were determined than for all the arrays investigated so far.

Pseudo 3 × 3 electrode array

[M 14] [P 14] In order to complete the possible design variations of circular arrays, a pseudo 3 × 3 array of electrodes was developed [29]. This allows quasi-unidirectional movement of droplets. Each movement occurs about a different pivot point. The mixing time was slower as compared with the 2 × 4 array, having the same total number of electrodes.

Overall comparison of all electrode arrays

[M 14] [P 14] To have an overall comparison of the results given in all the subsections above, a comparison of the mixing time for moving droplets on the different arrays were made (see Table 1.1). The mixing times are roughly halved on adding a second dimension, i.e. additional pivot points (e.g. on going from a 1 × 2 to a 2 × 2 array) [29]. The mixing times are also reduced by having more electrodes, i.e. additional translation steps (e.g. on going from a 1 × 2 to a 1 × 3 array). The best result, i.e. the lowest mixing time, is given by the 2 × 4 array mixer.

Table 1.1 Mixing times for liquid droplet mixing in various 1 × *N* and 2 × *N* array mixers at 16 Hz [29].

Mixer array	1 × 2	2 × 2	1 × 3	2 × 3	1 × 4	2 × 4
Mixing time (s)	16.8	10.0	12.1	6.1	4.6	2.9

1.2.8 Moving- and Oscillating-droplet Mixing by Dielectrophoresis

Most Relevant Citations

Peer-reviewed journals: [99–101]; proceedings contributions: [101].

In addition to using electrowetting (see Figure 1.42), electrophoretic and dielectrophoretic forces can be used for moving droplets [99]. Electrophoretic droplet movement depends on the application of large DC fields, which may pose problems for fluid systems such as suspensions. For dielectrophoretic operation, however, AC fields are sufficient, as for electrowetting.

Electrowetting forces depend largely on the cleanness of the substrates, since even small traces of impurities may notably change the wetting behavior [99]. This certainly can be controlled under laboratory conditions, but may experience limitations under 'dirty' real-world applications. Further limitations on the droplet volume are given by the geometric constraints of the electrode chamber which needs to be wetted at the floor and ceiling.

Dielectrophoretic forces act as body forces on the droplet, rather than being surface forces as in the case of electrowetting [99]. By proper design of electric field geometries and using non-wetting surfaces, the dielectrophoretic forces can be set into action. This allows two-dimensional translation of the droplets and droplet injection over a large range of volume scales for reagent titration.

Dielectrophoretic forces depend on the polarizibility of species, rather than on movement of charges [99]. This allows the movement of any type of droplet being immersed by a dielectrically distinct immiscible carrier medium. Since dielectric forces are generated by spatially inhomogeneous fields, no mechanical actuation is required. In addition to this, dielectrophoretic droplet movement benefits from the general advantages given by droplet microfluidic, i.e. discrete, well-known very small volumes, no need for channels, avoidance of dead volumes and more.

1.2.8.1 Mixer 15 [M 15]: Dielectrophoretic Droplet Micro Mixer

This dielectrophoretic droplet mixer, called a programmable fluidic processor, contains two rows of 32 pads for the electrodes; these rows being at the upper and lower edges of the substrate and connected to the 8 × 8 electrode array in the center of the chip [99]. The electrodes form a square matrix with an upper and lower half owing to the connectivity to the pads. The electrodes have a square shape.

Mixer type	Dielectrophoretic droplet micro mixer	Array format	8 × 8
Substrate material	Glass	Electrode width, length	30 μm, 30 μm; 100 μm, 100 μm; 300 μm, 300 μm
Electrode material	Gold on titanium	Passivation layer material and thickness	Fluoro-Pel®, ~0.5 μm
Total chip size	40 mm × 40 mm		

A two-dimensional array was patterned by standard photolithographic techniques on a substrate [99]. An open fluid reservoir was achieved simply by sealing the internal part of the device, including the array, with an O-ring. This open-face arrangement was for laboratory experiments, while the device has to be sealed for later practical uses. A hydrophobic coating served for electrical insulation. The device was connected to a computer-controlled switching circuit.

1.2.8.2 Mixer 16 [M 16]: Electrical Phase-array Panel Micro Mixer

From this device, several sub-versions were made. These so-called electrical panel devices were realized as three- and six-phase array devices and a nine (3 × 3) electrode dot device [100, 101]. The three- and six-phase array devices consist of a substrate with electrode arrays of a paste printed on using a silkscreen printmaking process. This substrate is covered with an insulating covering film.

Mixer type	Electrical phase-array panel device	Multi pitch length	0.5 mm, 0.75 mm, 1.0 mm, 2.0 mm
Substrate material	Polyester	Multi pitch width	0.2 mm
Substrate thickness	300 μm	Insulating covering film material	Teflon or polypropylene
Electrode material	Silver	Insulating covering film thickness	90 μm

1.2.8.3 Mixer 17 [M 17]: Electrical Dot-array Micro Mixer

This nine (3 × 3) electrode dot device comprises a four-layer printed circuit board [100, 101]. The first layer has a nine-phase electrode dot matrix. The entire electrode dot matrix consists of multiple 3 × 3 units. Each electrode has a hole in its center for connection to the electric circuits of the layers below. These layers were connected to external terminals. Sequential voltages were applied to the three-phase electrode columns and lines.

Mixer type	Electrical dot-array panel device	Electrode unit in the array	3 × 3
Layer material	Glass reinforced epoxy	Total electrode area	150 × 150 mm
Average pitch and width of first layer	1.0 mm, 0.5 mm	Insulating covering film thickness	90 μm

1.2.8.4 Mixing Characterization Protocols/Simulation

[P 15] The reservoir of the dielectrophoretic droplet mixer was filled with 1-bromododecane, a low-permittivity, low-viscosity, water-immiscible hydrocarbon [99]. AC signals of up to 180 V_{p-p} and frequencies between 5 and 500 kHz were formed using a programmable function generator and amplified by variable gain amplifiers.

The AC signal was distributed to an array of solid-state switches, operating under a microcontroller that was driven by a Lab software program. Images of the moved droplets were taken using a microscope and a CCD camera. A trans-illuminating quartz–halogen lamp was used for most viewing and a xenon lamp for fluorescence studies.

In addition to point-to-point movement of droplets, the dielectrophoretic droplet mixer could form droplets at pressurized orifices near electrodes [99]. In the latter, the fluid is normally kept by a 'hydrostatic holdoff', which is then overcome by the dielectrophoretic action.

The droplets were moved by switching electrodes on and off, creating a minimum of the electric field energy and exerting a lateral force on the droplet [99].

[P 16] Sequential voltages with six-phase rectangular profiles were applied [100, 101]. Following some first evaluations, it turned out that the sequence [+++ –] was the most efficient means for droplet transport. Most smooth droplet transports were found for voltages > 350 V. The upper limit was, however, fixed at 400 V, since otherwise cross-electrode discharge may lead to damage.

As surrounding liquids oils or a specially liquid, fluorinert, can be used [100]. For the actual experiments, a vegetable oil was chosen. It has a density lower than most aqueous solutions so that such droplets sink in the oil medium towards the panel surface. The droplets were applied using a normal pipet. Such prepared droplets are almost spherical on the polypropylene film. Often superhydrophobic materials were used as substrates.

1.2.8.5 **Typical Results**

Controlled droplet motion

[M 15] [P 15] In order to demonstrate the basic functionality of the dielectrophoretic fluid processor, some complex droplet motions were made (see Figure 1.43) [99]. As an example, one 0.18 nl and two 0.065 nl droplets were guided over an 8 × 8 array (30 μm electrodes) and one after the other combined with a 0.38 nl droplet. Spontaneous fusion was observed. In this way, a 0.69 nl droplet resulted, containing all the initial volumes. This droplet was finally moved over the array. A speed of > 1.8 mm s^{-1} was achieved.

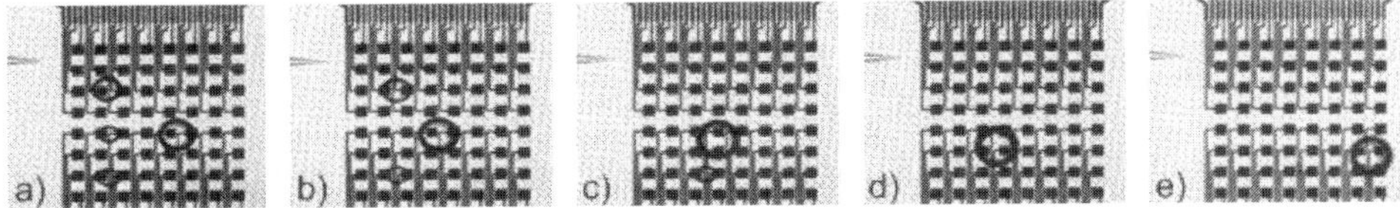

Figure 1.43 Multiple droplet motion and fusion on an 8 × 8 array fluid processor with 30 μm electrodes. (a) Four droplets of varying size, three on one line and one separate large droplet; (b) spontaneous fusion of the middle droplet on the line with the large droplet; (c) spontaneous fusion of the upper droplet on the line with the large droplet; (d) spontaneous fusion of the lower droplet on the line with the large droplet; (e) movement of the remaining large droplet to the edge of the array [99] (by courtesy of RSC).

Non-mechanical droplet metering

[M 15] [P 15] Discrete aliquots of the liquid were injected onto a reaction surface by means of dielectrophoretic forces generated by a fluid processor [99]. Micropipet injectors were immersed in the fluid reservoir and placed close to the electrodes. A vertical distance of 10–20 μm was maintained. By means of a micromanipulator, the lateral distance was adjusted. The pressure was controlled by a syringe pump.

The droplet injection was initiated by energizing the electrode nearest the injector [99]. By dielectrophoretic forces, fluid is drawn out of the pipet and forms a droplet, which then moves to the edge of the electrode. The dielectrophoretic forces have to be balanced properly with the interfacial tension forces to give uniformly sized droplets. 18 μm (3 pl) sized droplets; for example, were made in that way.

Droplet fusion and reaction

[M 15] [P 15] Two aqueous droplets in a liquid hydrocarbon were placed on two adjacent electrodes (30 μm) with a 30 μm distance [99]. By activation of intervening electrodes the droplets were brought together and gave spontaneous fusion. Fast mixing of the droplet contents followed this merging; using the reaction between fluorescein in dilute HCl solution and NaOH solution, the change in fluorescence intensity could be taken as rough measure of the mixing speed. In this way, the speed of the diffusion front was estimated to be about 170 μm s^{-1}.

Droplet-based fluorescence assay

[M 15] [P 15] As some sort of feasibility test for biochemical analysis, nine droplets of *o*-phthalaldehyde were placed using a micropipet on an electrode of an 8 × 8 array [99]. Varying quantities of a 0.5 g l^{-1} solution of bovine serum albumin were the metered towards these droplets from a second injector, the newly introduced droplets being 0.18 nl in volume (~70 μm). Such produced fused droplets had protein concentrations in the range 0.266–232 μg ml^{-1}. This corresponded well with the varying fluorescence intensities of the droplet observed at the maximum wavelength of the fluorescence spectra of the protein-bound *o*-phthalaldehyde, i.e. demonstrating that the reaction was achieved.

Droplet motion in electrode array devices with sequential voltage

[M 16] [P 16] Droplet transport could be achieved for frequencies of the sequential voltages in the range 0.5–3.0 Hz [100, 101]. An increase in the ratio of electrode width to pitch facilitated the droplet transport. Since only perpendicular transport can also be achieved, fluid guiding is necessary. This can be accomplished, e.g., by use of thin polymer films.

Droplet motion in electrode dot devices with sequential voltage

[M 17] [P 16] Using electrode dot devices, each droplet can be moved independently [100, 101]. This was evidenced by two droplets next to each other, where only one electrode was activated. Only the respective droplet moved and the other stayed where it was.

Chemical reaction in electrode dot devices with sequential voltage

[M 17] [P 16] A pH-indicator reaction was performed by merging droplets containing phenolphthalein and aqueous NaOH solution on an electrode dot device [100, 101]. The reaction immediately followed the merging and mixing of the droplet, as evidenced by the respective color change.

Biochemical reaction in six-phase electrode array devices with sequential voltage

[M 16] [P 16] The same findings as for the chemical reaction (see *Chemical reaction in electrode dot devices with sequential voltage*) were made for the luciferin–luciferase enzyme reaction with adenosine triphosphate on the six-phase electrode array device [100, 101]. The reaction rapidly followed the mixing, as evident from the luciferin luminescence of the droplet after merging.

Problems during chemical reaction in electrode array devices with sequential voltage

[M 16] [P 16] For reasons of depletion of reactants during a chemical reaction, as given, e.g., for the phenolphthalein pH change (see above), the surface tension of the respective droplets may change considerably [100, 101]. In the case of large voltages, this may lead to deformation of the droplets and finally to break-up of the droplet. Many small droplets were generated.

1.2.9 Bulge Mixing on Structured Surface Microchip

Most Relevant Citations

Peer-reviewed journals: [102].

In addition to using electrowetting effects for droplet movement and merging, surface-guided flow due to selective wettability can do a similar job [102]. Structured surfaces can form so-called 'liquid micro channels' without having any wall, besides the surface structure. These continuous, cylinder-like liquids can undergo fragmentation above a certain critical liquid load to bulges, structures similar to extended droplets. By having a proper geometry of the surface structures, the position of the bulges can be fixed and by approaching such structures close together, mixing between selected solutions can be performed at a selected place.

1.2.9.1 Mixer 18 [M 18]: Structured Surface Bulge Micro Mixer

Mixer type	Structured surface bulge micro mixer	MgF_2 layer thickness	20 nm
Substrate material	Hydrophobic silicone rubber or thiolated gold substrate	Surface structure width	About 30–40 μm *
Wettable surface structure	MgF_2	Surface structure length	Several hundred μm *

* Parameters are not specified in [102]; data are judged from scanning electron micrographs presented.

This device comprises a hydrophobic surface using the wettability pattern of hydrophilic stripes for a surface-guided flow, 'micro channels' [102]. As substrates, hydrophobic materials were employed. The hydrophilic stripes were generated by mask-through thermal vapor deposition of MgF_2 on a silicone rubber or thiolated gold substrate.

1.2.9.2 Mixing Characterization Protocols/Simulation

[P 17] The substrate was exposed to an atmosphere with 40% humidity at room temperature. By cooling to 5 °C the dew point was reached and water condensed on the wetting hydrophilic stripes. Such covered hydrophilic stripes can be interpreted as liquid micro channels with a channel wall in only one dimension [102].

1.2.9.3 Typical Results

Filling of the stripes/fluid instability and droplet break-up at large coverage

[M 18] [P 17] By depositing a small amount of water on the hydrophilic stripes the channels are shaped homogeneously as cylinder segments. The real shape of the cylinders depends on the contact angle. If subsequently more and more water is deposited on the stripes, the volume of the cylinders increases until suddenly a single bulge with a characteristic shape is formed (see Figure 1.44).

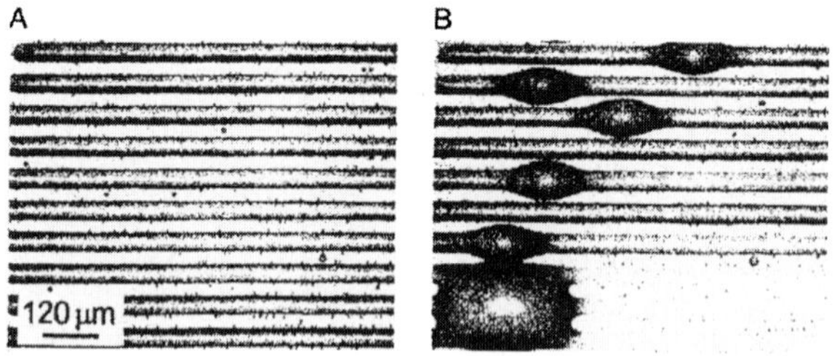

Figure 1.44 Stripes covered with water. (A) Low water coverage with constant cross section and small contact angle; (B) high water coverage with partially bulge formation and merging of two bulges of two neighboring channels [102] (by courtesy of AAAS).

Theoretical analysis of the fluid instability

[M 18] [P 17] In theoretical work, the surface channel instability was investigated [102]. It could be shown that the surface channel instability is different from the Rayleigh-plateau instability (compare [72], Chapter 3), which describes the decay of a fluid cylinder into a periodically droplet array. In contrast to the Rayleigh-plateau instability, the fluid cylinders attached to a surface form a single bulge instead of decaying into droplets. A detailed theoretical description of the experimental findings is given [102].

Bulge mixing at corners

[M 18] [P 17] The wettable pattern can be used as micro-fluidic chips or micro reactors. Pairs or multiples of different hydrophobic stripes can be filled, e.g., with

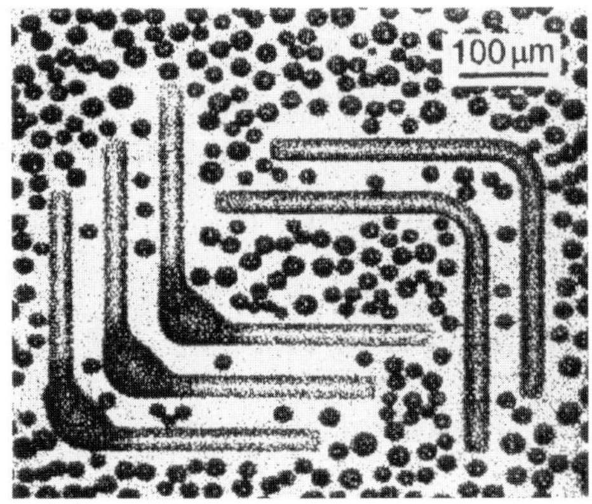

Figure 1.45 Surface micro channel domains with corners. The bulge prefers to sit at the corner because of maximum contact with the hydrophilic stripe. If the corners are close enough, two or more adjacent bulges will coalesce and form a fluidic bridge [102] (by courtesy of AAAS).

different reactants. By overloading with liquids, the bulges formed can coalesce and form bridges between two or more of these surface channels. By merging the bulges, mixing of the fluids is easily performed. This bridging position happens by chance, so to control the bridging position on a chip, the shape of the stripes used either changing the width or forming corners into the otherwise straight surface channels (see Figure 1.45) [102].

Solidification of the fluidic network

[M 18] [P 17] If the filled liquid channels and bridges can be fixed by, e.g., freezing, polymerization or sol–gel reactions, a mold insert could be formed which would allow one to transfer the fluidic network structure as 'real' channels into plastic or other materials [102].

1.2.10
Valveless Micropumping Mixing

Most Relevant Citations

Peer-reviewed journals: [103].

A powerful pumping principle is represented by PZT (Pb–Zr–Ti)-driven valveless micropumps which by operation at high frequency (kHz range) is said to be able to induce turbulence locally [103]. When using two feed inlets into a pumping chamber and one outlet, the pumping action can be transferred into mixing action owing to the turbulence achieved by the heavy flow motion.

1.2.10.1 Mixer 19 [M 19]: Valveless Micropumping Micro Mixer

This valveless micropump comprises a pair of diffuser elements and a chamber with oscillating diaphragm (see Figure 1.46) [103]. The two diffusers are located at the side of the chamber. By variation of the chamber volume owing to diaphragm displacement, fluid motion is generated. Flow resistance of the diffusers is directionally varying so that directed flow guidance is given. During pumping, work turbulence is induced.

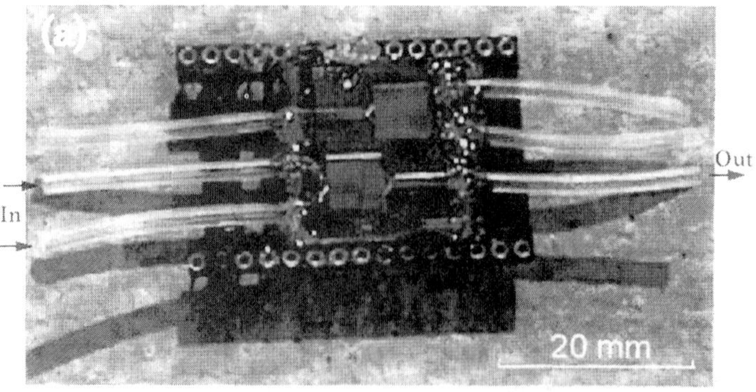

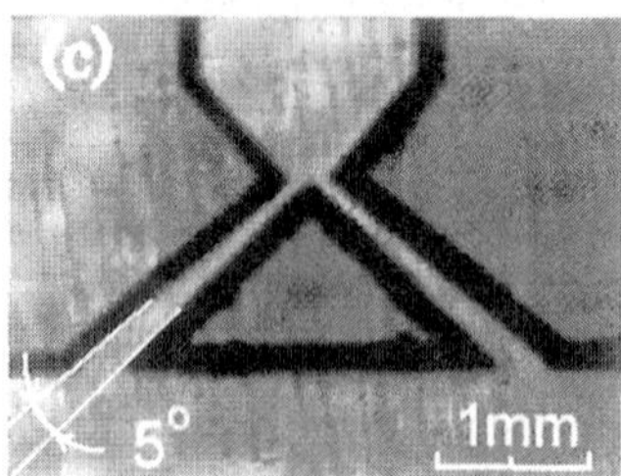

Figure 1.46 Valveless micropumping mixer with two mixing chambers. (a) Photograph of the total device; (b) details of the diffusers at the inlet; (c) details of the diffusers at the outlet [103] (by courtesy of Kluwer Academic Publishers).

HF isotropic etching of glass was used for making the diffuser and mixing-chamber structures [103]. The diaphragm was etched into silicon by standard silicon micromachining. Anodic bonding was used for sealing of these two structured plates. A PZT piece was glued on the diaphragm.

Mixer type	Valveless micro-pumping micro mixer	Diaphragm length, width, thickness	6 mm, 6 mm, 0.15 mm
Mixing chamber and diffuser material	Pyrex glass	Mixing chamber length, width, depth	6 mm, 6 mm, 0.04 mm
Diaphragm material	Silicon	Mixing chamber length, width, thickness	6 mm, 6 mm, 0.2 mm
Diffuser length, angle	1.5 mm, 5°	Total device size	20 mm × 20 mm

1.2.10.2 **Mixing Characterization Protocols/Simulation**

[P 18] A square-wave voltage (120 V peak-to-peak) was applied for excitation of the diaphragm to change the volume of the mixing chamber [103]. The pumping mixer was operated with water, methanol and ethanol. The flow rates were measured with a 70 μm glass micropipet and the pressures were determined using a miniature pressure meter.

1.2.10.3 Typical Results

Pumping performance

[M 19] [P 18] Pumping action up to 10 kHz was achieved for excitation by square-wave voltage [103]. Pumping ethanol, 15.6 μl min^{-1} as the maximum flow rate at a zero flow pressure of 2.16 kPa was achieved (870 Hz square wave). Operation was also achieved with sinusoidal and triangular excitation.

Mixing performance

[M 19] [P 18] For characterization of mixing performance, colored solutions were generated by dissolving various pigments in water, methanol and ethanol [103]. The formation of a homogeneous color was followed on mixing each of these solutions with the corresponding pure uncolored liquids. It was concluded that mixing is efficient, although no details on the mixing time and flow patterns created are given.

1.2.11 Membrane-actuated Micropumping Mixing

Most Relevant Citations

Proceedings contributions: [30].

1.2.11.1 Mixer 20 [M 20]: Membrane-actuated Micropumping Micro Mixer

This device (see Figure 1.47) achieves active mixing by fast periodical changes of a mixing-chamber volume via actuation of a thin membrane by a piezo disk [30]. Similar to the pumping action of membrane pumps, based on the same principle, fluid motion is achieved in this way. The design thus comprises two layers with the bottom part containing the inlet and outlet ports and with the top part forming the mixing chamber. The top side of the mixing chamber is the thin membrane, above which the piezo disk is located. The outlet port is placed in the center of the mixing chamber, while the two inlet ports are on opposite sides close to the channel wall.

The chip carrier contains electrical contacts for wire bonding of the piezo disk [30].

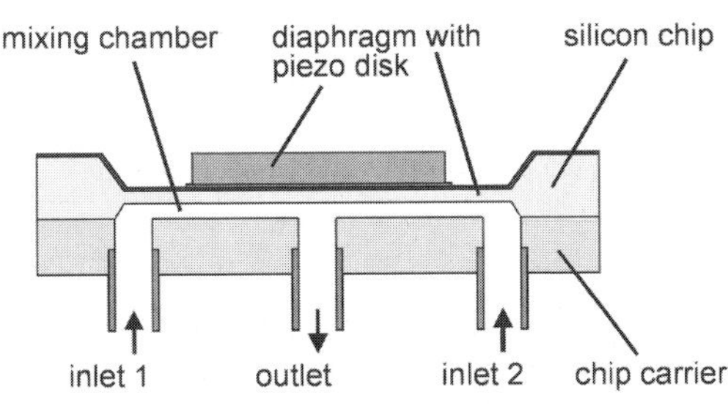

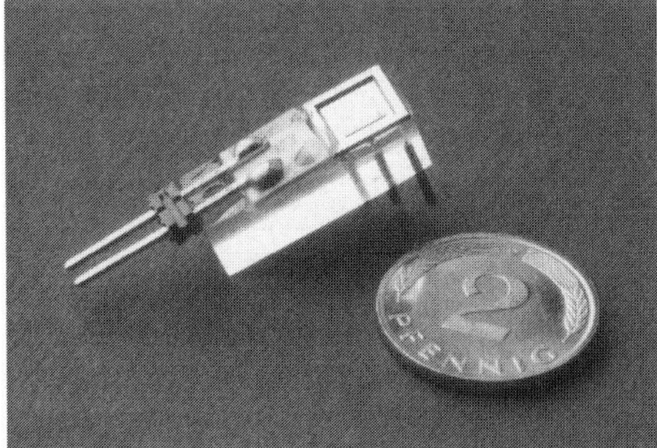

Figure 1.47 Schematic design of the membrane-actuated micropumping mixer (left) and photograph of the device (right) [30] (by courtesy of Kluwer Academic Publishers).

Standard silicon micromachining was applied for device manufacture using wet-chemical etching by KOH [30].

Mixer type	Membrane-actuated micropumping micro mixer	Membrane thickness	40 μm
Membrane and mixer material	Silicon	Mixing chamber length, width, depth	5.1 mm, 5.1 mm, 15 μm
Mixer top side layer material	Alumina	Chip carrier material	PPMA; PEEK
Total membrane area	5.1 mm × 5.1 mm	Total device size	7 mm × 7 mm

1.2.11.2 Mixing Characterization Protocols/Simulation

[P 19] Dilution-type mixing monitoring of a blue-dyed aqueous solution and water was carried out using video taping [30].

For testing the μTAS module, colorimetric pH detection with bromothymol blue indicator was used [30]. A stock solution was prepared at a concentration of 3 mg ml^{-1} in ethanol. Thereafter, this stock solution was diluted with water to 40 μg ml^{-1}. Later a water-soluble salt was employed. Detection was carried out at 610 nm, close to the maximum absorption of the dye.

1.2.11.3 Typical Results

Passive mode of the mixer

[M 20] [P 19] An investigation of the flow patterns in the mixer outlet was made by dilution-type dye imaging [30]. On contacting the two streams only in the mixer devices without any membrane actuation (static or passive case), a bi-layered system results with no obvious degree of mixing, as is to be expected.

Active mode with single downward stroke

[M 20] [P 19] An active single downward stroke yields a parabolic shock wave traveling along the mixer outlet and disturbing the laminar pattern, thus promoting mixing [30]. Maximal action is achieved when placing the outlet in the center of the mixing chamber, thus being below the maximal deflection of the membrane.

Active mode with many oscillating strokes

[M 20] [P 19] Setting multiple downward strokes yields superposition of many parabolic shock waves traveling along the mixer outlet [30]. In this way, a much more effective disturbance of the laminar pattern is given than for a single stroke only. Using a dilution-type mixing experiment, high dispersion of the dye is evident, which is indicative of a high mixing efficiency.

Amplitude and frequency

[M 20] [P 19] By colorimetric measurement the impact of the frequency and amplitude of actuation on the mixing efficiency was characterized (100 Hz;

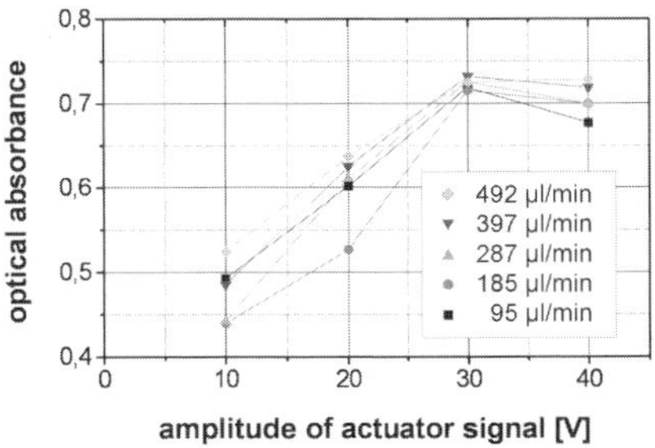

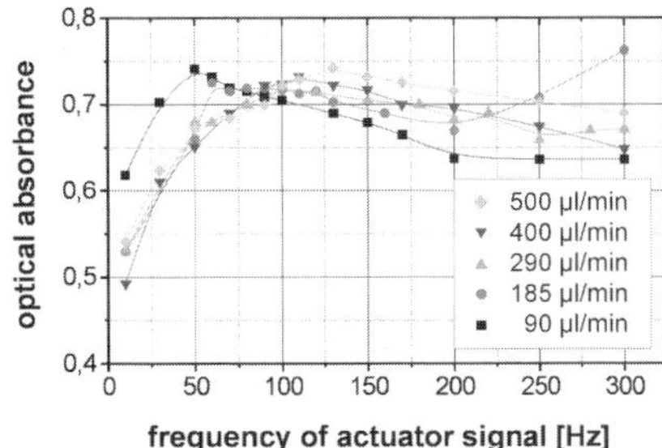

Figure 1.48 Mixing efficiency, determined by photometry, as a function of the amplitude (left) and frequency (right) of the actuating membrane [30] (by courtesy of Kluwer Academic Publishers).

95–492 µl min^{-1}) (see Figure 1.48) [30]. An optimum signal amplitude between 30 and 40 V was found corresponding to a total membrane stroke of 6–7 µm. For lower voltages, mixing was incomplete. At higher amplitude, no further improvement was noted. This is probably due to a too large deflection of the membrane touching the bottom of the mixing chamber and actually closing the outlet channel, hence massively disrupting the mixing process.

At most measured flow rates, mixing performance is nearby constant over the frequency of the actuator signal, actually giving a slight maximum (30 V) [30]. The differences between curves taken at various flow rates are due to flow-dependent variation of the perturbation zone distances in the mixer outlet.

Functioning within a µTAS module for colorimetric analysis

[M 20] [P 19] The µTAS module is made for performing colorimetric analyses as typically applied in cuvette tests, e.g. for on-site water analysis [30]. A continuous test replaces copious manual pipetting of the sampling volumes. The module consists of a micro flow restrictor, a micro mixer and an optical microcuvette for colorimetric analysis. The sample is injected by a conventional FIA (flow injection analysis) system. By close connection, a dead volume of only 2.2 µl is given.

Owing to the very small mixing chamber, fast changes by mixing can be induced, resulting in a fast response of the µTAS module, i.e. a large test-throughput frequency may potentially be achieved [30]. Several cycles were reported between baseline and active mixing measurements. The original state was reached perfectly after each measurement, which means, e.g., that no sample cross-contamination occurred. The results also prove that complete mixing is ascertained. In this way, a full pH calibration curve was repeatedly obtained.

1.2.12
Micro Impeller Mixing

Most Relevant Citations

Proceedings contributions: [32].

Impellers of all types dominate conventional mixing; numerous textbooks are available in this field (see e.g. [104, 105]). This is typically carried out in a batchwise manner. Since in many μTAS applications liquids in mixing chambers of comparatively extended volumes have to be mixed, it is not too far-fetched to develop miniaturized stirrers [32]. Claimed advantages of micro impellers are the possibility of matching the impeller diameter to the mixing volume, of performing large-area mixing, of effecting mixing on-demand (switch on/off) and the flexibility of the mixing approach, e.g. concerning the choice of liquids.

1.2.12.1 **Mixer 21 [M 21]: Impeller Micro Mixer**

This device contains micro impellers several tens of microns in diameter consisting of a cap, hub and two rotary blades which were made by electroplating and thin-film technology from the ferromagnetic material Permalloy (see Figure 1.49) [32]. These mixers are built in molded PDMS structures containing recesses as mixing chambers. Upon actuation with a standard magnetic stirrer, rotational speeds up to 600 rpm can be reached.

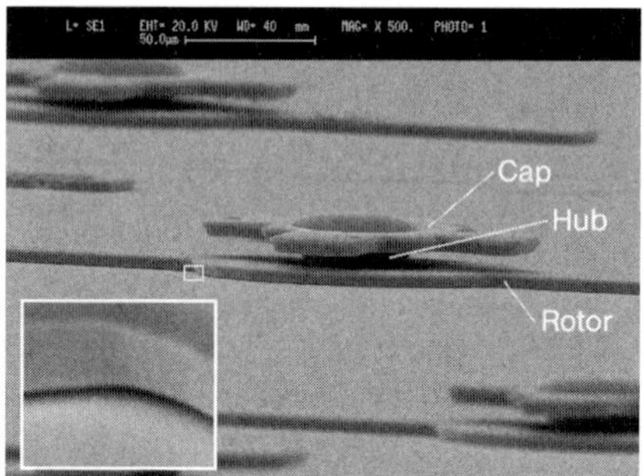

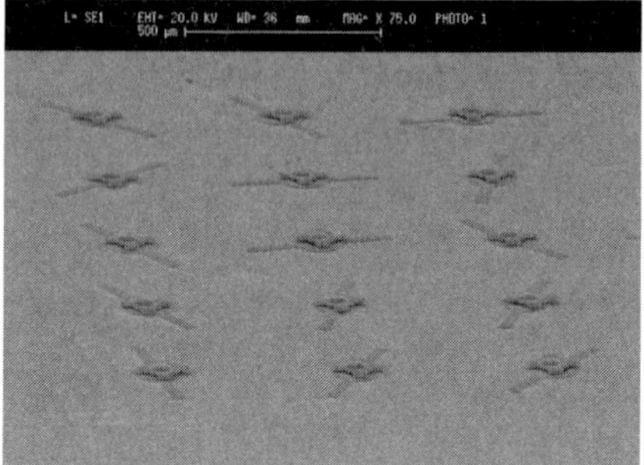

Figure 1.49 Scanning electron micrograph of a 5 × 3 micro-impeller array [32] (by courtesy of Kluwer Academic Publishers).

Mixer type	Impeller micro mixer	Rotational impeller speed	Up to 600 rpm
Mixer material	Permalloy	Exemplary chamber diameter, depth	2.5 mm, 40 μm
Impeller diameter	About 100 μm		

1.2.12.2 **Mixer 22 [M 22]: Ferromagnetic Sphere-chain Micro Mixer**

This device forms impeller-like structures by the assembly of several small objects such as spheres, instead of making an artificial micro impeller by micromachining [104, 105]. This has the advantage that comparatively cheap items can be used for building up the impeller. One way to do so is to use ferromagnetic spheres which rest in a mixing chamber. In the presence of a magnetic field, the spheres attract each other and form a chain. If a rotating magnetic field is applied the chain rotates, essentially like an impeller, and stirs the liquid in the chamber. The diameter of the spheres is chosen so that it is slightly smaller than the depth of the mixing chamber; the number of spheres is given by the simple geometric constraint that the length of the chain has to be slightly shorter than the chamber diameter (for circular chambers) or width (when being rectangular or square). The spheres are retained by a kind of frit structure at the chamber end. The spheres can be introduced before sealing of the mixing chamber or after by suspending them via the feed channels.

1.2.12.3 Mixing Characterization Protocols/Simulation

[P 20] Flow visualization was achieved by a dilution-type experiment [32].

A simulation model was used based on CFD-ACE [32].

1.2.12.4 Typical Results

Scanometric color index profiles at T-junction

[M 21] [P 20] Mixing near a T-junction (channel 750 μm wide and 40 μm deep) is complete at a 0.17 μl min^{-1} flow rate when using a micro impeller at 120 rpm, as demonstrated by scanometric color index profiles providing line concentration profiles [32].

Flow visualization in mixing chamber

[M 21] [P 20] Mixing of an extended circular mixing chamber (diameter 2.5 mm, depth 40 μm) is complete after 55 s when using a 3 × 3 array of micro impellers at 600 rpm, as demonstrated by dye flow visualization and color histograms [32].

Simulation of mixing compared to experimental results in an impeller array

[M 21] [P 20] CFD simulation results for micro impeller mixing in an extended circular mixing chamber (width 2 mm, length 2 mm, depth 50 μm) is in line with experimental results yielded under similar condition. Mixing is complete within less than 1 min when using a 3 × 3 array of micro impellers at 600 rpm [32].

Liquid mixing time

[M 21] [P 20] A liquid mixing time of 55 s was determined for mixing a volume of an extended circular mixing chamber (diameter 2.5 mm and depth 40 μm; impeller at 600 rpm) (see Figure 1.50) [32].

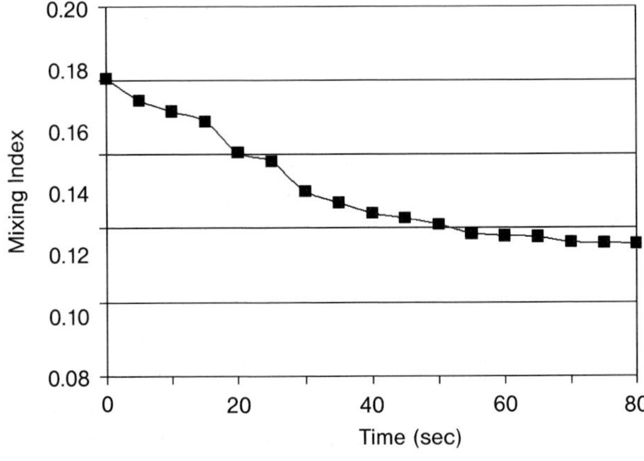

Figure 1.50 Liquid mixing time of micro-impeller mixing in a mixing chamber derived by plotting a mixing index yield by photometry versus time [32] (by courtesy of Kluwer Academic Publishers).

1.2.13 Magnetic Micro-bead Mixing

Most Relevant Citations

Proceedings contributions: [106]; patents: [107].

An integrated micro device containing a mixer was developed for biochemical experiments [106, 107]. For amplification reactions the handling of very small volumes (< 10 μl) of DNA and RNA under non-batch conditions was required. Thus, the key issue for design of the mixer was to minimize the volume of the components of the micro device, thereby, among other things, stimulating the development of a micro mixer with reduced dead volume.

1.2.13.1 Mixer 23 [M 23]: Magnetic Micro-bead Micro Mixer

This magnetic micro-bead mixer relies on a dynamic mixing principle, based on stirring of externally driven small particles in a small mixing chamber [106, 107]. Magnetic beads made of nickel were enclosed in this chamber. The externally stimulated bead motion resulted in rapid homogenization of the solutions to be mixed.

1.2.14 Rotating-blade Dynamic Micro Mixer

Most Relevant Citations

Proceedings contributions: [108].

A dynamic micro mixer has been proposed using curved blades mounted on a rotating shaft [108]. The system can be operated with and without a stator system. Advantages are seen in a more uniform flow distribution, since no feed splitting such as in interdigital micro mixers is necessary and the energy demand is said to be lower than in T-mixers. In this way, lamellae are generated in a very thin and narrow helical pattern. The mixer was characterized for its macro and micro mixing using a reaction sensitive to back-mixing. The power demand was also tested.

The above-mentioned results were presented in the framework of an oral presentation [108]. In the open literature there is no further information known to the authors.

1.3 Passive Mixing

1.3.1 Vertical Y- and T-type Configuration Diffusive Mixing

Most Relevant Citations

Peer-reviewed journals: [68]; proceedings contributions: [57, 69, 70].

Y- and T-type micro mixers resemble mixing tees which are used for simple, undemanding mixing tasks conventionally. Whereas the latter are typically operated under turbulent-flow conditions, their microstructured analogues use solely diffusion in laminar regimes for mixing, by virtue of decreasing the distances and enlarging the specific interfaces, respectively.

When mixing is done exclusively by diffusion, bi-layered flow typically is given. This bi-layer can be arranged vertically (with respect to the plane of mixer plate or the ground level of the mixing channel) or horizontally. In the first case, a two-level feed section is required which is placed on top of each other at the injection zone into the mixing channel. Horizontal injection can be realized more simply with regard to microfabrication; two branches of a T or a Y merge within one plane and are followed by the mixing channel.

In a simulation study, generic investigations on the impact of design details of Y- and T-type micro mixers of the mixing efficiency for gaseous mixing were made [57]. Similar findings were reported for liquid mixing in cross- and T-type micro mixers [57, 68–70].

1.3.1.1 Mixer 24 [M 24]: T-type Micro Mixer

A design of a T-type micro mixer was proposed with the same cross-sections in the feed branches and the mixing channel attached [57]. The T-shape was exactly matched, i.e. the branches are at a 90° angle to the mixing channel.

Mixer type	T-type micro mixer	Mixing channel width, depth, length	500 μm, 300 μm, 5 mm
Mixer material	Artificial design	Angle between branch and mixing channel	90°
Feed branch channel width, depth, length	500 μm, 300 μm, 5.5 mm		

1.3.1.2 Mixer 25 [M 25]: Y-type Micro Mixer

A design of a Y-type micro mixer was proposed with the same cross-sections in the feed branches and the mixing channel attached [57]. The Y-shape was matched, i.e. the branches are at 45° angle. An inverse Y-shape design with a –45° angle was also proposed.

Mixer type	Y-type micro mixer	Mixing channel width, depth, length	500 μm, 300 μm, 5 mm
Mixer material	Artificial design	Angle between branch and mixing channel	45°
Feed branch channel width, depth, length	500 μm, 300 μm, 5.5 mm		

1.3.1.3 Mixer 26 [M 26]: Y-type Micro Mixer with Venturi Throttle

A design of a Y-type micro mixer with a Venturi-type throttle (a tiny orifice which opens like a diffuser to the mixing channel) at the connection of feed branches and mixing channel was proposed [57]. The same cross-sections were assumed for the feed branches and the mixing channel attached.

Mixer type	Y-type micro mixer with Venturi throttle	Mixing channel width, depth, length	500 μm, 300 μm, 5 mm
Mixer material	Artificial design	Angle between branch and mixing channel	45°
Feed branch channel width, depth, length	500 μm, 300 μm, 5.5 mm	Venturi throttle diameter	160 μm

1.3.1.4 Mixer 27 [M 27]: Y-type Micro Mixer with Extended Serpentine Path

An Y-junction feeds a long serpentine channel with 18 bends, arranged on a square format [69]. A top plate was thermally bonded under pressure.

Milling was applied for device manufacture [69].

Mixer type	Y-type micro mixer with extended serpentine path	Mixing channel width, depth, length	200 μm, 200 μm, 40 mm
Mixer material	PMMA	Total plate dimensions	30 mm, 30 mm, 50 mm

1.3.1.5 Mixer 28 [M 28]: T-type Micro Mixer with Straight Path

For means of liquid mixing, a mixer device was designed with a simple T-structure with a straight mixing channel [68]. The width of the mixing channel is twice the width of the inlet channels to ensure undisturbed vertical bi-lamination. At the three ends of the mixing channel structure small square ports are formed which are connected to larger square inlet ports for flow guidance to the macro world.

Mixer type	T-type micro mixer with straight path	End port at channels	500 μm × 500 μm
Mixer material	Silicon/Pyrex glass	Inlet ports with end ports in the center	2000 μm × 2000 μm
Mixing channel widths	20, 40, 60, 100, 200 μm	Thickness of Pyrex top plate	500 μm
Inlet channel widths	10, 20, 30, 50, 100 μm	Material of bottom plate	Polystyrene
Mixing channel length	5.65 mm	Thickness of bottom plate	2 mm
Channel length of both inlets	3.0 mm	Hole diameter in bottom plate	1.58 mm
Channel depth	Various depths, all < 100 μm	Tubing material	PEEK

A silicon wafer was manufactured using three photolithographic and etch steps [68]. Etching was performed from both sides of the wafer so that conduits were produced. Vertical walls were achieved by applying a silicon dry etching technique. Thus, the channels had a rectangular cross-section. The etched silicon structure was anodically bonded to a Pyrex glass wafer. The array of micro mixers produced was then diced into pieces. Using an adhesive, the mixers were bonded to a polymer plate with fluid connectors.

1.3.1.6 Mixing Characterization Protocols/Simulation

[P 21] The mixing of gaseous methanol and oxygen was simulated. The equations applied for the calculation were based on the Navier–Stokes (pressure and velocity) and the species convection–diffusion equation [57]. As the diffusivity value for the binary gas mixture 2.8×10^{-5} m^2 s^{-1} was taken. The flow was laminar in all cases; adiabatic conditions were applied at the domain boundaries. Compressibility and slip effects were taken into account. The inlet temperature was set to 400 K. The total number of cells was ~17 000 in all cases.

As base case, the following scenario was considered [57]. The velocities of both fluids were set to 0.3 m s^{-1}, corresponding to a Peclet number of 8.08 in the mixing channel (see Figure 1.51). The mixing length of the base case amounts to 2.83 mm and the pressure drop is 14.6 Pa.

[P 22] Red and green pen inks were diluted and contacted in a Y-type flow configuration at 20 μl min^{-1} [69]. The colored flows were visualized by a color CCD

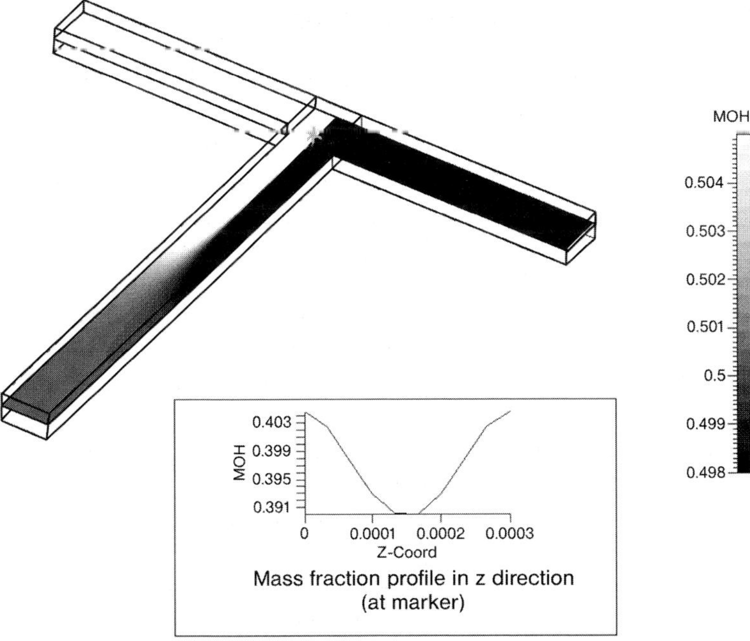

Figure 1.51 Methanol mass fraction for mixing in the T-micro mixer design with base case geometry and parameters [57].

camera with a magnifying lens using direct illumination of the micro chip from the camera. Color data were extracted from the images.

[P 23] Pumping was achieved by raising the pressure in stainless-steel liquid reservoirs through gas flow from a nitrogen gas cylinder [68]. Albeit maybe involving more technical expenditure than simply using pumps, this has the advantage of notably reduced pulsations. The pressure of the liquid just before entering the micro mixer was measured by a pressure transducer.

A commercial blue dye solution was mixed with colorless deionized water [68]. Images were taken by an optical microscope fitted with a video camera. The color images were converted to gray-scale images and the combined effect of red, green and blue (RGB) elements in a pixel of an image were analyzed. A computer program was written for automated RGB analysis. Using premixed solutions, some sort of calibration curve was obtained.

In analogy with this dilution-type mixing experiment, reaction-type experiments were undertaken so as to have complementary information, that is to get a more 3-D image of the mixing process [68]. The hydrolysis of dichloroacetyl phenol red by sodium hydroxide solution was studied, a reaction known to be completed within about 100 μs. The same RGB analysis and calibration curves as mentioned above were used.

[P 24] Computer simulations were carried out using the software Fluent 6 [68]. A 3-D solid model of the T-channel micro mixer was built and named in Gambit. The simulations were made solely for the zone of the T-junction, since for all other zones, including the downstream section of the mixing channel, laminar flow was assumed. Thus, a fine mesh of 173 000 brick elements could be used for the solid model.

[P 25] Confocal fluorescence microscopy was applied to generate 3-D profiles of the species concentration in a Y mixer [70]. A commercial non-fluorescent compound, Fluo-3 (5 mmol l^{-1}), forms a strong fluorescent complex with calcium ions; actually calcium chloride solution (1 mmol l^{-1}) was used.

A numerical analysis using FlumeCAD was made, solving the incompressible Navier–Stokes equation for the velocity and pressure fields [70]. The steady-state velocity field was then used in the coupled solution of three species transport equations (two reagents and one product). Further details are given in [70].

1.3.1.7 Typical Results

Fluid velocity variation – gas mixing

[M 24] [P 21] The inlet gas velocities of gaseous methanol and oxygen were varied from $1.0 \cdot 10^{-3}$ to 0.5 m s^{-1} (Pe range from 0.027 to 13.49) [57]. Mixing lengths in the range of 1–3 mm were determined. At small Pe, mixing occurs at the interface between the fluids in the T-junction. At larger Pe, mixing is completed in the mixing channel and requires the above-mentioned length scale.

Aspect ratio variation at constant width – gas mixing

[M 24] [P 21] The base case aspect ratio is 0.6. Applying the parameter set defined for the base case (see [P 21] above), it was found that an increase of the aspect ratio

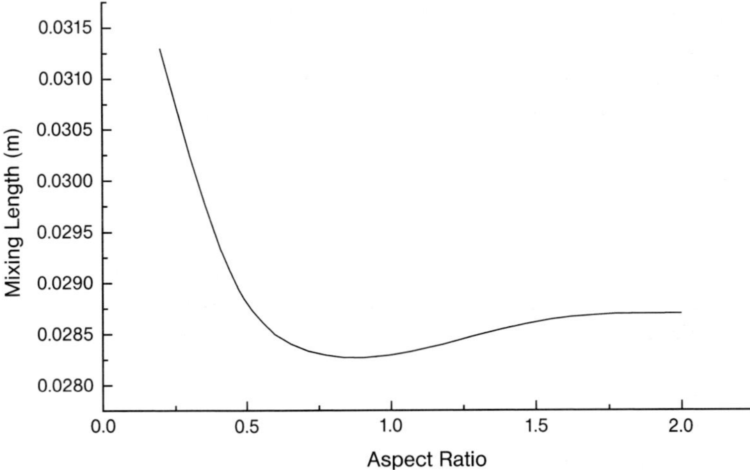

Figure 1.52 Mixing length as a function of the aspect ratio for a T-type micro mixer with a mixing channel width of 500 μm [57] (by courtesy of IOP Publishing Ltd.).

initially decreases steeply the mixing length, which is attributed to the effect of horizontal wall shear due to different viscosities and hence different velocity profiles (see Figure 1.52) [57]. Thereafter, a minimum exists at an aspect ratio of 0.8. At larger ratios the mixing length increases slightly and becomes invariant of the aspect ratio at a value of 1.5. At increasing aspect ratio, the effect of horizontal wall shear decreases, which leads to a symmetrical velocity profile and hence improved mixing. Diffusion becomes the dominant mixing mechanism at aspect ratios > 0.45. Mixing is further influenced by the different residence times of the two species and their longitudinal velocities, giving rise the minimum curve observed.

Aspect ratio variation at constant hydraulic diameter – gas mixing

[M 24] [P 21] The hydraulic diameter of the base case channel is 375 μm [57]. On increasing the aspect ratio at this constant hydraulic diameter, a monotonically decreasing mixing length is found. This is due to a decrease in the diffusion distances in the vertical to the flow. However, some of this mixing improving effect is counterbalanced by velocity gradients in the longitudinal direction, reducing to an extent the mixing efficiency.

Mixing angle variation – gas mixing

[M 25] [P 21] The impact of the mixing angle on the mixing efficiency and the pressure drop was investigated [57]. Physically speaking this is equivalent to transferring the T design into a Y or inverse Y design. The angle values are defined with respect to the horizontal, i.e. the feed branch axis of the T-mixer. Thus, the angle of the latter device is 0°, whereas the normal Y design has a value of 45°.

The variation of the mixing length, from 1.93 to 2.25 mm, is fairly small compared with the changes in the feed angle, from +45° to –65° (see Table 1.2) [57]. Therefore,

Table 1.2 Calculated mixing lengths and pressure drop for Y-type micro mixers with different mixing angles of the two feed branches [57].

Mixing angle (°)	*Mixing length (mm)*	*Pressure drop (Pa)*
+45°	2.03	3.54
+30°	1.93	3.57
+20°	1.93	3.59
+0°	1.96	3.64
–20°	1.97	3.68
–45°	2.12	3.71
–65°	2.25	3.76

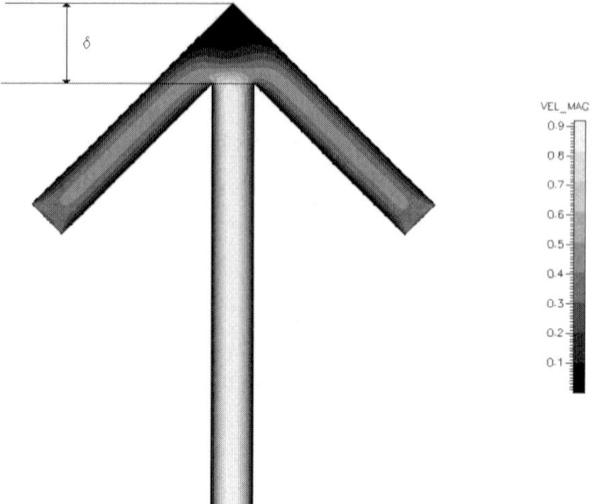

Figure 1.53 Velocity magnitude contours of gas flows in an inverse Y-type micro mixer with a –45° mixing angle at 0.3 m inlet velocities [57] (by courtesy of IOP Publishing Ltd.).

it can be concluded that the angle variation has only a minor impact and, vice versa, this parameter can be chosen rather freely, if other demands make this necessary, e.g. to reduce the footprint area of the mixing element via negative-angle feeds. In contrast, the latter devices have stagnant zones (see Figure 1.53), whereas positive-angle devices have not.

In addition, the angle variation has only a minor impact on the pressure loss, ranging from 3.57 to 3.76 Pa [57].

Venturi-type mixing: impact of throttle diameter – gas mixing

[M 26] [P 21] One means of speeding up mixing is based on putting an orifice directly after the T-junction [57]. Since this involves built-up of a large pressure drop, special designs for pressure recovery such as Venturi-type designs were introduced. It was aimed to exploit this effect also on the micro-scale. Using a throttle diameter of 160 μm (and otherwise the base-case parameters), differences in the results between 2-D and 3-D simulations were pointed out.

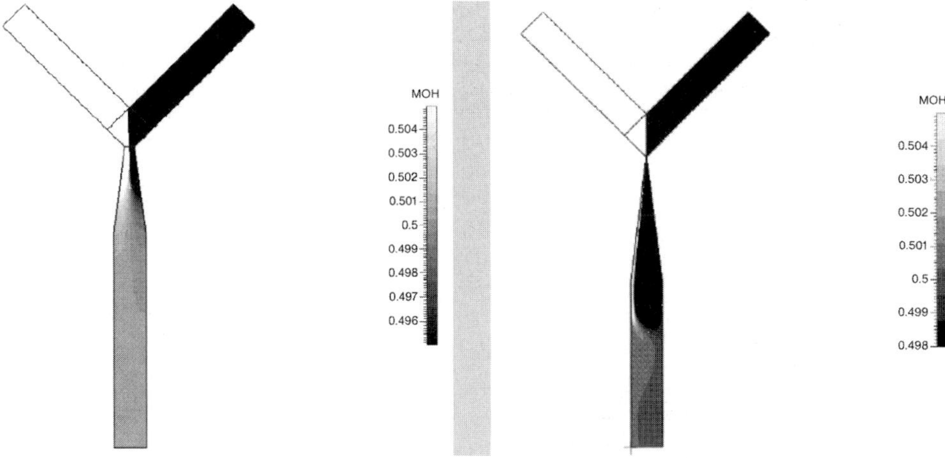

Figure 1.54 Methanol mass fraction contours in a 160 μm throttle at an inlet velocity of 0.3 m s^{-1} (left); methanol mass fraction contours in a 10 μm throttle at an inlet velocity of 5 m s^{-1} (right) [57] (by courtesy of IOP Publishing Ltd.).

Varying the throttle diameter from 10 to 500 μm has a considerable effect on the mixing length, as is to be expected (see Figure 1.54 left) [57]. In this way, the mixing length can be reduced from about 2 mm to < 0.1 mm. The actual shape of the corresponding curve is slightly sigmoidal with points of inflection owing to changes in velocity profiles. Below 10 μm the flow is in the slip regime.

The corresponding pressure drops ranged from 3.54 to 1631 Pa for the largest and smallest throttle sizes [57].

Venturi-type mixing: impact of inlet velocity – gas mixing

[M 26] [P 21] Only the 10 μm throttle is capable of completing gas mixing within the length of the mixing channel attached even at high gas inlet velocities [57]. Thus, for this design the inlet velocities were varied up to 5 m s^{-1}. For the latter velocity, a mixing length of 3.5 mm results. The corresponding pressure drops range from 1632 to $125 \cdot 10^3$ Pa [57].

An increase of the mixing length with increasing inlet velocity is found, displaying two points of inflection [57]. The first point at low velocity is associated with a core fluid moving at higher velocity within the throttle. The second, high-velocity inflection relates to the appearance of recirculations just behind the throttle (see Figure 1.54 right).

Use of color data – liquid mixing

[M 27] [P 22] The applicability of using red, green and blue color data to imagine bi-laminated flows from red and green solutions is discussed in [69]. Finally, the red color data were chosen for further investigations.

In this way, cross-sectional concentration profiles could be determined with fairly high accuracy [69].

Impact of surface roughness and bend shape – liquid mixing

[M 27] [P 22] The impact of surface roughness in micro channels and of bends on the mixing of a bi-laminated fluid was investigated [69]. The surface roughness of the mixing channel was varied from 0.66 to 38.5 μm by changing the milling speed. Cross-sectional concentration profiles were measured before and after passing a rough/right angle, smooth/right angle and smooth/round corner angle bends. A kind of mixing degree was defined from the ratio of the areas of the concentration profiles.

The highest mixing degree was found for the rough/right angle design, probably owing to secondary flows [69]. The two other designs behaved similarly.

Detailed visualization of flow patterns at the start and end of a T-channel – liquid mixing

[M 28] [P 23] The mixing times were compared for four T-channel micro mixers of different dimensions and thus hydraulic diameter at constant Reynolds number [68]. By means of a dilution-type experiment, completeness of mixing was judged from the positions along the mixing channel where a homogeneous color texture was achieved, after having confirmed that the same holds for a reaction-type experiment. The expected sequence of mixing times was found, i.e. the smaller the hydraulic diameter, the shorter is the mixing time. The quantitative data are given in Table 1.3.

Table 1.3 Liquid mixing times for T-channel micro mixers of varying hydraulic diameter [68].

Channel width, depth (μm)	*Hydraulic diameter of the micro mixer (μm)*	*Reynolds number*	*Mixing time (ms)*
200, 82	116	442	1.48
200, 51	81	467	0.97
100, 51	67	489	0.77
60, 51	55	–	–

The millisecond-mixing times are orders of magnitude faster than calculated times for diffusion based on Fick's law [68].

Detailed visualization of flow patterns at the start and end of a T-channel – liquid mixing

[M 28] [P 23] The liquid mixing process in the T-shaped micro mixer was investigated using both dilution-type and reactive-type imaging approaches [68]. In order to provide fine details, images were taken at the inlet T-junction and outlet region, both containing roughly half of the mixing channel separately.

Very distinct patterns were found depending on the flow regime [68]. For Reynolds numbers below 150, bi-laminated streams were found, displaying a low degree of color interpenetration and thus mixing. This is in line with the reaction experiments, showing virtually colorless solutions. For $Re = 150–400$, striations are found both in the dilution- and reaction-type experiments, i.e. layer-like fluidic compartments with either predominant dark or light color. This is due to cross-flow of the lamellae starting already at the T-junction. Here, new interfaces are created and mixing is intensified. The formation of striations is more pronounced in the dilution-type than the reaction-type experiments. This is indicative of segregation of the lamellae into compartments of still comparatively large volume. For *Re* > 400 (up to 500 being investigated), fairly homogeneous color textures result for both types of imaging.

Simulation of liquid flow at the T-junction – liquid mixing

[M 28] [P 23] The liquid flow was simulated for six different scenarios, involving symmetrical and asymmetric flow rates, equal and different viscosities and presence or absence of vertical velocity components [68]. A mixer device with a 100 μm channel was taken.

Scenario 1: For the symmetrical case (equal flow rates and viscosities), almost no penetration of the liquid layers into each other is achieved [68]. Symmetrical fluid trajectories along the two symmetry planes of the mixing channel result (see Figure 1.55).

Scenario 2: For the case with a vertical (z-) flow velocity component, one layer went slightly upwards, the other slightly downwards (actually the z-component was only 1% of the y-velocity along the channel axis) [68]. Upon collision, a swirling flow is formed and a transverse motion of species is induced, strongly enhancing mixing (see Figure 1.56).

Scenario 3: This effect is only given at high Reynolds number (489) and is not observed at low Reynolds number (121).

Scenario 4: For the asymmetric case, i.e. different flow rates (at the same viscosity), also penetration of species from one layer to the other was observed [68].

Scenario 5: For dissimilar viscosities, one having twice the value of the other, also penetration of the lower-viscosity species into the high-viscosity layer was observed [68].

Scenario 6: When the z-component was more intense, 10% of the y-velocity component instead of 1%, the highest degree of penetration from all six scenarios was reached [68].

The mixing efficiency was characterized by a dimensionless parameter, called the intensity of segregation, for the six scenarios [68]. This parameter is based on the distribution of the mass fractions of the species. The results given in Table 1.4 show the importance of the z-component of the flow velocity to induce penetration of species and thus to promote mixing.

Table 1.4 Dimensionless values for the intensity of mixing at the outlet face of the T-channel micro mixer for the six scenarios simulated [68].

Scenario	1	2	3	4	5	6
Intensity of mixing	0.36	0.71	0.34	0.35	0.55	0.83

The results show that small undulations such as inaccurate surface profiling of the micro channels, possibly inducing a z-component or leading to asymmetric flow rates, may considerably change mixing [68]. The same holds when asymmetric flow pumping is given for external reasons, e.g. improper action of pumping such as pulsations.

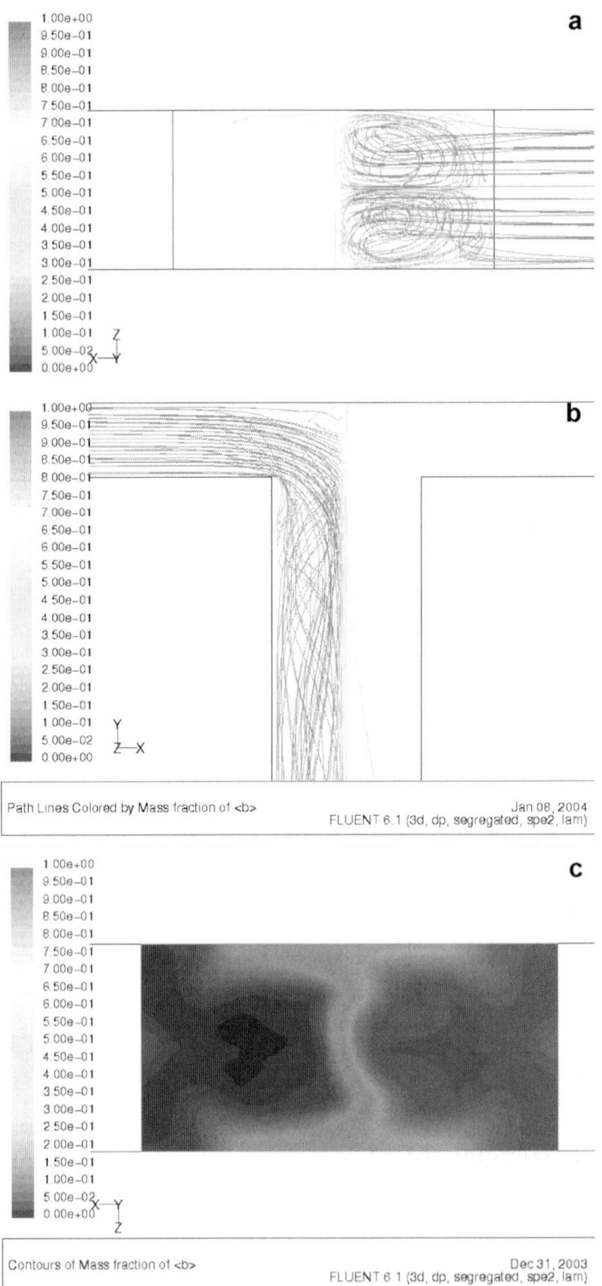

Figure 1.55 Scenario 1: symmetric flow rates lead to symmetric flow fields giving an overall two-layered fluid structure with hardly any species penetration and a low degree of mixing. (a) and (b) show fluid trajectories of one species for the cross-sectional area at the mixing channel front and the whole T-channel design; (c) gives a mass-fraction contour plot of the other species at the outlet face [68] (by courtesy of Elsevier Ltd.).

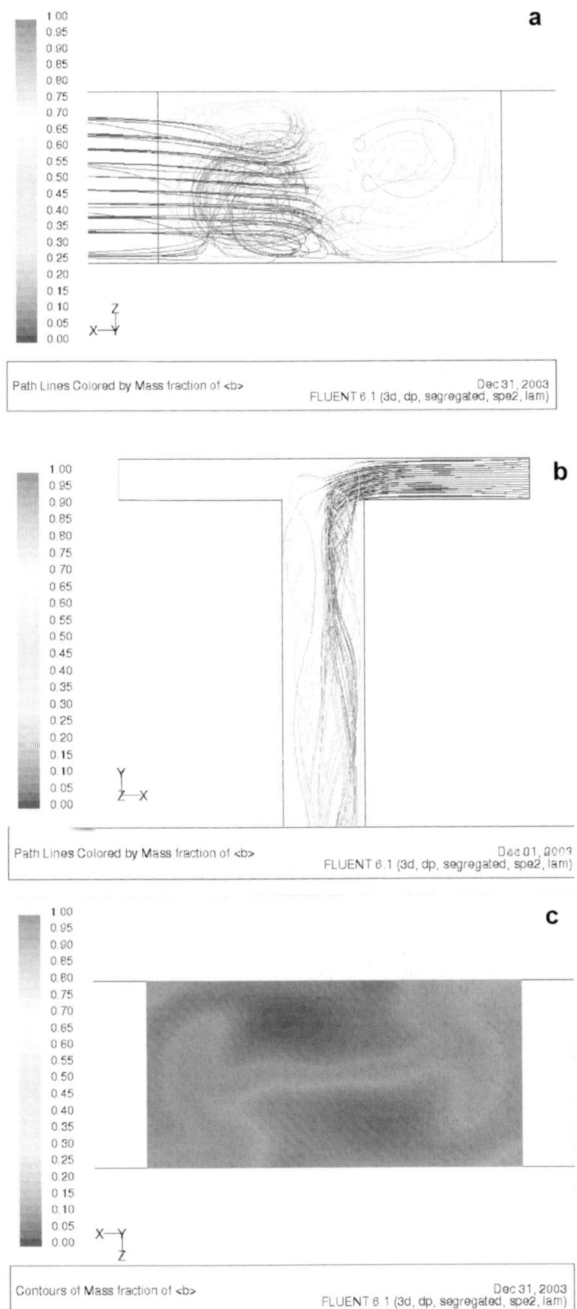

Figure 1.56 Scenario 2: vertical z-components of the flow lead to species penetration and induce secondary flows. (a) and (b) show fluid trajectories of one species for the cross-sectional area at the mixing channel front and the whole T-channel design; (c) gives a mass-fraction contour plot of the other species at the outlet face [68] (by courtesy of Elsevier Ltd.).

Relationship of pressure drop and Reynolds number – liquid mixing

[M 28] [P 23] According to laminar flow theory, the Reynolds number of the flow varies linearly with the pressure gradient [68]. This is only given for one T-channel mixer device presented above, whereas all others show progressively less increase in Reynolds number with increasing pressure gradient. This is due to the recirculations of the crossing flows mentioned above. While at low Reynolds number the viscous forces dominate, the inertia forces for the reasons given above come into play. This consumes part of the energy of the flow, which is used for mixing and decreases the flow velocity.

Based on these considerations, a schematic was proposed showing regions of vortices and of secondary cross flow in the T-junction by which the findings given above can be explained (see Figure 1.57) [68]. This is said to resemble secondary flow of Prandtl's first kind as a result of centrifugal force when fluid flows in curved path.

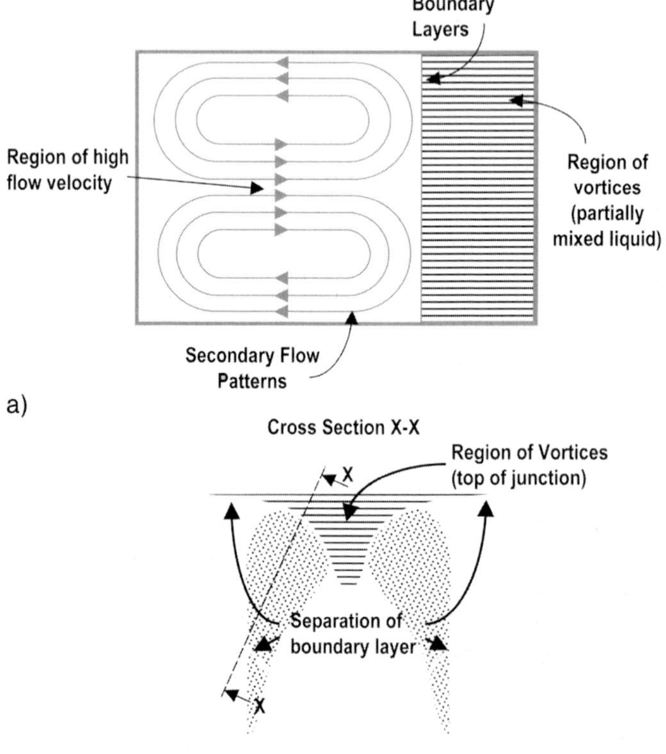

Figure 1.57 (a) Cross-sectional view of the image (b) at X–X showing the separation of a partially mixed zone and a recirculation zone, inducing cross flow for mixing. (b) Top view of the T-junction showing the respective distribution of the same zones [68] (by courtesy of Elsevier Ltd.).

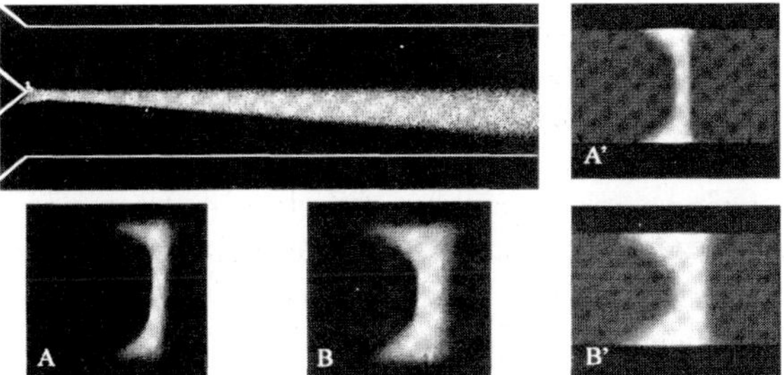

Figure 1.58 Experimental images of the lateral and cross-sectional profiles of a fluorescent species formed by mixing and subsequent immediate reaction. Left, experimental images; right, simulated images [70] (by courtesy of Kluwer Academic Publishers).

Cross-sectional concentration profiles in a Y micro mixer

(No details on mixer in [109]) [P 25] By use of confocal microscopy, cross-sectional concentration profiles were derived in a Y micro mixer (see Figure 1.58) [70]. At the top and bottom of the channel large fluorescent areas were found, while this region thinned in the channel center. The experimental images perfectly match the simulated concentration profiles.

Similar very good agreement was found for a quantitative mixing analysis when determining the diffusive widths at various locations along the mixing channel and for various flow rates [70]. Thus, the use of the confocal fluorescence technique for complex mixing analysis was demonstrated and compared with the potential of standard microscopy set-ups.

1.3.2 Horizontally Bi-laminating Y-feed Mixing

Most Relevant Citations

Peer-reviewed journals: [110].

The usually slow mixing characteristics of unfocused bi-lamination can be exploited for a purpose. It allows one the sensing of a reaction in a rather unmixed state and, by stopping the flow, to have short mixing reaching the reaction state [110]. Thus, by pulsing the flow rate, rapid switching between reactant and product flows can be achieved, e.g. to investigate the kinetics of a reaction. In one realized example (see below), only two streams were layered (bi-lamination), albeit fed by multi-channel feeds.

1.3.2.1 Mixer 29 [M 29]: Unfocused Horizontally Bi-laminating Y-feed Micro Mixer

An interdigital micro mixer with two multi-parallel feed lines and a rectangular mixing chamber was realized for pulsed/stopped-flow operation of reacting flows [110]. The feed lines are mirror-imaged inclined to the longitudinal axis of the mixing chamber and 'merge' to an alternate pattern just in front of the latter. Actually, only a bi-lamination is achieved since in each multi-channel feed, i.e. on one device level, only one fluid is guided. The parallel splitting serves to achieve a homogeneous distribution of the flow in the axial direction.

The material choice and hence the microfabrication was based on the specific needs of Fourier transform infrared (FTIR) spectroscopy, to have IR-transparent materials [110]. For this reason, CaF_2 disks are an essential part of the device. The main part of the device consists of two polymer layers sandwiching one metal layer which are encased by the two disks, acting as a kind of end plates.

In a first step, the negative working photoresist SU-8 is spin-coated on to the disk and soft baked [110]. The disk is then UV-exposed to pattern the bottom layer. A silver thin metal layer is thereafter evaporated. The metal layer is spin-coated with an AZ-type photoresist, dried, exposed and developed. In this way, the metal layer can be developed independently from the patterning of the SU-8 layer underneath. The metal layer is patterned by wet-chemical etching. As a next step, a second SU-8 layer is deposited, soft-baked and exposed. Top and bottom layers are now developed. After a hard bake, a second CaF_2 disk is attached to the stack and sealed by a light-curing epoxy resin.

Mixer type	Unfocused bi-lamination micro mixer	Metal layer material	Silver (evaporated)
Mixer material	Polymer SU-8	Metal layer thickness	2 μm
Polymer layer thickness	~10 μm	Number of feed channels	7
Inspection window material	CaF_2	Feed channel width, depth	100 μm, ~10 μm
Inspection window thickness, diameter	1 mm, 20 mm	Mixing channel width, depth	1 mm, ~22 μm

1.3.2.2 Mixing Characterization Protocols/Simulation

[P 26] Time-resolved FTIR spectroscopy was performed by operation of an infrared spectrometer in the rapid scan acquisition mode (see Figure 1.59) [110]. The effective time span between subsequent spectra was 65 ms. Further gains in time resolution can be achieved when setting the spectral resolution lower (here 8 cm^{-1}) or by using the step-scan instead of rapid-scan mode.

For liquid feed, a double-channel syringe pump with two 500 μl syringes, a high-speed pneumatic switching valve and PTFE tubing with an inner diameter of 254 μm was used [110]; 0.5 μm PTFE microfilters were set in front of the micro mixer to remove particles. The flow rate was set to 6 ml h^{-1}.

For repetition of stopped-flow shots, the control software of the spectrometer and the software for pump and valve operation were coupled [110]. A typical FTIR measurement section was as follows. After start of the pumping with opened valve, the FTIR spectrometer recorded five interferograms for the unmixed state, prior to reaction. Following a signal by the spectrometer, the valve switched to the stopped-flow mode. Now the liquid was standing in the mixing channel and mixing took place via diffusion. By continued FTIR scanning the course of reaction could then be followed. Now the valve was opened again and the sequence could be repeated as often as desired to obtain a sufficient signal-to-noise ratio.

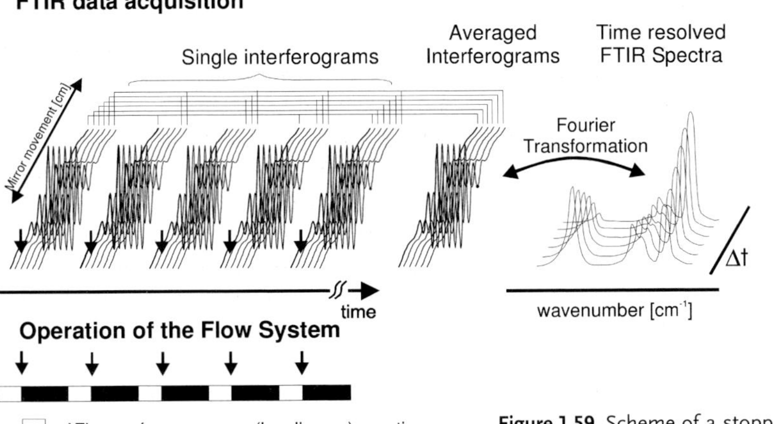

Figure 1.59 Scheme of a stopped-flow rapid scan measurement sequence [110] (by courtesy of RSC).

1.3.2.3 Typical Results

Velocity and pressure profiles by CFD simulations

[M 29] [P 26] Even velocity profiles are found throughout the whole rectangular mixing channel at a pressure drop of 0.8 bar [110]. This pressure drop is considerably lower than for a former, similar design with more complex feed.

Contour plot of mass fractions/mixing time/pre-mixing

[M 29] [P 26] A calculated contour plot of the mass fraction of one component with a diffusion coefficient of $1 \cdot 10^{-9}\ m^2\ s^{-1}$ is indicative of diffusive mixing [110]. It can be seen that completion of mixing is only expected at larger distances from

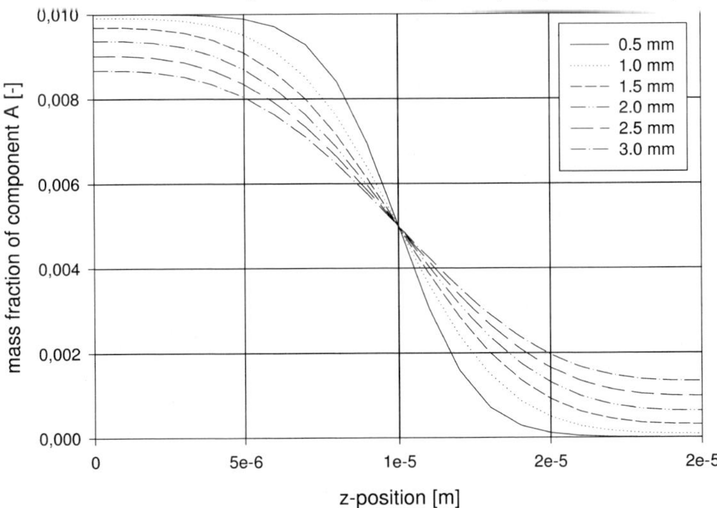

Figure 1.60 Contour plot of the mass fraction of one component with a diffusion coefficient of $1 \cdot 10^{-9}\ m^2\ s^{-1}$ across the longitudinal axis of the mixing channel at various distance from the feed section [110] (by courtesy of RSC).

the inlet. Pre-mixing takes place to a certain extent, which depends on the distance from the feed section (see Figure 1.60).

At a distance of 1.0 mm from the inlet, which corresponds to the center of the IR beam, a mixing time of about 100 ms is calculated for the stopped-flow mode (see *Neutralization model reaction*, below) [110]. This mixing time is < 50% compared with a previous, similar design with more complex feed.

Neutralization model reaction

[M 29] [P 26] The neutralization of acetic acid with sodium hydroxide is an extremely fast reaction which can be monitored by FTIR spectroscopy [110]. The speed of the reaction allows one to monitor mixing directly, since the reaction will always immediately follow the mixing course. It can be seen that the time for completion of mixing is about 100 ms, in good accord with the simulation results (see Figure 1.61 and *Contour plot of mass fractions/mixing time/pre-mixing*, above).

It also evident from the FTIR spectra that a certain degree of pre-mixing, due to diffusion in the continuous-flow mode, occurred, leading to a correspondingly large pre-reaction [110].

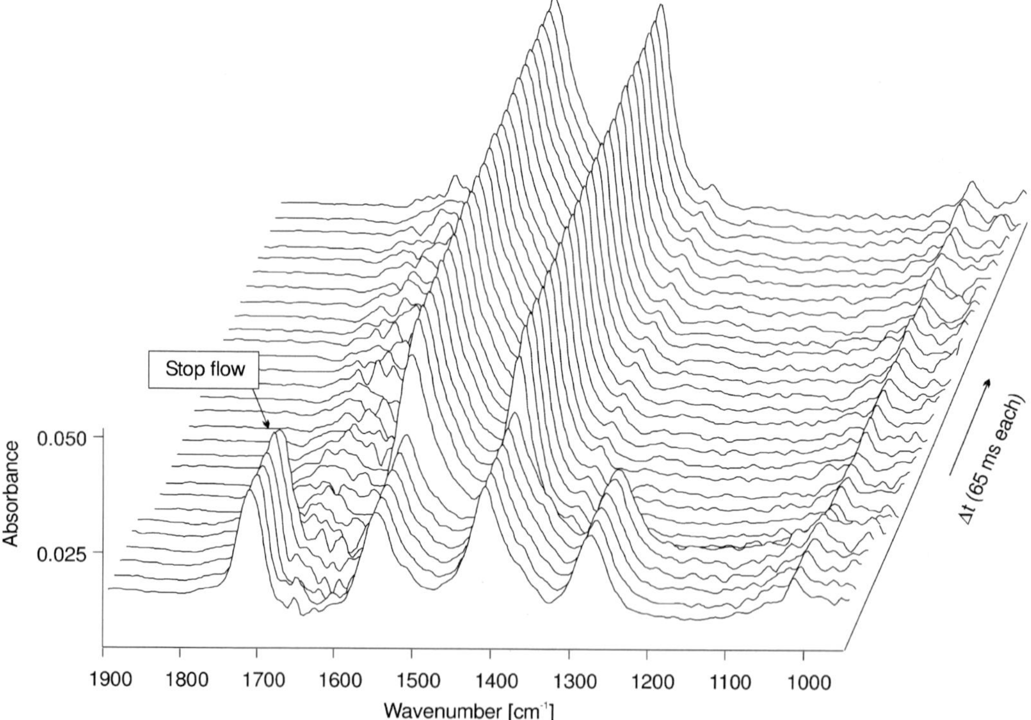

Figure 1.61 Stack plot of FTIR spectra at various time spans (each 65 ms) for the neutralization reaction between acetic acid and sodium hydroxide [110] (by courtesy of RSC).

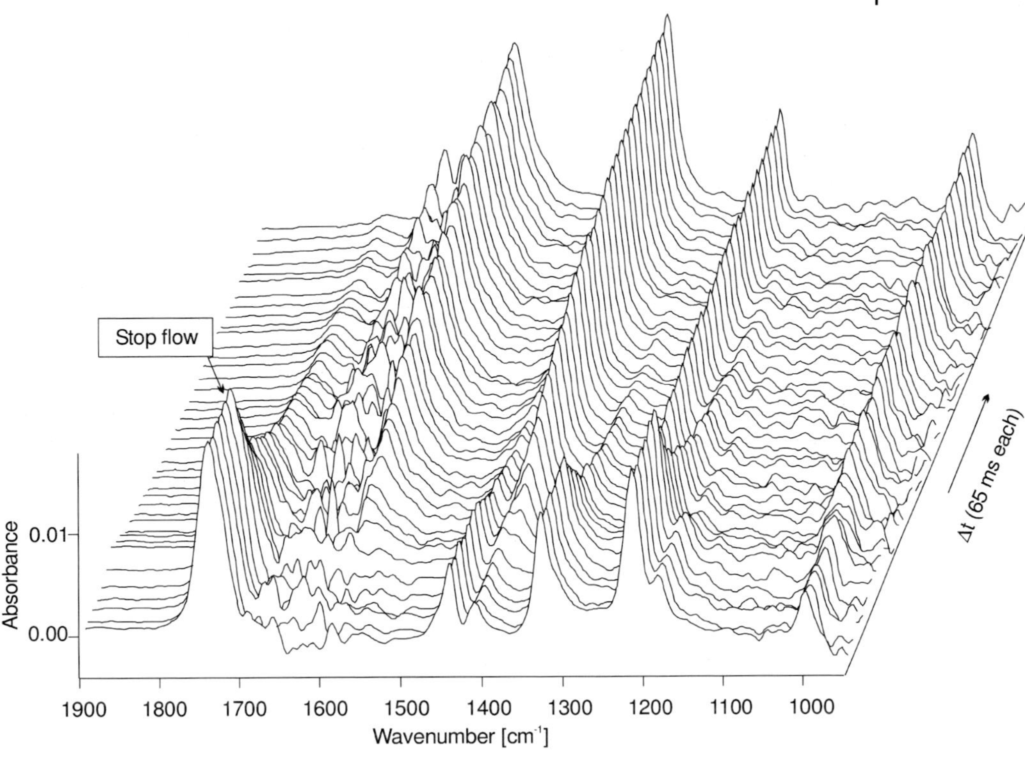

Figure 1.62 Stack plot of FTIR spectra at various time spans (each 65 ms) for the saponification reaction between methyl monochloroacetate and sodium hydroxide [110] (by courtesy of RSC).

Saponification model reaction

[M 29] [P 26] The saponification of methyl monochloroacetate with sodium hydroxide is a slow reaction which can be monitored by FTIR spectroscopy [110]. Owing to the slow reaction, despite pre-mixing no detectable reaction could be monitored by FTIR (compare with *Neutralization model reaction*). On stopping the flow, the reaction products chloroacetate and methanol appear (see Figure 1.62). After a few hundred milliseconds of reaction time, the reaction is completed.

Accessible FTIR spectral region

[M 29] [P 26] When using water as liquid (see description of the two reactions, above), the accessible spectral region to follow chemical reactions was determined as 950–1600 and 1700–3000 cm^{-1} [110]. Owing to the strong absorption of the bending vibration of the solvent water, the 1600–1700 cm^{-1} region cannot be used. This poses a limit for monitoring biological samples having amide bonds which absorb in the same spectral region.

1.3.3
Capillary-force, Self-filling Bi-laminating Mixing

Most Relevant Citations

Proceedings contributions: [111].

The development of this mixing concept orients on generating a portable tool for medical and environmental analysis, preferably done in field experiments on-site [111]. This means first of all eliminating pumps for feeding of the solutions to be mixed, since these are usually costly and heavy parts. In addition, ruggedness was a major desire, since many field samples contain particles or are contaminated, different from 'clean' laboratory probes. Hence the target was to find a simple self-filling principle which in turn more or less automatically is followed by reasonably fast mixing.

Capillary forces are a suitable means for manual self-filling of samples into minute channel cavities [111]. However, this usually allows one to fill in only one liquid. Therefore, a specially technique had to be developed which allows the filling of two liquids, based on a self-closing and self-re-opening mechanism.

1.3.3.1 Mixer 30 [M 30]: Capillary-force, Self-filling Bi-laminating Micro Mixer

Two adjacent micro channels of trapezoid shape are separated by a type of cantilever structure leaving a small gap which can permit fluid contact (see Figure 1.63) [111]. The left channel has a smaller cross-sectional area and is placed at a higher position than the other more voluminous channel.

The first solution, the sample taken in the field, is filled in the upper channel by using the capillary-force self-filling mechanism (see Figure 1.64) [111]. By capillary forces the cantilever is slightly bent and automatically closes the sample volume, leaving the second channel still unfilled. The second solution to perform the analysis is filled into the second channel, directly on-site or later, in the latter case still manually or also by using pumps.

The device actually made contained seven pairs of such adjacent channels with a small gap, placed in parallel in a quadratic format [111].

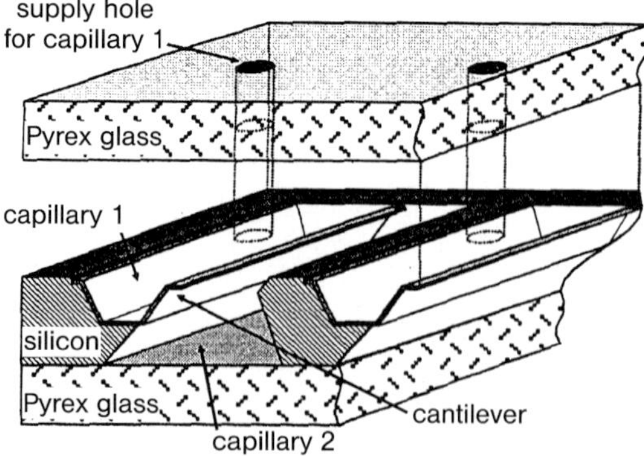

Figure 1.63 Schematic of the capillary-force, self-filling device [111] (by courtesy of Springer-Verlag).

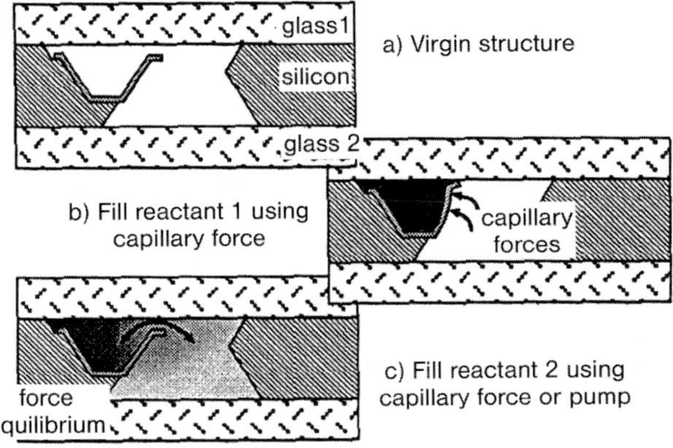

Figure 1.64 Schematic of the capillary-force, self-filling and self-closing mechanism, followed by re-opening, again owing to capillary forces. Thereafter, an interface is provided for mixing of the solutions, being previously encased [111] (by courtesy of Springer-Verlag).

It can be envisaged that, facing the target application in environmental and medical analysis, cleaning can be done by solvents or hot vapor [111]. Alternatively, disposable materials may be employed so that the device can be used as once-only tool. For medical analysis, the integration of a needle may be advisable.

The channels are anisotropically etched into silicon and are separated by a thin SiO_2 layer, the cantilever [111]. This microstructure, open to both sides, is closed by anodic bonding to Pyrex glass wafers.

Mixer type	Capillary-force, self-filling bi-laminating micro mixer	Gap for diffusion	20 μm
Mixer material	Silicon	Volume channel 1	0.22 μl
Cover material	Pyrex	Volume channel 1	0.86 μl
Channel width, depth, length for liquid 1	300 μm, 200 μm, 10 mm	Total size of the mixing chip	20 × 20 mm
Channel width, depth, length for liquid 2	500 μm, 300 μm, 10 mm	Pairs of adjacent channels	7
Cantilever plate thickness, length, height	2 μm, 10 mm, 245 μm		

1.3.3.2 **Mixing Characterization Protocols/Simulation**

[P 27] Red and blue comestible colors were chosen as test liquids [111].

Analysis was done using light microscopy [111]. Owing to the transparent glass cover, other spectroscopic techniques such as UV–Vis can also be applied.

1.3.3.3 Typical Results

Confirmation of self-filling and automixing principles

[M 30] [P 27] The self-filling principle was confirmed by introducing two colored solutions serially into the channels [111]. This also proved the self-closing mechanism of the cantilever, because no cross-over of liquids was observed. A gross change of color in the larger channel could be observed as the end result of diffusion of two colored solution into each other. Details of the diffusion process itself could not be observed owing to the limited contrast of the light microscopy observation.

Mixing Time

[M 30] [P 27] Based on general assumptions of diffusion in aqueous solutions, the mixing in the self-filling device is assumed to be of the order of seconds and minutes [111]. This is slow compared with other micro mixers, but having a fast mixing time is not a requirement for the target applications of the device, namely field analysis and medical applications. The time needed for mixing is more than given during sampling and transfer to the measuring equipment.

1.3.4 Cross-injection Mixing with Square Static Mixing Elements

Most Relevant Citations

Proceedings contributions: [71].

The insertion of small static mixing elements (SME) is common to achieve swirls and eddies in pipe flow, albeit usually not being turbulent [71]. The flow obstacles are fairly small compared with the pipe diameter, unlike typical packings of static mixers which fully cover the diameter of the channel. Such mixing elements provide abrupt changes in surface orientation to result in flow separation and subsequent eddy production.

1.3.4.1 Mixer 31 [M 31]: Cross-shaped Micro Mixer with Static Mixing Elements

The channel structure of the mixer is a simple cross, i.e. four channels which all merge at one junction [71]. A cross was preferred over a T-channel mixer since two interfaces instead of only one are initially created when the fluids are contacted. The top channel feeds one fluid, while the other fluid is injected via the left and right channels. The last, bottom channel functions as mixing and outlet zone. Squares, much smaller than the channel width, are positioned at the walls of this mixing channel and function as static mixing elements. The squares are positioned on alternate sides of the channels and at a distance corresponding to multiple square widths.

Three versions of the cross-shaped mixer were designed for simulation [71]. One version contained no SMEs, one was equipped with two SMEs on alternate sides of the mixing channel, and one with five SMEs, two of them in the inlet side channels directly at the cross and three on alternate sides of the mixing channel.

A micro fabrication of the designs mentioned above has not been realized so far; the designs were only taken for simulation [71].

Mixer type	Cross-shaped micro mixer with static mixing elements	Outlet channel width, depth	30 μm, 40 μm
Mixer material		Side channel width, depth	25 μm, 40 μm
Inlet channel width, depth	40 μm, 40 μm	Static mixing element width, length, depth	10 μm, 10 μm, 40 μm
Distance cross–second SME (Two-SME design)	75 μm	Distance cross–first SME (Two-SME design)	45 μm

1.3.4.2 Mixing Characterization Protocols/Simulation

[P 28] A 3-D solid model of the cross-shaped micro mixer is meshed to a sufficiently fine scale with brick elements of 2 μm for the simulations [71]. Simulation results were intended at very short time scales, e.g. in intervals of 50 μs, to verify the mixing patterns at the initial state after application of pressure. The numerical values of the mass fraction are taken to give quantitative measures of the mixing efficiency. The pre-processor fluidics solver and post-processor of ConventorWare™ were used for the simulations. The software FLUENT 5 was used for verification of these results, since the former software is so far not a widely established tool for fluid dynamic simulation.

A pressure of 2 bar was set at the inlets and of $2 \cdot 10^{-4}$ bar at the outlet of the micro mixer [71].

1.3.4.3 Typical Results

Comparison of cross-injection mixing without and with static mixing elements

[M 31] [P 28] Flow simulations revealed that swirled flow generating eddies can be achieved in a cross-shaped micro mixer with SME [71]. Comparing the distribution of the mass fractions at the mixing channel's outlet, it is evident that complete mixing can be achieved for the cross-shaped mixer with two SMEs, whereas the same overall channel structure without any SMEs, a simple straight channel, is still largely segregated, exhibiting a coiled three-layered fluid (see Figure 1.65).

Time evolution of the flow patterns in the cross-injection SME mixing at the initial stage

[M 31] [P 28] The time evolution of the flow patterns in the cross-shaped micro mixer with two static mixing elements was monitored by simulation at time intervals of 50, 150, 500 μs and 1 ms after application of pressure [71]. In addition to seeing the evolution of the swirling patterns, it was concluded from this analysis that at 500 μs a nearly homogeneous distribution of the mass fractions is given and at 1 ms this is indeed completed. Hence the theoretical mixing time of the mixer may be below 1 ms.

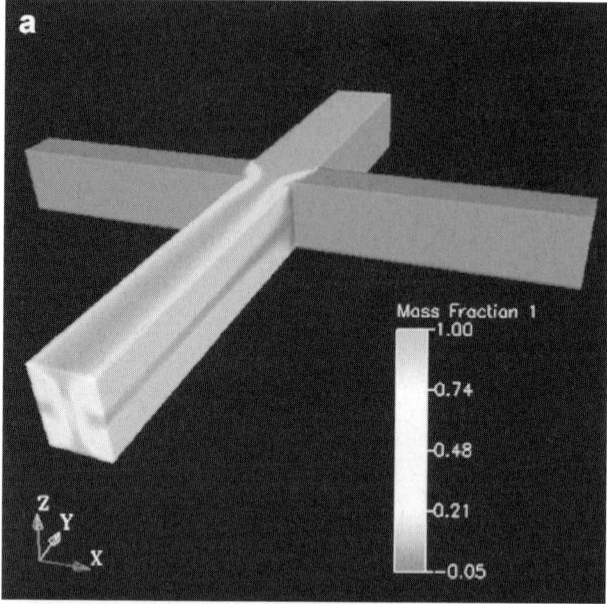

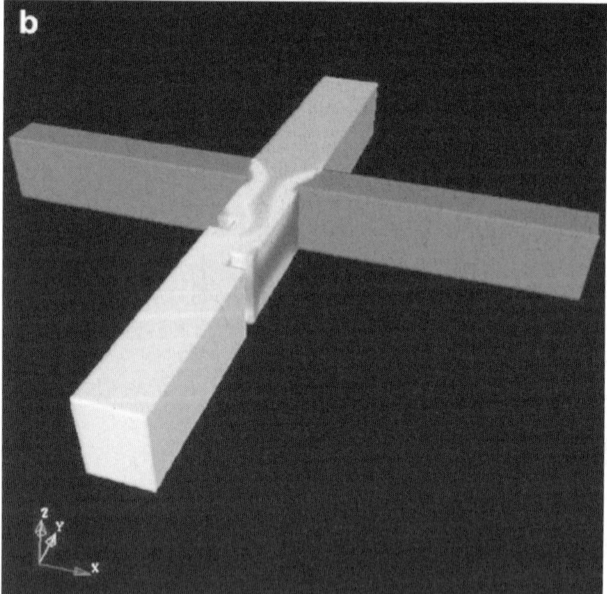

Figure 1.65 CFD simulations giving 3-D mass contour plots in the cross-channel structure for a design without and one with two static mixing elements. The completeness of mixing can be judged from the cross-sectional mass distribution at the outlet [71] (by courtesy of Elsevier Ltd.).

Verification of the ConventorWare™ software

[M 31] [P 28] The flow patterns simulated by the ConventorWare™ and FLUENT 5 software concerning the flow in the cross-shaped micro mixer with two static mixing elements the same; hence the predictability of the ConventorWare™ software was demonstrated [71].

Velocity vector and contour plots

[M 31] [P 28] Velocity vector plots show the separation of the boundary layer before and after approaching the static mixing element [71]. Backflow occurs in the separation region. By this unsteady reattachment of the flow, new interfaces are constantly generated, when the flow has to pass a series of such mixing elements.

The velocity contour plots show a higher velocity, a higher velocity gradient and rapid change of the direction of the velocity components in the proximity of the static mixing elements compared with the rest of the flow in the channel (see Figure 1.66) [71].

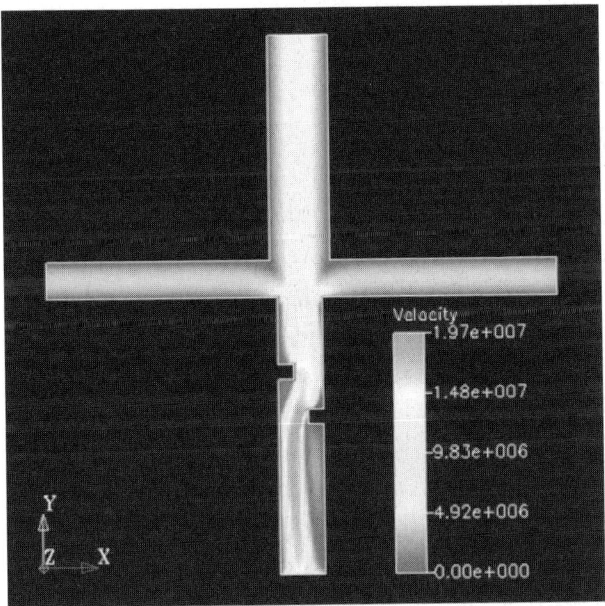

Figure 1.66 Velocity contour plots of the flow in the cross-channel mixer [71] (by courtesy of Elsevier Ltd.).

Pressure drop plots – intensity of segregation and velocity vector plots

[M 31] [P 28] Pressure drop plots show that the bulk of the pressure drop occurs in the mixing channel due to the action of the static mixing elements [71]. This means that the additional pressure drop has to be paid for by speeding up mixing, e.g. at the cost of reducing the overall flow rate. The decrease in flow rate is estimated to be 30–37% compared with flow through a straight channel of the same size. However, this loss in capacity is more than counterbalanced by the increased mixing

performance, as the calculated intensities of segregation and variation coefficient show. The latter are measures for the mixing efficiency. Here, the cross-shaped micro mixers with SMEs are much better than the same channel structure without static elements (see Table 1.5). The incorporation of five elements is still better than a two-SME cross-shaped mixer, having completion of mixing already after flow passage through the SME zone (without the need to further pass the residual open channel).

Table 1.5 Measures for mixing efficiency calculated from mass contour plots yielded by CFD simulation – benchmarking cross-shaped mixers with and without static mixing elements (SME) [71].

Flow configuration	*Intensity of segregation*	*Coefficient of variation*	*Flow rate ($\mu l\ s^{-1}$)*	*Mean flow velocity ($m\ s^{-1}$)*
No SME	0.3115	0.5343	11.9	9.92
2 SMEs	0.0211	0.1349	8.55	7.13
5 SMEs	0.0006	0.0210	7.45	6.21

1.3.5 Hydrodynamic Focusing Cross-Injection Mixing

Most Relevant Citations

Peer-reviewed journals: [112]; proceedings contributions: [113].

Ultra-fast diffusive mixing can only be achieved by an extreme reduction of the fluid layer thickness, being not more than several tens of nanometers [112]. Microfabricated devices alone are not adequate to generate such thin lamellae. Therefore, further compression of the micrometer thin lamellae, generated in a microstructure, has to be performed by hydrodynamic means. This is achievable by contacting two fluids of extremely different volume flows. Owing to volume conservation, the fluid of higher flow tends to form a thick lamella, while an ultrathin lamella of the other fluid results. Since the final thin lamella thickness is achieved after only a short entrance flow region, the whole process is termed 'hydrodynamic focusing'.

1.3.5.1 Mixer 32 [M 32]: Hydrodynamic Focusing Cross-injection Micro Mixer

In a cross-flow configuration, one stream with one fluid enters from the left and two streams carrying the same other fluid are fed from above and below [112]. The latter have much larger flow rates as the first stream so that this layer is hydrodynamically compressed when all three streams enter the outlet channel on the right side of the cross. The channel which carries the stream to be focused narrows to a kind of nozzle when approaching the T-junction

In order to analyze mixing phenomena occurring very close to the first fluid contact, i.e. directly behind the mixing element, a planar design completely covered by a transparent plate was chosen (see Figure 1.67) [112]. This permits characterization without any dead times, hence permitting observation of the entire mixing process.

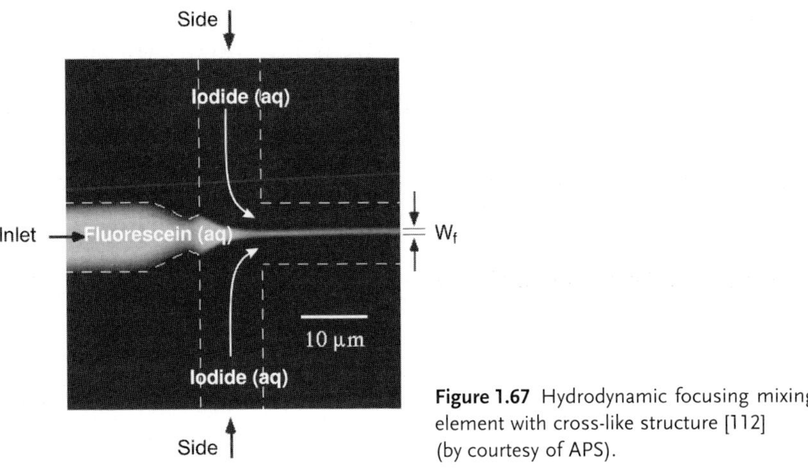

Figure 1.67 Hydrodynamic focusing mixing element with cross-like structure [112] (by courtesy of APS).

The mixers were realized by means of silicon micromachining using standard photolithographic techniques and a chlorine reactive ion etching process [112]. The structured silicon wafers were sealed with a thin coating of cured silicone rubber. This silicone layer could be withdrawn after use, allowing cleaning of the mixer. After repetition of coating, experiments could be continued using the cleaned structure. Four holes, for fluid connection of inlet and side flows and of the mixture to external sample reservoirs, were drilled into the backside of the wafer.

Mixer type	Hydrodynamic focusing cross-injection micro mixer	Channel width	10 μm
Mixer material	Silicon	Nozzle width	2 μm
Cover material	Glass		

1.3.5.2 Mixing Characterization Protocols/Simulation

[P 29] No details on the experimental results were given in [113].

[P 30] Sample dosing was achieved without using any pump. Instead, the liquids were fed by applying pressure, controlled by regulating the incoming flow of nitrogen gas, on the head of each reservoir [112]. The finite precision of pressure control directly influenced the minimum thickness of lamellae achievable.

Fluorescence monitoring was used for depicting the concentration changes due to the mixing process [112].

1.3.5.3 Typical Results

Thickness of compressed lamellae

[M 32] [P 30] The focusing width was controlled by varying the volume flows of the side and inlet flows [112]. Widths as small as 50 nm were measured, yielding a nearly instantaneous interdiffusion of the side flow into the inlet flow. However, it

should be mentioned that this is not true for the dispersion of the inlet flow within the overall mixing volume, i.e. the fluid of the side flow has to serve as an excess reagent.

After mixing, the time evolution of a subsequent reaction can be spatially separated, and the resolution is only determined by the flow velocity [112]. A resolution better than 1 μs μm^{-1} was achieved since laminar flow was even maintained at high velocities, owing to the small size of the channels.

Sample consumption

[M 32] [P 30] Owing to the small mixer dimensions, the sample consumption is low [112]. The volume flows of the focused reactant stream are typically of the order of nanoliters per second, which is more than three orders of magnitude lower than the typical rates needed for turbulent mixers. This is important, in particular, for applications in biochemistry, usually demanding the consumption of expensive samples.

Variation of jet width and mixing time by adjustment of feed pressures

[M 32] [P 29] Based on an analogy of the fluid flow with the current flow in a resistive circuit, the jet width was calculated for various pressures [113]. In the range of pressures investigated, jets from 1 down to 0.02 μm, i.e. 20 nm, width resulted. These calculated values are in excellent agreement with experimental data. Thus, mixing can be performed within microseconds with only nanoliter sample consumption.

[M 32] [P 30] It was found that hydrodynamic focusing was achievable in the mixer over a wide range of volume flow ratios, expressed as the ratio α of the corresponding pressures, defined as follows:

$$\alpha = \frac{P_s}{P_i} \tag{1.1}$$

where P_s is the pressure of the side flow channel and P_i refers to that of the inlet flow channel [112].The minimum and maximum values of α, limiting the range of focusing, were determined as $\alpha_{min} = 0.48$ and $\alpha_{max} = 1.28$ [112]. Increasing α results in a narrowing of the focused stream. Below the lower limit, the inlet flow enters the side channels, instead being directed to the outlet channel. Above the upper limit, the same phenomena occur for the side flow, hence streaming into the inlet flow channel. Both flow reversals were independent of the overall pressures P_s and P_i.

The experimental data for α were compared with theoretical values calculated by means of analogy considerations to electric flow (Ohm's law) [112]. Simple circuit models based on a network of resistors were applied to simulate the cross-type configuration chosen. It could be shown that the calculated values for α_{min} and α_{max} were in excellent agreement with the experimental data.

Visualization of the hydrodynamically focused lamellae by fluorescence

[M 32] [P 30] The mixing process was visualized by epifluorescence and confocal microscopy images (see Figure 1.68) [112]. A bright inlet flow, labeled with a

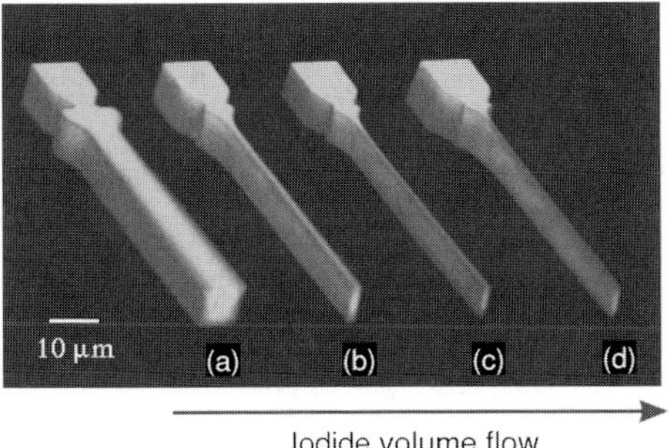

Figure 1.68 Visualization of hydrodynamic focusing of a fluid layer by means of fluorescence imaging [112] (by courtesy of APS).

fluorescent dye, was mixed with non-fluorescent buffer side flows. The light emission of excited molecules of a focused fluorescein solution was quenched by a solution containing iodide ions. Thus, the time evolution of iodide diffusion into the fluorescein solution was revealed as a change in the fluorescence intensity.

Lamellae of the fluorescein solution larger than 1 μm could be directly imaged [112]. In addition, a non-imaging approach, based on a fluorescence technique, was applied to analyze even smaller lamellae. This technique allowed one to determine widths as small as 50 nm. Data obtained at a flow of 5 nl s^{-1} showed that the mixing process was completed at best after 10 μs. The experiments consumed extremely small sample volumes, needing only 25 nl of fluorescein solution and 1 μl of the buffer solution per measurement.

1.3.6 Geometric Focusing Bi-laminating Mixing

Most Relevant Citations

Peer-reviewed journals: [114]; proceedings contributions: [115, 116].

The focusing bi-laminating micro mixer was realized in the framework of the development of a flow injection analysis (FIA) system [114, 115]. The mixer is placed downstream of the two-fold injection of sample and reagent streams into the carrier. Thereafter, the mixed stream enters a reaction chamber and finally passes a detector. An easy integration is required, as the mixing element is part of an integrated system which has the minimization of size as one issue.

FIA is a chemical analysis method based plug-wise injection into a carrier stream [114–116]. These plug samples can be further manipulated, e.g. by reaction to compounds better detectable by a detector. The sample consumption can be reduced via miniaturization; however, then usually laminar-flow conditions are given so that micro mixers are needed which are efficient in that regime.

The analysis target was the detection of ammonia in aqueous solutions, e.g. in surface water [114–116]. The Berthelot reaction scheme was employed to monitor ammonia, by chemical conversion to indophenol blue, using a chlorination step first followed by coupling of two phenol moieties. The absorption of the dye was measured by a photometric-type experiment. Full conversion by efficient mixing is mandatory for a good analysis.

Design targets for the application were to have a total system volume < 1 cm^3 and to achieve one sample throughput per minute [115]. The reaction time for the Berthelot reaction is about 40 s so that only 20 s are left for all other process steps. This involves the injection of the sample, mixing and an optical absorption measurement. Therefore, mixing has to be fast, requiring only a few seconds.

1.3.6.1 Mixer 33 [M 33]: Geometric Focusing Bi-laminating Micro Mixer

This micro mixer is based on a bi-laminated stream which has to pass a considerably narrowed flow passage having the function of reducing the diffusion distance and thereby speeding up liquid mixing [114–116].

A central demand for the mixer is the very accurate setting of the channel width. For this reason, a microfabrication technique was chosen that is known to achieve very smooth and ideally-vertical channel walls, namely the reactive ion etching (RIE) technique [114–116]. The fluidic connections are made from the bottom by RIE. Anodic bonding serves for interconnection.

Mixer type	Geometric focusing bi-laminating micro mixer	Initial bi-laminating channel width, depth, length	300 μm, 200 μm, 1.65 mm
Mixer material	Silicon/Pyrex	Focusing channel width, depth, length	100 μm, 200 μm, 1.65 mm

1.3.6.2 Mixing Characterization Protocols/Simulation

[P 31] Liquid mixing times were calculated based on assuming diffusion as the only mixing mechanism and considering Fick's law which takes into account the diffusion constant and the diffusion distance [114].

For judging completion of mixing experimentally, a flow guidance with two injection inlets and two outlets was used [114]. A phenol solution was injected via one inlet port; the completion of mixing was indicated when both outlets gave the same phenol concentration. The concentration was determined by conductivity measurement. Mixing efficiency was defined as the ratio of the concentrations in the outlets. Thus, this parameter ranges from 0 to 1 for no to complete mixing if one stream is charged with the target species and the other not at all.

The experimental set-up used Hamilton syringes for liquid feed [114]. Two syringes pump and another two extract solution from the mixer. The pressure drop was measured differentially, by determining the pressure at one inlet and outlet, respectively.

1.3.6.3 Typical Results

Calculated liquid mixing time

[M 33] [P 31] The liquid mixing time was calculated for phenol ($D = 0.89 \cdot 10^{-9}$ m^2 s^{-1}) in an aqueous solution using the bi-laminating focusing mixer [114, 116]. For example, with a reduction in the channel width from 300 to 100 μm the time can be reduced from 35 to 4 s.

Calculated focusing channel length

[M 33] [P 31] For the bi-laminating focusing mixer, the length required for liquid mixing was calculated [114, 116]. For a flow of 6 μl min^{-1} and a channel height of 200 μm, a mixing length of 1.65 mm results.

Calculated pressure drop

[M 33] [P 31] The pressure drop was calculated for a flow of 6 μl min^{-1} at various micro channel widths [114]. For channel widths in the range 50–300 μm, the pressure drop decreases from about 1000 to 300 Pa.

Experimental liquid mixing time

[M 33] [P 31] The liquid mixing time was measured for phenol ($D = 0.89 \cdot 10^{-9}$ m^2 s^{-1}) in an aqueous solution using the bi-laminating focusing mixer [114]. At flow rates exceeding 60 μl min^{-1} virtually no mixing occurred, whereas nearly complete mixing was achieved below 1 μl min^{-1}.

1.3.7
Bi-laminating Microfluidic Networks for Generation of Gradients

Most Relevant Citations

Peer-reviewed journals: [117].

Gradients of diffusible substances having chemo-attractant or chemo-repellent properties play an important role in, e.g., biological pattern formation, morphogenesis, angiogenesis and axon path finding. The generation of such gradients in solution, even with a complex shape, divergent from the normal parabolic-shaped concentration profiles, is possible by using networks of micro channels designed to control diffusive mixing of substances [117].

1.3.7.1 Mixer 34 [M 34]: Bi-laminating Microfluidic Network

Several fluid manipulation steps occur within the 'Christmas tree', including splitting, combining and mixing of the different fluids. While keeping the number of inlets low, the splitting increases the number of streams with different concentration of the fluids forming a concentration profile in the broad outlet channel. One fluidic device described is a three-input/nine-output network, i.e. the three input fluid streams are split into four sub-streams which are mixed in the serpentine channel array (shown as a zigzag line in Figure 1.69). The used flow rates ensure a complete diffusive mixing. Subsequently, after flow through the four serpentine channels, the fluids form again inputs which are split into five sub-streams entering a second array of serpentine channels. Downstream of the microfluidic network all sub-streams are merged forming concentration profiles of different shape in the outlet channel [117].

The microfluidic devices, often called 'Christmas tree' owing to the pattern of the channels, were fabricated from polydimethylsiloxane (PMDS) using common micro technologies (see Figure 1.69). A mask is formed by using a high-resolution 3300 dpi printer to transfer the CAD pattern to a transparent mask substrate.

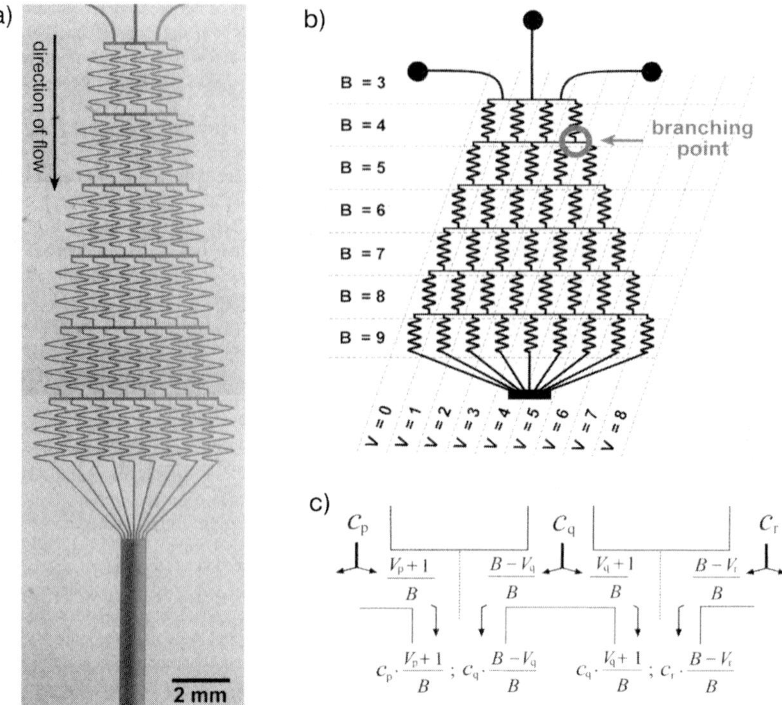

Figure 1.69 (a) Schematic of a 3-input/9-output 'Chrismas tree' microfluidic network ([117]). (b) Schematic for explanation of the mathematical solution of the microfluidic network. (c) Schematic demonstration of the derivation of equations governing the splitting ratios at the branching points. The dotted lines indicate the boundary between the two combined streams. The concentrations at the end of the serpentine channels can be calculated by multiplying the concentration of the incoming streams (cp, cq, cr) with the corresponding splitting ratio $[(V_p + 1)/B, (B - V_q)/B, (V_q + 1)/B$ and $(B - V_r)/B]$ [117] (by courtesy of APS).

A minimum feature size of ~20 μm is achievable with this method. The mask pattern is transferred into SU-8 photoresist coated on a silicon wafer using 1 : 1 contact photolithography and, after some subsequent process steps, a negative master of the channel network is achieved. Positive replicas can be achieved by molding the master with PMDS again. After punching the inlet and outlet holes, the PMDS surface and the glass cover plate are treated in air plasma. Putting both parts together, an irreversible tight seal is formed.

1.3.7.2 Experimental Characterization Protocols/Simulation

[P 32] The concentration profiles are generated by permuting the order at the inlets of fluorescein solutions (fluorescein in 100 mM $NaHCO_3$ buffer, pH 8.3, green) with different concentrations (100, 50 and 0%) and a solution of tetramethylrhodamine ethyl ester in ethanol. The overall flow velocity ranges from 800 μm s^{-1} to 1.2 mm s^{-1} [117].

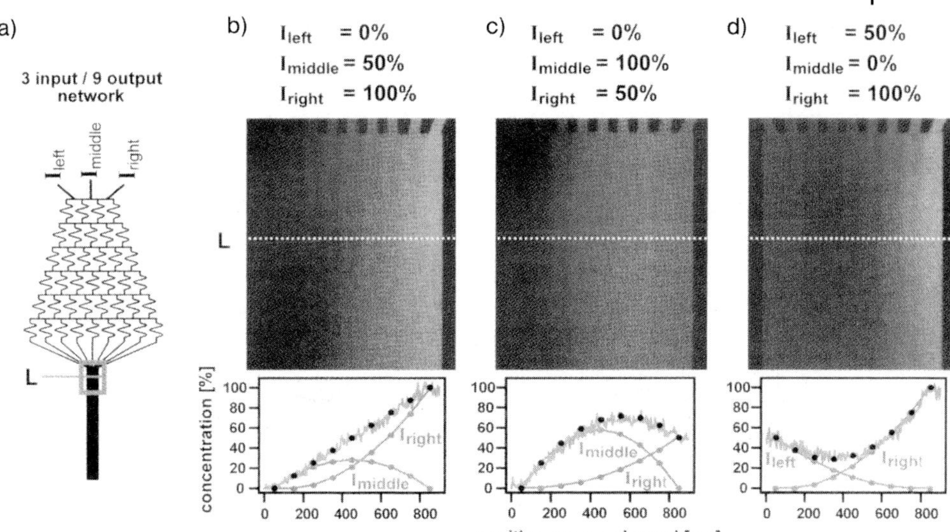

Figure 1.70 (a) Schematic of a 3-inlet/9-outlet microfluidic network. (b) Linear and (c, d) parabolic gradients of fluorescein in solution. The inlet concentrations are indicated by I_i. The plots show the fluorescence intensity profile across the broad outlet channel. The theoretically calculated concentration profiles and the contributions of the individual inlets are marked • [117] (by courtesy of ACS).

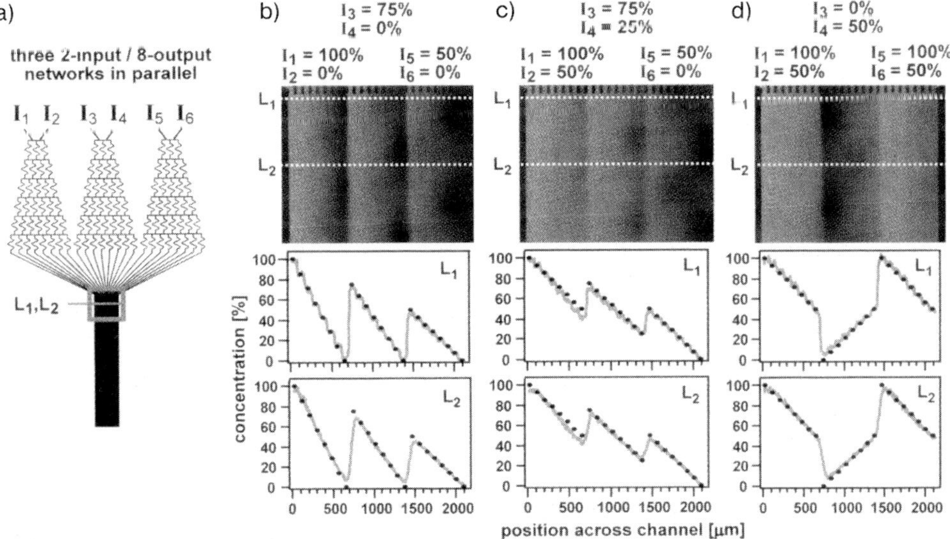

Figure 1.71 (a) Schematic of a 2-inlet/8-outlet microfluidic network in parallel. (b) Linear and (c, d) parabolic gradients of fluorescein in solution. The inlet concentrations are I_i. The plots show the fluorescence intensity profile across the broad outlet channel at the beginning of the channel and 800 μm downstream. The theoretically calculated concentration profiles are marked • [117] (by courtesy of ACS).

1.3.7.3 Typical Results

Linear and parabolic gradients

[M 34] [P 32] Linear and parabolic gradients can be generated at the outlet of the microfluidic network. In all cases, calculated concentrations [see Figure 1.70 (•)] are in a good agreement with the experimentally observed concentration profiles.

Periodic gradients

[M 34] [P 32] Periodic gradients can be formed by combining, in parallel, three networks each having two inlets. Since in each of these networks independently different linear concentration profiles can be generated, a variety of saw-tooth gradients can be established when the individual linear profiles are brought together in the broad outlet channel (see Figure 1.71) [117].

Increasing the number of inlets from two to three at the individual networks extends the accessible range of profiles. It is possible to form symmetric gradients of three parabolic parts or, by using preformed linear gradients, mixed profiles consisting of linear and parabolic parts are accessible [117].

Superpositioned gradients

[M 34] [P 32] Superposition of gradients can be generated by feeding the inlets of a two two-inlet/eight-outlet parallel network with different substances, e.g. of fluorescein (green) and tetramethylrhodamine ethyl ester (red). In both branches of the network an individual saw-tooth profile is generated. When both saw-tooth profiles merge in the outlet channel, superimposed periodic profiles are formed. Overlapping gradients can be used for comparing and quantifying the role of competing gradients in chemotaxis [117].

1.3.8 Bifurcation Multi-laminating Diffusive Mixing

Most Relevant Citations

Peer-reviewed journals: [42]; micro machining: [118]; analytical application/kinetic studies: [119].

Multi-lamination mixing relies on the generation of an alternating arrangement of thin fluid compartments, multi-lamellae, which are then mixed by diffusion. Multi-lamination is realized by alternating feed arrangements (type *A–B–A–B–* ...), the outlets of which direct into a flow-through mixing chamber and thus create the multi-lamellae pattern. The aim is to generate sufficiently small fluid compartments so that steep concentration gradients result which give fast mixing by diffusion. Most common feed schemes for interdigital mixing are bifurcation branching (see below) and interdigital branching (see Section 1.3.9).

Bifurcation branching achieves equidistribution by flow symmetry (see Figure 1.72). The outlet arrangement is basically identical with the interdigital feed concept. While the bifurcation concept does not demand the build-up of an additional pressure barrier (see the interdigital structures in Section 1.3.9), besides the pressure drop for the flow passage itself, it poses much higher demands on the structural precision and the absence of fouling or deposits within the channels, as both will break the ideal splitting geometry of bifurcation.

Bifurcation structures are self-replicating patterns which serially branch channels until a multitude is reached. In this way, a main stream can be split into many sub-streams just relying on flow symmetry. These sub-streams can be contacted in an interdigital-type arrangement with the sub-streams of a second feed and then be guided into a main channel or into an inverse bifurcation structure.

1.3.8.1 Mixer 35 [M 35]: Bifurcation Multi-laminating Micro Mixer

One inlet channel is split by a bifurcation structure into 16 sub-channels. On the backside of the same wafer, an identical bifurcation structure is made (see Figure 1.72) [42]. These sub-streams are redirected via wafer-through nozzles close to the end of the bifurcation structure on the other side. Here, the sub-streams are guided through short channels which lie next to the ends of the bifurcation structure. Two such alternate channels merge repeatedly to yield finally an inverse bifurcation structure which ends in a broad channel. Different from the first bifurcation structure having the same width all along the flow path, in the inverse structure the width increases constantly owing to the increasing number of streams. In this way, the lamellae thickness of the fluid is kept constant. The broad outlet channel is very long to ensure completion of mixing by diffusion.

Microfabrication is effected by etching a wafer from both sides which contains through-holes for the fluid connectors [42]. Two transparent cover plates close the open structures and also have through-holes. A glass capillary was attached to the three-plate chip. The holes in the cover plate were made by sandblasting.

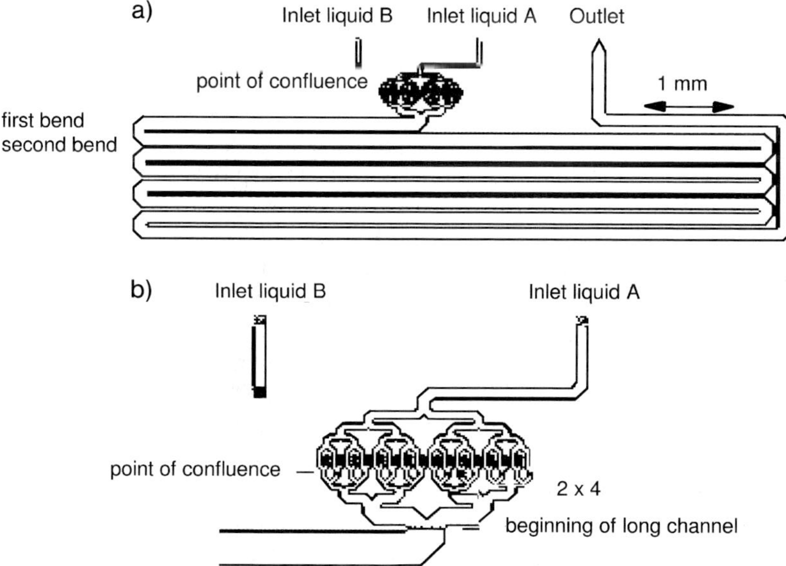

Figure 1.72 Schematic of the mixing element of the bifurcation laminating micro mixer [42] (by courtesy of RSC).

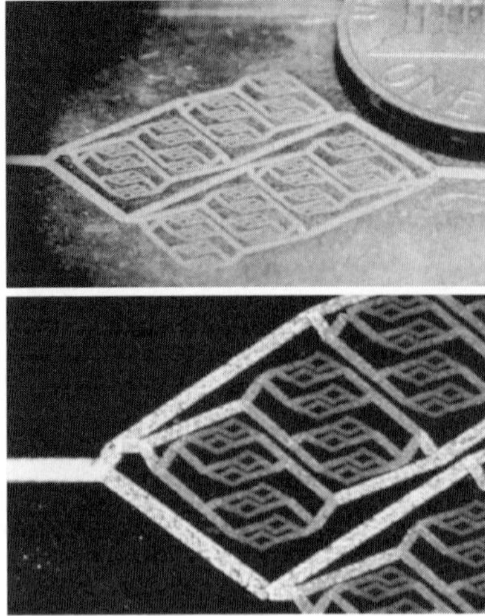

Figure 1.73 Biomimetic vasculatures [118] (by courtesy of RSC).

Mixer type	Bifurcation multi-laminating micro mixer	Channel width	49–58 μm
Mixer material	Silicon	Capillary material	Fused silica
Cover material	Glass	Capillary outer diameter, inner diameter	375 μm, 50–150 μm
Number of bifurcation branches	16	Outer size of the chip	5 mm × 10 mm
Nozzle (lamellae) width	20 μm	Total internal volume	600 nl

The fabrication of bifurcation structures was also achieved by a high-brightness diode-pumped Nd:YAG laser direct write method [118]. A PDMS structure was realized in one step and tested for mixing function by fluorescence imaging. The method provides rapid prototyping of master structures and is said to have high flexibility. Furthermore, it allows one to realize a bifurcation structure with multi-width multi-depth micro channels, i.e. a 3-D object. Such structures are similar to biomimetic vasculatures whose channel diameters change according to Murray's law, which states that the cube of the radius of a parent vessel equals the sum of the cubes of the radii of the daughters (see Figure 1.73). In this way, micro channel networks similar to physiological vascular systems may be fabricated, having all the favorable properties of the latter. First steps in this directionare described in [118].

1.3.8.2 Mixing Characterization Protocols/Simulation

[P 33] Reactive imaging was used for mixing and flow-pattern analysis [42]. Fluorescence quenching (using a photomultiplier tube) offers a very sensitive method to determine the completion of mixing.

Syringe pumps (25 μl syringes) were used for feeding the streams [42]. The flow is visualized via a microscope using a video camera. For quantitative measurements, the set-up is further modified.

For flow visualization, fluorescence solutions contained 40 μmol l^{-1} fluorescein and 100 μmol l^{-1} rhodamine B [42]. For the quenching experiments, 20 μmol l^{-1} fluorescein and 2 mol l^{-1} potassium iodide in borate puffer were used.

1.3.8.3 Typical Results

Ratio of dead volumes

[M 35] [P 33] The calculated dead volume referred to the inlet volumes (about 36 nl), which is small compared with the total internal mixer volume (600 nl) [42].

Liquid mixing time

[M 35] [P 33] The liquid mixing time for 95% mixing is about 15 ms [42]. Further mixing requires more time, also because the lamellae close to the wall are thicker than the internal ones.

Flow asymmetry

[M 35] [P 33] The lamellae close to the channel walls are larger than those in the interior [42]. In addition, a velocity profile exists, the lamellae at the walls moving much slower than the internal ones. Thus, the mixing of these different lamellae is for two reasons completed at a very different length. For high degrees of mixing, therefore, the diffusion of the boundary layers becomes dominant.

Quantitative mixing judgement

[M 35] [P 33] By fluorescence quenching, the degree of mixing at various stages of the flow path could be determined (see Figure 1.74) [42]. The data in the long outlet channel are in good agreement. In contrast, mixing in the flow passages beforehand is slower than expected. This is explained by the greater striation thickness at the point of confluence. From there, thinning of the lamellae occurs rapidly, consequently promoting mixing. At the point of confluence of the inverse bifurcation structure, the slower mixing is explained by the many layers being close to the channel walls. Owing to passing a curved flow passage (bend), the lamellae are, in addition, compressed in an asymmetric way, again slowing the mixing.

Time-resolved FTIR spectrometry for protein conformation kinetic study

[M 35] [protocol see [119]] A protein conformation kinetic study of the small protein ubiquitin was performed both in the continuous and in a stopped-flow mode at low reactant consumption [119]. The bifurcation mixer was used prior to an IR flow cell for data monitoring. The change of conformation from native to the A-state was followed when adding methanol under low pH conditions to the protein solution. In the continuous mode, long data acquisition could be made and the reaction time was determined by the flow rate and the volume interconnecting zone between the mixer and IR flow cell, which was small, but not negligible. In the stopped-flow mode, the reaction time resolved was dependent on the time resolution of the FTIR instrument.

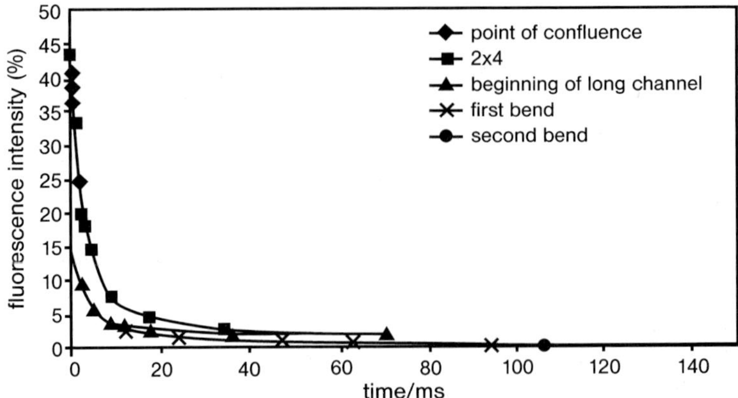

Figure 1.74 Decay of fluorescence intensity at selected points in the bifurcation laminating mixing element. These data are taken as a measure for spatially judging the mixing efficiency [42] (by courtesy of RSC).

1.3.9
Interdigital Multi-laminating Diffusive Mixing (Normal and Focusing)

Most Relevant Citations

Peer-reviewed journals: [20, 36–39, 41, 67]; proceedings contributions: [34, 40, 120, 121]; patents: [122, 123].

Interdigital feeds, termed in analogy with the respectively arranged electrode structures, provide multiple outlet ports with alternately arranged fluids (type *A–B–A–B*, etc.) [34, 67]. The typical architecture of interdigital feeds comprises a large reservoir from which many equal sub-stream channels branch ending in nozzle-type outlets, for each fluid. The equidistribution to these is given by applying a pressure barrier, i.e. making the hydraulic diameter of the sub-stream channel much smaller than that of the reservoir.

When the multi-lamellae pattern is vertical to the feed reservoir layers, this is named vertical multi-laminating. When both are oriented in the same way, this is the horizontal variant.

In a vertically multi-laminating variant of the interdigital principle, the two arrays of the multiple sub-streams, belonging to two fluids, are overlaid in such a way that the above-mentioned alternate arrangement results (see Figure 1.75) [34, 67]. This usually requires the reservoirs to be placed in two different layers of the microstructured device. The sub-stream channels typically start in the two different layers and merge in a common layer.

The direction of the feed flows before approaching the mixing chamber in the vertical multi-laminating variant can be counter-, cross-flow or co-flow, the selection of which is mainly governed by microfabrication needs (see Figure 1.75). In the case of counter-flows, the streams have to be re-directed before entering the mixing channels. In this way, co-flow entering is achieved.

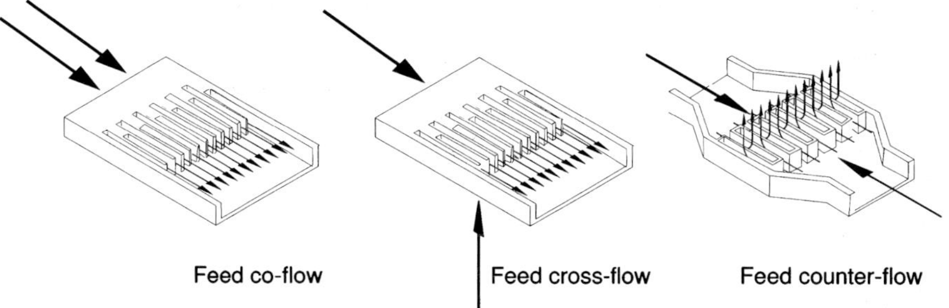

Figure 1.75 Vertically multi-laminated interdigital feeds. For the horizontal variant, counter-, cross- and co-flow injection schemes are given (large arrows). The alternating streams enter the mixing channel in all cases as co-flows (small arrows) [39].

In a horizontally multi-laminating variant, plates, typically with multi-channel arrays, are stacked on *A–B–A–B*-type manner (see Figure 1.75) [40, 41]. Here, the complete multi-channels feed one lamellae, while in the vertical variant this is done by a single channel. In this way, fairly large throughputs can be achieved. The outlets of the multi-channel arrays of the vertical variant may have straight or oblique directions (see Figure 1.81) [40, 65]. In the latter case, it is thought that the fluids keep this direction for a distance so that swirling flow patterns may be generated.

Considering the width of typical interdigital outlets, which are of the order of several tens of microns, it stands to reason that multi-lamination on its own is a rather slow mixing process, typically needing liquid mixing times of the order of seconds (see Figure 1.76 and, e.g., [34]).

This may be adequate for μTAS applications, where stream velocities are low and hence residence times are large. For most chemical and chemical engineering applications, however, faster mixing times are required. Though today's micro-

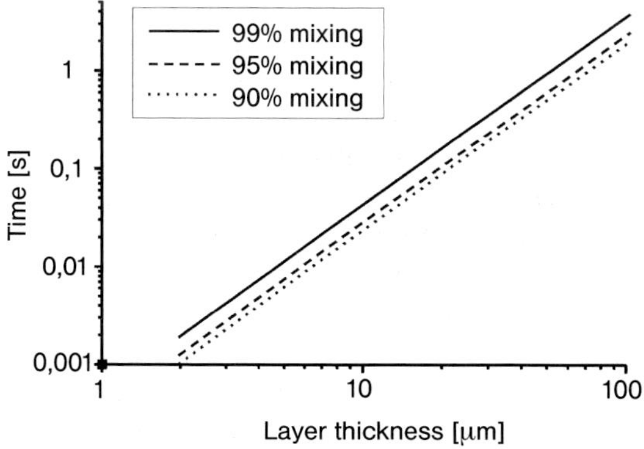

Figure 1.76 Typical mixing times calculated for a range of diffusion distances, physically corresponding to the lamellae width, assuming diffusion to be the only mixing mechanism [34] (by courtesy of AMI).

fabrication potentially allows one to manufacture outlet widths at the micron level (and below), this is neither cost-efficient nor practical facing throughput or fouling–stability issues. An alternative concept for mixing speed-up utilizes the well-known hydrodynamic focusing concept [112] (see Section 1.3.5). In this way, fluid lamellae are compressed by external forces, e.g. stream constraints. Favorably, this can be achieved by posing geometric constraints [114–116]. This principle, performed first for bi-laminated streams [114–116] (see Section 1.3.6), has been transferred to interdigital multi-laminated devices [20, 35, 37, 124]. The result of geometric focusing is the generation thinner lamellae which mix faster.

Although the compression of lamellae follows a hydrodynamic principle, it was decided to term this approach 'geometric focusing' throughout this chapter, to distinguish this design-based method better from the processing-based hydrodynamic focusing relying on setting very different pressures for the inlet streams.

Geometric focusing via triangular-shaped focusing chambers using horizontal co-flow injection schemes has become state-of-the-art for interdigital mixers (see Figure 1.77) [20, 35, 37, 124].

Initially, geometries other than triangles, such as sections of an arc, were employed, mainly for fabrication reasons [36]. These had a too extreme focusing ratio resulting in inhomogeneous lamellae formation and lamellae tilting and winding at high *Re* [20]. Thereafter, triangular focusing chambers were introduced, with focusing ratios of the order of several tens [20, 35, 37, 124]. Based on these first achievements and a semi-analytical optimization study, a novel focusing design of much larger focusing ratio was developed with the targets to have regular focusing with equally spaced lamellae at a liquid mixing time of only a few milliseconds [20, 37]. In addition, it was demanded that the mixing should be completed within

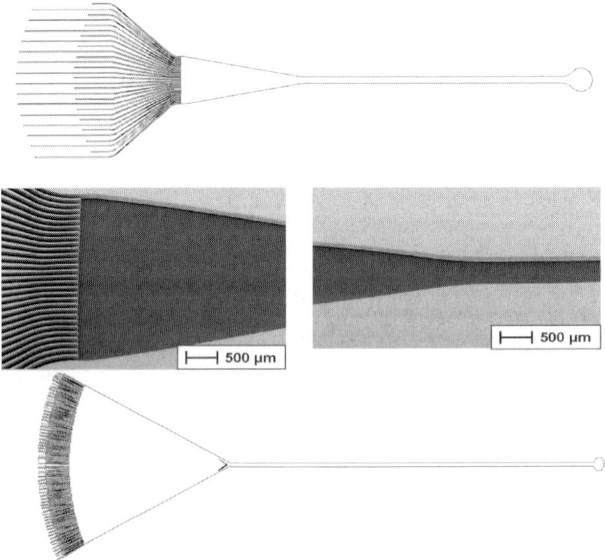

Figure 1.77 A triangular focusing geometry for multi-lamellae flows (source IMM).

the micro device, and not in the attached tubing [20, 37]. The novel, optimized design was termed *SuperFocus* (see [M 44]). A second-generation version of the *SuperFocus* mixer was developed with the issue of throughput enhancement into the range of several hundred liters per hour, considering an aqueous system [39].

The focusing–interdigital approach is comparatively simple to scale up, as this means only increasing the number of lamellae within one plane (equalling-up or internal numbering-up). In addition, the flow rate can be further increased simply by enlarging the height of the fluid compartments in the vertical direction, i.e. having a larger structural depth of the flow-through mixing chamber.

The interdigital feed with flow compression (geometric focusing) can have additional flow expansion which introduces a jet in a flow-through chamber [20, 36, 37].

1.3.9.1 Mixer 36 [M 36]: Unfocused Interdigital Multi-laminating Micro Mixer with Co-flow Injection Scheme (I), 'Rectangular Mixer'

This micro mixer simply creates multi-laminated streams via an interdigital feed arrangement, but does not rely on geometric focusing of these streams by geometric constraints as triangular mixers [M 43] do [20] (see also [39]). Mixing is performed in a mixing chamber, which has a rectangular format from top view, giving the name of the device, 'rectangular mixer'.

The rectangular mixer is composed of four plates (see Figure 1.78) [20, 39, 124]. The second plate contains the complete interdigital feed for one fluid (reservoir + multi-channel feed), the multi-channel feed of the other fluid (without reservoir), the focusing section and the mixing channel. The third plate comprises two rows of multiple holes serving as conduits to the underlying fluid reservoir. Fluid distribution is achieved via these holes. A first plate serves as cover.

The microstructures are fabricated by photoetching of a specially glass (see Figure 1.79) [20, 124]. Such microstructured plates are joined by soldering. The joint plate stack is held in a frame.

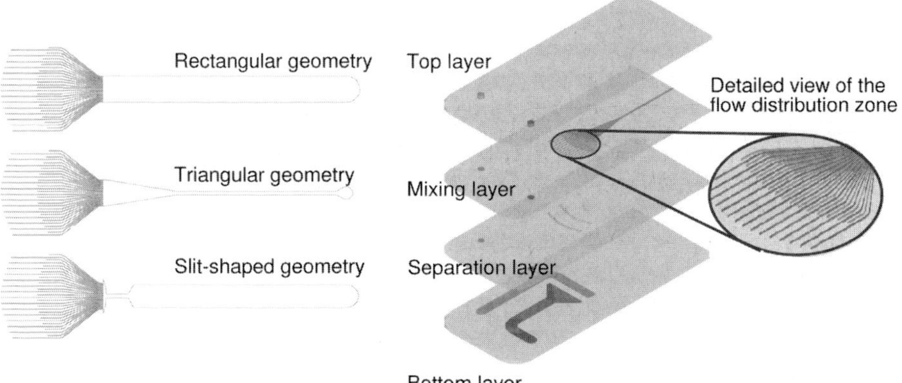

Figure 1.78 Exploded view of a glass micro mixer (right) and design variations of the mixing zone (left) [124] (source IMM).

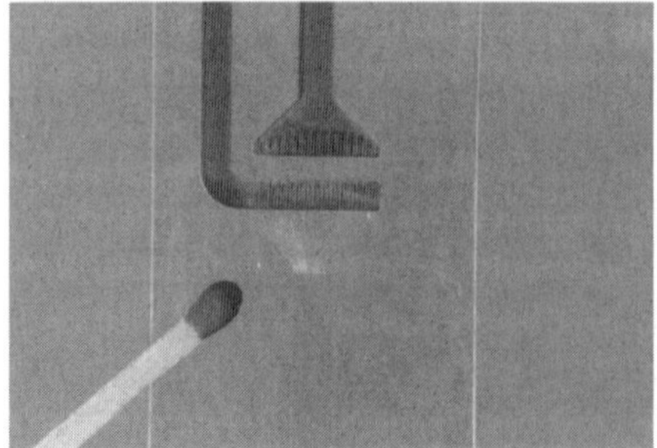

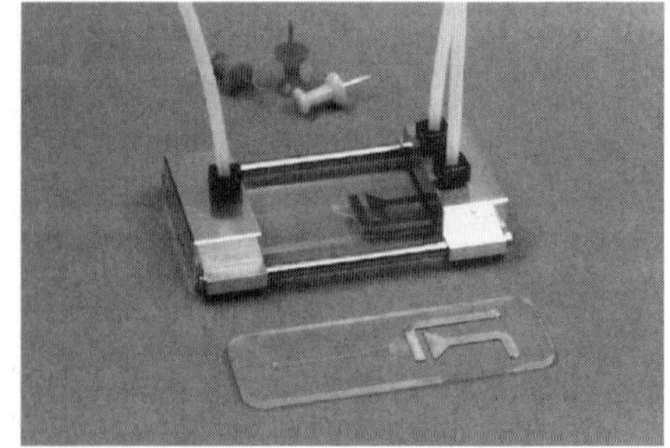

Figure 1.79 Microstructured four-plate glass stack of the interdigital micro mixer (left) and assembled device with steel frame (right) [124] (source IMM).

Mixer type	Interdigital rectangular micro mixer with co-flow injection scheme	Feed channel width and depth, width of separating walls	60 μm, 150 μm, 50 μm
Mixer material	Specially glass (Foturan®)	Number of micro channels	2 × 15
Mixer frame material	Stainless steel/ aluminum	Mixing chamber length, width, height	27.4 mm, 3.25 mm, 150 μm
Device outer dimensions: length, width, thickness	76 mm, 26 mm, 2.3 mm		

1.3.9.2 Mixer 37 [M 37]: Interdigital Vertically Multi-laminating Micro Mixer with Co-flow Injection Scheme (II)

This device uses an alternate arrangement of streams (see Figure 1.80) which are much thinner compared with their width, i.e. the diffusion distances were set deliberately large [67]. The corresponding diffusion-based mixing was named lateral mixing. For comparison, a so-called vertical mixing device was made (see Figure 1.80). Here, the thin layers are overlaid so that the diffusion distances are kept small, rather than placing them side by side.

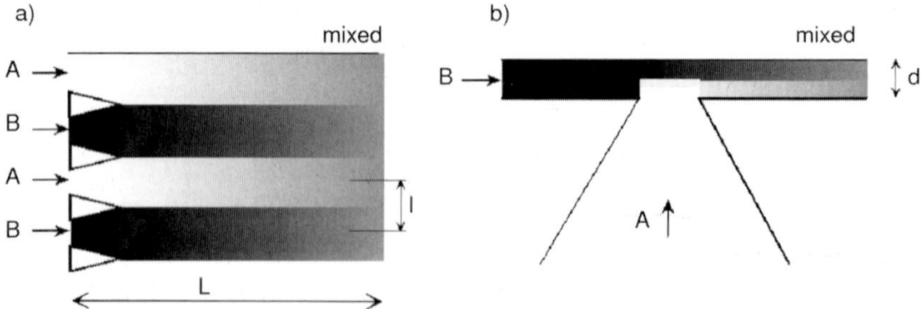

Figure 1.80 Schematic of the flow guidance and the mixing principle of micro mixers for (a) lateral view of lamination and (b) vertical view of lamination [67] (by courtesy of IOP Publishing Limited).

In practice, the above-mentioned design issues are realized by having a large reservoir within a multi-plate rig on which the mixing device is mounted and screwed using seals [67]. At the top of this reservoir several orifices form through-holes to an interdigital structure for each liquid. The orifices of one liquid are arranged in-line, perpendicular to the liquid flow. From each orifice a separate feed channel originates und guides to a rectangular flow-through chamber. The orifices of the other liquid are placed in-line, at a distance to the first line of orifices. The position of each orifice is chosen so that it is placed in between two feed channels carrying the other liquid, yielding an *A–B–A–B*-type pattern. Behind this second row of orifices, multiple, parallel channels lead to a large rectangular flow-through chamber where multi-lamellae mixing takes place.

Micro fabrication was made by conventional silicon wet etching. Sealing was achieved by anodic bonding to Pyrex glass [67].

Mixer type	Interdigital vertically multi-laminating micro mixer with co-flow injection scheme	Feed channel width, depth, length for liquid 2	10 μm, 5 μm, 0.335 mm
Mixer material	Silicon	Mixing chamber width	2.21 mm
Cover material	Pyrex	Total chip size	7.5 × 4 mm
Number of micro channels liquids 1 and 2	14, 13	Rig material	PVC
Feed channel width, depth, length for liquid 1	40 μm, 5 μm, 2.335 mm	Rig sealing material	Silicone rubber
Fin distance between the feed channels	70 μm		

1.3.9.3 Mixer 38 [M 38]: Interdigital Horizontally Bi-laminating Micro Mixer with Cross-flow Injection Scheme, Reference Case to [M 37]

This device performs a bi-lamination using two fluid layers of basically the same dimensions as the device reported above, [M 37] (see Figure 1.80) [67]. In contrast, however, these layers are not laterally oriented, but are superposed in the vertical direction. The device serves mainly as a benchmark for the performance of [M 37].

Fluid feed is realized by feeding one liquid through one rectangular through-inlet, [67]. This flow is entering a flow-through chamber and fills this volume. At a distance to the first inlet, the second inlet is introduced from below via 42 rectangular orifices so that the lower half of the channel depth is filled with the other liquid stream, the upper half being composed of the first liquid. The orifices are grouped into four rows, two with 10 orifices and two with 11 orifices. From there, a bi-laminated flow is passed through the chamber, which now serves for diffusion mixing. The whole flow-through chamber has a lower depth as compared with its width.

The device is mounted and screwed using seals on a multi-plate rig on which the mixing device is located [67].

Micro fabrication was effected by conventional silicon wet etching. Sealing was achieved by anodic bonding to Pyrex glass [67].

Mixer type	Interdigital horizontally bi-laminating micro mixer with cross-flow injection scheme, reference case to [M 37]	Mixing chamber depth, width	5 μm, 2.21 mm
Mixer material	Silicon	Total chip size	7.5 × 4 mm
Cover material	Pyrex	Rig material	PVC
Number of micro channels liquids 1 and 2	1 : 1	Rig sealing material	Silicone rubber

1.3.9.4 Mixer 39 [M 39]: Interdigital Horizontally Multi-laminating Micro Mixer with Co-flow Injection Scheme

This device achieves a horizontal orientation of the lamellae by alternate vertical stacking of plates, which carry one fluid each. This is different from the interdigital micro mixers described above having a vertical *A–B*-type orientation [40, 41].

Vertical orientation is preferably achieved by placing multi-channel plates as a stack in such a way that the plates carry the fluids alternately [41]. For this purpose, two sets of plates have to be manufactured which typically have a mirror-imaged design to be fed from two oppositely placed fluidic ports. The stack-like design is amenable to numbering-up to operate even many thousands of channels in parallel and large throughputs [65].

Such multi-platelet stacks were realized with tilted, straight (V-type) and curved (P-type) multi-channel arrays (see Figure 1.81) [41]. The V-type device has slanted outlets so that neighboring jets may collide; the P-type has parallel streams.

The microstructures were realized by a special, advanced turning technique [41]. The platelet stacks were diffusion bonded for sealing and encased in a housing by

P-Mixer: curved channels

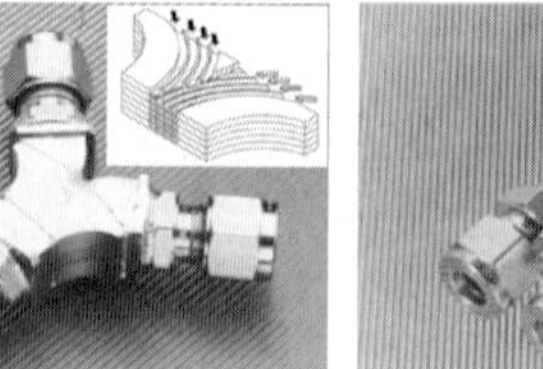

V-Mixer: straight channels

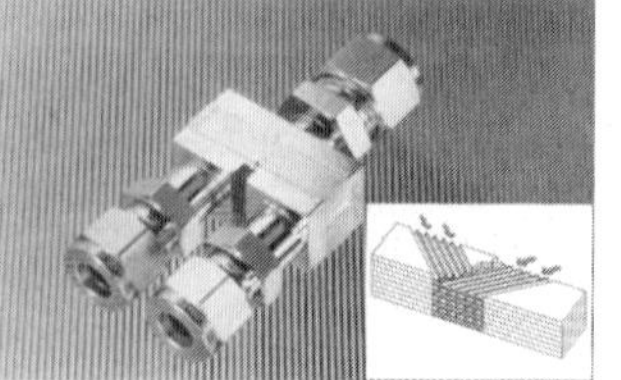

V-Mixer with minimized dead volume

Figure 1.81 Two different horizontally multi-laminating mixers with different co-flow schemes; curved-channel design (P-type, left), straight-channel design (middle). In addition, a part of a V-type mixer with minimized dead volume is shown [Pfeifer et al., Chem. Ing. Tech. 76, 5 (2004) 607].

electron beam welding. The housing also contains a mixing chamber typically of the same cross-section as the stack.

The flow in the feed channels is laminar, as is to be expected [41]. Owing to the size of the mixing chamber and the high fluid velocities, liquid mixing is expected to have a fast transition from laminar to turbulent. Evidence for this is given below. Different from conventional micro mixers, the primary vortices are in the range of about 100 μm, hence smaller than usual. Thus, mixing is different and should be faster.

(a) Version described in [41].

Mixer type	Interdigital horizontally multi-laminating micro mixer (I)	Feed channel depth, width	70 μm, 100 μm
Mixer material	Stainless steel	Wall width	30 μm
Housing material	Stainless steel	Number of micro channels	2500

(b) Version described in [40] (see Figure 1.82).

Mixer type	Interdigital horizontally multi-laminating micro mixer (II)	Number of channels per passage	480
Mixer material	Stainless steel	Platelet thickness	200 μm
Housing material	Stainless steel	Width of micro-structured stack	7.9 mm
Feed channel depth, width, length	100 μm, 100 μm, 20 mm	Width of mixer chamber	11.2 mm
Number of channels per platelet	40	Height of mixer device	4.8 mm
Number of platelets per passage	12		

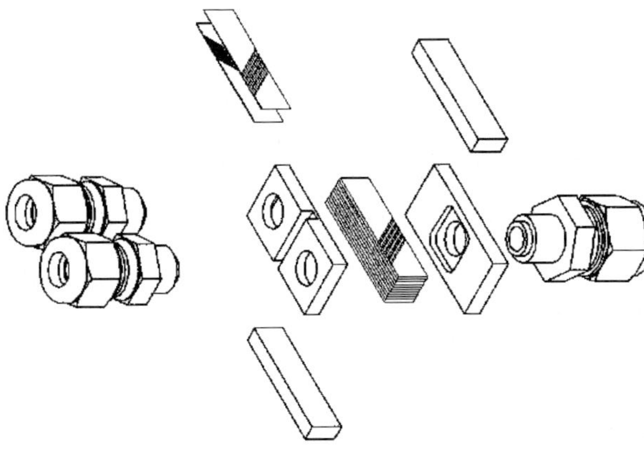

Figure 1.82 Exploded view of a horizontally multi-laminating mixer with a multiple-late stack as central feed element, having multiple parallel channels on each plate [Pfeifer et al., Chem. Ing. Tech. 76, 5 (2004) 607].

1.3.9.5 Mixer 40 [M 40]: Interdigital Vertically Multi-laminating Micro Mixer with Counter-flow Injection Scheme – '3-D Slit Mixer'

This device is made from a microstructured inlay with an interdigital element which is inserted in a recess of a two-piece housing (see Figure 1.83) [36].

The inlay has a counter-flow configuration consisting of many parallel channels fed by one reservoir each (see Figure 1.84). The parallel channels branch in an interdigital structure. The shape of the channels was not straight, but was of meandering nature, mainly for fabrication reasons to allow the realization of high-aspect-ratio structures. From there the sub-streams are redirected at 90° into a flow restriction zone, called 'slit', owing to its initial cross-sectional shape. This area is rapidly decreased, since the vertical extension resembles the shape of a small section of an arc [20]. Thereby, hydrodynamic focusing takes place, yielding lamellae compression to speed up diffusion. The slit conduit is connected to tubing. This tubing itself can consist of two cylinders with expanding flow cross-section or simply one cylinder, roughly of the same hydraulic diameter as the end of the slit. In the case of expanded tubing, it is known that a jet forms at sufficiently high flow velocities and possibly jet mixing could assist diffusion mixing. The extent of the jet mixing so far has not been quantified. The fluid connectors are all set side-by-side on the top part of the housing. The bottom part contains the recess.

The inlay is typically made by the LIGA technique, a combination of deep X-ray lithography, electroforming and molding, which gives very precise and steep microstructures [36]. Here, the electroformed structure made in metal was taken as the end product; no molding step was applied. Alternatively, advanced silicon etching (ASE) can be applied to achieve structures of virtually the same quality in silicon. Specially inlays were made in stainless steels including high-alloyed

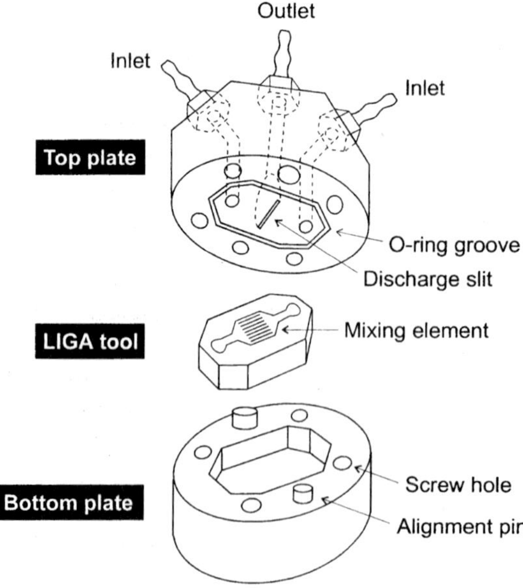

Figure 1.83 Exploded view of the interdigital mixer device [36] (by courtesy of ACS).

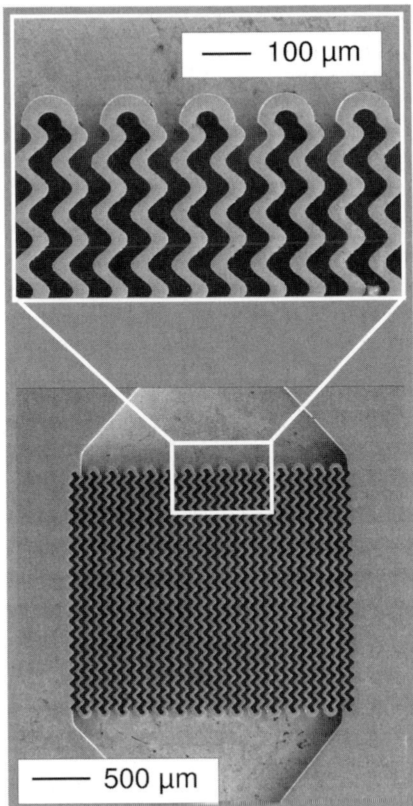

Figure 1.84 SEM image of the interdigital mixer element [36] (by courtesy of ACS).

modifications such as Hastelloy by using µEDM. To save machining time, linear channels of reduced aspect ratio were made instead of the periodically curved ones. By means of laser ablation, similar structures were made in polymers.

The housing was fabricated by precision engineering, including drilling, milling and µEDM.

In a very early version, the interdigital elements presented above were made in glass by photoetching and joined by a thermal bonding process [125, 126]. In this way, transparent structures were achieved. The mixer was part of an integrated microfluidic system composed of many photoetched glass plates which were thermally bonded [126]. Heating and cooling channels and delay loops were placed besides two mixing elements connected in series. Instead of having tiny slits, larger square openings were put above the interdigital elements, probably at the price of reduced mixing. Characterization of such system has not been reported in the open literature.

Later, the same fabrication and interconnection process was used with an interdigital design, being more adapted to the basic elements, flat plates [20, 124]. Instead of having a complex 3-D channel guidance, the flows always follow the plate horizontally and only change from one plate to another at certain location.

Mixer type	Interdigital vertically multi-laminating micro mixer with counter-flow injection scheme	Slit width	60 μm
Mixer inlay material	Metals such as nickel or silver; to a lesser extent: silicon, stainless steels, polymers	Tubing diameter	50 μm
Mixer housing material	Stainless steel	Second tubing diameter (specially version)	500 μm
Feed channel width, depth, length (two versions)	40 μm, 300 μm, 1 mm 25 μm, 300 μm, 1 mm	Sealing material	O-ring polymer seals
Number of micro channels for one feed (two versions)	15, 18	Total size (diameter, height)	20 mm, 16 mm

1.3.9.6 Mixer 41 [M 41]: Interdigital Vertically Multi-laminating Micro Mixer with Counter-flow Injection Scheme, 10-fold Array

This device is based on a numbered-up design of the cross-flow interdigital elements of [M 40] [36]. The basic parts of this device are essentially the same, a micro-structured inlay and a two-piece housing (see Figure 1.85, bottom). Ten interdigital mixing elements are operated in parallel in a star-like arrangement (see Figure 1.85, top). The feed distribution of one fluid is achieved from tubing to a ring-like channel structure in the top housing part. The other fluid is fed to the center of the inlay and spread in a star-like manner. The laminated streams are guided through a dodecagon-shaped slit and collected via a through-hole in the slit channel.

Fabrication is carried out in the same way as reported for [M 40].

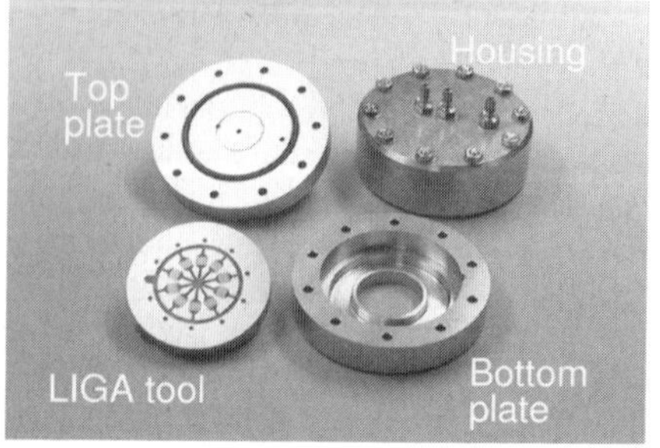

Figure 1.85 Single and 10-fold array interdigital mixer inlays (left) and assembled and disassembled 10-fold array micro mixer device (right) [36] (by courtesy of ACS).

Mixer type	Interdigital vertically multi-laminating micro mixer with counter-flow injection scheme, 10-fold array	Number of micro channels for one feed and one element (two versions)	15, 18
Mixer inlay material	Metals such as nickel or silver	Slit width	350 μm
Mixer housing material	Stainless steel	Second tubing diameter (one version only)	500 μm
Feed channel width, depth, length (two versions)	40 μm, 300 μm, 1 mm 25 μm, 300 μm, 1 mm	Sealing material	O-ring polymer seals

1.3.9.7 Mixer 42 [M 42]: Interdigital Vertically Multi-laminating Micro Mixer with 'Slit-type' Focusing – 'Plane Slit Mixer'

This slit-type interdigital micro mixer is a planar version of the device [M 40] (see also [M 41]) mentioned above, because it comprises an identical focusing zone [20, 37, 124]. In slit-type focusing zone use, the multi-laminated streams are compressed at large focusing ratio by geometric constraints [20]. Then, the mixing channel with the multi-laminated streams is connected to a mixing chamber of much larger cross-section where a jet is formed surrounded by two eddies at sufficiently large flow velocity.

The construction and microfabrication are identical with those of the rectangular micro mixer [M 36] (see above) [20, 37, 124].

Mixer type	Interdigital vertically multi-laminating micro mixer with 'slit-type' focusing	Number of micro channels	2 × 15
Mixer material	Specially glass (Foturan®)	Slit depth in steel housing	60 μm
Mixer frame material	Stainless steel/ aluminum	Mixing channel: length, width, height	19.4 mm, 500 μm, 150 μm
Device outer dimensions: length, width, thickness	76 mm, 26 mm, 2.3 mm	Focusing factor	6.5
Slit-type chamber: initial width, focused width, depth, focusing length, expansion width, expansion length, expansion angle	4.30 mm, 500 μm, 150 μm, 300 μm, 2.8 mm, 24 mm, 126.7°		

1.3.9.8 Mixer 43 [M 43]: Interdigital Vertically Multi-laminating Micro Mixer with Triangular Focusing (I)

Triangular micro mixers use hydrodynamic focusing of multi-laminated streams by geometric constraints [20, 39, 124]. The multi-channel feed section with the two reservoirs and pressure barriers is initially set on two plate levels. The feed channels are combined to the interdigital structure on one plate level in a co-flow injection scheme. In a triangularly shaped focusing section, a linear decrease in the cross-section is achieved. In standard triangular mixers, focusing ratios, i.e. the amount of the lamellae compression, are small (about 2–10); a specially version, the so-called SuperFocus, is designed to have very high focusing ratios up to 200 [20, 37, 39, 121].

The construction and microfabrication are identical with those of the rectangular micro mixer [M 36] (see above) [20, 124].

Mixer type	Interdigital vertically multi-laminating micro mixer with triangular focusing	Number of micro channels	2 × 15
Mixer material	Specially glass (Foturan®)	Mixing chamber: length, initial width, height, opening angle	8 mm, 3.2 mm, 150 μm, 20°
Mixer frame material	Stainless steel/aluminum	Mixing channel length, width, height	19.4 mm, 50 μm, 150 μm
Device outer dimensions: length, width, thickness	76 mm, 26 mm, 2.3 mm	Focusing factor	6.5
Feed channel width and depth, width of separating walls	60 μm, 150 μm, 50 μm		

1.3.9.9 Mixer 44 [M 44]: Interdigital Vertically Multi-laminating Micro Mixer with Optimized Triangular Focusing – 'SuperFocus'

The SuperFocus micro mixer is a specially version of the triangular mixer [M 43] with a very high focusing ratio (see Figure 1.86) [20, 37, 39, 121]. High ratios a/b (a = arc of inlet channel ends; b = mixing channel width) are achieved by increasing both the inlet (a) and outlet (b) diameter of the focusing zone. Owing to this fact, the number of inlet channels of the SuperFocus mixer is notably increased compared with the triangular device [M 43]. The inlet channels are made larger to diminish fouling effects, when SuperFocus mixers are designed for large liquid throughputs. The focusing length c should be kept small, in order to have a small residence time in the focusing zone, but not too small, since otherwise lamellae tilting and twisting may occur (see *Lamellae twisting in multi-lamination patterns*).

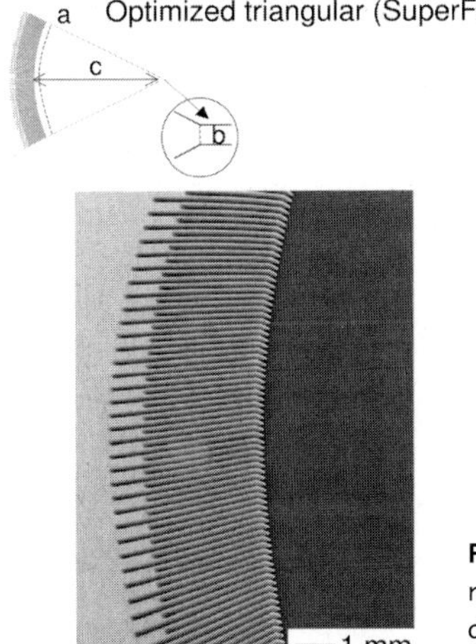

Figure 1.86 Design of an optimized triangular interdigital micro mixer, termed SuperFocus, and SEM image showing detail of the corresponding feeding zone [20] (by courtesy of AIChE).

(a) Glass version [20, 37, 39, 121]

The glass version of the SuperFocus mixer is composed of three layers containing similar structures to the standard triangular mixer [M 43] (see above) [20, 37, 39, 121]. Photoetching of a specially glass is used for manufacture.

Mixer type	Interdigital vertically multi-laminating micro mixer with optimized triangular focusing – SuperFocus	Number of micro channels	2 × 62
Mixer material	Specially glass (Foturan®)	Mixing chamber: length, initial width, height, opening angle	22 mm, 19.84 mm, 500 μm, 50°
Mixer frame material	Stainless steel/ aluminum	Mixing channel: length, width, height	50 mm, 500 μm, 500 μm
Device outer dimensions: length, width, thickness	100 mm, 26 mm, 5.2 mm	Focusing factor	39.7
Feed channel width and depth, width of separating walls	100 μm, 500 μm, 60 μm		

(b) Steel version [39]

The steel version comprises a two-piece housing, on request with an inspection window. Thin-wire erosion is used for microfabrication of the tiny cogs which served as inlet structures to achieve the multi-lamination (see Figure 1.87). High-speed

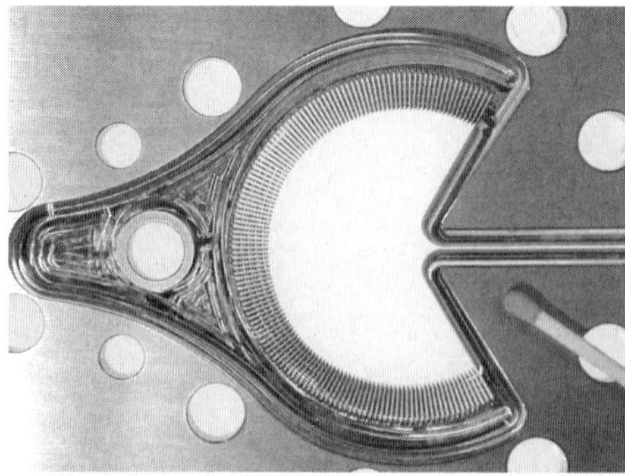

Figure 1.87 Central part of the steel SuperFocus mixer, a plate with the interdigital cogs, machined by μEDM (source IMM).

milling is used for manufacture of the fluid distribution structures which are large reservoirs. These feeds are arranged on opposite sides of a plate, comprising in its center the focusing and mixing zones (see Figure 1.87). This plate is inserted in a two-piece housing and sealed with graphite gaskets. The feeds are open to every second cog, which forms a channel after assembly within the housing. In this way an interdigital feed channel arrangement is created.

Mixer type	Interdigital vertically multi-laminating micro mixer with triangular optimized focusing – SuperFocus	Number of micro channels	2 × 69
Mixer and housing material	Stainless steel	Mixing chamber: length, initial width, height, opening angle	20 mm, 82.96 mm, 5 mm, 240°
Device outer dimensions: diameter, thickness	140 mm, 35 mm	Mixing channel: length, width, height	38 mm, 500 μm, 5 mm
Feed channel width and depth, width of separating walls	360 μm, 5 mm, 250 μm	Focusing factor	165.9

1.3.9.10 Mixer 45 [M 45]: Interdigital Vertically Multi-laminating Micro Mixer with Triangular Focusing Zone (II)

This interdigital micro mixer with a triangular focusing zone was realized consisting of a main channel and five feed channels for each fluid [120]. The two sets of feed channels are interdigitated, end at the beginning of the main channel and have the shape of half a star. At the end of the channels, ports are arranged on the section of an arc. There is an inner arc for one fluid and an outer arc for the other. The sets of ports are fed by one connector each. The main mixing channel has a short focusing region at the beginning, albeit of low focusing ratio.

Microfabrication was made using the SU-8 technique [120]. In particular, a two-layer manufacture without intermediate resist development was chosen to create two-levelled channel structures. SU-8 was brought on to a silicon substrate and UV irradiated twice, then dissolution of part of the resist was performed. For sealing, a special technique was developed. An SU-8 layer coated on a quartz or Pyrex substrate was pressed against the exposed resist structure. By baking and UV light exposure through the transparent top plate, both SU-8 layers are merged to one body, yielding a capped micro channel. If desired, the silicon substrate can be removed by wet-chemical etching.

Mixer type	Interdigital vertical multi-laminating micro mixer with triangular focusing zone	Feed channel width	50 μm
Mixer material	SU-8 Polymer	Mixing channel width	1 mm

1.3.9.11 Mixer 46 [M 46]: Interdigital Vertically Multi-laminating Micro Mixer with Flow-re-directed Focusing Zone

In this version of a focusing interdigital mixer, one feed is realized via through-holes in the top plate of a device [34]. The reservoir is placed below the top plate; contacting with the other fluid is achieved on the same level. The holes are framed by a U-type wall so that the flow is allowed to pass via the opening of this ‘U’ structure (see Figure 1.88). The other fluid is passed through a conduit outside the frame. The frames are arranged in arrays with a short distance to each other, so shaping this conduit. The groups of such linear arrays form a V-like superstructure.

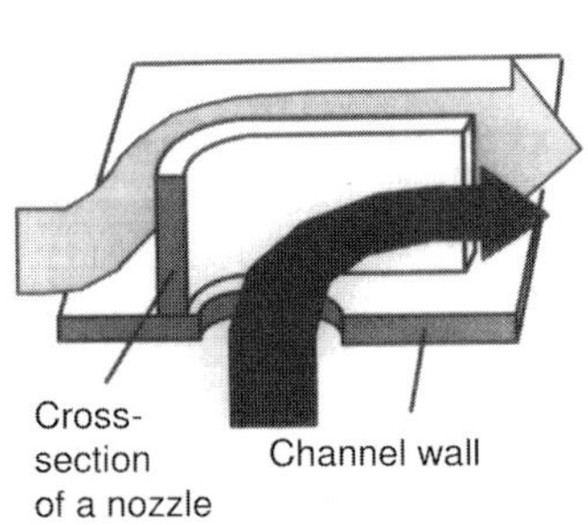

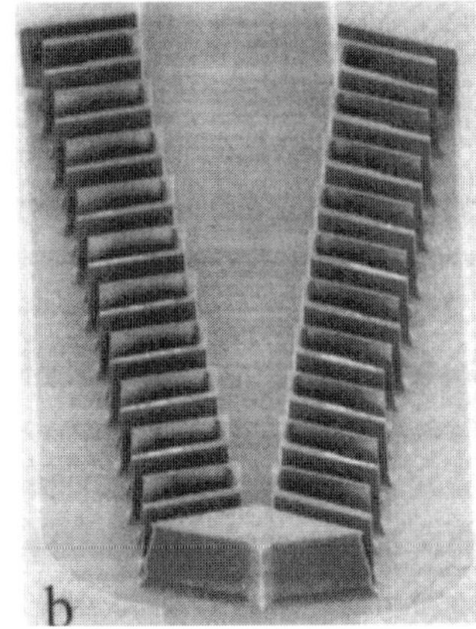

Figure 1.88 ‘U’-type frame with hole and opening forming an injection-nozzle structure (left) for an interdigital multi-laminating feed of V-like superstructure (right) [34] (by courtesy of AMI).

In the interior of the 'V' the reservoir of the one fluid is given (the other being below as mentioned above). Outside the 'V', two triangular-shaped mixing chambers are formed with a cross-sectional expansion in the flow direction in such a way that the lamellae thickness is kept constant. The flow is re-directed within these mixing chambers nearly perpendicular to the injection via the 'U'-type openings. In this way, geometric focusing of the flow is achieved, decreasing the lamellae width and promoting mixing. The two flow-through chambers merge, as do the corresponding multi-lamellae streams.

Microfabrication was effected by anisotropic reactive ion etching (RIE) using a plasma [34]. Back-side structures were realized by wet-chemical etching. A transparent plate is bonded on top of the microstructures.

Mixer type	Interdigital multi-laminating micro mixer with flow-re-directed focusing zone	Channel depth	50 μm
Mixer material	Silicon	'U' frame width, length	~40 μm, ~80 μm
Cover material	Glass	Lamella initial thickness	~30 μm

No further details on geometries are given in [34].

1.3.9.12 Mixing Characterization Protocols/Simulation

[P 34] Dilution-type dye imaging with phenol red indicator and dilute hydrochloric acid solution was applied and monitored by video taping [34]. Details on flow rates were not given.

[P 35] A reaction system with two competitive parallel reactions was used for mixing characterization [36]. The Dushman reaction involves the mixing of iodate, iodide and sodium acetate in one solution and a strong acid such as sulfuric acid or hydrochloric acid in another solution. If mixing is fast, the neutralization of the acid and the base dominates as the faster reaction. The redox reaction of iodide and iodate then is a slow process; nearly no iodine is formed as the redox product.

For slow, bad mixing there will be local acid and base excesses [36]. The excess of the acid can promote the redox reaction mentioned above. This acid-catalyzed reaction is much faster than the redox reaction without acid. Thus measurable contents of iodine are formed. These quantities can be detected photometrically using a UV–Vis spectrometer.

The reaction was originally used for characterizing batch mixing [78]. Here, highly concentrated drops were added to large volume in a vessel having a lower concentration [36]. For a micro-reactor operation using typically 1 : 1 ratios of the solutions to be mixed, the concentration had to be increased relative to the batch-mixing protocol to achieve comparable sensitivity. Later, an optimized micro-reactor protocol was developed using a boronic acid buffer to avoid post-reaction of the mixed volume to iodine which was present for the original protocol [79]. Thereby, the prior existing flow-dependent error of the determination of the mixing efficiency was eliminated.

[P 36] Reactive-type dye imaging with a 10% ammonia solution in water and a solution of phenolphthalein in ethanol–water (5 g in 500 ml/500 ml) were contacted to yield an intense, pink color after mixing [67]. Transparent tubing was used for estimating the volume flow by measuring the flow speed of moving bubbles, inserted deliberately. The pressure was set hydrostatically in the range 3.5–9 kPa.

[P 37] A competitive parallel reaction was used for analyzing mixing efficiency in azo-type reaction between 1- and 2-naphthol and diazotized sulfanilic acid was chosen (see Scheme 1.1) [41]. This was mainly done owing to the speed of the reaction. Large differences are given for the rate constants, as long as complete mixing is achieved. Thus the ratio of the isomers is indicative of the mixing efficiency.

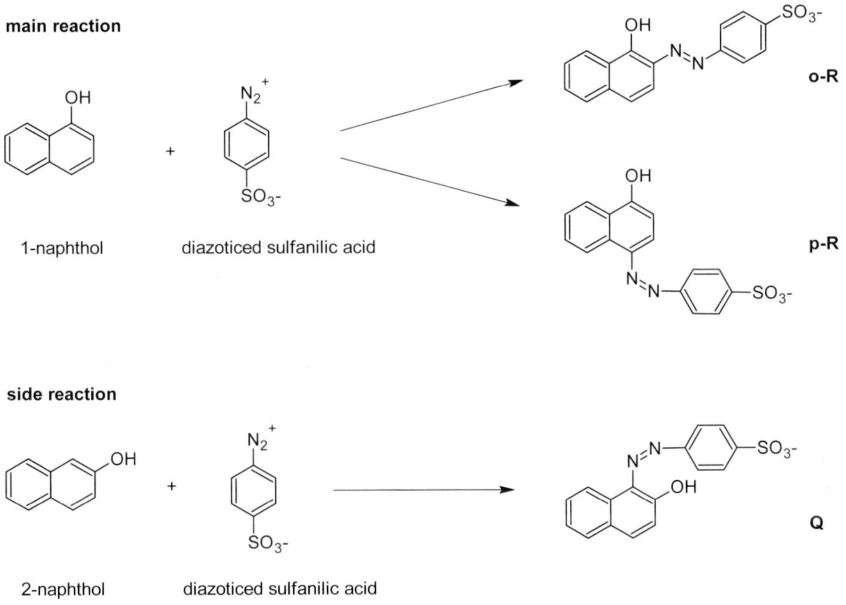

Scheme 1.1 Reaction scheme for the azo reaction with 1- and 2-naphthol with diazotized sulfanilic acid.

A mathematical correlation between the selectivity of the azo reaction and the scale of mixing by laminar and turbulent flow is described in [41].

A modified azo reaction protocol is given in [122, 123].

Computational fluid dynamics were used to describe the flow which undergoes a fast transition from laminar (at the fluid outlets) to turbulent (in the large mixing chamber) [41]. Using the commercial tool FLUENT, the following different turbulence models were applied: a $k\varepsilon$ model, an RNG–$k\varepsilon$ model and a Reynolds-stress model. For the last model, each stream is solved by a separate equation; for the two first models, two-equation models are applied. To have the simulations at practical times, the computing was done for part of the design using symmetry correlations.

[P 38] Gas mixing in a flow-through mixing chamber was followed by sampling at defined places in the front of the mixer feed outlets [40]. This was achieved by a motor-driven capillary, with an outer diameter of 30 μm and an inner diameter of 20 μm, which automatically scans a pre-defined path. Larger capillaries may disturb the flow and give false results. The position of the camera was checked with a CCD camera. A typical scan was a lateral movement over the full extension of the mixing chamber followed by a slight change in the vertical direction and scanning laterally in the reverse way and so on. The velocity of the capillary was restricted to 2.5 μm s^{-1}, basically owing to have a sufficient response time of the analysis system, using a mass spectrometer.

The fluid temperature was kept at 25 °C. Argon–nitrogen and helium–nitrogen mixing was investigated at throughputs from 1.25 to 5.0 l min^{-1}, corresponding to 4.34 to 17.36 m s^{-1} [40].

[P 39] Dilution-type mixing uses blue-colored water and pure water solutions as two feeds [20, 39]. The flow pattern is visualized and mixing is indicated by giving intermediate blue colors. Photographic imaging was applied.

Aqueous solutions with high concentrations of the commercial dye 'water blue' (5 g l^{-1}, $6.25 \cdot 10^{-3}$ M) and pure water were fed through the interdigital micro mixers [7, 20]. These blue-colored solutions provide excellent contrast owing to the high solubility and large extinction coefficient of the dye. They are, in particular, suitable to image multi-laminated systems arranged parallel to the direction of observation and multi-phase systems such as gas/liquid and liquid/liquid (results not shown here; see for instance [127]). However, when assuming a layer structure horizontal or tilted with respect to the observer, it stands to reason that one cannot distinguish between a real mixed system and a layered fluid structure.

One disadvantage of using the dye 'water blue' imaging has to be mentioned [7, 20]. Owing to the high molecular weight of this dye, diffusion is certainly different from low-molecular weight species such as the iron ions. However, when using organic dyes, this feature is inherent to most molecules thereof, belonging to condensed extended aromatic systems.

[P 40] A reaction-type mixing uses iron trichloride and sodium rhodanide solutions as two feeds [20, 39]. A brownish solution results with the product iron rhodanide. Photographic imaging was applied.

Mixing of uncolored iron ion (Fe^{3+}) and rhodanide (SCN^-) solutions (81.3 g l^{-1}, 0.5 M $FeCl_3$, and 40.5 g l^{-1}, 0.5 M NaSCN), resulting in the formation of the respective brownish complex, turned out to provide reasonable contrast and is free from any plugging phenomena [20]. Different from the 'water blue' solutions (see [P 39]), potentially providing information both on the fluid layer formation at the very beginning and on the course of mixing, the iron rhodanide system only displays the completion of mixing, linked to the color reaction. Thereby, mixing even in complex fluid systems, e.g. tilted lamellae, can be visualized.

[P 41] Semi-analytical calculations were performed for determining mixing quality [39]. Assuming uniform lamella formation and the absence of lamellae tilting, and

neglecting the velocity distribution in the channel, the convection-diffusion problem of mixing can be translated into a pure diffusion problem. Assuming periodic boundary conditions, the problem can be solved by modeling the diffusion process over the domain of a single lamella. A diffusion coefficient of 10^{-9} m^2 s^{-1} was assumed.

The numerical results are based on the solution of the incompressible Navier–Stokes equation:

$$\frac{\partial v_i}{\partial t} + (v_j \cdot \nabla_j)\, v_i = -\frac{1}{\rho}\nabla_i p + \frac{\eta}{\rho}\nabla^2 v_i\,,\ \nabla_i v_i = 0 \tag{1.2}$$

and a convection–diffusion equation for the concentration field:

$$\frac{\partial c}{\partial t} + (v_i \cdot \nabla_i)\, c = D\nabla^2 c \tag{1.3}$$

by means of the finite-volume method. V_i is the fluid velocity, ρ and η its density and viscosity, D the binary diffusion constant and c denotes concentration, respectively. For pressure–velocity coupling, the SIMPLEC algorithm was used. The simulations were done with the commercial flow solver CFX4 from AEA Technology.

A mixing residual was defined as given in [37] (see [39]):

$$r = \frac{1}{A}\int_S \left| c(x,t) - \frac{1}{2} \right| da \tag{1.4}$$

The initial value is 0.5 for concentration values of 0 and 1.

[P 42] A further extension of the above-mentioned approach [P 41] for design optimization is described in [38]. Here, the convection–diffusion problem was also reduced to a pure diffusion problem. Further details can be found in [38].

The CPU times needed for these calculations were < 1 min [38].

[P 43] A defined illumination was established by using a light-guide system with two goose necks for quantitative in-line monitoring of concentration profiles through transparent micro mixers [20]. Owing to problems with achieving a constant fixed illumination with the existing equipment, a new calibration was made for each experimental run (typically lasting a day) with each mixer by using standard solutions of the dye with known concentration. The images of both calibration and experimental solutions were converted to gray-scale format. Thereby, the concentration distribution was gathered when analyzing the images taken under fixed illumination with images of the calibration solutions. Using imaging software, concentration profiles along the channel cross-section were obtained.

1.3.9.13 **Typical Results**

[M 36] [P 39] A regular multi-lamination pattern composed of 30 lamellae is found for a rectangular interdigital mixer [20, 39, 124]. Within the limits of a photographic image, no notable deviations are observed. No mixing is observed for residence times ranging from 4.3 to 1140 ms.

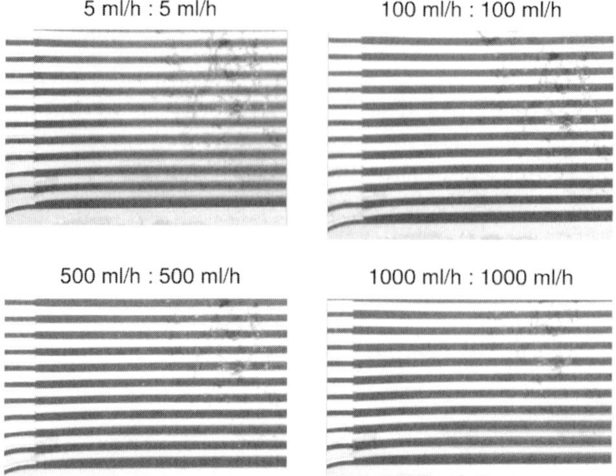

Figure 1.89 Multi-lamination flow patterns in the rectangular interdigital micro mixer for various flow rates [20] (by courtesy of AIChE).

The flow pattern remains the same for total flow rates from 10 to 2000 ml h^{-1}, corresponding to residence times of 4.3–1140 ms (see Figure 1.89). [20]. Thus 'pure' laminar flow without any secondary flow patterns applies.

[M 43] [P 39] A regular multi-lamination pattern composed of 30 lamellae is found for a triangular interdigital mixer [20, 39]. Within the limits of a photographic image no notable deviations are observed.

At a low volume flow rate of 20 ml h^{-1}, separated light and dark sections become increasingly smaller when passing through the triangular zone and reveal a more homogeneous color profile after entering the rectangular outlet zone. This is confirmed by detailed analysis of the concentration profile (see Figure 1.90).

Moreover, Figure 1.90 shows that not all lamellae have absolutely equal thickness, as opposed to the observations in Figure 1.89. The thickness decreases from the interior towards the exterior. An analysis of the lamellae thickness at various cross-sectional positions was derived by precise measurement of the respective high-

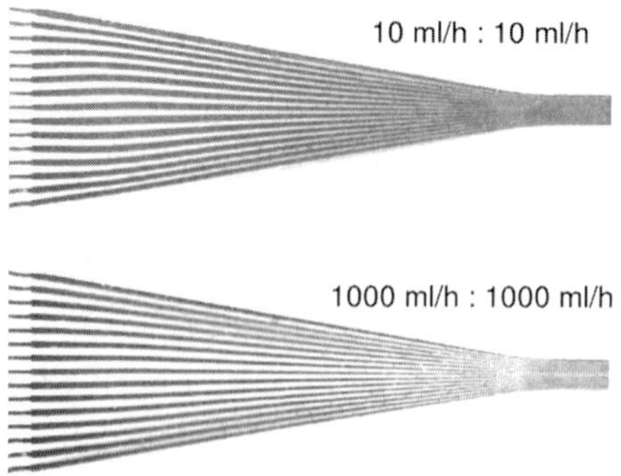

Figure 1.90 Multi-lamination flow patterns superposed by geometric focusing in the triangular interdigital micro mixer at two flow rates [20] (by courtesy of AIChE).

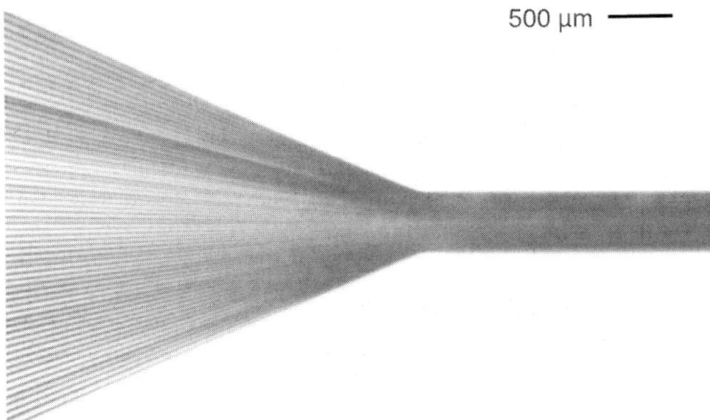

Figure 1.91 Multi-lamination flow pattern superposed by focusing in the SuperFocus interdigital micro mixer, visualized by rhodanide reactive imaging (4 : 1 l h^{-1}) [20] (by courtesy of AIChE).

contrast image at high magnification. The interior lamellae are up to 50% larger than those near the wall. This deviation is probably caused by the parallel orientation of the inlets, i.e. the various lamellae flows have different angles with respect to the channels' direction, rather than being guided in the same direction. Thereby, the lamellae width becomes slightly dependent on the channel position.

[M 44a] [P 40] A fairly regular multi-lamination pattern composed of 124 lamellae (see Figure 1.91) is found for a SuperFocus interdigital mixer (glass version; small arc of interdigital feeds) [20, 39, 121]. Small deviations in lamellae thickness obviously lead to the different share of mixing observed visually, i.e. parts of the lamellae are darker than the rest. It seems that the interior lamellae have less mixing, i.e. are lighter in color.

The fairly uniform color formation in the mixing channel demonstrates that mixing is close to completion [20]. A total flow of 8 l h^{-1} at a pressure drop of 2.5 bar can be achieved owing to the parallel feed of many inlet channels.

[M 44a] [P 40] A regular multi-lamination pattern composed of 138 lamellae is found for a SuperFocus interdigital mixer (glass version; large arc of interdigital feeds) [39, 121]. Deviations are found only for some outer lamellae leading to the different share of mixing. Owing to the dilution-type experiment, these areas have more diffuse colors. The reason is unclear at present; it may be due to lamellae tilting (albeit this is not in line with simulations) or may reflect the velocity profile.

When using an asymmetric flow ratio (5 : 1) for still better flow visualization, the regularity of the multi-laminated pattern is the most striking feature (see Figure 1.92) [39].

Multi-lamination patterns for focused, re-directed streams

[M 46] [P 34] A multi-lamination pattern was evident in the central and inner parts of the flow when entering a triangular focusing section with re-directed flow (~ 90°) [34]. The outer part, i.e. close to the channel walls, was disordered, probably owing to flow maldistribution.

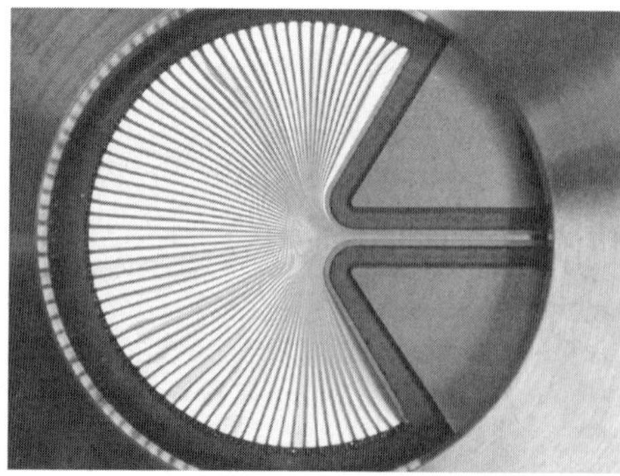

Figure 1.92 Regular arrangement of liquid lamellae in the focusing chamber of the SuperFocus micro mixer (steel version; large arc of interdigital feeds). For better flow visualization, an asymmetric flow ratio (5 : 1) was chosen, setting the dyed water solution at a lower flow rate [39].

Multi-lamination patterns for lateral multi-lamination

[M 37] [P 36] The onset of mixing for a lateral multi-lamellae flow was demonstrated by pH-driven color formation using an indicator solution (7 kPa; 1.2 μl min^{-1}) [67]. Two channels of the multi-lamellae flow were blocked so that also non-dyed zones were visible. Applying higher flow rates led to incomplete mixing. The fact that complete mixing could be only achieved at comparatively long residence times in the order of one second or so can be explained by the rather large diffusion distances, about 100 μm, provided by the first-generation device.

Simulations predict a high degree of mixing for the conditions of the experiment [67]; however, the visual inspections so far only vaguely confirm this.

Multi-lamination patterns for vertical bi-lamination – reference case

[M 38] [P 36] Using a reactive visualization experiment based on pH-driven color formation, the homogeneous texture of the color of the mixed flow at the end of the mixing chamber suggests that mixing was not far from completion for a vertical multi-lamellae flow (3.5 kPa; 0.9 μl min^{-1}) [67].

Some deviations from ideal are also arise. Directly next to some outer orifices color formation is much more intense, probably owing to the longer residence time of the flow next to the walls. The same is applies to nearly all orifices in the fourth row. However, both color formations vanish shortly after a rising so that a colorless flow follows in each case, yielding color only at considerable distance at the end of the mixing chamber.

Simulations predict a high degree of mixing for the conditions of the experiment [67]; however, only the visual inspections so far can vaguely confirm this.

In-line concentration monitoring of multi-laminated streams

[M 36] [P 39] The concentration profiles of a multi-laminated (unfocused) stream along the cross-sectional axis of a rectangular interdigital micro mixer were determined by a special illumination technique (see Figure 1.93) [20]. These profiles were taken at a distance to the interdigital outlets within the rectangular mixing chamber attached. Within one plot, which was taken at one flow rate and one residence time, minor differences can be seen, e.g. small statistical variations in dye intensity and lamellae thickness or systematic differences such as thicker

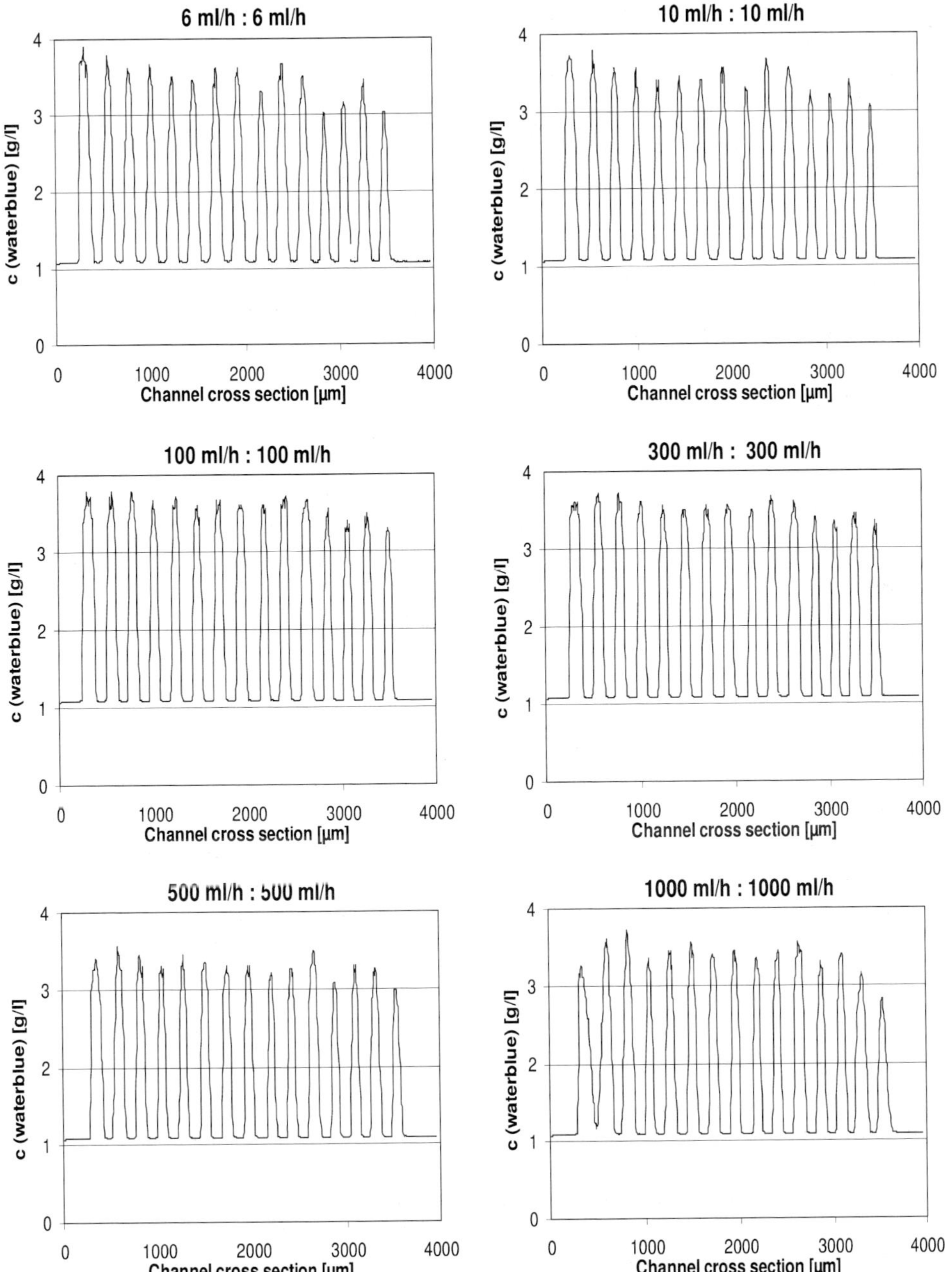

Figure 1.93 Cross-sectional concentration profiles for multi-lamination flow patterns in the rectangular interdigital micro mixer for various flow rates [20] (by courtesy of AIChE).

lamellae at the channel walls. The different plots are virtually the same, apart from minor deviations. The latter may result from statistical fluctuations, e.g. pulsating pumps (see the inner lamellae), or systematic deviations, e.g. by surface or flow distribution effects.

Hence, the main information is that the in-line concentration profiles confirm the uniform multi-lamellae formation over a favorable large range of flow rates given already by photographic flow-pattern imaging (see *Multi-lamination patterns*) [20].

[M 43] [P 39] The concentration profiles of the triangular interdigital mixer differ from those of the rectangular mixer for two reasons [20]. First, the degree of mixing is higher owing to the geometric focusing effect. At total volume flows of 12 and 20 ml h^{-1}, long residence times and more flattened profiles are obtained compared with those of the rectangular mixer. The profile at 12 ml h^{-1} indicates a higher degree of mixing than that at 20 ml h^{-1} due to the longer mixing time. Moreover, both profiles show strong deviations in concentration between outer and inner lamellae (curved profile), in particular for one of the outer lamellae next to the mixing chamber wall. The origin for the curved profile is not totally clear but may be correlated with small bending of the cover of the mixing chamber, thereby slightly changing the optical path over the cross-section.

The second major difference of the flow patterns of the triangular to the rectangular mixers that more complex concentration profiles are found which have more peaks than the number of generated lamellae [20]. This becomes evident at volume flows ≥ 200 ml h^{-1}.

Combined hydrodynamic and geometric focusing of multi-laminated streams

[M 36] [P 39] Hydrodynamic focusing was achieved in a rectangular interdigital mixer by setting the two flow rates of the liquids at different levels [20]. In this way, one set of lamellae can be thinned considerably. The corresponding flow pattern are described as a function fo the ratio of the individual flow rates of the two liquids.

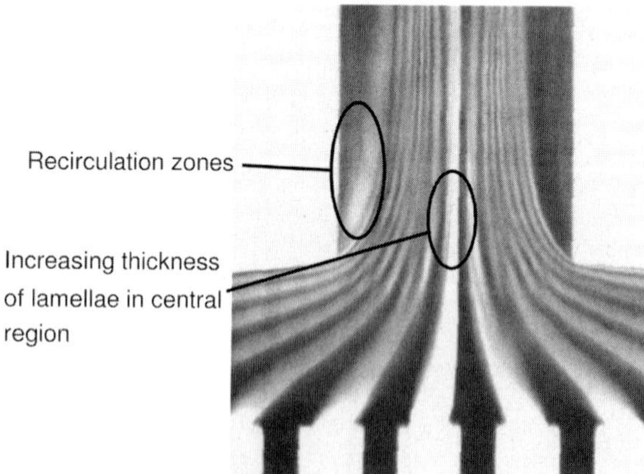

Figure 1.94 Multi-lamination flow pattern superposed by focusing in the slit-shaped interdigital micro mixer. Blurred zones indicate lamellae tilting and winding within the entire focusing zone [20] (by courtesy of AIChE).

[M 43] [P 39] For a triangular interdigital mixer, the possibility of exerting hydrodynamic focusing was demonstrated, i.e. the enlarging of lamellae thickness for one liquid at the expense of thinning the other by having a higher flow rate for the first liquid [20, 124]. Since a triangular device is applied, the hydrodynamic effect is combined with geometric focusing in this way.

[M 42] [P 39] For a slit-type interdigital mixer the focusing performance was also characterized [20]. Owing to the moderately high focusing ratio in combination with the short focusing length, the thickness of the lamellae was considerably different. The interior lamellae are much thicker than the exterior ones. Dead zones are situated on both sides of the multi-laminated stream in the adjacent mixing channel. These dead zones are enriched with one fluid, thus are far from the final mixture.

The most striking feature, however, is that blurred zones are found in the entire focusing region with the exception of the central part (see Figure 1.94). Here, simulation results show that lamellae tilting and winding occurs.

Mixing time and mixing length of non-focusing and focusing interdigital mixers

[M 42] [M 43] [M 44] [M 36] [P 41] The mixing performances of several types of interdigital mixers were compared at a fixed flow rate (100 ml h^{-1}) – the non-focusing rectangular and the focusing triangular, slit-type and SuperFocus mixers (see Figure 1.95) [37] (see also [121]). The mixers were compared using semi-analytical calculations under certain simplified assumptions (see [37] and [P 41]). As expected, the performance of the rectangular device is worse than those of all focusing mixers. The difference is smaller when comparing mixing length than mixing time owing to the increase in flow velocity for the focusing mixers (which increases the mixing length). The SuperFocus mixer is still considerably better than the triangular mixer owing to the higher focusing ratio. For the slit-shaped mixer, only diffusion within the interior smaller lamellae was taken into account in the semi-analytical calculations. Thus, the performance is close to that of the SuperFocus mixer. However, other regions in the slit-shaped mixer will give slower mixing characteristics.

A mixing time of about 5 ms for 95% mixing is determined for the SuperFocus mixer [37] (see also [121]). This corresponds to a mixing length of about 0.6 mm at a volume flow rate of 100 ml h^{-1}. This means that complete mixing can be achieved even for flow rates of several l h^{-1} in SuperFocus mixers, needing not more than a 20 mm mixing channel length for flow rates up to 4 l h^{-1}.

[M 44a] [P 40] In order to demonstrate that fast mixing at high flow rates can be achieved in the SuperFocus mixer, experimental and theoretical derived mixing residuals were compared (see Figure 1.96) [37] (see also [121]). The experimental data were collected from on-line concentration monitoring by a spectrometric analysis of calibrated photographic images using a gray-scale analysis of the originally colored images. Albeit the experimental and theoretical data sets do not match perfectly, there is reasonable agreement, proving the assumption that SuperFocus mixers can achieve fast mixing also at l h^{-1}-flow rates. The experimental mixing length needed for 95% mixing (~5 ms) at 4 l h^{-1} amounts to about 35 mm.

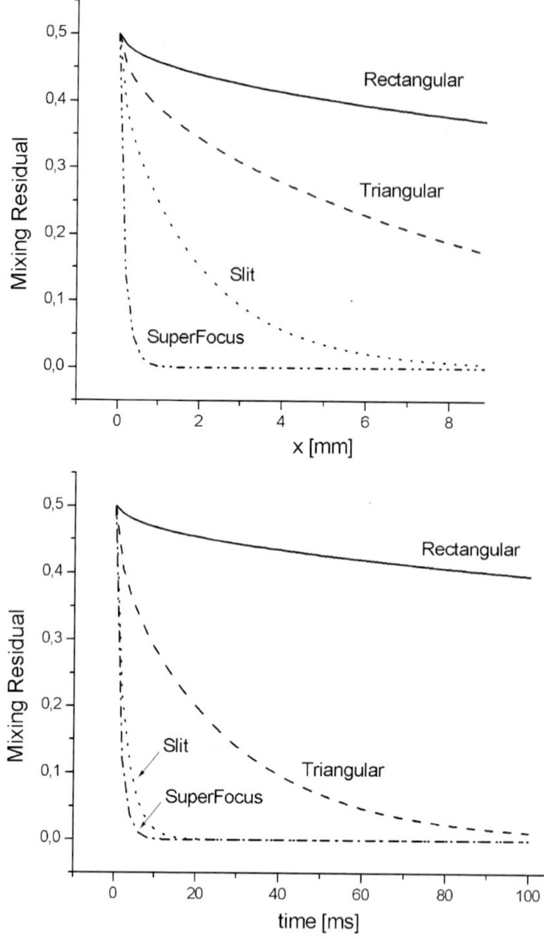

Figure 1.95 Mixing residual as a function of distance (top) and time (bottom) [37] (by courtesy of AIChE).

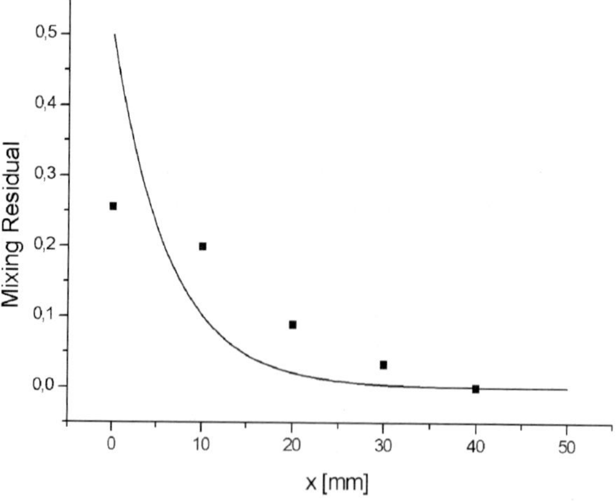

Figure 1.96 Comparison of experimental (squares) and theoretical (solid line) mixing residuals as a function of the length of the channel flow passage [37] (by courtesy of AIChE).

Optimization of focusing interdigital mixers by analytical modeling

[M 44a] [P 42] An intrinsic problem with the focusing concept is that a share of the flow passage has to be taken for the focusing zone in addition to the mixing zone [38] (see also [37]). Accordingly, additional residence time is required and part of the mixing sets in already in the mixing zone. Thus, the mixing performance as specified in *Liquid mixing time – benchmarking*, deliberately ignoring the focusing zone effects to circumvent a too complex analysis, is actually worse, to an extent which has to be specified.

In [38], an optimization strategy is shown for reducing the focusing zone effects with regard to residence time and mixing. The main measure for optimization is to change the depth of the focusing zone (in conjunction with changing the depth of the mixing channel attached). Therefore, the residence time was plotted as a function of the chamber depth in the focusing zone for various pressures. The quantity pressure was chosen, since it reflects other parameters of the SuperFocus mixers as a sum, regarding the focusing and mixing channel zones. Depending on the parameters chosen, residence times of the order of several tens of milliseconds are predicted. With increasing chamber depth, the residence time within the focusing chamber has a smaller relative share, which is counter-intuitive at first sight; however, this relates to the assumption of a constant pressure which is required for structural changes of the mixing channel which overcompensate the effect of changing the chamber depth. Above a chamber height of 500 μm, the residence times become constant. At higher pressures, shorter residence times are yielded over the whole range of chamber depths investigated.

Similar behavior was observed when calculating the mixing residual [38]. The percentage of mixing can be reduced to about 10% with a sufficiently deep focusing chamber. At small depths, the amount of mixing before entering the mixing channel increases drastically. This relates directly to the residence time, enhancing diffusion.

The focusing angle also has a considerable impact on the share of the residence time in the focusing zone and the respective share of mixing [38]. Larger focusing angles decrease the residence time and mixing, e.g. down to 20 ms at 100°.

The CPU times needed for these calculations were < 1 min [38].

Jet flow patterns created from multi-lamination patterns

[M 42] [P 39] In a slit-type interdigital micro mixer, first multi-lamination patterns are created [20, 124]. Then, the multi-laminated stream is introduced into a chamber of larger cross-section. A jet is formed and induces two eddies in the dead zones adjacent to the initial channel (see Figure 1.97). Mixing is not completed within the chamber, as evidenced by the non-uniform color distribution. The contribution of the jet mixing to the total mixing is unclear.

The jet formation is only observed above a certain threshold flow rate [20]. At flow rates of 500 ml h^{-1} jet formation takes place, whereas a multi-laminated pattern is found at 10 ml h^{-1} (see Figure 1.98). In the latter case, thus a simple defocusing of the stream occurs. At still higher flow rates (2 l h^{-1}), the texture of the liquid in the large mixing chamber is homogeneous, i.e. the jet cannot be seen visually any longer. Already at the beginning of the channel attached to the chamber, the multi-

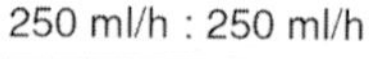

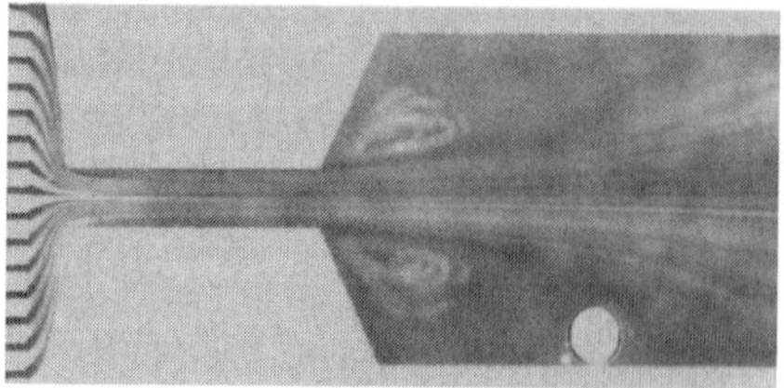

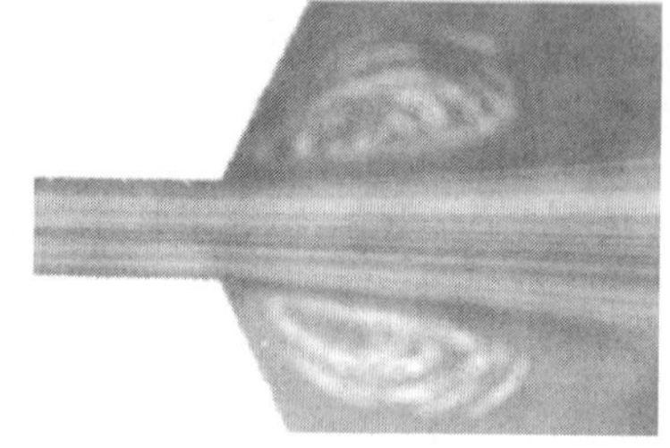

Figure 1.97 Jet formation by reopening of the cross-sectional flow area for a multi-lamination stream which was before focused in the slit-shaped interdigital micro mixer [20] (by courtesy of AIChE).

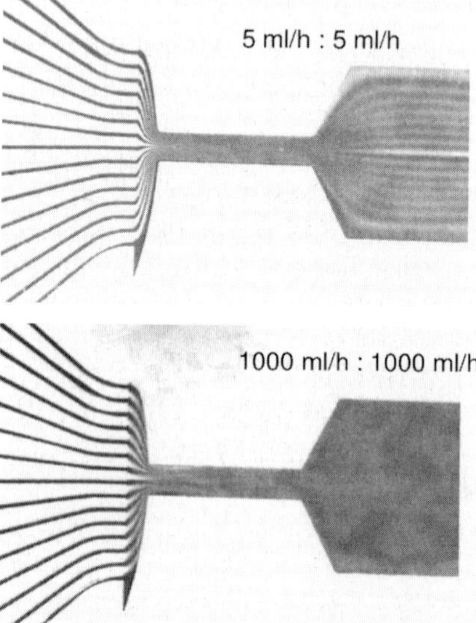

Figure 1.98 Pure multi-lamination flow pattern (top, 10 ml h^{-1}) and superposed by focusing and reopening/jet formation (bottom, 2 l h^{-1}) in the slit-shaped interdigital micro mixer visualized by blue-colored water dilution imaging [20] (by courtesy of AIChE).

lamination pattern of the focusing zone breaks up and after some distance the color of the solution is fairly homogeneous. This may be due to mixing; however, it could also be caused by tilting of lamellae. In any case, this demonstrates that the degree of dispersion of the two liquids is increased.

In a decisive experiment it was judged whether mixing or only dispersion by tilting has taken place (see Figure 1.99) [20]. By reactive rhodanide imaging it was found that at 2 l h^{-1} no mixing occurred, as demonstrated by the absence of color formation due to reaction within the jet area. In the dead zones nearby, having much longer residence times, color formation was intense. The validity of this analysis is finally

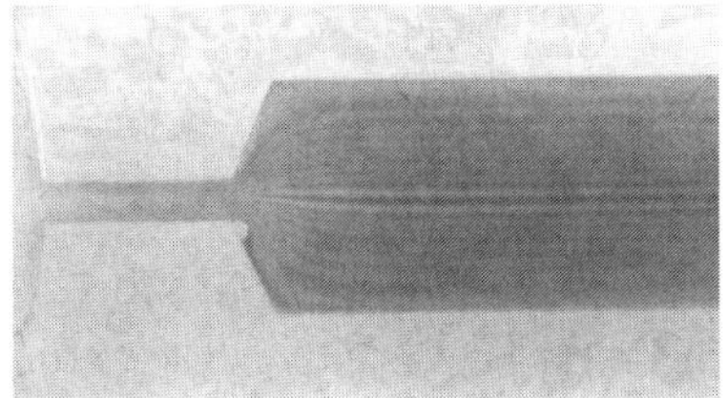

Figure 1.99 Pure multi-lamination flow pattern (top, 10 ml h^{-1}) and superposed by focusing and reopening/jet formation (bottom, 2 l h^{-1}) in the slit-shaped interdigital micro mixer visualized by rhodanide reactive imaging [20] (by courtesy of AIChE).

demonstrated by the fact that dilution and reactive images gave equal information at a low flow rate of 10 ml h^{-1}. Here, indeed, mixing took place and lamellae tilting was absent. Hence, the results also show that there is a need for the two mixing characterizations (dilution and reactive) in order to obtain complementary information to have information at a 3-D level on the mixing process.

Applicability of CFD simulation to diffusion mass transfer – numerical diffusion

[M 44a] [P 40] Numerical errors which are due to discretization of the convective terms in the transport equation of the concentration fields introduce an additional, unphysical diffusion mechanism [37]. Especially for liquid–liquid mixing with characteristic diffusion constants of the order of 10^{-9} m^2 s^{-1} this so-called numerical diffusion (ND) is likely to dominate diffusive mass transfer on computational grids.

There are couple of measures that can taken in order to minimize ND [37]. Higher order discretization schemes such as the QUICK scheme reduce the numerical errors. Furthermore, ND depends strongly on the relative orientation of flow velocity and grid cells. ND can be minimized by choosing grid cells with edges parallel to the local flow velocity.

For the example of the slit-type interdigital micro mixer, it was shown that the higher order differencing scheme (QUICK scheme) better describe the flow pattern of the multi-lamellae than a first-order upwind scheme [37]. In the latter case, gray areas between the black and white encoded lamellae are found in the focusing zone and are due to numerical diffusion, but not correlated with a physical diffusive mixing process. However, the QUICK scheme also shows gray areas due to numerical diffusion in the mixing channel. Hence, such CFD simulations cannot properly give a quantitative description of the mixing process, but can only be used for a qualitative judgement. For quantitative analysis, semi-analytical calculations were applied instead.

Figure 1.100 Characterizing mixing by diffusion of a dye in an aqueous solution (left) and by a chemical reaction yielding a colored product (right) in the slit-shaped interidigital micro mixer; viewing direction is the flow direction [37] (by courtesy of AIChE).

Lamellae twisting in multi-lamination patterns

[M 42] [P 39] [P 40] In the focusing zone of the slit-type interdigital micro mixer, having an extreme focusing ratio, lamellae twisting and winding are observed [37] (a first hint is given in [124]). This is evidenced by having diffuse colors up to uniform color formation for dilution-type experiments, while finding no color formation for reaction experiments (see Figure 1.100). Both results seem to be contradictory at first sight. However, the latter definitely means that mixing is absent. Thus, for the first result, often indicative of mixing, a new explanation needs to be given. The dilution–color formation can be explained by having both colored and uncolored zones in the optical axis under investigation (see Figure 1.101). The result is a 'mixed' optical spectrum, but not relying on a mixed fluid system, but rather characterizing a dispersed one.

The underlying process was described by CFD simulations, taking into account the experiences made for reducing the numerical diffusion (see *Applicability of CFD simulation to diffusion – numerical diffusion*). Owing to inertial forces stemming from the velocity distribution in the channel, tilting occurs at low *Re* and transforms to lamellae winding at *Re* larger than about 50. Details on the lamellae winding are given in Figure 1.101.

Figure 1.101 Orientation of liquid lamellae in the mixing channel of the slit-shaped interdigital micro mixer; viewing direction is the flow direction [37] (by courtesy of AIChE).

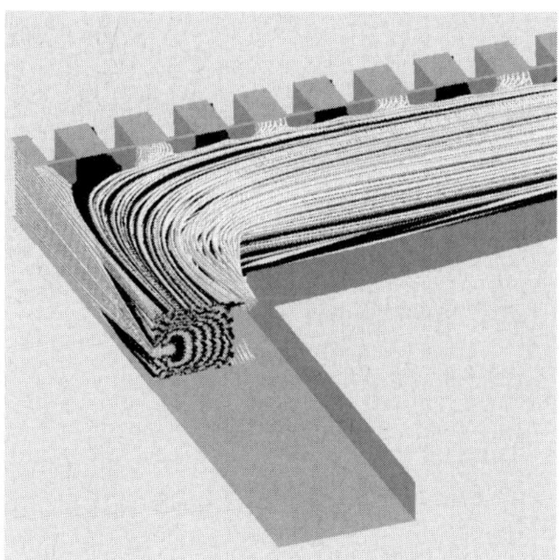

Figure 1.102 Streamline pattern in the slit-shaped interdigital micro mixer at $Re = 2160$ [37] (by courtesy of AIChE).

At still higher *Re* (2160), the lamellae become highly intertwined, as evidenced by the streamline pattern (see Figure 1.102) [37].

Liquid mixing time – benchmarking

[M 44a] [P 41] The liquid mixing time was calculated to 3.5 ms for a SuperFocus mixer creating lamellae about 4 μm thick (see Table 1.6) [39] (see also [121]). This calculation neglects the premixing in the focusing region and the respective residence time.

For an experimental validation of the mixing time, see *Liquid mixing length – benchmarking*, below. For calculation of the share on mixing of the focusing zone, see *Share of mixing of focusing zone relative to mixing zone*, below.

[M 43] [M 36] [P 41] The liquid mixing time was calculated to be 59 ms for a standard triangular mixer creating lamellae about 110 μm thick (see Table 1.6) [39]. For comparison, the liquid mixing time of the rectangular mixer having 100 μm thin lamellae is 2.48 s.

[M 43] [M 36] [P 39] The liquid mixing time was measured to be about 100 ms for a standard triangular mixer and about 3 s for the rectangular mixer [124]. The differences from the calculated liquid mixing times given above may be due to inaccuracies of judging the completion of mixing by visual inspection of dyed solutions.

Table 1.6 Comparison of mixing performance of three interdigital micro mixers with 8 l h^{-1} aqueous streams.

Type of mixer	*Mixing length for mixing residual of 0.05* (m)*	*Mixing time for mixing residual of 0.05* (s)*
Rectangular	11.30	2480
Triangular	1.74	59
SuperFocus	0.031	3.5

* For the definition of the mixing residual, see [39].

Liquid mixing length – benchmarking

[M 44a] [P 41] At a total flow rate of 8 l^{-1}, the liquid mixing length was calculated to be 3.1 cm for a SuperFocus mixer creating lamellae about 4 μm thick (see Table 1.6) [39] (see also [121]). This calculation neglects the premixing in the focusing region and the respective mixing length.

[M 44a] [P 40] At a total flow rate of 4 l h^{-1}, a liquid mixing length of about 3 cm was measured for a SuperFocus mixer creating lamellae about 4 μm thick (mixing residual of 0.05) [39]. These experimental results correspond reasonably to calculation results (see *Liquid mixing time – benchmarking,* above) and generally confirm the fast, millisecond mixing properties of the device.

[M 43] [M 36] [P 41] At a total flow rate of 8 l h^{-1}, the liquid mixing length was calculated to be 1.74 m for a standard triangular mixer creating lamellae about 110 μm thick (see Table 1.6) [39]. For comparison, the liquid mixing length of the rectangular mixer having 100 μm thin lamellae is 11.30 m.

Share of mixing of focusing zone relative to mixing zone

[M 44a] [P 42] For focusing interdigital mixers, part of the residence time after lamellae contacting is needed to pass the focusing chamber. Here, mixing starts under changing conditions which is not desired and actually prolongs the real mixing time. Therefore, a reduction of the time needed for flowing through the focusing chamber is an important optimization function of the SuperFocus mixer [38]. This can be achieved by setting a different geometry, i.e. chamber depth, in the focusing zone as compared with the mixing zone.

Whereas the share of mixing in the focusing zone can be as high as 60% for a depth of 0.3 mm, it can be reduced to about 12% for a 0.5 mm depth at a certain flow rate, equivalent to a pressure drop of 4.5 bar (see Figure 1.103) [38]. Above 0.5 mm the share remains constant. The pressure drop, i.e. the flow rate, also has an influence on the mixing share, e.g. a decrease in pressure from 4.5 to 0.5 bar increases the mixing share from about 12 to 30%.

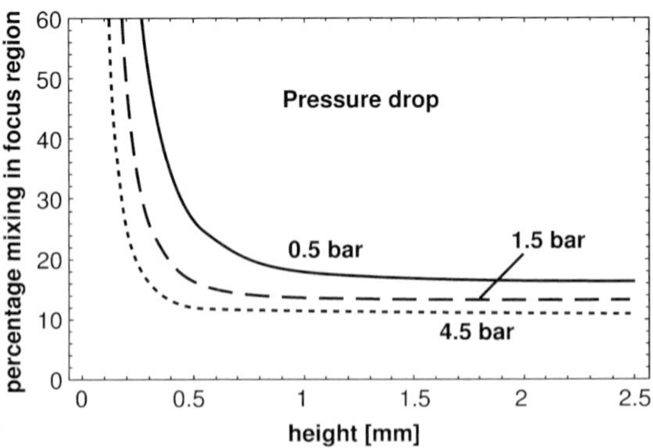

Figure 1.103 Share of mixing in the focusing zone relative to the mixing zone as a function of the focusing zone depth. The impact of the design details, expressed as pressure drop, is also given (full line, 0.5 bar; long dashed line, 1.5 bar; short dashed line, 4.5 bar) [38] (by courtesy of Elsevier Ltd.).

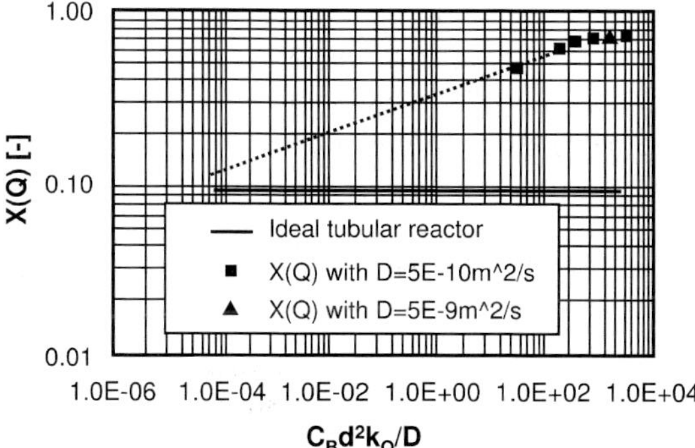

Figure 1.104 Selectivity of an azo-type parallel reaction for testing mixing efficiency as a function of the dimensionless number Π [41] (by courtesy of Elsevier Ltd.).

Competitive parallel reactions – mixing process under laminar flow

[M 39] [P 37] Using an azo-type competitive reaction, the mixing efficiency could be determined via the selectivity [41]. Using a P-type micro mixer, laminar flow mixing could be investigated (see Figure 1.104). The selectivities measured are far from the ideal behavior of a tubular reactor.

The diffusion length was varied from 10 to 100 μm; for the whole range the selectivities did not indicate good mixing [41]. This shows that pure laminar diffusion cannot compare with the ideal tubular reactor. Hence, such mixers may not be adequate for fast to very fast reactions, when side reaction selectivities are considered.

[M 39] [P 37] Using an azo-type competitive reaction, the selectivities were compared for the P- and V-type micro mixers having straight and oblique fluid injection, respectively [41]. In this way, laminar- and turbulent-flow mixing achieved by vertical interdigital microstructured mixers can be compared. The selectivities of the turbulent V-type mixer are better to some extent as compared with the P-type device; however, neither approaches the characteristics of the ideal tubular reactor. The micro devices, however, are better than a conventional jet mixer.

The difference between the microstructured mixers can be understood when considering the fact that the crossing of the streams yields additional shearing which provides new interfaces for mixing [41]. Concentration differences are reduced much faster in this way.

The latter is confirmed by CFD; perpendicular-oriented velocity components are quickly compensated downstream [41]. After about 800 μm, parallel flow of the streams is attained as for the P-type mixer.

V-mixer – comparison with conventional nozzle mixer

[M 39] [P 37] Using an azo-type competitive reaction, the selectivities were compared for the V-type micro mixers and conventional nozzle mixers with and without internals as a function of a specific power parameter [122, 123]. Considerably lower amounts of consecutive products, indicative of good mixing, are formed. All mixers show increasing mixing efficiency on enhancing the power.

Turbulent energy for P-mixer and V-mixer

[M 39] [P 37] As is to be expected from the above-mentioned results, the turbulent energy is higher for the V-mixer than for the P-mixer, as CFD simulations prove [41]. These findings correlate with the analysis of the velocity components, also given above.

Applicability of turbulent models

[M 39] [P 37] The Reynolds–stress model describes best the experimental findings out of three turbulent models investigated (see Figure 1.105) [41]. Then, the model was used for predictions of the mixing efficiency as determined by an azo-type parallel reaction. It was found that the wall thickness has no major influence, whereas the channel depth, as expected, has an influence, affecting the shearing.

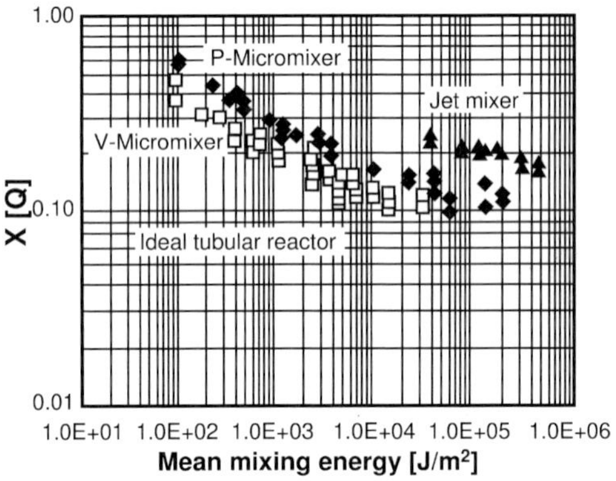

Figure 1.105 Experimental (azo-type parallel reaction) and calculated selectivities by the Reynolds–stress model [41] (by courtesy of Elsevier Ltd.).

Gas mixing by vertical interdigital feeds

[M 39] [P 38] Mixing argon and nitrogen in a vertical interdigital feed arrangement yields periodical, alternating (multi-lamination) concentration patterns, when analyzed at a short distance of 50 μm to the feed outlet [40]. The concentration profile matches the feed outlet architecture. At a larger distance of 550 μm, mixing has taken place to a large extent, i.e. the amplitude of the periodical concentration profile is decreased. For the lamellae close to the outlet, mixing is slightly less advanced as compared with the interior. After a mixing length of 800 μm, complete mixing was attained.

Mixing degree of gas mixing

[M 39] [P 38] The degree of mixing of argon–nitrogen was determined as a function of the mixing length [40]. As expected, an increase in mixing is found. On enhancing the volume flow from 1.25 to 5.0 l min^{-1}, the degree of mixing at a given mixing length is smaller. Using species of faster diffusion such as He yields higher degrees of mixing at a given mixing length (see Figure 1.106). In most cases, complete mixing is achieved after an 800 μm passage.

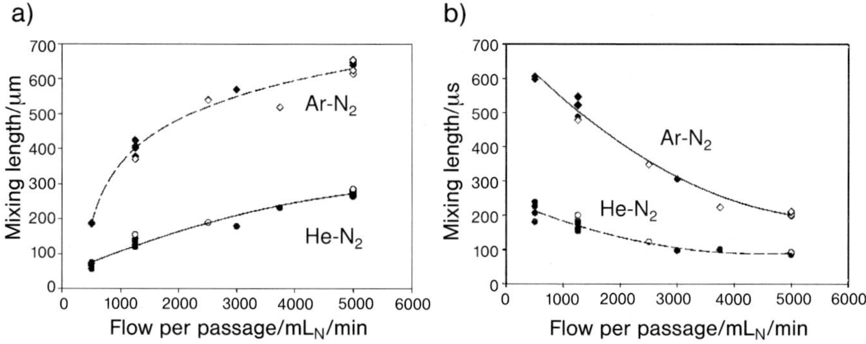

Figure 1.106 Mixing length and mixing time for the V-type micro mixer with vertical interdigital arrangement for two gas systems (Ar-N_2 and He-$N_{2)}$ to be mixed [Pfeifer et al., Chem. Ing. Tech. 76, 5 (2004) 607].

Shear stress for angled injection during gas mixing

[M 39] [P 38] For angled gas-stream injection, it was detected that mixing is enhanced towards the (hypothetical) colliding point (see Figure 1.106) [40]. This is due to shear stress creating new interfaces for mixing. This leads to a discontinuity in the respective mixing degree–distance plots. For a 90° injection, the collision point was at 200 µm, the distance between the flow outlets also being 200 µm.

Benchmarking to standard laboratory mixing tool

[M 41] [P 35] The mixing efficiency of an interdigital micro mixer array was compared with that of mixing tees and heavily stirred vessels regarding to liquid mixing [36]. Although the comparison is more of a qualitative nature, as the flow conditions and other parameters are not equal, it gives within these limits insight into the general feasibility of using a micro mixer. Using a competitive reaction method, the micro mixer turned out to be superior to all benchmarks. The turbulent operated mixing tee was the closest reference, the same tool under laminar conditions had the worst mixing efficiency.

Design details – achievement of vertical lamination by 90° feed turn

[M 41] [P 35] The focusing zone of a specially interdigital micro mixer, termed 'slit' owing to its shape, with counter-flow and then 90° redirected feed turn, serves for flow restriction and achieves vertical lamination, if the correct geometry is chosen [36]. This is even more demanding when the focusing zone is a ring-like zone fed by multiple mixing elements instead being a single 'slit', which is the case for a 10-fold mixing array. It was shown that an decrease in slit width from 350 to 200 µm notably increases the mixing efficiency.

Design details – improving the feed channel geometry

[M 41] [P 35] The inlet flow distribution has an effect on the mixing performance in the sense that deviations from ideal feed have their influence on the gross mixing efficiency [36]. For an interdigital 10-fold mixer array, the cross-section of the inlet

channel was varied with the aim of adjusting the pressure barrier which serves for flow equilibration. A positive effect on the mixing efficiency was seen.

Flow rate dependence

[M 41] [P 35] The mixing efficiency in the slit-shaped micro mixer is constant for a comparatively large volume flow range (0.5–3 l h^{-1}) [36]. Only at flow rates lower than 0.5 l h^{-1} is considerably reduced performance attained.

Comparison of single-element and array devices – validity of the numbering-up concept

[M 40] [M 41] [P 35] The performances of a single-element and array micro mixer devices of interdigital feed type were compared [36]. Comparable mixing efficiencies were found, the single device being slightly better, as expected. The array mixer allows one to perform good mixing at high throughput which the single device cannot reach.

Impact of lamellae width on the mixing performance

[M 40] [M 41] [P 35] For interdigital micro mixers, the widths of the feed channel and correspondingly of the lamellae width were varied [36]. The reactive mixing characterization approach used, based on competing reactions, was able to reflect the expected increase in mixing efficiency through faster diffusion. For both single-element and array micro mixer devices improved mixing was detected on reducing the channel width from 40 to 25 μm.

Luciferin/luciferase reaction to monitor multi-lamellae formation/interdigital structures for biochemical applications

[No mixer specified here; see [128]] [no protocol] Interdigital flow configurations having a triangular focusing zone were part of an integrated biochemical system, the so-called Micro Work Bench [128]. A planar feed architecture with two rows of nozzles gave an alternate feed arrangement. An interdigital element with two outer lamellae from one fluid and one inner lamella of the other feed was realized and a respective 'numbered-up' element. As test reaction, the bioluminescence of the enzymatic reaction of firefly luciferase with ATP solution was employed, resulting in a change of fluorescence intensity, which marks the flow pattern. The thus derived patterns only roughly resembled the expected multi-lamellae flow patterns, probably for reasons of insufficient flow distribution. The fluorescence intensity decreased, indicating some mixing effect, but not to zero owing to the formation of thick in addition to thin lamellae.

Interplay reaction kinetics and multi-lamination – a generic analysis

[No details on mixer] [no protocol] Recently, the interplay between reaction kinetics and multi-lamination has been theoretically analyzed for the first time [129]. Selected types of reactions, based on different scenarios of the elemental main, side and consecutive reactions, were defined which are common in organic synthesis. For these reactions simple, but nonetheless valuable, kinetic equations were assumed. For the multi-lamination mixing also selected scenarios were taken, including small

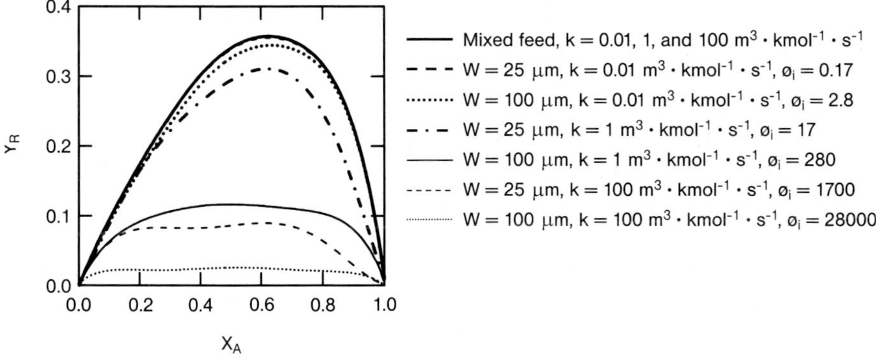

Figure 1.107 Relation between yield of a product R and conversion of a reactant A for different rate constants and lamination widths for one selected scenario of elemental reaction (two reactants A + B form R, while B can react with R as well in a consecutive reaction to the consecutive product S). W: lamellae width; k: rate constant; ϕ_i: ratio of reaction rate to diffusion rate [129] (by courtesy of Elsevier Ltd.).

lamellae widths in multi-laminated flow configurations and large lamellae widths in bi-laminated flows. Only diffusive mixing was assumed for reasons of simplicity. Then, it was calculated what might happen if such a selected reaction is undergone in such a lamellae configuration. It was found that the choice of lamellae has a strong effect on the reaction course and hence on selectivity and yield (see Figure 1.107). For some processes, thick lamellae are the preferred choice. Based on such conclusions, optimized design configurations were proposed in which part of the flow, containing mainly side products, was removed under laminar-flow conditions, leaving a purer stream with enhanced main product share. Thus, an in-line separation technique was proposed for increase in product purity.

1.3.10
Interdigital Concentric Consecutive Mixing

Most Relevant Citations

Peer-reviewed journals: [130]; proceedings contributions: [66]; contributions within books: [82].

In the section above, it was mentioned that microstructured platelet stacks are used for interdigital multi-lamination [40, 65]. Here, the feeds are oriented parallel towards a large mixing chamber. A variation of this principle arises, when a consecutive injection (instead of parallel) of the alternate fluid feeds is made. For this purpose, it is advised to set the mixing chamber in the center of the platelet and to benefit from a homogeneous concentric injection. Thus the corresponding platelets carry a larger breakout in their interior and smaller conduits for fluid feed in their outer region. Separation platelets (with breakout) and gaskets have to be inserted between the fluid carrying platelets in order to achieve separation between the two feed fluids. The fluid carrying platelets for the two feeds can be different in shape [66] or rely on the same design [130]. The concept of consecutive injection is in particular amenable to an internal numbering-up, since stacking of thin plates on guide pins is very facile way to create hundreds (or thousands, on request) of alternate feeds.

The large diameter of the breakout and the high flow rates typically applied suggest strongly that mixing is performed under turbulent conditions [130]; final evidence in favor has not been given so far. Experimental findings on the mixing quality, however, again strongly support this assumption (see *Mixing quality vs. pressure)*. At low flow rates, initial simulations, relying only on a coarse grid, hint that a circle segment fluid distribution is achieved, rather than achieving the more finely dispersed onion skin-like pattern.

1.3.10.1 **Mixer 47 [M 47]: Interdigital Consecutive Micro Mixer, StarLam300**

This interdigital concentric consecutive microstructured mixer is composed of a housing with a cylindrical recess in which a platelet stack is inserted (see Figure 1.108) [130]. The device can be operated up to 100 bar and at a temperature of 600 °C.

The platelets contain a star-like breakout (see Figure 1.109), giving the device name of the series, StarLam [130]. From six ports the fluids are guided to the platelet center; a breakout of smaller diameter in the second type of plate (non-feed carrying) actually defines the whole flow conduit. Twelve holes are placed on a circular arc next to the platelet boundary.

Mounting of the device is facilitated, since the assembly of the many platelets to a stack is achieved by open accessible guide pins, where the platelets are threaded and thereby become pre-oriented; demounting is done in the same way (see Figure 1.110) [130]. The thread platelet stack can be uncompressed along the guide pins and thus facile cleaning is achieved. Another means to encounter fouling is

Figure 1.108 Interdigital concentric consecutive microstructured mixers have the potential for use as chemical production tools, as evidenced by their comparatively large outer dimensions and owing to an internal numbering-up of the microstructured platelets. Left, front: StarLam300; right, back: StarLam3000 [130].

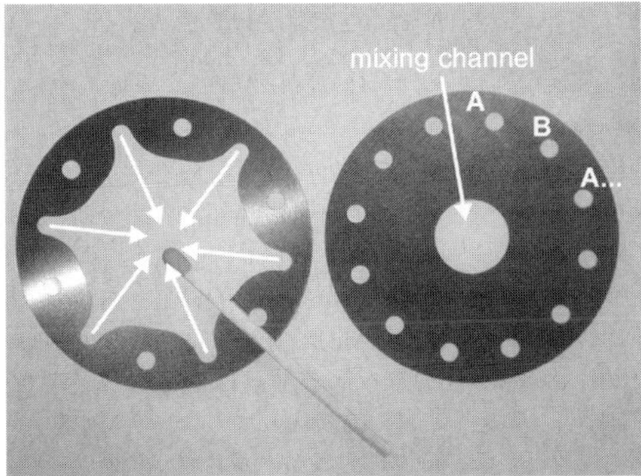

Figure 1.109 Platelet with star-shaped breakout for fluid carriage into the mixing chamber and feed conduits (left) and separation platelet with conduits for feed distribution to the platelets (right) [130].

Figure 1.110 Uncompressed stack of five platelets mounted on two guide pins and arrows indicating the fluid distribution [130].

initially to diminish it by variable adaptation of the micro spaces, i.e. choosing platelets of thicknesses ranging from 50 to 250 μm.

The housing was made using conventional milling, turning and drilling [130]. The two platelets were realized by laser cutting with a YAG laser at 1064 nm, which allows the manufacture of all stainless steels and alloys including those customary in trade, including Hastelloy, titanium and tantalum. The design is amenable to punching or wet-chemical etching.

Mixer type	Interdigital concentric consecutive micro mixer StarLam300	Number of platelets required for forming a stack	2
Housing and platelet material	Stainless steel	Platelet material	1.4571/X 6 CrNiMo Ti 17 12 2/ASTM 316 Ti
Housing dimensions (without connectors)	ϕ 40 mm × 65 mm	Number of platelets	2 × 65
Fluidic connections	Swagelok	Platelet thickness	100 μm
Mixer platelet stack dimensions	ϕ 22 mm × max. 17 mm	Seals material	Graphite; Viton; Chemraz

1.3.10.2 Mixer 48 [M 48]: Interdigital Consecutive Micro Mixer, StarLam3000

The StarLam3000 is the large-capacity version of its smaller counterpart, the StarLam300. The construction and fabrication of the StarLam3000 are identical with those of the StarLam300 (see [M 47]), only the design specifications vary (see below).

Mixer type	Interdigital concentric consecutive micro mixer StarLam3000	Number of platelets required for forming a stack	2
Housing and platelet material	Stainless steel	Platelet material	1.4571/X 6 CrNiMo Ti 17 12 2/ASTM 316 Ti
Housing dimensions (without connectors)	ϕ 100 mm × 150 mm	Number of platelets	2 × 113
Fluidic connections	Swagelok	Platelet thickness	250 μm
Mixer platelet stack dimensions	ϕ 56 mm × max. 57 mm	Seals material	Graphite; Viton; Chemraz

1.3.10.2 Mixing Characterization Protocols/Simulation

[P 44] Pumping for determination of the highest flow rates and pressure drops was realized using a normal water conduit (28 mm copper pipe with 1 inch ball valve), a water meter customary in trade (metering precision up to 5000 l h^{-1}) and a stopwatch [130]. The pressure drops were determined using a simple analog pressure gage (scale up to 4 bar).

[P 45] For mixing characterization, three different types of commercial membrane pumps were used depending on the capacity regime (high, thousands of l h^{-1}; medium, hundreds of l h^{-1}; tens of l h^{-1}).

An original protocol for mixing characterization using the competitive iodate/iodide reaction (Dushman reaction) is given in [36]. Especially for mixing at long residence times, the mixing results are superposed by a post-reaction which thus falsifies to an extent the overall result. Meanwhile, an optimized protocol using a borate buffer has been described, yielding a stable, alkaline mixture without post reaction [79]. An aqueous solution containing 0.0319 mol l^{-1} potassium iodide, 0.00635 mol l^{-1} potassium iodate, 0.0909 mol l^{-1} sodium hydroxide and 0.0909 mol l^{-1} boric acid needs to be mixed with 0.015 mol l^{-1} sulfuric acid [79].

1.3.10.3 Typical Results

Maximum flow rate

[M 47] [M 48] [P 44] The current maximum flow rate of a small-scale interdigital concentric mixer (StarLam300) is 1000 l h^{-1} at about 3 bar pressure drop [130]. For the large-scale counterpart (StarLam3000) even 3000 l h^{-1} at only 0.7 bar is determined. Extrapolating the flow rate-pressure relationship further to 10 bar yields a flow rate of nearly 2000 l h^{-1} for the StarLam300 and nearly 10 000 l h^{-1} for the

StarLam3000, i.e. 10 $m^3 h^{-1}$. An older StarLam version is reported to achieve about 300 l h^{-1} at 8 or 12 bar, depending on specification.

This approaches the performance of conventional static mixers, closing the gap between this smallest class of industrially used apparatus and micro mixers for analytical purposes [130]. Thus, a toolbox of microstructured mixers is available for the whole range of flow rates. This box comprises interdigital multi-laminating with a few ml h^{-1} capacity and split-and-recombine micro mixers (caterpillar type) up to 100 l h^{-1} as well as interdigital concentric microstructured mixers (StarLam type) series achieving 300, 1000 and up to 3000 l h^{-1}, dependent on the respective sub-version.

Maximum quality vs. flow rate

[M 47] [M 48] [P 45] Two interdigital concentric microstructured mixers of small and large scale (StarLam type) were compared for their mixing efficiency [130]. The efficiency was determined by a competitive reaction approach and is inversely proportional to a measured UV absorption signal, i.e. the extinction, relating to one reaction product, iodine. The small StarLam300 reaches extinctions as low as 0.1 at maximum flow rate, which is considered to correlate with a good mixing efficiency. The best value known for laboratory micro mixers (still smaller and lower throughput) is 0.01 [22]. The mixing performance of an older small-scale StarLam version even comes very close to this best performance of 0.01, which equals that of the caterpillar mixer at high flow rate.

The large-scale interdigital concentric microstructured mixer (StarLam3000) gives a mixing efficiency which is average, equivalent to an extinction of 0.55 (see Figure 1.111) [130]. The highest flow rate applied in these mixing-efficiency measurements did by far not correspond to the maximum possible flow rate of the StarLam3000, since for reasons of sufficient chemical stability only medium-capacity pumps could be chosen; the maximum flow rate, however, was determined using water only. Thus, it makes sense to extrapolate the mixing efficiency of StarLam3000 further to still higher flow rates. Performing this up to 1000 l h^{-1}, a similar good performance of the large version to that for its small-scale counterpart (StarLam300) is obtained.

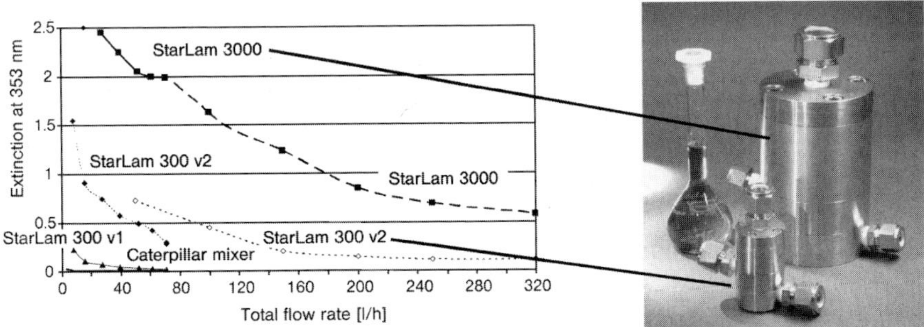

Figure 1.111 Flow rate dependence of the Dushman mixing quality of the interdigital concentric consecutive microstructured mixers of the StarLam series and other micro mixers taken for benchmarking [130].

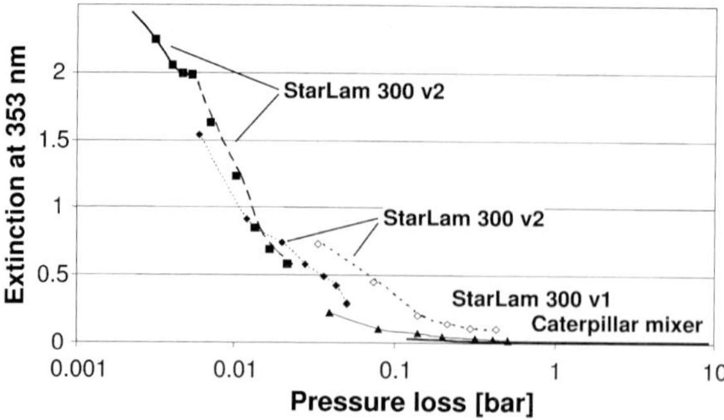

Figure 1.112 Pressure-drop dependence of the Dushman mixing quality of the interdigital concentric consecutive microstructured mixers of the StarLam series and other micro mixers taken for benchmarking [130].

Mixing quality vs. pressure

[M 47] [M 48] [P 45] If the Dushman mixing efficiency of different microstructured mixers is plotted versus the pressure drop, all data sets orient on one curve (see Figure 1.112) [130]. This means that all these devices have the same decreasing relationship of pressure drop with respect to mixing efficiency, which is typical for turbulent operated static mixers. This demonstrates that achieving respectively high flow rates for each device is a prerequisite for good mixing performance. This is further evidence for the toolbox mixer concept, built from grouped devices, mentioned above.

Mixing quality vs. normalized flow

[M 47] [M 48] [P 45] To obtain a more general view, the flow rate can be normalized as the ratio of measured to maximum flow rate [130]. An analysis of the so-determined mixing qualities allows one to judge which percentage range of flow rates gives good mixing quality, if a threshold value for the latter is defined. The current measurements of StarLam300 and StarLam3000, which are only available up to 35% normalized flow at present owing to pumping limits, show that there is continuity when considering the percentage flows. This is further corroborated when considering the small-scale predecessor version of StarLam300. Already at 7% normalized flow rate, good mixing efficiency is achieved.

1.3.11
Cyclone Laminating Mixing

Most Relevant Citations

Proceedings contributions: [109, 131, 132].

Cyclone mixers give a rotational flow field [109]. The corresponding formation of vortex patterns is another way of laminating and focusing streams. It is hoped that by folding of the vortices, thinning of the lamellae can be achieved, with an increase of residence time in the mixer [131]. A full rotation should halve the lamellae width. Hence the optimization parameter may be to have as many rotations as possible.

1.3.11.1 Mixer 49 [M 49]: Cyclone Laminating Micro Mixer, Tangential Injection (I)

This cyclone mixer comprises a spherical mixing chamber with 16 equispaced nozzles which introduce the feeds tangentially at high (m s^{-1}) fluid velocity [109] (see also [132]).

Microstructuring was effected by dry reactive ion etching (DRIE) [109, 132]. Two microstructured wafers were sandwiched between two powder-blasted wafers with through-holes and feed-rings.

Mixer type	Cyclone laminating micro mixer, tangential injection	Cover material	Pyrex
Mixer material	Silicon		

More information on the geometry was not given in [109, 132].

1.3.11.2 Mixer 50 [M 50]: Cyclone Laminating Micro Mixer, Tangential Injection (II)

This cyclone mixer comprises a cylindrical mixing chamber with many nozzles, arranged on the circular arc of the chamber (see Figure 1.113) [131]. At the the center of the floor of the chamber an outlet hole collects the multi-laminated streams.

Figure 1.113 Cyclone mixer design with tangential inlets only.
Left: photograph of a glass platelet carrying a cylindrical mixing chamber surrounded by tangential inlets.
Right: SEM image showing a detail of the same platelet [131] (source IMM).

A steel version with small mixing chamber depth and a glass version with a deeper chamber were realized. In the glass version, the nozzle feeds are located at the top of the mixing chamber, mainly for fabrication reasons, which has an extended depth. The nozzles perform fluid injection at a small angle, in the direction of the intended tangential flow. The nozzles are flat channel-type structures with strongly decreasing width so that the fluid passes through a tiny opening into the cylindrical mixing chamber.

(a) Microstructuring was effected by the LIGA process in the case of the steel version to yield a metal replica containing the cyclone-flow element [131]. This microstructured plate was inserted as an inlay in a recess of a steel housing consisting of two parts which were connected by pressure fitting using seals. The housing also contained fluid connectors.

Mixer type	Cyclone laminating micro mixer, tangential injection	Mixing chamber diameter	5 mm
Mixer material	Metal (inlay); stainless steel (housing)	Mixing chamber depth	150 μm
Nozzle width	40 μm	Outlet channel diameter	500 μm
Nozzle number	16		

(b) For the glass version, photoetching of glass and subsequent thermal bonding were applied [131]. The plate architecture was complex, i.e. as many as seven plates were needed to achieve the overall flow guidance. The thus-formed glass stack was inserted in a steel or aluminum frame with fluid connectors.

Mixer type	Cyclone laminating micro mixer, tangential injection	Mixing chamber diameter	5 mm
Mixer material	Glass	Mixing chamber depth	1150 μm
Nozzle width	40 μm	Outlet channel diameter	500 μm
Nozzle number	16	Number of micro-structured plates	7

1.3.11.3 Mixer 51 [M 51]: Cyclone Laminating Micro Mixer, Cross-flow Injection

This cyclone mixer has a similar fluidic arrangement as [M 50] and was similarly fabricated, but performs the nozzle feed differently [131]. Alternating arranged vertical and tangential inlets are used, instead of having only tangential ones. Thereby, it is intended to collide jets at roughly a 90° angle in order to achieve fragmentation of the jets to smaller bodies for improved mixing. This mixer was mainly designed for multi-phase operation, in particular for gas/liquid contacting. Owing to the large space requirements of such feeding architecture, the number of nozzles is less than for [M 50]. Eleven plates were needed for the overall fluid architecture. A specially version for slurry operation was also made.

(a) Standard version

Mixer type	Cyclone laminating micro mixer, cross-flow injection	Mixing chamber diameter	5 mm
Mixer material	Glass	Mixing chamber depth	1150 μm
Nozzle width, liquid and gas side	50 μm, 30 μm	Outlet channel diameter	500 μm
Nozzle number	2 × 8	Number of micro-structured plates	7

(b) Specially version for slurry operation

Mixer type	Cyclone laminating micro mixer, cross-flow injection	Mixing chamber diameter	10 mm
Mixer material	Glass	Mixing chamber depth	1150 μm
Nozzle width, liquid and gas side	50 and 150 μm (two sets), 30 μm	Outlet channel diameter	500 μm
Nozzle number	2 × 8	Number of micro-structured plates	11

1.3.11.4 **Mixing Characterization Protocols/Simulation**

[P 46] Mixing speed was studied by finite element modeling tools [109].

Deionized water and 0.1 mM fluorescein solution were contacted in a dilution-type experiment (1 bar injection pressure) [109].

1.3.11.5 **Typical Results**

Extension of rotational flow field in mixer outlet

[M 49] [P 46] The extension of the rotational flow field in the mixer outlet was visualized (see Figure 1.114) (inlet speed 10 m s^{-1}; Re_{max} 200; 10 bar) [109].

Impact of mixer geometry on total rotation

[M 49] [P 46] The total rotation was given as a function of the dimensionless geometric parameter ratio, namely the chamber diameter to nozzle width and the chamber diameter to height ratios, at various heights of injection [109]. The total rotation is notably dependent on the height of injection. Fluid volumes entering at half-height undergo the largest rotation. The mixing chamber geometry has a much smaller influence, albeit minimizing the width and optimal diameter to height ratios slightly improve the rotation.

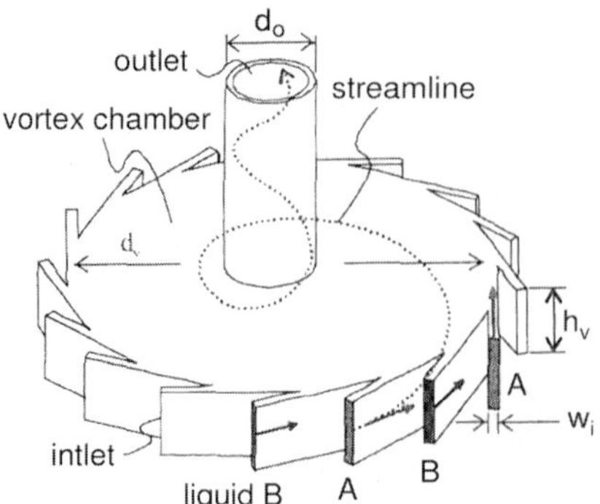

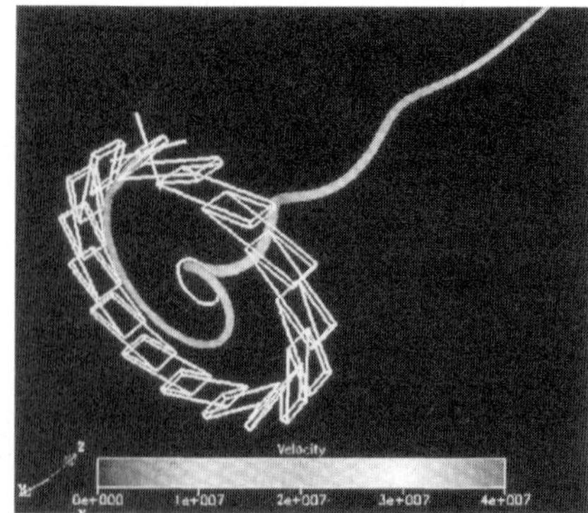

Figure 1.114 Schematic of the cyclone mixer design (left) and streamline indicating the rotational flow field in the cyclone micro mixer [109] (by courtesy of Kluwer Academic Publishers).

Vortex flow patterns

[M 49] [P 46] The formation of vortex flow patterns was confirmed both by simulation and experimentally [109] (see also [132]). While the simulation suggests spirally wound fluid flow, the first experiment yields only a direct straight flow into the center outlet. However, this difference might be merely due to the fact that very initial results are presented in [109].

[M 50] [P 39] At low total flow rates, star-shaped profiles, i.e. the streams go straight from the nozzles to the center outlet hole, are visible, which indicates reasonable flow distribution (see Figure 1.115) [131]. The top image reveals a larger dark-colored fraction than the transparent one. Since large-area mixing is not likely for the short residence times in the chamber, this indicates that the fluid compartments are folded towards the optical observation axis. The top image then sums up optical information from different fluid layers, giving rise to an integral color. At higher flow rates, the streams take a more tangential direction, now folding the layers in the vertical direction which gives the typical vortex structure. At still higher flow rates, images of nearly homogeneous color texture are obtained, which again is seen to be due to layer folding and not to mixing (for reasons of the short residence time).

The images of the experimental patterns resemble with fairly large detail simulated flow images [131]. For example, fine structures such as edged bending at the inlet are found on both types of images. In the experiment, however, the image is not totally symmetrical. About one-quarter of the top view of the chamber has nearly homogeneous color, which is likely due to flow maldistribution, i.e. one fluid is fed here at some higher flow rate and occupies a much larger space.

[M 51] [P 39] For sole liquid flows, this cyclone mixer also reveals vortex-type flow patterns [131]. By increasing the volume flow, the extent of spiral winding of the liquid layers increases as in the observations made for the tangential-flow device previously (see Figure 1.115). Correspondingly, at Reynolds numbers of a few hundred, hydrodynamic focusing of liquid lamellae sets in, by which time the lamellae width becomes considerably reduced. The homogeneous texture of the

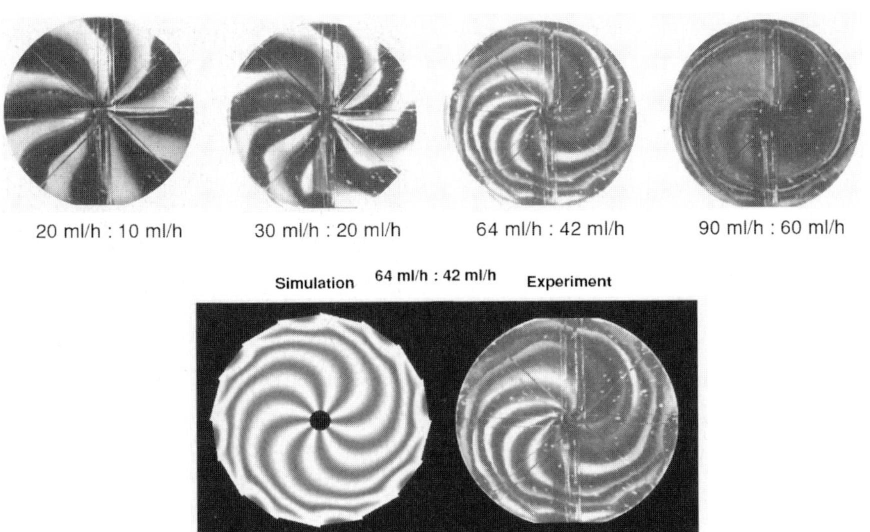

Figure 1.115 Top: photographs of vortex-flow pattern formation at various flow rates when contacting water and dyed water (acid blue) solutions in the tangential-flow cyclone micro mixer. Bottom: comparison of the experimentally derived image of one vortex pattern with the predicted image by CFD simulation [131] (source IMM).

liquid flow in the chamber suggests a high degree of mixing; however, simulations predict that this is due to tilting of the lamellae.

Similar vortex patterns were also found for gas/liquid flows [131], which is outside the scope of this book, however.

1.3.12 Concentric Capillary-in-capillary and Capillary-in-tube Mixing

Most Relevant Citations

Peer-reviewed journals: [133] (see also [134]).

Capillary-in-capillary mixers use simple base structures, namely capillaries, for fluid feed and as mixing chambers [133, 134]). The very simple idea is to insert a smaller capillary into a larger one under strict control of the distance of these two objects. Albeit the capillaries themselves then may have macro dimensions, micro-space conditions may be achieved by proper adjustment of their relative diameters and, in addition, by adjustment of the following mixing chamber, i.e. thereby creating thin fluid compartments. These compartments can be two circles of an arc of similar diameter, if a cylindrical flow obstacle is placed in the larger capillary in the mixing section and the two fluids need to be passed through the free distance between the obstacle and the large capillary's wall [133, 134]. If this is not the case, one fluid forms an inner cylindrical stream surrounded by the outer shell of the other fluid, provided that laminar conditions pertain.

Capillary-in-capillary mixers were used for electrospray ionization mass spectrometry (ESI-MS), which allows one to perform on-line kinetic studies for a wide range of applications in chemistry, bioorganic chemistry, isotope exchange experiments and enzymology, just to name a few [133]. ESI-MS is a method alternative to the traditionally employed quench-flow techniques with off-line analysis.

1.3.12.1 **Mixer 52 [M 52]: Capillary-in-capillary Micro Mixer**

A laboratory-built continuous-flow mixing set-up was constructed around a mixer made from two concentric capillaries. The capillary of larger diameter was placed in a three-way union; a sleeve was inserted at one end of the union to allow insertion of the smaller capillary. The inner capillary was closed by a rapid curing, self-priming polymer glue. Upstream a notch is cut into the side of the inner capillary so that the respective fluid has a conduit to contact the other. Both fluids have to flow along the passage until the glue plug end with very narrow diameter, by which fast completion of mixing is achieved. Then, the mixed flow was directly introduced into an analytical instrument, which was here a mass spectrometer [133].

Two syringe pumps fed the mixer, the inner capillary directly from behind, the outer capillary being fed by the third port of the three-way union, in a vertical direction to the other feed [133]. A Delron block was then positioned at the end of the outer capillary, providing a connection to a mass spectrometer.

The mixing volume of the capillary-in-capillary mixer is adjustable and can be controlled automatically, i.e. it is not necessary to have a set of various capillaries and to exchange them to match the single conditions of each new experiment [133].

Mixer type	Capillary-in-capillary micro mixer	Distance notch–glue plug, i.e. mixing length	2 mm
Inner and outer diameters of large capillary	182 μm, 356 μm	Mixing volume and dead (mixing) time	2 nl, 8 ms
Length of outer capillary	13 cm	Glue material	Self-priming polyimide
Inner and outer diameters of small capillary	100 μm, 167 μm	Sleeve material	Flexon
Distance between outer and inner capillaries, i.e. mixing width	8 μm		

1.3.12.2 **Mixer 53 [M 53]: Capillary-in-tube Micro Mixer**

This double-pipe mixer contained as connection module a sort of T-piece as an outer shell to connect and fix three outer capillaries for two inlet flows and one outlet flow [134]. One tube of relatively large diameter was passed through two openings of the T-frame, so that the tube end was at the level of the end of the

T-frame. The tube end was glued to seal it and injection into the tube was made from the side through the third opening of the T-frame. In the center of the tube a small-diameter capillary was inserted; the second fluid was fed through this flow conduit. Since this small capillary had a shorter length in the outlet section than the large tube, the respective fluids were contacted from this point.

The double-pipe mixer was designed and so far only used for contacting and reacting immiscible fluids [134]. The respective flow-pattern maps were derived and annular and slug flows as well as complete spread of the inner-tube fluid were identified as distinct regimes. Since in this chapter only miscible liquids are concerned, no protocol and no results are given for the mixer below. However, the device is mentioned, since it could in principle be used also for mixing miscible fluids.

Mixer type	Capillary-in-tube micro mixer	Inner diameter of large tube	3000 μm
Tube material	Glass	Inner and outer diameters of small capillary	1588 μm, 307/607/878 μm

1.3.12.3 **Mixing Characterization Protocols/Simulation**

[P 47] The capillary-in-capillary mixer was especially developed for millisecond time-resolved studies by ESI-MS. Since this ranges involves an application and characterization of the mixer itself was not performed, for further information on the experimental details the reader is referred to [133].

1.3.12.4 **Typical Results**

Application to ESI-MS analysis

[M 52] [P 47] The capillary-in-capillary mixer proved functionality for millisecond time-resolved studies by ESI-MS [133]. The experiments were performed in two modes of operation: in a 'spectral mode' with recording of entire mass spectra and in a 'kinetic mode' where the intensity of selected ion signals can be monitored as a function of the average reaction time. This enabled new means of resolving kinetic data, i.e. to measure reliably first-order rate constants up to at least 100 s^{-1}. This performance is four times better than for reported ESI-MS experiments.

By control experiments with a standard commercial stopped-flow instrument, it was checked that true kinetic information was derived, i.e. that there are no mixing effects any more, which limit the applicability of the device [133].

1.3.13 Droplet Separation-layer Mixing

Most Relevant Citations

Peer-reviewed journals: [39, 135, 136]; proceedings contributions: [53, 137]; chapter in encyclopedia: [138]. See also [139] for a conceptual and fabrication study of a micro dosage device generating two droplets from different sources which can be merged.

Separation-layer micro mixers are specially tools for mixing solutions which react fast or tend to foul otherwise [39, 53, 135–138]. The most prominent example of such processes is probably the generation of particles by immediate precipitation, as e.g. for calcium carbonate formation. Separation-layer mixers thus overcome the limits of normal micro mixers, which tend to clog under such conditions.

Separation-layer mixers comprise a specially interdigital feed which inserts either a miscible or immiscible layer between the solutions to be mixed, i.e. a multi-layer feed with a three-layer periodicity is generated [39, 53, 135–138]. In the case of miscible solutions, the separation layer constitutes most often the solvent itself. By this measure, mixing is 'postponed' to a further stage of process equipment or at least not initiated in close vicinity of the interdigital feed. Accordingly, the solutions to be mixed are fed in a fairly defined, dispersed state to a later stage of processing, but not as a mixed state on a molecular level. It is common laboratory experience that other types of miniaturized process equipment such as tube reactors/heat exchangers do not suffer from precipitation to the same extent as mixers with their tiny nozzles do.

Two groups of separation-layer mixers were described, one having a regular interdigital feed in a planar structure and the other having a concentric interdigital feed [39, 53, 135–138]. The planar structures create multi-lamination patterns, with every third layer being the separation layer. The fluid-compartment architecture of the concentric devices composed of an inner stream is surrounded by several rings of an arc.

The separation-layer technique benefits from the unique feature of micro mixers, such as to operate in a laminar flow regime [135]. By the absence of convective recirculation patterns, at least close to the inlet, the separation layer remains as a barrier between the solution to be mixed, as long as it is not passed by molecules owing to diffusive transport.

The concentric and planar feeds can be operated in two modes, allowing either drop- or stream-like injection of liquids in mixing chambers attached or in any other processing device such as a reaction tube [53]. Drop-like injection can be used to create segmented flows, either gas/liquid and immiscible liquid/liquid. Apart from performing multi-phase reactions, such segmented flows may be used to achieve plug-flow operation with one reacting and one inert segment. This combines effective recirculation patterns for good mixing and defined setting of residence time. Again, among other uses, this may be particularly beneficial for precipitations.

Alternatively to mixing of the precipitating reactants in a droplet, these may be injected via two separate nozzles in a micro chamber where reaction takes place and possibly further downstream transport is initiated. Although such devices have not been published to the best of our knowledge, a conceptual and fabrication study of a micro dosage device gives a very schematic idea of how such configurations may look and how they may be realized. Here, configurations are proposed with two nozzles generating two droplets from different sources which merge in a chamber below [139].

1.3.13.1 Mixer 54 [M 54]: Concentric Separation-layer Interdigital Micro Mixer

The concentric separation-layer micro mixer is constructed as an assembly of stacked plates for feed supply with three tubes, performing lamination for mixing, set into one another (see Figure 1.116) [39, 53, 136–138] (see also [135]). The tubes are inserted into a frit. The three feed lines are each connected to a tube. In this way, a tri-layered concentric fluidic system is achieved. Besides mixing three solutions, a major application for the device is to separate the two fluids to be mixed by a separation layer, usually being the solvent of the two solutions. This is to delay the mixing process in order to avoid unwanted fouling problems at the mixer outlet. This is particularly valuable for spontaneous precipitation reactions which are the main field of application of the mixer.

The steel plates are made by milling and the PEEK tubes by turning and milling [39].

Mixer type	Concentric separation-layer interdigital micro mixer	Tube inner diameters	1.5 mm, 2.5 mm, 3.4 mm
Tube material	PEEK	Tube outer diameters	2.0 mm, 3.0 mm, 4.0 mm
Housing material	Stainless steel	Tube lengths	28.50 mm, 21.75 mm, 15.00 mm
Device outer dimensions: length, width, thickness	41 mm, 41 mm, 24 mm	Fluid layer thickness (from interior to outside)	1.5 mm, 200 μm, 200 μm

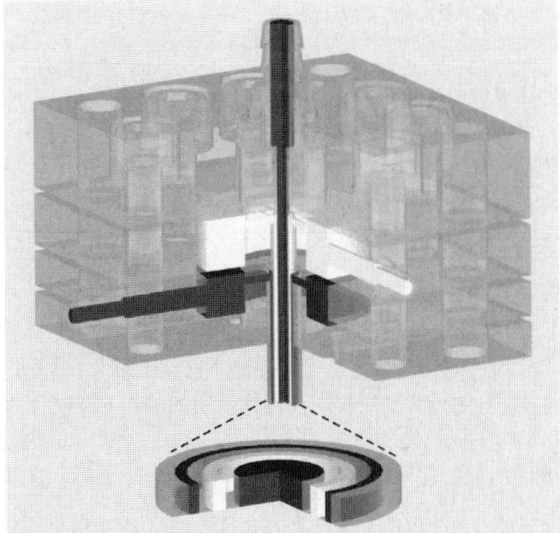

Figure 1.116 Schematic of the construction of the concentric separation-layer micro mixer and magnified concentric fluid layer arrangement when leaving the mixer [53] (source IMM).

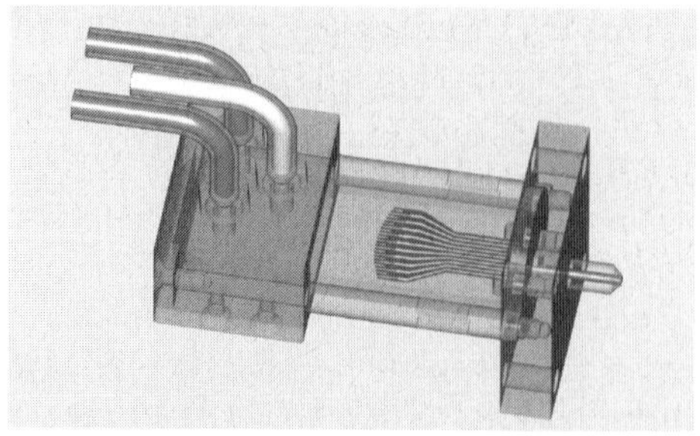

Figure 1.117 Schematic of the construction of the stacked separation-layer micro mixer [53] (source IMM).

1.3.13.2 **Mixer 55 [M 55]: Planar Separation-layer Interdigital Micro Mixer**

The planar separation layer mixer is constructed as an assembly of stacked plates held in a frame which comprises the fluid connectors (see Figure 1.117) [39, 53, 136–138] (see also [135]). The interdigital feed structure is virtually the same as for normal interdigital micro mixers (see Section 1.3.9) with the exception that, owing to the feed of three fluids, three reservoirs and three multi-channel feeds in three plates are needed. Such interdigital feeding structure generates alternately arranged lamellae of the three liquids. The mixing chamber is rectangular from the feed inlet until close to the outlet and then becomes tapered to fit better to the round fluid adapters normally used.

The glass plates were manufactured by photoetching of a specially glass and were irreversibly bonded by a thermal soldering process [39].

Mixer type	Planar separation-layer interdigital micro mixer	Feed channel width, depth	60 μm, 500 μm
Tube material	Glass	Fin width	50 μm
Frame material	Stainless steel, aluminum	Fluid layer thickness	110 μm
Device outer dimensions: length, width, thickness	80 mm, 41 mm, 41 mm	Mixing chamber width, depth, length	10 mm, 500 μm, 20 mm

1.3.13.3 **Mixing Characterization Protocols/Simulation**

[P 48] Flow patterns may be visualized simply by photographic imaging using light reflection, if precipitates are generated during the mixing process [39]. This was performed for droplets generated at a concentric separation-layer micro mixer. The calcium carbonate reaction may be used as a fast precipitation following immediately the mixing process.

[P 49] CFD simulations were made for monitoring the flow patterns within a droplet which is generated at a concentric separation-layer micro mixer [39]. Diffusion–convection equations of two user scalars have to be solved in addition to the corresponding equation for the volume fraction of the fluids within a multiphase CFD simulation.

The computation of free surface flows was done by means of volume tracking and relies on the solution of an advection equation for an additional scalar field with simultaneous solution of the Navier–Stokes equation [135]. Surface tension is of major importance owing to the large surface-to-volume ratio of the small droplets. CFX4.4 software was applied. The so-called 'surface-sharpening' algorithm was applied, which ensures global mass conservation of the phases. The calculation of mixing is done by solving an additional convection–diffusion equation for species concentration. As for many other studies, numerical diffusion can provide a non-physical mechanism which adds to the real mixing. High-order discretization schemes are employed to suppress the numerical diffusion. For discretization of the convective transport terms a higher order upwind scheme with a flux limiter was used. The mesh cell number had to be limited to 50 000 owing to constraints of present workstations.

[P 50] The fouling and plugging sensitivity was tested under real-case conditions [53]. Three test reactions were used to test the fouling sensitivity, namely the quaternization of an amine, the forced precipitation of calcium carbonate and amide formation of an acid chloride (see below). In most cases, high concentrations of the reactant solutions were used in the range 10^{-2}–1 mol l^{-1}, which commonly are applied in laboratory research and chemical production. It was proved that these reactions cannot be carried out using standard micro mixers, such as the IMM interdigital micro mixers. In the last two cases, plugging within a few seconds or less occurs, naturally depending on the experimental protocol applied.

The separation-layer micro mixer was mounted about 2 cm above a funnel-shaped glass element which was connected to a glass tubular reactor, not being cooled [53]. The end of the tube was set about 2 cm above a glass beaker collecting the solutions. All experiments were made using the micro mixer and the mixer–tubular reactor set-up only.

Quaternization of 4,4′-bipyridyl with ethyl bromoacetate

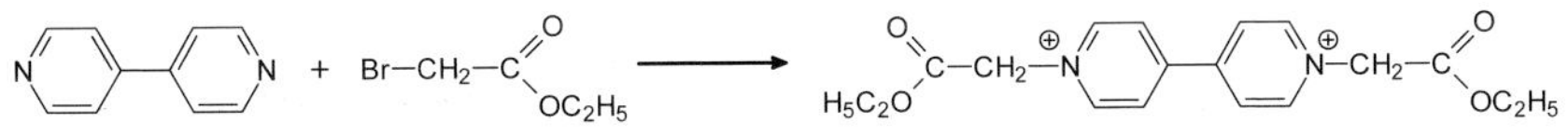

This reaction results in the precipitation of nearly the entire amount of the product if apolar, non-protic solvents are chosen [53]. This particle formation, however, needs a certain induction time, dependent on solvent and temperature.

Layers of 4,4′-bipyridyl (0.3 mol l^{-1} in dichloromethane) and ethyl bromoacetate (0.3 mol l^{-1} in dichloromethane) and a separation layer of dichloromethane are fitted into each other by means of the concentric separation mixer [53]. The reaction temperature is 22 °C. The reaction solution is inserted as droplets or a continuous stream either directly or via the tubular reactor in the beaker. The precipitate solution yielded is passed through a frit and the remaining solid is washed with dichloromethane and dried at elevated temperature and weighed. The quaternized product is characterized by NMR spectroscopy.

For further details of the experiments see, [53] or [72].

Formation of calcium carbonate

$$Ca^{2+} + 2\ CO_3^{2-} \rightarrow CaCO_3$$

The formation of calcium carbonate is an inorganic reaction which also results in immediate precipitation [53].

Layers of calcium nitrate (40 mmol l^{-1} in water; $CaNO_3 \cdot 4\ H_2O$), potassium carbonate (40 mmol l^{-1} in water) and a separation layer of water are fitted into each other by means of the concentric separation mixer [53]. The reaction temperature is 22 °C. The reaction solution forms droplets in a dodecane reservoir and inserted as such a segmented flow in the tubular reactor [137, 138].

For further details of the experiments, see [53] or [72].

Amide formation from acetyl chloride and *n*-butylamine

$$CH_3{-}C(=O)Cl + H_2N{-}(CH_2)_3{-}CH_3 \xrightarrow[Et_3NH^{\oplus}Cl^{\ominus}]{Et_3N} CH_3{-}C(=O){-}NH{-}(CH_2)_3{-}CH_3$$

This reaction results more or less immediately after contact of the two reactant solutions with precipitation [53]. The reaction time is of the order of a few tens of milliseconds, if mixing is adequate.

Layers of acetyl chloride (0.79 mol l^{-1}) in tetrahydrofuran (THF), *n*-butylamine (0.80 mol l^{-1}) and triethylamine (0.80 mol l^{-1}) in THF, and a separation layer of THF are fitted into each other by means of the concentric separation-layer mixer [53]. The reaction temperature is 22 °C. The reaction solution is inserted as droplets or a continuous stream either directly or via the tubular reactor ina beaker containing water. With rigorous stirring, hydrolysis of the acid chloride and hence stopping of the reaction are achieved. The phases are separated and the aqueous phase is extracted with THF. The combined THF phases are dried over Na_2SO_4. After filtration, the THF solvent is evaporated at 25 mbar. The remaining amide product is characterized by FTIR spectroscopy.

In a second experiment, higher concentrations were applied: acetyl chloride (0.198 mol l^{-1}), *n*-butylamine (0.200 mol l^{-1}) and triethylamine (0.200 mol l^{-1}) [53].

1.3.13.4 **Typical Results**

Flow patterns in concentrically multi-layered droplet mixers

[M 54] [P 48] Separation-layer micro mixers with concentric multi-layered outlets can be operated in a droplet-forming mode [53] If fast precipitating solutions are contacted in this way with a solvent layer for initial separation, the part of the droplet close to the tube outlets remains transparent, which demonstrates that a tri-layered system still exists with the two reacting solutions not being intermixed, as evidenced by calcium carbonate formation in aqueous solutions as described in [39, 136]. At the droplet end cap the layers collide and circulation flow sets in. As a result, mixing is achieved and precipitation occurs. The circulation patterns are visualized by the particle trajectories.

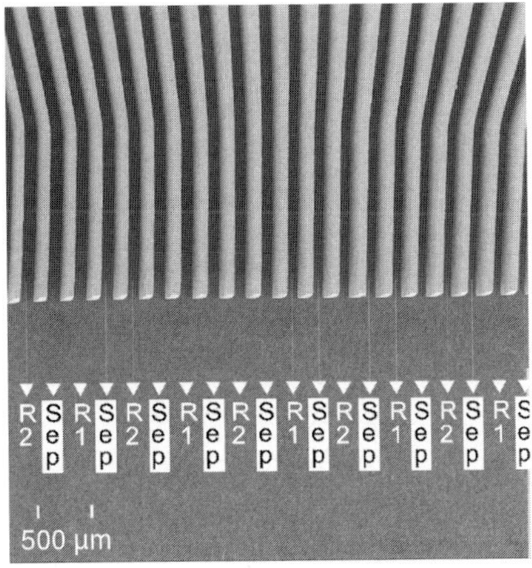

Figure 1.118 SEM Micrograph of the planar interdigital separation-layer micro mixer (left). The multi-lamination pattern is indicated. R1, R2 and Sep refer to reactant layer 1, reactant layer 2 and separation layer, respectively. Photograph of iron and rhodanide layers each separated by a water layer flowing through the glass separation mixer (right). The multi-laminated flow in the mixing chamber is nearly transparent, hence no mixing occurred. Close to the outlet and more intense in the droplet attached discoloration due to mixing and subsequent generation of the reaction product is visible ([53]; source IMM).

Flow patterns in planar multi-layered stream mixers

[M 55] [P 40] Separation-layer micro mixers with planar multi-layered outlets are usually operated in continuous streaming mode. When using the iron–rhodanide reactive approach, no color formation occurs for most of the mixing chamber, which indirectly proves that nearly ideal multi-lamination flow patterns were formed [53, 136] (see Figure 1.118). This also demonstrates that no recirculation patterns such as for the concentric device arise. Hence mixing can be much better controlled and the 'postponing effect' is more defined than with concentric layer mixer. On the otherhand, fouling still can occur in the planar device, as the fluids are encased in a miniature mixing chamber, whereas the droplets of the concentric mixer are only exposed to a fluidic environment (no walls).

Only at the end of the mixing chamber is the formation of the colored product iron rhodanide visible (see Figure 1.118) [53]. A droplet formed at the outlet (for low volume flow rates) demonstrates even better the delay with regard to mixing.

Simulated equivalent diameters of droplets

[M 54] [P 49] A validation of the volume-tracking method, applied for droplet simulation, was performed by comparing the simulated equivalent diameters of droplets with data derived from analytical correlations [135]. Assuming a nozzle diameter of 1 mm and taking water and dodecane as dispersing and continuous

liquids, respectively, the analytical correlations yield values of 3.7–4.1 mm for the droplet diameter. The CFD simulations predict droplet diameters of 3.6 and 3.8 mm for flows of 10 and 40 mm s^{-1} ($Re = 10$ and 40), respectively.

Hence reasonable agreement of the CFD simulations and analytical correlation is achieved, demonstrating that the latter can be applied for droplet simulation studies [135].

Two-fluid droplet generation and mixing – close contact

[M 54] [P 49] The formation of a droplet from two aqueous solutions and the respective mixing was simulated [135] (see Figure 1.120). The inner circular liquid has 400 μm diameter and is initially separated by 100 μm from the outer annular fluid, being 150 μm thin. A series of snapshots, the concentrations given in grayscale, show the interfacial development, i.e. the flow pattern within the droplet (the diffusivity of the liquid molecules was set to zero).

A large toroidal eddy evolves and a strong upward flow in the center of the droplet is observed [135]. A more detailed image is provided by the velocity distribution and reveals that actually two toroidal eddiesexists, a small one below and a large one above the stagnant interface between the upwards and downwards flows (see Figure 1.119). The mixing cannot be judged precisely owing to the numerical diffusion; however, good mixing may be assumed, especially in the droplet cap, on observing the interfacial stretching.

Two mechanisms were proposed for the strong upward and downward flows [135]. The first was based on interplay of the liquid–liquid and solid–liquid surface tensions and the gravitational and inertia forces. The second is correlated with the neck formation, the droplet break-up and the retossing of the fluid.

At the inlet the flow is almost linear and a vertical interface between the two fluids exists [135]. For this reason, fouling of the inlet is predicted and this agrees with the experimental findings.

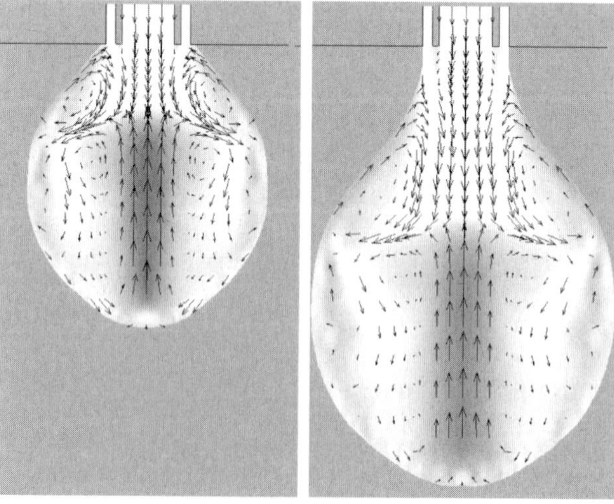

Figure 1.119 Velocity distributions for $t = 0.289$ and 0.6 s. For better visualization the vertical velocity component is encoded with gray tones: dark gray (white) corresponds to the maximum upward (downward) flow velocity [135].

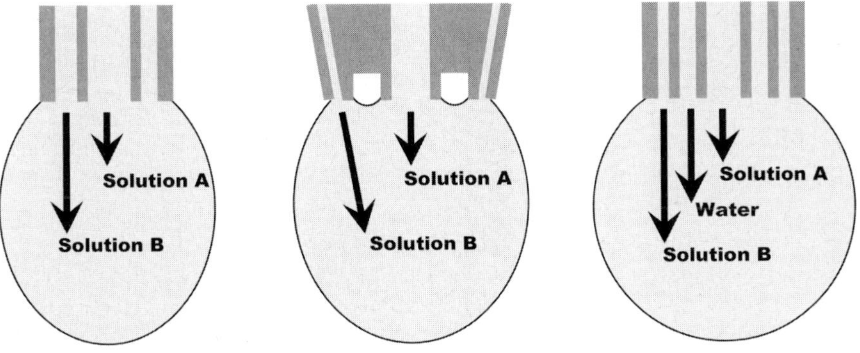

Figure 1.120 Schematic design of the various mixing nozzles analyzed: two-fluid droplet close contact (left); two-fluid droplet separated by annulus (middle); three-fluid droplet with separation layer (right) [135].

Two-fluid droplet generation and mixing – fluids separated by annulus

[M 54] [P 49] Since the results mentioned above indicate that the low configuration with two closely positioned liquids is prone to fouling, simulations with a new configuration were made, placing an annulus between the two liquids [135] (see Figure 1.120). The contact of the two-fluid mixture with the nozzle walls is minimized, i.e. the wall is wetted by one fluid only. Thus, fouling should be reduced compared with the configuration with both liquids in close contact. However, the whirls are less, i.e. mixing is expected to be less.

Three-fluid droplet generation and mixing – separation layer

[M 54] [P 49] Simulations predict an irrotational flow pattern close to the inlet on inserting a separation-layer liquid between the two liquids to be mixed (see Figure 1.121; see also Figure 1.120). [135] (see Figure 1.121; see also [136]). Thus, this concept may be adequate to prevent fouling by keeping precipitation zones away from the tiny mixer structures. The color encoded fluid distributions suggest a low degree of mixing, which, however, is uncertain, as numerical diffusion affected most settings of the simulations.

The relative velocity distributions for the initial droplet formation exhibit a strong upward flow with toroidal eddies [135] (see also [136]). When approaching the droplet pinch-off, however, the eddies mostly disappear and only adjacent up- and downward flows are found.

Experimental findings corroborate qualitatively the predictions of the simulations [135] (see also [136]). Photographs were made with a transmitted light microscope for solutions giving a white precipitate of calcium carbonate. The precipitate scatters the light and can be detected with high resolution already with low quantities formed. In order to observe fine streaks of the precipitates (and not clouds covering most of the droplet volume), the relative flow of the separation-layer liquid was set high and the photographs were taken at an early stage of the droplet formation. Close to the inlet only a transparent solution was found, indicating absence of mixing here which is compatible with an irrotational flow pattern. At the droplet cap, a two-times folded

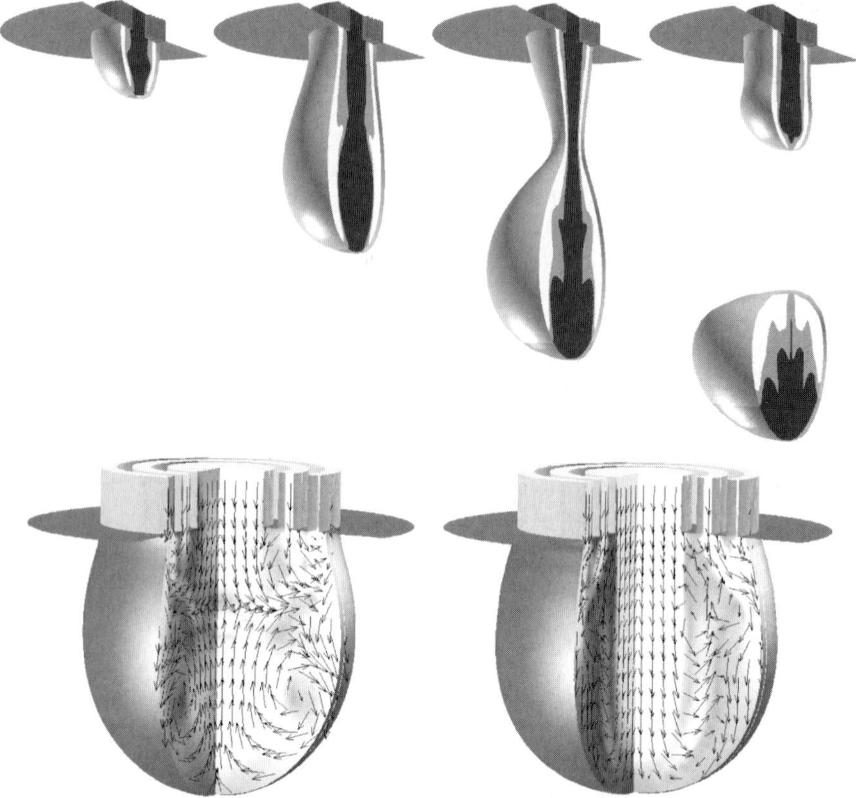

Figure 1.121 Distributions of one reacting solution, the separation-layer solution and another reacting solution in black, gray and white, respectively, at time steps (from left to right) $t = 0.03$, 0.20, 0.27 and 0.30 s (top). All inlet velocities were 100 mm s^{-1}. The dimensions of the separation-layer mixer are specified in [135]. The velocity distributions are given as well (bottom). Left: t = 0.07 s; right: t = 0.76 s. Further specifications are given in [135].

eddy is found at each side of the droplet, which proves that mixing occurred. The eddies rotate, however, in the opposite direction, as predicted by the simulations. This may be due to small misalignments of the streams in the experiment breaking the perfect rotational symmetry which is assumed in the simulations.

A further experimental proof relies on the fact that stable operation with precipitating solutions can be maintained over periods of 8 h [135].

Robustness of the planar separation-layer mixer concept

[M 54] [M 55] [P 40] Preliminary findings indicate that the mixing chamber walls of the planar separation-layer micro mixer may have a negative impact on fouling, particularly during the start-up phase and after pumping fluctuations [53]. This was not observed for the concentric mixer which does not rely on a mixing chamber; the only wall present is the mixer surface.

Robustness of the concentric separation-layer mixer concept

[M 54] [P 50] For the concentric separation-layer mixer detailed investigations on the robustness of the reactor concept were undertaken [53]. The operation time until plugging was monitored for three well-known fast-precipitating reactions from inorganic and organic chemistry.

For the quaternization of 4,4′-bipyridyl with ethyl bromoacetate, stable operation for at least 2 h, sometimes ranging up to 8 h, was achieved [53]. The yield amounted to 75%. This yield is in the same order as for laboratory-batch operation.

For the formation of calcium carbonate, stable operation for 8 h could be achieved [53].

For amide formation from acetyl chloride and *n*-butylamine, stable operation of at least 1 h could be achieved [53]. Particularly advantageous was the setting of the flow rates to 5 : 250 : 5 ml h^{-1}, which could be operated for 3 h. The yields achieved were between 87 and 100%. The lower yields were obtained at high total flow rates.

Parametric studies on the impact of total flow rate, relative flow rates and concentration on the flow pattern within droplets

[M 54] [P 50] High-resolution images of the droplets at the concentric separation-layer micro mixer display details on their flow patterns when using a reactive precipitation approach. The images were taken just before rupture, i.e. refer to the most advanced mixing state during droplet formation. Thereafter, the droplets fall owing to gravity and are mixed by recirculation owing to the droplet rotation [53].

The images show a transparent and turbid portion within the droplet, indicating the absence and presence of precipitated particulates (see Figure 1.122) [53].

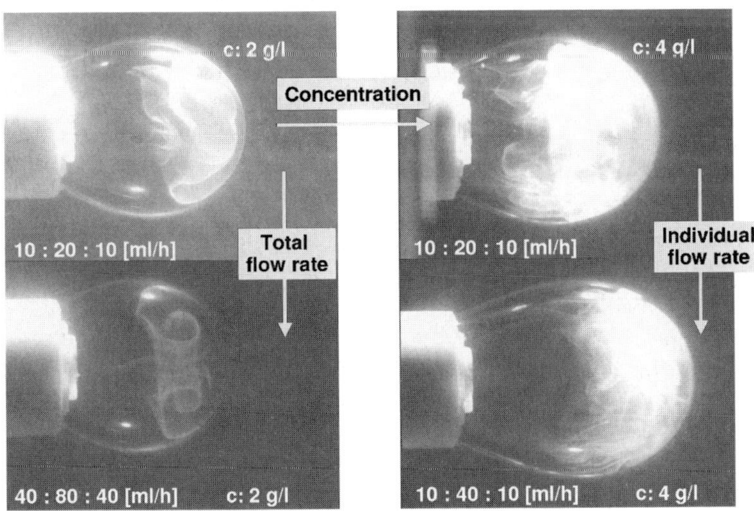

Figure 1.122 Photographs documenting the extent and location of calcium carbonate precipitation in droplets generated by the concentric separation layer mixer. The photos were rotated by 90° to allow a better comparison ([53]; source IMM).

On increasing the concentration of the precipitating solution from 2 to 4 g l^{-1}, the portion of the transparent section at the inlet is enlarged relative to the turbid portion containing the precipitated particulates in the bubble cap [53]. In particular, no particle strings entangling both portions are visible any longer.

On increasing the relative flow rate from 10 : 20 : 10 ml h^{-1} to 10 : 40 : 10 ml h^{-1} at 4 g l^{-1}, the portion of the transparent section at the inlet is enlarged relative to the turbid portion containing the precipitated particulates in the bubble cap [53]. In particular, no particle strings entangling both portions are visible any longer.

On increasing the total flow rate from 10 : 20 : 10 ml h^{-1} four-fold, the frequency of droplet formation is enhanced [53]. Accordingly, the residence time before rupture is reduced. This is evident by the lower degree of precipitation due to less mixing. These thin, streamline-like precipitation lines give an idea of the underlying flow patterns which are wound at the bubble cap owing to the interplay of up and down streams.

[M 54] [P 48] CFD simulations for the flow in the separation-layer micro mixer predict a stable, almost irrotational flow pattern in the inlet region, which is in line with the experimental findings of a transparent region mentioned above [39]. This pattern is maintained until the droplet end cap. Changes only occur when the droplet breaks up and falls, inducing rotational flow.

Fouling prevention under fast precipitating conditions

[M 54] [P 48] The ability of the separation-layer concept to prevent fouling even under severe conditions was demonstrated for two fast precipitating reactions, bisquaternization of 4,4′-bipyridyl and amide formation from acetyl chloride [39]. Stable operation for a few hours, and in selected cases longer, was achieved.

1.3.14
Split-and-recombine Mixing

Most Relevant Citations

Peer-reviewed journals: [7, 43]; proceedings contributions: [125, 126, 140–142]; patents: [143, 144].

Split-and-recombine (SAR) micro mixers use channel splitting and recombination [7, 43, 125, 126, 140–142]. Thereby, an originally bi-laminated flow is first divided into two sets of two lamellae, each of which is guided separately in branches of the channel. During this flow passage, the cross-sections of the streams are changed in such a way that their combination yields the original cross-section. This is done repeatedly, typically 5–10 times, until 2^n lamellae are achieved (n being the number of SAR cycles). In this way, a flow system with channel/two-channel elements as repeat unit is created. The SAR devices differ in the exact design of these repeat units which, e.g., may be of fork-like design.

For the first step, the lamellae division, a splitting plane, basically acting like a knife, can be useful [7, 140], not only for precise cutting but also for avoiding friction forces to deform the lamellae. SAR mixers were described with [7, 140] and without a splitting plane [43, 125, 126, 141, 142].

SAR mixing can also be achieved by in-channel structures [7]. The splitting here is achieved by spatially separated flow guidance of split lamellae within the channel. Elevating and descending structures can serve this purpose [7].

SAR mixing in the truest sense is only possible for very low Reynolds numbers, typically < 100 [7]. At other regimes, secondary flow superposes the SAR patterns. In terms of mixing, this may even be beneficial as a faster mixing time results. However, the typical SAR flow patterns cannot be identified by flow monitoring, so that, e.g., the design cannot be optimized by these simple means.

1.3.14.1 Mixer 56 [M 56]: Möbius-type Split-and-recombine Micro Mixer

The design of this SAR device was oriented on the so-called Möbius band, which is a twisted structure (see Figure 1.123) [141]. Two horizontal fluid layers are separated in the center. Thereafter, both double layers are turned by 90° in the same direction so that two vertically laminated systems are achieved. Then, the layers are joined to give a four-lamellae system.

The actual system consisted of two parallel rows with 10 SAR elements each [141]. The channels underwent a linear change of the width and depth to turn and reshape the flow. The mixer was an element that was intended to be integrated in a system comprising a mixing unit, reaction channel and an optical detector with crossed cylindrical lenses and optical fibers.

Microfabrication was done by means of excimer laser ablation in a polymer substrate [141]. This substrate was covered by a thin polymer sheet by thermal bonding under compression.

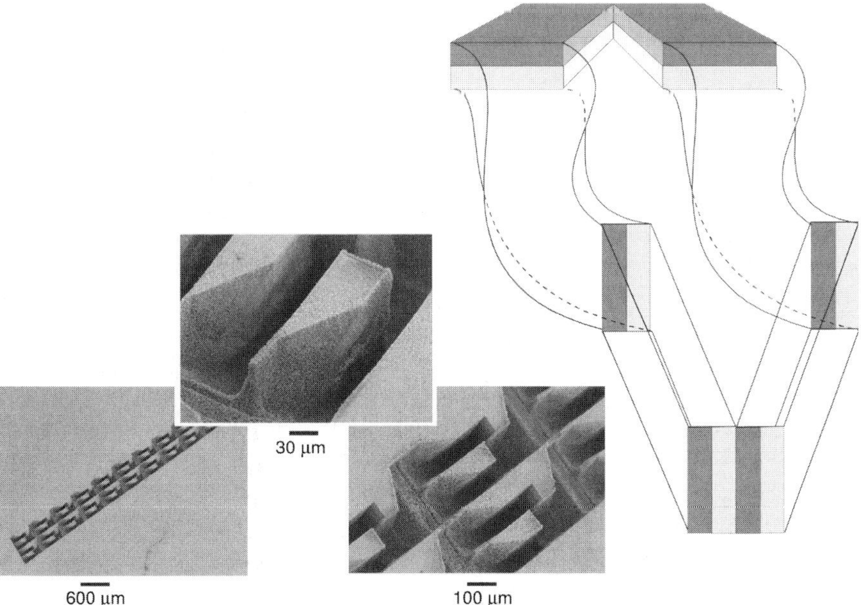

Figure 1.123 Schematic of the SAR process in a Möbius-type SAR micro mixer [141] (by courtesy of Kluwer Academic Publishers).

Mixer type	Möbius-type SAR micro mixer	Top layer thickness	250 μm
Mixer material	Polycarbonate	Channel width	275 μm
Plate thickness	1 mm	SAR element length	500 μm
Top plate material	PMMA		

1.3.14.2 Mixer 57 [M 57]: Möbius-type Split-and-recombine Micro Mixer with Fins

Another Möbius-type mixer is described in [142]. Again, a two-fold 90° rotation is said to be the basic principle, doubling the number of layers for laminar flow. After the first rotation, the laminated fluids are split and thereafter recombined, restoring the original geometry.

A channel with a series of diagonal fins performs the first 90° turn [142]. Then, the channel is split symmetrically into two channels, each having the same series of diagonal fins. Thereafter, the two channels are recombined by a junction being mirror-imaged to the first split. The original channel structure is restored. From here, the SAR step can be repeated multiple times. Experimentally it was found that seven fins are adequate for performing a correct 90° turn.

The microstructure was realized by a dry-film photoresist technique and based on established techniques from printed circuit board technology [142]. Dry resists are available as thin films, e.g. of thickness 50 or 100 μm. The resist films are encased in other polymer materials which are later removed. The resist films can be deposited on various base materials such as silicon or polymers giving mechanical stability. Lamination is carried out with a roller laminator. Then, exposure is made and spray development without any solvents follows. The process steps can be repeated at multi-laminated structures. Closed structures can be made in this way. Finally, the resist is cured.

Mixer type	Möbius-type SAR micro mixer with fins	Fin width, depth	50 μm, 200 μm
Mixer material	VACREL®	Internal mixer volume	6.5 μl
Plate material	PC-board glass-reinforced base material FR4 with 35 μm Cu layer	Fin number (per rotation)	7
Plate thickness	1.6 mm	Number of mixing stages	4
Channel width, depth	600 μm, 200 μm		

1.3.14.3 Mixer 58 [M 58]: Fork-element Split-and-recombine Micro Mixer

The design of this mixer is based on a series of fork-shaped elements (see Figure 1.124) [43, 144]. Initially, two parallel feed channels are placed so that one

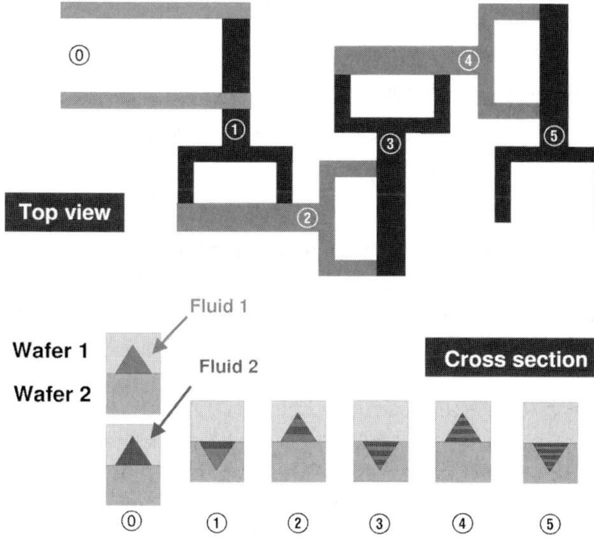

Figure 1.124 Schematic of the SAR process using a series of fork-like elements [43] (by courtesy of IOP Publishing Ltd.).

channel is attached to the end of a fork element. The other is close to the T-junction of the fork. Thereby, the fluid layers are superposed. For reasons of simplicity, a lateral bi-laminated system is supposed. The flow is then guided through the fork like an inverse tee so that splitting into two sub-streams is achieved. The two end channels of the fork are attached to the next fork in the same way as for the first two channels. Each fork is turned by 90° in an anti-clockwise manner. By this repeated splitting process, the interface between the fluids is enhanced so that mixing should be promoted. For laminar flow, it is expected to have a multi-layered fluid after several SAR passages in the ideal case. The channels were of triangular shape owing to the fabrication technique chosen (see below).

Wet-chemical etching with KOH solutions was used for microfabrication [43]. The structures were etched until the natural etch stop was reached. Bonding at 400 °C was carried out after an oxygen plasma treatment of the silicon wafer. Stainless-steel tubes were inserted in openings generated by sawing. These tubes led into transparent polymer tubing. An epoxy resin served for closing open parts in the whole system. A micropump was operated at a constant flow rate of 10 μl h^{-1}. The homogeneity of the fluid at the mixer outlet was investigated by using an optical microscope.

Mixer type	Fork-element SAR micro mixer	Number of mixing elements	5, 10, 15, 20
Mixer material	Silicon	Channel widths at surface	150 μm, 200 μm
Cross-sectional area	1600 μm^2	SAR element length, width	720 μm, 500 μm

1.3.14.4 **Mixer 59 [M 59]: Stack Split-and-recombine Micro Mixer**

This SAR device, named a stack mixer, is composed of two plates which both contain microstructures [125, 126]. By face-to-face positioning of these microstructures, a micro channel network yielding the SAR path is formed.

The two inlets feed an initial, rectangular channel, thus generating the two fluids forming adjacent lamellae with a vertical boundary (see Figure 1.125) [125, 126]. The lamellae are split horizontally by forming two sub-channels out of the inlet channel. The new channels have half the initial channel depth at same channel width. These channels are set then on the same level, at the plate interface, but still separated. Then the channels approach to each other until they merge. Thereafter, the channel width is halved and the depth doubled, restoring the initial channel geometry. This process can be repeated many times; actually five elements were placed in series in the device realized.

The actual device consists of two parallel rows of five mixing elements each [125, 126]. The lateral displacement and recombination of the sub-channels are done via 90° angles. In contrast, the variation of channel depth and width is done in a continuous fashion, yielding a curved horizontal and vertical structure. In particular, the curved floor structure is demanding, since a real 3-D and not just a multi-level microfabrication process is needed.

The mixer was integrated in a system composed of a four-plate device consisting of a top plate with fluid connectors, the stack mixer, a reaction channel and an optical detector using fiber optics [126]. The optical detector using fibers and cylindrical lenses to guide and focus the light and was made by LIGA for reasons of high demands on structural precision to avoid losses of light.

Microfabrication was achieved by the so-called laser-LIGA process [126] based on the replication of a master made by pulsed excimer laser ablation [125, 126]. Then, electroforming and injection molding processes followed to result finally in a replicated polymer structure identical with the master, but being amenable to mass production [126]. The plates were joined by thermal compression by heating slightly above the glass transition temperature and applying gentle pressure.

No details of the geometry specifications are given in [125, 126]. The tabulated specifications were taken from SEM images in [126].

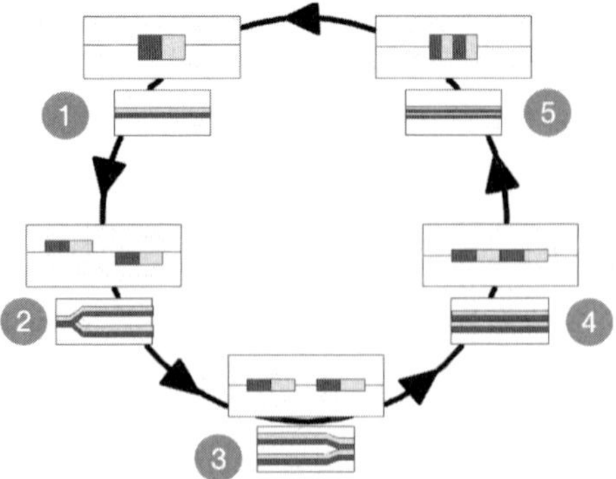

Figure 1.125 Schematic of the splitting and recombination process [125].

Mixer type	Stack SAR micro mixer	Number of parallel rows	2
Mixer material	PMMA, polycarbonate	Initial and halved channel width	120 μm, 60 μm
Number of mixing elements	5	Mixing element length	600 μm

No characterization is provided in [125].

1.3.14.5 Mixer 60 [M60]: Up-down Curved Split-and-recombine Micro Mixer

This eight-step SAR mixer was proposed with a split of one channel into two and a subsequent recombination per element, termed caterpillar mixer [7]. Since the channel is split along the horizontal axis, vertical changes of the split channels for recombination have to be undertaken. In addition, a change of the cross-section of the merged geometry has to be made to restore the original channel geometry. The vertical displacement of the split channels is achieved by smooth up and down steps, yielding a continuously curved flow structure (not step-like).

The fluidic network is formed by the assembly of two identically microstructured plates [7]. The horizontal splitting layer is inserted between the two plates as well as seals. A steel housing serves for compression and provides the fluidic connections.

The mixer plates were fabricated out of a transparent material to allow for optical inspection of the mixing process [7]. The plates were structured by milling. The unruffled surface of the channel was evident by optical and SEM inspection, revealing a continuous change of brightness. The horizontal splitting layer was realized from a thin sheet by means of pulsed sublimation cutting using an Nd:YAG laser. The graphite gaskets were made by pulsed laser cutting

As a predecessor device, a caterpillar mixer with step-like internal structure was fabricated; without horizontal splitting layer and no separation of the streams [66].

Mixer type	Up-down curved SAR micro mixer	Splitting layer material	Stainless steel
Mixer material	PMMA	Splitting layer thickness	100 μm
Number of SAR steps	8	Gasket material	Graphite
Channel width, depth, total length (all 8 steps)	2 mm, 4 mm, 96 mm		

The same mixing geometry scaled by a factor of 0.5 was realized in stainless steel [7].

1.3.14.6 Mixer 61 [M 61]: Multiple-collisions Split-and-recombine Micro Mixer

This SAR mixer, based on multiple collisions of one stream, was described by industry in a patent [143]. The same company has established the first production plant with microstructured devices which was described in the literature, the devices being microstructured mixers [145].

One stream is split into two curved sub-streams which simply merge and then instantly split again [143]. The mixing is performed by several crosses in a row which are connected, with two channels for inlet and two for outlet. This process is repeated many times. Typical dimensions are not given and the fabrication is also not specified. Since no testing is reported, which is not surprising for an industrial proprietary method, no further information on this mixer is given below.

1.3.14.7 Mixer 62 [M 62]: Separation-plate Split-and-recombine Micro Mixer

This micro mixer extends the SAR concepts mentioned above, the sequential lamination being actively supported by the use of a separation plate for flow splitting (see Figure 1.126) [140]. As a result of the splitting, two sub-channels are formed, initially at the same height level within the device. Then, one of these sub-channels undergoes a downward movement in a lower level of the device. The two flow channels are so sandwiched, the first carrying the two fluids separate from each other. At certain locations there are conduits within the sandwich to achieve flow recombination on one of the levels of the sandwich. This procedure is repeated many times.

Channel structures are etched on two plates which are later positioned face-to-face to give the overall fluid structure [140]. In the region where the channels overlap, they are separated by the separation plate defined by an etch stop layer. The channel covered by this structured plate was generated by underetching in the <100> direction through slits in the plate. The micro mixer is assembled from a silicon and a glass wafer connected by anodic bonding.

Mixer type	Separation-plate SAR micro mixer	Width of channels for inlet branches	150 μm
Mixer material	Silicon/glass	Separation-plate thickness	5 μm
Maximum channel width	300 μm	Slit width	15 μm
Channel depth	30 μm		

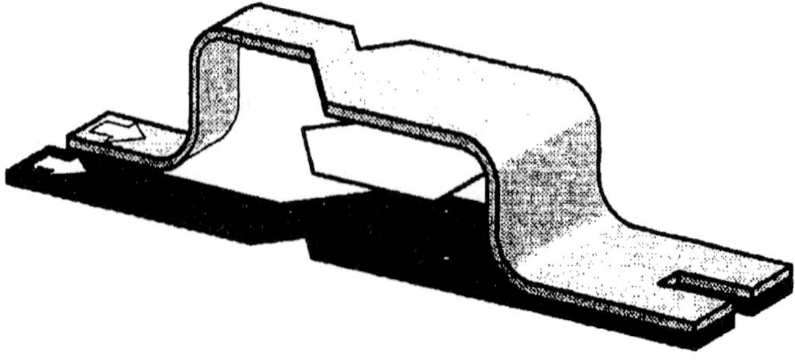

Figure 1.126 Schematic of a SAR mixing concept using a separation plate [140] (by courtesy of IEEE).

1.3.14.8 **Mixing Characterization Protocols/Simulation**

[P 51] The reduction of a solution with potassium permanganate in alkaline ethanol was used as a test reaction [141]. The course of this reaction can be simply followed by visual observation of the color changes. The reactant solution is purple, the intermediate manganate is green and the final product, manganese dioxide, is yellow to brown. One drawback relates to the precipitation of the product, which needs a cleaning step with sulfuric acid for dissolution.

[P 52] A double-syringe pump was used for liquid feed [142]. An iodine–starch solution was mixed with a photographic fixer solution in a ratio of 1 : 3.5. Thereby, the intense blue color changed to pale blue. The mixing process was followed by means of a stereo microscope.

[P 53] Mixing to the following four fluid systems was investigated [43]: (a) aqueous solutions of hydrochloric acid (1 M) and methyl orange; (b) water and air; (c) oil and air; (d) water and oil.

[P 54] The mixing experiments were performed at relatively low *Re*. For this purpose, 85% glycerol–water solutions were used, with dynamic viscosity and density of about 100 mPas and 1.2 kg l^{-1}, respectively [7]. Total flow rates in the range 0.2 l h^{-1} ($Re = 0.22$) and 2 l h^{-1} ($Re = 2.2$) were applied. For the given dimensions, CFD simulation showed that for *Re* above about 15, corresponding to a total flow rate of 13.5 l h^{-1}, secondary flow induced by inertial forces had a notable effect. Thus the applied flow rates were well below the critical value and pronounced secondary flow effects were not observed.

The combination of two complementary flow-pattern imaging approaches allows for reliable experimental quantization of the mixing quality [7]. First, mixing of a transparent and a dyed sub-stream is optically inspected by means of transmitted light microscopy. Second, the use of two transparent sub-streams of yellowish iron ion (Fe^{3+}) and transparent rhodanide (SCN^-) solutions which form a deep colored complex after mixing allows one to identify those regions where the first method mistakenly suggests good mixing due to layered lamellae configurations.

For initial simulations, a structured grid comprising 220 000 hexahedral cells was used [7]. Using the commercial solver CFX4 the Navier–Stokes equation was solved and simultaneously a convection equation for a user scalar describing a non-diffusive tracer. To minimize the discretization error, the tracer was solely used for lamellae visualization and hence the diffusivity was set to zero. A higher order differencing scheme (QUICK) was used for discretization and the SIMPLEC algorithm was applied for pressure–velocity coupling. Liquid properties of water were chosen.

[P 55] For details of the simulation approach see [140].

Flow visualization experiments were made using a pH-indicator reaction between phenol red and an acidic solution resulting in a color change. Images were recorded by a light microscope [140].

1.3.14.9 Typical Results

Theoretical analysis of mixing via symmetric and asymmetric multi-lamination

[M 62] [P 55] The mixer design was based on a detailed theoretical analysis of flow and diffusion phenomena in the laminar regime for idealized channel geometry [140]. The time evolution of so-called asymmetric and symmetric concentration profiles was first compared for stationary conditions, i.e. without any flow.

In the symmetric case, three inner layers B–A–B of equal thickness are surrounded by two layers A of half thickness, yielding an overall layer sequence A–B–A–B–A [140]. In the asymmetric case, all layers A and B have equal thickness, but the outer layers belong to two different fluids. Thus, the sequence refers to A–B–A–B. The time required to achieve a homogeneous concentration profile, i.e. the mixing time, is much longer for the asymmetric lamination.

In addition, dynamic studies were performed including moving fluids. Generally, similar results were obtained [140]. Different from the stationary case, the width of the inner lamellae is decreased relative to the outer layers owing to the parabolic flow profile.

Completion of mixing

[M 57] [P 52] For a Möbius-type four-stage SAR mixer complete mixing was achieved at flow rates up to 285 $\mu l\ min^{-1}$ [142].

Relationship of residence time to reaction species for a consecutive reaction

[M 56] [P 51] In a laminar, diffusive mixing system, residence time relates to the course of mixing. Therefore, flow rates were varied when using a Möbius-type SAR mixer to vary the residence time and hence the product spectra of a consecutive reaction [141]. Reactant, intermediate and final product were all colored so that the mixing and reaction course could be directly followed by visual observation. The reduction of permanganate finally to manganese dioxide was investigated.

At fast flow rates, no color change was observed in the SAR mixer [141]. At lower flow rates, the formation of the green manganate was evident. At still lower flow rates, the brown manganese dioxide was observed. The exact flow rates were not specified.

The example shows that by control of the residence time the yield of a consecutive product can be maximized and that intermediates may be isolated at high concentration, if parameter settings are optimal [141].

Separation-plate SAR: visual observation

[M 62] [P 55] The mixing process was observed by flow visualization using light microscopy. A pH reaction between phenol red and an acid yielded a color change [140]. Thereby, it was possible to resolve zones with incomplete mixing in the neighborhood of the contacting zone, in particular those referring to the uneven lamination caused by the slits. A homogeneous mixture was formed in the outlet channel. Complete mixing in the whole cross-section of the outlet channel was obtained in 100–300 ms. The flow range applied was 0.01–0.1 $\mu l\ s^{-1}$, which corresponds to a highly viscous flow with a Reynolds number < 1. The mixing times measured were not dependent on the flow rates.

Fork SAR: miscible liquids – visual observation

[M 58] [P 53] The mixing process was investigated by a pH-driven color reaction [43]. A complete change of the color was detected at the mixer outlet, i.e. mixing was assumed to be complete. This was found for SAR mixers with five and 20 mixing elements, hence a low number of mixing steps seems to be sufficient.

Fork SAR: gas–water contacting – visual observation

[M 58] [P 53] Air and water were contacted in a five-step fork-like micro mixer [43]. Bubbles of 200–500 μm size resulted. The mixing in 10 mixing elements resulted in even smaller bubbles below 100 μm. Coalescence led to the formation of larger bubbles within a few minutes.

Fork SAR: gas–oil contacting – visual observation

[M 58] [P 53] Oil and water were contacted in a five-step fork-like micro mixer [43]. Bubbles down to 100 μm in size resulted, giving the whole fluid system a nearly white optical appearance. The foams were more stable than the water-based foams. About 1 h was needed until all bubbles were removed from the oil phase. When using a 20-element mixer, the oil-based foams were stable for 2–3 days.

Fork SAR: oil–water contacting – visual observation

[M 58] [P 53] Oil and water were contacted in a five-step fork-like micro mixer [43]. Small water droplets were achieved in this way, which soon gave a segmented oil and water flow. When using a 15-step fork-like micro mixer, smaller droplets resulted giving an opaque system of emulsion nature. The separation into two separate phases took 3–4 days.

Increase in interfacial stretching for split–recombine mixing – lyapunov exponent

[M 60] [P 54] In the framework of chaotic convection, the mixing performance is commonly characterized by interfacial stretching:

$$\lambda(t) = \lim_{L_0 \to 0} \frac{L(t)}{L_0} \tag{1.5}$$

where L_0 and $L(t)$ denote the characteristic dimensions of the interfacial area at $t = 0$ and at finite time t, respectively [7]. In the case of a 2-D incompressible flow, the diffusive mass transport, which determines the mixing performance, depends quadratically on the interfacial stretching λ. Chaotic flows ensure particularly efficient mixing since they imply an exponential increase of stretching over time.

Accordingly, the finite-time Lyapunov exponent σ may be defined via

$$\lambda(t) \approx e^{\sigma t} \tag{1.6}$$

Although the repeated application of the SAR principle leads to a highly regular lamellae pattern, it also exhibits an exponential increase in interfacial area [7]. The corresponding stretching factor is given by

$$\lambda(t) = (2^n - 1) \approx 2^{u\,t/l} \tag{1.7}$$

where n, u and l denote the number of SAR steps, the mean velocity and the length of one SAR unit, respectively [7]. According to above-mentioned equation, the SAR mixer has a finite-time Lyapunov exponent of $\sigma = \ln 2(u/l)$. Thus, even in the case of a highly regular flow pattern a positive finite time Lyapunov exponent can be achieved – a characteristic feature of chaotic advection. Whereas chaotic advection generally induces regions of regular flows and poor mixing besides regions of high chaoticity where mixing predominantly occurs here, in the case of an ideal SAR multi-lamination, spatial homogeneous mixing is obtained.

Since the final lamella dimension does not depend solely on the channel width, but also on the number of SAR steps, thorough mixing can be achieved under moderate pressure drops [7]. A linear total pressure drop $\Delta p = n\, p_0$, where p_0 denotes the pressure drop of one SAR step, leads to an exponential decrease in the lamella dimension $L_n = L_0/2^n$.

Simulation of cross-sectional flow patterns without splitting plane and for non-separated SAR flows

[M 60] [P 54] SAR flow splitting can be performed using split channels or done in-line in one channel. Concerning the quality of flow splitting in the latter case, CFD simulations were performed on the example of a so-called caterpillar micro mixer [7].

For $Re \approx 1$, the SAR principle is imperfectly realized when aiming at separating the flow in-line in one channel (without splitting of the channel structure itself) [7]. Owing to internal friction, the horizontal splitting taking place in the first half of the geometry does not preserve the initial lamination (see Figure 1.127). This leads to an S-shaped lamellae configuration at the center of the geometry. A similar problem occurs in the recombination step (lower row in Figure 1.127). Again, owing to the internal friction the lamellae configuration is changed, ending up with a more pronounced S-shaped structure rather than a multi-lamellae configuration.

In the case of high Re, the situation is different. Inertial forces induce a secondary flow which causes strong tilting and entanglement of the lamellae and, hence, a considerable increase in interfacial area is achieved [7].

Figure 1.127 Cross-sections taken from a CFD simulation for $Re = 1$, displaying the lamellae reshaping in a SAR caterpillar mixer. The initial configuration is shown in the small cross-section at upper left. The other cross-sections in the upper row show the splitting at $x = 0.6$ and 1.2 mm. The cross-sections in the lower row show recombination at $x = 1.8$ and 2.4 mm at the end of the first cell [7] (by courtesy of RSC).

Simulation of cross-sectional flow patterns with splitting plane and for separated SAR flows

[M 60] [P 54] The impact of having truly separated flows was shown by CFD simulations [7]. An essential part of the flow splitting, besides having split channels which recombine later, is a splitting plane which cuts the flow like a knife. These design aspects are central parts of an optimized SAR caterpillar mixer.

Cross-sectional views for such a design depicting the lamellae configurations for $Re = 3.45$ were determined (see Figure 1.128) [7]. According to the CFD simulation results, an almost prefect lamellae configuration is reached after passing one SAR step. The same holds true if a series of three SAR steps is simulated. Thus, by consequent implementation of the above design rules, a vertically aligned multi-lamination pattern is achieved.

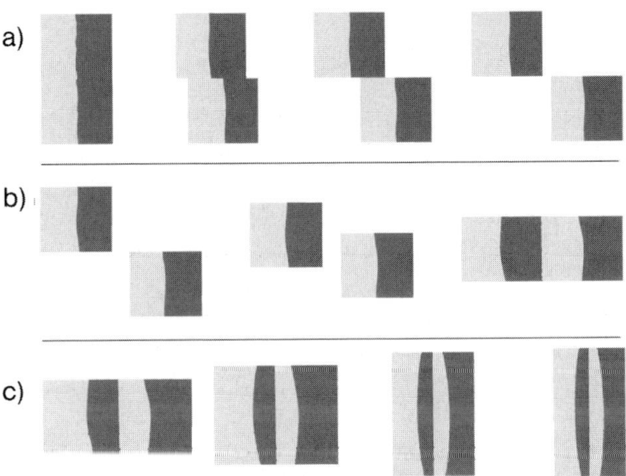

Figure 1.128 Cross-sectional views of lamellae configurations within a SAR step for $Re = 3.45$. (a) Splitting; (b) rearrangement of sub-streams and recombination; (c) reshaping [7] (by courtesy of RSC).

For increased flow rates, however, the CFD simulations show more and more deviations from an ideal SAR multi-lamination pattern [7]. Since inertial forces come into play, a secondary flow pattern is superposed on the SAR velocity profile of the creeping flow regime. Streamlines seeded at the initial lamellae interface in top view for various Reynolds numbers. The lamellae pattern right at the outlet for the same set of Reynolds numbers were also given. Further simulations showed that for Re above ~15 the center lamellae are thinned out until they detach from the top and bottom walls for $Re \approx 30$ (see center image of Figure 1.128).

Description of inertia forces which distort SAR flows – the Dean number

[M 60] [P 54] An analysis of the role of secondary flows which distort the SAR flows owing to inertia forces was carried out by simulations [7]. This analysis was based on using the Dean number as a measure of how to achieve ideal SAR flows, following

some assumptions. The lower the Dean number, the more ideal should be the SAR flow.

It comes out that large channel widths and heights imply low Dean numbers [7]. From this argument, the SAR mixer should be realized in macroscopic dimensions. On the other hand, in order to achieve fast mixing by diffusion, the dimension of final lamellae should be of the order of microns and a compact mixer format implies a moderate number of SAR steps, of the order of 10. Hence, an inlet width of a few millimeters is appropriate; the precise value, however, depends on the requirements of the particular process of the mixing application. Concerning the channel height, it is suggested that one should also have a large dimension in order to achieve high volume flows under low Dean numbers. However, depending on the fluid properties of the lamellae, instabilities may develop under certain conditions. The susceptibility of the lamellae to instabilities decreases with smaller channel dimensions. Again, a channel height of a few millimeters is a reasonable compromise.

Flow pattern analysis – dilution-type imaging

[M 60] [P 54] Dilution-type flow pattern analysis was performed for a caterpillar SAR mixer (see Figure 1.129) [7]. At a total flow rate of $0.2\,\mathrm{l\,h^{-1}}$, highly regular lamellae patterns are observed up to the eighth SAR step. According to the low Reynolds number of 0.22, secondary flow effects are suppressed and homogeneous mixing is to be expected.

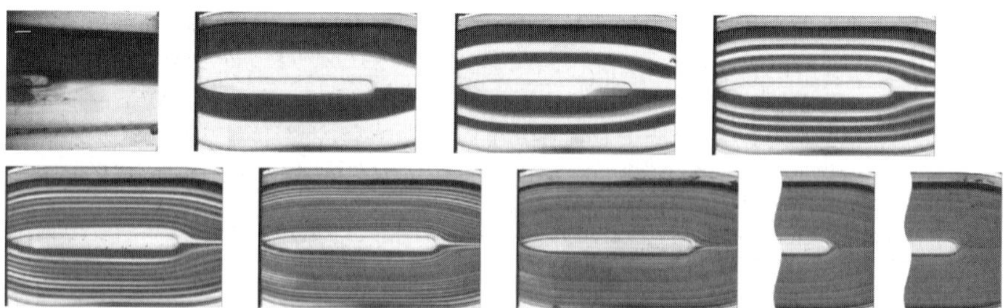

Figure 1.129 Optical inspection of multi-lamination in the SAR mixer. The dyed (blue-colored water) and transparent lamellae of an 85% glycerol–water solution are shown in dark and light gray, respectively. The applied total volume flow rate of $0.2\,\mathrm{l\,h^{-1}}$ corresponds to $Re = 0.22$ [7] (by courtesy of RSC).

Flow pattern analysis – reaction-type imaging

[M 60] [P 54] In addition to the dilution-type imaging, the flow patterns in the SAR caterpillar mixer were also analyzed by using the iron–rhodanide reaction, a reactive approach (see Figure 1.130) [7]. At a total flow rate of $0.2\,\mathrm{l\,h^{-1}}$, the two almost transparent solutions give in the first SAR step the formation of the brown iron–rhodanide complex at the fluid interface. The increasing numbers of lamellae become evident from the increasing number of dark-colored interfaces. By mixing via diffusion, a uniform distribution of the brown iron–rhodanide complex is derived in the last mixing step. Mixing seems to be completed after eight mixing steps.

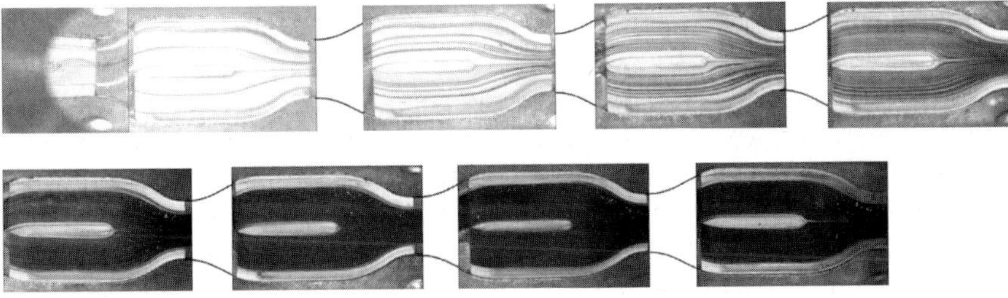

Figure 1.130 Optical inspection of mixing in the SAR mixer. The applied total volume flow rate of 0.2 l h^{-1} is the same as for the experiment shown in Figure 1.129. Starting from a bi-lamination of yellowish iron ion (Fe^{3+}) and transparent rhodanide (SCN^{-}) solutions, homogeneous mixing is achieved in the eighth mixing step, indicated by the deep brown color of the iron–rhodanide complex formed [7] (by courtesy of RSC).

Flow pattern analysis – comparison between simulation and experiment
[M 60] [P 54] Good agreement between simulated and experimental flow patterns was found [7]. A simulated streamline pattern for *Re* = 0.22 was superimposed on a photograph image of the flow pattern for the first SAR step. Besides the global SAR performance, both findings indicate a certain curvature of the lamella interface, owing to the increasing width of the dark colored interface after half of the flow element passage.

Analysis of performance loss due to non-ideal geometries caused by fabrication needs
[M 62] [P 55] Fluid dynamic simulations revealed the influence of cross-flow through the slits in the separation plate which were introduced only for reasons of microfabrication, i.e. underetching the plate to connect in- and outlet micro channels [140]. It could be shown that the present design suffered from the problem that about 30% of the total flow was fed through the slits, leading to uneven lamination. Further analysis demonstrated that this effect is reduced either using thinner slits, i.e. increasing the pressure loss, or combining several mixing elements in one cascade.

1.3.15
Rotation-and-break-up Mixing

Most Relevant Citations
Peer-reviewed journals: [146].

A number of micro mixers use secondary or rotational flows, which are, e.g., created by in-channel flow structures, to stretch and fold fluids. The mixing approach here superposes the rotation by a break-up step, which basically is a splitting step [146]. This was done based on the analysis of elementary mixing steps and their corresponding transfer to low Reynolds number mixing.

1.3.15.1 **Mixer 63 [M 63]: Rotation-and-break-up Micro Mixer (I)**

This device has serial in-channel mixing elements composed of convex and concave units [146]. The mixer was made in two versions with short units (200 μm) and long segments (400 μm).

Microstructuring was effected using a replica molding method [146]. A negative photoresist (SU-8) was spin-coated as a thin film on a silicon wafer and a pattern was generated using a photo chrome mask. A prepolymer was spin-coated on to this structure, cured and separated from the master. In this way, a positive structure was yielded in PDMS. Two such positive structures with different designs were bonded by a self-align method using methanol after oxygen plasma treatment.

Mixer type	Rotation-and-break-up micro mixer	Channel width, depth, length	100 μm, 50 μm, 4 mm
Mixer material	PDMS	Length of in-channel units	200 μm or 400 μm

1.3.15.2 **Mixer 64 [M 64]: Rotation-and-break-up Micro Mixer (II)**

This device has similar structures to [M 63], but provides a more complex texture of their units (see Figure 1.131) [146]. Five mixing units are placed laterally within the channel instead of only one in the case of [M 63]. The next row of five units is

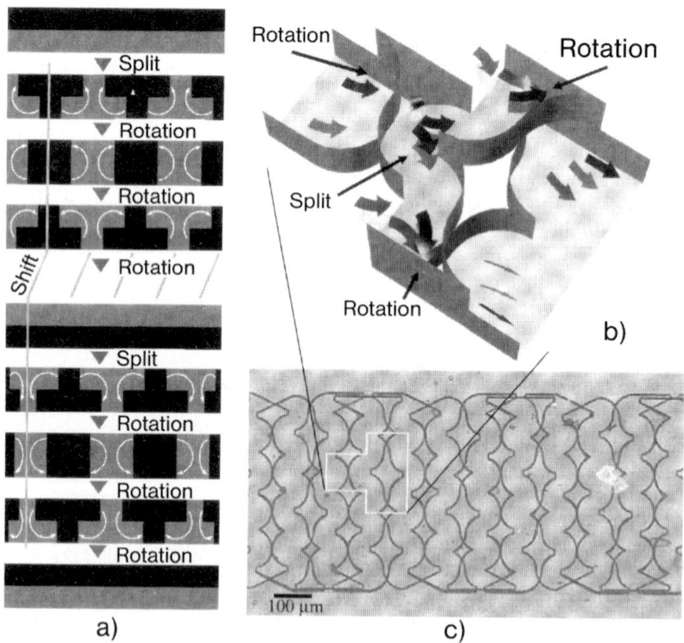

Figure 1.131 Mixer for flow rotation and break-up. (a) Cross-sectional images of the micro channel during one mixing cycle; (b) schematic of the flow treatment involving splitting (break-up), rotation and merging steps; (c) optical micrograph of the mixer [146] (by courtesy of IOP Publishing Ltd.).

displaced by half the lateral extension of the mixing units. In this way, a complex flow pattern is yielded which performs a splitting (break-up) action and leads to rotation of such split flows.

Fabrication of the device was performed as described for [M 63].

1.3.15.3 **Mixing Characterization Protocols/Simulation**

[P 56] Mixing was analyzed by fluorescence flow visualization using a confocal scanning microscope [146]. Water and aqueous solutions (0.02 w.-%) were mixed using syringe pumps at flow rates of 3, 30 and 150 μl min^{-1}.

The standard deviation of the fluorescence mixing intensity was taken as a measure for judging mixing efficiency [146]. The intensity of the pixels was monitored and corrected by using flat and dark field images. All data processes were conducted using a commercial image data processor. The Sobel edge algorithm was applied.

Simulation was done using the CFD-ACE+ program considering the geometry of the element with the two convex and concave units [146]. By applying periodic boundary conditions, designs with repeated mixing steps could be modeled. Particle traces were generated by integrating numerically through the discrete velocity field using an adaptive Runge–Kutta scheme.

1.3.15.4 **Typical Results**

CFD imaging of the flow rotation

[M 63] [P 56] Rotation of the flow is given both at high and low Reynolds numbers, as shown by CFD simulations (see Figure 1.132) [146]. The distortion of the interfaces is more pronounced at higher Reynolds number. When analyzing the wrapping of the interface, it becomes evident that striations were observed only for high Reynolds number flow. For low Reynolds number flow, only the position of the flow changed and the interface was not distorted, since no convective acceleration was given.

Experimental imaging of the flow rotation

[M 63] [P 56] The experimental findings are in accordance to the predictions made by the CFD simulations (see *CFD Imaging of the Flow Rotation*) [146]. Mixing is much better at high (50) than at low (1) Reynolds number for the reasons given above.

A mixer with short units gives better mixing performance than a mixer with long unit [146].

Experimental imaging of the flow rotation with break-up

[M 64] [P 56] The addition of a break-up step in addition to the flow rotation gives positive moments for mixing (see Figure 1.133) [146]. This leads to an increase in the interfacial area by stretching and folding and producing striations. At low *Re*, fluids were broken up and small fragments of blobs were generated. Besides the increase in interfacial area, steep concentration gradients were generated, which speed up diffusion.

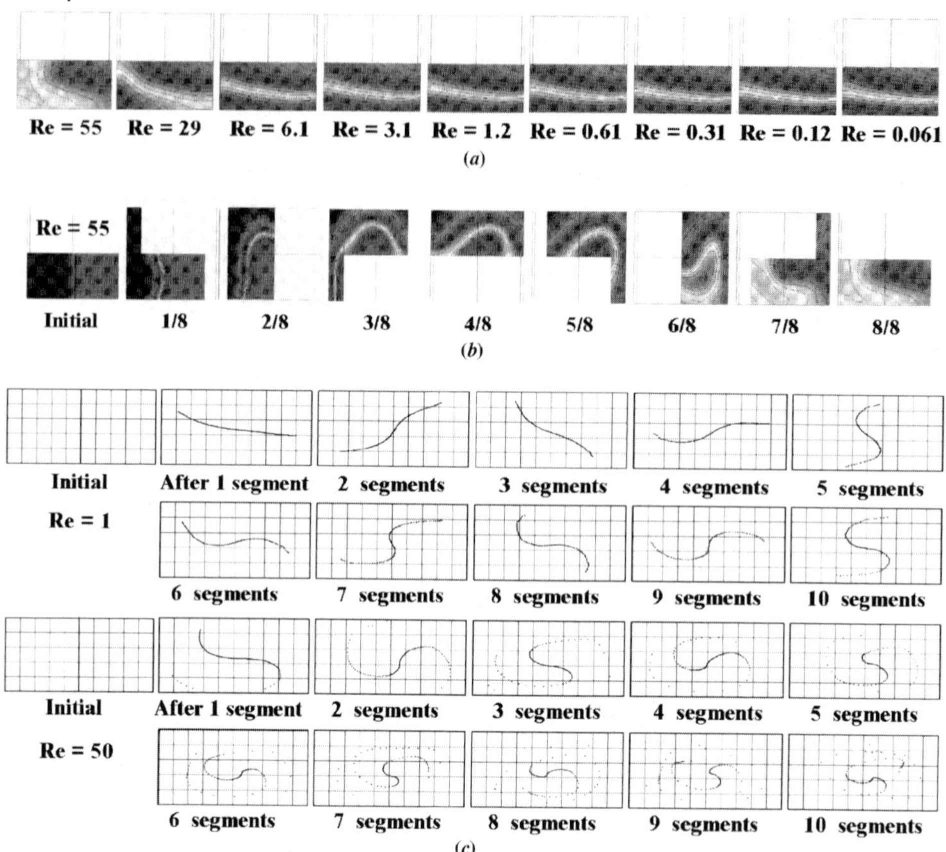

Figure 1.132 (a) Concentration differences and imaging the distortion of the interface between two fluids for different *Re*. At low *Re*, the interface is not much deformed. (b) Wrapping of the interface between two fluids. Striations are only given at high *Re*. At low *Re*, only the position of the fluids was changed. (c) Consecutive development of the interface when passing the fluid through the mixer segments [146] (by courtesy of IOP Publishing Ltd.).

Reduction of mixing lengths by fluid break-up

[M 64] [P 56] More than 70% mixing is achieved for a mixing length of only 4 mm for $Re = 1$, 10 and 50 when using rotation and break-up [146]. In contrast, for pure diffusion complete mixing is given after 1, 10 and 50 m for the same *Re*. Accordingly, the rotation and break-up have a considerable impact.

Optimal Reynolds number

[M 64] [P 56] There is an optimal *Re* for a mixer which is based on flow rotation and break-up [146]. For the design given here, stretching/folding, break-up and diffusion occur at $Re = 10$, yielding the most efficient mixing for the cases studied. At $Re = 1$, stretching and folding hardly occur, and only break-up and diffusion are found. At $Re = 50$, stretching and folding and also break-up are relevant; owing to the high flow velocity, diffusion is not completed.

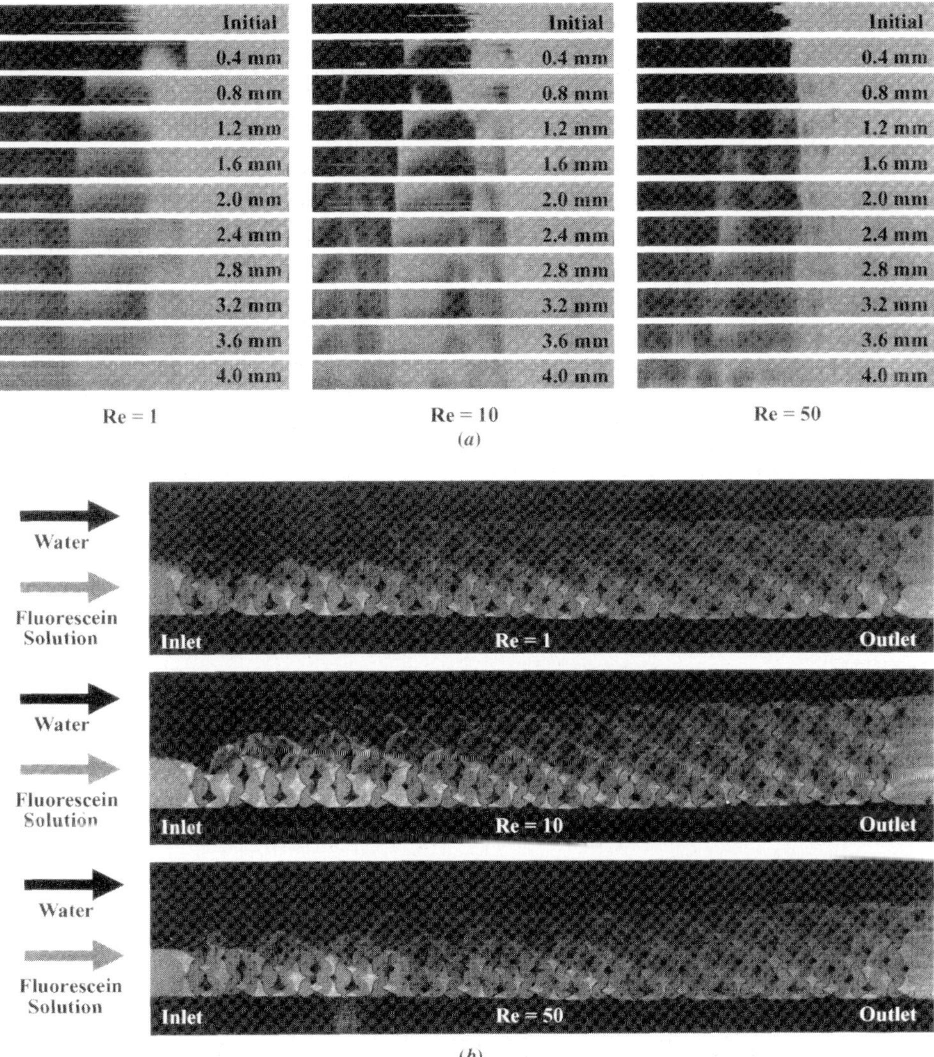

Figure 1.133 Fluorescence micrographs obtained in a mixer using flow rotation and break-up. (a) Cross-sectional images showing the break-up which produces the smaller fragments of blobs. Thereby, the interfacial area is increased and large concentration gradients are provided. (b) Top of the mixer, clearly showing the propagation of one fluid into the other [146] (by courtesy of IOP Publishing Ltd.).

1.3.16
Micro-plume Injection Mixing

Most Relevant Citations

Proceedings contributions: [51, 147] (see also [148]).

By feed of a fluid through a nozzle array, which is a plate with many tiny holes, so-called micro-plume injection into a micro channel can be achieved [51, 147]. Typically, the micro channel's floor is perforated in a section in this way and a closed-channel fraction follows for completion of mixing. Large specific interfaces can in principle be achieved depending on the nozzle diameter. This mixing concept benefits from conceptual simplicity and fits well to existing MEMS techniques. Furthermore, it consumes less footprint area and therefore does not create much dead space, which is one of the prime requirements during μTAS developments.

It stands to reason that plumes are only formed under certain hydrodynamic conditions, e.g. ratios of flow rates of the liquids. Otherwise, simple bi-lamination with comparatively low specific interface may occur. In addition, a flow maldistribution within the array may occur for certain conditions, i.e. most flow passes the first row of nozzles at the expense of the residual holes. So far, there is, to the best of our knowledge, no detailed report on modeling these aspects or an experimental proof where indeed plume fluid structures are visualized; only gross characterization of the mixing was given (see below).

A micro-plume injection mixer as described above was also part of a microfluidic system composed of a mixer with an array of nozzles as through-holes, stirring part and reactor with integrated heater [148]. The main reason for choosing this mixing concept was to achieve integration into a simple straight channel overall structure with minimal dead space. The holes (10 μm in diameter with 30–100 μm depth) were fabricated in macroporous silicon by electrochemical etching. The system was intended for use in biochemical analysis; no details on specific applications or testing were given in the reference.

1.3.16.1 Mixer 65 [M 65]: Micro-plume Injection Micro Mixer

The central part of this device is a square mixing chamber which contains a sieve bottom through which a multitude of fluid bulbs, 'micro plumes', are injected into a main stream (see Figure 1.133) [51, 147]. The idea is to enlarge the fluid interface and to speed up diffuse mixing. The fluid of the main stream is fed through one channel which enters at half chamber width and leaves on the opposite chamber side again through a channel at half chamber width. The second fluid is fed through a channel oriented at 90° relative to the other feed channel and also leads to the chamber at half width. This channel is on a lower level than the other feed and the outlet channels and leads to a reservoir beneath the mixing chamber.

Bulk micromachining was applied for structuring a silicon wafer [51]. The sieve and the mixing chamber as through-structures were obtained by etching from both sides. The wafer was covered from both sides with Pyrex glass by anodic bonding. A light-reflecting silicon layer of the same size was deposited on the Pyrex plate by

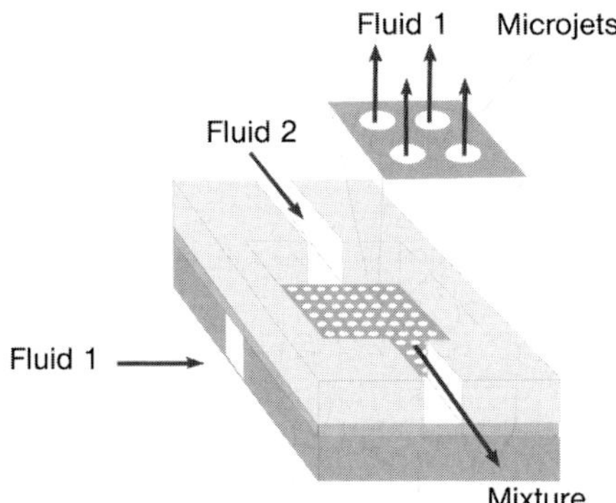

Figure 1.134 Schematic of the micro-plume injection mixer design (newly drawn following [51]).

thin-film technology and structured by reactive ion etching. Inlet and outlet ports were drilled into the top glass wafer.

Mixer type	Micro-plume injection mixer	Hole depth of the sieve	10–30 μm
Mixer material	Silicon	Distance between the holes of the sieve	100 μm
Top plate material	Pyrex	Number of holes in the sieve	400
Hole diameter of the sieve	10–30 μm	Mixing chamber dimensions	2.2 × 2 × 0.33 mm

1.3.16.2 Mixing Characterization Protocols/Simulation

[P 57] A fluorescent dye was injected through the sieve of the micro-plume injection mixer [51]. The changes in fluorescence intensity thereafter in the mixing chamber were monitored by using a fluorescence microscope.

In addition, visible absorption measurements using an optical-fiber setup were performed directly in the mixing chamber [51]. The optical path and hence the signal intensity could be substantially increased by passing the light through the long axis of the mixing chamber using multiple reflections between the mixer bottom plate and the thin-film coated layer. By 90° reflection at the beginning and end of the optical reflection zone, the signal is introduced and passed via fibers, respectively.

1.3.16.3 Typical Results

Numerical analysis of mixing time

[M 65] [P 57] By numerical analysis of the diffusion process, it was shown that the multiple injection of the micro-plumes leads to notable speed-up of mixing compared with a single-point injection, as expected [51]. The latter corresponds in a wider sense to a mixing tee flow configuration. Mixing is therefore improved for all flow rates; the best results are given at higher flow rates. These simulation results were taken as input for the design of the mixer.

Top-view fluorescence mixing detection

[M 65] [P 57] The fluorescence measurements proved complete mixing within a few seconds [51].

In-line visible mixing detection

[M 65] [P 57] Visible absorption measurements proved complete mixing at a flow rate of about 1 μl s^{-1} [51].

1.3.17 Slug Injection Mixing

Most Relevant Citations

Proceedings contributions: [149].

1.3.17.1 Mixer 66 [M 66]: Segmented-flow Micro Mixer

In this device, two liquids are injected using minute conventional valves with two injectors into a mixing channel arranged as a serpentine structure [149]. The switching rate of the valves is 1 Hz.

Fabrication is effected by precision milling of PMMA laminates which are sealed by diffusion bonding [149].

Mixer type	Segmented flow micro mixer	Mixer internal volume	10 μl
Material	PMMA	Injector width, depth	100 μm, 30 μm
Channel width, depth, length	1000 μm, 200 μm, 100 mm	Total size of the chip	5 cm × 4 cm × 0.5 cm

1.3.17.2 Mixing Characterization Protocols/Simulation

[P 58] Dilution-type mixing was accomplished with blue and yellow solutions which were image-processed and then yielded black and white colors, the mixture being gray [149].

Pressure-driven operation was used for fluid feed [149].

1.3.17.3 Typical Results

Mode of flow injection – impact on mixing

[M 66] [P 58] When two flow injectors were constantly open, the two flows injected into a micro channel formed a bi-laminated pattern [149]. For alternative injection, equal alternate slugs, i.e. a segmented flow, were found. Upon flow transport, mixing by diffusion occurs and the color of the plugs is 'smeared'. The mixing was claimed to be rapid; however, no time scale is given in the reference. A concentration profile over an arbitrary channel length gives an idea of how the course of mixing develops.

1.3.18 Secondary Flow Mixing in Zig-zag Micro Channels

Most Relevant Citations

Peer-reviewed journals: [59]; proceedings contributions: [150, 151].

The continuous change of flow direction in zig-zag channels can induce secondary flow patterns at sufficiently high *Re*, which besides diffusion can act as a mixing mechanism. By means of recirculation patterns, material is transported transverse to the flow direction and improves the mixing.

1.3.18.1 Mixer 67 [M 67]: Y-type Micro Mixer with Zig-zag or Straight Channel

All mixers described in this sub-section have a micromachined Y-type contactor (90° angle between the two feed channels) attached to a micro channel where mixing occurs (see Figure 1.135) [59].

Mixer type	Y-type micro mixer with a zig-zag or straight channel	Zig-zag channels (periodic step 1, 2, 4, 8): effective length	2828 μm
Mixer material	PET	Reference case A (straight channel): width, length	141 μm, 2000 μm
Periodic step	0, 100, 200, 300, 400, ∞ μm	Reference case B (straight channel): width, length	100 μm, 2828 μm
Ratio of periodic step to channel width (meandering ratio)	0, 1, 2, 4, 8, ∞	Reference case C (straight channel): width, length	100 μm, 2000 μm
Zig-zag channels (periodic step 1, 2, 4, 8): width, length	100 μm, 2000 μm		

The first group of devices has a long zig-zag micro channel attached, with 90° angles for each fluid turn [59]. Four devices differing in the periodic step were made, with equal width and linear length of the micro channel (see Figure 1.136). To classify these mixers, the ratio of the periodic step to the channel width was introduced. In the following, this will be termed the meandering ratio. As reference

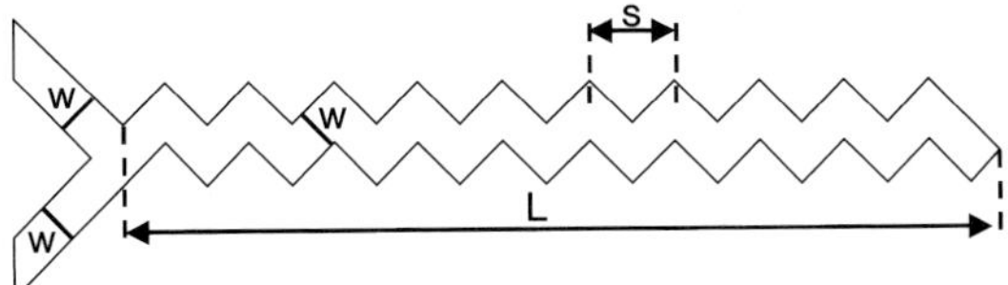

Figure 1.135 Characteristic dimensions of a microfluidic mixing element with Y-type contactor attached to a zig-zag channel: *w*, width of the micro channel; *s*, linear length of the periodic step; *L*, linear length of the zig-zag micro channel [59] (by courtesy of ACS).

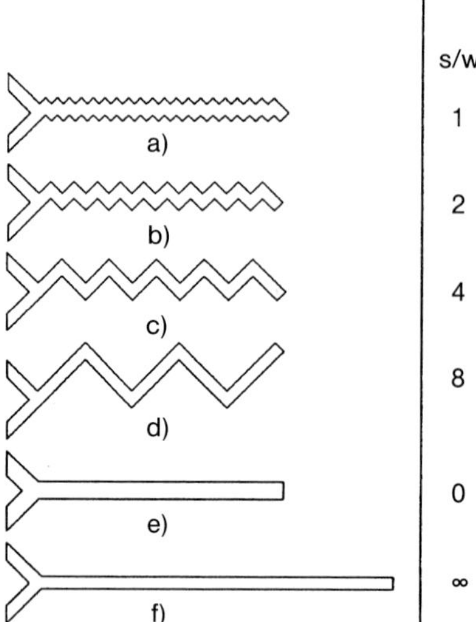

Figure 1.136 Microfluidic mixing elements with Y-type contactor attached to a zig-zag channel having different geometries in the meandering section. (a) $s/w = 1$; (b) $s/w = 2$; (c) $s/w = 4$; (d) $s/w = 8$; (e) $s/w \to 0$; and (f) $s/w \to \infty$ [59] (by courtesy of ACS).

cases, three straight channel devices were made, representing the asymptotic cases with the periodic step approaching zero or infinite (see Figure 1.136). Accordingly, the linear length of these two devices changes.

The micro devices were fabricated from a polyethylene terephthalate (PET) substrate using 193 nm ArF excimer laser ablation [59]. Microstructures produced in this way were thermally sealed by a lamination machine.

1.3.18.2 **Mixer 68 [M 68]: T-type Micro Mixer with Zig-zag or Straight Channel**

Two sets of bend and straight channels were designed, with different hydraulic diameter, thus called 'micro' and 'mini' [151]. All contained a T-type contactor at the beginning.

(a) The 'mini' channels were made by traditional precision engineering machining [151].

Mixer type	T-type micro mixer with a zig-zag or straight channel; 'mini'	Length between two bends (for zig-zag mixer only)	1 μm
Mixer material	Acrylic plastic	Bend length to hydraulic diameter	2.5 (zig-zag), 250 (straight)
Micro channel width, depth, length	300 μm, 600 μm, 100 mm	Number of bends	80

(b) The 'micro' channels were fabricated by silicon micro machining [151]. A glass cover was bonded to allow visual observation.

Mixer type	T-type micro mixer with a zig-zag or straight channel; 'micro'	Length between two bends (for zig-zag mixer only)	2.5 μm
Mixer material	Silicon/glass	Bend length to hydraulic diameter	57 (zig-zag), 114 (straight)
Micro channel width, depth, length	180 μm, 25 μm, 25 mm (zig-zag) or 5 mm (straight)	Number of bends	10

1.3.18.3 Mixing Characterization Protocols/Simulation

[P 59] Mixing was simulated based on a model considering the laminar mixing of species along a 2-D section of a micro channel [59]. The following assumptions were made: the flow profile is constant along the depth axis; variations in concentration do not modify the viscosity and density of the fluid; and the channel walls are smooth; the wall surface tension forces are neglected. The momentum and mass transport equations are solved in two steps, as described in detail in [59]. For the hydrodynamic calculations, a parabolic Poiseuille profile is assumed at both inlet boundaries. The fluid velocity is fixed to 0 along the wall boundaries (no slip conditions), and a free condition is assumed at the outlet boundary.

As mixing efficiency, the ratio between minimal and maximal concentrations at the outlet cross-section is defined [59].

[P 60] The observation of the flow patterns was performed using a microscope with a video camera [151]. Flow feed was achieved by hydrostatic pressure. A pH-indicator reaction with bromothymol blue and sodium hydroxide solution was applied, resulting in a dark blue color. Experiments were performed at flow rates of 50, 900 and 4000 μl min^{-1} for the 'mini' channels and 60 and 6000 μl min^{-1} for the 'micro' channels.

1.3.18.4 Typical Results

Mixing efficiency as a function of the meandering ratio

[M 67] [P 59] The mixing efficiency was calculated as a function of the ratio of periodic step to channel width, i.e. of the degree of meandering of the channel, having discrete values of 1, 2, 4, and 8 [59]. For low *Re* of 0.26 (fluic velocity: $1.3 \cdot 10^{-3}$ m s^{-1}) a monotonic increase in the mixing efficiency with increasing geometric ratio was found, from 65 to 84% (see Figure 1.137). For *Re* of the order of 1000 times higher (267; fluid velocity: 1.3 m s^{-1}), a more pronounced increase up to a maximum at a geometric ratio of 4 was found, from then slightly decreasing up to a ratio of 8. The maximum efficiency was near 99%; it was still about 88% at a ratio of 8.

For reference case A (largest width), a low mixing efficiency of 29% is found, as expected [59]. On decreasing this parameter from 141 to 100 μm (reference case B), consequently the mixing efficiency increases to 72%. A further increase in mixing length to the value of the effective length (2828 μm; reference case C) even further increases the mixing efficiency to 86%.

The results show that zig-zag channels may have superior mixing performance to geometrically straight channels [59]. This increase in performance is gained, however, only at large *Re* and large ratios of periodic step to width (i.e. moderate meandering).

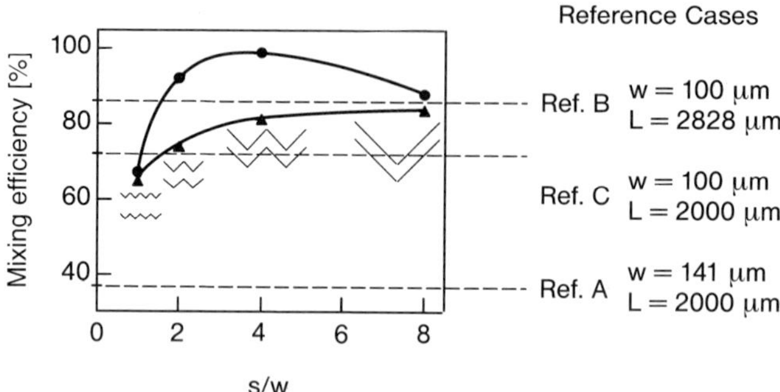

Figure 1.137 Mixing efficiency versus the meandering ratio s/w at $Pe = 2600$ and at two different *Re* for a set of zig-zag and straight-channel micro mixers. $Re = 0.26$ (triangles); $Re = 267$ (circles) [59]; w = channel width, L = channel length (by courtesy of ACS).

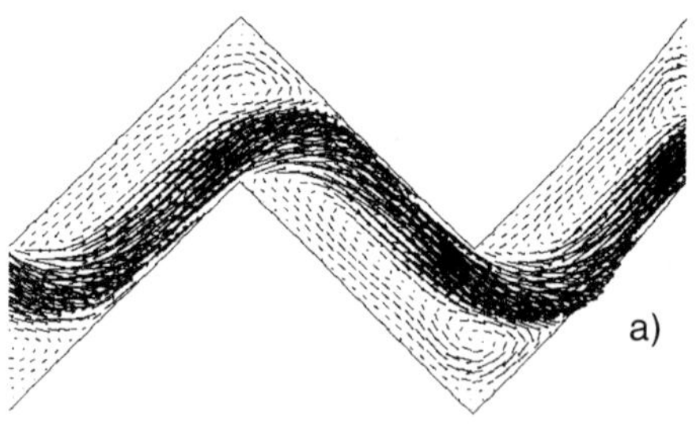

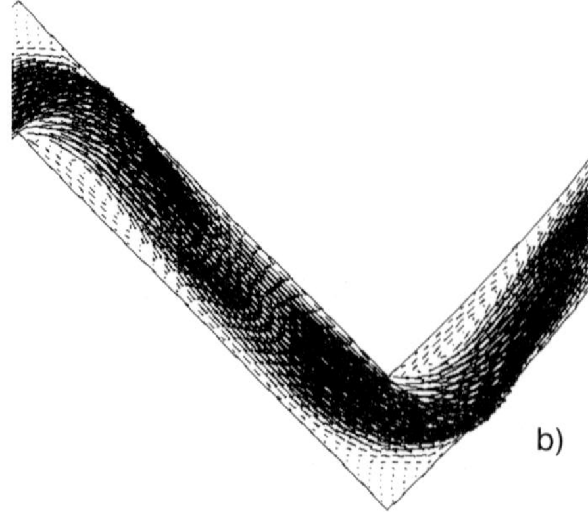

Figure 1.138 Flow patterns in two zig-zag micro mixers at $Re = 267$, given as velocity vectors. (a) $s = 400$ μm; (b) $s = 800$ μm [59] (by courtesy of ACS).

Hydrodynamics at different meandering ratios

[M 67] [P 59] Fluid dynamic simulations show that the mixing efficiency is influenced by recirculation patterns which are exhibited in the corners of the zig-zag channels, i.e. at the 90° fluidic turns (see Figure 1.138) [59]. Recirculations induce a transversal component of the velocity and thus transport material from the interface to the walls. At a meandering ratio of 4, this pattern is fully developed and expands into the entire arm of the channel after each angle. At a ratio of 8, recirculation is fixed to the corner only. In this view, it is understandable that for the latter flow geometry lower mixing efficiency (88%) is found than for the meandering ratio 4 case (near 99%).

Mixing efficiency as a function of the Reynolds number

[M 67] [P 59] At a constant Peclet number of 2600, the mixing efficiency was calculated for different *Re* in the range 0.26–267 [59]. Up to $Re = 80$, the mixing efficiency stays almost constant at 81% (see Figure 1.139). Here, only diffusion causes mixing. For $Re > 80$, a pronounced increase in mixing efficiency takes places, which is due to the induction of flow circulation patterns (see *Hydrodynamics at different meandering ratios*).

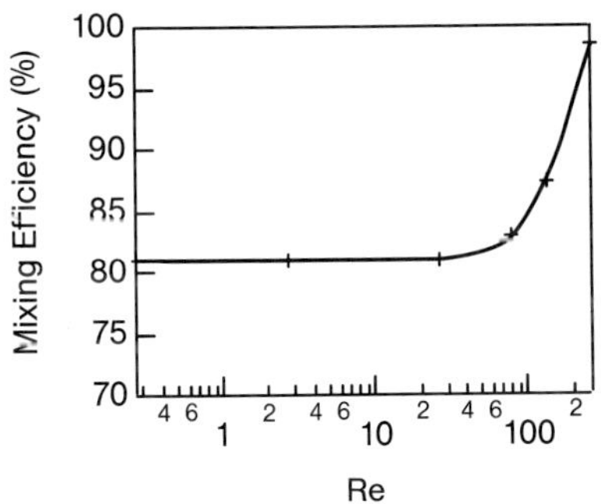

Figure 1.139 Mixing efficiency versus *Re* number for one zig-zag micro mixer, with $s = 400$ μm, at $Pe = 2600$ [59] (by courtesy of ACS).

Maximum velocity development in zig-zag channels

[M 67] [P 59] The ratio of the maximum velocity in a zig-zag channel normalized by the maximum velocity in a straight channel was calculated as a function of the number of zig-zag turns for various meandering ratios [59]. Up to about six turns, this ratio increases as a consequence of developing more recirculation zones which restrict the flow path for the 'non-circulated' flow and thus increases the corresponding maximum velocity.

When the outlet maximum velocity (after eight turns) at a meandering ratio of 4 is plotted versus *Re*, a curve similar to the dependence of the mixing efficiency on *Re* is yielded [59]. At $Re \approx 80$, the outlet maximum velocity increases considerably, whereas at low *Re* it remains nearly constant.

Mixing efficiency at various diffusion constants

[M 67] [P 59] The mixing efficiency was calculated for various diffusion constants, ranging from $2 \cdot 10^{-7}$ to $10 \cdot 10^{-7}$ m^2 s^{-1} [59]. A meandering ratio of 4 was assumed

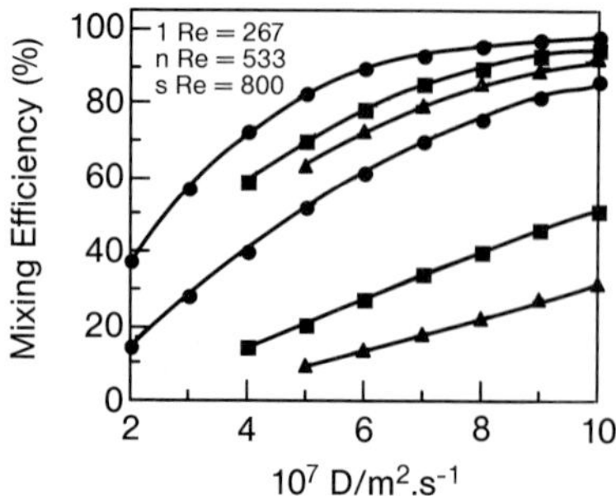

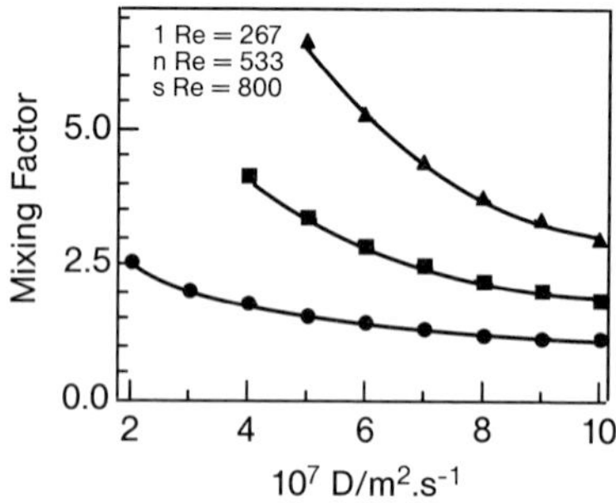

Figure 1.140 Mixing efficiency versus diffusion constant for one zig-zag micro mixer (lines) with s = 400 μm and one straight-channel micro mixer (dashed lines), reference case B, at three different *Re* values (top). Mixing factor, the ratio between the mixing efficiency of comparable zig-zag and straight-channel micro mixers, versus diffusion constant (bottom) [59] (by courtesy of ACS).

for the zig-zag channel and three *Re* values were taken into account, 267, 533 and 800. For comparison, the same analysis was performed with a straight channel of comparable dimensions, again at the same *Re* values. As expected, the efficiency increases in all cases with larger diffusion constants (see Figure 1.140). The performance of the zig-zag channel at all three *Re* values is better than the best performance of the straight channel. Also as to be expected, the difference between the mixing efficiencies at the three *Re* values is remarkable in the linear case. This is simply due to the respective differences of residence times and diffusion widths. Concerning the zig-zag case, the relative differences between the same curves are much smaller, which is due to the compensating effect of recirculation.

A mixing factor was introduced, defined as the ratio of the mixing efficiency of the zig-zag and the straight channels under otherwise equal conditions [59]. The mixing factor decreases with increasing diffusion constant, showing that the zig-zag channels are most effective for solutes with low diffusion constant. It is also confirmed that the largest differences, up to a value of about 6.5, are given for operation at high *Re* (800). Here, the relative contribution of recirculation patterns to mixing is the highest.

Experimental mixing efficiency as a function of the flow rate

[M 67] [P 59] The mixing efficiency was derived experimentally in a device with a meandering ratio of 8, i.e. a periodic step of 800 μm [59]. The channels were 100 μm wide and 48 μm deep with a linear distance of 2000 μm. An aqueous buffer solution and a buffer solution with fluorescein were contacted; the fluorescence intensity was measured at the end of the channel. Owing to the known problems of biasing

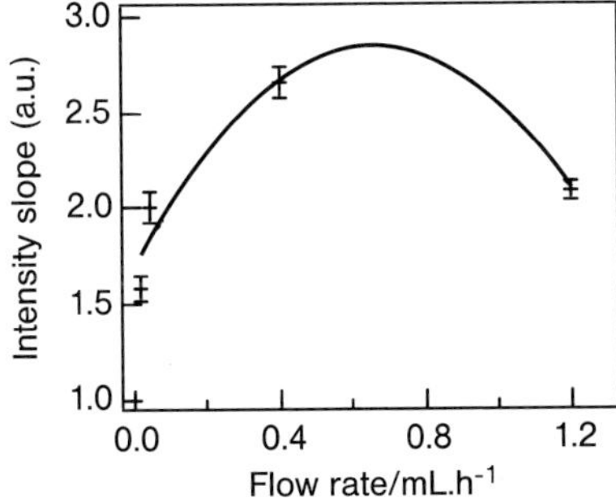

Figure 1.141 Slope of the fluorescence intensity versus flow rate for one zig-zag micro mixer with $s = 800$ μm and $w = 100$ μm. The slope is proportional to the mixing efficiency. In a dilution-type experiment, an aqueous fluorescein solution is mixed with an aqueous solution [59] (by courtesy of ACS).

for such dilution experiments, i.e. to an overestimation of mixing efficiency as a result of limited resolution, the mixing efficiency was estimated from the slope of the intensity curves, rather than taking these values directly (see Figure 1.141).

The analysis of the slope of the intensity as a function of the flow rate (from 0.04 to 1.2 ml h^{-1}) is qualitatively in accordance with the theoretical predictions from the mixing-efficiency calculations [59]. The slope is highest at the lowest flow rate, owing to the large residence time. It decreases when the flow rate is increased, but increases again above a certain critical flow rate. The corresponding increase in mixing efficiency is caused by recirculation flows, as is known from the calculations and the hydrodynamic studies discussed above (see *Mixing efficiency as a function of the meandering ratio* and *hydrodynamics at different meandering ratios*). The respective critical *Re* is 7; this, however, differs considerably, by an order of magnitude, from the theoretical *Re* of 80, derived from the calculations. The origin of this difference is not clear. It is speculated that it may be caused by non-smooth surface profiles of the micro channels due to re-deposition processes of material during the excimer laser manufacturing. Such attached particles are flow obstacles and also contribute to mixing.

Comparison with straight/zig-zag channels

[M 68] [P 60] At a very early stage of worldwide activities in the field of mixing with micro mixers, an experimental comparison was made between the mixing performance of straight and zig-zag channels [151]. The investigations covered mini and micro channels.

The 'mini' channels had a width of 300 μm and a depth of 600 μm and 80 curves were placed on a passage of 100 mm, yielding a length from bend to bend of 1 mm [151]. The curved channels showed very different behavior depending on the flow rate and correspondingly the Reynolds number. At $Re = 1.85$ and a flow rate of 50 μl min^{-1}, mixing was similar to that of a straight channel of the same total length. Bi-laminated patterns were visible and a weakly developed mixing was found close to the interface. The zig-zag structure had no impact, acting only as a dead zone.

At $Re = 33$ and a flow rate of 900 μl min^{-1}, the flow pattern changed to secondary flow with recirculation [151]. The mixing was completed at the end of the channel, as found by visual inspection. This was much further improved at $Re = 148$ and a flow rate of 4000 μl min^{-1}; here complete mixing was achieved after a passage of

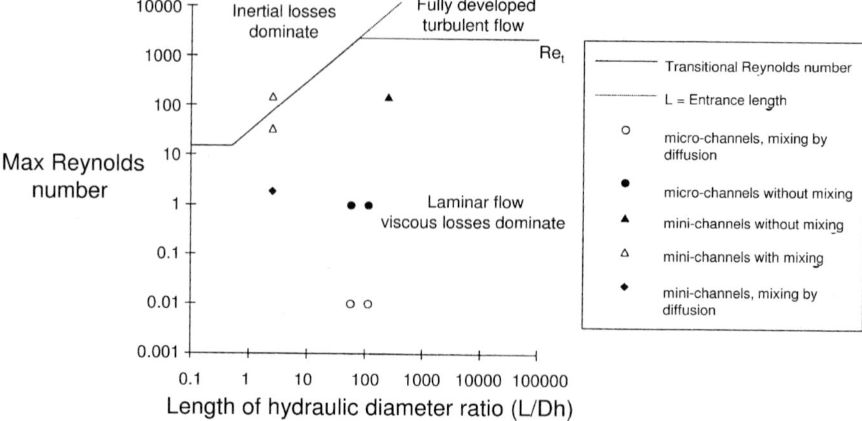

Figure 1.142 A kind of flow-pattern map for zig-zag micro mixers of various dimensions correlating their Reynolds numbers with their hydrodynamic regimes as a function of their ratio of bend length to hydraulic diameter [151] (by courtesy of Kluwer Academic Publishers).

about one-third of the channel length. For the same conditions, no observable mixing was found for a straight channel of the same length.

The bended 'micro' channels had a width of 180 μm and a depth of 25 μm and a reduced length (25 mm) compared with the 'mini' channels [151]. The flow in such channels was characterized at two very low *Re* (1.0 and 0.1) and compared with the flow in straight channels under some hydrodynamic conditions. In all four cases, undisturbed laminar flow was found. Mixing was only detectable at $Re = 0.1$ owing to diffusion mixing at a much prolonged residence time. At $Re = 1$, no mixing could be detected.

Based on these experiments, a kind of flow-pattern map was proposed describing a region of laminar flow where viscous losses dominate, an intermediate region with secondary flow where inertial losses dominate (albeit still not turbulent) and a region of fully developed turbulent flow (see Figure 1.142) [151]. The transitional Reynolds number from the pure laminar to the secondary-flow regime increases with the ratio of bend length to hydraulic diameter.

Comparison with straight/column-filled channels – use in micellar electrokinetic chromatography

[No details on mixer] [no protocol] By means of fluorescence imaging (buffer and fluorescein solutions), the flow patterns in the following microstructured elements are derived: in a straight channel, a straight channel with a row of rectangular columns in the center, and a zig-zag channel with 90° turns only [150]. For details on the mixers, see the original literature. The most homogeneous color texture was given for the zig-zag channel. The straight channel gave a bi-laminated flow; a similar finding was made for the column–row mixer, where the columns actually act more as a separation wall than to induce secondary flow and to improve mixing.

The zig-zag mixer was then used as a pre-column reactor to derivatize biogenic amines such as histamine and tyramine with *o*-phthaldialdehyde and to detect the corresponding products with micellar electrokinetic chromatography [150]. A separation of four amines, histamine, tyramine, putrescine and tryptamine, was demonstrated in that way.

1.3.19
Mixing by Helical Flows in Curved and Meander Micro Channels

Most Relevant Citations
Peer-reviewed journals: [47, 152, 153]; proceedings contributions: [49].

The generation of secondary 'helical' flows in suitably curved channels, known as Dean vortices, is not an entirely new fluidic phenomenon discovered at the micro scale; actually it was found already also for wound tubings of conventional diameter (see a summary in [152]). In-depth studies concerning Dean vortices in curved channels were made in the framework of various applications such as filtration, heat exchange, friction and mixing.

When guiding fluids through curved channels, the maximum in the velocity profile is displaced towards the outer channel wall and Dean vortices form (as reported, e.g., in [152]). The latter are typically characterized by two counter-rotating vortices above and below the symmetry plane of the channel coinciding with its plane of curvature. Fluid is transported outwards in this plane by means of centrifugal forces. By recirculation, back transport along the channel walls is induced.

A Dean number of ~140 is a kind of threshold value [47, 152]. For lower values, two counter-rotating vortices are found, whereas for higher values, two additional counter rotating vortices appear which are close to the center of the outer channel wall. Means to achieve this are changes in the flow velocity, the hydraulic diameter and the radius of curvature.

Not only can simple helical flows be induced, but also chaotic flow patterns of alternating helical flows can be achieved [152]. For the latter purpose, a switch between two flow patterns has to be achieved by either consecutively changing the geometry parameters (hydraulic diameter or radius of curvature) or altering the Reynolds number such that the Dean number changes via its threshold value. Accordingly, improved mixing can be achieved by periodic alterations of flow patterns in simple curved channels, laid within planar geometry. The corresponding type of micro mixer can be based on an alternative change of curvature, i.e. a meander design [152].

Helical flows are not only found for 2-D mixer designs. In analogy with findings for macro-scale alternating helical coils, microstructured 3-D designs were proposed [49, 153]. For reasons of limitations of today's micro fabrication, not real helices were made, but easier to fabricate structures such as complex micro channels based of L-shaped elements arranged in a 3-D fashion.

1.3.19.1 Mixer 69 [M 69]: Curved Channel Micro Mixer
This refers to a generic design of a curved channel as given in [152] (see Figure 1.143).

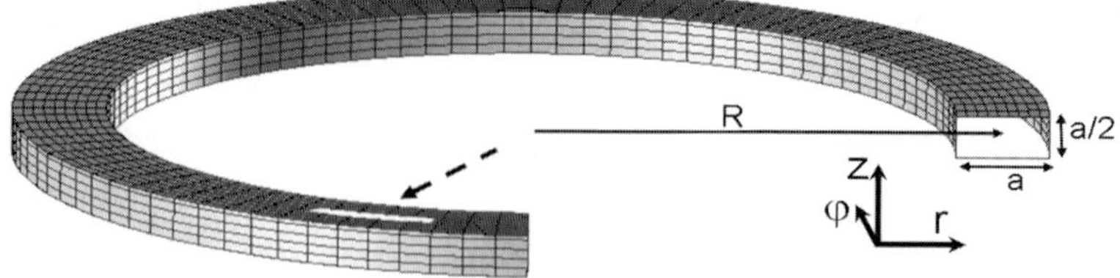

Figure 1.143 Model geometry of the curved square channel used for simulating helical flows. Only half the geometry is shown due to reflection symmetry; *a* and *R* denote the channel dimension and the radius of curvature, respectively [152].

Mixer type	Curved channel micro mixer	Channel width	200 μm
Mixer material	Not realized, simulation work only	Radius of curvature	1 mm

1.3.19.2 **Mixer 70 [M 70]: Meander Channel Micro Mixer**

Here, micro mixers with a repeatedly curved, i.e. meander, design are considered [152].

Version (a). This refers to a generic design of a meander channel as given in [152] (see Figure 1.144).

Mixer type	Meander channel micro mixer	Channel width	200 μm
Mixer material	Not realized, simulation work only	Radius of curvature	1 mm

Version (b). This design of meander mixer was actually realized. It refers to a structure with a large number of mixing elements which was micro-machined by precision milling in a plastic material [47]. Irreversible sealing of the channel was accomplished by insertion of a thin PMMA foil via solvent bonding.

Mixer type	Meander channel micro mixer	Number of mixing elements	20
Mixer material	PMMA	Thickness of PMMA foil for sealing	100 μm
Channel width, depth	1 mm, 1 mm	Inlet connectors	Plastic tubing of about 1.4 mm inner diameter
Channel length	322 mm	Outlet connector	1/8 inch steel tube connector

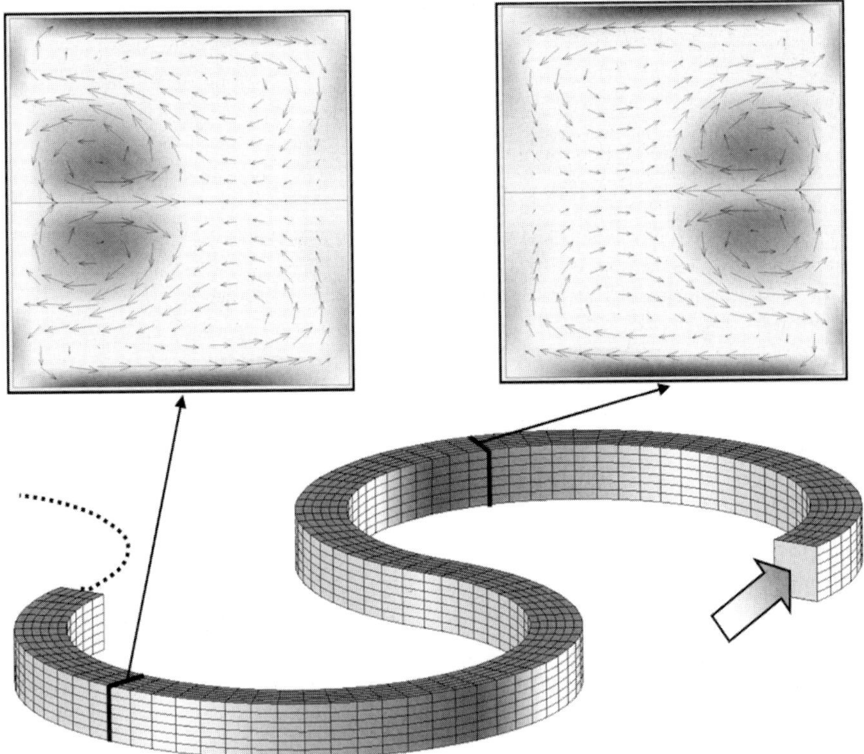

Figure 1.144 Top, cross-sectional view of secondary flows taken in flow direction; bottom, sketch of a meander channel build up from circular segments [152].

1.3.19.3 Mixer 71 [M 71]: 3-D L-shaped Serpentine Micro Mixer

A tortured, 3-D flow serpentine path was achieved by complex connection of various L-shaped microstructured elements [49, 153]. Such a structure is a miniature analog of alternating macro-scale helical coils that are known to give chaotic flows. Basically, the flow exposes multiple 90° turns instead of being guided along bendt structures. The fluids are introduced via a T-structure.

The L-shaped structures can simply be made by conventional silicon micro machining. The microstructured wafer was covered by a glass slip using a thin layer of a silicone adhesive [153].

Mixer type	3-D L-shaped serpentine micro mixer	Long and short axis of L-element	700 μm, 400 μm
Mixer material	Silicon/glass	Channel width, depth	300 μm (at the surface), 150 μm

1.3.19.4 Mixing Characterization Protocols/Simulation

[P 61] The numerical simulations were based on the solution of the incompressible Navier–Stokes equation and a convection–diffusion equation for a concentration field by means of the finite-volume method [152]. The Einstein convention of summation over repeated indices was used. For pressure–velocity coupling, the SIMPLEC algorithm and for discretization of the species concentration equation the QUICK differencing scheme were applied. 'Hybrid' and the 'central' differencing schemes referred to velocities and pressure, respectively (commercial flow solvers CFX4 and CFX5).

Because numerical errors due to discretization of a convective term introduce an additional, unphysical diffusion mechanism, termed numerical diffusion (ND), the diffusion coefficient D was set to zero [152]. The resulting concentration fields nonetheless are indicative of the distribution of a solute within the micro channel volume. In this way, convective patterns can be derived for the redistribution of the liquid transverse to the flow direction. Accordingly, the stretching, tilting and thinning of liquid lamellae can be followed.

[P 62] A Lagrangian particle tracking technique, i.e. the computation of trajectories of massless tracer particles, which allows the computation of interfacial stretching factors, was coupled to CFD simulation [47]. Some calculations concerning the residence time distribution were also performed. A constant, uniform velocity and pressure were applied at the inlet and outlet, respectively. The existence of a fully developed flow without any noticeable effect of the inlet and outlet boundaries was assured by inspection of the computed flow fields obtained in the third mixer segment for all Reynolds numbers under study.

A two-step procedure was used for numerical computation of the mixing performance [47]. First, the velocity and pressure fields were derived by solving the Navier–Stokes equations and the equation of mass conservation for an incompressible fluid. In a second step, trajectories of mass less particles were computed by streamline integration of the velocity field.

The commercial flow solver CFX4, relying on the finite volume method (FVM), was applied to solve the velocity field [47]. The SIMPLEC algorithm and the QUICK differencing scheme were used for pressure–velocity coupling and for discretization of the velocity fields, respectively. For an accelerated convergence the algebraic multigrid (AMG) iterative method was applied.

Two strategies of particle tracking were used. As a first strategy, tracers were distributed along the 'interface' of the two fluids (virtual, since miscible) for computation of interfacial stretching factors [47]. Then an iterative method for computing interfacial stretching was established.

In a second approach, the paths of a large number of tracers, which were distributed uniformly over a certain volume element, were followed [47]. The time at which the tracers penetrate out of the element was taken. By such an iterative procedure the residence-time distribution was determined.

[P 63] A reactive-type flow visualization method was used for quantification of mixing [47] (citing a protocol described in detail in [20] and given in [P 40]). Colorless solutions of iron(III) nitrate and sodium rhodanide form a colored compound,

Fe(III) rhodanide. The local concentration of the colored compound is determined photometrically from the digitized images of the flow inside the channel.

Two continuously working double-action syringe pumps were used for liquid pumping [47]. The pulsations of the step motors were damped by a 20 ml gas ballast. Overall flow rates ranged from 200 to 2000 ml h^{-1}. The corresponding Dean numbers tested ranged from 35 to 351. Digital video equipment with planar illumination from a fluorescence light source from below was used for flow monitoring.

It can be seen that even if the residence time for the $K = 141$ flow is about a factor of 4 smaller than for $K = 35$ at the same position, mixing proceeded to a higher degree for the larger Dean number, as is clearly visible especially when comparing the images recorded at position 2.

The Lambert–Beer law was applied for the determination of the mixing performance, the local concentration c being measured by photometry via the relation

$$\ln\frac{\Phi_0}{\Phi} = \varepsilon\, c\, l \qquad (1.8)$$

where Φ (Φ_0) denotes the photometric brightness obtained from the digitized images with finite (zero) rhodanide concentration, ε is the extinction coefficient, c is the $Fe(SCN)_3$ concentration in the dye solution and l is the layer thickness.

[P 64] Phenolphthalein in ethanol and sodium hydroxide in ethanol were contacted in the serpentine mixer [153]. Microscopy observations were used for qualitative judging of mixing. Cross-sectional imaging was obtained by focusing of the microscope on the vertical segments of the channel. Apart from this qualitative analysis, quantitative information was derived by normalized average red dye intensity analysis from the microscope images.

1.3.19.5 **Typical Results**

Flow patterns of helical flow, depicted by velocity vectors

[M 70a] [P 62] Computational flow simulation of the secondary flow, depicted by velocity vectors, was performed for Dean numbers of 10 and 100 [47]. The helical flow is weak for the smaller Dean number. The center of rotation is located close to the midpoint of the patch. For a Dean number of 100, a notable increase in the relative strength of the helical flow is observed: the center of the vortex is shifted towards the outer channel wall.

At a still higher Dean number of 200, the flow pattern changes considerably (see Figure 1.145) [47]. An additional counter-rotating vortex appears is found, yielding a 4-vortex flow pattern. The new vortex is located close to the outer channel wall. On following the flow-pattern evolution along the flow path the four-vortex pattern, it is observed that it is well developed already at an early stage.

Species concentration and velocity fields along the flow passage at Dean numbers above and below the threshold value

[M 69] [P 61] For a curved channel design, species concentration and velocity fields of the secondary flow were given (see Figure 1.146) [152]. These fields each were

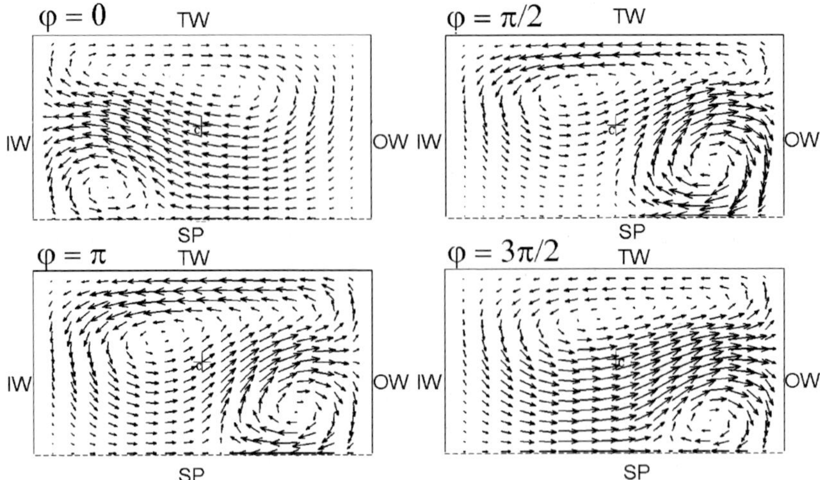

Figure 1.145 Evolution of the four-vortex helical flow in a mixing element for $K = 200$ [47].

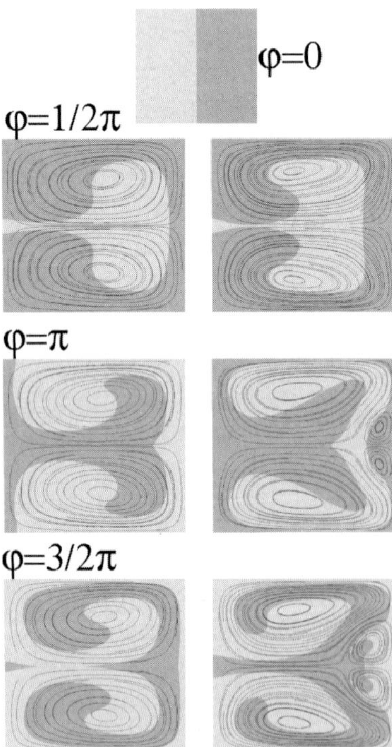

Figure 1.146 Species concentration (encoded in gray) given for cross-sections at the inlet, outlet and two intermediate positions for two Dean numbers, $K = 150$ (left) and 300 (right). The initial condition is shown on the left, i.e. two lamellae. Additionally, the velocity fields of the secondary flow are shown [152].

determined at various channel cross-sections, at the inlet, outlet and at two intermediate positions. Two Dean numbers, $K = 150$ and 450, were considered. Starting from two initial lamellae, it was found that the velocity profiles were at the beginning qualitatively identical, thus giving similar concentration distributions. Already at an early stage in the flow passage, an additional pair of counter-rotating vortices develop for $K = 450$. These vortices are found close to the center of the outer channel wall. In this way, one lamella is incompletely displaced, leaving a certain fraction. At a later stage in the flow passage, a well developed, steady velocity profile with four vortices is achieved.

Vorticity of helical flows

[M 69] [P 61] In a curved channel, helical flows can be produced with four vortices, composed of two times two types, a small and large one (see Figure 1.147) [152]. The total vorticity of the small vortices was integrated over the relevant part of the cross-section. It is found that these vortices start to develop at Dean numbers around 200. The strongest increase in the vorticity is observed at Dean numbers between 300 and 400.

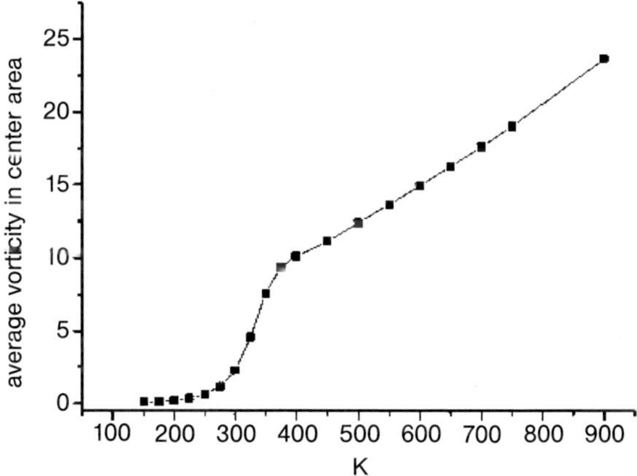

Figure 1.147 Total secondary vorticity as function of the Dean number [152].

Interface stretching

[M 70a] [P 62] As an initial situation, the fluid interface is set as a vertical straight line across the inlet [47]. An iterative method was employed for interface tracking owing to the high interfacial stretching. In this way, the seeding density was adjusted to eliminate interface self-crossing during particle tracking and to improve the accuracy of interface stretching calculations. The tracer numbers referring to Dean numbers $K = 10$, 100 and 200 were about 3000, 10000 and 56000, respectively.

As interface stretching factor, the interface length at a certain position divided by the initial interface length is defined [47]. Such stretching factors were given as a function of the number of segments of a meander mixer (see Figure 1.148). At a Dean number $K = 10$ nearly no stretching takes places. In contrasta, exponential

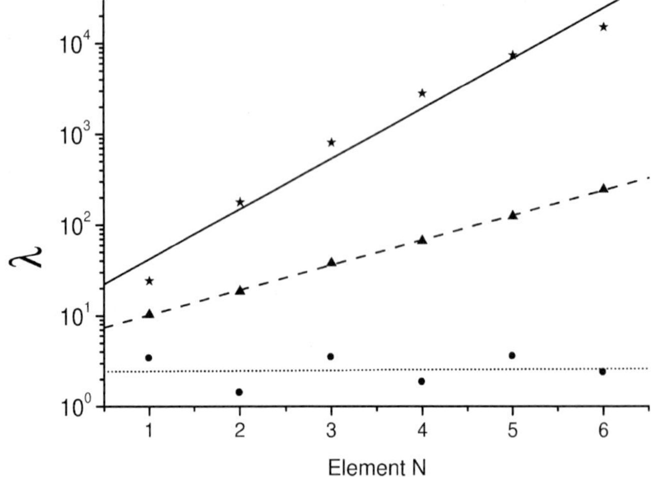

Figure 1.148 Interface stretching factor vs. element number for different Dean numbers: (•) $K = 10$; (▲) $K = 100$; (★) $K = 200$ [47].

growth of the interface is observed for $K = 100$ and 200. A positive Lyapunov exponent is given, confirming chaotic mixing at large Dean numbers.

The flow pattern switch at $K = 100$ differs from that at $K = 200$ [47]. In the first case, the center of rotation of the two Dean vortices changes when the sign of curvature is altered, similar to the situation reported in [154]. At the higher Dean number, a four-vortex pattern of pronounced asymmetry is given; a meandering flow has here a significant impact on the resulting change of the flow pattern.

At $K = 10$, the helical flow is nearly symmetric and a meandering flow has hardly any impact on the interface stretching; actually the initial flow situation is restored at certain positions (see Figure 1.149) [47]. At $K = 100$, an asymmetric two-vortex system experiences interfacial stretching similar to the 'blinking-vortex' flow pattern reported in [154]. At $K = 200$, the appearance of the two new vortices leads to even more substantial changes of the superposed flow pattern, giving large increases in interfaces. Later investigations confirmed that a four-vortex vortices system is yielded already at $K = 150$.

Residence time distribution

[M 70a] [P 62] Simulated residence time distributions (RTDs) were obtained for $K = 200$ at various positions for a meander mixer and compared with RTDs simulated for Poiseuille and for plug flow in a straight square channel with the same cross-sectional dimension (see Figure 1.150) [47]. The RTDs in the meander mixer come close to those given by plug flow, but are much narrower than those of Poiseuille flow. The RTD in the meander mixer is the better, the more downstream the flow is investigated. This is due to the homogenization of the flow by chaotic flow patterns.

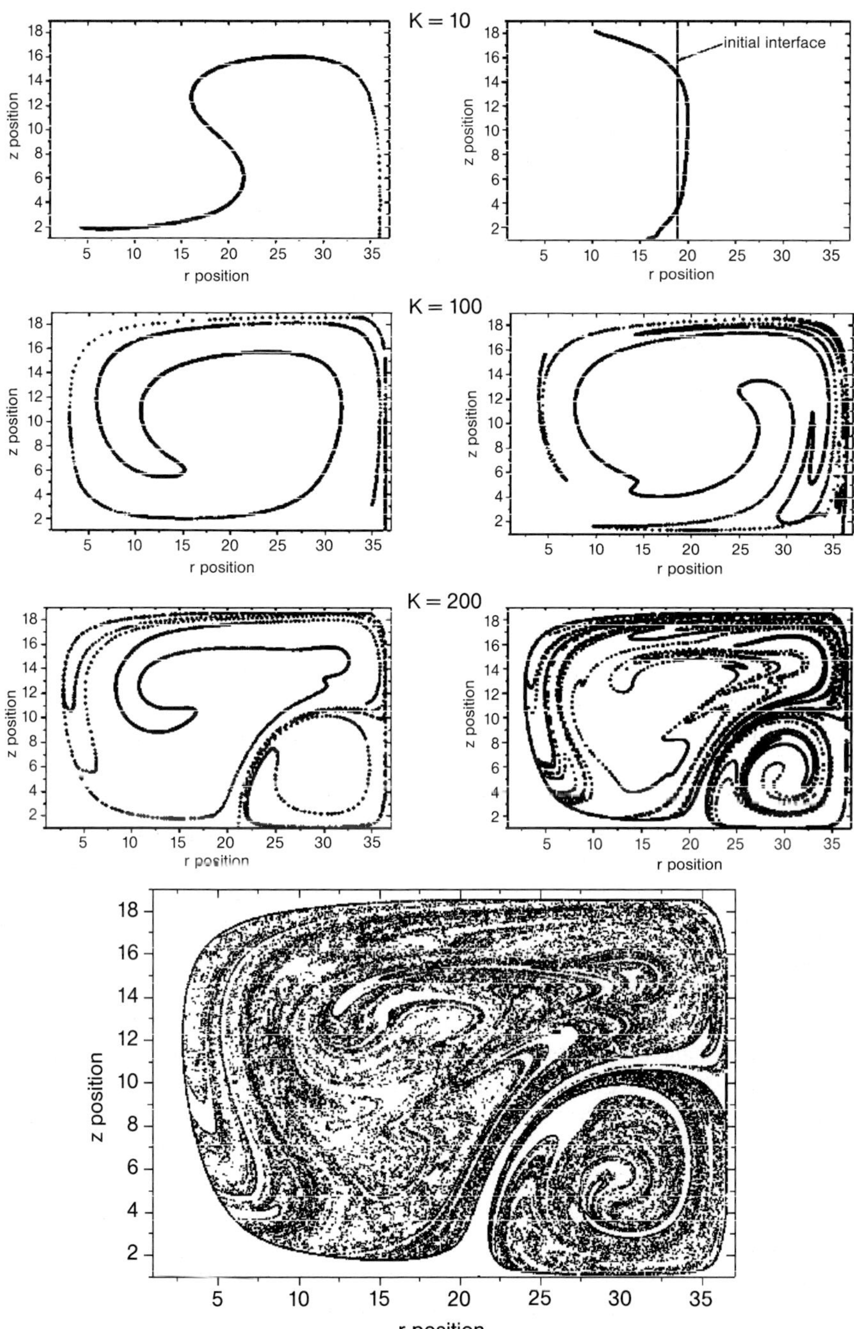

Figure 1.149 Interface shapes obtained from particle tracking for $K = 10$ (upper row), $K = 100$ (middle row) and $K = 200$ (lower row). The left column shows the shapes after the first mixing element and the right column those after the second mixing element. The last image at the bottom refers to the shape after the sixth mixing element at K = 200 [47].

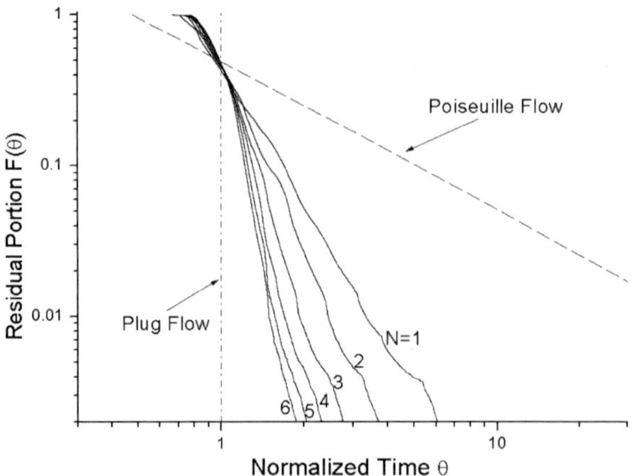

Figure 1.150 Residual portion of particles remaining in the meander mixer for $K = 200$ [47].

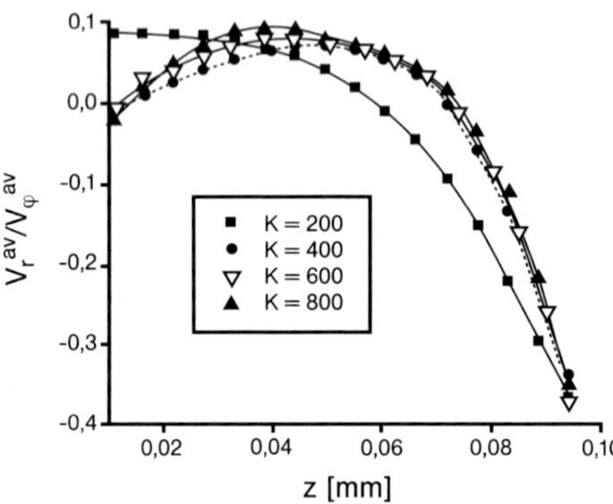

Figure 1.151 Relative transverse velocities averaged over a certain zone for Dean numbers between $K = 200$ and 800 [152].

Benchmarking of helical flow strength in curved and bas-relieved channels

[M 70a] [P 61] The strength of the helical flow in curved channels was compared with that in bas-relief structured micro channels, considering the averaged relative transverse velocity for a series of Dean numbers (see Figure 1.151) [152]. The curve yielded at $K = 200$ is distinctively different from those obtained at $K = 400$, 600 and 800, which equal each other. This is due to the different flow patterns; at low Dean numbers only one pair of vortices is given, whereas two pairs of vortices develop at larger Dean numbers, yielding curves with an intermediate maximum of the relative transverse velocity.

The relative transverse velocity of the curved channel corresponds well with data published for bas-relief structured channels [152]. Hence, also the very simple design of a curved channel can lead to efficient mixing.

Periodic switching of helical flow in curved channels – chaotic advection

[M 70a] [P 61] It is known that stirring and chaotic mixing can be achieved by an unsteady potential flow [154]. This can also be utilized for helical flows in curved

channels [152]. For this purpose, the geometric parameters need to be repeatedly changed to alter the Dean number. In this way, a periodic switch between the flow regimes above and below a threshold value of the Dean number, i.e. between two and four counter-rotating vortices, is achieved.

Hence alternating flows can also be achieved by repetitive induction of entrance flow effects [152]. This relies to a constant re-direction of flows from one curved channel into another.

The sum of both effects, the flow-pattern switch and the entrance flow effects, is achieved when the sign of curvature is simply altered [152]. The respective design is that of a meander channel. At large Dean numbers, chaotic flow can be induced in this way, as evidenced by cross-sectional views of different segments of the meander channel.

Qualitative judgement of mixing – visual inspection

[M 70b] [P 63] At a Dean number $K = 141$, video images show improved color formation for reactive imaging in the meander micro mixer as compared with $K = 35$, which proves more advanced mixing (see Figure 1.152) [47]. This is particularly remarkable, since at the higher Dean number the residence time for the flow is about a factor of 4 shorter than for the $K = 35$ at the same position. Thus, the theoretically predicted higher vorticity (see *Vorticity of helical flows* and *Interface stretching*) indeed results in experimentally confirmed better mixing.

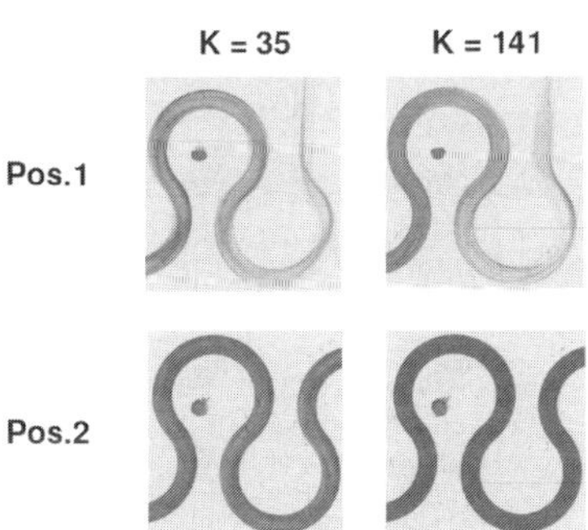

Figure 1.152 Micro photographs of the mixing patterns in the meander micro mixer for different Dean numbers and positions [47].

Quantitative judgement of mixing – reactive imaging analyzed by photometry

[M 70b] [P 63] Photometric concentration profiles were obtained for mixing of 50 mM reactant solutions at 25 °C and $K = 246$ in a meander micro mixer close to the inlet (see Figure 1.153) [47]. At the start of the mixing process only one lamella is visible, corresponding to about 10 mM reaction product. Further downstream, the next profiles provide a multitude of lamellae, which represent between 2 and 24 mM product. At still further positions in the meander channel, more homogeneous flow patterns are yielded and the product content is increased to between 15 and 28 mM. Finally, the concentration increases to values close to 50 mM which indicates complete mixing.

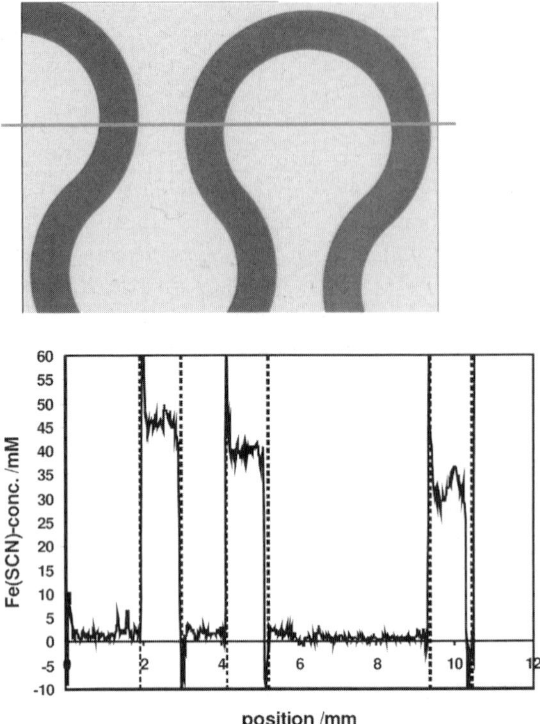

Figure 1.153 Top: micro photograph of the flow patterns obtained from the mixing and reaction taken close to the inlet of the mixer. The horizontal line denotes the cut over which the concentration was determined. Bottom: resulting photometric concentration profile. The dashed lines enclose the channel regions [47].

Mixing time – reactive imaging analyzed by photometry

[M 70b] [P 63] The relationship between mixing time and Dean number was determined using a meander micro mixer and a reactive-imaging approach [47]. For Dean numbers ranging from $K = 105$ to 141, the mixing time is considerably reduced. This correlates with the predicted changes in flow patterns by simulation (see *Vorticity of helical flows* and *Interface stretching*). The relative strength of the helical flow, as compared with the mean flow velocity, increases with K. In addition, the 'blinking vortex' principle becomes more and more effective, because the centers of the vortices are continuously shifted towards the outer channel wall. As a result of both effects, chaotic advection arises.

The results obtained on mixing time also give experimental evidence for the existence of a threshold value around $K = 140$, which is associated with a change in vorticity [47]. At this point, the helical flow pattern switches from a two- to the four-vortex fluid system. Besides qualitatively confirming such a threshold value, the experiments are also in very good quantitative accord with the value predicted by the simulations.

Above $K = 141$, a steady but comparatively small reduction in mixing time was measured [47]. This confirms that the four-vortex pattern is relatively stable for Dean numbers above the threshold value. The increase in interfacial stretching obviously is now more compensated by the reduction in residence time, as for the cases reported above.

In this regime at $K = 141$, mixing times below 50 ms are reached [47]. For the geometry chosen, a Dean number of 140 corresponds to a Reynolds number of about 313. Hence the critical transition for meander (Dean) micro mixers is almost an order of magnitude smaller than the critical Reynolds number for the laminar-to-turbulent transition in straight channels. Hence meander micro mixers can achieve fast mixing in a regime where mass transfer in straight channels is usually slow and dominated by diffusion. Considering their ease of fabrication, meander micro mixers can be considered a simple, but efficient means and design.

Flow patterns in L-shaped micro mixer
[M 71] [P 64] Flow cross-sectional images were taken in the 3-D L-shaped micro mixer using a reactive approach to characterize mixing [153]. Initially, a bi-laminated system is observed, then the interface becomes more and more stretched and elongated, and finally the streams intertwine deeply and mixing results from this enlarging of specific interface.

Concentration monitoring in L-shaped micro mixer
[M 71] [P 64] A sort of concentration monitoring was achieved for the 3-D L-shaped micro mixer giving a quantitative analysis of the mixing [153]. A normalized average intensity clearly shows the speeding up of mixing at higher *Re*. At $Re = 70$, mixing is complete within 20 ms.

Mixing and bacteria capture in L-shaped micro mixer
[M 71] [details of protocol in [49]] L- and C-shaped serpentine PDMS micro mixers use chaotic advection [49]. Numerical results regarding the degree of mixing at different flow rates and at different locations were determined. The degree of mixing actually increased with increasing flow rate (0.05–0.4 ml min^{-1}), despite the reduction of the residence, more than compensated by increasing secondary-flow mixing. This is corroborated by experimental findings, showing superior performance at high Reynold number.

A biological application, bacteria capture by mixing of blood/bacteria samples with magnetic beads, was carried out in the L-shaped micro mixer [49]. A high capture efficiency of 99% was obtained at a short time of 0.15 s. The blood cells and bacteria remained intact after the mixing process, evidencing the low shear strain field of the flow.

1.3.20
Distributive Mixing with Traditional Static Mixer Designs

Most Relevant Citations
Peer-reviewed journals: [2].

A straightforward idea is to 'shrink' the designs of conventional static mixers, i.e. to keep the geometry exactly by simply decreasing all dimensions with the same reduction factor [2]. The static mixers are basically pipes with in-line elements which serve as flow obstacles. They thus perform distributive mixing, e.g. splitting, rearranging and recombining or stretching and folding of the flow. The miniaturized versions are expected to induce a comparable action. This looks at first sight similar to the split-and-recombine (SAR) mixers. However, the distributive mixing described here results in much more complex flow patterns, not comparable to the lamellae patterns obtained at low Reynolds number of SAR devices. At high Reynolds numbers, both mixing approaches may behave similarly, but this is not known in detail at present.

1.3.20.1 Mixer 72 [M 72]: Intersecting Elements Microstructured Mixer

This mixer resembles conventionally widely used Sulzer™ and Koch™ static mixers which have an intersecting design, a series of rigid elements. A similar miniaturized design with four in-line mixing elements was realized [2]. Each mixing element is made by multiple bars placed at ±45° to yield intersecting channels. The mixing elements result from their predecessor by reflection and a 90° rotation.

The CAD design generated the geometry by performing Boolean operations on simple volumic primitives [2]. The microstructure was realized by microstereolithography. This technique is based on a layer-by-layer light-induced polymerization of a liquid resin, a pre-polymerized solution. Actually, about 1800 layers of 5 μm thickness were superimposed in this way (5 h manufacturing time).

Mixer type	Intersecting elements micro mixer	Mixing element: width (tube diameter), length	1200 μm, 1200 μm
Mixer material	Polymer	Number of bars	24
Number of mixing elements	4	Angle of bars	±45°

1.3.20.2 Mixer 73 [M 73]: Helical Elements Micro Mixer

This mixer resembles the conventionally widely used Kenics™ static mixer having a short-helix design with alternately arranged right- and left-handed elements. The fluid is stretched and folded. A similar miniaturized design with six in-line mixing elements was realized [2]. Each helix element is aligned at 90° from the previous one and has a twist angle of 90°. Right- and left-handed elements are alternately arranged within the tube.

Mixer type	Helical micro mixer	Number of mixing elements	6
Mixer material	Polymer	Mixing element: width (tube diameter), length	1200 μm, 900 μm

1.3.20.3 Mixing Characterization Protocols/Simulation

[P 65] Pumping was achieved by pressuring water with nitrogen [2]. The pressure was measured with a pressure gauge. The flow rate was determined gravimetrically. Degassed, dionized, filtered water was used as fluid. Before the experiments, the mixers were primed to remove resting air bubbles.

Qualitative, numerical simulations were performed with the commercial tool FLUENT-5 to evaluate mixing efficiency [2]. The simulations were oriented on concepts employed for conventional 3-D static mixing. The micro-mixer geometries were laid out using the GAMBIT predecessor as well as the meshing of surfaces and volumes and the specification of boundary conditions. Entrance and exit sections were also simulated.

1.3.20.4 Typical Results

Volume flow–pressure drop relation

[M 72] [M 73] [P 65] A linear dependence of the pressure drop on the volume flow rate was observed both for the intersecting and the helical microstructured mixers (see Figure 1.154) [2]. The pressure drop of the helical device is lower than that of the intersecting mixer. A flow rate of about 1 l h^{-1} was found for the helical mixer at a pressure of about 150 mbar, whereas the intersecting mixer gives 0.75 l h^{-1} at about 250 mbar.

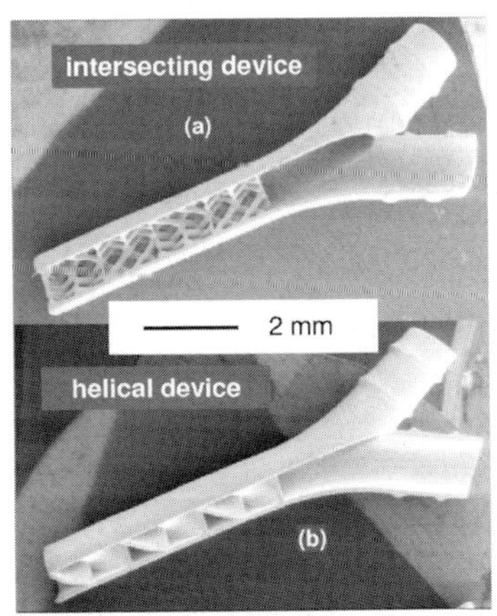

Figure 1.154 Cut view of micro mixers manufactured by microstereolithography.
(a) Intersecting channel device;
(b) helical-element device [2] (by courtesy of RSC).

Cross-sectional velocity profiles

[M 72] [M 73] [P 65] The analysis of cross-sectional velocity profiles (water as fluid; $Re = 12$) shows that the intersecting structures have intricate gradient fields near the bars of the internals, while the helical device displays entrance and exit effects over more than one-quarter of the flow field (see [155] e.g. for fluid flow through macroscopic helical static elements) [2].

Mixing efficiency

[M 72] [M 73] [P 65] When judging mixing efficiency by particle trajectories (water as fluid; $Re = 12$), it is evident that the intersecting device performs manifold splitting

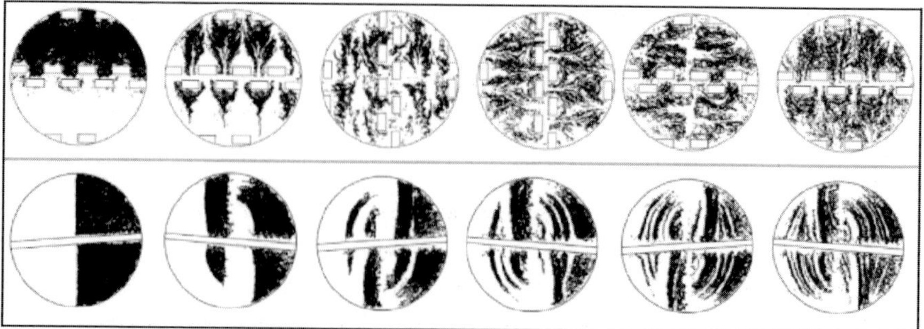

Figure 1.155 Particle-tracking imaging aiming at visualization of the mixing process ($Re = 12$). The location of 65 000 particles is given at various locations at the beginning of the structured in-line elements. Top, intersecting mixer; bottom, helical mixer [2] (by courtesy of RSC).

and recombining of the flow, yielding a fine-dispersed system at the end, i.e. achieving a good mixing efficiency (see Figure 1.155). In contrast, flow stretching and folding are found for the helical device, resulting in a coarsely textured fluid, which means less efficient mixing.

1.3.21 Passive Chaotic Mixing by Posing Grooves to Viscous Flows

Most Relevant Citations

Peer-reviewed journals: [44, 45, 152, 156, 157]; micro fabrication: [118]; proceedings contributions: [142, 158].

Grooves oriented at an oblique angle on the ground level of a micro channel are known to induce transverse flows by using a steady axial pressure gradient [44]. These grooves pose an anisotropic resistance to viscous flows, mainly in the orthogonal direction. In this way, an axial pressure gradient is built up and generates a mean transverse component of the flow. The flow originates at the floor structures and circulates back across the top of the channel, giving helical streamlines for the full flow.

The functioning of grooves was proven under pressure-driven [44] and electro-osmotic flow [156] conditions and both cases are described below.

1.3.21.1 Mixer 74 [M 74]: Non-grooved Channel – Reference Case

As a reference case, this straight channel design without any grooves was taken [44]. In the following, one typical set of design specifications is given.

Mixer type	Non-grooved micro channel	Channel width, depth	200 μm, 70 μm
Mixer material	Polydimethyl-siloxane		

1.3.21.2 Mixer 75 [M 75]: Oblique, Straight-grooved Micro Mixer (I)

The main feature of this class of mixers is patterned grooves on the floor of the micro channel [44]. In the simplest version, the array of grooves presents a repeating, periodic sequence of the base structure. In the following, one typical set of design specifications is given.

Microfabrication was done by means of photolithography using an SU-8 photoresist [44]. In a two-step procedure first the negatively shaped channel structure was established and thereafter the groove pattern was generated. Molds of these structures were made in PDMS, yielding the positive structure. Sealing was achieved by exposure of the PDMS structure to a plasma for 1 min and subsequent coverage with a glass cover slip.

Mixer type	Oblique, straight-grooved mixer	Channel width, depth	200 μm, 70 μm
Mixer material	Polydimethyl-siloxane	Angle of groove tilting	45°
Ratio groove structure height to channel depth	0.2		

1.3.21.3 Mixer 76 [M 76]: Oblique, Asymmetrically Grooved Micro Mixer – Staggered Herringbone Mixer (SHM)

In this more advanced version of the patterned grooved mixer, more complex patterns are created which have a repeated sequence of mirror-imaged sub-arrays, typically consisting of bendt grooves (see Figure 1.156) [44]. These are known to create a sequence of rotational and extensional local flows, i.e. the shape of the grooves is varied as a function of the channel length. A prominent design of this class is the staggered herringbone mixer (SHM). Here, the positions of the centers of rotation and of the up and down wellings (local extensional flow) of the transverse flow are exchanged. In the following, one typical set of design specifications is given.

Microfabrication was effected in the same way as described for the [M 75] device [44].

Mixer type	Oblique, asymmetrically-grooved mixer: staggered herringbone mixer (SHM)	Channel width, depth	200 μm, 77 μm
Mixer material	Polydimethyl-siloxane	Angle of groove tilting	45°
Ratio groove structure height to channel depth	0.23		

The fabrication of staggered herringbone structures was also achieved by a high-brightness diode-pumped Nd:YAG laser direct write method [118]. A PDMS

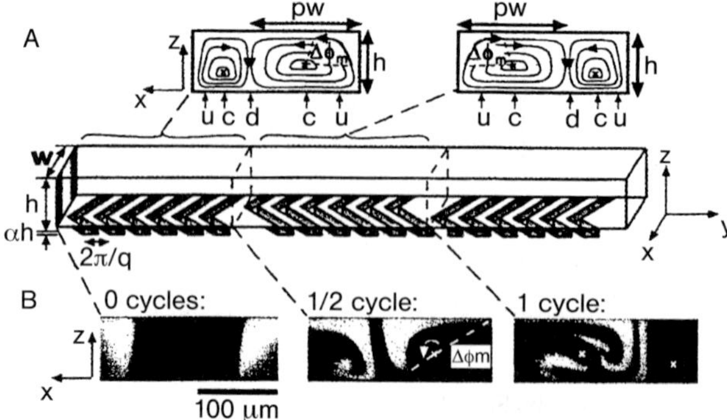

Figure 1.156 (A) Design of the staggered herringbone micro mixer, displaying one-and-a-half cycle of the floor structures (SHM). Streamlines are also given. For details of the symbols used, which describe the rotational and extensional flow, see [44]. (B) Confococal micrographs of vertical cross-sections of the channel given in (A) [44] (by courtesy of AAAS).

structure was realized in one step and tested for mixing function by fluorescence imaging. The method provides rapid prototyping of master structures and is said to have high flexibility.

1.3.21.4 Mixer 77 [M 77]: Oblique, Straight-grooved Micro Mixer (II)

This device contains in a T-type micro channel oblique, straight grooves, termed slanted wells in the original publication (see Figure 1.156) [156]. The grooves have the same purpose as given for the devices discussed below, namely to induce lateral transport to promote mixing.

The T-shape was imprinted into a polymer substrate yielding a trapezoid cross-sectional shape [156]. By excimer laser fabrication, a series of slanted wells beginning close to the T-junction were made. This microstructure was then sealed by a polymer cover.

Mixer type	Oblique, straight-grooved mixer	Channel width at top and bottom, depth	72 μm, 28 μm, 31 μm
Mixer material	Polycarbonate	Number of grooves	4
Cover material	Polyethylene terephthalate glycol	Cross-sectional groove width	14 μm
Inlet and outlet (mixing) channel width	511 μm, 100 μm	Spacing between the grooves	15.2 μm

1.3.21.5 Mixer 78 [M 78]: Diagonal-grooved Micro Mixer

A mixer, which is part of a sensor system, has diagonal grooves on the channel bottom [142]. This creates a lateral flow enlarging the interfacial area. No indication is given whether this measure causes the flow to become chaotic.

Mixer type	Diagonal grooved mixer	Plate thickness	1.6 mm
Mixer material	VACREL®	Channel width at top and bottom, depth	600 μm, 100 μm, 170 mm
Plate material	PC-board glass-reinforced base material FR4 with 35 μm Cu layer	Groove width, depth	200 μm, 50 μm

1.3.21.6 Mixing Characterization Protocols/Simulation

[P 66] Equal streams of 1 mM solutions of fluorescein-labeled polymer (polyethylenimine, molecular weight 500 000) in water–glycerol mixtures (0 and 80% glycerol) and a clear solution were injected into the channel [44]. The flow was achieved by compressed air at constant pressure. Imaging was achieved by applying a confocal fluorescence microscope.

[P 67] Simulations were made following experiments made previously [156]. Therein 0.11 mM Rhodamine B solutions in 20 mM carbonate buffer were mixed with the same carbonate buffer. For the buffer solution, the physical properties of water were approximated. For Rhodamine B, a diffusion coefficient of $2.8 \cdot 10^{-6}$ cm^2 s^{-1} was taken. Electroosmotic flow was applied for liquid transport. For all of the walls in the domain the electroosmotic (EO) mobility was set to $3.4 \cdot 10^{-4}$ cm^2 V^{-1} s^{-1}, which corresponds to a zeta potential (ζ) of –44.1 mV. The electric field in the outlet channel was 1160 V cm^{-1}. The Reynolds number was 0.22. The electric field strength was set low in order to decrease diffusive (pre-)mixing prior to the groove structure.

A 3-D model geometry was generated using the CFD-ACE+ v6.6 software package [156]. Both steady-state and transient mixing simulations were undertaken, the latter describing the mixing of plugs. As plug size 100 μm was assumed, being initially located 50 μm before the series of grooves. The plug mixing was investigated for a straight channel, without a T-junction so as to avoid the band broadening which would have occurred for such a flow configuration. Further details of the plug experiments are given in [156].

[P 68] A double-syringe pump was used for liquid feed [142]. An iodine–starch solution was mixed with a photographic fixer solution at a ratio of 1 : 3.5. Thereby, the intense blue color changed to a pale blue. The mixing process was followed by means of a stereo microscope.

1.3.21.7 Typical Results

Completion of mixing

[M 78] [P 68] For a diagonal-groove mixer, complete mixing was observed by reactive-type imaging for flow rates up to 1.3 μl min^{-1} [142]. For comparison, complete mixing is given for a channel without grooves only up to 0.12 μl min^{-1}. Hence the diagonal grooves allow for a 10-fold increase in flow rate at similar mixing performance.

Impact of Reynolds number on flow

[M 75] [M 76] [P 66] For grooves of small extension relative to the channel diameter, it was found that a variation of *Re* (in the range $Re < 1$) does not change the form of the flow, i.e. the shape of the trajectories [44]. Even at Re < 100 the form of the flow is qualitatively maintained.

Modeling the relationship between groove geometry and flow form

[M 75] [M 76] [P 66] Experimentally found dependences of the average rate of rotation on the geometry of the grooves can be described by a simple model [44].

Groove asymmetry

[M 75] [M 76] [P 66] Symmetrical grooves produce non-chaotic flows; at the optimized asymmetrical groove geometry most of the cross-sectional area is filled with chaotic flow [44].

Groove depth

[M 77] [P 67] Lateral transport is enhanced on increasing the groove depth at constant channel depth, as evidenced using a four-groove structure angled at 45° (fluid transport given by electroosmotic field) [156]. For groove depths exceeding 50 μm (at a channel depth of 50 μm), no additional impact is given, however; the mixing effect is constant from hereon.

This behavior can be correlated with the magnitude of the *z*-component of the electric field, E_z (see Figure 1.157) [156]. This increases with groove depth and is spatially maximal at the ends of the slanted grooves.

Groove angle

[M 77] [P 67] The angle of the grooves, relative to the channel long axis, was varied from 15 to 90° for a four-grooved structure (fluid transport given by electroosmotic field; 50 μm deep grooves) [156]. The concentration profiles yielded by exposing the flow to 90° perpendicular grooves show no effect of lateral mixing, as expected. The bi-laminated pattern remains nearly undisturbed. For all other angled grooves, lateral mass transport is evident. The smaller the angle, the larger is the effect. This is explained by electric-potential plots as a function of the relevant structural dimensions. As the groove angle decreases, the difference in potential between the grooves' ends increases. This leads to a larger electric field and to enhanced electroosmotic flow.

The disadvantage of small-angled grooves, however, is that they consume a much larger footprint area so that for practical solutions a compromise between lateral-mixing capacity and area of the mixing element may has to made [156].

Fluid folding vs. fluid stretching

[M 77] [P 67] For groove angles as small as 15°, fluid folding, finally leading to recirculation patterns, is observed, whereas at larger angles only fluid stretching occurs (see Figure 1.158 and *Groove angle*, above) [156].

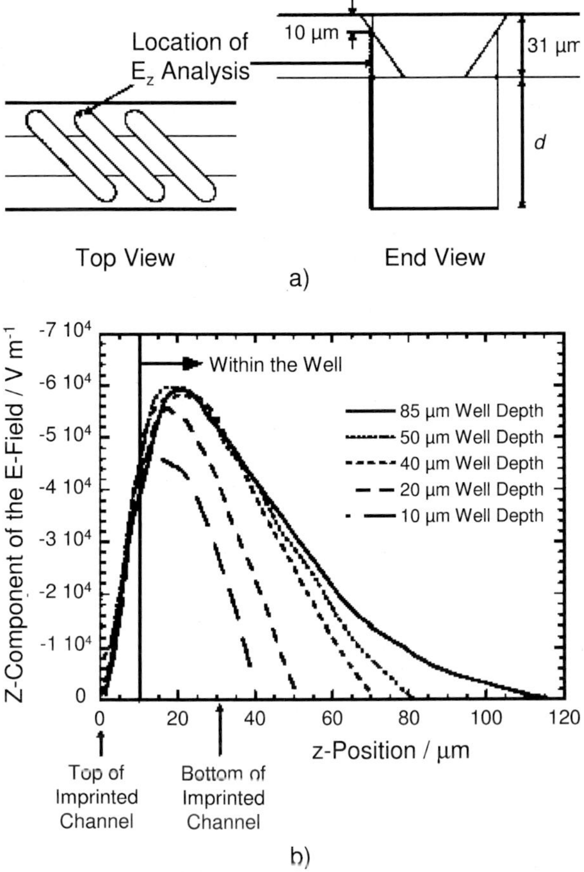

Figure 1.157 Analysis of the magnitude of the z-component of the electric field E_z. (a) Depiction of the location of the analysis; (b) E_z as a function of the depth of the grooves for various absolute groove depths [156] (by courtesy of RSC).

Wall EO mobility variation

[M 77] [P 67] The wall EO mobility is dependent on the manufacturing process of the grooves [156]. It is known, for example, that laser ablated surfaces have larger EO mobility than imprinted surfaces. For different ratios of the groove-(laser ablated)-to-channel-(imprinted) mobilities, their effect on the concentration profiles was investigated. Only at ratios of about 3.0 is enhanced lateral transport observed, evident from the presence of fluid folding and recirculation instead of fluid stretching only (groove angle, 45°; groove depth, 50 µm). Such high ratios are above the experimental data obtained for the naked surfaces so far; hence surface modifications will be needed (e.g. deposition of polyelectrolyte multi-layers) to increase the ratio further. If this is done, the ratio of the groove-to-channel mobilities will be a second means of increasing lateral transport and hence chaotic mixing under the action of an electroosmotic field, in addition to the variation of the groove angle (see *Groove angle*, above).

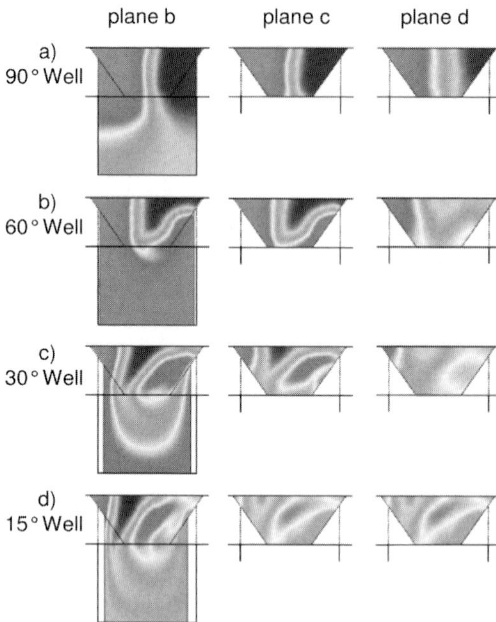

Figure 1.158 Concentration profiles of Rhodamine B imaged on various planes of the four-grooved channels for different groove angles at a constant groove depth of 50 μm [156] (by courtesy of RSC).

Transient mixing of plugs

[M 77] [P 67] When mixing plugs, it is desired to have fast mixing similar to steady flows; however, in addition, the axial dispersion should be minimized, since the corresponding increase in plug length should be kept minimal [156].

As to be expected, the axial dispersion is larger at increased groove depth, because the grooves somehow act like dead zones and the residence time within the grooves is longer (groove angle: 15°) [156]. Too shallow grooves give rise to insufficient lateral mixing.

Concentration profiles as a function of time, similar to time–age residence time plots, give quantitative information on the degree of axial dispersion (see Figure 1.159) (groove angle, 15°; groove depth, 20 μm) [156]. This was done for ratios of the groove-to-channel mobilities of 1 and 2. The axial dispersion in both cases is low, as a comparison of the curves obtained by guiding flow through grooved channels and a blank channel (without groove internals) reveals. The residence-time plots of the grooved structures are only to a small extent broader (< 10%) as compared with the undisturbed flow. The axial broadening of the flow at a ratio of the groove-to-channel mobilities of 2 is lower, which is explained by the larger lateral flux. It was confirmed by concentration plots that for both grooved-channel structures mixing is effective; the standard deviation of the concentration across the channel is reduced by 72%.

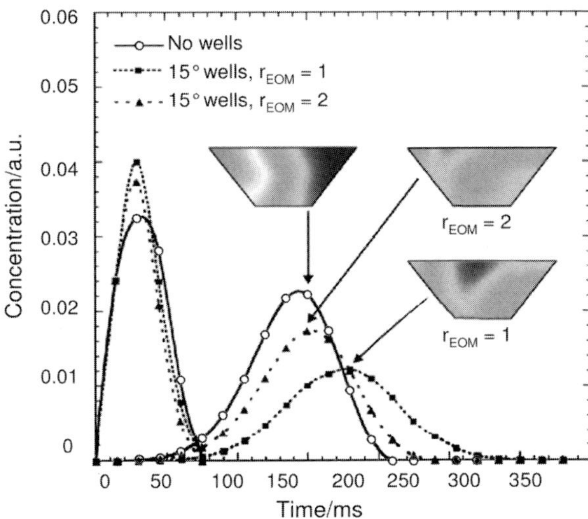

Figure 1.159 Concentration profiles of an injected plug of Rhodamine B traveling over a four-groove structure in a micro channel at two different groove-to-channel mobilities r_{EOM} and for an unstructured channel as reference case. This plot describes the axial broadening of the pulse signal. In addition, cross-sectional concentration profiles are given to analyze the corresponding impact on mixing [156] (by courtesy of RSC).

Twisting flow

[M 75] [P 66] 3-D twisting flow is found for a mixer design with an array of multiple obliquely oriented, straight grooves [44].

Benchmarking of non-grooved, oblique grooved and staggered herringbone designs

[M 74] [M 75] [M 76] [P 66] Confocal fluorescence imaging of the cross-sections of the flow in the micro channel were used for judging the mixing quality. For a simple channel without any grooves, almost no mixing is detected (see Figure 1.160) [44]. For the mixer design with an array of multiple obliquely oriented, straight grooves incomplete mixing is found, even after a 30 mm flow passage. The staggered herringbone mixer gives good mixing, even at high *Pe* ($9 \cdot 10^5$). The fluorescence images show an increase of fluidic filaments, i.e. of specific interfaces, the more cycles are passed.

Quantified mixing efficiency

[M 75] [M 76] [P 66] The standard deviation of the fluorescence intensity of cross-sectional flow images was taken as a measure of the mixing efficiency (see Figure 1.160) [44]. The SHM mixer performs well over a large range of *Pe*. The mixing length required for 90% mixing increases by less than a factor of 3 from the lowest to the highest *Pe*. The straight-grooved mixer and the non-grooved channel have much reduced performance.

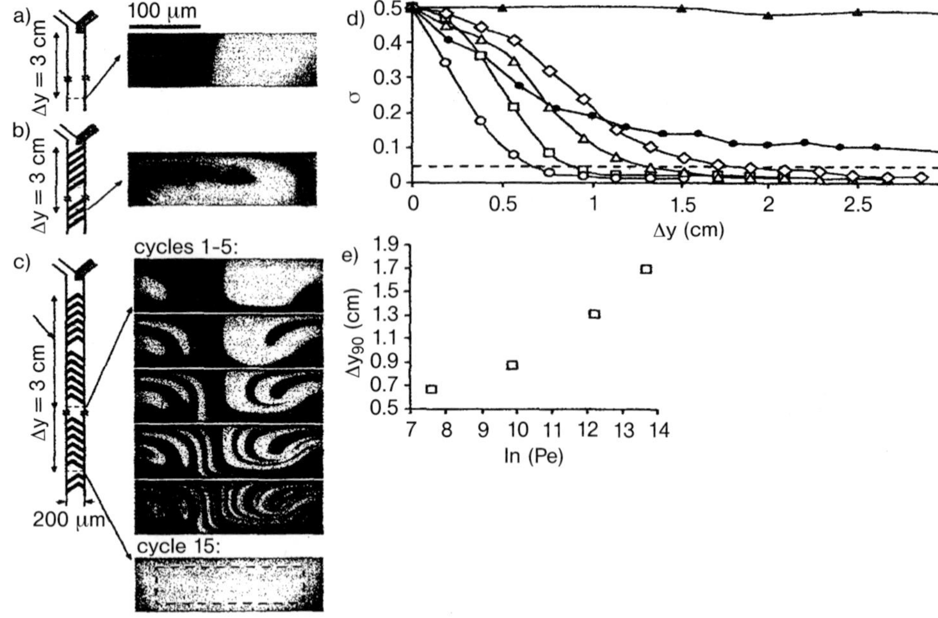

Figure 1.160 (A) Confocal micrograph of the vertical cross-section of the channel for a non-grooved channel. (B) Same type of image for the oblique, straight-groove design. (C) Same type of images for the oblique, asymmetric groove design (staggered herringbone mixer, SHM). (D) Standard deviation, a measure of mixing efficiency, versus mixing length for various *Pe*: SHM (open symbols), (○) $Pe = 2 \cdot 10^3$; (□) $Pe = 2 \cdot 10^4$; (△) $Pe = 2 \cdot 10^5$; (◇) $Pe = 9 \cdot 10^5$; (▲) oblique, straight-rigded channel; (●) non-grooved channel. (E) Mixing length for 90% mixing as a function of *Pe*. For further details of this figure, see [44] (by courtesy of AAAS).

Impact of Peclet number

[M 76] [P 66] The mixing length for the staggered herringbone mixer increases linearly with ln(*Pe*) (see Figure 1.160) [44].

Mixing lengths for a real-case example

[M 74] [M 76] [P 66] The mixing length for different micro channels was calculated for the mixing of a protein solution in aqueous buffer (molecular weight 10^5; diffusion constant 10^{-6} cm^2 s^{-1}; velocity 1 cm s^{-1}) [44]. For $Pe = 10^4$, a simple, non-grooved channel requires 100 cm for mixing completion, whereas 1 cm is sufficient for the SHM. If the velocity is increased by a factor of 10 ($Pe = 10^4$), the mixing length of the SHM device amounts to 1.5 cm. A non-grooved channel will need 10 m for the same mixing task.

Axial dispersion

[M 74] [M 76] [P 66] Axial dispersion, an important parameter, e.g., in pressure-driven flows in liquid chromatography, is significantly reduced for the SHM-type

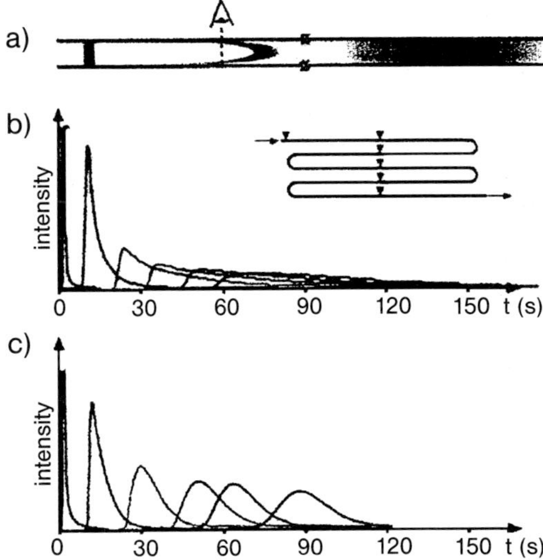

Figure 1.161 Axial dispersion with and without SHM. (a) Schematic showing the dispersion of a plug in Poiseuille flow. (b) Unstirred Poiseuille flow in a non-grooved channel (depth, 70 μm; width, 200 μm; $Pe \approx 10^4$). (c) Stirred Poiseuille flow in a staggered herringbone mixer (SHM) of the same design [44] (by courtesy of AAAS).

mixing as compared with non-grooved flows (see Figure 1.161) [44]. In addition, the dispersion curves for the non-grooved flow are notably asymmetric, which is due to a low-dispersed central flow and a highly dispersed flow close to the walls caused by shear forces. In contrast, the flow in the SHM device is initially asymmetric, but becomes symmetrical the more groove arrays are passed.

Simulation of helical flows for channel with oblique ridges – relative transverse velocity *[M 74] [M 76] [P 66]* The flow of a channel with a structured bottom wall with dimensions corresponding to the oblique-ridge structured channel examined experimentally [44, 45] was simulated. For the twisted, helical flow, the ratio of the average velocities in the y- and x-directions was calculated as function of the vertical position in the channel (z-coordinate). The relative transverse velocity exhibits an almost linear increase inside the grooves, a steep increase above the grooves, changes sign and flattens from about half the channel height to the top wall, there approaching a maximum. The maximum value of 0.052 yielded by the simulation is in good accord with the experimental result of 0.06 [44, 45].

Simulation of helical flows for channel with oblique ridges – relative structure height *[M 76] [P 66]* The relative structure height was varied from 0.01 to 0.76 [44]. The corresponding maximum relative transverse velocities calculated, i.e. the boundary values at the top, correspond well to the experimental values of α. Also, good agreement between the numerical results and an analytical approximation was achieved.

Simulation of helical flows for channel with oblique ridges – single- and double-sided structured channels

[M 76] [P 66] The relative transverse velocities of single- and double-sided structured channels were compared, each having oblique ridges [44]. For the double-sided structured channel, a symmetric curve with a point of reflection was obtained, as expected. Both curves almost coincide except for the regions close to the top of the channel. Owing to the correlation of the absolute transverse velocity and the increase in the interfacial area, an improved mixing performance when using double-sided structured channels can be predicted compared with solely single-sided channels.

This is also evident when considering the convective entanglement of two initially vertical liquid lamellae. For the double-sided structured channel, a considerable increase in the lamellae entanglement and an enlarged interfacial surface area are observed.

Simulation of helical flows for channel with oblique ridges – Reynolds number flow regime

[M 76] [P 66] The relative transverse velocity depends only vaguely on the Reynolds number for oblique ridged channels [44]. This is surprising at first sight, since the absolute transverse velocity within the oblique ridges is considerably enhanced in this way.

Investigations on band broadening and mixing

[No details on the mixer] [no protocol] The mixing of a fluorescent dye with a transparent solution was performed in a mixer with alternate arranged oblique ridges to obtain data on band broadening and mixing [158]. The ridges were manufactured by excimer-laser ablation in a polymer. Fluorescence intensity profiles were thus derived and compared with results obtained for an equivalent channel structure without the ridges and with theoretically derived perfect mixing. The measured profiles were close to the perfect mixing and very different from those obtained without ridge mixing, which resembled bi-laminated structures. Hence, the results are in line with the findings reported above.

1.3.22 Chaotic Mixing by Twisted Surfaces

Most Relevant Citations

Peer-reviewed journals: [46].

Chaotic particle motion can occur for two-dimensional velocity fields which are time dependent and for three-dimensional velocity fields [46]. For induction of such complex fields, one option is to guide the flow close to structured surfaces, e.g. placed on a channel's bottom, starting from a T-type flow inlet configurations.

1.3.22.1 Mixer 79 [M 79]: Twisted Surface Micro Mixer

Four T-type mixer designs were investigated which differ in the type of structuring of the mixing channel of the T-structure (see Figure 1.162) [46]. As a reference

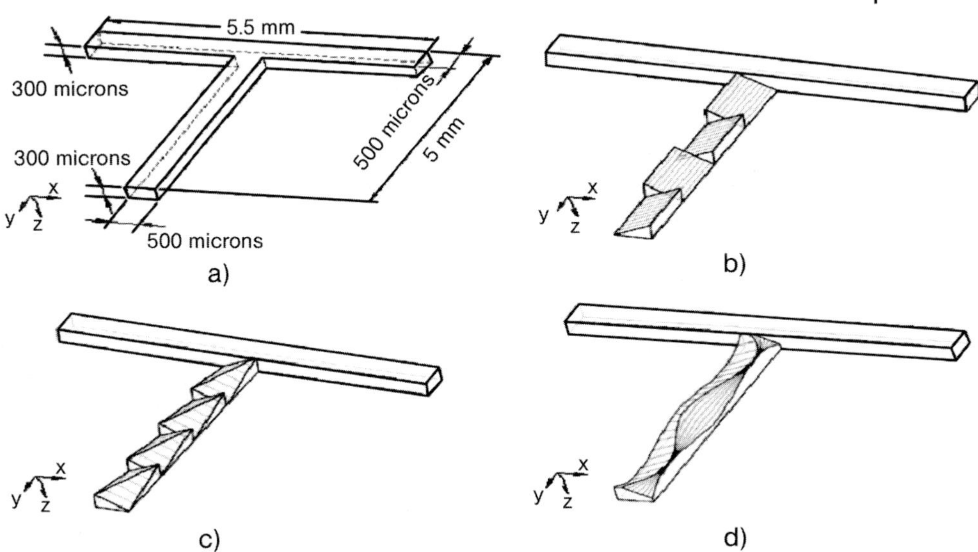

Figure 1.162 Design of four T-type mixers, three of them containing twisted microstructured internals [46] (by courtesy of RSC).

case, a channel without any internals was used. Three mixer designs with twisted internals were developed, with inclined, oblique and wave-like structures on the channel's surface. The idea is that by interaction with the flow, chaotic regimes are induced passively. For this reason, the structures are arranged in an alternately periodic manner so that the flow repeatedly sways around the structures.

Mixer type	Twisted surface micro mixer	Feed channel width, depth, length	300 μm, 500 μm, 5.5 mm (whole T with two feeds)
Mixer material	Simulation study only Polymers such as PMMA and PDMS may be suitable	Mixing channel width, depth, length	300 μm, 500 μm, 5 mm

1.3.22.2 **Mixing Characterization Protocols/Simulation**

[P 69] Gas mixing by diffusion and convection was investigated. Simulations were performed by CFD-ACE software on a personal computer [46]. The finite-element method and three-dimensional unstructured grids were used for solving the governing equations, which are the continuity equation, momentum conservation (Navier–Stokes) equations and species convection–diffusion equation. Laminar flow and adiabatic conditions were assumed. The SIMPLEC method was adopted for pressure–velocity coupling. The total number of elements ranged from 9000 to 16 000 depending on the complexity of the internal structure investigated.

1.3.22.3 Typical Results

Mixing mechanism – diffusion vs. convection

[M 79] [P 69] By calculation of the Peclet number (*Pe*), the ratio of mass transport by convection relative to diffusion was analyzed [46]. For the range of velocities investigated (0.5–2.5 m s^{-1}), *Pe* values > 2 were found, which is indicative of the dominance of convection for all simulations carried out that are described below. For instance, $Pe = 6.75$ is obtained for a velocity of 0.5 m s^{-1}.

T-type mixing without internals – impact of velocity

[M 79] [P 69] Mass contour fraction plots within the T-type mixer design for oxygen mixing at different inlet velocities were calculated [46]. At an inlet velocity of 0.5 m s^{-1}, mixing is completed after about two-thirds of the flow passage. Only a low degree of mixing, and virtually bi-laminated streams, are found for an inlet velocity of 2.5 m s^{-1}.

T-type mixing with internals – mass contour fractions

[M 79] [P 69] Mass contour fraction plots were simulated for gas mixing in three T-type mixer designs with internals (four sub-sections) at an inlet velocity of 2.0 m s^{-1} [46]. It turned out that the inclined design gave the best results.

A mixing length of 5.64 mm for 99% mixing was found for the inclined mixer, whereas a length of 7.90 mm was required for an unstructured channel [46]. Hence a positive effect for enhancing mixing is given, albeit the absolute magnitude is not very large, being of the order of 30%.

This is in line with tracks of the fluid flow giving the sway of the flow [46]. The largest amplitude is given for the inclined mixer, hence 'disturbing' to the largest extent the flow which is expected to result in improved mixing.

Mixing with the inclined mixer with various sub-sections – mass contour fractions

[M 79] [P 69] Mass contour fraction plots were simulated for gas mixing in the inclined T-type mixer design with various numbers of sub-sections (2–12) at an inlet velocity of 2.0 m s^{-1} and at a fixed mixing channel length of 4.5 mm [46]. A plot of the mixing length (for 99% mixing) as a function of the number of sub-sections exhibits a minimum-type dependence (see Table 1.7). The fastest mixing

Table 1.7 Calculated mixing lengths for the inclined mixer with different numbers of sub-sections (at an inlet velocity of 2.0 m s^{-1}) [46].

Number of sub-sections at fixed length of 4.5 mm	*Length per section (mm)*	*Mixing length (mm)*
2	2.2500	6.44
4	1.1250	5.62
6	0.7500	5.46
8	0.5625	5.64
10	0.4500	5.71
12	0.3750	6.26

is obtained for 6 sub-sections. The difference in mixing lengths between the best and the worst structures amounts to about 15%. For more than six sub-sections, the structures obviously stand too close to have an impact on the flow, i.e. the structuring does not result in swaying any longer.

1.3.23 Chaotic Mixing by Barrier and Groove Integration

Most Relevant Citations

Peer-reviewed journals: [3, 58].

The 'chaos screw' is a mixing device developed for chaotic mixing for polymer extrusion processes [58]. The channel for the molten polymer is defined by the interstice between the extruder shell and the screw root, two adjacent flights, and a barrel wall, similar to a flight, but of smaller height. The device has macroscopic dimensions and operates at $Re << 1$. The barriers are so placed that barrier-free zones are followed by barrier-containing zones; the flow field consequently changes. By use of the barrier structures, a hyperbolic point is caused in the flow field. The flow field changes depending on the location along the down-channel direction in an alternating manner, creating chaotic flow.

Based on these considerations, a planar version of the 'chaos screw' was designed with micron-sized dimensions, the barrier embedded micro mixer [3, 58].

1.3.23.1 Mixer 80 [M 80]: Barrier-embedded Micro Mixer with Slanted Grooves

In this barrier-embedded micro mixer, oblique grooves, placed at the channel's floor, and barrier structures at the channel ceiling replace the flights and barriers of the 'chaos screw' [58]. The placement of the structures, however, is not essential for the functions. The positions at the floor and ceilings may change without loss of performance; two-sided structures are expected to exhibit improved mixing over single-sided structures. The barriers are placed periodically so that barrier-free zones intersect.

In the no-barrier zone, the cross-sectional flow field helical flow is induced by the grooves and shows non-linear rotation with only one elliptic point [58]. In the barrier zone, a spatially periodic perturbation on the helical flow is imposed and thereby two co-rotating flows form, characterized by a hyperbolic point and two elliptic points. By periodic change of the two flow fields, a chaotic flow can be generated.

This passive chaotic mixing concept is achieved with only minor modification of the flow channel, i.e. placement of only small obstacles, so that a low pressure loss for a given mixing task is expected [58].

The flow field of the barrier-embedded micro mixer has one hyperbolic point which considerably increases the specific interfacial fluid area, leading to rapid mixing by stretching and folding [58]. This is different to the staggered herringbone micro mixer with oblique grooves changing alternately the angle of tilting. Here, counter-rotating flows having one parabolic point with a lesser degree of interfacial area gain.

The mixer is composed of two plates. Two SU-8 masters were fabricated by conventional photolithography: one for the bottom channel with the slanted grooves by two-step photolithography and the other for the top with the barrier structures by one-step photolithography [58]. The photolithographic processes can be made simpler when both grooves and barriers are put on the bottom plate. To prevent irreversible bonding of the PDMS replicas, having plasma-oxygen treated surfaces, before proper alignment, methanol was used as a surfactant.

Mixer type	Barrier-embedded micro mixer with slanted grooves	Groove tilt angle	45°
Mixer material	Polydimethyl-siloxane	Mixing channel width, depth, length	240 μm, 60 μm, 21 mm
Groove depth, length	100 μm, 9 μm	Barrier width, height	30 μm, 40 μm

For comparison, a simply slanted grooved micro channel (without barriers) and a T-shaped channel (without grooves and barriers) were fabricated using the same technologies, but with fewer steps for the photolithography [58].

1.3.23.2 **Mixer 81 [M 81]: Barrier-embedded Micro Mixer with Helical Elements**

This barrier-embedded micro mixer was made with helical static elements, resembling the internals of a Kenics mixer, and barrier structures at the channel ceiling [3].

Fluid flow through macroscopic helical static elements (see, e.g., [4, 64] for static mixers) has been well studied by many groups (see, e.g., [155]). In previous studies miniaturized helical elements (without barriers) were realized [2]. Here, their mixing performance was inferior to that with intersecting static elements, resembling Sulzer packings. However, the latter reduced considerably the throughput by high-pressure drop. Hence, it was envisaged to develop a mixer with helical static mixing elements with good mixing efficiency and reasonable throughput. For this reason, the helical elements were equipped with barriers.

Mixer type	Barrier-embedded micro mixer with helical elements	Number of helical elements	6
Mixer material	Photopolymer resin SL-5410	Number of barriers per helical element	2
Total number of layers for the stereolithography	272	Mixing channel diameter, length	1 mm, 9.4 mm
Layer thickness mixer part	40 μm	Barrier thickness, height, length	100 μm, 150 μm, 413 μm
Helical element thickness, length	100 μm, ~1.64 mm	Cross-section of square inlets	1000 μm × 1000 μm

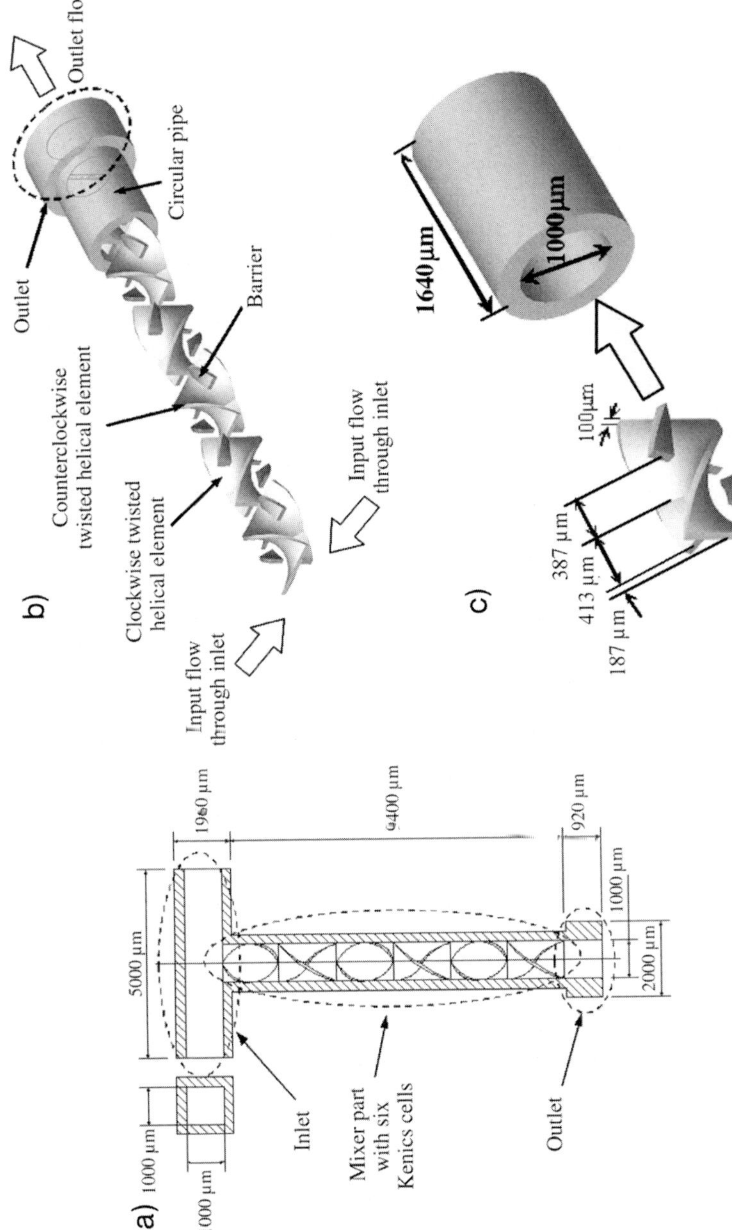

Figure 1.163 Schematics of (a) the top view of the Kenics micro mixer with detailed dimensions, (b) the 3-D view of the barrier-embedded micro mixer and (c) the 3-D view of one mixing element with detailed dimensions [3] (by courtesy of IOP Publishing Ltd.).

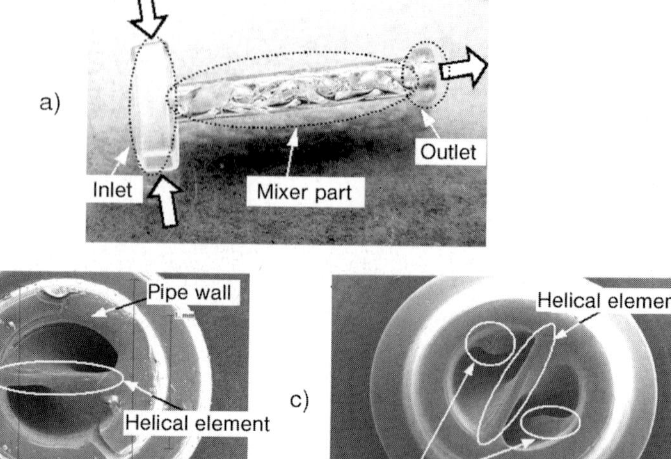

Figure 1.164 (a) Complete view of the barrier-embedded helical micro mixer; (b) Kenics cell with a helical element only; (c) Kenics cell with a helical element and two barriers [3] (by courtesy of IOP Publishing Ltd.).

The static elements were internals of a circular T-pipe. The mixer realized had Kenics cells with helical elements twisted by 180°, giving a fully 3-D structure (see Figure 1.163) [3]. Two barriers were placed on the wall of the pipe. Counterclockwise- and clockwise-rotated helical elements were alternately arranged. Four barriers were introduced in the pipe wall in each Kenics cell. For comparison, pipes with only helical elements (no barriers) and empty pipes (no internals) were also fabricated.

Microfabrication was achieved by micro-stereolithography using a CW Ar^+ laser (see Figure 1.164) [3].

1.3.23.3 Mixing Characterization Protocols/Simulation

[P 70] The mixing performance along the channel length was characterized by a pH-indicator reaction using phenolphthalein (0.31 mol l^{-1}) and NaOH (0.33 mol l^{-1}) in 99% ethanol solution [58]. Images were taken using a stereoscopic microscope and a graphic grabber system. A white lamp served as illumination system.

As a second imaging approach, confocal laser scanning microscopy was applied to monitor the cross-sectional mixing of Rhodamine B solutions (99% ethanol) and pure 99% ethanol [58]. Laser scanning over the entire cross-sectional area was performed and at various locations along the channel.

1.3.23.4 Typical Results

Reaction imaging of mixing along the channel passage

[M 80] [P 70] The mixing performance of three mixing devices was characterized by a pH-indicator reaction using phenolphthalein: a T-channel, slanted grooved

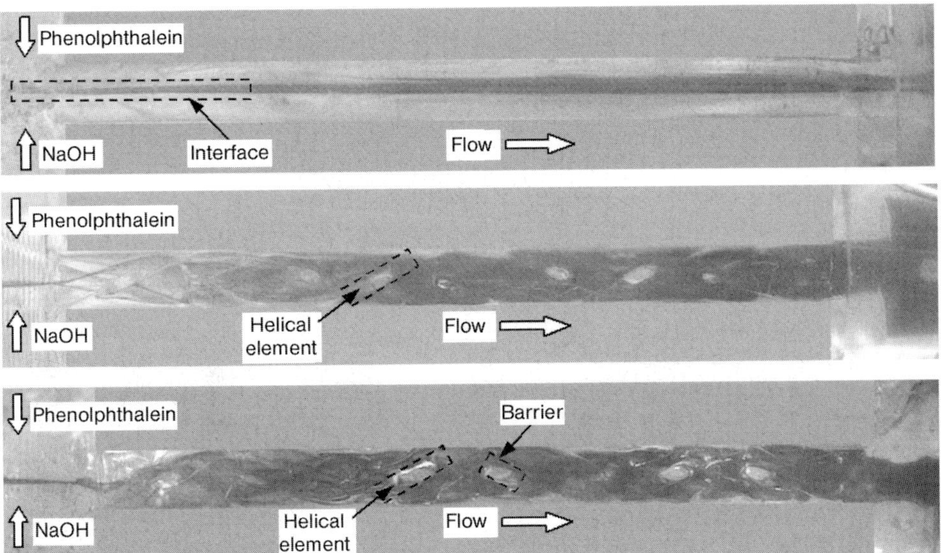

Figure 1.165 Images of the color formation as measure for mixing using a reactive visualization approach. (a) Circular pipe; (b) helical micro mixer; (c) barrier-embedded helical micro mixer [3] (by courtesy of IOP Publishing Ltd.).

and barrier-embedded micro mixer [58]. The color formation of the indicator was imaged at three locations, referring to the start of contacting the solutions, the center of the mixing channel and its end.

The T-shaped mixer yields color formation only at the interface of the bi-laminated flow which becomes clearly visible at about half the channel length [58]. The other two mixers show already at the beginning of the channel notable mixing, which then results in a fairly homogeneous color after half of the channel passage (which does not necessarily mean approaching 100% mixing). From direct visual observation, albeit not evident in the images, it is stated that the barrier-embedded micro mixer has a deeper color formation than the slanted grooved one, i.e. potentially may have a higher degree of mixing.

[M 81] [P 70] A microscopy-image analysis of the color formation due to a reactive approach reveals that the micro mixer with helical elements and barriers gives a better performance than the micro mixer with helical elements but without barriers and a reference pipe structure without either helical elements or barriers (see Figure 1.167) [3]. The pipe gives the expected profile with two colorless fluid compartments on top and at bottom, separated by a colored interface. This is indicative of the absence of any swirling, secondary flow.

Quantification of the degree of mixing

[M 80] [P 70] The intensity of the color in the channel was monitored for quantitative analysis of the degree of mixing [58]. Images were taken over 11 regions and a normalized intensity was derived.

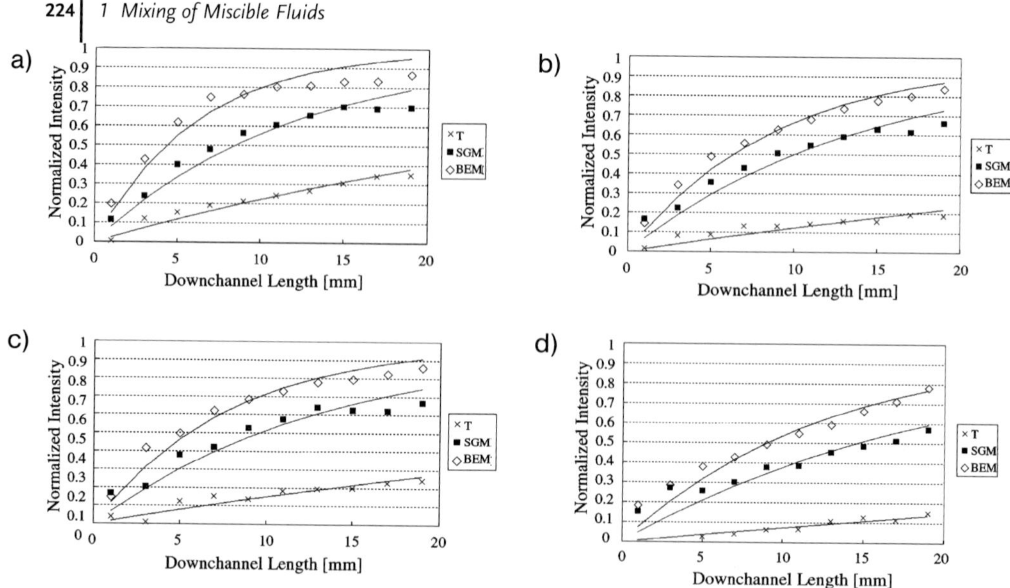

Figure 1.166 Normalized average intensity changes along the down-channel direction of three micro-mixer devices with barrier-embedded, slanted grooved and T-channel structures at (a) $Re = 0.228$, (b) 0.457, (c) 0.685 and (d) 2.28 [58] (by courtesy of IOP Publishing Ltd.).

A plot of the normalized intensity versus the down-channel length clearly shows that the mixing efficiencies of the devices have the following sequence: barrier embedded mixer > slanted grooved mixer >> T-channel mixer (see Figure 1.166) [58]. When operating at larger Reynolds numbers (up to 2.28), the residence time is decreased, since operation is done at constant length. Hence, the measured mixing performances are lower compared with the initial experiment at $Re = 0.228$. The relative degree of mixing performance is lower for the T-channel mixer than for the other two grooved mixers. In the latter case, the helical flow circulation is intensified at higher *Re*, partly compensating the effect of the reduction of the residence time.

[M 81] [P 70] A plot of the normalized intensity versus the down-channel length clearly shows that the mixing efficiencies of the devices have the following sequence: barrier-embedded helical mixer > helical mixer >> pipe mixer (see also Figure 1.165) [3]. The relative degree of mixing performance is lower for the pipe mixer than for the other two helical mixers. This is most evident for the three plots with $Re \geq 13.96$.

Cost for enhancing mixing performance by barrier structures

[M 80] [P 70] From prior experience with the 'chaotic screw' extruder, it was estimated that the presence of the barrier structures in the barrier-embedded mixer accounts for about a 10% reduction in flow rate as compared with the slanted grooved mixer [58]. Thus, a large mixing improvement is given at the cost of only a small reduction in flow rate.

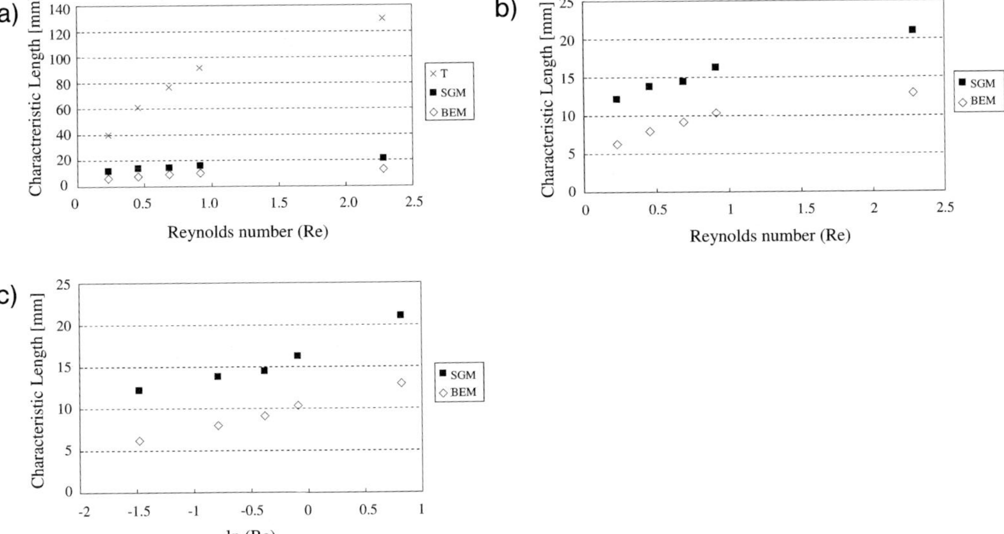

Figure 1.167 (a) Characteristic mixing length of all three micro devices and (b) detailed view of the mixing length of the barrier embedded and slanted grooved micro mixers only, as a function of the flow regime. (c) Ln(*Re*) plot of the latter data set [58] (by courtesy of IOP Publishing Ltd.).

Characteristic mixing length

[M 80] [P 70] A characteristic mixing length was defined [58]. Such a mixing length is two times shorter for the barrier-embedded mixer than the slanted grooved mixer for the whole flow range observed (see Figure 1.167). The mixing length is about 4–10 times shorter than for the T-channel mixer. The T-device exhibits a strong increase in the mixing length with increasing *Re*, as this means a reduction in the residence time. The slanted grooved and barrier-embedded mixers, in contrast, exhibit only a slightly increasing mixing length for *Re* ranging from 0.228 to 2.28. Actually, a logarithmic increase in mixing length with *Re* is observed.

[M 81] [P 70] The characteristic mixing length is much shorter for the barrier-embedded helical mixer and the helical mixer than the pipe mixer [3]. The barrier-embedded helical mixer is slightly better than the helical mixer; hence, the effect of the barriers isevident, albeit not strong. The characteristic mixing length of 2–3 mm of the barrier-embedded helical mixer increases only slightly Reynolds numbers ranging from 7 to 28.

Cross-sectional mixing

[M 80] [P 70] Cross-sectional images of the flow patterns, gained by confocal microscopy, demonstrate the interfacial stretching and hence are indicative of mixing in the barrier-embedded mixer [58]. At the non-barrier entrance structure clearly the helical flow induced by the slanted grooves is evident. In the barrier

section, the flow rotation is further clearly improved. At the exit the image is homogeneous, which is indicative of at least large fluid segregation, if not demonstrating completion of mixing.

1.3.24 Distributive Mixing by Cross-sectional Confining and Enlargement

Most Relevant Citations

Proceedings contributions: [159].

The mixer discussed below is part of a ferrofluidic microsystem [159]. Ferrofluids can be used for fluid displacement by magnetic action using an external permanent magnet source. Ferrofluids are superparamagnetic liquids.

1.3.24.1 Mixer 82 [M 82]: Distributive Micro Mixer with Varying Flow Restriction

Two micro mixer designs were developed, both containing confining and expanding structures for distributive mixing by interface enlargement [159]. The confining and expanding structure can be the micro channel itself, which repeatedly changes its cross-section or internals such as columns which are flow obstacles.

The first design has a slightly curved serpentine channel comprising square columns. The second design is an arrow-shaped channel [159]. Circular holes, two before and one after the mixing channel, serve as fluid reservoirs for both designs.

The mixers were fabricated by deep reactive ion etching (DRIE) into silicon [159]. The silicon structure was anodically bonded to a glass wafer.

Mixer type	Distributive micro mixer with varying flow restriction	Mixing channel width, depth, length	1000 μm, 30 μm, ~8 mm
Mixer material	Silicon	Feed channel width, depth	400 μm, 30 μm
Top plate material	Glass	Reservoir diameter (fluids to be mixed)	1.5 mm
Top plate thickness	0.5 mm	Reservoir diameter (mixed fluids)	2 mm

1.3.24.2 Mixing Characterization Protocols/Simulation

[P 71] For fluid transport, ferrofluid magnetic displacement by using a commercial hydrocarbon-based ferrofluid was applied [159]. This fluid is compatible with several aqueous dyed solutions. Miniature permanent magnets were used.

The mixing of dyed fluids was observed by microscopy [159].

1.3.24.3 Typical Results

Mixing pattern observation

[M 82] [P 71] Observations on the course of mixing for two mixer designs were made by imaging the complete channel with flow-confining structures [159]. The

solutions to be mixed are moved by ferrofluid displacement using an external magnet. No details were given; only the finding of the onset of mixing was stated.

1.3.25
Time-pulsing Mixing

Most Relevant Citations

Peer-reviewed journals: [26]; proceedings contributions: [48].

Time pulsing can be achieved by simply varying the flow rates of the streams in two inlets of a mixer [26] or by using more intricate flow structures such as the superposition of unsteady cross-flow injection and a steady flow in a main channel [48].

By simply varying the flow rates in the inlet channels periodically with time, mixing can be enhanced, without any need for further moving parts or complex fluid architectures (see Figure 1.168) [26]. A low-frequency sinusoidal flow rate is superimposed upon a steady flow rate. For steady-state inlet velocities of 1.0 mm s^{-1}, time-dependent inlet velocities of the form 1.0 + 7.5 sin $(5 \cdot 2 \pi t)$ mm s^{-1} were used, where t is being the time.

Time pulsing can be characterized by the Strouhal number (St), which is a dimensionless parameter describing the ratio of the flow characteristic time scale (L/V) to the pulse time period ($1/f$) [Eq. (1.8)] [26], where L is the hydraulic diameter, V the average velocity in the outlet channel and f the pulsing frequency. For a pulsing frequency of 5 Hz, a Strouhal number of 0.375 derives.

$$\mathrm{St} = (f\,L) \,/\, V = (L/V) \,/\, (1/f) \tag{1.9}$$

Another paper describes the stretching and folding of material lines yielded by simulation and experimental imaging, induced by time-pulsing mixing via unsteady cross-flow injection in a steady-flow main channel [48].

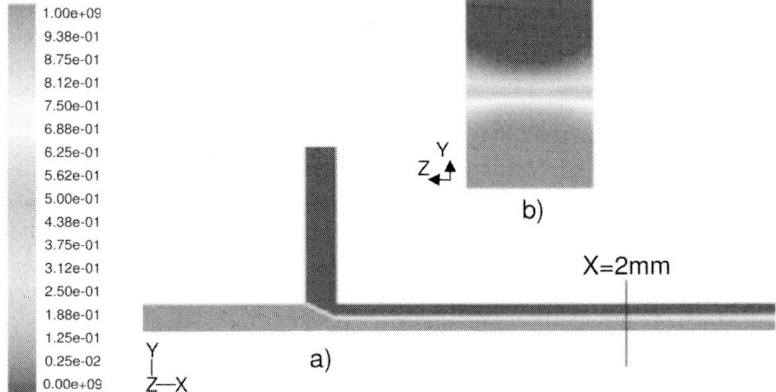

Figure 1.168 Inlet mean velocity as a function of time in the control case (dashed line) and in the biased sinusoidal pulsing case (solid line) [26] (by courtesy of RSC).

1.3.25.1 Mixer 83 [M 83]: Time-pulsing Cross-flow Micro Mixer (I)

[M 83] This time-pulsing device comprises a cross-flow channel arrangement with three channel branches, specifically consisting of one main channel on to which a perpendicular channel is so attached that a smaller part of the main channel serves as inlet and the larger residual part is the mixing flow-through chamber [26]. The perpendicular channel then has the function of a second inlet. Both the inlet and outlet channels have the same size. These dimensions were assumed to be typical for many μTAS applications and are also thought to be amenable to mass production, e.g. by means of injection molding. Furthermore, devices made in this way can be fed by pumping, both mechanically and electrically. Both options can undergo periodic switching of the feed flows.

A numerically controlled mill fabricated trenches and through holes at the ends of the trenches in a thin, colorless, polyacrylic acid plate [26]. This thin plate was bonded to a more rigid, transparent polycarbonate base plate with UV-curing optical adhesive. The through holes in the plate were connected to holes in the base plate, thereby forming one conduit. The latter holes did not penetrate the base plate, but rather were connected to cross-type arranged borings. Thereby, fluid connections could be attached to the side of the base plate.

A glass cover slip bonded to the acrylic plate enclosed the trenches, forming in this way the micro channels [26]. (18 and 25 gage) Syringe needles were light press fitted into these cross-holes. Bonding the joints with epoxy ensured good seals and mechanical integrity.

Mixer type	Time pulsing cross-flow mixer	Channel width, depth	200 μm, 120 μm
Mixer material	Polyacrylic acid/glass	Channel length before, after confluence	1.25 mm, 3 mm

1.3.25.2 Mixer 84 [M 84]: Time-pulsing Cross-flow Micro Mixer (II)

This device consists simply of a T-type junction where two pressure-driven flows are contacted (see Figure 1.169) [48]. After a short passage, such a bi-laminated stream is exposed to injections from both sides by adjacent channels, which yields a cross-flow configuration. From there on, the injected flow passes a long main channel and finally reaches a reservoir.

The switching-on of the transverse flow leads to bending of the material line when the flow passes the intersection at the cross-flow configuration (see Figure 1.170) [48]. After switching off of the transverse flow, folding of the material line takes place owing to the parabolic flow profile. This enhances the interfacial area and speeds up mixing, in a traditional way. Moreover, chaotic flow can be achieved by multiple repetition of the transverse injection structures and hence multiple stretching and folding.

The micro mixer was fabricated from two plates by standard MEMS technology, using deep reactive ion etching (DRIE) [48]. Anodic bonding is used for sealing the plates.

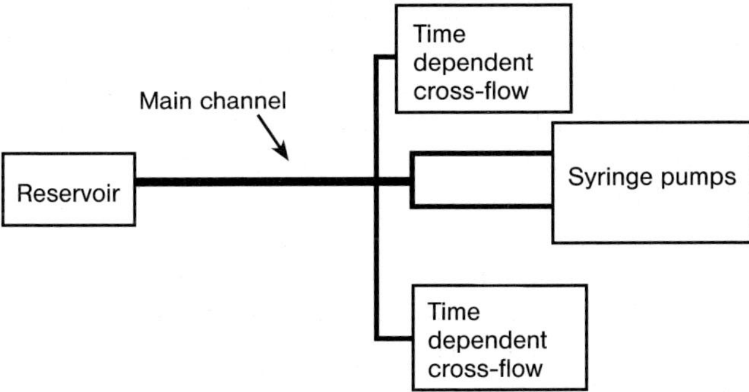

Figure 1.169 Schematic of the time pulsing cross-flow mixer [48].

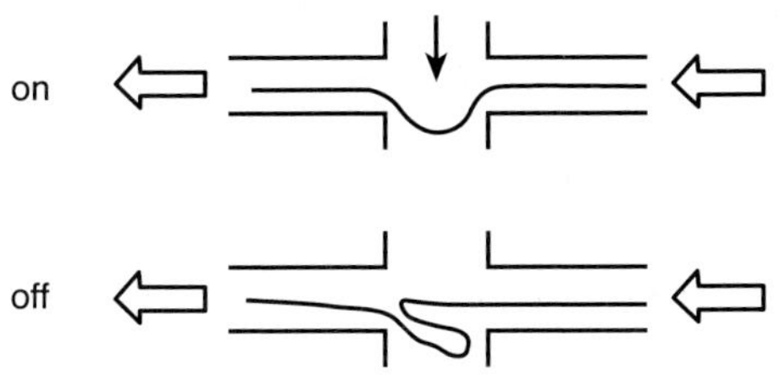

Figure 1.170 Schematic of the working principle of the mixer: stretching and folding of material lines are achieved by periodic perturbation of the main flow with adjacent flows [48] (by courtesy of Springer-Verlag).

Mixer type	Time pulsing cross-flow mixer	Top plate	Pyrex
Mixer material	Silicon		

No details on the dimensions of the mixer are given [48].

1.3.25.3 **Mixing Characterization Protocols/Simulation**

[P 72] Detailed simulations and two feasibility experiments were carried out [26].

For both the simulations and the experiments, aqueous solutions were fed at room temperature with a time-averaged mean velocity of 1.0 mm s^{-1} ($Re = 0.3$) [26]. For reasons of simplicity, the same solutions were fed through both inlets for the simulations. In the experiments, a dye dissolved in water was mixed with pure water; the nature of the dye was not disclosed. Mass fractions were recorded. From this the weighted standard deviation was calculated, giving the degree of mixing when being normalized by an average value of the mass fraction (for the exact definitions see [26]).

The kinematic viscosity considered in the simulations was that of water. Here, as diffusion constant D a value of 10^{-10} m^2 s^{-1}, a typical value for diffusion of small proteins in aqueous solutions, was taken.

The modeling considered 1.25 mm of the channel before the confluence and 3 mm of the channel after the confluence with a plane of symmetry at half the channel depth. The computational domain was discretized with structured hexahedral meshes, the size of the cells being 10 μm long. Diffusion was modeled.

Tests were conducted with peristaltic pumps [26]. Two signals from –1.2 to +1.2 V DC control the pumping from maximum reverse to maximum forward flow. The control signal comes from the center taps of potentiometers used as voltage dividers. A function generator sends a sinusoidal signal to one end of both voltage dividers and a power supply fixes the other ends of the voltage dividers to specified voltages. Mixing is observed using a standard microscope and images were taken on a color video camera.

[P 73] Syringe pumps were used for the liquid feed of the main and adjacent streams [48]. The injection flow was provided by a sinusoidal mode.

Sucrose or glycerol solutions were contacted with the same solutions containing a fluorescent tracer [48]. Both streams were fed with the same flow rate. Fluorescence microscopy was used for flow monitoring.

A two-dimensional kinematic simulation was performed, assuming Pouseuille flow in both the main and adjacent channels.

1.3.25.4 **Typical Results**

Constant flow rate for both inlets

[M 83] [P 72] When contacting two feeds in the cross-flow mixers with constant inlet flow rates of 1.0 mm s^{-1}, a tri-lamellae flow pattern is generated, as expected ($Re = 0.3$) [26]. Two broader outer lamellae are found separated by a thin, diffuse layer. Both lamellae remain totally unchanged, i.e. contain only the material which they originally possessed, and are thus unmixed.

At the interface a small diffusive region exists where mixing takes place. This region is broader on the top and bottom of the channel, i.e. is smallest at half channel height. This is due to the higher velocity in the center of the channel, reducing the residence time.

The overall degree of mixing for the whole channel is therefore low, being limited to only 12% [26]

Increase in constant flow rate for both inlets

[M 83] [P 72] On increasing the constant inlet flow rate from 1.0 to 8.5 mm s^{-1} (from $Re = 0.3$ to 2.55), a tri-lamellae flow pattern is again generated [26]. Mixing here is even decreased compared with the 1.0 mm s^{-1} case, since residence time is reduced and only diffusion takes place. The degree of mixing is 8% in absolute terms, hence is 34% less than in the 1.0 mm s^{-1} case.

The flow rate of 8.5 mm s^{-1} corresponds to the maximum flow rate induced by period switching of the flow [26].

Pulsed flow for the perpendicular inlet

[M 83] [P 72] The inlet flow rate in one feed channel was set constant at 1.0 mm s^{-1} and the flow rate of the other inlet was varied periodically in time with 1.0 + 7.5 sin ($5 \cdot 2 \pi t$) mm s^{-1} [26]. The flow pattern is again of tri-lamellae type with two outer unmixed and an inner diffuse, mixed layers; however, the diffuse mixed zone is now much larger than in the constant flow case (see Figure 1.171 and *Constant flow rate for both inlets*).

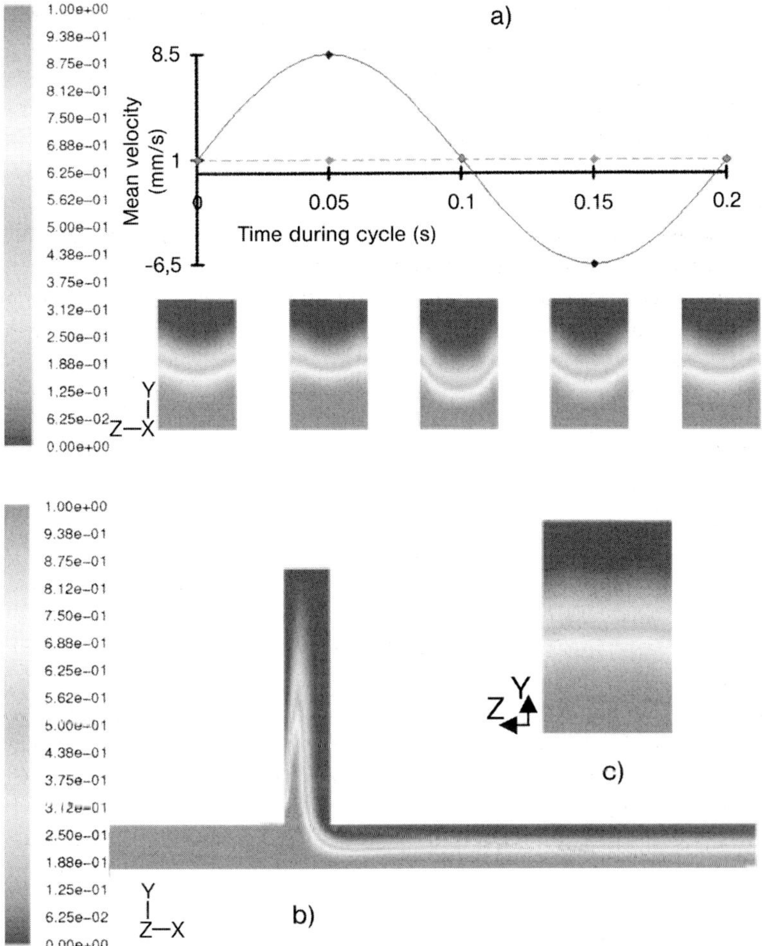

Figure 1.171 Numerical simulation results obtained with a pulsed flow from the perpendicular inlet. (a) Mean velocity as a function of time in the inlet (dashed line) and in the perpendicular inlet (solid line). Contour levels of the mass fraction of one liquid in the *YZ*-plane cross-section taken 0.25 mm downstream of the confluence at various times marked on the previous curves are also given. (b) Contour levels of the mass fraction of one liquid in the *XY*-plane at half the channel depth at the first time in the cycle shown in (a). The channels in (b) are clipped short in the *X*-direction for more compact stacking. (c) Contour levels of the mass fraction of one liquid in the *YZ*-plane cross-section taken 2 mm downstream of the confluence at the first time shown in (a) [26] (by courtesy of RSC).

As a consequence of the time pulsing of one inlet flow rate, one of the feed lamellae and the diffuse zone penetrate also the inlet region of the other fluid, i.e. generating a tri-lamellae peak here. The diffuse interface is now no longer symmetric with regard to the channel width (as for constant flows), but is curved and changes with time. Thereby, material transport is done, the diffuse zone becomes broader and mixing is enhanced. The degree of mixing is now 22%, being 79% larger than for constant flows. As expected, the degree of curvature of the interface is largest near the point of confluence, e.g. as images taken at 0.25 mm evidence, and becomes small downstream. At a 2.0 mm distance from the point of confluence, nearly no time-dependent behavior can be observed any longer.

If twice the amplitude is taken, the degree is unchanged at 22% [26]. If the standard pulsing is done in a channel half as deep, the degree of mixing is again nearly unchanged, 21%.

Variation of amplitude of time pulsing

[M 83] [P 72] If the amplitude for time pulsing is doubled, the degree of mixing remains unchanged, being 22% in both cases [26]. The increase in amplitude amounts to a maximum flow rate of 17.0 mm s^{-1} instead of 8.5 mm s^{-1} used formerly.

Variation of channel depth

[M 83] [P 72] If the channel depth is halved at otherwise the same time pulsing conditions, the degree of mixing remains nearly unchanged, being 22 and 21% for full-channel and half-channel depth, respectively [26].

Pulsed flow from both inlets in-phase

[M 83] [P 72] When pulsing both inlet flow rates in-phase, the degree of mixing is only 19%, being 13% lower than for one-inlet-flow pulsing (see *Pulsed flow for the perpendicular inlet*) [26]. The two fluids are basically flowing side-by-side, albeit moved forwards and backwards. The interface is not stretched extensively by this action.

Pulsed flow from both inlets with 90° phase difference

[M 83] [P 72] When pulsing both inlet flow rates with a phase difference of 90° (amplitude and frequency being the same), the degree of mixing is notably increased to 59%, being more than doubled as compared with one-inlet-flow pulsing (see Figure 1.172 and *Pulsed flow for the perpendicular inlet*) [26]. One inlet flow is operated with $1.0 + 7.5 \sin(5 \cdot 2\pi t)$ mm s^{-1} and the other with $1.0 + 7.5 \sin(5 \cdot 2\pi t + \pi/2)$ mm s^{-1}.

The flow pattern is again of lamellae type with a much more pronounced degree of diffuse, mixed layers (see *Constant flow rate for both inlets*) [26]. As a consequence of the time pulsing of both inlet flow rates, both inlet streams also penetrate the other inlet region periodically. The diffuse interface is now twice folded, resulting in a mixed island deeply penetrating into the other liquid. Good mixing occurs already very close to the point of confluence and persists in the outlet channel.

Pulsed flow from both inlets with 180° phase difference

[M 83] [P 72] When pulsing both inlet flow rates with a phase difference of 180° (amplitude and frequency being the same), the degree of mixing is notably increased to 56%, being more than doubled compared with one-inlet-flow pulsing (see Figure 1.173 and *Pulsed flow for the perpendicular inlet*) [26]. One inlet flow is operated

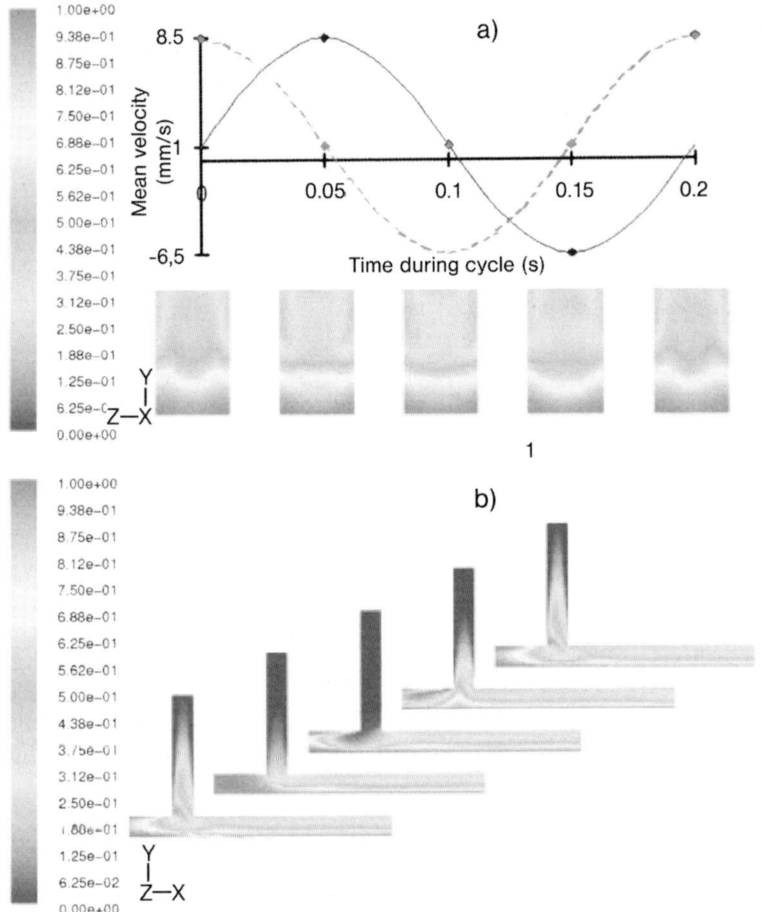

Figure 1.172 Numerical simulation results obtained with a pulsed flow from both inlets, one flow being at 90° phase difference to the other. (a) Mean velocity as a function of time in the inlet (dashed line) and in the perpendicular inlet (solid line). Contour levels of the mass fraction of one liquid in the *YZ*-plane cross-section taken 0.25 mm downstream of the confluence at various times marked on the previous curves are also given. (b) Contour levels of the mass fraction of one liquid in the *XY*-plane at half the channel depth at the first time in the cycle shown in (a). The channels in (b) are clipped short in the *X*-direction for more compact stacking [26] (by courtesy of RSC).

with $1.0 + 7.5 \sin(5 \cdot 2\pi t)$ mm s^{-1} and the other with $1.0 + 7.5 \sin(5 \cdot 2\pi t + \pi)$ mm s^{-1}.

However, the mixing degree is not higher than for 90° phase pulsing, but actually slightly lower [26]. Nonetheless, 180° phase operation has distinct advantages. Only by this operation is a constant, time-invariant outlet flow achieved, which is important when the mixed flow has to enter a downstream microfluidic element.

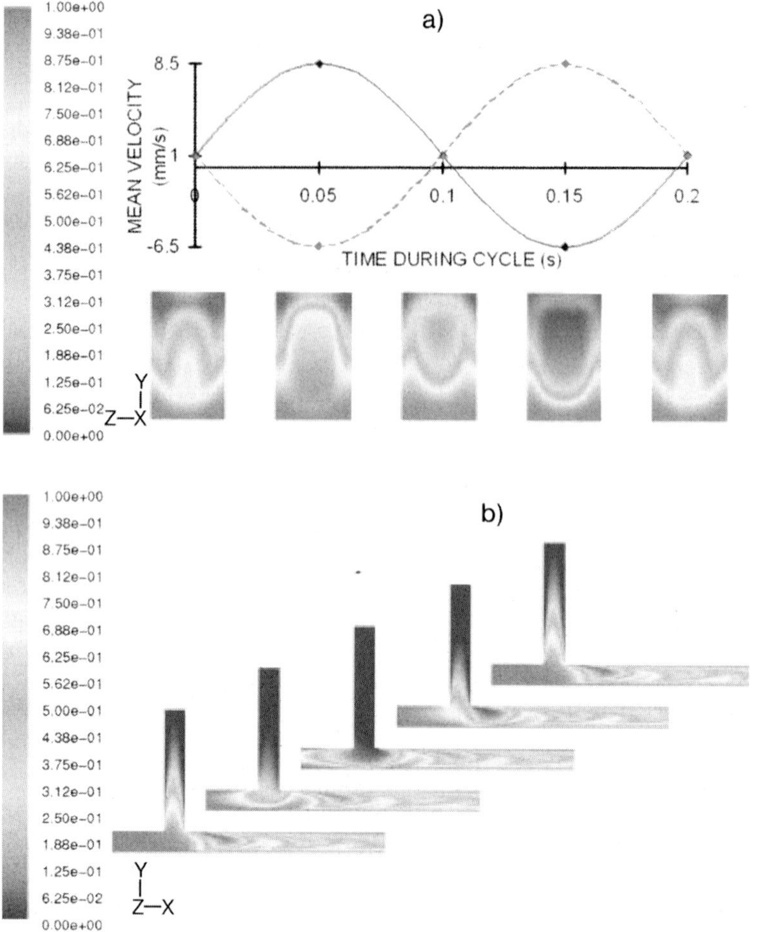

Figure 1.173 Numerical simulation results obtained with a pulsed flow from both inlets, one flow being at 180° phase difference to the other. (a) Mean velocity as a function of time in the inlet (dashed line) and in the perpendicular inlet (solid line). Contour levels of the mass fraction of one liquid in the *YZ*-plane cross-section taken 0.25 mm downstream of the confluence at various times marked on the previous curves are also given. (b) Contour levels of the mass fraction of one liquid in the *XY*-plane at half the channel depth at the first time in the cycle shown in (a). The channels in (b) are clipped short in the *X*-direction for more compact stacking [26] (by courtesy of RSC).

Here, one does not desire variation of the flow rate, as this may affect the performance of this and further microfluidic elements downstream. The individual flow rates are time dependent; owing to their interrelationship the sum is a constant over time.

The flow pattern is no longer of lamellae type (see *Constant flow rate for both inlets*), but rather consists of alternate puffs of the two liquids [26]. Each puff has the form of a hairpin-like structure, since it travels at the velocity of the flow, which depends on the distance from the center of the channel. In this way, hairpins become more elongated the more they travel downstream. The hairpins slowly dissipate as they travel downstream owing to diffusion. The diffuse interface is now significantly folded; the composition of the mixed island, however, varies with time, having a greater content of the one or the other liquid in a periodic manner.

Pulsed flow from both inlets with irrational ratio of frequencies

[M 83] [P 72] When pulsing both inlet flow rates with an irrational ratio of frequencies, a lesser degree of mixing of only 52% compared with the 90° phase time pulsing with the same frequency is obtained (mixing degree 56%) [26]. The different pulsing frequencies are sometimes undesirably close in phase, leading to reduced mixing. In a closed system (not a flow-through system), time pulsing at different frequencies may lead to a different situation, giving rise to improved results.

Comparison experiment with numerical simulation results

[M 83] [P 72] For constant flows at both inlets, the same lamellae-type flow pattern is found [26]. For time pulsing with 180° phase difference, the experimental image also roughly resembles the computer simulation. To have a realistic image made by the latter, several images taken at various depths of the channel were overlaid, since the experimental image consists of optical data gathered along the optical path which is parallel to the channel depth. In particular, the penetration of one liquid into the other is evident.

Simulation of the bending of material lines

[M 84] [P 73] Simulation were undertaken to reveal the impact of the perturbation amplitude on the material line folding [48]. At small amplitude, only weak oscillations are observed. With increasing perturbation, the oscillation amplitude becomes larger. At still larger amplitude, chaotic behavior is observed, correlated with multiple folding. An extremely convoluted periodic flow pattern is formed downstream, promoting mixing.

Intricate periodically bendt patterns result downstream in the main channel [48].

Determination of the degree of chaoticity

[M 84] [P 73] The degree of chaos was determined by calculating the Lyapunov exponent, a measure for material line stretching [48]. Within the parametric limits of the simulation study made, the highest Lyapunov exponent was 0.1. On adding a further adjacent channel, this parameter can be increased to 0.4.

Experimental flow visualization

[M 84] [P 73] The intricate periodically bendt patterns downstream in the main channel, predicted by the simulation, are confirmed by experimental flow visualization (see Figure 1.174) [48]. This is supported by the parabolic velocity profile along the channel, promoting convolute interface structures. The profiles flatten downstream.

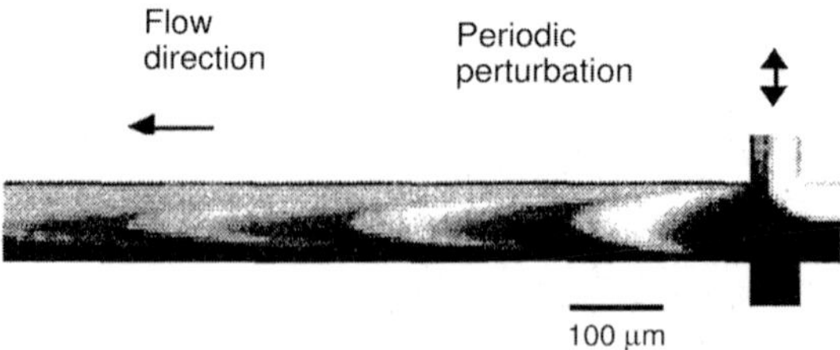

Figure 1.174 Experimental image of the intricate, periodic flow patterns generated by the oscillating injection via adjacent channels into a main stream [48] (by courtesy of Springer-Verlag).

1.3.26
Bimodal Intersecting Channel Mixing

Most Relevant Citations

Peer-reviewed journals: [160, 161].

Capillary electrochromatography column (CEC) operation in the sub-15 nl volume range demands mixers that have an order of degree smaller internal volumes, i.e. a few hundred picoliters or less. The internal volumes of today's micro mixers are typically in the range of a few hundred nanoliters, and hence are not suitable. Therefore, a new mixing concept was developed that was oriented on regular packed-bed mixing (see Figure 1.175) [160]. Here, the goal is to achieve transverse mixing along the flow path of fluids flowing around the obstacles. Highly regular micromachined packed beds can also be realized; however, it still seems problematic to achieve sufficient transverse mass transfer. This is particularly due to heterogeneities in the flow path of existing packed beds. An improvement in this situation was seen in promoting convection streams by so-called trans-channel coupling, which means the deliberate formation of voids in the packed bed which are due to convective transport besides the diffusive transport that occurs in the interstices of the packed bed's objects. These voids are naturally not located linearly downstream (as otherwise all the flow will take this option), but should meander along the residual packed-bed volume. In this way, a main stream is split repeatedly into smaller side streams, which after passage merge again with the main stream.

The bimodal intersecting channel micro mixer is the micromachined version of such a packed-bed-with-voids approach (see Figure 1.175) [160]. It has two types of channels, larger ones for the main convective transport and many smaller ones for decreasing the diffusion path so that efficient mixing can take place (see Figure 1.176). Liquid transport generally is achieved by electroosmotic flow (EOF) action. In the course of the flow, many repeated sequences of flow splitting and recombination are given. The mixer has two domains with micro channels of very different hydraulic diameter, thus being *bimodal*, which *intersect* and of different mixing mechanism; some regions are dominated by EOF transport and diffusive mixing (large channels), others are ruled by convective transport (small channels).

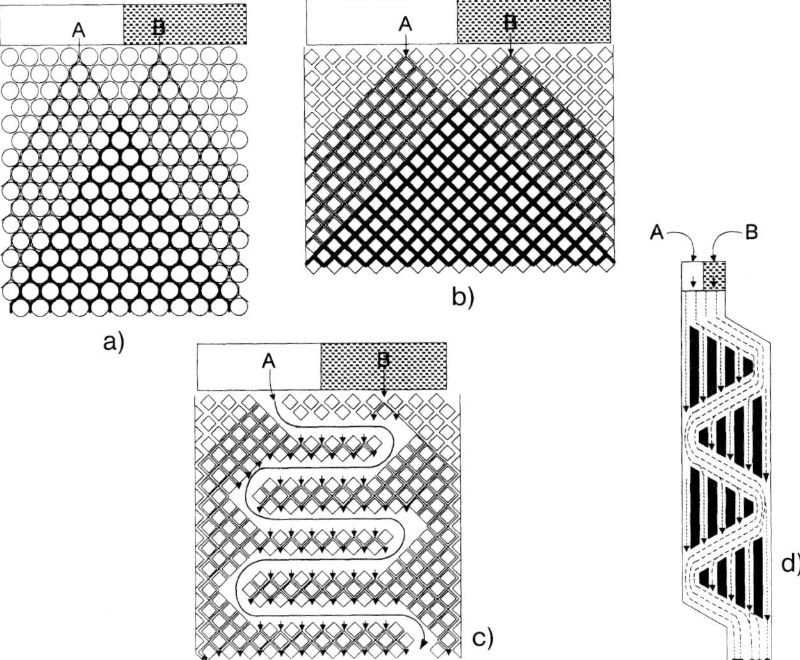

Figure 1.175 (a) 2-D image of a packed bed composed of densely packed spherical particles. Two confluent liquid streams (A and B) are injected at distant points on top of the packed bed and move statistically downwards. At a certain point they merge and mixing takes place by diffusion at constantly recreated and renewed interfaces. (b) A similar flow arrangement with square bed objects; however, this time a micromachined packed bed is shown. (c) Artificial new structure of a packed bed where some objects of the packed bed have been removed. Now, flow conduits exist which are intended to promote lateral mixing. (d) Bimodal intersecting channel micro mixer, which is a micromachined design resembling the idea depicted in (C) [160] (by courtesy of ACS).

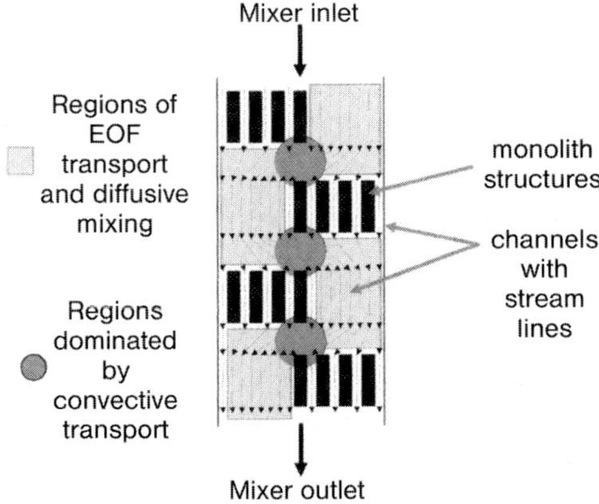

Figure 1.176 A suggestion as to how the flow might develop in the bimodal intersecting channel micro mixer. Distinct zones with small and large channels serve to promote mixing by diffusive and convective transport, respectively [160] (by courtesy of ACS).

1.3.26.1 Mixer 85 [M 85]: Bimodal Intersecting Channel Micro Mixer

This micro device contains one zig-zag-like main micro channel. In the direction of the flow, four smaller channels intersect this main flow path [160]. These smaller channels have all the same width, but differ in length. The two confluent streams enter the micro mixer in parallel fashion.

The micro mixer was made by standard photolithographic and etching methods on an ultra-flat quartz substrate [160]. The microstructure created in this way was shielded by a cover plate by thermal bonding. This cover plate contained access holes for fluid supply and withdrawal. Internal surfaces of the chip were activated to have a high density of silanol groups needed for the EOF transport.

Mixer type	Bimodal intersecting channel micro mixer	Large micro channel: width, depth	27 μm, 10 μm
Mixer material	Quartz	Small micro channel: width, depth	5 μm, 10 μm
Flatness of quartz substrate	< 3 μm across the whole wafer surface	Total volume of the mixer	~100 pl
Whole mixer structure: width, length, depth	100 μm, 200 μm, 10 μm		

1.3.26.2 Mixing Characterization Protocols/Simulation

[P 74] The liquids were moved by electroosmotic flow using a computer-controlled power supply [160]. Analog voltages were generated by an analog power output card. The voltage output was set by a laboratory-written LabVIEW program. The flow rate was set to 0.25 mm s^{-1}.

Mixing was characterized by an optical microscope with an epifluorescence attachment [160]. A 10^{-4} M fluorescein solution in sodium phosphate buffer (10 mM, pH 7, with trace of methanol) was applied. A filter cube was used for fitting the excitation and emission characteristics, allowing selective passages of the radiation. A CCD digital camera collected real-time fluorescence images.

For simulation, a 3-D random walk algorithm was developed to study diffusion-controlled mixing phenomena [160]. Several assumptions were made, i.e. only EOF carries out fluid transport, only neutral and point-like analytes are present and the transport in each dimension is fully independent. An elastic collision mechanism was applied for molecule-wall collisions. The analyte was introduced as a stream of 200 molecules ms^{-1}.

[P 75] A static enzyme assay experiment was carried out using a stopped-flow method [161]. This is commonly used for monitoring reaction kinetics. β-Galactosidase was used as model enzyme to convert the substrate fluorescein mono-β-D-galactopyranoside (FMG) via hydrolysis into fluorescein. As buffer solution 10 mM potassium phosphate at pH 7.2 with 1 mM ascorbic acid was used to minimize photobleaching. The enzymatic reaction is accompanied by a change in fluorescence intensity which can be monitored with a microscope.

1.3.26.3 Typical Results

Studies on diffusive mixing in a channel without the micro mixer

[M 85] [P 74] In order to demonstrate the need for a micro mixer, theoretical studies were undertaken to show how slow diffusion in a simple micro channel is, under the flow parameters applied for the micro mixer and in a channel of dimensions similar to those for the purposes envisaged [160]. The information was given as a 2-D plot on the spatial evolution of concentration within the micro channel along the flow axis. Although both fast and slow diffusion constants (10^{-6} and 10^{-8} $cm^2\ s^{-1}$ were assumed, in no case was complete mixing found for a flow passage of 1 mm. These theoretical findings were confirmed by control experiments with bi-laminated channel flows, demonstrating that even after a 3 mm flow passage (100 and 20 μm channel width and depth, respectively) mixing is not completed, as expected for diffusion in such dimensions.

It was further calculated how long micro channel have to be to ensure complete mixing [160]. It was found that depending on the diffusion constant assumed this can take several millimeters of flow passage. A mixing length of a few hundred micrometers, as desired, was only valid for a high diffusion constant of 10^{-6} $cm^2\ s^{-1}$ and a very short mixing channel width of 10 μm. The latter was regarded as feasible concerning today's micromachining capabilities, but as rather impracticable. Therefore, the need for incorporating a micro mixer was demonstrated.

Microscopy images/fluorescence intensity profiles

[M 85] [P 74] Microscopy images and fluorescence intensity profiles were recorded for bi-laminated streams (not having passed the micro mixer) and streams which leave the outlet of the micro mixer (see Figure 1.177) [160]. In the first case, two separated areas are found, divided by a thin interface. This hydrodynamic informa tion corresponds well to the fluorescence intensity profile having two distinct zones, showing the maximum and zero concentration of the dye separated by an ~25 μm thin mixed interface. For a stream mixed by the micro mixer, a rather uniformly colored microscopy image and even fluorescence intensity profile were found, which indicates that complete mixing had occurred. Hence, the basic functioning of the mixer could be proven.

Static enzyme assay – stopped flow method

[M 85] [P 75] In order to determine reaction kinetics by a stopped-flow method, the mixing step within a microfluidic system was improved [161]. Traditionally, an efficient external T-channel mixer is used for this purpose. Now, the bimodal intersecting channel micro mixer was integrated into the system. The mixer has the following advantages over the current state of the art:

- mixing volume is ≤ 100 pl
- transport time through the mixer is 0.5–1 s
- axial distance through the mixer is 400 μm
- volume of the entire channel network is ~6 nl = consumption of enzyme solution.

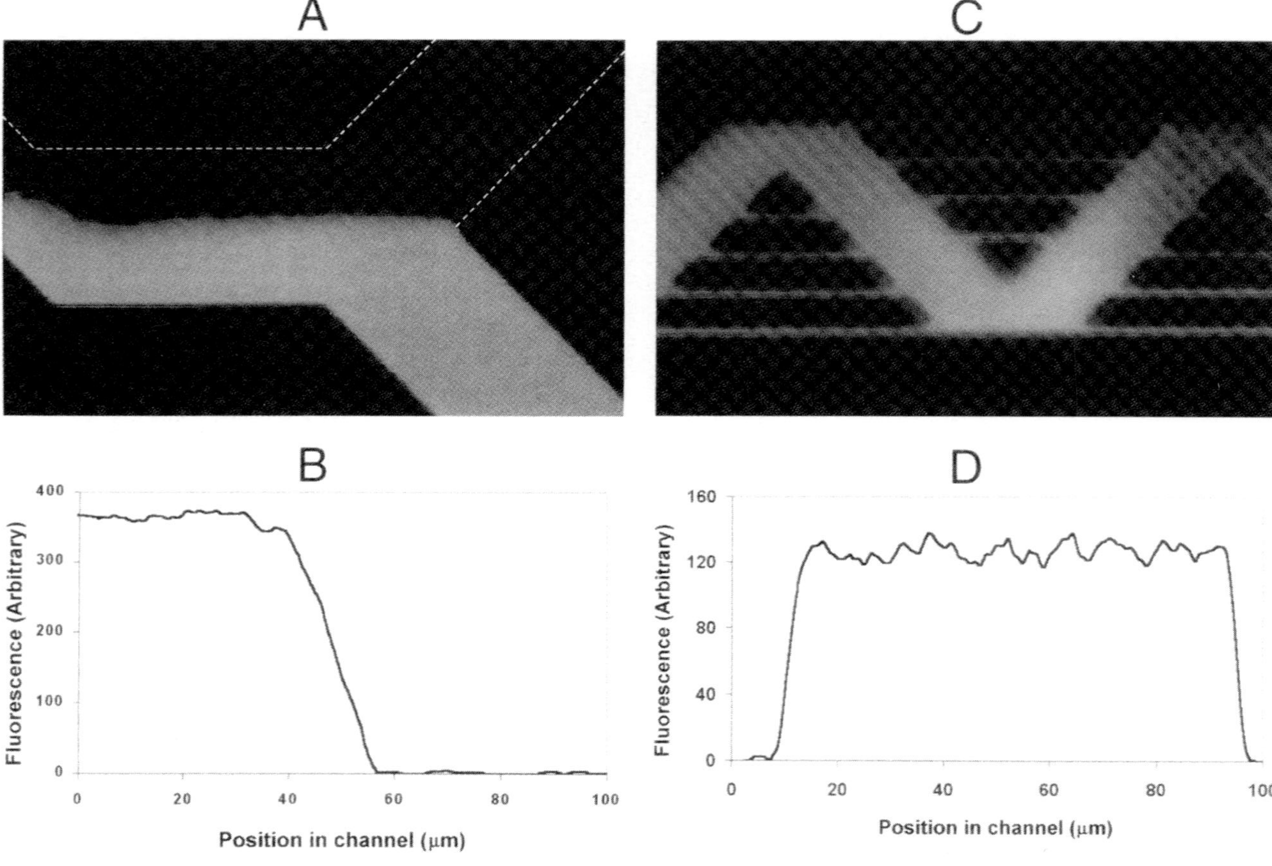

Figure 1.177 (A) Two solutions (fluorescein–buffer and buffer) merge in a bi-laminated configuration into a larger channel. A distinct interface is indicative of slow mixing only. (B) A fluorescence intensity profile over the cross-section of the bi-laminated flow of (A) was taken. Two discrete areas compare well with the findings of the flow pattern. (C) Microscopy image of the flow at the outlet of the micro mixer. Visual inspection suggests intense mixing. (D) A fluorescence intensity profile over the cross-section of the mixer outlet flow of (C) was taken. The even profile suggests completion of mixing [160] (by courtesy of ACS).

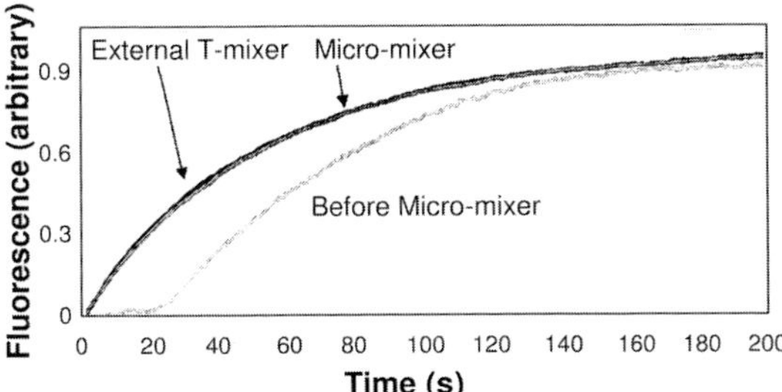

Figure 1.178 Stopped-flow enzymatic assays when using an external T-channel micro mixer, the bimodal intersecting channel micro mixer and the flow before the micro mixer with only diffusion mixing [161] (by courtesy of ACS).

Other advantages concerning variation of the ratio of the substances to be mixed, flushing of the reaction mixture, rinsing capabilities and recycle times are also mentioned [161].

By fluorescence monitoring of the concentration profile along the channel width in a special microfluidic chip, it could be shown that mixing times may be of the order of several tens of seconds for low molecular weight molecules and of several hundred seconds for high molecular weight molecules in aqueous solutions [161]. This fits expectations from simple calculations on the rate of diffusion. Therefore, a micro mixer is required.

It could be shown that the bimodal intersecting channel micro mixer clearly improves the mixing performance. The corresponding axial concentration profile is virtually flat, i.e. mixing has led to the same concentration within the channel [161]. Without the use of the micro mixer more or less the initial concentration profile is found, i.e. a two-sectioned profile with two constant concentration values due to the bi-laminated structure.

When performing the stopped-flow enzymatic assay, the substrate fluorescein mono-β-D-galactopyranoside (FMG) is converted via hydrolysis into fluorescein using the enzyme β-galactosidase (see Figure 1.178) [161]. By monitoring the accompanying change in fluorescence intensity, it was found that the corresponding yield increase for fluorescein over time is identical with that observed when using an external T-channel mixer. Since the latter is a validated and commonly used device, this means optimal functioning of the novel intersecting channel micro mixer. It was also demonstrated that processing which did not make use of this micro mixer results in a poorer performance, i.e. lower yield per unit time.

1.3.27 Micro-bead Interstices Mixing

Most Relevant Citations

Peer-reviewed journals: [162].

The mixing results of the bimodal intersecting mixer show that microstructured analogs of packed-bed structures can mix efficiently. Small micro channels resemble here the interstices of the packed bed. It was therefore logical to look also at the mixing properties of small packed beds themselves. Small spherical polymer particulates, so-called micro beads, are commercially available and can be placed in a section of a micro channel forming a miniature packed bed [162].

1.3.27.1 **Mixer 86 [M 86]: Micro-bead Interstices Micro Mixer**

A microfluidic system was made consisting of a Y-type channel structure and a packed bed of micro beads in a zone in the center of the outlet channel [162]. At the end of the micro-bead bed a weir is placed. The outlet channel then reopens to a larger flow zone which has several parallel micro channels, splitting the main flow into many sub-streams. These micro channels serve as spatially addressable detection lanes.

The device was fabricated using standard lithographic and replica molding methods [162].

Mixer type	Micro-bead interstices mixer	Width of detection lanes	50 μm
Mixer material	PDMS	Number of detection lanes	10
Inlet channel width	100 μm	Micro-bead material	Polystyrene
Outlet channel width	100 μm	Micro-bead diameter	15 μm
Depth of all channels	23 μm	Depth of weir	7–12 μm

1.3.27.2 **Mixing Characterization Protocols/Simulation**

[P 76] The micro beads were introduced into the channel by means of hand pumping in less than 30 s [162].

Mixing was monitored by fluorescence imaging [162]. For this purpose, a 50 μM fluorescein solution in Tris–HCl (pH 7.4) buffer and pure buffer were pumped through the two inlet channels at a flow rate of 0.5 μl min^{-1}.

To analyze the mixing, fluorescence micrographs were taken and also a cross-sectional quantitative analysis of the mixing intensity was performed [162]. A reaction analysis with a two-step reaction was also made. Glucose in buffer was first oxidized to gluconic acid and hydrogen peroxide using the enzyme biotin-labeled glucose oxidase fixed to the micro beads, which were coated with streptavidin. As a follow-up reaction, hydrogen peroxide and amplex red reacted to give fluorescent resorufin using the enzyme horseradish peroxidase.

1.3.27.3 **Typical Results**

Flow patterns without micro-bead bed

[M 86] [P 76] Flow-pattern analysis without micro beads was performed to ensure that laminar-flow properties are maintained throughout the whole microfluidic system, especially concerning the weir conduit [162]. Fluorescence images taken at the Y-contact, at the empty channel and at the parallel detection lanes prove that bi-

laminated patterns are found (one layer being fluorescent, the other not). This is indicative of laminar flow without any secondary flows. This also shows that mixing is weak in the absence of the micro-bead packing, which is to be expected, however.

A quantitative data analysis of the fluorescence intensity in the 10 detection lanes revealed strong intensity in five lanes, one weak signal in one lane and four lanes without any detectable fluorescence, which is in line with a bi-laminated pattern fed to a multi-channel architecture [162].

Flow patterns with micro-bead bed

[M 86] [P 76] The flow patterns with micro beads give a completely different fluorescence texture compared with the experiment without beads (see *Flow patterns without micro-bead bed,* above) [162]. While the initial situation is the same, i.e. a bi-laminated structure after the Y-contact, nearly homogeneous flow patterns are found starting with the micro-bead zones and in the region of the parallel detection lanes (see Figure 1.179). Besides simple visual inspection, this is corroborated by quantitative data. The fluorescence intensity in the 10 detection lanes is virtually the same.

Reaction performance of the micro-bead bed

[M 86] [P 76] The above-mentioned findings of the mixing performance of the micro-bead bed fluidic system suggest fast and complete mixing. Therefore, fast reactions should be carried out without any delay. In order to prove this, a biochemical reaction, the oxidation of glucose in buffer to gluconic acid and hydrogen peroxide, was studied [162]. In a additional micro-bead zone, fluorescent resorufin is generated by reaction of hydrogen peroxide and amplex red using the enzyme horseradish peroxidase.

Fluorescence images at the start and end of the micro-bead zone reveal that within only a short distance after entering the micro-bead bed, fluorescence is observed, whereas there is no fluorescence before [162]. A quantitative fluorescence analysis confirms this visual observation.

1.3.28
Recycle-flow Coanda-effect Mixing Based on Taylor Dispersion

Most Relevant Citations

Peer-reviewed journals: [163]; proceedings contributions: [55]

The use of the Coanda effect is based on the desire to have a second passive momentum to speed up mixing in addition to diffusion [55, 163]. The second momentum is based on so-called transverse dispersion produced by passive structures, which is in analogy with the Taylor convective radial dispersion ('Taylor dispersion') (see Figure 1.180 and [163] for further details). It was further desired to have a flat ('in-plane') structure and not a 3-D structure, since only the first type can be easily integrated into a μTAS system, typically also being flat. A further design criterion was to have a micro mixer with improved dispersion and velocity profiles.

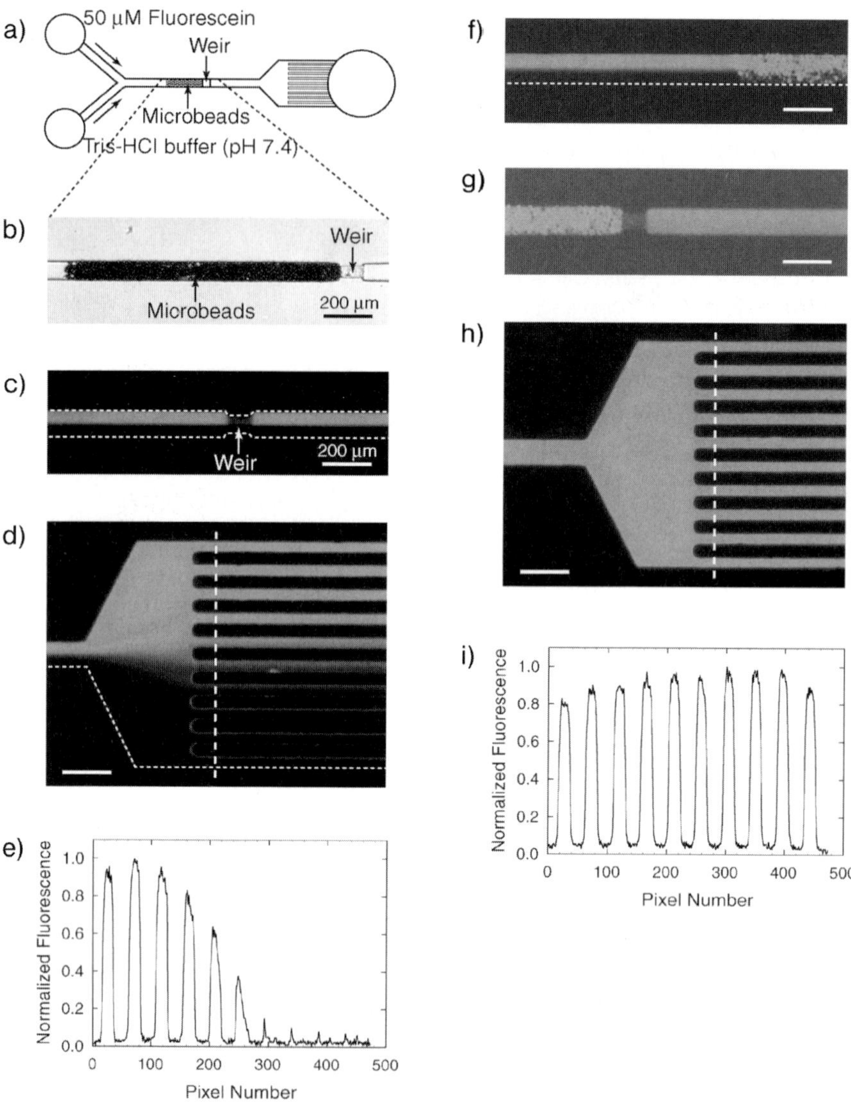

Figure 1.179 (a) Schematic of the microfluidic device with a zone of micro beads. (b) Optical image of the micro beds packed in the micro channel. (c) Fluorescence micrograph of the weir in the channel in the absence of micro beads. (d) Fluorescence micrograph of the 10-lane detection zone in the absence of micro beads. (e) Fluorescence intensity line scans in the 10-lane detection zone in the absence of micro beads, as indicated by the dashed line in (d). (f) Fluorescence micrograph of the channel zone before the micro beads in the presence of micro beads. (g) Fluorescence micrograph of the mixed stream after leaving the micro bead zone. (h) Fluorescence micrograph of the 10-lane detection zone in the presence of micro beads. (i) Fluorescence intensity line scans in the 10-lane detection zone in the presence of micro beads, as indicated by the dashed line in (h) [162] (by courtesy of ACS).

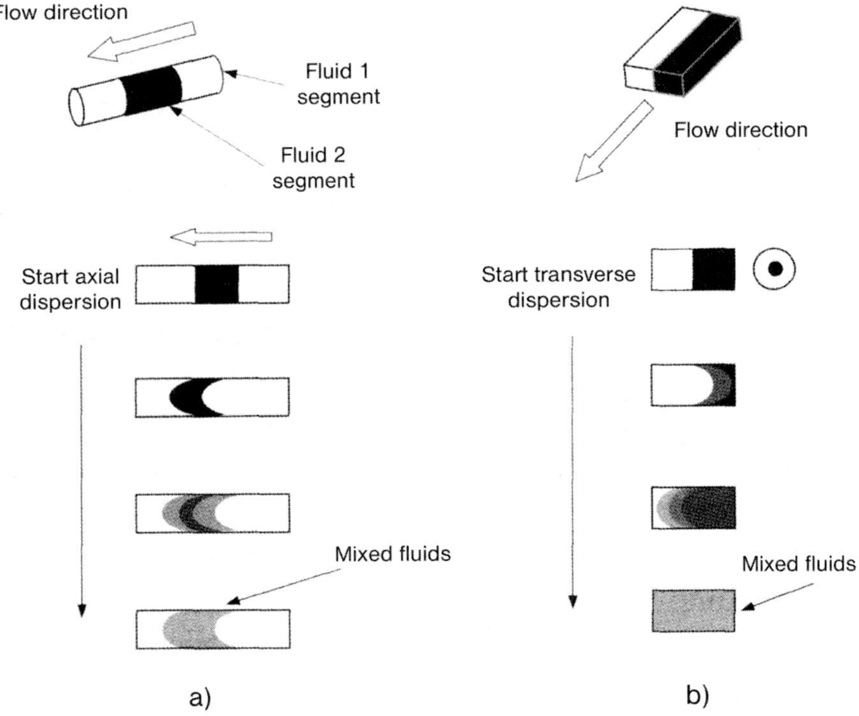

Figure 1.180 (a) Mixing in a capillary tube by Taylor dispersion. (b) Design concept of the micro mixer using a Tesla structure to exploit the Coanda effect [163] (by courtesy of RSC).

Relying on the knowledge of flow pattern generated by the Coanda effect in-plane micro valves and micro pumps, it was envisaged to transfer this technique to a micro-mixer device [55]. In the latter case, using a Tesla structure (see [163] for geometric details), the flow is redirected and collision of streams occurs.

1.3.28.1 Mixer 87 [M 87]: Coanda-effect Micro Mixer with Tesla Structures

In the Coanda-type mixer, the fluids are first bi-laminated in a T-type configuration and then pass a so-called Tesla structure (see [55] for geometric details) which comprises angled surfaces (so-called wing structure) [55]. By flowing along the latter, splitting and redirection of the flow are achieved, which leads to a kind of collision (see Figure 1.181). While one stream passes the major angled passage, the neighboring stream approaches both this major passage and a smaller secondary passage, set in a Y-type flow configuration. This stream splits into two sub-streams according to the different pressure losses of both passages. The flow of both passages is so oriented that collision results. Thus, a larger stream with predominantly one fluid collides with a smaller stream of the pure other liquid. Mixing is said to take place by turbulence.

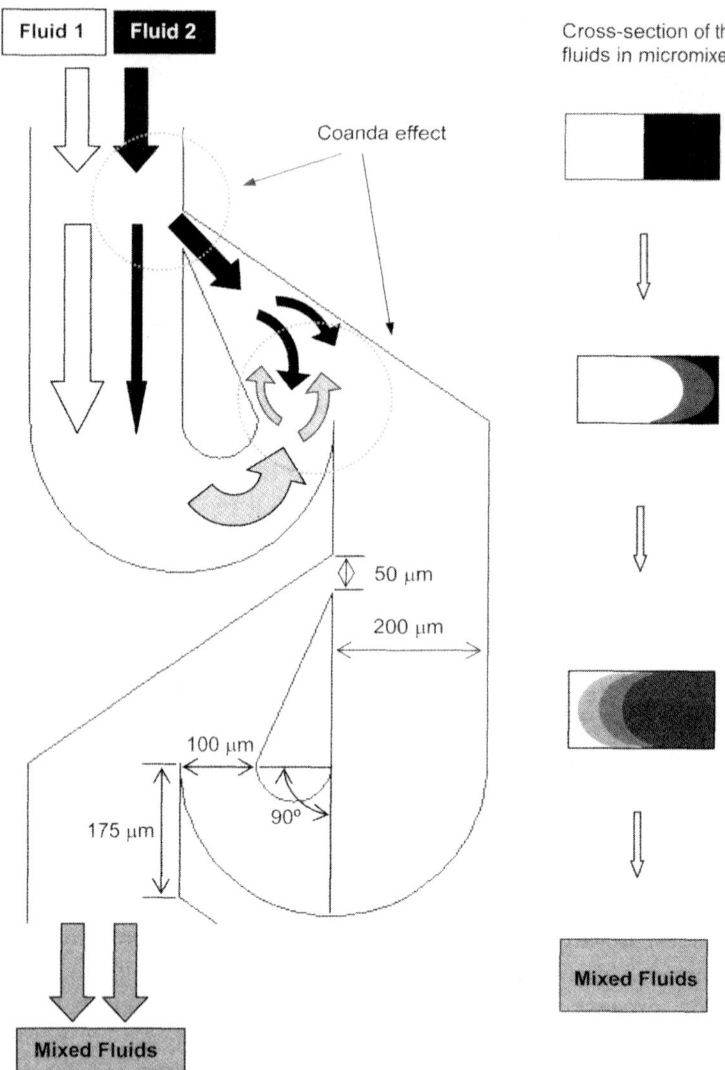

Figure 1.181 Schematic of the action of the Coanda effect on the fluid flow, expressed by major stream directions [163] (by courtesy of RSC).

The Tesla structure is repeated many times in a row so that a sequential mixing is achieved [55, 163].

Microfabrication was achieved by UV-lithography with a negative photoresist SU-8 2035 based on a nickel plate [163]. By means of electroforming of the exposed and developed resist structure, a nickel mold was prepared (see Figure 1.182). By hot embossing, the final polymer mixer structures were prepared. Holes for fluid connectors were drilled. The microstructured substrate was joined with a blank polymer plate by thermal fusion bonding.

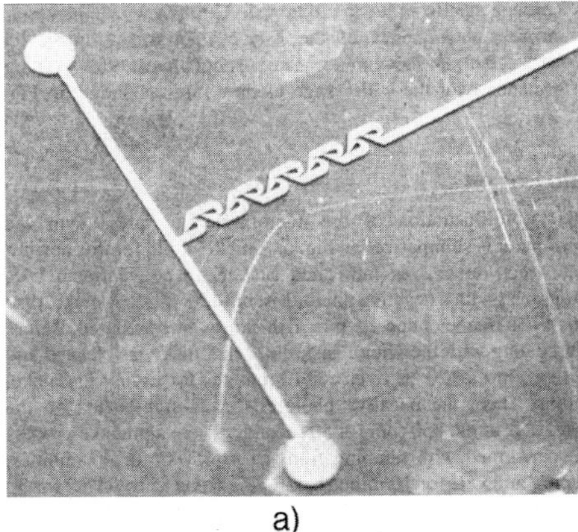

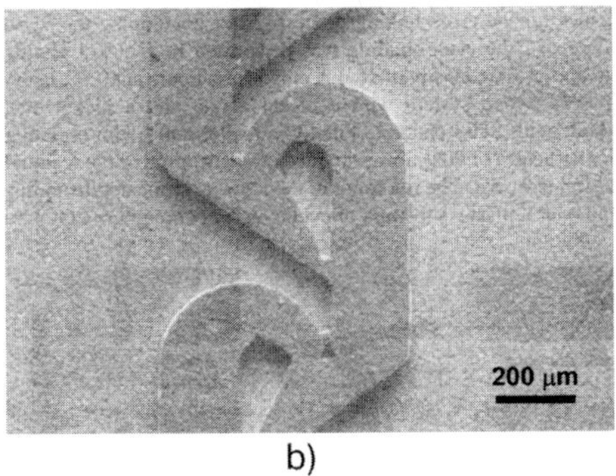

Figure 1.182 Nickel mold for producing the Coanda micro mixer. (a) Microphotograph of the whole device; (b) scanning electron micrograph of the underlying Tesla flow structure [163] (by courtesy of RSC).

Mixer type	Coanda-effect micro mixer with Tesla structures	Angle of turn for the main passage	90°
Mixer material	PDMS; cycloolefin copolymer (COC)	Length of the passage with combined streams before next Tesla structure	175 µm
Opening diameter of main passage, initial, final	200 µm, 100 µm	Channel depth	110 µm
Opening diameter of side passage	50 µm		

1.3.28.2 Mixing Characterization Protocols/Simulation

[P 77] Simulation was done using CFD-ACE+ [55, 163]. The physical properties of water were assumed. Discretization with structured grids of 15 µm length was used. The first-order upwind scheme and conjugate gradient with preconditioning solver were applied.

[P 78] Blue and yellow deionized water streams were mixed in a dilution-type experiment using a syringe pump [55]. A CCD camera recorded the changes in color in the mixer.

[P 79] A quantitative analysis of mixing performance was made by use of pH analysis [163]. Solutions of different pH were mixed and the corresponding changes in pH were collected from four outlet ports. Then, 0.05 M potassium biphthalate buffer (pH 4, 25 °C) and 0.05 M potassium phosphate monobasic–sodium hydroxide buffer (pH 7, 25 °C) were mixed, fed by a syringe pump. The pH changes were monitored with a commercial pH sensor. A reference curve with the sensor was taken in advance.

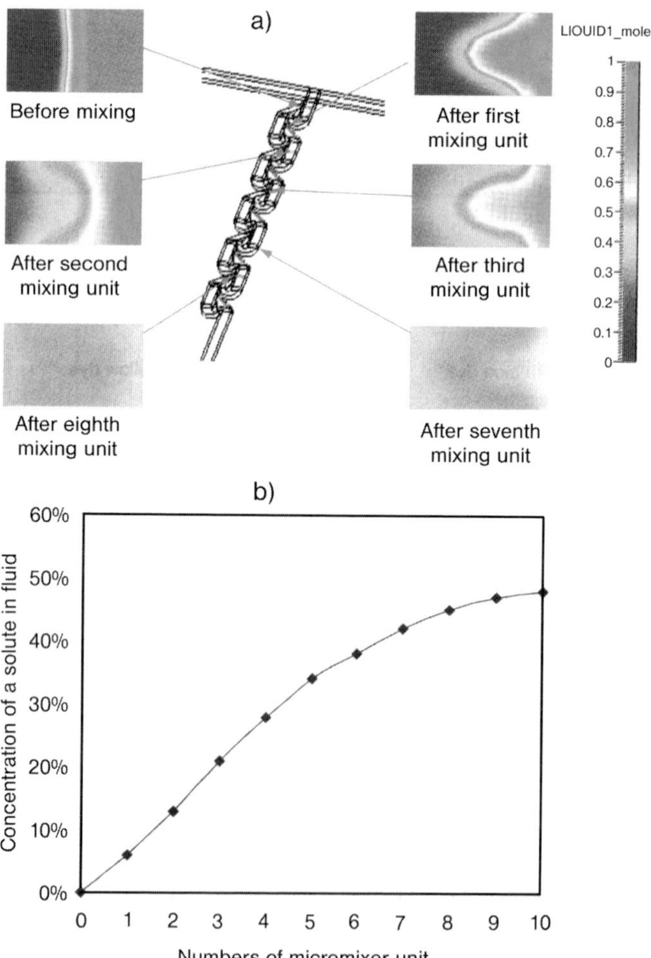

Figure 1.183 (a) Simulation results for the Coanda micro mixer at a flow rate of 50 μl min^{-1}. Cross-sections of the concentration patterns for various mixing units are given [163] (by courtesy of RSC).
(b) Quantitative concentration development along the mixing flow path.

1.3.28.3 **Typical Results**

Impact of the Coanda structure on mixing

[M 87] [P 77] [P 78] Simulations show that the largest contribution to the mixing process is exerted by the Coanda structure and not by diffusion owing to the lamination beforehand. A dilution-type experiment with blue and yellow dyes reveals the formation of a central mixed zone in the initial units, as evident from the green color. Further, downstream the degree of mixing is further increased (see Figure 1.184) [55].

Effectiveness of transverse dispersion

[M 87] [P 77] Simulations show that, as intended, transverse dispersion is achieved at the Coanda structure and mixing is completed after passing four Tesla structures (see Figure 1.183 and 1.184) [163].

Comparison with mixers of other designs

[M 87] [P 77] [P 78] Three designs were compared for their mixing performance with the Coanda micro mixer, a Coanda design without wing, a T-type mixer and a ring-type mixer [55] (see also [163]). An analysis of flow patterns at 50 μl min^{-1} by a dilution-type experiment clearly reveals the onset of mixing for the 'real' Coanda structure, whereas all three other designs do not give any visual hint of mixing.

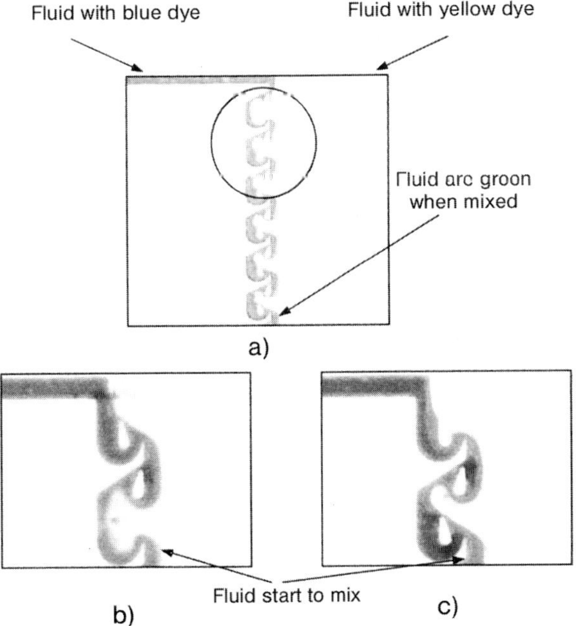

Figure 1.184 Dilution-type experiment using blue- and yellow-colored streams for visualization fo the flow patterns during mixing. (a) Flow-pattern image of the entire mixer device. (b) Flow-pattern image of the first two mixing units for 1 μl min^{-1}. (c) Flow pattern image of the first two mixing units for 100 μl min^{-1} [55] (by courtesy of Kluwer Academic Publishers).

In addition, simulation results predict that after only two mixing stages complete mixing is achieved [55, 163].

Mixing performance as a function of flow rate

[M 87] [P 78] The mixing performance of the Coanda micro mixer is nearly constant over a large range of volume flows, from 1 to 100 μl min^{-1} [55]. The small deviation from ideal is maximum at a flow rate of about 10 μl min^{-1}. The T-type mixer instead shows a notable decrease in mixing efficiency with flow rate, simply owing to the reduction in residence time.

For a flow rate of 1 μl min^{-1}, CCD-camera monitoring of a dilution-type experiment showed a homogeneous texture of the images, when the flow had passed five mixing elements [163]. Albeit this does not necessarily indicate completeness of mixing, it illustrates that a certain degree of dispersion was achieved.

Quantitative pH analysis of mixing performance as a function of flow rate

[M 87] [P 79] By injection of solutions of different pH, quantitative analysis of mixing efficiency was performed (see Figure 1.185) [163]. It is found that the Coanda micro

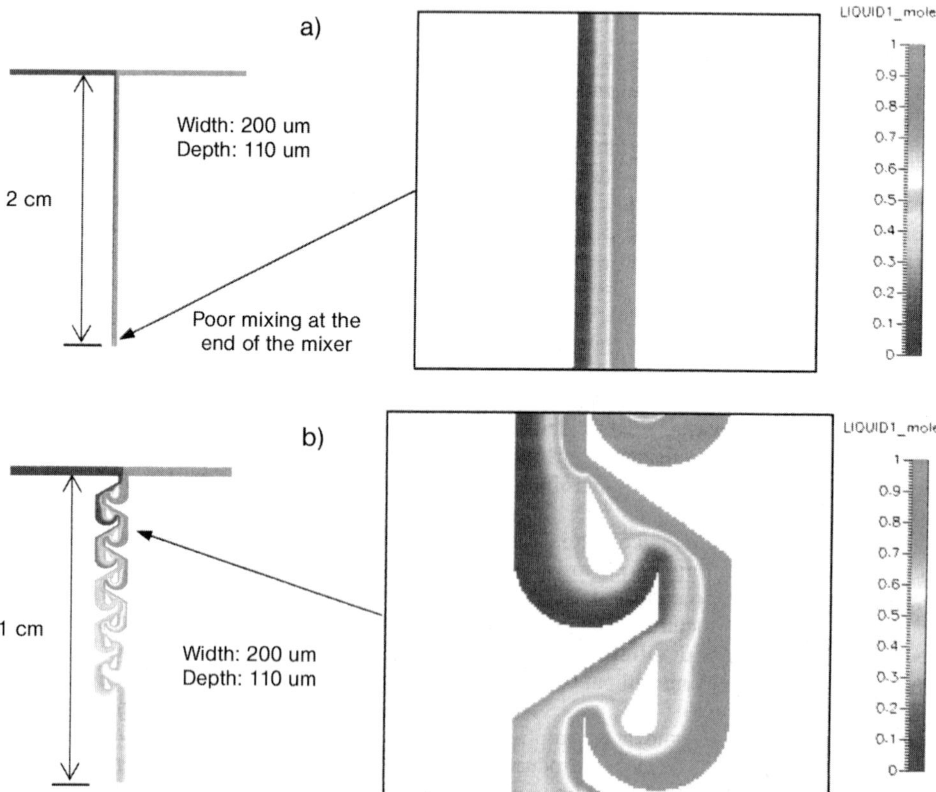

Figure 1.185 Comparison of flow patterns for T-channel and two Coanda micro mixers, revealing the degree of mixing [163] (by courtesy of RSC).

mixer displays a nearly flat curve of mixing efficiency as a function of mixing length, whereas the T-channel mixer shows a decrease, owing to the reduction in residence time and to basing diffusion as the only mixing mechanism. The use of 10 instead of five mixing elements gives better results. Both curves are not far from the ideal behavior, which is a totally flat curve, i.e. the mixing efficiency does not change with flow rate.

[M 87] [P 77] These experimental results are similar to predictions made by simulation on the same comparison [163]. The only small difference between experiment and simulation is that the real performance of the Coanda mixer is somewhat lower than the predicted one, as expected. This requires design optimization.

1.3.29
Recycle-flow Mixing Based on Eddy Formation

Most Relevant Citations

Peer-reviewed journals: [56].

Recycle mixers have a mixing chamber which contains, in addition to the fluid inlet and outlet, additional outlets and re-inlets which build up a recycle loop to withdraw fed fluid and to insert it back to the chamber [56]. Thus a portion of the fluid passes simply through the mixer, while another portion re-enters. The effect is not only prolongation of the residence time, but also, and this is more important, a generation of secondary flows, e.g. eddies, when the primary jet is contacted with the re-entering jets. This generates new interfaces and enhances mixing. The ratio of through and re-entering streams is governed by the pressure losses of both flow paths so that the overall mixing principle is a passive one.

Recycle reactors equipped with pumps are widely used in the chemical industry to achieve high conversion and to reduce the reactor volume for a given conversion [56].

1.3.29.1 Mixer 88 [M 88]: Recycle-flow Micro Mixer

The mixing chamber of the recycle-flow mixer has an inverse V-shape with the inlet at top (at the 'V-corner') and two outlets at the branches of the 'V' (see Figure 1.186) [56]. The two outlets are recombined to one outlet. The width of the 'V'-type mixing chamber is 10 times larger than the inlet and outlet widths in order to allow the formation of secondary flows. At the branches of the 'V', two side channels for recycle flow are also placed. Via two 90° turns the flow re-enters the mixing chamber at the top part, adjacent to the inlet. Five mixing elements were connected in series to ensure complete mixing. The combined outlet of one mixing element serves as inlet for the next. Two large, square-like ports placed in line fed the series of mixing elements; the same type of port served as the device outlet behind the elements.

Conventional photolithography with a negative photoresist SU-8 on a silicon wafer was used for manufacture [56]. PDMS pre-polymer and a curing agent were mixed and poured on this master. The PDMS replica was separated and inlet and outlet holes were punched into the replica. The PDMS microstructure was bonded to glass slides by oxygen plasma treatment.

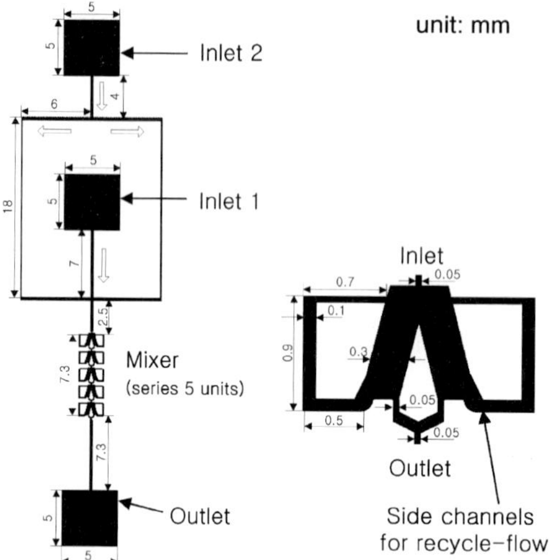

Figure 1.186 Schematic of the recycle-flow micro mixer with five mixing elements [56] (by courtesy of Transducer Research Foundation).

Mixer type	Recycle-flow micro mixer	Outlet channel width	50 μm
Mixer material	PDMS/glass	Side channel width initial, at re-entry	100 μm, 50 μm
Number of mixing units in series	5	Length of V-type mixing chamber and of side channel	900 μm
Total depth (for all channels)	100 μm, 150 μm, 200 μm	Mixing chamber inlet width, side-arm width	500 μm, 300 μm
Inlet channel width	50 μm	Total device size	5 cm × 1.5 cm

1.3.29.2 **Mixing Characterization Protocols/Simulation**

[P 80] The simulation was carried out with CFD-ACE+ [56]. The diffusion, velocity and pressure were calculated in the user scalar module using a finite element method and three-dimensional structured grids. Quantitative analysis of mixing efficiencies was performed.

Mixing was characterized by dilution-type imaging using ink and pure water solutions [56]. As a reactive approach, color imaging was performed by a pH-indicator reaction, merging phenolphthalein solution (0.01 M in water–ethanol) and aqueous NaOH solution (0.3 M).

1.3.29.3 **Typical Results**

Recycle flow ratio and flow pattern

[M 88] [P 80] For an overall channel depth of 150 μm, CFD simulations were performed as a function of flow rates and Reynolds number [56]. At low $Re = 7$

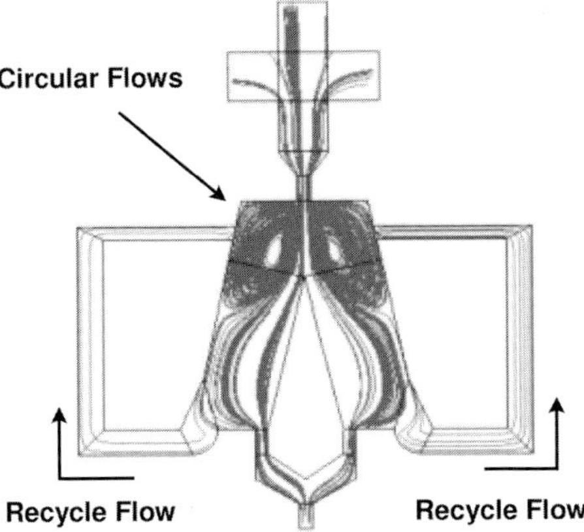

Figure 1.187 Simulated flow pattern of the first element of the recycle-flow micro mixer ($Re = 14$; 150 μm) [56] (by courtesy of Transducer Research Foundation).

(0.05 ml min^{-1}), the flow split into the recycle channel, but ran in the opposite direction, as expected. At low $Re = 14$ (0.1 ml min^{-1}), the correct flow passage into the recycle loop was observed. A detailed flow pattern of such operation depicts eddy formation on both sides when the two re-entering streams collide with the primary inlet stream (see Figure 1.187). The amount of the recycle to primary inlet flow was low, amounting to 0.2%. At low $Re = 42$ (0.3 ml min^{-1}), this share increased to 3%. The results were explained by the relative pressure drops of outlet and inlet streams.

Quantitative mixing analysis

[M 88] [P 80] The simulated mixing percentage in the recycle-flow micro mixer increased with increasing flow rate and Reynolds number ($Re = 7$–42; 0.05–0.3 ml min^{-1}) [56]. The efficiencies also increased with increasing number of mixing elements. For $Re = 28$ and 42, nearly complete mixing was achieved after passage of five mixing elements. The difference between the latter two flow regimes was only obvious for a passage of two mixing elements and then was equal.

The depth of the mixing chamber also had a notable influence [56]. A large depth of 200 μm gave much improved performance, while the simulated mixing percentage was nearly the same for depths of 150 and 100 μm.

Dilution- and reaction-type flow imaging – CFD flow simulation

[M 88] [P 80] Good agreement was found for flow patterns predicted by CFD simulation and a dilution-type experiment using ink for flow visualization ($Re = 14$; 150 μm) [56]. At the outlet fully mixed profiles are found for both simulation and experiment.

Similar findings were made in the reactive experiment. Since now initially only the interfaces are colored, fine details of the flow patterns can be visualized ($Re = 14$; 150 μm) [56]. The results confirm the existence of two eddies at the re-entering flow at the top of the mixing chamber (see Figure 1.188).

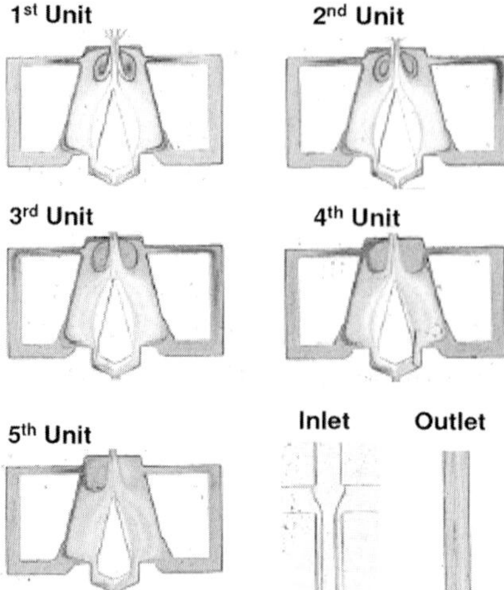

Figure 1.188 Microscope images of the color formation due to a reactive characterization of mixing in the five mixing elements of the recycle-flow micro mixer ($Re = 28$; 150 μm). Phenolphthalein and NaOH solutions were mixed [56] (by courtesy of Transducer Research Foundation).

1.3.30 Cantilever-valve Injection Mixing

Most Relevant Citations
Proceedings contributions: [33].

The combination of pumping action and mixing, as given e.g. for the mixer pumps, is sometimes used in the chemical industry. Thus, attempts at mixing in micro spaces were described which make use of flow energy, which is anyway available within miniature integrated systems. For example, micro pumps or valves create fluid motion; part of this energy can be transferred into mixing action.

1.3.30.1 Mixer 89 [M 89]: Cantilever-valve Injection Micro Mixer

This micro mixer is based on using a cantilever-plate flapper valve for injection of volume into a main stream and performing mixing in a subsequent flow-through chamber (see Figure 1.189) [33]. This design is realized by sandwiching two microstructured wafers, the bottom wafer containing the valve and two ports and the upper wafer with the mixing chamber.

The operation of the mixer/valve is as follows. A sample flow passes continuously through the chamber [33]. At certain times, small reagent volumes are injected via the valve and diffusive mixing occurs. As a first guess, it is thought that the injected volume gives an underneath layer in the mixing chamber, similar to a bilayer configuration. Thereafter, the injection mode is stopped and the mixed reagent flow passes downstream to another functional element.

The relative flow rates determine the mixing ratio and pressure at reagent and sample ports [33]. Accordingly, the device is passively pressure-actuated.

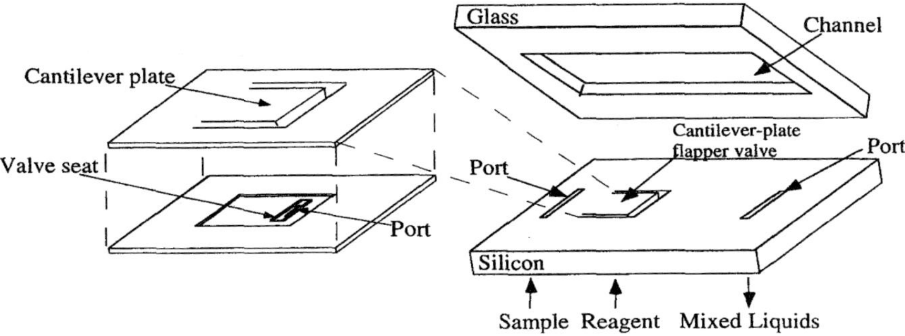

Figure 1.189 Schematic exploded views of the cantilever-valve injection mixer [33] (by courtesy of Kluwer Academic Publishers).

Microfabrication is based on standard silicon micro machining protocols using wet-chemical etching/deep reactive ion etching and oxide protection layers [33]. On one wafer the valve seat and the ports are made in this way. This microstructured silicon wafer is fusion bonded with an SOI (silicon-on-insulator) wafer with a thin silicon overlayer forming the cantilever. After removal of the bulk silicon from this sandwich, it is joined by anodic bonding with a glass wafer, which is structured using a three-part masking scheme.

Mixer type	Cantilever-valve injection mixer	Distance fin-to-fin (underneath the cantilever), fin width	70 μm, 5 μm
Mixing chamber material (top plate)	Glass	Distance from cantilever to fins below	2.3 μm
Valve and port material (bottom plate)	Silicon	Distance fin-to-cantilever end (underneath the cantilever)	515 μm
Cantilever width, length, thickness	800 μm, 600 μm, 3 μm	Flow-through mixing chamber width, depth	940 μm, 70 μm

1.3.30.2 **Mixing Characterization Protocols/Simulation**

[P 81] The investigation is based on the dependence of the quantum yield of the fluorescent dye fluroescein on pH [33]. It is high at large pH (alkaline media) and low at small pH (acidic media). Hence a basic solution of fluorescein is mixed with an acidic solution. In this way, a fluorescence turn-off length (FTOL) is defined, which, however, is not equal to the mixing length, but may give an idea of it.

Simulations of three-component time-dependent diffusion were made based on two slightly different models using Matlab software [33]. Basically, fluid layer thicknesses are predicted, which determine diffusion distances. In this way, the H^+ and OH^- concentrations are revealed which can be related back to the pH. By the known fluorescence intensity–pH relationship, the quantum yield is thus given.

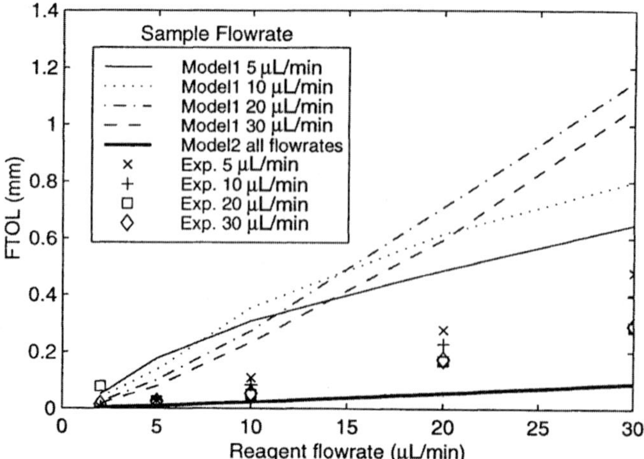

Figure 1.190 Comparison of simulated and experimentally determined fluorescence turn-off lengths (FTOL) [33] (by courtesy of Kluwer Academic Publishers).

1.3.30.3 **Typical Results**

[M 89] [P 81] The experimental turn-off lengths for fluorescence quantum yield decay are between the two predictions based on different models (see Figure 1.190) [33]. Hence the models give a rough estimate. From these data, mixing times can be extracted.

The data also show that mixing is probably better than expected for diffusion as the only mechanism [33]. This is to be expected, since owing to shearing of the layers it is likely that secondary motion is induced. Further, the measurements show that the FTOL and the mixing length are not necessarily equal.

1.3.31
Serial Diffusion Mixer for Concentration Gradients

Most Relevant Citations

Peer-reviewed journals: [164].

The generation of defined concentration gradients in a flowing stream can have many practical analytical uses, e.g. for drug screening (see e.g. [164]). Continuous concentration gradients have a continuously decreasing concentration along the channel, whereas discontinuous gradients give a spontaneous change in concentration at a certain locations and have near-constant concentration in between, similar to a step-like profile. By use of specially fluidic networks, both types of concentration gradients can be realized. One solution is based on the multiple serial contacting of a main stream with side flows at certain, spatially confined locations. For a short time, typically, mixing can occur here via diffusion. Between the contact zones, the main stream flow is 'undisturbed', i.e. no material from other sources is introduced, and only concentration equilibration within the main channel is achieved.

To realize such interactions between many streams in a complex fluidic network, excellent pressure control over all inlets is required, because minute differences may result in undesired breakthroughs or even changes in the flow direction [164].

1.3.31.1 Mixer 90 [M 90]: Serial-diffusion Micro Mixer for Concentration Gradients

A micro channel network is composed of four channel systems, each two representing a symmetric, mirror-imaged design (see Figure 1.191) [164]. Two channels approach each other in a Y-type configuration and come close, do not merge, but are separated by a dam-like structure, restricting the interface to a very small area. The dam structure is due to special application demands for biological cell immobilization, as the fluidic network is part of a μTAS chip, and is not essential for the functioning of the network to work for concentration gradient generation. After the side-to-side arrangement of the two main channels, the flow is redirected twice, yielding partly two separate channels again. Finally, the side-to-side arrange-

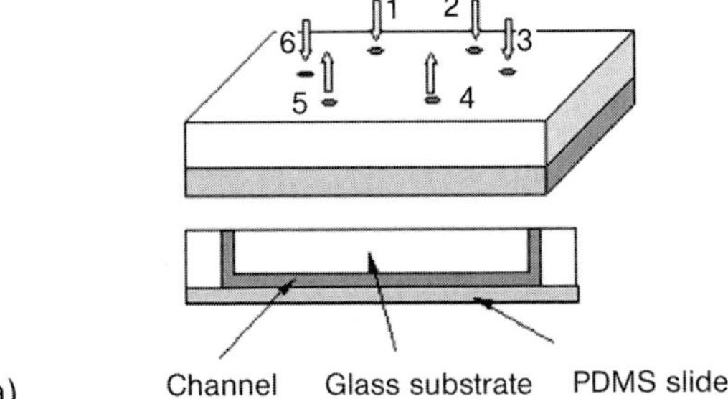

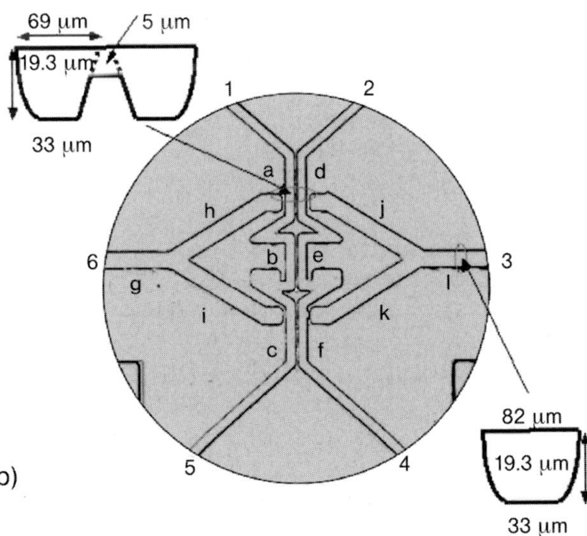

Figure 1.191 (a) Construction of the microfluidic chip and (b) schematic of the microfluidic network for feeding analyte and buffer solutions [164] (by courtesy of RSC).

ment is given again with a dam structure and thereafter the channels are separated by an inverse-Y-type structure.

The two symmetrical side channels are arranged like a Y-configuration, which attaches at its ends two times each the main channels [164]. At these intersections, mixing takes place via diffusion.

The microfluidic structure was manufactured in glass by classical photolithography and wet-chemical HF etching methods [164]. Holes for fluid connection were drilled. A polymer membrane was prepared in-house and served as a top plate to give a glass–polymer sandwich chip. Pressure sealing was applied.

Mixer type	Serial-diffusion mixer for concentration gradients	Channel depth at dam structure	5 μm
Fluidic network material (bottom plate)	Glass	Side channel width at top and bottom of the channel, depth	82 μm, 33 μm, 19.3 μm
Cover material (top plate)	PDMS elastomer	Total size of the chip	5 × 4 × 0.5 cm
Main channel width at top and bottom of the channel, depth	69 μm, 33 μm, 19.3 μm	Inner diameter of connectors	3 mm

1.3.31.2 Mixing Characterization Protocols/Simulation

[P 82] Dilution-type mixing was accomplished with the fluorescent dyes acridine orange (0.01% solution in 20 mM in TE buffer; see below) or trypan blue (prepared in 0.85% saline) contacted with buffer solution (TE buffer: 10 mmol l^{-1} Tris–HCl, pH 7.4, 1 mmol l^{-1} EDTA, pH 8.0) [164]. Images were taken by a laser scanning confocal microscope. Profiling data analysis was employed along detection lines.

Numerical calculations using MATHEMATICA software were made based on a theoretical model which assumes flow distribution in circular pipes under laminar conditions as described by the Bernoulli equation and applies an electrical circuit model based on Ohm's law [164].

1.3.31.3 Typical Results

Continuous concentration gradients

[M 90] [P 82] As a symmetrical flow mode, an operation is described in which the fluorescent dye and buffer solutions are guided in the two main channels side-by-side [164]. Buffer solution is introduced from the two side channels and can mix with the main channel at the two intersections.

In this way, a discontinuous concentration gradient can be built up, as evident from the presence of three distinct concentration plateaus measured by the fluorescence intensity (see Figure 1.192) [164]. For the last mixing step, fluorescence intensity monitoring was possible along the whole flow path from the intersection point (start of mixing) until a considerable passage downstream (completion of

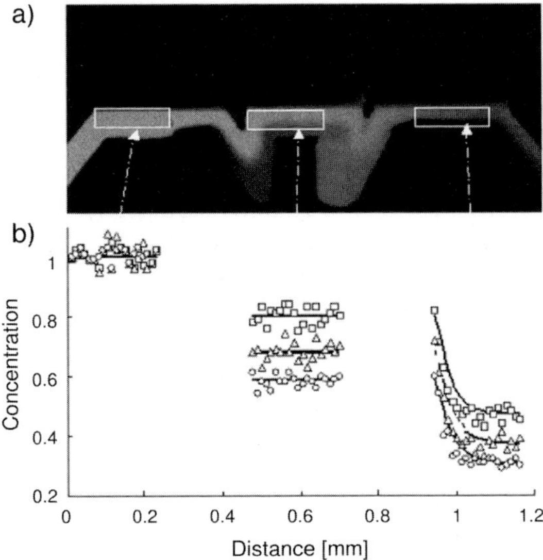

Figure 1.192 (a) Fluorescence micrograph in the main channel of the microfluidic network. (b) Discontinuous, step-like concentration profiles generated in this way [164] (by courtesy of RSC).

mixing). Clearly, the decay of the fluorescence signal due to the increased dilution of the buffer solution within the main channel was visible. Finally, a plateau of the concentration was reached, since mixing was completed.

Numerical simulations match the experimental findings perfectly [164].

The exact shape of the concentration gradient can be controlled by adjusting the pressure of the main stream to the side stream [164]. The pressure of the buffer solution from the side has a key function here.

Mixing of fluorescent dyes of different molecular weight

[M 90] [P 82] As expected, the decay of intensity is not equal for the two different fluorescent dyes having different molecular masses and hence different diffusion coefficients [164]. In a sense, this indicates also that diffuse mixing is effective and is a major contribution to mixing. However, some deviations from diffusion as the only mixing mechanism seem to take place, since the theoretical models are not fully able to describe the experimental behavior. This indicates the presence of other secondary-flow mechanisms besides diffusion.

Asymmetric flow mode

[M 90] [P 82] As asymmetric flow mode an operation is described for which the fluorescent dye is guided in one main channel and the other main channel is left empty [164]. Buffer solution is introduced from the two side channels and can mix with the main channel at the two intersections. In this way, asymmetric pressures are generated between the symmetrical inlets.

A large concentration gradient was so established by having a mixing protocol different to the symmetrical flow mode (for details, see [164]). Simulation results basically agree with the experimental findings.

1.3.32 Double T-junction Turbulent Mixing

Most Relevant Citations

Peer-reviewed journals: [6].

Multi-laminating devices may yield millisecond liquid mixing times. However, mixing devices are needed for the sub-millisecond range, e.g. as measuring tools for studying fast consecutive reactions and processes. Such studies are, in particular, important in the field of quench-flow and stopped-flow analysis, e.g. to monitor protein folding, where a time resolution of < 1 ms is required [6]. Therefore, a sudden start and stop of processes and reactions by intimate mixing has to be ensured. Moreover, the volume of the mixing chamber, which usually is fairly large for macroscopic mixing devices, has to be strongly reduced in order to minimize dead volumes and, hence, dead times.

1.3.32.1 Mixer 91 [M 91]: Double T-junction Micro Mixer

The mixing device comprises a micro channel system with three inlet and one outlet ports, which resembles two mixing tees connected in series (see Figure 1.193) [6]. The outlet of the first mixing tee forms one inlet of the second mixing tee, and the other inlet is fed by an external source. In the second mixing tee, another fluid is introduced and mixed with the mixture of the first two fluids.

Basically, this geometry provides a residence time channel of defined length separated by two mixing tees [6]. In the idealized case of indefinitely short mixing, this configuration would enable one immediately to start a reaction and to control the reaction time by variation of flow rate or length of flow passage before stopping the reaction by a second mixing process. In practice, a major part of the volume of the residence time channel will, even if good mixing is ensured, be demanded by mixing issues rather than by needs of reaction (as proven by the results below). As the mixing mechanism turbulent mixing was used, as this is said to result in fast mixing. For this reason, flow velocities of the liquid streams had to be used which are unusually high for microfluidic devices.

Mixer type	Double T-junction micro mixer	Sub-channel width, length	400 μm, ~2 mm
Mixer material	Silicon	Outlet hole diameter	750 μm
Cover material	Glass	Total size of the mixing chip	10 mm × 10 mm
Number of sub-channels	6		

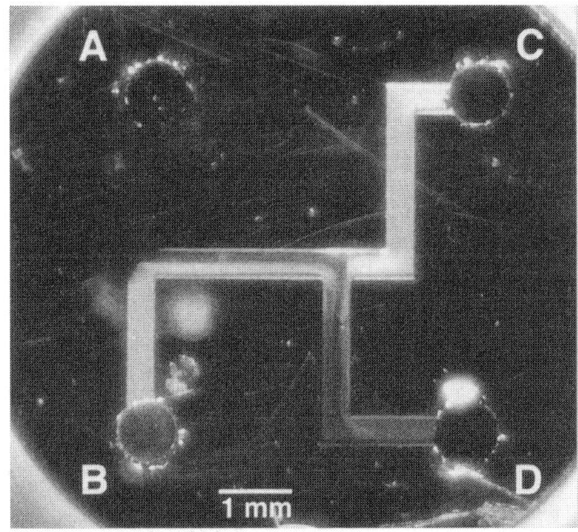

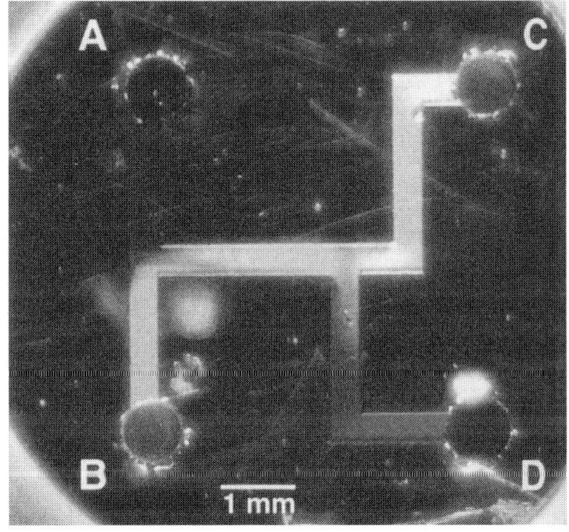

Figure 1.193 Schematic representation of the double mixing tee-type arrangement of a micro mixer for sub-millisecond quench-flow analysis [6] (by courtesy of ACS).

The micro channel system was fabricated by standard silicon micromachining via etching of a silicon wafer with potassium hydroxide using thermal oxide as an etch mask [6]. The double mixing tee configuration consists of six micro channels. For fluid connection, an outlet hole was drilled into the silicon chip. The chip was anodically bonded to a glass slide with three inlet holes, clamped in a holder and, thereby, connected to a commercially available quench-flow instrument.

1.3.32.2 **Mixing Characterization Protocols/Simulation**

[P 83] In feasibility experiments, the time scale of mixing was determined and compared with that of the reactions to be investigated [6]. The first and second mixing processes were followed by flow visualization using two-fold color changes of a pH indicator-containing solution. In a first step, mixing with an acid resulted in a color change. Second, the original color was re-established by subsequent addition of a base.

As a real reaction, the basic hydrolysis of phenyl chloroacetate was investigated in a kind of quench-flow analysis experiment [6].

In order to vary residence times and mixing performance, flow rates in the range 1.8–9.0 l h^{-1} were applied, corresponding to Reynolds numbers from 1000 up to 8000 [6].

1.3.32.3 Typical Results

Ensuring well-developed turbulence and high mixing quality

[M 91] [P 83] The mixing device was tested as a measuring tool for studying fast consecutive reactions and processes [6]. This concerned applications in quench-flow and stopped-flow analysis, where a time resolution of < 1 ms is required.

A pH-indicator reaction was used first for flow visualization. A fully developed color profile, proving high mixing quality, along the channel axis of the mixer was only observed for flow rates as high as 2.7 l h^{-1} [6]. For smaller flow rates, e.g. 0.7 l h^{-1}, only incomplete mixing resulted.

Feasibility of dual fast mixing for quench-flow analysis

[M 91] [P 83] On the basis of the above-mentioned results, a flow range of 1.8–9.0 l h^{-1} was chosen for the investigation of the basic hydrolysis of phenyl chloroacetate as a test reaction for quench-flow analysis [6]. This reaction is sufficiently fast to follow the mixing. Chemical reaction data for the hydrolysis of phenyl chloroacetate showed a mixing dead time of 100 μs. Hence, the feasibility of initiating and quenching reactions with time intervals down to 110 μs in the micro mixer could be proven.

1.3.33
Jet Collision Turbulent or Swirling-flow Mixing

Most Relevant Citations

Peer-reviewed journals: [39, 54, 136, 165, 167]; chapter in encyclopedia: [138]; proceedings contributions: [53, 137, 166]; patents: [52].

Impinging jets are one way to generate large interfaces by ‘converting’ fluids from large reservoirs into multiple small-diameter micro plumes [39, 52–54, 136, 137, 166, 167]. The various impinging jet techniques differ in the way in which they exploit such freshly generated interfaces. Impinging jets can be contacted in a mixing chamber or in free space. They may be guided into air or gas media or introduced into a liquid (which was filled from prior jets). The jets may be directed without offset [39, 136], i.e. collide, or with off-set [54], to penetrate deeply into a liquid and by flowing aside the next jet to create eddies which further enlarge the interface.

By collision of jets, in particular at high fluid velocity, mixing can be accomplished. At high velocities, turbulent mixing can be achieved [39, 52–54, 136, 137, 166, 167]. Even under laminar conditions, collision mixing may be adequate, e.g. when no other type of micro mixer may be suitable and, nevertheless, microfluidics is chosen as the processing concept [39, 136]. This is the case, e.g., when performing precipitation reactions which can be mixed using jet collision in a free-flow configuration [39, 136, 165].

Jets can be collided frontal, i.e. at a 180° angle [52]. Here, the energy input is maximal. The fluid can be virtually 'atomized' in this way and has to be surrounded by a mixing chamber to be re-collected. In a Y-type flow configuration, a 'softer' contact can be achieved [39, 136]. The two jets simply merge to a third one. Under unfavorable conditions, however, a separated, bi-laminated fluid system is achieved.

1.3.33.1 **Mixer 92 [M 92]: Frontal-collision Impinging Jet Micro Mixer, 'MicroJet Reactor'**

[M 92] A special frontal-collision mixer merges two high-velocity jets in a small mixing chamber which is a recess in a solid metal piece (see Figure 1.194) [52, 166]. The jets are introduced via fluid connectors from opposite sides of the chamber and need a very accurate alignment, so as not to be displaced for merging. Owing to the very high velocities, being close to or even above supersonic speed, misalignment may cause destruction of the wall material by cavitation effects. In some versions of the device, massive spheres are introduced which prevent this material damage by being hit by the fluid stream and protecting so the wall beneath.

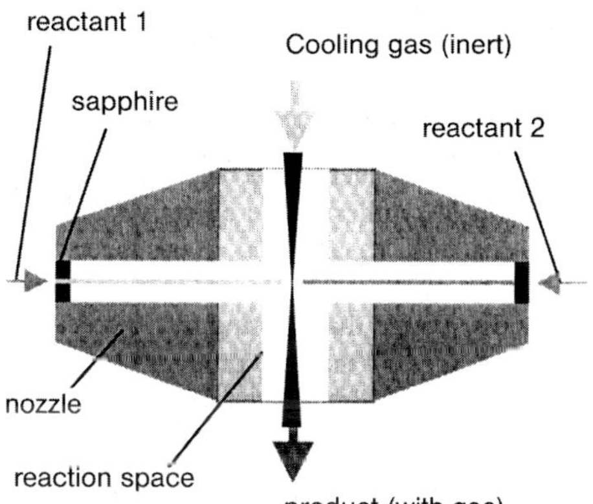

Figure 1.194 Schematic of a commercial frontal-collision mixer [166] (by courtesy of VDE-Verlag).

After jet merging, the material is brought out by an air stream, which redirects the merged stream by 90° into an outlet tubing [52, 166]. This is particularly done because a major application task of the mixer is to perform precipitation reactions. In the latter case, the mixed fluid is sprayed as small droplets or particles. If the droplets still contain solvent, nanoparticles can be produced by evaporation.

No details on the geometry of the device and the fabrication were published.

1.3.33.2 **Mixer 93 [M 93]: Y-Type Collision Impinging Jet Micro Mixer**

Impinging jet micro mixers merge two liquid jets in a Y-like configuration at a point outside the flow-guiding element [39, 136]. The latter contains the two fluid connectors on the side and two tiny outlet nozzles. The nozzles are angled in such a way that the jets merge. The impinging jet element can be inserted into a housing, e.g. serving as a mixing chamber, enabling operation under an inert gas atmosphere, and maintaining temperature control.

Impinging jet elements were constructed as a cylindrical block comprising two feed tubes which become smaller towards the outlet (see Figure 1.195) [39, 136].

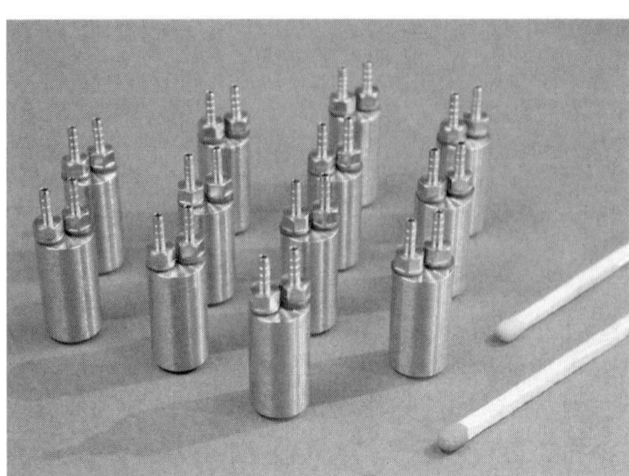

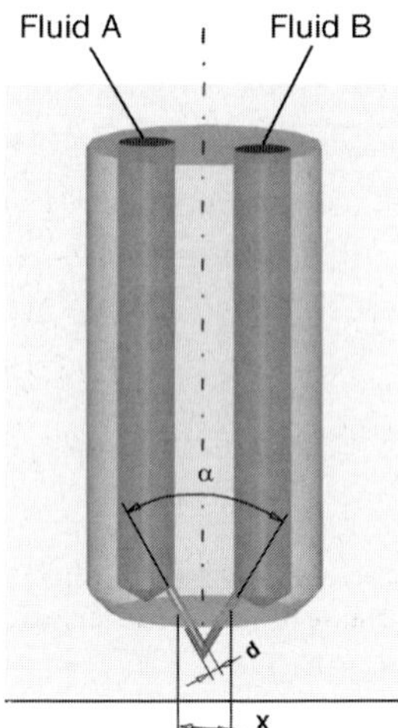

Figure 1.195 Photograph of the impinging jet micro mixer (left) and schematic of the design of the device (right) [53] (source IMM).

Accordingly, the main characteristics of these mixers are the diameter of the outlet boring, the interspaces between the borings and the angle defined by the orientation of borings (relative to the normal) [39, 136]. For one study, nine impinging jet micro mixers were made which differ in these specifications, listed in Table 1.8.

Table 1.8 List of parameters specifying the different impinging jet micro mixers: diameter of the outlet boring *d*, the interspaces between the borings *x* and the angle *α* defined by the orientation of borings ([53]; source IMM).

Jet mixer No.	*Diameter of outlet boring, d (μm)*	*Interspaces between the borings, x (mm)*	*Angle α defined by the orientation of borings (°)*
1	1000	3	60
2	500	2	45
3	500	2	60
4	500	2	90
5	500	3	45
6	500	3	60
7	500	4	45
8	350	3	45
9	350	4	60

The jet mixers were made of Hastelloy and manufactured by die sinking, a special variant of the μ-EDM technique using small cylindrical electrodes [39, 136].

1.3.33.3 **Mixer 94 [M 94]: Impinging Jet Array Micro Mixer**

Arrays with multiple oblique impinging jets are generated via 3-D channel networks which feed fluid from a reservoir via the outlet nozzles of the network into a mixing chamber (see Figures 1.196 and 1.197) [54]. Perforated plates contain such arrays.

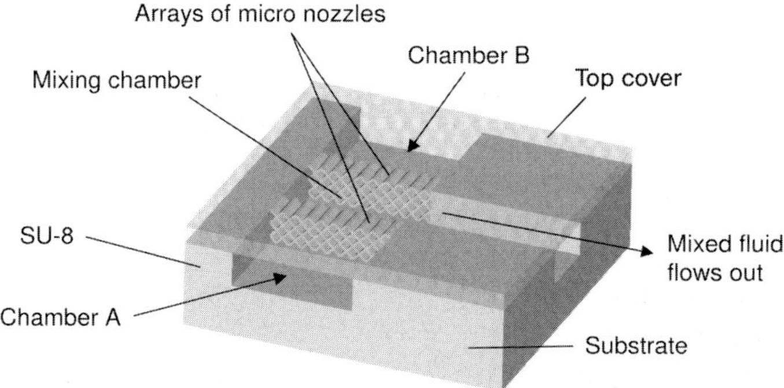

Figure 1.196 Schematic design of the impinging-jet array micro mixer [54] (by courtesy of IOP Publishing Ltd.).

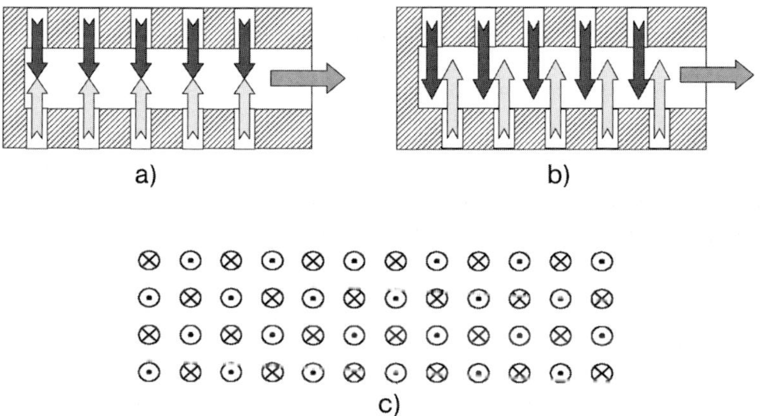

Figure 1.197 Schematic of the flow guidance in the impinging-jet array micro mixer. (A) Without offset; (B) with offset; (C) cross-sectional view of the jet orientation under offset conditions. The two symbols denote jets which have a 90° and a –90° direction relative to the plane of the paper [54] (by courtesy of IOP Publishing Ltd.).

These plates are inserted in a T-channel structure in a still thicker plate, directly at the T-junction at the ends of the two feed channels. In this way, a rectangular mixing chamber is formed in the outlet channel of the T-structure, having multi-nozzle arrays on both sides. The arrays can be positioned with and without offset to collide the jets directly and to let them flow aside. In the latter case, it is thought that eddies are produced at sufficiently high velocity of the jets.

The micro mixer was fabricated by UV lithography of SU-8 [54]. Two lithography masks were employed, one for the inlet and outlet channels and sidewalls and the other for making the array of micro nozzles. A non-conventional tilted lithographic method was used. A thick layer of SU-8 was exposed at 45° and –45° as well as other angles. A special resist development technique for the small and deep nozzle structures also had to be explored.

Mixer type	Impinging-jet array micro mixer	Nozzle length	300 μm
Mixer material	SU-8 resist	Nozzle angle (relative to array front)	28°
Distance between two nozzles	1000 μm	Mixing chamber width (i.e. array distance) depth, length	210 μm, 1000 μm, 5000 μm
Cross-sectional area of the nozzles	70 × 70 μm		

1.3.33.4 Mixing Characterization Protocols/Simulation

[P 84] A jet of acetyl chloride (0.197 mol l^{-1}) in THF at a flow rate of 1000 ml l^{-1} and a jet consisting of *n*-butylamine (0.200 mol l^{-1}) and triethylamine (0.200 mol l^{-1}) in THF at a flow rate of 1000 ml h^{-1} are generated by an impinging jet mixer [53]. Both jets merge in a Y-type flow configuration. The reaction temperatures are 22 °C. The reaction solution was inserted as droplets or a continuous stream either directly or via the tubular reactor in a beaker containing water. By rigorous stirring, hydrolysis of the acid chloride and hence stopping of the reaction are achieved. The phases are separated and the aqueous phase is extracted with THF. The combined THF phases are dried over Na_2SO_4. After filtration, the THF solvent is evaporated at 25 mbar. The remaining amide product is characterized by FTIR spectroscopy.

[P 85] Two set-ups comprising the above-mentioned mixer type were tested, both being micro mixer–tubular reactor configurations [53]. These set-ups differ in the type of tube used. A first set-up was equipped with a spirally wound steel tube, a geometry commonly found in chemical applications (due to savings of space). The vertically mounted tube was 3.30 m long and of 4 mm inner diameter. In contrast, the second set-up comprised a 0.75 m long straight glass tube of 0.3 mm inner diameter. Both set-ups used a laboratory-made housing in which the impinging jet mixer was inserted. The housing chamber was tapered like a funnel in order to collect the liquid mixture and to introduce it directly into the tubular reactor attached, without any wakes. Four inspection windows around the housing allowed judging of the jet formation and the extent of fouling. Feeds into the housing optionally allowed the insertion of inert or cooling/heating gas or an immiscible liquid as heat transfer medium.

[P 86] A solution with a commercial fluorescent dye (1.2 mM) and deionized water were fed by syringe pumps into the micro mixer [54]. A mercury lamp illuminated the mixing chamber. Filters were used to select between the emission light and the reflected light. A microscope with a digital camera was used for flow monitoring.

1.3.33.5 **Typical Results**

Robustness test for the impinging-jet mixer – fouling sensitivity during precipitations
[M 93] [P 84] Different types of flow are found for the impinging jet mixers, dependent on the flow rate (see Table 1.9 and Figure 1.198) [53]. At low flow rates, the two fluids merge immediately after leaving the outlets, wet the mixer surface and finally form droplets. At increasing flow rate, the frequency of the droplets increases and results in merging to a jet. For flow rates next to the droplet regime, the jets formed are often not stable with regard to the axial position, but rather are moving along an angle to the normal. This was termed 'wobbly jet', whereas a stable jet is at a standstill in the normal position. When the two initial jets join at a distance to yield a third jet, this is referred as 'Y-type'. 'T-type' merging is characterized by immediate contact after leaving the outlets (similar to the droplet formation). Thereby, only one jet is generated. Increasing the flow rate further yields a broadening of the jet, first being termed 'fan-shaped' and thereafter 'fanned-out' when broadening is intense.

Table 1.9 Types of flow for the different impinging-jet micro mixers, as defined in Table 1.8 ([53]; source IMM).

Jet mixer No.	*600 ml h^{-1}*	*1000 ml h^{-1}*	*1400 ml h^{-1}*	*1800 ml h^{-1}*	*2200 ml h^{-1}*	*2600 ml h^{-1}*	*3000 ml h^{-1}*
1	Droplets	Droplets	Droplets	Droplets	Droplets	Stable jet, T-type	Stable jet, T-type
2	Droplets	Droplets	Wobbly jet, T-type	Wobbly jet, T-type	Stable jet, T-type	Fan-shaped jet, T-type	Fan-shaped jet, T-type
3	Droplets	Droplets	Wobbly jet, T-type	Stable jet, T-type	Stable jet, T-type	Fan-shaped jet, Y-type	Fanned-out jet, Y-type
4	Droplets	Droplets	Droplet/jet-transition	Wobbly jet, T-type	Wobbly jet, T-type	Stable jet, T-type	Fanned-out jet, T-type
5	Droplets	Wobbly jet, T-type	Stable jet, T-type	Stable jet, T-type	Stable jet, Y-type	Fan-shaped jet, Y-type	Fan-shaped jet, Y-type
6	Droplets	Wobbly jet, T-type	Stable jet, T-type	stable jet, T-Type	Stable jet, T-type	Stable jet, T-type	Fan-shaped jet, Y-type
7	Droplets	Droplets	Wobbly jet, T-type	Wobbly jet, T-type	Wobbly jet, T-type	Stable jet, Y-type	stable jet, Y-Type
8	Droplets	Droplets	Wobbly jet, Y-type	Stable jet, Y-type	Fan-shaped jet, Y-type	Fan-shaped jet, Y-type	Fanned-out jet, Y-type
9	Droplets	Droplet/jet-transition	Wobbly jet, T-type	–*	–*	–*	–*

* No measurement was possible owing to a too high pressure loss using the existing equipment.

The preferred type of flow concerning the operation of fouling-sensitive reactions is the stable Y-type jet [53]. Mixing is here performed at a position distant from the mixer outlets. Since the mixer surface ideally is only wetted separately by the reactant solutions, the precipitation is kept a 'safe' distance from the micro channels. With regard to fouling, any operation resulting in direct merging of the streams is not

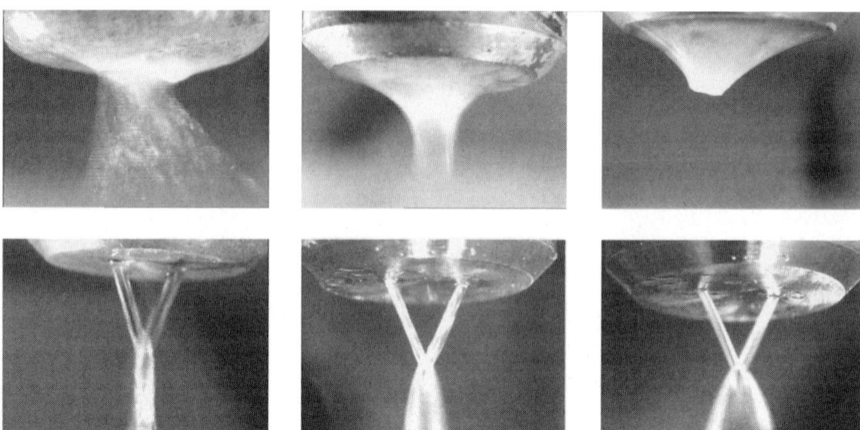

Figure 1.198 Types of flow for different impinging jet mixers, using the acetyl chloride (THF) and *n*-butylamine/triethylamine (THF) solutions ([53]; source IMM). Top left: wobbly jet, T-type (initial flow before plugging); total flow rate 200 ml h^{-1}, mixer type 3. Top middle: stable jet, T-type (initial flow before plugging); 2000 ml h^{-1}, mixer type 6. Top right: device plugged after some minutes of T-type jet operation; 2000 ml h^{-1}, mixer type 2. Bottom left: stable jet, Y-type; 2000 ml h^{-1}, mixer type 5. Bottom middle: fan-shaped jet, Y-type; 2000 ml h^{-1}, mixer type 7. Bottom right: fanned-out jet, Y-type; 3000 ml h^{-1}, mixer type 7.

favored, since the mixed solution wets the micro channels, usually leading rapidly to plugging. The advantages of stable versus wobbly and normally thin versus fanned-out jets currently are not foreseen as clearly.

The results show that devices with tiny openings seem to give the most preferred flow patterns [53]. For instance, device 8 with openings of 350 μm diameter has a favorably large Y-type flow regime, ranging from 700 to at least 1500 ml h^{-1}. The results indicate that a distance of 3 mm and an angle of 45° are particularly suitable for the generation of Y-type flow, as evidenced by the respective large range of flow rates of mixers 5 and 8. In contrast, at higher angles the collision point of the two jets is closer to the mixer surface, hence wetting and correspondingly T-type jets are more likely.

Mixing quality determined by competing reactions approach

[M 93] [P 84] A chemical method based on competitive reactions was used for the analysis of mixing quality [53]. An analysis of the mixing quality of the impinging jet mixers 1–9 (see Table 1.8) in the flow rate range from 300 to 1500 ml h^{-1} reveals a decrease in absorption for all curves, and an increase in mixing quality, with increasing flow rate until a constant, low level is approached (see Figure 1.199). Obviously, the penetration of the jets is intensified at high flow rates, generating larger specific interfaces for mixing.

Jet mixers 8 and 9 with the smallest openings of 350 μm (see Table 1.8) give the best mixing, whereas mixer type 1 with 1000 μm borings performs worst [53]. Even

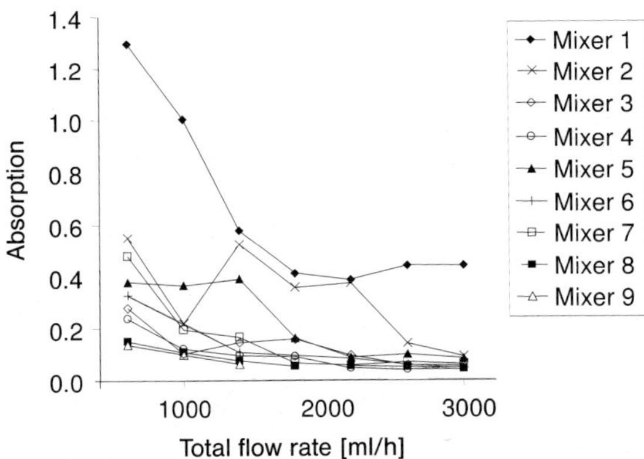

Figure 1.199 Determination of the mixing quality of the different impinging jet micro mixers using a competing reactions approach based on measuring UV–Vis absorption ([53]; source IMM).

at the largest flow rate measured, the performance of the latter device is lower than that of most of the other mixers operated at low flow rate. Regarding the influence of the interspaces, there is no clear indication of better mixing with mixers 5 and 6 ($x = 3$ mm) than mixers 2–4 ($x = 2$ mm) or 7 ($x = 4$ mm). Accordingly, mixing in Y-type flow mode is not necessarily superior to T-type mixing.

Moreover, the comparison of the mixing performance of the grouped mixers 5/6 and 8/9 allows one to judge the influence of the angle α [53]. It was found that the mixing quality at 60° is better than that at 45°. This can be explained by the more frontal collision of the jets at 60°.

Mixers 2–4, in contrast to all other devices, do not exhibit a continuous decrease in absorption, but rather a maximum of absorption at a flow rate of about 700–900 ml h^{-1} [53]. This is most prominent for mixer 2 having an angle of 45°. In view of the hydrodynamic investigations summarized in Table 1.9, this corresponds to the change from the droplet to the T-type wobbly jet regime. This relationship is further confirmed by the small plateau of mixing quality of mixer 7, also having this change of flow.

Robustness test for the impinging-jet mixer – fouling sensitivity during precipitations *[M 93] [P 84]* As a test for fouling sensitivity, amide formation from acetyl chloride and *n*-butylamine was carried out using different impinging jet mixers at various total flow rates [53]. Depending on the flow pattern generated, the extent of fouling varied widely. In the case of droplet or T-type jet formation, processing suffered severely from plugging of the mixer outlets. For Y-type jets, fouling is at least significantly reduced, in some cases totally absent.

It turned out that one impinging jet mixer design (two 350 μm openings, separated by 3 mm, and inclined to each other by 45°), giving stable Y-type flows, was particularly advantageous [53].

Fouling sensitivity during precipitations for two set-ups with different tube geometry *[M 93] [P 85]* Two set-ups with two different kinds of tubing, straight and spirally wound, were tested for their fouling sensitivity, each comprising a impinging-jet micro mixer [53]. When performing amide formation in the set-up with the spirally wound steel tube, plugging occurred after only 42 s operating time (see Table 1.10). In contrast, the set-up with the straight glass tube could be used for a much longer period, at least 20 min. This indicates only a minimal time, since the operation was deliberately stopped owing to the large amount of chemicals and solvents already consumed.

Table 1.10 Operating times achieved with the different set-ups with various process parameters [53].

Set-up	*Dimensions*	*Experimental protocol*	*Operation time/ observations*
Mixer/funnel/tube linear	Mixer type 8, 0.75 m long tube with 0.3 mm i.d.	0.200 mol l^{-1}, 2000 l h^{-1}	20 min*, Et_3NHCl lumps + gas bubbles
Mixer/funnel/tube linear	Mixer type 8, 0.75 m long tube with 0.3 mm i.d.	0.400 mol l^{-1}, 2000 l h^{-1}	38 min*, extensive Et_3NHCl lump formation + gas bubbles
Mixer/funnel linear, tube connector bendt, tube linear	Mixer type 8, 0.30 m long bend connector with 0.3 mm i.d., 0.75 m long tube with 0.3 mm i.d.	0.400 mol l^{-1}, 2000 l h^{-1}	30 s, system plugged owing to extensive Et_3NHCl lump formation
Mixer/funnel linear, tube spirally wound	Mixer type 8, 3.30 m long tube with 4 mm i.d.	0.200 mol l^{-1}, 2000 l h^{-1}	42 s, system plugged

* The operation was stopped intentionally, i.e. much longer processing is possible.

In addition, the hydrodynamics were monitored. The main features were gas bubble formation (HCl) and particulate formation (Et_3NHCl) and agglomeration, both due to the reaction. The flow in the glass tube seemed to be rather undisturbed [53]; only from time to time, bubble formation due to HCl gas evolution and passing of Et_3NHCl lumps were observed. These bubbles and lumps moved with the liquid mixture and were rinsed out of the tube and hence did not behave as obstacles which could cause a breakdown of the flow.

In a further run, the reactant concentration was doubled (see Table 1.10) [53]; 0.395 mol l^{-1} acetyl chloride in THF and 0.400 mol l^{-1} *n*-butylamine and 0.400 mol l^{-1} Et_3N in THF were processed in the second set-up with the straight tube at a total flow rate of 2000 ml h^{-1}. Although extensive precipitation of Et_3NHCl was observed, still the respective lumps were carried out of the tube. After 38 min of operation, no plugging was observed.

When processing in a set-up with a short, curved flow element (0.3 m long bendt Teflon tube of 0.3 mm inner diameter) between the funnel and straight tubular reactor plugging occurred after only 30 s (see Table 1.10) [53]. Hence the insertion of curved flow passages is detrimental, even for only short paths.

Accordingly, the results show that a linear arrangement of the whole set-up is crucial for the avoidance of fouling [53]. Then, even highly concentrated solutions entrapping solid lumps and gas bubbles can be passed through tubes of diameter of much less than 1 mm. Thus, means of building a properly working set-up were derived to handle fast and heavily precipitating reactions, while still using micro-space operation.

Two arrays of oblique impinging jets in frontal collision and in offset orientation

[M 94] [P 86] By means of fluorescence imaging, it could be shown that mixing is only completed in a downstream section below the mixing chamber and the nozzle-array section [54]. This is explained by the comparatively low interfacial enlargement on direct collision.

In contrast, injection under offset conditions, where the jets slide along each other, shows completion of the mixing already in the mixing chamber within 1 1 s (see Figure 1.200) [54]. This is explained by the generation of eddies by guiding the jets in different velocity directions.

It was further found that the impinging-jet array mixer with narrower mixing chambers, i.e. shorter gap between the jets, give better mixing [54]. This is thought to be due to the higher possibility of complete penetration to the other side of the mixing chamber. Lastly, smaller sized and consequently more jets in total have more efficient mixing, as expected.

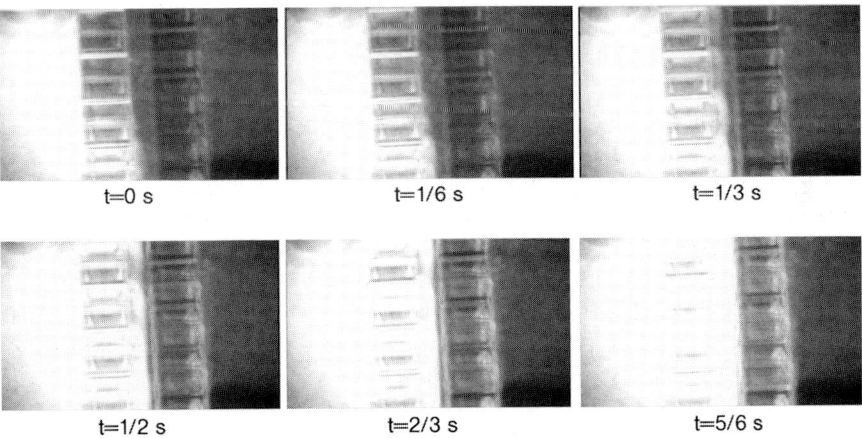

Figure 1.200 Dilution-type fluorescence imaging for visualization of the mixing process in the impinging-jet array micro mixer under offset conditions for the arrays at various times [54] (by courtesy of IOP Publishing Ltd.).

References

1 Kresta, S. M., Krebs, R., Martin, T., The future of mixing research, *Chem. Eng. Technol.* **2004**, *27*, 208–214.

2 Bertsch, A., Heimgartner, S., Cousseau, P., Renaud, P., Static micromixers based on large-scale industrial mixer geometry, *Lab Chip* **2001**, *1*, 56–60.

3 Kim, D. S., Lee, I. H., Kwon, T. H., Cho, D.-W., A barrier embedded Kenics micromixer, *J. Micromech. Microeng.* **2004**, *14*, 1294–1301.

4 Streiff, F. A., Rogers, J. A., Don't overlook static mixer reactors, *Chem. Eng.* **1994**, *6*, 77–82.

5 Oroskar, A. R., Vanden Bussche, K., Towler, G., Scale up vs numbering up – can miniaturization change the rules in chemical processing, in *Proceedings of the VDE World Microtechnologies Congress, MICRO.tec 2000* (25–27 Sept. 2000), VDE Verlag, Berlin, EXPO Hannover, **2000**, 385–392,

6 Bökenkamp, D., Desai, A., Yang, X., Tai, Y.-C., Marzluff, E. M., Mayo, S. L., Microfabricated silicon mixers for submillisecond quench flow analysis, *Anal. Chem.* **1998**, *70*, 232–236.

7 Schönfeld, F., Hessel, V., Hofmann, C., An optimized split-and-recombine micro mixer with uniform 'chaotic' mixing, *Lab Chip* **2004**, *4*, 65–69.

8 Reyes, D. R., Iossifidis, D., Auroux, P. A., Manz, A., Micro total analysis systems: 1. Introduction, theory, and technology, *Anal. Chem.* **2002**, *74*, 2623–2636.

9 Auroux, P. A., Iossifidis, D., Reyes, D. R., Manz, A., Micro total analysis systems: 2. Analytical standard operations and applications, *Anal. Chem.* **2002**, *74*, 2637–2652.

10 Ehrnström, R., Miniaturization and integration: challenges and breakthroughs in microfluidics, *Lab Chip* **2002**, *2*, 26N–30N.

11 Manz, A., Becker, H., *Microsystem Technology in Chemistry and Life Science, Topics in Current Chemistry*, Vol. 194, Springer-Verlag, Heidelberg, **1998**.

12 Knapp, M. R., Kopf-Sill, A., Dubrow, R., Chow, A., Chien, R.-L., Chow, C., Parce, J. W., Commercialized and emerging lab-on-a-chip applications, in Ramsey, J. M., van den Berg, A. (Eds.), *Micro Total Analysis Systems*, Kluwer, Dordrecht, **2001**, 7–9,

13 van den Berg, A., Lammerink, T. S. J., Micro Total Analysis Systems: Microfluidic Aspects, Integration Concept and Applications, in Manz, A., Becker, H. (Eds.), *Topics in Current Chemistry*, Vol. 194, Springer Verlag, Heidelberg, **1998**, 21–49.

14 Ramsey, J. M., Miniature chemical measurement systems, in Widmer, E., Verpoorte, E., Banard, S. (Eds.), *Proceedings of the 2nd International Symposium on Miniaturized Total Analysis Systems, Analytical Methods and Instrumentation, Special Issue μTAS '96*, Basel, **1996**, 24–27.

15 van den Berg, A., Bergveld, P., *μ-TAS: Miniaturized Total Chemical Analysis Systems*, Kluwer, Dordrecht, **1995**.

16 Veser, G., Experimental and theoretical investigation of H_2 oxidation in a high-temperature catalytic microreactor, *Chem. Eng. Sci.* **2001**, *56*, 1265–1273.

17 Kestenbaum, H., Lange de Olivera, A., Schmidt, W., Schüth, H., Ehrfeld, W., Gebauer, K., Löwe, H., Richter, T., Synthesis of ethylene oxide in a catalytic microreactor system, *Stud. Surf. Sci. Catal.* **2000**, *130*, 2741–2746.

18 Kestenbaum, H., Lange de Olivera, A., Schmidt, W., Schüth, F., Ehrfeld, W., Gebauer, K., Löwe, H., Richter, T., Silver-catalyzed oxidation of ethylene to ethylene oxide in a microreaction system, *Ind. Eng. Chem. Res.* **2000**, *41*, 710–719.

19 Haas-Santo, K., Görke, O., Schubert, K., Fiedler, J., Funke, H., A microstructure reactor system for the controlled oxidation of hydrogen for possible application in space, in Matlosz, M., Ehrfeld, W., Baselt, J. P. (Eds.), *Microreaction Technology – IMRET 5: Proc. of the 5th International Conference on Microreaction Technology*, Springer-Verlag, Berlin, **2001**, 313–320.

20 Hessel, V., Hardt, S., Löwe, H., Schönfeld, F., Laminar mixing in different interdigital micromixers – Part I: Experimental characterization, *AIChE J.* **2003**, *49*, 566–577.

21 Gravesen, P., Branjeberg, J., Jensen, O. S., Microfluidics – a review, in *Proceedings of Micro Mechanics Europe, MME '93* (September 1993), Neuchatel, Switzerland, **1993**, 143–164.

22 Yang, Z., Matsumoto, S., Goto, H., Matsumoto, M., Maeda, R., Ultrasonic micromixer for microfluidic systems, *Sens. Actuators A* **2001**, *93*, 266–272.

23 Liu, R. H., Yang, J., Pindera, m. Z., Athavale, M., Grodzinski, P., Bubble-induced acoustic micromixing, *Lab Chip* **2002**, *2 (3)*, 151–157.

24 Liu, R. H., Lenigk, R., Druyor-Sanchez, R. L., Yang, J., Grodzinski, P., Hybridization enhancement using cavitation microstreaming, *Anal. Chem.* **2003**, *75*, 1911–1917.

25 Oddy, M. H., Santiago, J. G., Mikkelsen, J. C., Electrokinetic instability micromixing, *Anal. Chem.* **2001**, *73*, 5822–5832.

26 Glasgow, I., Aubry, N., Enhancement of microfluidic mixing using time pulsing, *Lab Chip* **2003**, *3*, 114–120.

27 Niu, X., Lee, Y.-K., Efficient spatial–temporal chaotic mixing in microchannels, *J. Micromech. Microeng.* **2003**, *13*, 454–462.

28 Qian, S., Bau, H. H., A chaotic electroosmotic stirrer, *Anal. Chem.* **2002**, *74*, 3616–3625.

29 Paik, P., Pamula, V. K., Fair, R. B., Rapid droplet mixers for digital microfluidic systems, *Lab Chip* **2003**, *3*, 253–259.

30 Woias, P., Hauser, K., Yacoub-George, E., An active silicon micromixer for µTAS applications, in van den Berg, A., Olthuis, W., Bergveld, P. (Eds.), *Micro Total Analysis Systems*, Kluwer, Dordrecht, **2000**, 277–282.

31 West, J., Karamata, B., Lillis, B., Gleeson, J. P., Alderman, J., Collins, J. K., Lane, W., Mathewson, A., Berney, H., Application of magnetohydrodynamic actuation to continuous flow chemistry, *Lab Chip* **2002**, *2*, 224–230.

32 Lu, L.-H., Ryu, K. S., Liu, C., A novel microstirrer and arrays for microfluidic mixing, in Ramsey, J. M., van den Berg, A. (Eds.), *Micro Total Analysis Systems*, Kluwer, Dordrecht, **2001**, 28–30.

33 Voldman, J., Gray, M. L., Schmidt, M. A., Liquid mixing studies with an integrated mixer/valve, in Harrison, J., van den Berg, A. (Eds.), *Micro Total Analysis Systems*, Kluwer, Dordrecht, **1998**, 181–184.

34 Branebjerg, J., Larsen, U. D., Blankenstein, G., Fast mixing by parallel multilayer lamination, in Widmer, E., Verpoorte, E., Banard, S. (Eds.), *Proceedings of the 2nd International Symposium on Miniaturized Total Analysis Systems, Analytical Methods and Instrumentation, Special Issue µTAS '96*, Basel, **1996**, 228–230.

35 Floyd, T. M., Losey, M. W., Firebaugh, S. L., Jensen, K. F., Schmidt, M. A., Novel liquid phase microreactors for safe production of hazardous specially chemicals, in Ehrfeld, W. (Ed.), *Microreaction Technology: 3rd International Conference on Microreaction Technology, Proc. of IMRET 3*, Springer-Verlag, Berlin, **2000**, 171–180.

36 Ehrfeld, W., Golbig, K., Hessel, V., Löwe, H., Richter, T., Characterization of mixing in micromixers by a test reaction: single mixing units and mixer arrays, *Ind. Eng. Chem. Res.* **1999**, *38*, 1075–1082.

37 Hardt, S., Schönfeld, F., Laminar mixing in different interdigital micromixers – Part 2: Numerical simulations, *AIChE J.* **2003**, *49*, 578–584.

38 Drese, K., Optimization of interdigital micromixers via analytical modeling – exemplified with the SuperFocus mixer, *Chem. Eng. J.* **2004**, *101*, 403–407.

39 Löb, P., Drese, K. S., Hessel, V., Hardt, S., Hofmann, C., Löwe, H., Schenk, R., Schönfeld, F., Werner, B., Steering of liquid mixing speed in interdigital micromixers – from very fast to deliberately slow mixing, *Chem. Eng. Technol.* **2003**, *27*, 340–345.

40 Zech, T., Hönicke, D., Fichtner, M., Schubert, K., Superior performance of static micromixers, in *Proceedings of the 4th International Conference on Microreaction Technology, IMRET 4* (5–9 March 2000), AIChE Topical Conf. Proc., Atlanta, GA, **2000**, 390–399.

41 Ehlers, S., Elgeti, K., Menzel, T., Wiessmeier, G., Mixing in the offstream of a microchannel system, *Chem. Eng.* Proc. **2000**, *39*, 291–298.

42 Bessoth, F. G., deMellow, A. J., Manz, A., Microstructure for efficient continuous flow mixing, *Anal. Commun.* **1999**, *36*, 213–215.

43 Schwesinger, N., Frank, T., Wurmus, H., A modular microfluidic system with an integrated micromixer, *J. Micromech. Microeng.* **1996**, *6*, 99–102.

44 Stroock, A. D., Dertinger, S. K. W., Ajdari, A., Mezic, I., Stone, H. A., Whitesides, G. M., Chaotic mixing in microchannels, *Science* **2002**, *295*, 647–651.

45 Stroock, A. D., Dertinger, S. K. W., Whitesides, G. M., Ajdari, A., Patterning flows using grooved surfaces, *Anal. Chem.* **2002**, *74*, 5306–5312.

46 Jen, C.-P., Wu, C.-Y., Lin, Y.-C., Wu, C. Y., Design and simulation of the micromixer with chaotic advection in twisted microchannels, *Lab Chip* **2003**, *3*, 73–76.

47 Jiang, F., Drese, K. S., Hardt, S., Küpper, M., Schönfeld, F., Helical flows and chaotic mixing in curved micro channels, *AIChE J.* **2004**, *50* (9), 2297–2305.

48 Lee, Y.-K., Deval, J., Ho, C. M., Tabeling, P., Chaotic mixing in electrokinetically and pressure driven micro flows, in Matlosz, M., Ehrfeld, W., Baselt, J. P. (Eds.), *Microreaction Technology – IMRET 5: Proc. of the 5th International Conference on Microreaction Technology*, Springer-Verlag, Berlin, **2001**, 185–191.

49 Liu, R. H., Ward, M., Bonnano, J., Ganser, D., Athavale, M., Grodzinski, P., Plastic on-line chaotic micromixer for biological applications, in Ramsey, J. M., van den Berg, A. (Eds.), *Micro Total Analysis Systems*, Kluwer, Dordrecht, **2001**, 163–164.

50 Solomon, T. N., Mezic, I., Uniform resonant chaotic mixing in fluid flows, *Nature* **2003**, *425*, 376–380.

51 Miyake, R., Lammerink, T. S. J., Elwenspoek, M., Fluitman, J. H. J., Micromixer with fast diffusion, in *Proceedings of the IEEE-MEMS '93* (Feb. 1993), Fort Lauderdale, FL, **1993**, 248–253.

52 Penth, B., *Method and device for carrying out chemical and physical processes*, WO 00/61275, **1999**.

53 Werner, B., Donnet, M., Hessel, V., Hofmann, C., Jongen, N., Löwe, H., Schenk, R., Ziogas, A., Specially suited micromixers for process involving strong fouling, in *Proceedings of the 6th International Conference on Microreaction Technology, IMRET 6* (11–14 March 2002), AIChE Pub. No. 164, New Orleans, **2002**, 168–183.

54 Yang, R., Williams, J. D., Wang, W., A rapid micro-mixer/reactor based on arras of spatially impinging micro-jets, *J. Micromech. Microeng.* **2004**, *14*, 1345–1351.

55 Hong, C.-C., Choi, J.-W., Ahn, C. H., A novel in-plane passive micromixer using coanda effect, in Ramsey, J. M., van den Berg, A. (Eds.), *Micro Total Analysis Systems*, Kluwer, Dordrecht, **2001**, 31–33.

56 Jeon, M. K., Kim, J.-H., Yoon, H. J., Noh, J., Kim, S. H., Yoon, E., Park, H. G., Woo, S. I., Design and characterization of a passive recycle-micromixer, in *Proceedings of the 7th International Conference on Micro Total Analysis Systems* (5–9 Sept. 2003), Squaw Valley, CA, **2003**, 109–112.

57 Gobby, D., Angeli, P., Gavriilidis, A., Mixing characteristics of T-type microfluidic mixers, *J. Micromech. Microeng.* **2001**, *11*, 126–132.

58 Kim, D. S., Lee, S. W., Kwon, T. H., Lee, S. S., A barrier embedded chaotic micromixer, *J. Micromech. Microeng.* **2004**, *14*, 798–805.

59 Mengeaud, V., Josserand, J., Girault, H. H., Mixing processes in a zigzag microchannel: finite element simulations and optical study, *Anal. Chem.* **2002**, *74*, 4279–4286.

60 Hardt, S., Drese, K. S., Hessel, V., Schönfeld, F., Passive micro mixers for applications in the micro reactor and μTAS field, in *Proceedings of the 2nd International Conference on Microchannels and Minichannels* (17–19 June 2004), Rochester, NY, **2005**, in print.

61 Hessel, V., Löwe, H., Micro mixers – a review on passive and active mixing priciples, *Chem. Eng. Sci.* **2005**, in print.

62 Hessel, V., Löwe, H., Mixing principles of micro mixers – passive and active mixing, in Wang, Y., Holladay, J. (Eds.), *Microreactor Technology and Process Intensification, Microreactor Technology and Process Intensification Symposium of the 226th ACS National Meeting* (7–11 Sep. 2003), New York, **2005**, in print.

63 Jensen, K. F., Smaller, faster chemistry, *Nature* **1998**, *393*, 735–736.

64 Bayer, T., Himmler, K., Hessel, V., Don't be baffled by static mixers, *Chem. Eng.* **2003**, *5*, 2–9.

65 Schubert, K., Brandner, J., Fichtner, M., Linder, G., Schygulla, U., Wenka, A., Microstructure devices for applications in thermal and chemical process engineering, *Microscale Thermal Eng.* **2001**, *5*, 17–39.

66 Löwe, H., Ehrfeld, W., Hessel, V., Richter, T., Schiewe, J., Micromixing technology, in *Proceedings of the 4th International Conference on Microreaction Technology, IMRET 4* (5–9 March 2000), AIChE Topical Conf. Proc., Atlanta, GA, **2000**, 31–47.

67 Koch, M., Chatelain, D., Evans, A. G. R., Brunnschweiler, A., Two simple micromixers based on silicon, *J. Micromech. Microeng.* **1998**, *8*, 123–126.

68 Wong, S. H., Ward, M. C. L., Wharton, C. W., Micro T-mixer as a rapid mixing micromixer, *Sens. Actuators B* **2004**, *100*, 359–379.

69 Kawazumi, H., Tashiro, A., Ogino, K., Maeda, H., Observation of fluidic behavior in a polymethylmethacrylate-fabricated microchannel by a simple spectroscopic analysis, *Lab Chip* **2002**, *2*, 8–10.

70 Greiner, K. B., Deshpande, M., Gilbert, J. R., Design analysis and 3D measurement of diffusive broadening in a Y-mixer, in van den Berg, A., Olthuis, W., Bergveld, P. (Eds.), *Micro Total Analysis Systems*, Kluwer, Dordrecht, **2000**, 87–90.

71 Wong, S. H., Bryant, P., Ward, M., Wharton, C., Investigation of mixing in a cross-shaped micromixer with static mixing elements for reaction kinetics studies, *Sens. Actuators B* **2003**, *95*, 414–424.

72 Hessel, V., Hardt, S., Löwe, H., *Chemical Micro Process Engineering – Fundamentals, Modeling and Reactions*, Wiley-VCH, Weinheim, **2004**.

73 Skelton, V., Greenway, G. M., Haswell, S. J., Styring, P., Morgan, D. O., Warrington, B. H., Wong, S., Micro reaction technology: synthetic chemical optimization methodology of Wittig synthesis enabling a semi-automated micro reactor for combinatorial screening, in *Proceedings of the 4th International Conference on Microreaction Technology, IMRET 4* (5–9 March 2000), AIChE Topical Conf. Proc., Atlanta, GA, **2000**, 78–88.

74 Bourne, J. R., Crivelli, E., Rys, P., 289. Chemical selectivities by mass diffusion. V. Mixing-disguised azo coupling reactions, *Helv. Chim. Acta* **1977**, *60*, 2944–2956.

75 Bourne, J. R., kut, O. M., Lenzner, J., Maire, H., An improved reaction system to investigate micromixing in high-intensity mixers, *Ind. Chem. Res.* **1992**, *31*, 949–958.

76 Bourne, J. R., kut, O. M., Lenzner, J., Maire, H., Kinetics of the diazo coupling between 1-naphthol and diazotized sulfanilic acid, *Ind. Chem. Res.* **1990**, *29*, 1761–1765.

77 Roessler, A., Rys, P., Selektivität mischungsmaskierter Reaktionen: Wenn die Rührgeschwindigkeit die Produktverteilung bestimmt, *Chemie in unserer Zeit* **2001**, *35*, 314–322.

78 Villermaux, J., Falk, L., Fournier, M.-C., Detrez, C., Use of parallel competing reactions to characterize micromixing efficiency, *AIChE Sym. Ser.* **1991**, *88*, 286, 6.

79 Panic, S., Loebbecke, S., Tuercke, T., Antes, J., Boskovic, D., Experimental approaches to a better understanding of mixing performance of microfluidic devices, *Chem. Eng. J.* **2004**, *101*, 409–419.

80 Loebbecke, C. S., Schweikert, W., Tuercke, T., Antes, J., Marioth, E., Krause, H., Applications of FTIR microscopy for process monitoring in silicon microreactors, in *Proceedings of the VDE World Microtechnologies Congress, MICRO.tec 2000* (25–27 Sept. 2000), VDE Verlag, Berlin, EXPO Hannover, **2000**, 789–791.

81 Loebbecke, S., Antes, J., Tuercke, T., Boskovich, D., Schweikert, W., Marioth, E., Schnuerer, F., Krause, H. H., Black box microreactor? Possibilitiess for a better control and understanding of microreaction processes by applying suitable analytics, in *Proceedings of the 6th International Conference on Microreaction Technology,*

IMRET 6 (11–14 March 2002), AIChE Pub. No. 164, New Orleans, **2002**, 37–38.

82 Ehrfeld, W., Hessel, V., Löwe, H., *Microreactors*, Wiley-VCH, Weinheim **2000**.

83 Jähnisch, K., Hessel, V., Löwe, H., Baerns, M., Chemistry in Microstructured Reactors, *Angew. Chem. Int. Ed.* **2004**, *43*, 406–446.

84 Hessel, V., Löwe, H., Micro chemical engineering: components – plant concepts – user acceptance: Part I, *Chem. Eng. Technol.* **2003**, *26*, 13–24.

85 Pennemann, H., Watts, P., Haswell, S., Hessel, V., Löwe, H., Application of micro reactors: benchmarking for analytical applications and lab-scale synthesis, *Org. Proc. Res. Dev.* **2004**, 422–439.

86 Jensen, K. F., Microreaction engineering – is small better? *Chem. Eng. Sci.* **2001**, *56*, 293–303.

87 Fletcher, P. D. I., Haswell, S. J., Pombo-Villar, E., Warrington, B. H., Watts, P., Wong, S. Y. F., Zhang, X., Micro reactors: principle and applications in organic synthesis, *Tetrahedron* **2002**, *58*, 4735–4757.

88 Haswell, S. T., Watts, P., Green chemistry: synthesis in micro reactors, *Green Chem.* **2003**, *5*, 240–249.

89 Gavriilidis, A., Angeli, P., Cao, E., Yeong, K. K., Wan, Y. S. S., Technology and application of microengineered reactors, *Trans. IChemE.* **2002**, *80/A*, 3–30.

90 de Mello, A., Wootton, R., But what is it good for? Applications of microreactor technology for the fine chemical industry, *Lab Chip* **2002**, *2*, 7N–13N.

91 El Moctar, A. O., Aubry, N., Batton, J., Electro-hydrodynamic micro-fluidic mixer, *Lab Chip* **2003**, *3*, 273–280.

92 Lee, C.-Y., Lee, G.-B., Fu, L.-M., Lee, K.-H., Yang, R.-J., Electrokinetically driven active micro-mixers utilizing zeta potential variation induced by field effect, *J. Micromech. Microeng.* **2004**, *14*, 1390–1398.

93 Oddy, M. H., Santiago, J. G., C., M. J., Electrokinetic instability micromixer, in Ramsey, J. M., van den Berg, A. (Eds.), *Micro Total Analysis Systems*, Kluwer, Dordrecht, **2001**, 34–36.

94 Choi, J.-W., Hong, C.-C., Ahn, C. H., An electrokinetic active mixer, in Ramsey, J. M., van den Berg, A. (Eds.), *Micro Total Analysis Systems*, Kluwer, Dordrecht, **2001**, 621–622.

95 Wilson, C. F., Wallace, M. I., Morishima, K., Simpson, G. J., Zare, R. N., Coupled Electrorotation of Polymer Microspheres for Microfluidic Sensing and Mixing, *Anal. Chem.* **2004**, *74*, 5099–5104.

96 Bau, H. H., Zhong, J., Yi, M., A minute magneto hydro dynamic (MHD) mixer, *Sens. Actuators B* **2001**, *79*, 207–215.

97 Paik, P., Pamula, V. K., Pollack, M. G., Fair, R. B., Electrowetting-based droplet mixers for microfluidic systems, *Lab Chip* **2003**, *1*, 28–32.

98 Pollack, M. G., Shederov, A. D., Fair, R. B., Electrowetting-based actuation for integrated microfluidics, *Lab Chip* **2002**, *2*, 96–101.

99 Schwartz, J. A., Vykoual, J. V., Gascoyne., R. C., Droplet-based chemistry on a programmable microchip, *Lab Chip* **2004**, *4*, 11–17.

100 Taniguchi, T., Torii, T., Higuchi, T., Chemical reactions in microdroplets by electrostatic manipulation of droplets in liquid media, *Lab Chip* **2002**, *2*, 19–23.

101 Torii, T., Taniguchi, T., T., N., Higuchi, T., Chemical reaction in microdroplets, in *Proceedings of the 6th International Conference on Microreaction Technology, IMRET* 6 (11–14 March 2002), AIChE Pub. No. 164, New Orleans, **2002**, 192–201.

102 Gau, H., Herminghaus, S., Lenz, P., Lipowsky, R., Liquid morphologies on structured surfaces: from microchannels to microchips, *Science* **1999**, *283*, 46–49.

103 Yang, Z., Goto, H., Matsumoto, M., Yada, T., Micro mixer incorporated with piezoelectrically driven valveless micropump, in Harrison, J., van den Berg, A. (Eds.), *Micro Total Analysis Systems*, Kluwer, Dordrecht, **1998**, 177–180.

104 Zlokarnik, M., Judat, H., *Rührtechnik – Theorie und Praxis*, Springer-Verlag, Berlin, **1999**.

105 Tatterson, G. B., *Scaleup and Design of Industrial Mixing Processes*, McGraw-Hill, New York, **1994**.

106 Schmidt, K., Ehricht, R., Ellinger, T., McCaskill, J. S., A microflow reactor with components for mixing, separation and detection for biochemical experiments, in EHRFELD, W., RINARD, I. H., WEGENG, R. S. (Eds.), *Process Miniaturization: 2nd International Conference on Microreaction Technology, IMRET 2, Topical Conf. Preprints, AIChE*, New Orleans, **1998**, 125–126.

107 SCHMIDT, K., MCCASKILL, J. S., *Schaltbarer dynamischer Mikromischer mit minimalem Totvolumen*, DE 19728520, IMB Institut für Molekulare Biotechnologie e.V. Jena, **1997**.

108 WALZEL, P., LANGER, G., New dynamic micro mixer for viscous fluids, in *Proceedings of the 27th International Exhibition-Congress on Chemical Engineering, Environmental Protection and Biotechnology, ACHEMA* (19–24 May 2003), DECHEMA, Frankfurt, **2003**, 21.

109 BÖHM, S., GREINER, K., SCHLAUTMANN, S., DE VRIES, S., VAN DEN BERG, A., A rapid vortex micromixer for studying high-speed chemical reactions, in RAMSEY, J. M., VAN DEN BERG, A. (Eds.), *Micro Total Analysis Systems*, Kluwer, Dordrecht, **2001**, 25–27.

110 HINSMANN, P., FRANK, J., SVASEK, P., HARASEK, M., LENDL, B., Design, simulation and application of a new micromixing device for time resolved infrared spectroscopy of chemical reactions in solution, *Lab Chip* **2001**, *1*, 16–21.

111 SEIDEL, R. U., SIM, D. Y., MENZ, W., ESASHI, M., A capillary force filled automixing device, in EHRFELD, W. (Ed.), *Microreaction Technology: 3rd International Conference on Microreaction Technology, Proc. of IMRET 3*, Springer-Verlag, Berlin, **2000**, 506–513.

112 KNIGHT, J. B., VISHWANATH, A., BRODY, J. P., AUSTIN, R. H., Hydrodynamic focusing on a silicon chip: mixing nanoliters in microseconds, *Phys. Rev. Lett.* **1998**, *80 (17)*, 3863–3866.

113 BAKAJIN, O., CARISON, R., CHOU, C. F., CHAN, S. S., GABEL, C., KNIGHT, J., COX, T., AUSTIN, R. H., Sizing, fractionating and mixing of biological objects via microfabricated devices, in HARRISON, J., VAN DEN BERG, A. (Eds.), *Micro Total Analysis Systems*, Kluwer, Dordrecht, **1998**, 193–198.

114 VEENSTRA, T. T., LAMMERINK, T. S. J., ELWENSPOEK, M. C., VAN DEN BERG, A., Characterization method for a new diffusion mixer applicable in micro flow injection analysis systems, *J. Micromech. Microeng.* **1998**, *9*, 199–202.

115 VEENSTRA, T. T., LAMMERINK, T. S. J., VAN DEN BERG, A., ELWENSPOEK, M. C., A mixer based on diffsion-mixing designed for the MAFIAS-Project, in *Proceedings of the Sensor Technology in the Netherlands: State of the Art* (2 March 1998), Enschede, Netherlands, **1998**, 263–267.

116 VEENSTRA, T. T., LAMMERINK, T. S. J., VAN DEN BERG, A., ELWENSPOEK, M. C., Characterization method for a new diffusion mixer applicable in micro flow injection analysis systems, in *Proceedings of the Micromechanics Europe (MNE)* (3 June 1999), Ulvik, Norway, **1999**, 186–189.

117 DERTINGER, S. K. W., CHIU, D. T., JEON, N. L., WHITESIDES, G. M., Generation of gradients having complex shapes using microfluidic networks, *Anal. Chem.* **2001**, *73*, 1240–1246.

118 LIM, D., KAMOTANI, Y., CHO, B., MAZUMDER, J., TAKAYAMA, S., Fabrication of microfluidic mixers and artificial vasculares using a high-brightness diode-pumped Nd:YAG laser direct write method, *Lab Chip* **2003**, *3*, 318–323.

119 KAKUTA, M., HINSMAN, P., MANZ, A., LENDL, B., Time-resolved Fourier transform infrared spectrometry using a microfabricated continuous flow mixer: application to protein conformation study using the example of ubiquitin, *Lab Chip* **2003**, *3*, 82–85.

120 JACKMAN, R. J., FLOYD, T. M., GHODSSI, R., SCHMIDT, M. A., JENSEN, K. F., Integrated microchemical reactors fabricated by both conventional and unconventional techniques, in *Proceedings of the 4th International Conference on Microreaction Technology, IMRET 4* (5–9 March 2000), AIChE Topical Conf. Proc., Atlanta, GA, **2000**, 62–70.

121 HESSEL, V., DIETRICH, T., FREITAG, A., HARDT, S., HOFMANN, C., LÖWE, H., PENNEMANN, H., ZIOGAS, A., Fast mixing in interdigital micromixers achieved by means of extreme focusing, in *Proceedings of the 6th International Conference on Microreaction Technology, IMRET 6* (11–14 March 2002), AIChE Pub. No. 164, New Orleans, **2002**, 297–305.

122 Schubert, K., Bier, W., Linder, G., Seidel, G., Menzel, T., Koglin, B., Preisigke, H.-J., *Method and device for performing chemical reactions with the aid of microstructure mixing*, WO 95/30476, Bayer, **1994**.

123 Schubert, K., Bier, W., Linder, G., Seidel, D., Menzel, T., Koglin, B., Preisigke, H.-J., Herrmann, E., *Verfahren und Vorrichtung zur Durchführung chemischer Reaktionen mittels Mikrostruktur-Mischung*, EP 0758981, Bayer, **1994**.

124 Herweck, T., Hardt, S., Hessel, V., Löwe, H., Hofmann, C., Weise, F., Dietrich, T., Freitag, A., Visualization of flow patterns and chemical synthesis in transparent micromixers, in Matlosz, M., Ehrfeld, W., Baselt, J. P. (Eds.), *Microreaction Technology – IMRET 5: Proc. of the 5th International Conference on Microreaction Technology*, Springer-Verlag, Berlin, **2001**, 215–229.

125 Hessel, V., Ehrfeld, W., Möbius, H., Richter, T., Russow, K., Potentials and realization of micro reactors, in *Proceedings of the International Symposium on Microsystems, Intelligent Materials and Robots* (27–29 Sept. 1995), Sendai, Japan, **1995**, 45–48.

126 Möbius, H., Ehrfeld, W., Hessel, V., Richter, T., Sensor controlled processes in chemical microreactors, in *Proceedings of the 8th Int. Conf on Solid- State Sensors and Actuators, Transducers '95 -Eurosensors IX* (25–29 June 1995), Stockholm, **1995**, 775–778.

127 Löb, P., Pennemann, H., Hessel, V., g/l-Dispersions in interdigital micromixers with different mixing chamber geometries, *Chem. Eng. J.* **2004**, *101*, 75–85.

128 Fujii, T., Hosokawa, K., Shoji, S., Yotsomoto, A., Nojima, T., Endo, I., Development of a microfabricated biochemical workbench – improving the mixing efficiency, in Harrison, J., van den Berg, A. (Eds.), *Micro Total Analysis Systems*, Kluwer, Dordrecht, **1998**, 173–176.

129 Aoki, N., Hasebe, S., Mae, K., Mixing in microreactors: effectiveness of lamination segments as a form of feed on product distribution for multiple reactions, *Chem. Eng. J.* **2004**, *101*, 323–331.

130 Werner, B., Hessel, V., Löb, P., Mischer mit mikrostrukturierten Folien für chemische Produktionsaufgaben, *Chem. Ing. Tech.* **2004**, *76*, 567–574.

131 Hardt, S., Dietrich, T., Freitag, A., Hessel, V., Löwe, H., Hofmann, C., Oroskar, A., Schönfeld, F., Vanden Bussche, K., Radial and tangential injection of liquid/liquid and gas/liquid streams and focusing thereof in a special cyclone mixer, in *Proceedings of the 6th International Conference on Microreaction Technology, IMRET* 6 (11–14 March 2002), AIChE Pub. No. 164, New Orleans, **2002**, 329–344.

132 van den Berg, A., Integrated micro- and nanofluidics: silicon revisited, in Ramsey, J. M., van den Berg, A. (Eds.), *Micro Total Analysis Systems*, Kluwer, Dordrecht, **2001**, 207–209.

133 Wilson, D. J., Konermann, L., A capillary mixer with adjustable reaction chamber volume for millisecond time-resolved studies by electrospray mass spectrometry, *Anal. Chem.* **2003**, *75*, 6408–6414.

134 Tagaki, M., Maki, T., Miyahara, M., Mae, K., Production of titania nanoparticles by using a new microreactor assembled with same axle dual pipe, *Chem. Eng. J.* **2004**, *101*, 269–276.

135 Schönfeld, F., Rensink, D., Simulation of droplet generation by mixing nozzles, *Chem. Eng. Technol.* **2003**, *26*, 585–591.

136 Schenk, R., Hessel, V., Werner, B., Ziogas, A., Hofmann, C., Donnet, M., Jongen, N., Micromixers as a tool for powder production, *Chem. Eng. Trans.* **2002**, *1*, 909–914.

137 Schenk, R., Donnet, M., Hessel, V., Hofmann, C., Jongen, N., Löwe, H., Suitability of various types of micromixers for the forced precipitation of calcium carbonate, in Matlosz, M., Ehrfeld, W., Baselt, J. P. (Eds.), *Microreaction Technology – IMRET 5: Proc. of the 5th International Conference on Microreaction Technology*, Springer-Verlag, Berlin, **2001**, 489–498.

138 Schenk, R., Hessel, V., Jongen, N., Buscaglia, V., Guillemet-Fritsch, S., Jones, A. G., Nanopowders produced using microreactors, in *Encyclopedia of Nanoscience and Nanotechnology*, Vol. 7, **2004**, 287–296.

139 Adamschik, M., Schmid, P., Maier, C., Hinz, M., Seliger, S., Hofer, E. P., Kohn, E., Micro dosage controlled micro reactor based on CVD-diamond, in *Proceedings of the 4th International Conference on Microreaction Technology, IMRET 4* (5–9 March 2000), AIChE Topical Conf. Proc., Atlanta, GA, **2000**, 114–120.

140 Branebjerg, J., Gravesen, P., Krog, J. P., Nielsen, C. R., Fast mixing by lamination, in *Proceedings of IEEE-MEMS '96* (12–15 Feb. 1996), San Diego, CA, **1996**, 441–446.

141 Mensinger, H., Richter, T., Hessel, V., Döpper, J., Ehrfeld, W., Microreactor with integrated static mixer and analysis system, in van den Berg, A., Bergfeld, P. (Eds.), *Micro Total Analysis Systems*, Kluwer, Dordrecht, **1995**, 237–243.

142 Svasek, P., Jobst, G., Urban, G., Svasek, E., Dry film resist based fluid handling components for µTAS, in Widmer, H. M., Verpoorte, E., Barnard, S. (Eds.), *Proceedings of the 2nd International Symposium on Miniaturized Total Analysis Systems, Analytical Methods & Instrumentation, Special Issue µTAS '96*, Basel, **1996**, 78–80.

143 Koop, U., Schmelz, M., Beirau, A., *Mikromischer*, DE 19746583, Merck, **1997**.

144 Schwesinger, N., Frank, T., *Device for mixing small quantities of liquids*, DE 96911955, Merck, **1995**.

145 Krummradt, H., Kopp, U., Stoldt, J., Experiences with the use of microreactors in organic synthesis, in Ehrfeld, W. (Ed.), *Microreaction Technology: 3rd International Conference on Microreaction Technology, Proc. of IMRET 3*, Springer-Verlag, Berlin, **2000**, 181–186.

146 Park, S.-J., Kim, J. K., Park, J., Chung, S., Chung, C., Chang, J. K., Rapid three-dimensional passive rotation micromixer using the breakup process, *J. Micromech. Microeng.* **2004**, *14*, 6–14.

147 Fluitman, J. H., van den Berg, A., Lammerick, T. S., Micromechanical components for µTAS, in van den Berg, A., Bergfeld, P. (Eds.), *Micro Total Analysis Systems*, Kluwer, Dordrecht, **1995**, 73–83.

148 Yotsumoto, A., Namakura, R., Shoji, S., Wada, T., Fabrication of an integrated mixing/reaction micro flow cell for µTAS, in Harrison, J., van den Berg, A. (Eds.), *Micro Total Analysis Systems*, Kluwer, Dordrecht, **1998**, 185–188.

149 Tracey, M. C., Cox, T. I., Davis, J. B., Microfluidic mixer employing temporally interleaved liqiuid slugs and parabolic flow, in Ramsey, J. M., van den Berg, A. (Eds.), *Micro Total Analysis Systems*, Kluwer, Dordrecht, **2001**, 141–141.

150 Ro, K. W., Lim, K., Hahn, J. H., PDMS Microchip for precolumn reaction and micellar electrokinetic chromatography of biogenic amines, in Ramsey, J. M., van den Berg, A. (Eds.), *Micro Total Analysis Systems*, Kluwer, Dordrecht, **2001**, 651–562.

151 Branebjerg, J., Fabius, B., Gravesen, B., Application of miniature analyzers: from microfluidic components to µTAS, in van den Berg, A., Bergfeld, P. (Eds.), *Micro Total Analysis Systems*, Kluwer, Dordrecht, **1995**, 141–151.

152 Schönfeld, F., Hardt, S., Simulation of helical flows in microchannels, *AIChE J.* **2004**, *50*, 771–778.

153 Liu, R. H., Yang, J., Lenigk, R., Bonanno, J., Grodzinski, P., Self-contained, fully integrated biochip for sample preparation, polymerase chain reaction amplification, and DNA microarray detection, *Anal. Chem.* **2004**, *76*, 1824–1831.

154 Aref, H., Stirring by chaotic advection, *J. Fluid Mech.* **1984**, *143*, 1–21.

155 Ujhidy, A., Nemeth, J., Szepvölgyi, J., Fluid flow in tubes with helical elements, *Chem. Eng. Proc.* **2003**, *42*, 1–7.

156 Johnson, T. J., Locasio, L. E., Characterization and optimization of slanted well designs for microfluidic mixing under electroosmotic flow, *Lab Chip* **2002**, *2*, 135–140.

157 Aubin, J., Fletcher, D. F., Bertrand, J., Xuereb, C., Characterization of the mixing quality in micromixers, *Chem. Eng. Technol.* **2003**, *26*, 1262–1270.

158 Johnson, T. J., Ross, D., Gaitan, M., Locascio, L. E., Laser modification of channels to reduce band broadening or to increase mixing, in Ramsey, J. M., van den Berg, A. (Eds.), *Micro Total Analysis Systems*, Kluwer, Dordrecht, **2001**, 603–604.

159 Perez-Castillejos, R., Esteve, J., Acero, M. C., Plaza, J. A., Ferrrofluidics for disposable microfluidic systems, in Ramsey, J. M., van den Berg, A. (Eds.), *Micro Total Analysis Systems*, Kluwer, Dordrecht, **2001**, 492–494.

160 He, B., Burke, J., Zhang, X., Zhang, R., Regnier, F. E., A picoliter-volume mixer for microfluidic analytical systems, *Anal. Chem.* **2001**, *73*, 1942–1947.

161 Burke, B. J., Regnier, F. E., Stopped-flow enzyme assays on a chip using a microfabricated mixer, *Anal. Chem.* **2003**, *75*, 1786–1791.

162 Seong, G. H., Crooks, R. M., Efficient mixing and reactions with microfluidic channels using microbead-supported catalysts, *J. Am. Chem. Soc.* **2002**, *124*, 13360–13361.

163 Hong, C.-C., Choi, J.-W., Ahn, C. H., A novel in-plane passive microfluidic mixer with modified Tesla structures, *Lab Chip* **2004**, *4*, 109–113.

164 Yang, M., Yang, J., Li, C.-W., Zhao, J., Generation of concentration gradientby controlled flow distribution and diffusive mixing in a microfluidic chip, *Lab Chip* **2002**, *2*, 158–163.

165 Donnet, M., Bowen, P., Jongen, N., Lemaitre, J., Hofmann, H., Schreiner, A., Jones, A. G., Schenk, R., Hofmann, C., Successful scale-up from millilitre batch optimization to a small scale continuous production using the segmented flow tubular reactor: example of calcium carbonate precipitation, *Chem. Eng. Trans.* **2002**, *1*, 1353–1358.

166 Penth, B., New non-clogging micro-reactor for chemical processing and nano materials, in *Proceedings of the Micro.tec 2000, VDE World Microtechnologies Congress Expo 2000* (25–27 Sept. 2000), VDE Verlag, Berlin, **2000**, 401–405.

167 Schönfeld, F., Rensink, D., Simulation of droplet generating by mixing nozzles, *Chem. Eng. Technol.* **2003**, *26* (5), 585–591.

2
Micro Structured Fuel Processors for Energy Generation

2.1
Outline and Definitions

This chapter deals with development work in the field of micro structured fuel processors, which convert various fuels to hydrogen for fuel cells and other power generation modules.

2.1.1
Power Range and Applications

The equivalent electrical power output of fuel processors, which are currently under worldwide investigation, covers a wide range from less than 1 W to several megawatts. However, owing to their miniaturization, potential micro structured fuel processors seem to be most competitive with conventional technology in all application fields which are restricted by space demands, namely mobile and residential applications. This limits their upper power equivalent to about 100 kW. Portable fuel cell systems promise to be the first commercial market for fuel cells according to a market study performed by Fuel Cell Today in July 2003 [1]. According to the same report, the number of systems built, being mostly prototypes, has increased dramatically to more than 3 000 in 2003. Most of these systems have used to date PEM fuel cells. Low-power fuel processors (1–250 W) compete with both conventional storage equipment such as batteries and simpler fuel cell-based solutions such as direct methanol fuel cells (DMFC). The systems dedicated to cellular phones and laptops developed by Toshiba, NEC and Smart Fuel Cell are based on DMFC technology [2]. However, other companies such as Casio and Motorola have developed micro structured fuel processors for methanol steam reforming (see Section 2.7).

Fuel cell systems for residential applications are commonly developed for the generation of power and heat, which increases their overall efficiency considerably, as even low-temperature off-heat may be utilized for hot water generation. For mobile applications one needs to distinguish between systems dedicated to supplying the drive train and auxiliary power units (APU), which generate either additional energy for vehicles (e.g. air conditioning and refrigerator systems of trucks) or work as standalone systems for power supply. The application fields of

Chemical Micro Process Engineering. V. Hessel, H. Löwe, A. Müller, G. Kolb.

ISBN: 3-527-30998-5

APUs are manifold. They reach from the small scale such as traffic lights or standalone measurement instruments over the medium scale such as laptops to the larger scale such as the automotive and leisure industry (e.g. boats and caravans). In the last couple of years the worldwide research and development focus has moved from the drive trains to APUs, which is the less challenging task concerning system dynamics and start-up time.

2.1.2
Overall Assembly

A fuel processor is a complex device which converts conventional fuels such as gasoline or a renewable fuel such as ethanol into hydrogen. During the conversion process, which is the reforming step, carbon dioxide and various by-products are formed, the most important of them being carbon monoxide in concentrations up to 16 vol.% and more depending on the fuel and the reforming process. Conventional proton exchange membrane (PEM) fuel cells are the most common technology applied for mobile applications. This fuel cell type cannot tolerate carbon monoxide in the long term at trace concentrations exceeding 10 ppm [3], and therefore gas purification devices need to be switched after the reformer. The gas purification may be done via three different main technologies:

- conversion of the carbon monoxide (corresponding reactions are water-gas shift, preferential oxidation and methanation; see below)
- membrane separation of the hydrogen from the other reformer products
- pressure swing adsorption.

So far, micro structured fuel processor systems seem to be limited to the first two technologies. Figure 2.1 shows a general flow scheme of a fuel processor with heterogeneously catalyzed reactors for gas purification. Devices shown in dashed lines are not mandatory for the system.

From this flow scheme derives the following list of peripheral equipment necessary to run a fuel processor:

- fuel and water tanks
- supply devices such as pumps and compressors
- dosing equipment (flow controllers, valves)
- heat exchangers (not shown)
- evaporators
- temperature controllers
- sensors
- insulation/housing (not shown).

This section will provide information about micro structured reformer reactors, gas purification devices and catalytic burners, the last also in combination with an evaporator, for fuel processors. However, the specific problems related to the peripheral equipment will not be discussed in depth.

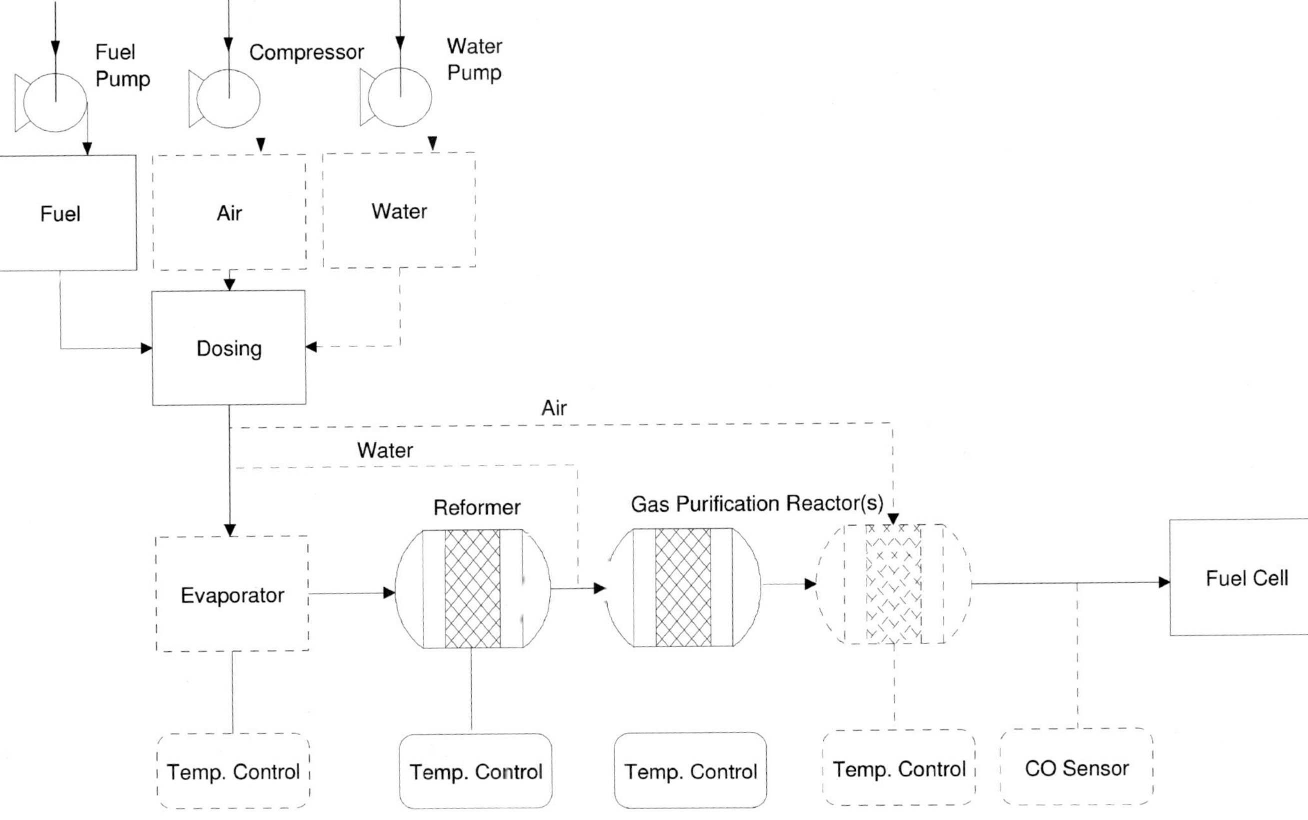

Figure 2.1 General flow scheme of a fuel processor with catalyzed gas purification.

2.1.3
Definitions

The overall efficiency of a fuel processor is commonly defined as the ratio of the lower heating value (*LHV*) of the hydrogen produced to the *LHV* of the fuel consumed:

$$\eta = \frac{LHV\ (H_2)}{LHV\ (\text{fuel})} \tag{2.1}$$

The overall efficiency of a fuel processor/fuel cell system is commonly defined as the ratio of the fuel cell stack power output *P* to the *LHV* of the fuel:

$$\eta = \frac{P\ (\text{fuel cell})}{LHV\ (\text{fuel})} \tag{2.2}$$

All composition data provided as percentages [%] in this chapter are based on gaseous compositions unless indicated otherwise and can be considered as molar percentages.

2.2
Factors Affecting the Competitiveness of Fuel Processors

The most critical characteristics of a fuel processor are the following:

- costs
- efficiency
- start-up time
- size
- weight
- responsiveness to load changes
- lifetime.

They will be discussed below with consideration of the advantages and drawbacks of micro structured devices.

2.2.1
Costs

According to studies by Arthur D. Little Inc. [4] and targets set by the US Partnership of New Generation Vehicles (PNGV), the future cost for a drive train fuel processor needs to be limited to US$ 30 kW^{-1}. Another study even limits the total cost of the whole drive train below the current value of a diesel turbo engine (US$ 50 kW^{-1}) [5]. However, this ambitious goal is far from being reached and certainly does not apply for residential applications [4] and APU systems, where values higher by one order of magnitude and more could be acceptable depending on the application. The target costs for total residential systems ranges around US$ 1500, but today's actual investment costs are about 15 000–20 000 US$ [6]. The costs of the fuel processor should contribute only about 12–15% to the overall system costs [4]. The fuel processor costs depend on the following main factors:

- fabrication technique
- type of fuel
- catalyst cost
- production quantity.

Cheap fabrication techniques feasible for mass production need to be applied for fuel processors. The various techniques available for micro structured fuel processors will be discussed in Section 2.9.

The type of fuel employed has a significant impact on the overall fuel processor cost. Applying fuels such as methanol with missing distribution system would generate a considerable surplus of cost. As an example, to displace 10% of the existing gasoline demand of the United States with methanol would generate a US$ 65 billion cost (status: 1998) [4], which charges the corresponding 25 million vehicles with approximately US$ 65 kW^{-1} cost. This is a strong argument for the utilization of fuels with existing infrastructure that has already been amortized. Especially for first niche applications, the infrastructure will play a dominant role. On the other hand, sustainability of power systems is a strong political argument, of course, which will drive the development of energy supply systems in the future more and more into the direction of renewable sources, especially for mobile systems, where carbon dioxide sequestration seems to be a less viable option.

Catalyst cost may play a significant role in the overall fuel processor cost and may reach values as high as 38% [4]. Here micro structured devices offer significant possibilities for cost reduction owing to the improved mass transfer in small channel systems, which allow for a higher efficiency of the catalyst.

Finally, the production quantity of fuel processors has a strong impact on costs. Different to most other applications discussed in the other chapters of this book, fuel processor fabrication is heading for high production quantities. Starting with some hundred to a thousand devices per year for first and niche applications, hundreds of thousands of fuel processors are the final goal. Increasing the production quantity from 100 to 10 000 devices reduces the fabrication cost by an order of magnitude [4].

2.2.2 Efficiency

System efficiency is a major aspect determining the competitiveness of fuel processors. The theoretically possible fuel processor efficiency decreases with increasing number of carbon atoms of the fuel molecule. This effect will be discussed in depth in Section 2.4.

The efficiency becomes more and more a crucial issue with increasing system size. Systems running at elevated pressure suffer from so-called parasitic power losses mainly stemming from the compressor energy demand [7]. For systems exceeding ~10 kW electrical power output capacity, the application of expander turbines for regaining the compression energy is a viable option. Consequently, the number of heat exchangers required for a specific fuel processor depends strongly on its overall power equivalent.

Applying compact micro structured heat exchangers allows for improvements in system efficiency at low cost concerning overall system size. In Section 2.5, micro structured reformer concepts will be presented which directly combine chemical reaction flow-paths of exothermic reactions with either heat-exchanging channels or endothermic reactions in an integrated device. Another opportunity for raising the overall system efficiency is the application of micro structured polymer heat exchangers which make use of low-temperature off-gas energy at high efficiency of the heat-exchanging process.

Few data are publicly available on fuel processor efficiency. However, a typical value of 77.3% was determined over the New European Drive Cycle (NEDC) for a fuel cell drive system supplied by methanol [8]. Johnson-Matthey reports even 89% for its HotSpot® module combined with a gas purification system [9].

2.2.3
Start-up Time

Another critical issue affecting end-users' acceptance is the start-up time of the fuel processor. This is affected mostly by the time demand for heating to the operating temperature. Three main options to get the individual devices of the fuel processor to the appropriate temperature exist:

- electrical power
- direct heating with a (start-up) burner
- indirect heating with a start-up burner.

Electrical heating requires power supply by an interim storage device, i.e. a battery. Even though batteries exist as buffer devices in most fuel cell system concepts, their size would need to increase considerably to meet the demands for start-up. Therefore, battery power is a less viable option especially for hydrocarbon reforming systems, where high operating temperatures of the reformer exceeding 600 °C need to be achieved.

Another option is direct heating through the flow path of the fuel by applying catalytic or conventional burners, which may be either part of the power supply system of the fuel processor or additionally installed devices (start-up burner). By direct contact of the combustion off-gas with the various devices, rapid start-up of even larger scale fuel processors seems to be feasible [10]. However, a major drawback is the restriction to catalyst systems which can tolerate hot combustion off-gases.

Indirect heating by a start-up burner through a dedicated flow path avoids the direct contact of the catalysts with combustion off-gases. Here micro structured channel systems offer unique possibilities at low overall pressure drop of the start-up system.

Generally, the start-up energy demand depends on the weight and heat capacity of the devices under consideration.

After the start-up heating procedure, stable operation of the fuel processor needs to be achieved. Micro structured devices allow for rapid system stabilization due to the fast dynamic response at the low residence times applied.

2.2.4 Size

A fuel processor size target of 600 W dm^{-3}, which is still an ambitious goal for most systems already existing (or under development), was given for 2000 by the US PGNV group [4]. The compactness of a fuel processor is mainly determined by the efficiency of both catalytic conversion processes and peripheral devices such as heat exchangers. Micro structured reactors allow for improved catalyst effectiveness and micro structured heat exchangers work at a high efficiency/volume ratio, thus leading to a substantial size reduction potential.

2.2.5 Weight

The weight of fuel processing devices is determined by their size and by the material employed. Therefore, a trade-off is necessary for each system and its individual components comparing the benefits of size reduction achieved by applying micro structured devices against the surplus of weight resulting from the application of metal foils normally made of stainless steel rather than ceramic honeycombs or conventional packed beds. The fuel processor weight target for 2 000 of 800 W kg^{-1} provided by the US PGNV group [4] is, like the groups size target, still ambitious.

2.2.6 Responsiveness to Load Changes

Generally, load changes from 5 to 100% should be tolerated by a fuel processor [11]. However, this is of course a more crucial issue for drive train applications than for APUs. Micro structured devices offer benefits concerning the load-change behavior of fuel processors owing to the small fluid volumes applied [12, 13].

2.2.7 Lifetime

The lifetime demand for automotive and portable applications is generally set to values around 5000 h, depending on the application, which is one order of magnitude lower than the demand of stationary (residential) applications.

2.3
Design Concepts of Micro Structured Reactors for Fuel Processing Applications

Depending on the targets heading for, various design concepts of micro structured reactor have been realized:

- Testing reactors: These are normally externally or electrically heated and may carry either single plates, a sandwich of plates or stacks of micro structured plates. They serve mostly for catalyst development purposes and to prove the feasibility of design concepts.
- Chip-like reactors: These devices utilize the potential of miniaturization and cheap mass production applying silicon as construction material. Future applications are mostly low-power systems (below a few watts).
- Integrated reactors: One type of integrated reactor is micro structured heat exchanger/reactor concepts, which may work as cross- or counter-flow reactors. Another type couples endothermic and exothermic reactions in two separate flow paths normally operated in the co-current mode. Both reactor types are designed as prototype components of future fuel processors for mobile applications.

In Sections 2.4–2.6 micro structured testing reactors for reforming, combustion and gas purification purposes and chip-like devices are presented; Section 2.7 is dedicated to integrated reactors and micro structured fuel processor concepts and prototypes.

2.4
Micro Structured Test Reactors for Fuel Processing

The general stoichiometry for the conversion of hydrocarbon or alcohol fuels $C_nH_mO_p$ can be written as follows [11]:

$$\begin{aligned} &C_nH_mO_p + x\,(O_2 + 3.76\,N_2) + (2\,n - 2\,x - p)\,H_2O\,(l) \\ &\rightarrow n\,CO_2 + \left(2\,n - 2\,x - p + \frac{m}{2}\right)H_2 + 3.76\,N_2 \end{aligned} \tag{2.3}$$

It was demonstrated that this general equation is valid for an overall fuel processor for both steam reforming (SR) and autothermal reforming (ATR) provided that the fuel combustion necessary to run the endothermic steam reforming process is taken into consideration [11]. It is even valid for the partial oxidation (POx), as soon as the water that is necessary in addition to run the subsequent water-gas shift (WGS) reactors is added to the system balance. Assuming the thermoneutral point, both ATR and SR are balanced concerning energy supply and demand. The molar oxygen to fuel (O_2/fuel) ratio x_0 at this point may be calculated for each fuel $C_nH_mO_p$ according to the following equation [11]:

$$x_0 = 0.312\,n - 0.5\,p + \frac{\Delta H_{\text{f, fuel}}}{\Delta H_{\text{f, water}}} \tag{2.4}$$

where ΔH_f is the heat of formation of fuel or water. The O_2/fuel ratio needs to be distinguished from the O_2/C ratio and from the O/C ratio, which is the atomic ratio of all feed oxygen atoms (which may partially stem from the fuel in case of alcohols) to the carbon atoms of the fuel. The value of O/C needs to be two in order to achieve full conversion.

It may be calculated that the theoretically possible efficiency of the fuel processor (see Section 2.1.3) decreases from methanol (96.3%) to gasoline (90.8%) owing to the increasing water demand of larger fuel molecules [11]. This is a consequence of the LHV definition, which does not take into consideration the heat of condensation of the water. However, the condensation energy of excess water may well be utilized in certain fuel processor systems, especially those designed for residential applications (hot water generation), thus raising the overall system efficiency. In some fuel processor concepts, the hydrogen (and for high-temperature fuel cells even carbon monoxide) not consumed by the fuel cell is fed back as an additional energy source, hence shrinking the fuel demand. This can lead to values of the fuel processor efficiency exceeding 100%, especially in steam reforming systems. However, this is merely a value resulting from the definition of fuel processor efficiency, and the overall efficiency of systems designed like that is not necessarily higher than for other concepts.

The reactor temperature required to prevent coke formation varies considerably for the different processes. Table 2.1 summarizes the values calculated assuming thermodynamic equilibrium for 2,2,4-trimethylpentane reforming. Generally, the coking tendency increases in the following order at constant O/C ratio: SR > ATR > POx. These calculations demonstrate that at steam to carbon ratios (S/C) > 2 and reaction temperatures > 600 °C, which is very common for hydrocarbon fuel processors, coke seems to be an unstable species especially under the conditions of steam reforming.

Table 2.1 Reactor temperature required to prevent coke formation for 2,2,4-trimethylpentane [11].

Reaction	*Feed composition*	*O/C ratio*	*Temperature (°C)*
Partial oxidation	$C_8H_{18} + 4\,(O_2 + 3.76\,N_2)$	1	1 180
Autothermal reforming	$C_8H_{18} + 2\,(O_2 + 3.76\,N_2) + 4\,H_2O$	1	1 030
Steam reforming	$C_8H_{18} + 8\,H_2O$	1	950
Autothermal reforming	$C_8H_{18} + 4\,(O_2 + 3.76\,N_2) + 8\,H_2O$	2	575
Steam reforming	$C_8H_{18} + 16\,H_2O$	2	225

In addition to coke formation and formation of carbon monoxide, the most familiar by-product is methane, which may be formed from carbon monoxide via the methanation reaction:

$$3\,H_2 + CO \rightarrow H_2O + CH_4 \qquad \Delta H_R = -253.7\ \text{kJ mol}^{-1} \tag{2.5}$$

However, most fuel cell systems can tolerate methane concentrations up to at least 1% in the reformate, no special purification reactions are required. In contrast, hence, removing small residual amounts of carbon monoxide from pre-purified reformate applying the methanation reaction may be considered as an alternative to the preferential oxidation of carbon monoxide, provided that the CO concentration is low enough to have no significant impact on the hydrogen yield. However, no applications of methanation for CO clean-up in micro structured devices appear to have been reported, hence the issue is not discussed in depth. Finally, during hydrocarbon reforming all hydrocarbon species (saturated and unsaturated) smaller than the feed molecule may be formed.

This section will focus on published work dealing with reactor development, catalyst development and investigations elucidating the mechanisms and kinetics of reforming in micro structured channel systems. Integrated concepts of fuel processors will be presented in Section 2.7.

An early proposal for applying coated catalyst systems in reforming applications not yet heading for mobile fuel cell systems was made about some 20 years ago by Ramshaw [14]. By testing various ceramic and metallic monoliths available for automotive exhaust systems, Ratnasamy et al. [15] demonstrated that reforming reactions such as autothermal reforming of methane benefit from channel dimensions in the sub-millimeter range. An increase in conversion from 80 to 98% was found on decreasing the cell size of monoliths from 1.6 to 0.8 mm under identical operating conditions (GHSV 212 000 h^{-1}, S/C 2.5, O/C 0.83).

2.4.1
Methanol Steam Reforming (MSR)

The overall equation for methanol steam reforming is:

$$CH_3OH + H_2O \rightarrow CO_2 + 3\,H_2 \qquad \Delta H_R = 48.8 \text{ kJ mol}^{-1} \tag{2.6}$$

The reaction may be considered as a combination of the endothermic decomposition of methanol:

$$CH_3OH \rightarrow CO + 3\,H_2 \qquad \Delta H_R = 89.2 \text{ kJ mol}^{-1} \tag{2.7}$$

Followed by the exothermic water-gas shift reaction:

$$CO + H_2O \rightarrow CO_2 + H_2 \qquad \Delta H_R = -40.4 \text{ kJ mol}^{-1} \tag{2.8}$$

An alternative reaction mechanism was proposed by Takahashi et al. [16]:

$$CH_3OH \rightarrow HCHO + H_2 \qquad \Delta H_R = 92 \text{ kJ mol}^{-1}$$

$$HCHO + H_2O \rightarrow HCOOH + H_2 \qquad \Delta H_R = 28.3 \text{ kJ mol}^{-1}$$

$$HCOOH \rightarrow CO_2 + H_2 \qquad \Delta H_R = -14.8 \text{ kJ mol}^{-1} \tag{2.9}$$

Methanol is the fuel which may be converted at the lowest reaction temperatures, normally not exceeding 300 °C. Hence relatively small amounts of carbon monoxide are generated. Figure 2.2 shows the CO concentration theoretically expected in the product flow according to a DuPont patent [17]. Thus, assuming a water/methanol molar ratio of > 2 and a reaction temperature of 300 °C, not more than 1.2% of CO will be present in the feed. This reduces the workload of the subsequent gas purification steps. Water-gas shift is therefore obsolete for fuel processors applying methanol steam reforming.

Palo et al. [18] found surprisingly low carbon monoxide concentrations, not higher than 1.2%, at reaction temperatures as high as 375 °C with their proprietary catalyst. This was ascribed to the short residence times applied (50–100 ms). Under these conditions, assuming the reaction mechanism proposed by Takahashi et al. shown above, carbon monoxide could only be formed by the reverse water-gas shift reaction, which is known to be slower than the reforming reaction. This is the case especially for catalyst systems with low activity towards water-gas shift. Holladay et al. [19] compared the performance of the same proprietary catalyst with that of a Cu/Zn catalyst which produced a higher carbon monoxide concentration of 3.1% in the reformate.

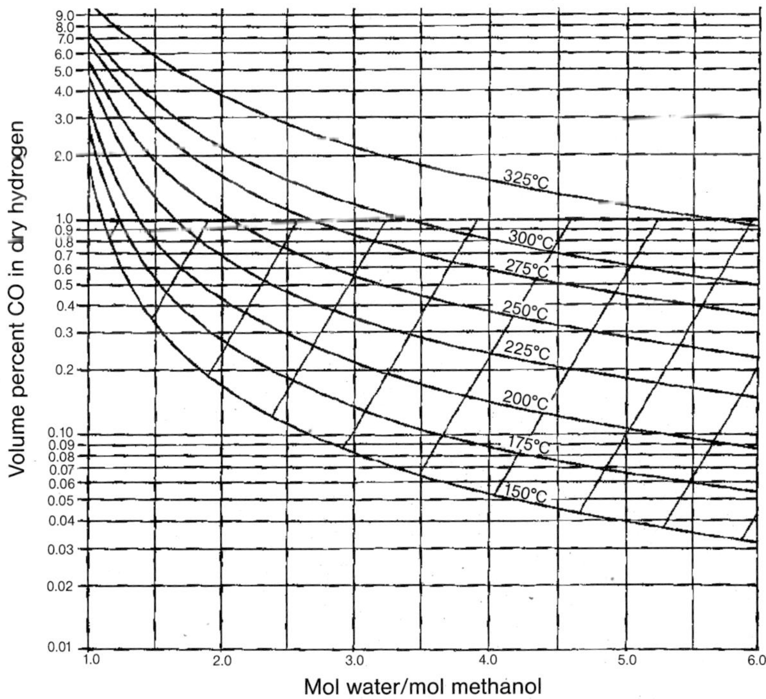

Figure 2.2 Theoretical carbon monoxide concentration obtained by methanol steam reforming as a function of the molar water/methanol feed ratio and reaction temperature [17].

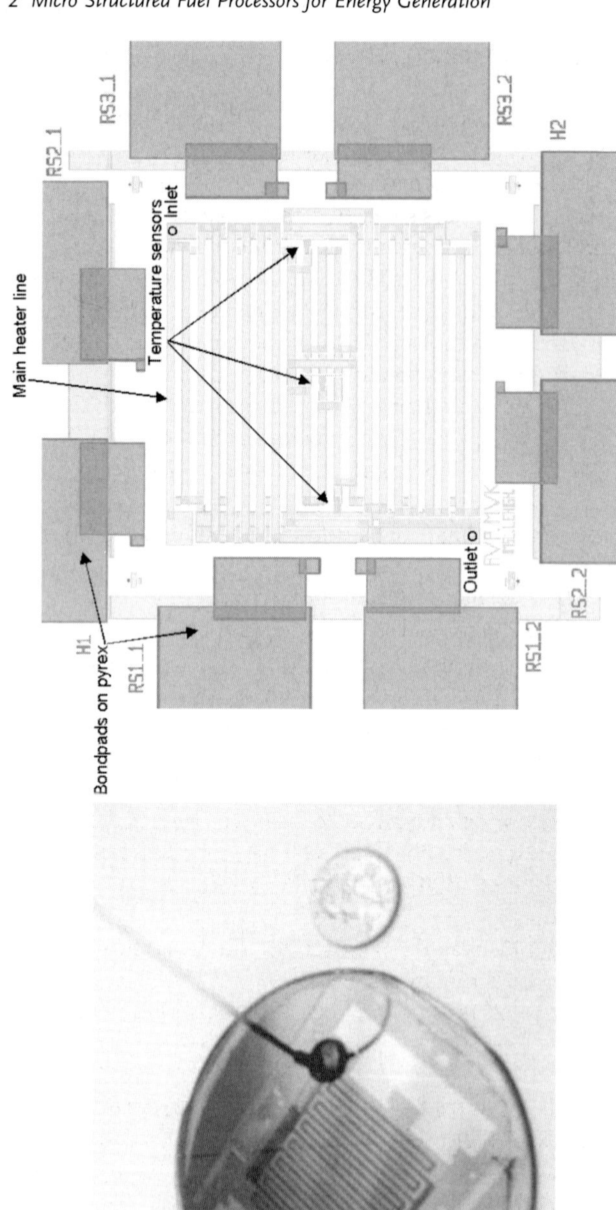

Figure 2.3 Photograph of the mounted device and superimposed view of the photomasks which shows the functionality [21] (by courtesy of SAGE Publications).

2.4.1.1 Methanol Steam Reforming 1 [MSR 1]: Electrically Heated Serpentine Channel Chip-like Reactor

Pattekar et al. [20] made use of a silicon chip for hydrogen production from methanol steam reforming. The chip-like reactors as a future part of a fuel processor/fuel cell system were regarded as an alternative to portable power supply such as batteries. The reactor housing was made of stainless steel, electrically heated and sealed with graphite. The micro channel was a long serpentine of 1 000 µm width and 230 µm depth fabricated by photolithography and KOH etching. Cu catalyst was sputtered to a thickness of 33 nm on to the chip. Preliminary simulations revealed a non-uniform temperature distribution in the reactor housing, pointing to the importance of proper insulation especially in lower power systems.

At a feed composition of 76 mol% methanol in steam, less than 7% conversion was achieved at 250 °C. Selectivity to carbon monoxide was higher than that to carbon dioxide and 7 mol% hydrogen was found in the product [20].

2.4.1.2 Methanol Steam Reforming 2 [MSR 2]: Electrically Heated Parallel Channel Chip-like Reactor

Later, Pattekar and Kothare [21] presented a silicon reactor fabricated by deep reactive ion etching (DRIE). It carried seven parallel micro channels of 400 µm depth and 1 000 µm width filled with commercial Cu/ZnO catalyst particles (from Süd-Chemie) trapped by a 20 µm filter, which also was made by DRIE, in the reactor. The reactor was covered by a Pyrex™ wafer applying anodic bonding. Details of the reactor are shown in Figure 2.3.

The catalyst was introduced as a suspension into the ready-made reactor. Resistance devices for temperature sensing and a meander for heating, both made of platinum, were incorporated into the reactor, which had overall dimensions of 40 × 40 × 3 mm. A pressure drop of approximately 1.6 bar was calculated for the whole reactor at the design point.

A deviation not exceeding 10% was determined for the flow through each individual channel by numerical simulation. Methanol conversion of 88% was achieved at 200 °C reaction temperature at a water/methanol molar feed composition of 1.5 [21].

2.4.1.3 Methanol Steam Reforming 3 [MSR 3]: Electrically Heated Stack-like Reactor

Pfeifer et al. [22] used a stack-like reactor heated by cartridges for methanol steam reforming. The aim of this work was catalyst development heading for a micro structured reformer supplying a 200 W fuel cell for small-scale mobile applications. Up to four plates carrying 80 channels 64 mm long and 100 µm wide and deep could be introduced into the reactor. To elucidate the dynamic behavior of their micro structured testing reactors, three electrically heated reformers of [MSR 3] type with housings made of both stainless steel and copper were set up by Pfeifer and co-workers [13, 23]. They were designed for a hydrogen output sufficient to feed a 200 W fuel cell (Figure 2.4) [13].

Numerical calculations revealed that isothermal operation was feasible throughout a catalyst layer as thick as 60 µm in a micro structured reactor under the conditions

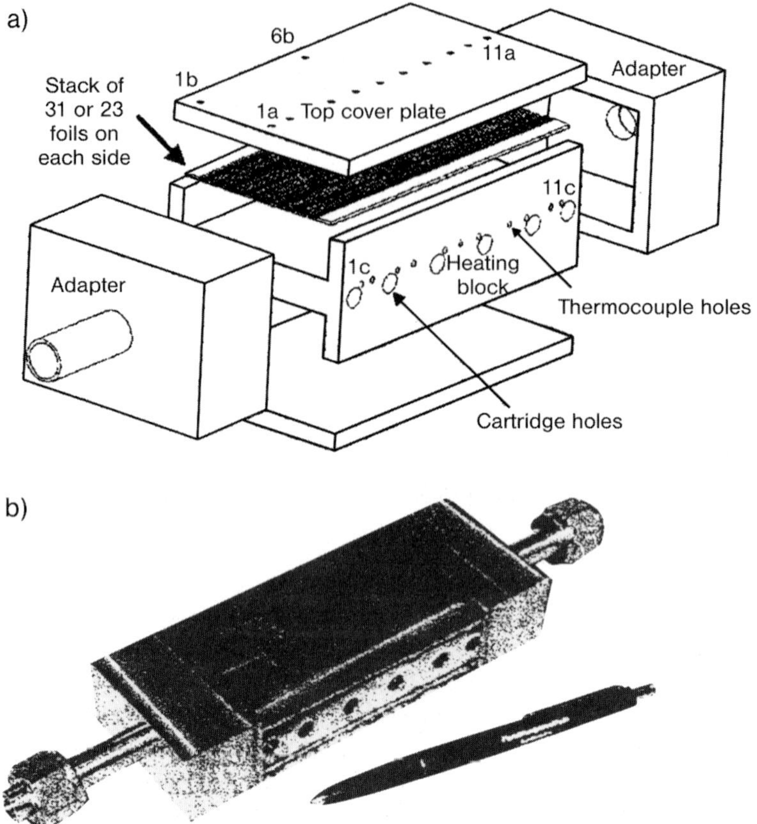

Figure 2.4 (a) Schematic exploded view of an electrically heated micro reformer and (b) a photograph of the mounted and welded device [13] (by courtesy of Springer VDI Verlag).

of fast heating and cooling. The characteristics of the devices are summarized in Table 2.2. In two of the reactors, additional fins were introduced over the whole length of the foils to improve the heat transfer throughout the reactor perpendicular to the flow direction.

The catalyst (Pd/ZnO) was introduced by wash coating prior to the mounting, which was performed by electron beam welding. Calcination and reduction of the catalyst were performed after the welding procedure. The total heating power of the six heating cartridges was 1.5 kW. Bores were introduced for temperature determination at various positions. The feed inlet temperature was set to 140 °C.

At 310 °C reaction temperature, a pressure of 1.25 bar and an S/C ratio of 1.9, 80% methanol conversion was achieved in the stainless-steel reactors compared with > 90% for the copper reactor. The carbon monoxide concentration determined in the reformate was 0.5 vol.%. Lower temperature gradients were found for the copper reactor (7 K) than the stainless-steel device (18 K). Non-stationary measurements were carried out at the reformer reactors. The reaction followed the

Table 2.2 Characteristics of three different electrically heated reactors ([MSR 3] type) [13].

	Type RS1	*Type RS2*	*Type RC*
Material foil	$AlMg_3$	$AlMg_3$	Copper
Material Housing	Steel 1.4301	Steel 1.4301	Copper
Channel width/depth (μm)	100/100	100/100	100/150
Additional heat transfer fins	3 (2 mm width)	None	3 (2 mm width)
Number of foils	62	62	46
Foil length/width (mm)	100/50	100/50	100/50
Reaction volume (cm^3)	17.3	19.8	19.3
Catalyst mass (g)	13.5	13.8	13.2
Total reactor weight (g)	1 900	1 907	2 136

temperature immediately. It took 60–70 s for the stainless-steel reactors to reach 90% of the final conversion on switching from bypass to reactor on stream. The chemical equilibration was calculated to take place even faster, within 10 s. Dimethyl ether was found as a by-product at concentrations up to several hundred ppm exclusively in the stainless-steel reformer.

2.4.1.4 Methanol Steam Reforming 4 [MSR 4]: Externally Heated Stack-like Reactor

Reuse et al. [24] applied a reactor carrying micro structured plates for methanol steam reforming over commercial copper-based low-temperature water-gas shift catalyst from Süd-Chemie. The reactor took up 20 plates made of FeCrAl alloy of size 20 mm × 20 mm × 0.2 mm. The channel size was 200 μm × 100 μm (Figure 2.5). The catalyst was conditioned by oxygen and hydrogen treatment.

Kinetic expressions were determined for both a tubular fixed-bed reactor containing 30 mg catalyst particles and the micro reactor coated with the catalyst particles. The coating was performed by filling the channel system with a suspension of the catalyst in 2-propanol and subsequent removal of the organic solvent by heat treatment at 500 °C [24].

Figure 2.5 Photograph of the externally heated stack-like reactor [24] (by courtesy of Springer Verlag).

Table 2.3 Reaction order of fixed-bed and micro reactors determined by Reuse et al. for methanol steam reforming [24].

Parameter	*Fixed bed*	*Micro channels*
m	0.70 ± 0.02	0.70 ± 0.1
n	0.1 ± 0.04	0.0 ± 0.1
o	-0.1 ± 0.1	-0.2 ± 0.1
k	$7.8 \cdot 10^{-5} \pm 0.9 \cdot 10^{-5}$	$4.8 \cdot 10^{-5} \pm 0.6 \cdot 10^{-5}$

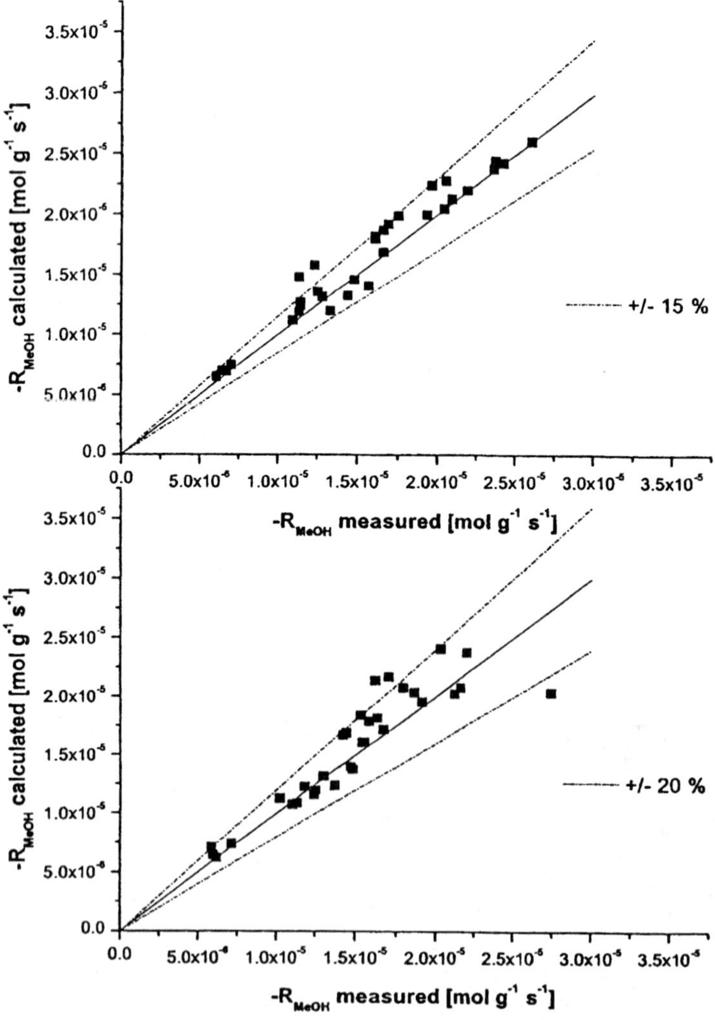

Figure 2.6 Parity plot of measured and experimentally determined reaction rates of methanol steam reforming [24] (by courtesy of Springer Verlag).

The experiments were carried out at a pressure of 1.5 bar and a flow rate of 80–270 $Ncm^3\ min^{-1}$. At 200 °C no deactivation of the catalyst was observed. As the rate of reaction was found to show a linear dependence on the residence time, differential conditions were assumed for the measurements. Because of the determined high activation energy of 56 $kJ\ mol^{-1}$, mass transport limitations were excluded. A power law kinetic expression of the following form was determined for methanol steam reforming:

$$-R_{CH_3OH} = k_0\, e^{-\frac{E_a}{RT}}\, p^m_{CH_3OH}\, p^n_{H_2O}\, p^o_{H_2} \tag{2.10}$$

Similar values were found for the reaction order in both systems at a lower rate of reaction for the micro channels as shown in Table 2.3 [24].

The inhibition by hydrogen was obviously more pronounced in the micro channels. Without hydrogen in the feed, the reaction rate was on average 34% higher for the coated catalysts. The kinetic expression described the reaction rate experimentally observed with an error of < 15% for the packed bed and < 20% for the micro channels (see Figure 2.6) [24].

2.4.1.5 **Methanol Steam Reforming 5 [MSR 5]: Electrically Heated Stack-like Reactor**

Cominos et al. [25, 26] developed a stack-like micro reactor for testing catalysts under methanol steam reforming conditions for PEM fuel cells (Figure 2.7). The stainless-steel device had outer dimensions of 75 mm × 45 mm × 110 mm and took up a stack of 5–15 plates carrying micro channels of 500 μm width and 350 μm depth with a length of 50 mm. The feed distributed between the channels and plates from a common inlet region. Thorough simulation work using the software CFD-ACE+ proved that the flow distribution deviated by < 2% for both the channels of one plate and for the individual plates.

Self-made Cu/Zn catalysts were prepared by introducing γ-alumina wash coats with an average thickness of 10 μm as carrier material into the micro channels. Their BET surface area was determined as 72 $m^2\ g^{-1}$ and the average pore diameter was 45 nm. The active components, Cu and Zn, were introduced by wet impregnation at two different load levels (8 and 16 wt.%) and weight ratios (3 : 1 and 1 : 1) [26].

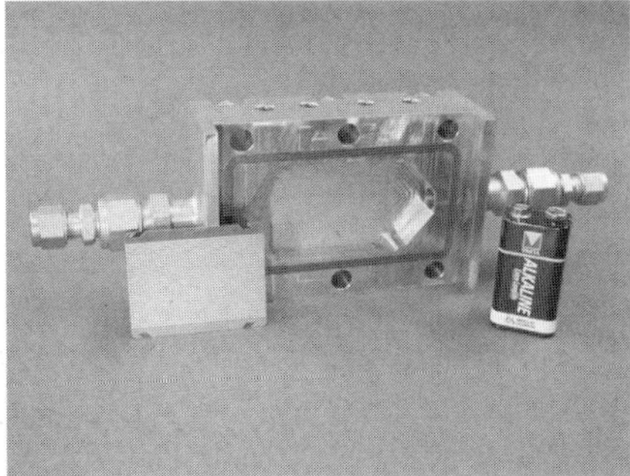

Figure 2.7 Close-up view of the experimental set-up and stack-like reactor [25] (source IMM).

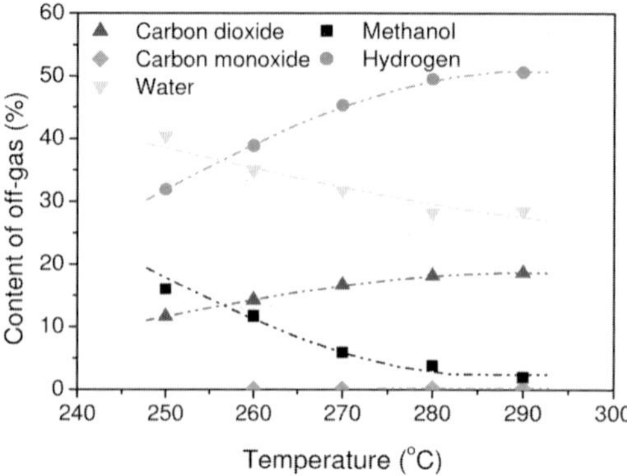

Figure 2.8 Content of the product gas at a flow rate of 500 Ncm^3 min^{-1} using 15 plates in reactor [MSR 5] [26] (by courtesy of Taylor & Francis Ltd.).

The experiments were carried out at residence times between 200 and 100 ms and total flow rates between 500 and 900 Ncm^3 min^{-1} using five coated plates. A 15% decrease in conversion was found when the flow rate was increased from 500 to 800 Ncm^3 min^{-1}, leading to a 36% relative decrease in the hydrogen content of the product. This was attributed to the slow kinetics of the reverse water-gas shift reaction. Increasing the temperature from 200 to 275 °C increased the conversion from 37 to 65%. At temperatures exceeding 250 °C, carbon monoxide formation started [26].

When 15 plates were introduced into the reactor, 80% conversion was achieved at a 290 °C reaction temperature, a residence time of 600 ms and an S/C ratio of two, resulting in a product containing more than 50% hydrogen and 0.25 vol.% CO (see Figure 2.8), which pointed to the need for an additional gas purification device to remove the CO before the product enters the fuel cell. From these results a power output of the testing device of 40 W and a power density of 1.7 kW Ndm^{-1} were calculated.

2.4.1.6 **Methanol Steam Reforming 6 [MSR 6]: Electrically Heated Screening Reactor**

Kolb et al. [27] developed a novel modular screening reactor for testing up to 10 catalyst carrier plates. The reactor was made of stainless steel and heated by heating cartridges (see Figure 2.9). The housing contained drawers in which the micro structured plates were placed. By variation of the end-caps of the reactor, two operational modes were possible: parallel screening of up to 10 plates and serial operation of up to 10 identically coated catalyst plates. The latter gave the opportunity of modifying the residence time via the catalyst mass under identical flow conditions. This reactor had dimensions of 70 mm × 55 mm × 64 mm; the micro structured channels were 300 μm deep and 500 μm wide with a length of 50 mm. The functionality of the reactor concerning cross-talk between the individual channels in the parallel operation mode was verified.

Ziogas et al. [28] performed catalyst screening with this reactor with catalysts coatings, which were made of various base aluminas such as corundum, boehmite and γ-alumina. Testing of Cu/Cr and Cu/Mn catalysts based on the different coatings for methanol steam reforming revealed differences in activity which were ascribed

Figure 2.9 Micro structured and modular screening reactor developed at IMM [25] (source IMM).

to both the different surface area and the morphology of the carrier material. The functionality of the serial operation mode of the reactor was proven for the same reaction system, indicating mass transfer limitations at higher flow rates.

2.4.1.7 Development of Catalyst Coatings for Methanol Steam Reforming in Micro Channels

After discussing the various test reactors, some work focusing more on catalyst development is presented below.

A commercial $CuO/ZnO/Al_2O_3$ catalyst was coated on quartz and fused silica capillaries by Bravo et al. [29] for methanol steam reforming and compared with packed-bed catalysts. The coatings had a thickness of 25 μm and showed 97% conversion and 97% selectivity towards carbon dioxide at 230 °C reaction temperature, a water/methanol molar feed composition of 1.1 and a space velocity of 45 kg_{cat} s mol^{-1} (methanol).

Pfeifer et al. [22] used a stack-like reactor ([MSR 3]) for testing of various CuO/ZnO and PdZnO catalyst systems:

- CuO/ZnO (1 : 1) catalyst based on nanoparticles sintered at 450 °C for 5 h; the BET surface area of the plates coated with this catalyst was 9.3 $m^2 g^{-1}$ and a maximum of the mesopore distribution was found at 100 nm; the layers had an average thickness of 20 ± 10 μm at the channel bottom.
- CuO/ZnO (1 : 1) catalyst based on nanoparticles sintered at 550 °C.
- $CuO/ZnO/TiO_2$ (39 : 39 : 22) catalyst based on nanoparticles sintered at 450 °C for 5 h.
- Pd/ZnO (1 : 99) catalyst based on ZnO nanoparticles which were impregnated with Pd acetate, calcined at 450 °C and reduced at 500 °C for 5 h in 1 vol.% H_2.
- $CuO/ZnO/TiO_2$ (25 : 2 : 3) catalyst based on TiO_2 nanoparticles which were impregnated with $Cu/ZnNO_3$ and calcined at 550 °C for 5 h.

The nanoparticles had an average particle size (APS) between 20 and 60 nm and BET surface areas between 17 and 58 $m^2 g^{-1}$, which decreased with increasing calcination temperature, reaching values below 1 $m^2 g^{-1}$ at 750 °C. To verify the

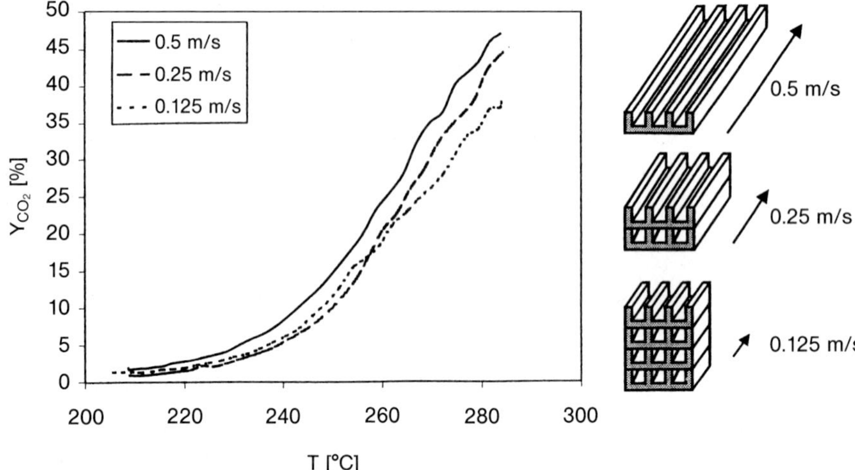

Figure 2.10 Carbon dioxide yield from methanol steam reforming vs. temperature at a constant residence time of 125 ms and different channel geometries, from Pfeifer et al. [22] (by courtesy of Springer Verlag).

absence of diffusion limitations, the modified residence time was varied, with one, two and 4 micro structured platelets of decreasing length L, $L/2$ and $L/4$. On decreasing the flow velocity u at the same time from u to to $u/2$ and $u/4$, the modified residence time $\tau = m_{cat}/Q$, m_{cat} being the catalyst mass and Q the total flow rate, remained unchanged, thus allowing direct comparison of the experiments. Figure 2.10 shows that no mass transport limitations were found for the reaction system.

For the standard experiments a 2 : 1 molar water/methanol mixture was fed together with 60 vol.% helium at a pressure of 3 bar into the reactor with a hydrodynamic residence time of 250 ms. The reaction started at 230 °C and showed maximum carbon dioxide yields between 235 and 250 °C. The best carbon dioxide yields were found for the catalyst based on TiO_2 impregnated with CuO/ZnO.

The Pd catalyst showed higher activity and less deactivation at 285 °C, but also higher CO yields, which exceeded the thermodynamic equilibrium for the water-gas shift reaction. It was assumed that the Pd could not form alloys with the Zn and therefore propagated CO formation in its elemental form. The activity of the Pd catalyst could be stabilized by small amounts of oxygen added to the feed, which was attributed to removal of coke. For the impregnated TiO_2 catalyst, CO yields exceeding the water-gas shift equilibrium were also found. Only the CuO/ZnO catalysts showed low carbon monoxide conversion at or (for the pre-reduced sample) even below the water-gas shift equilibrium, but at low degrees of conversion. Copper in its oxidized form is known to show low CO formation. However, under the reducing atmosphere the copper is likely to be in a mixed oxidation state, which propagates carbon monoxide formation.

Pfeifer et al. [23] focused on Pd/PdZn/ZnO systems for methanol steam reforming. In the same reactor ([MSR 3]) the formation of a Pd/Zn alloy at higher reduction temperatures was identified as the crucial step. Additionally, pre-impregnation of the ZnO particles with palladium before the coating procedure decreased the carbon monoxide selectivity by a factor of three. However, the CO concentration still exceeded the equilibrium of the water-gas shift reaction, which was not the case for a pelletized catalyst of the same type and therefore attributed to an interaction between the metal of the micro structured carrier and the catalyst. Chemisorption measurements with hydrogen and CO revealed a poor dispersion of < 10% for PdZn. TPO measurements revealed slow destruction of the PdZn alloy in air at temperatures exceeding 200 °C, and therefore diffusion of oxygen into the reactor was seen as a possible reason for the increased CO selectivity. However, copper-based catalysts are known to be much more sensitive to oxidation.

In a later study, Pfeifer et al. [30] prepared Pd/Zn catalysts by both pre- and post-impregnation of wash-coated zinc oxide particles with palladium and compared their performance in methanol steam reforming. The catalytic performance of the samples was tested at a 250 °C reaction temperature, 3 bar pressure, a S/C ratio of two and 250 ms residence time. The WHSV amounted as 0.3 Ndm^3 (min g_{cat})$^{-1}$. The thickness of the coatings was calculated to 20 μm. The formation of the PdZn alloy was proven to occur at temperatures exceeding 200 °C by XRD measurements.

For a standard sample containing 10% Pd, the stability of the catalyst against an oxidative atmosphere at temperatures up to 200 °C was verified by XRD. The effect of the parameters of catalyst preparation on palladium dispersion was investigated. Changing the temperature of the first calcination step of the pre-impregnated samples from 150 to 250 °C did not change the palladium dispersion. The temperature of the second calcination step (applied in both preparation routes) had an impact; the metal dispersion decreased above 450 °C. However, as the TOF was found to increase with increasing crystal size of PdZn, no deterioration of the catalytic activity was observed as both effects compensated each other. Variation (from 1 to 5%) of the hydrogen concentration during the final reduction step did not affect the dispersion, but maximum dispersion was found at 400 °C reduction temperature.

Finally, palladium dispersion was found to decrease asymptotically from 9% (1 wt.% palladium content) to 0.5% (20 wt.% palladium content). The effect of carbon monoxide concentrations exceeding the equilibrium values of the water-gas shift reaction was studied in depth. With increasing time on-stream and deactivation of the catalyst (containing 1% palladium), the carbon monoxide concentration exceeded the equilibrium up to 18-fold. Even higher values were found for lower reaction temperatures.

Pre-impregnation increased the carbon monoxide yields compared with post-impregnation as determined for the samples containing 10 wt.% palladium. The origin of the large amounts of carbon monoxide was finally attributed to interaction of palladium with the metal foils during the post-impregnation procedure, which was proven by the preparation of powder catalyst (no coating). These samples showed carbon monoxide concentrations below the equilibrium of the water-gas shift reaction. Isolated Pd(0) was considered to form the excess carbon monoxide. To assess the effect of the water-gas shift reaction on the reaction system, carbon

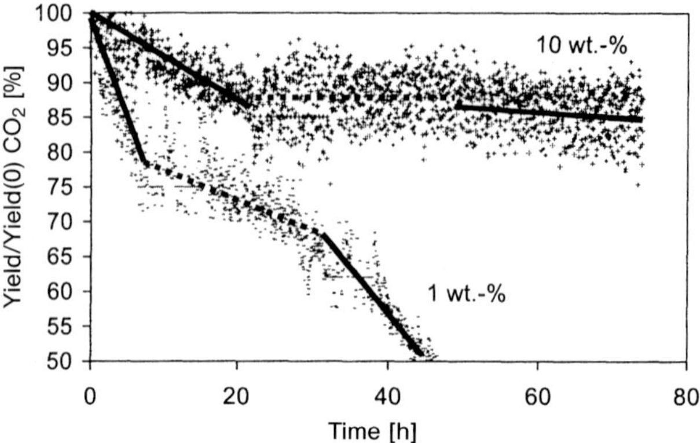

Figure 2.11 Ratio of initial yield to yield = $f(t)$ for CO_2 for 10 and 1 wt.% Pd catalyst vs. time on-stream at 285 °C before (continuous line) during (dotted line) and after (continuous line) air addition to the reaction mixture [30] (by courtesy of Elsevier Ltd.).

monoxide and carbon dioxide were added to the feed. They left the reactor unconverted, the water-gas shift species ratio was not affected and the reaction was therefore regarded as negligible under the experimental conditions applied.

At a 1% palladium loading, the initial activity of the pre-impregnated samples was much lower (25% conversion) than that of the post-impregnated samples (80% conversion). For both preparation routes, the highest activity was determined for the samples containing 10 wt.% palladium, which were also most stable against deactivation. At this loading level, the pre-impregnated sample even showed slightly superior activity. The addition of 2.5% air to the feed slowed the deactivation of the samples and, even after turning off the air, deactivation was slower than before (see Figure 2.11). Variation of the catalyst mass (layer thickness) almost linearly affected the conversion, but deactivation increased almost exponentially with decreasing catalyst mass.

Chin et al. [31] studied Pd/ZnO catalysts for methanol steam reforming, heading for a 10–50 W micro structured fuel processor. The catalysts under investigation contained 4.8, 9.0 and 16.7 wt.% Pd deposited on ZnO powder by impregnation. The catalysts were thoroughly characterized by thermogravimetry (TG), TPR, XRD and TEM.

It was found that a PdZn alloy disperses on the ZnO matrix under the conditions of methanol steam reforming. This alloy was regarded as the origin of the low carbon monoxide selectivity observed. In contrast, the presence of metallic Pd was thought to be the origin of high carbon monoxide selectivity.

Experiments were performed in 4 mm diameter quartz tubes carrying 200 mg of catalyst powder. A 1 : 1 weight ratio of water/methanol was fed to the reactor at a GHSV of 36 000 h^{-1}, corresponding to a short hydrodynamic residence time of 100 ms. The catalysts were reduced in 10% hydrogen in nitrogen at 350 °C prior to activity testing (Figure 2.12).

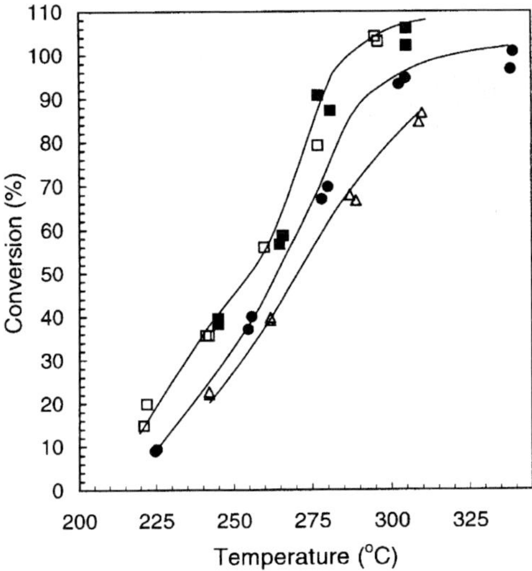

Figure 2.12 Methanol conversion vs. temperature for steam reforming of methanol. (△) 4.8, (●) 9.0 and (□) 16.7 wt.% Pt on ZnO (both filled and open squares); 0.1925 g catalyst, 100 ms contact time, 36 000 GHSV, $H_2O/C = 1.8$ and 1 atm [31] (by courtesy of Elsevier Ltd.).

Experimental results supported the assumption that this temperature was necessary to gain the PdZn alloy on the catalyst surface. No catalyst deactivation was detectable during the experiments. At 300 °C full conversion was achieved at a 100 ms residence time [32] and 5% and lower carbon monoxide selectivity. First order kinetics were determined, revealing $7.04 \cdot 10^{13}\ h^{-1}$ for the pre-exponential factor and 92.8 kJ mol^{-1} for the activation energy.

Later, Chin et al. [33] continued their work on Pd/ZnO catalysts. They stated that the major drawback of applying copper-based catalysts for mobile fuel processors was the sintering mechanisms at temperatures exceeding 270 °C and the pyrophoricity of copper-based catalysts. The latter is a safety issue regarding accidental exposure of the catalyst to air.

A detrimental effect of excess nitric acid on the ZnO support was observed, resulting in a reduced ZnO particle size and losses of surface area. This excess was present in the palladium nitrate solution that was applied for catalyst impregnation. Additionally it was found that the PdZn alloy was not only formed during the initial reduction step but also in situ in the hydrogen-rich reaction mixture of methanol steam reforming.

The [MSR 6] reactor type (see below) was applied for methanol steam reforming over $Cu/CeO_2/Al_2O_3$ catalysts by Men et al. [34, 35]. Wash coating of the alumina was performed, followed by subsequent impregnation steps with ceria and copper salt solutions. At 250 °C reaction temperature and a water/methanol molar ratio of 0.9, the copper/ceria atomic ratio was varied from 0 to 0.9, revealing the lowest conversion for pure ceria and a sharp maximum for a ratio of 0.1 (see Figure 2.13).

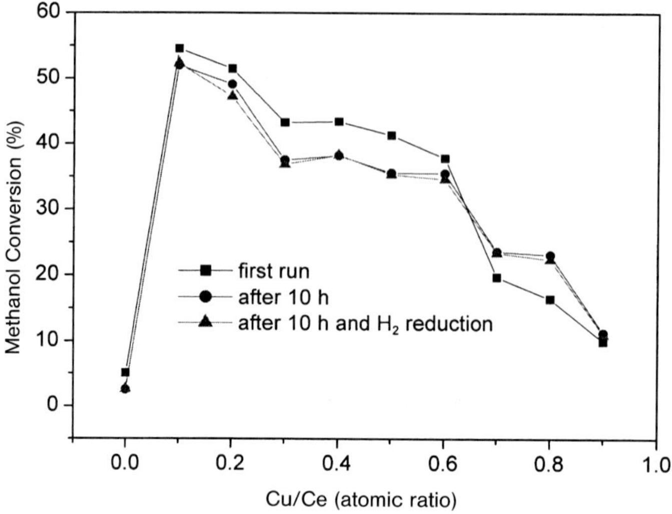

Figure 2.13 Influence of copper loading on catalytic activity over $Cu/CeO_2/\gamma\text{-}Al_2O_3$ catalysts [35] (by courtesy of Elsevier Ltd.).

The carbon monoxide selectivity was well below 2% for all samples. As a by-product, substantial amounts of dimethyl ether were found for all samples; the highest selectivity of 23% was detected over pure ceria. Only traces of another by-product, methyl formate, were measured. The dimethyl ether formation was attributed to separate dehydration of the methanol on the alumina surface.

However, the steam reforming activity of the catalyst was attributed to the copper/ceria metal/support interface. The improved dispersion of copper on ceria was supported by XPS measurements and both XPS and SEM–EDX measurements suggested the enrichment of copper and ceria on the alumina surface. Excess copper, however, is known to form bulk particles which do not contribute substantially to overall catalyst activity. For low copper loadings, an enhanced reducibility of copper was regarded as possible origin of the higher activity of the corresponding catalysts. However, Cu^+ was regarded as the most favorable oxidation state for methanol adsorption at this specific catalyst type.

2.4.2
Autothermal Methanol Reforming

The overall equation [Eq. (2.3)] for autothermal methanol reforming is as follows for 300 °C:

$$4\,CH_3OH + 3\,H_2O + 0.5\,O_2 \rightarrow 4\,CO_2 + 11\,H_2 \qquad \Delta H_R = +0\text{ kJ mol}^{-1} \qquad (2.11)$$

It is a combination of exothermic steam reforming and endothermic partial oxidation applying a stoichiometric feed ratio which allows for an overall zero energy balance. As the exothermic reaction is faster, a hot-spot is very common in autothermal reactors, which is manifested in the patented HotSpot® design of Johnson Matthey [9].

The benefits of applying micro structured reactors for autothermal reforming are manifold. Besides the well-known narrowing of the residence time distribution and the low pressure drops, hot-spot formation may well be reduced owing to the axial heat transfer of the wall material.

2.4.2.1 Autothermal Methanol Reforming 1 [AMR 1]: Micro Structured Autothermal Methanol Reformer

Chen et al. [36, 37] developed a $Pt/CeO_2/ZrO_2$ wash-coated catalyst and tested its performance on micro structured plates made of both stainless steel and aluminum. The micro channels of the stainless-steel plates were 170 µm deep, 500 µm wide and 30 mm long. Five plates carrying 48 channels each were incorporated into the reactor. The micro channels of the aluminum plates were 400 µm deep, 800 µm wide and 30 mm long. Four plates carrying 38 channels each were incorporated into this second reactor. A single micro structured platelet is shown in Figure 2.14.

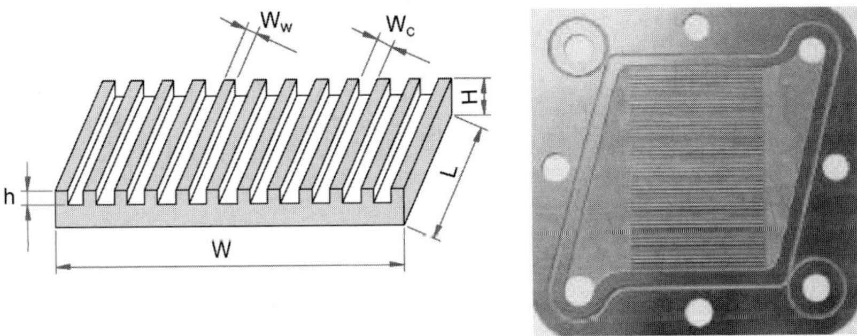

Figure 2.14 Photograph of a single microstructured platelet (right) [38] (by courtesy of ACS).

The catalyst coating was reduced in a mixture of 10% hydrogen in nitrogen for 2 h at 400 °C prior to testing. The S/C ratio was 1.2 and the molar oxygen to methanol ratio was 0.3 and higher for the experiments, which were performed at reaction temperatures between 300 and 450 °C and a GHSV of 7170 h^{-1}.

At this residence time, only 50% of the methanol feed could be converted. A reaction temperature as high as 400 °C and an increased oxygen/methanol ratio of 0.36 was necessary to achieve full conversion of the feed. This high reaction temperature resulted in a high carbon monoxide concentration in the reformate between 10.5 and 11% at 400 °C, which slightly decreased with increasing oxygen to methanol ratio. No further increase in the carbon monoxide content was observed on further increasing the reaction temperature to 450 °C.

2.4.2.2 Autothermal Methanol Reforming 2 [AMR 2]: Micro Structured String Reactor for Autothermal Methanol Reforming

Strings of brass wires with a diameter of 200–400 µm were applied by Horny et al. (Figure 2.15) [39]. The resulting voids thus had a hydraulic diameter in the region

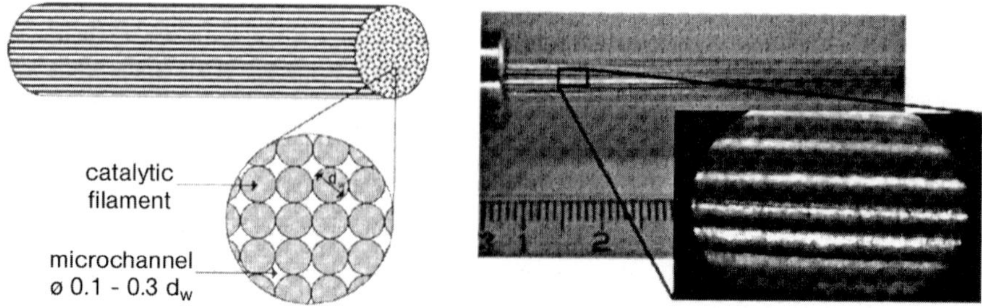

Figure 2.15 Schematic and photograph of the microstructured string reactor [39] (by courtesy of ACS).

of 100 μm. As methanol reforming is generally operated at relatively low reaction temperatures well below 350 °C, the advantages of brass, namely its high thermal conductivity (120 W mK^{-1}) and its composition (here 20–37% copper, 63–80% zinc), which is similar to the $Cu/ZnO/Al_2O_3$ catalyst commonly applied for methanol steam reforming, could be exploited. The catalyst was prepared by forming an aluminum/metal alloy on the outer surface of the brass. By acid or basic leaching of the aluminum, specific surface areas in the range between 30 and 55 $m^2\ g^{-1}$ could be achieved. During acid leaching mainly zinc was dissolved at increasing mass loss with increasing leaching time, whereas both zinc and aluminum were removed by basic leaching. In the latter case 2–3% mass losses were found independent of the duration of the leaching process.

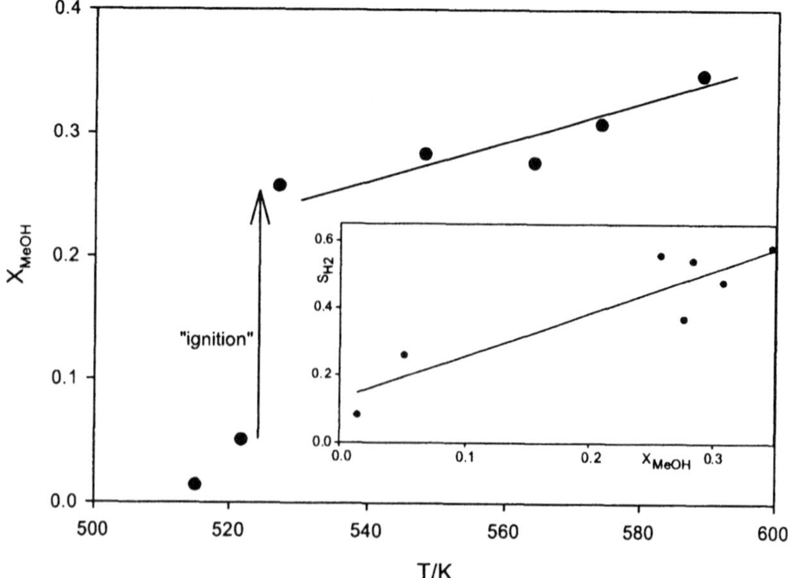

Figure 2.16 Methanol conversion as a function of temperature; hydrogen selectivity vs. methanol conversion for autothermal methanol reforming [39] (by courtesy of ACS).

For steam reforming, superior performance of the samples prepared by acid leaching concerning activity and stability was determined. A sample free of alumina showed no activity in steam reforming, emphasizing the need for alumina as active phase. The samples prepared by basic leaching showed rapid deactivation under steam reforming conditions. At 290 °C reaction temperature and a feed composition of 9 vol.% methanol and 11% water, which corresponds to a S/C ratio of 1.2, 65% methanol conversion could be achieved at 99% hydrogen selectivity for the CuZn37 samples treated by acid leaching for 20 min. Partial oxidation experiments were performed on the same sample.

For an O_2/CH_3OH molar ratio of 0.4 at 245 °C reaction temperature, 62% conversion was achieved and a hydrogen selectivity of about 82% was found. During experiments performed in the autothermal mode, below 250 °C reaction temperature low conversion (below 5%) of methanol was determined mainly due to total oxidation. At higher reaction temperatures, 'ignition' occurred, raising the methanol conversion above 25% at full oxygen conversion (see Figure 2.16). Up to 35% methanol conversion was achieved along with a hydrogen selectivity of 60%. Carbon dioxide selectivity was determined as 98%.

2.4.2.3 Catalyst Development for Methanol Decomposition

Maki et al. [40] developed a nickel dispersed carbon membrane catalyst about 100 µm thick. Between the membrane plates, the gas flow occurred in gaps of various thickness between 200 and 1 500 µm. The plates were mounted into a stack-like testing device for methanol decomposition to carbon monoxide and hydrogen.

A flow rate from 10 to 30 $Ncm^3\ min^{-1}$ of methanol vapor was fed into the reactor. Compared with a fixed-bed reactor, 85% conversion of the feed could be achieved at 380 °C reaction temperature, which was 60 K higher than for the fixed bed. However, the yield of carbon monoxide amounted to 40%, which was almost double the value found for the fixed bed, where carbon dioxide and methane yields were significantly higher owing to the increased rate of consecutive methanation and water-gas shift reactions at longer residence times. The maximum hydrogen yield of 1.5 mol $(mol_{CH3OH})^{-1}$ was found at 280 °C reaction temperature and a longer residence time (achieved by increasing the number of plates in the stack), which corresponded to 93% conversion.

2.4.3 Hydrocarbon Reforming

2.4.3.1 Methane Steam Reforming

Considering methane steam reforming [see Eq. (2.12)] as a large-scale process, Xu and Froment found that only the outer 2 mm of the catalyst pellets actually participates in the reaction [41], thus theoretically allowing for a two orders of magnitude reduction in catalyst volume. However, the well-known pressure drop limitations have prevented practical applications in the industrial field so far.

$$CH_4 + 2\ H_2O \rightarrow CO_2 + 4\ H_2 \qquad \Delta H_R = +212.5\ kJ\ mol^{-1} \qquad (2.12)$$

2.4.3.2 Development of Catalyst Coatings for Methane Steam Reforming in Micro Channels

Find et al. [42] developed a nickel-based catalyst for methane steam reforming. As material for the micro structured plates, AluchromY® steel, which is an FeCrAlloy (see Section 2.10.7) was applied. This steel forms a thin layer of alumina on its surface, which is less than 1 μm thick. This layer was used as an adhesion interface for the catalyst. Its formation was achieved by thermal treatment of micro structured plates for 4 h at 1 000 °C.

The catalyst itself was based on a nickel spinel ($NiAl_2O_4$) for stabilization. The active nickel was introduced as surplus of the stoichiometric content of the spinel to the catalyst slurry. The content of active nickel in the final catalyst could be adjusted via the pH during the precipitation. By XRD, α-alumina was identified as an additional phase in case the nickel was incompletely incorporated into the spinel. The sol–gel technique was then used to coat the plates with the catalyst slurry. Good catalyst adhesion was proved by mechanical stress and thermal shock tests.

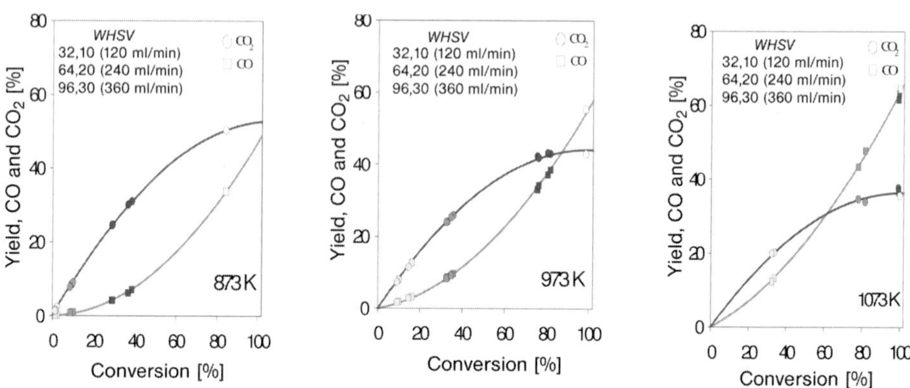

Figure 2.17 Conversion vs. yield plots for steam reforming with an $Ni/NiAl_2O_3$ catalyst [42] (by courtesy of J. Find).

Catalyst testing was performed in packed beds at a S/C ratio of three and reaction temperatures between 527 and 750 °C. The feed was composed of 12.5 vol.% methane and 37.5 vol.% steam, balance argon. At 700 °C reaction temperature and a space velocity of 32 h^{-1}, conversion rates close to the thermodynamic equilibrium could be achieved. With increasing WHSV, the point of equal carbon dioxide to carbon monoxide selectivity was shifted to higher temperatures (Figure 2.17). In other words, CO_2/CO ratio of one was always achieved at about 90% conversion. During 96 h of operation, the catalyst showed no detectable deactivation, in contrast to its commercial nickel-based counterpart.

2.4.3.3 Hydrocarbon Reforming 1 [HCR 1]: Micro Structured Monoliths for Partial Methane Oxidation

Mayer et al. [43] used an electrically heated micro heat exchanger, a micro mixer and a honeycomb reactor, all developed by Karlsruhe Research Center (Forschungszentrum Karlsruhe), for the partial oxidation of methane:

$$2\,CH_4 + 3\,O_2 \rightarrow 2\,CO + 4\,H_2O \qquad \Delta H_R = -471.7\ \text{kJ mol}^{-1} \qquad (2.13)$$

The authors provided various drivers for the application of micro channels, namely the safe operation of the explosive mixture [44], higher surface to channel volume compared with conventional ceramic monoliths and finally smaller pressure drop compared with packed beds or porous solid foams.

The reaction takes place at temperatures around 1 000 °C, a pressure of 25 bar and residence times in the order of few to several hundred milliseconds. Conventionally the process is performed with metal-loaded ceramic supports. From calculations, hot-spot formation up to 2320 °C might occur, which was expected to be minimized by the metallic catalyst.

Rhodium was chosen as reactor material which was also the catalytically active species. It has a high thermal conductivity of 120 W mK^{-1}. Channels 200 μm wide and 5 mm long were fabricated for the honeycomb reactor by mechanical micro machining of pure rhodium foils of 220 μm thickness. By wire-EDM, 60 μm wide channels 130 μm deep were produced. A total of 23 foils carrying 28 channels each were sealed by electron beam and laser welding, achieving a honeycomb which was pressure resistant up to 30 bar. During diffusion bonding, which was followed as an alternative route for sealing, large angled grin boundaries were formed. The honeycomb structure was then put into a ceramic holder carrying a heating wire of 150 W heating power (Figure 2.18). The maximum operation temperature of the reactor was 1 200 °C. The feed was mixed in the micro mixer, preheated in the heat exchanger to 300 °C, fed to the reactor and subsequently quenched in conventional equipment.

A feed of 1–10 $Ndm^3\ min^{-1}$ composed of 60 vol.% methane, 30 vol.% oxygen and 10 vol.% argon for the standard experiments on partial oxidation was passed through the reactor at an operating pressure between ambient and 25 bar. The activity increased with TOS up to 20 h. At 650 °C reaction temperature, mostly

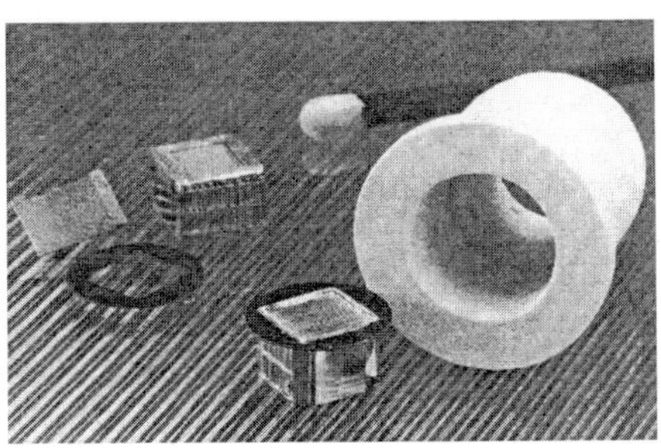

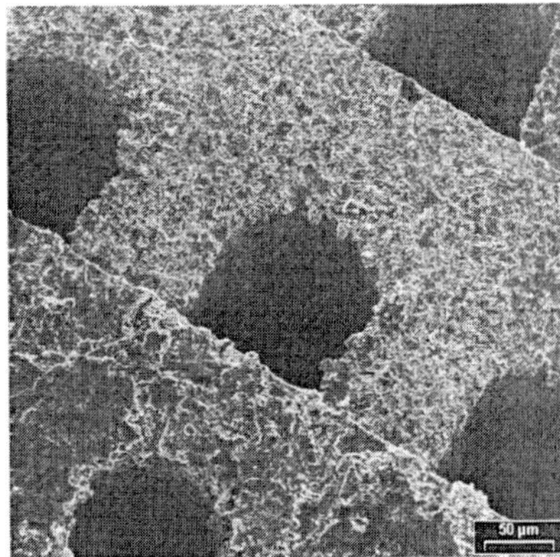

Figure 2.18 Photograph of the dismantled reactor (left) and SEM image of the front surface of the Rh foil stack fabricated by wire-EDM [42] (by courtesy of J. Find).

water and carbon dioxide were produced at 10% conversion. After ignition of the reaction at catalyst temperatures between 550 and 700 °C, temperature exceeding 1 000 °C were achieved within 1 min and mainly carbon monoxide and hydrogen were formed; 62% methane conversion and 98% oxygen conversion were achieved at 1190 °C for the fully activated catalyst.

Owing to the short residence time and the high reaction temperature, high selectivity towards hydrogen (78%) and carbon monoxide (92%) were found under standard testing conditions at a low carbon dioxide selectivity of 8%, which decreased with increasing temperature. However, neither the hydrogen nor carbon monoxide selectivity reached the thermodynamic equilibrium, which was attributed to the short residence times and operation in the kinetic regime rather than reverse homogeneous reactions taking place in the cooling zone of the set-up.

At about 1 140 °C reaction temperature, full oxygen conversion was always achieved when the CH_4/O_2 ratio was decreased from 2 to 1.5, but methane conversion increased from 45 to 96%. Carbon monoxide selectivity remained almost unchanged at 90%, but hydrogen selectivity increased from 75 to 83%. These effects were assumed to stem from the increased heat generation by enhanced methane combustion leading to a hot-spot at the reactor inlet.

Increasing the GHSV from $2 \cdot 10^5$ to $1.2 \cdot 10^6\ h^{-1}$ led to a 10% drop in methane conversion and deteriorated hydrogen and carbon monoxide yields, which were at least partially attributed to mass transport limitations. To check for mass transport limitations, the channel dimensions were varied from 60 to 120 μm width at a constant depth of about 130 μm. No effect of the channel diameter on conversion and selectivity was found at 900 °C.

At 1 090 °C reaction temperature and flow rates exceeding 3 $Ndm^3\ min^{-1}$, thermoneutral conditions were achieved and no external energy supply was necessary to maintain the reaction temperature. No deactivation of the catalyst could be observed during 200 h TOS. However, 0.1 wt.% material was lost from the monolith, which was attributed to sublimation effects [45].

To check for the effect of feeding air instead of pure oxygen, the inert gas content was increased at constant GSHV, which led to only a slight deterioration of the hydrogen and carbon monoxide selectivity owing to the reduced heat generation. Thermodynamic equilibrium could not be achieved, likely because of the short residence time of < 1 ms. Homogeneous reactions might well have taken place as the residence time before the quenching section was 30 times higher than that in the reaction zone.

Increasing the pressure from 1 to 12 bar at a constant reaction temperature of 1 200 °C, constant GHSV of $1.17 \cdot 10^6\ h^{-1}$ and constant CH_4/O_2 ratio of two led to lower conversion and selectivity towards hydrogen and carbon monoxide (see Figure 2.19). At a pressure of 12 bar a selectivity of 3% towards ethylene was found.

The influence of pressure was investigated by using air instead of oxygen and a 20 mm long honeycomb. Again, GHSV ($1.17 \cdot 10^6\ h^{-1}$), CH_4/O_2 (1.75) and reaction temperature (1 100 °C) were kept constant. On increasing the pressure from 1.5 to 20 bar, methane conversion dropped from 90% to 77% and selectivity towards CO and hydrogen deteriorated. In addition, higher pressure led to soot formation in

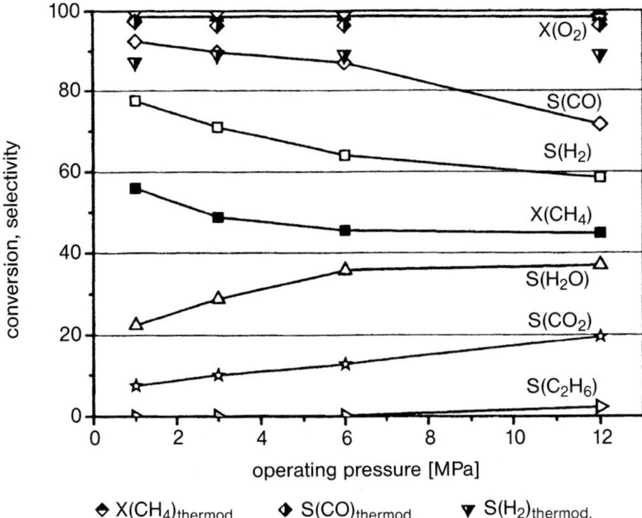

Figure 2.19 Conversion and selectivity vs. pressure for partial oxidation of methane at 1200 °C reaction temperature and an O/C ratio of 1; half-filled symbols indicate calculated thermodynamic equilibrium values [44] (by courtesy of ACS).

the section after the honeycomb and before the quench. It was proposed to control the heat formation by adding methane downstream in the reaction zone by nozzles to avoid product back-diffusion.

2.4.3.4 Hydrocarbon Reforming 2 [HCR 2]: Partial Methane Oxidation Heat Exchanger/Reactor

Early work by Tonkovich et al. [46] dealt with a heat exchanger/reactor containing catalyst powder for the partial oxidation of methane for distributed hydrogen production. The intention was to run the reaction safely in a micro structured reactor owing to the short residence times applied and the improved heat removal avoiding hot-spot formation.

A 5 wt.% rhodium catalyst was applied on mesoporous silica. The reactor with a total size of 7 cm × 3.8 cm × 0.4 cm consisted of nine plates each carrying 37 channels 35 mm long of 254 μm width and a large depth of 1.5 mm to take up the catalyst powder. In front of the reactor a heat exchanger was installed as preheater supplied with energy from a heating fluid. Downstream the reactor the products were quenched in another heat exchanger by cooling fluid. All three devices were integrated into a stack and sealed with nickel gaskets (Figure 2.20) [46].

A feed of 100 $Ncm^3 min^{-1}$ methane and 50 $Ncm^3 min^{-1}$ oxygen was introduced into the reactor at a pressure loss of < 2.5 mbar. The residence time of the reaction was 50 ms. 60% conversion was achieved along with a high carbon monoxide selectivity of 70% at 700 °C reaction temperature. Owing to the short residence times applied, no coke formation was observed and carbon monoxide selectivity was higher than expected from the thermodynamic equilibrium [46].

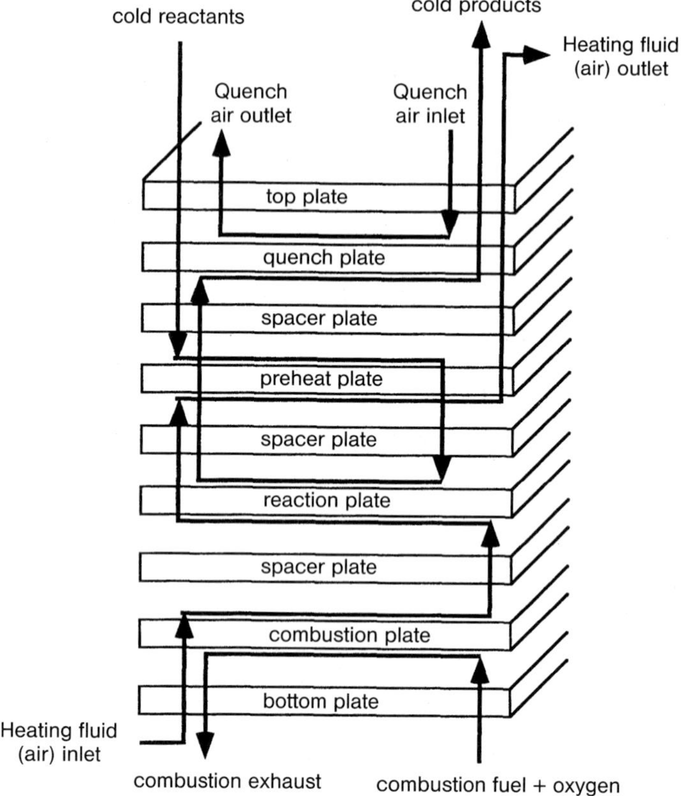

Figure 2.20 Schematic of flow paths [46] (by courtesy of AIChE).

2.4.3.5 Hydrocarbon Reforming 3 [HCR 3]: Micro Structured Autothermal Methane Reformer

Kikas et al. [47] operated a micro structured autothermal reformer for methane in the conventional and reverse flow mode. Owing to the reverse flow conditions, a more uniform temperature profile was achieved in the reactor (Figure 2.21).

In the unidirectional flow mode, 70–80% selectivity towards hydrogen could be achieved along with 55–60% conversion at a methane/oxygen ratio between 0.7 and 0.9. In the reverse flow mode, the hydrogen selectivity could be improved by 5% while maintaining the degree of conversion [47].

2.4.3.6 Hydrocarbon Reforming 4 [HCR 4]: Compact Membrane Reactor for Autothermal Methane Reforming

Kurungot et al. [48] developed a novel membrane material and a catalytic membrane reactor for the partial oxidation of methane. The driver of the development was the fact that rates of reforming reactions are much higher compared with the low permeability of conventional palladium membranes [49]. Silica was previously recognized as a low-cost alternative to palladium [50]. Additionally, the conventional

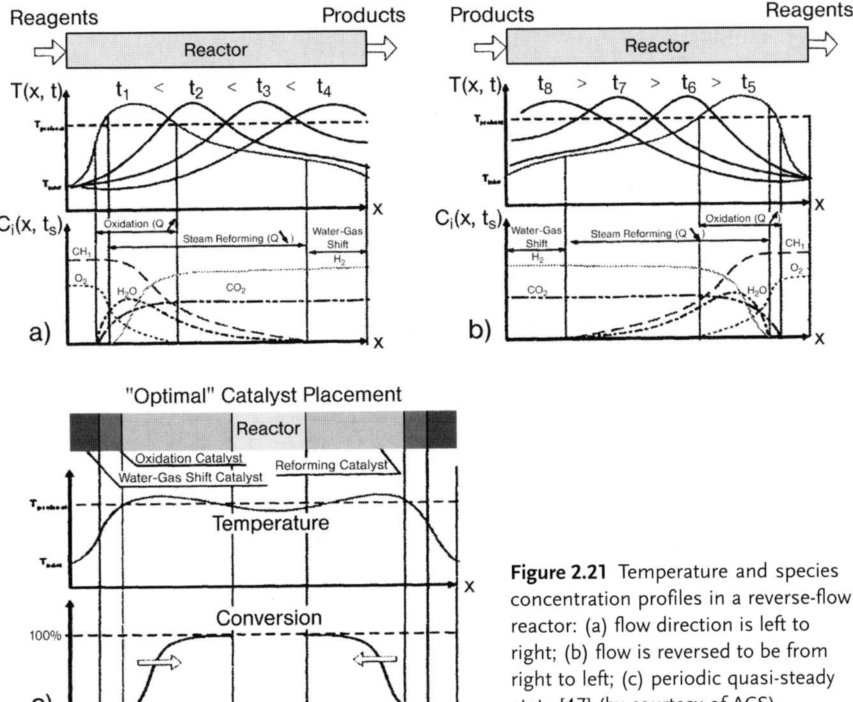

Figure 2.21 Temperature and species concentration profiles in a reverse-flow reactor: (a) flow direction is left to right; (b) flow is reversed to be from right to left; (c) periodic quasi-steady state [47] (by courtesy of ACS).

membrane reactor approach of a packed catalytic bed integrated with a hydrogen-selective membrane has limitations owing to the temperature gradients appearing in the catalyst zone.

The novel idea was to coat the reforming catalyst underneath the silica membrane. Rhodium catalyst on a γ-alumina carrier was regarded as a good candidate with sufficient mechanical stability owing to the low metal content. The catalyst was put between a support tube made of corundum and the silica membrane. The catalyst layer was deposited twice on the corundum tube (0.1 μm pore size) from a boehmite sol containing 1 wt.% rhodium by dip coating. In this way, a 1.5 μm catalyst layer with a surface area of 284 $m^2\ g^{-1}$ and an average pore diameter of 4.3 nm was achieved. Subsequently, the silica membrane (9 μm thickness) was deposited out of a silica sol by dip coating. After each coating calcination was performed at 600 °C [51].

The permeation of hydrogen and methane through the membrane was investigated, revealing increased permeability for hydrogen on raising the temperature, while the methane permeability remained at a low level. Between 100 and 525 °C, the separation factor (see Section 2.6.3) increased from 7.5 to 31. The hydrothermal stability of the membrane was verified at 525 °C for 8 h for a feed composed of 18% hydrogen, 18% methane and 74% steam. It revealed a decrease of the separation factor from 31 to 26 [51].

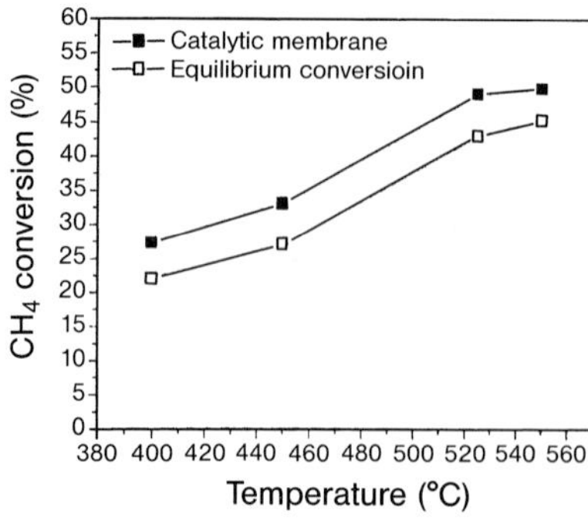

Figure 2.22 Effect of reaction temperature on methane conversion vs. temperature at S/C 3.5 and WHSV 6 Ndm^3 $(min\ g_{cat})^{-1}$[51] (by courtesy of Elsevier Ltd.).

Methane and air were fed to the reactor with an O/C ratio of 1.0 and additional steam at different S/C ratios. The reaction was performed under atmospheric pressure and a nitrogen flow was maintained on the permeate side. The results obtained with the membrane were compared with others generated with the same reactor type without a catalytic membrane in the temperature range 400–575 °C, which is low for methane reforming [51].

At 525 °C reaction temperature, S/C 3.5 and a WHSV of 4 Ndm^3 $(min\ g_{cat})^{-1}$, equilibrium conversion could be exceeded by 37% owing to the continuous removal of hydrogen from the mixture of reactands. Lowering the S/C ratio from 3.5 to 2.5 under these conditions decreased the conversion considerably, from 58.7 to 34.2% [51].

At higher WHSV, carbon formation was assumed owing to the color change of the membrane from yellow to gray. At S/C 3.5 and a WHSV of 6 Ndm^3 $(min\ g_{cat})^{-1}$, conversion increased from 25% to 50% between 400 °C and 560 °C reaction temperature. It exceeded the equilibrium conversion over the whole temperature range by about 10% (see Figure 2.22) [51].

2.4.3.7 Hydrocarbon Reforming 5 [HCR 5]: Sandwich Reactors Applied to Propane Steam Reforming

Kolb et al. [52] applied small externally heated sandwich-type reactors for catalyst screening for propane steam reforming. Two plates of 2 mm thickness were attached to each other and bonded by laser welding. The reactors where 41 mm long and 10 mm wide carrying 14 channels each, which were 25 mm long, 500 μm wide and 250 μm deep on each plate, thus resulting in a total channel cross-section of 500 μm × 500 μm when mounted.

The amount of wash coat which was deposited in the testing reactors was in the same range, between 14 and 17 mg, for the rhodium, platinum and palladium samples tested. The platinum sample was calcined after impregnation at a lower temperature of 450 °C, all other samples at 800 °C. The reason for this will be explained below. The content of the active noble metal was around 5 wt.%. All noble metal-containing samples were laboratory-made catalysts. A commercial α-alumina-based catalyst containing 14 wt.% Ni was added for comparison, as nickel catalysts are applied in industrial steam reforming [52].

Steam and propane were fed to the reactor with a low steam to carbon ratio (S/C) of 1.4, which corresponded to 5 Nml min^{-1} propane and 21 Nml min^{-1} steam balance nitrogen. The experiments were performed with three temperature ramps of 1 h each at 450, 550 and 650 °C [52].

The nickel-containing sample showed no detectable activity up to a reaction temperature of 650 °C within the short residence times applied for the tests. Increasing the reaction temperature to 750 °C led to a conversion of 6% for this sample. However, the catalyst was mostly active for propane dehydrogenation (44% selectivity) and had a selectivity of 28% towards both carbon dioxide and carbon monoxide [52].

The activity of the noble metal-based catalysts was already significant at 450 °C. The highest activity was found for the platinum catalyst. However, this sample was exclusively active in propane dehydrogenation at 450 °C. Additionally, the sample produced some methane at temperatures exceeding 450 °C. The palladium sample was less selective towards the dehydrogenation reaction at all temperatures applied, which is in line with results from the literature [53]. No methane formation was detected and the water-gas shift activity was lowest of all noble metal-based samples. However, its steam reforming activity was higher than that of the platinum sample. The rhodium sample was active exclusively for propane steam reforming at 550 °C. The platinum and palladium catalysts were deactivated significantly within 1 h even at 650 °C, which was attributed to coke formation. Owing to stabilization of the rhodium sample observed at 650 °C and the favorable selectivity pattern of this catalyst, it was considered the most promising candidate from these first experimental results [52].

A bimetallic Rh/Pt catalyst showed at 450 and 550 °C very similar activity to the rhodium sample but along with a less favorable selectivity pattern. Owing to the much less pronounced deactivation of this catalyst at 450 and 550 °C, higher activity was found at 650 °C. At this temperature, the sample had the highest activity and the best selectivity pattern of all samples under investigation so far. The least propene and no methane were formed, likely owing to the improved dispersion of the rhodium on platinum crystals. Hence, it may be assumed that addition of platinum to the Rh/Al_2O_3 wash coat both minimized its short-term deactivation under the experimental conditions applied and was beneficial for oxygen transport to the adsorbed carbon species [52].

It has already been mentioned that all rhodium-containing samples repeatedly showed a regain of activity when finally heated to 650 °C during the experimental protocol. The reversible adsorption of monoatomic carbon species might be an explanation for this observation [52].

Ceria-containing samples were prepared that showed a lower tendency towards coke formation, which was most pronounced for an $Rh/Pt/CeO_2$ catalyst. It showed no measurable deactivation for all reaction temperatures applied in the test protocol. Hence this catalyst showed the best performance of all the samples discussed above. For this $Rh/Pt/CeO_2$ catalyst, which was identified as the best catalyst regarding activity, selectivity and deactivation by coke formation, the effect of increasing the S/C ratio was determined at temperatures between 650 and 750°. These measure-

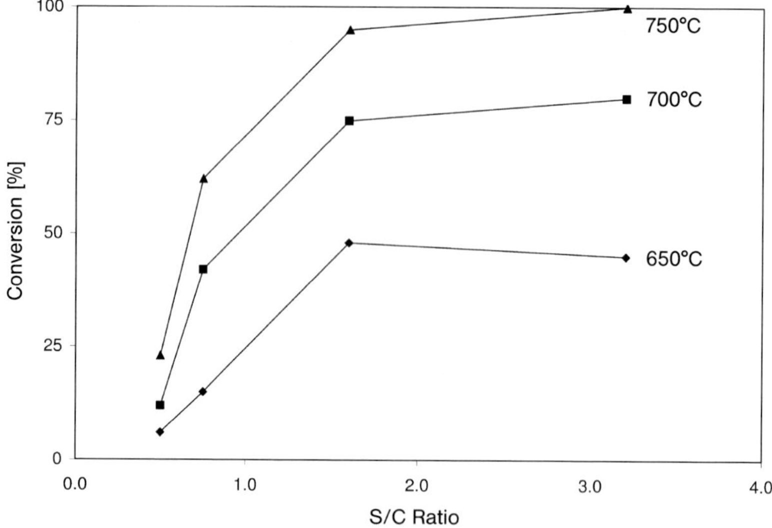

Figure 2.23 Conversion vs. S/C ratio at various reaction temperatures of propane steam reforming over $Rh/Pt/CeO_2$ catalyst [52] (by courtesy of Elsevier Ltd.).

ments were performed by keeping the water molar fraction constant and modifying the propane molar fraction in the feed. The total flow rate was kept constant at 100 Nml min^{-1} by balancing the nitrogen flow. The increase in propane conversion with increasing S/C ratio is shown for various temperatures in Figure 2.23 [52].

A strong effect of oxygen on platinum dispersion was found for both laboratory-made and commercial platinum catalysts. After treatment in air at 800 °C, almost the entire activity of the samples was lost for both types of samples, which was attributed to recrystallization of the platinum. However, this does not apply for the $Rh/Pt/CeO_2$ catalyst, where almost no difference in activity could be found. A sample treated at 800 °C showed an initial conversion of 33% at 450 °C reaction temperature compared with 40% for the sample pretreated at 450 °C tested at the same reaction temperature. At higher temperatures, both samples converted the propane completely. Not only the activity but also the selectivity pattern were very similar over the whole temperature range investigated.

To check the stability and performance of the laboratory-made $Rh/Pt/CeO_2$ catalyst, an undiluted mixture of propane and steam at an S/C ratio of 2.3 was fed to the reactor at a total flow rate of 120 Nml min^{-1} which corresponds to a residence time of 7 ms in the channel system under normalized conditions. After 6 h, full conversion was achieved with exclusive formation of carbon monoxide and carbon dioxide as carbonaceous species (see Figure 2.24); 62 mol% hydrogen, 7 mol% carbon dioxide and 12 mol% carbon monoxide were formed, ca. 20 mol% of water remaining; 160 Nml min^{-1} hydrogen was produced with 13.2 mg catalyst (including the carrier). Hence the turnover frequency amounted to ca. 700 Nl h^{-1} or 63 g h^{-1} of hydrogen per gram of catalyst per hour or 1 575 g of hydrogen per g of rhodium per hour.

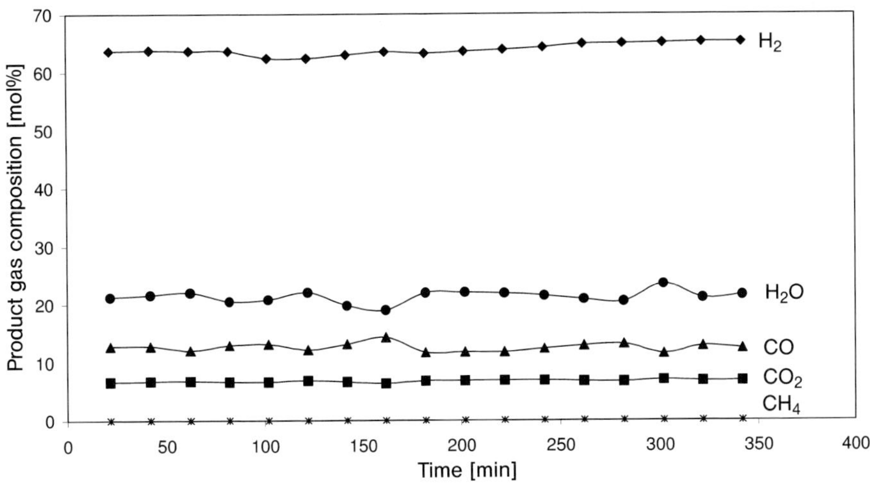

Figure 2.24 Product gas composition in undiluted feed determined with a $Rh/Pt/CeO_2$ catalyst (No. 13); 100 Nml min^{-1} flow rate, S/C 2.3 [52] (by courtesy of Elsevier Ltd.).

2.4.3.8 Hydrocarbon Reforming 6 [HCR 6]: Micro Structured Monoliths for Partial Propane Oxidation and Autothermal Reforming

Two reaction mechanisms for partial propane oxidation exist in the literature. One of them proposes that the reaction starts with catalytic combustion followed by reactions of a lower rate, namely steam reforming, CO_2 reforming and water-gas shift [54]. Aartun et al. [55] investigated both reactions. The other mechanism proposes that the partial oxidation reaction occurs directly at very short residence times [56], which are easier to achieve in the micro channels.

$$2\ C_3H_8 + 6\ O_2 \rightarrow 6\ CO + 8\ H_2 \qquad \Delta H_R = -229\ \text{kJ mol}^{-1} \qquad (2.14)$$

Metallic monoliths made of both rhodium ([HCR 1]) and FeCrAlloy (72.6% Fe, 22% Cr and 4.8% Al ([HCR 3]) carrying micro channels of 120 μm × 130 μm cross-section at various length (5 and 20 mm) were applied. The monoliths were prepared of micro structured foils by electron beam welding. After bonding, the FeCrAlloy was oxidized in air at 1 000 °C for 4 h to form an α-alumina layer, which was verified by XRD. Its thickness was determined as < 10 μm by SEM/EDX. The alumina layer was impregnated with rhodium chloride and alternatively with a nickel salt solution. The catalyst loading with nickel (30 mg) was much higher than that with rhodium (1 mg) (see Table 2.4). The amount of rhodium on the catalyst surface was determined as 3% by XPS.

Partial oxidation was carried out at an O_2/C ratio of ca. 0.63. GHSV amounted to 12.6 ms. The highest hydrogen yield was found for the FeCrAlloy monolith impregnated with rhodium at reaction temperatures exceeding 600 °C. Below this temperature, total oxidation dominated. Full conversion of propane was achieved at a reaction temperature as high as 1 000 °C at 58% hydrogen selectivity (see Figure 2.25).

Table 2.4 Properties of the microstructured reactors tested [55].

	Reactor			
	1	***2***	***3***	***4***
Material	Rhodium	FeCrAlloy	FeCrAlloy	FeCrAlloy
H × W × L (mm)	5.5 × 5.6 × 20	5.5 × 5.6 × 20	5.5 × 5.6 × 20	5.5 × 5.6 × 20
Channels	676	676	676	676
Channel dimensions (μm)	120 × 130	120 × 130	120 × 130	100 × 120
Geometric surface area of the channels (cm^2)	67.5	67.5	67.5	59.5
Porosity	0.34	0.34	0.34	0.26
Residence time (ms)	12.6	12.6	12.6	9.7
Impregnated with	–	Rh	Ni	–

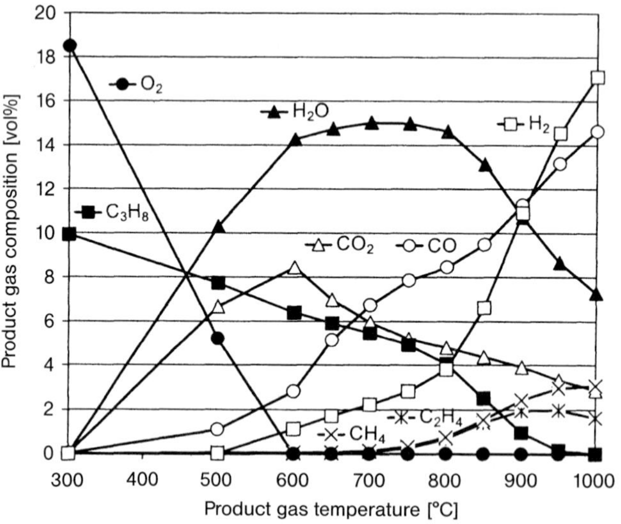

Figure 2.25 Product composition vs. product gas temperature for partial oxidation of propane in an $Rh/Al_2O_3/FeCr$ alloy reactor [55] (by courtesy of ACS).

The rhodium monolith showed a lower hydrogen selectivity of 50% at the same temperature and residence time. For the FeCrAlloy monolith impregnated with nickel, full conversion was achieved at 900 °C, but at a lower hydrogen yield [55].

The beneficial effect of the rhodium catalyst was proved by experiments at a reactor which was coated only with alumina. However, full conversion was also achieved at 1 000 °C, but along with inferior selectivity. Decreasing the residence time was beneficial for both conversion and hydrogen selectivity at the reactor impregnated with rhodium. This was explained by a temperature increase due to an increased rate of the combustion reaction, which in turn propagated the reforming reactions.

During autothermal reforming, the FeCrAlloy monolith impregnated with rhodium also performed best. The feed composition was set to an S/C ratio of one and an O/C ratio of two. Again, full conversion could not be achieved below 1000 °C. Owing to the water addition, an increased hydrogen selectivity of 87% was observed at that temperature. The total amount of carbon monoxide and hydrogen was higher for the same propane feed rate, which confirms, that not only water-gas shift but also reforming took place to a greater extent.

The reactor impregnated with nickel showed inferior performance again. Deactivation was observed, which was assumed to originate from coking, sintering, oxidation of the nickel or even losses of volatile nickel species. With increasing temperature, enhanced formation of by-products, namely methane and ethane, was observed in the reformate both under partial oxidation conditions and in the autothermal mode, which was attributed to thermal cracking.

2.4.3.9 Catalyst Development for the Autothermal Reforming of Isooctane and Gasoline in Micro Structures

Schwank et al. [57] developed a nickel catalyst on a $Ce_{0.75}Zr_{0.25}O_2$ support, coated it on metal foams and applied this micro structured porous catalytic system for the autothermal reforming of isooctane and simulated gasoline (Figure 2.26). The catalyst contained 5–15 wt.% nickel, had a BET surface area of 30 m^2 g^{-1} and showed a nickel dispersion of < 2%. The metal foams were 500 μm thick layers of FeCrAlloy, carrying 24 pores cm^{-1}. The foams were pretreated at 800 °C in air for 24 h to generate an oxide layer for better adhesion of the coating. Wet impregnation was applied as coating procedure.

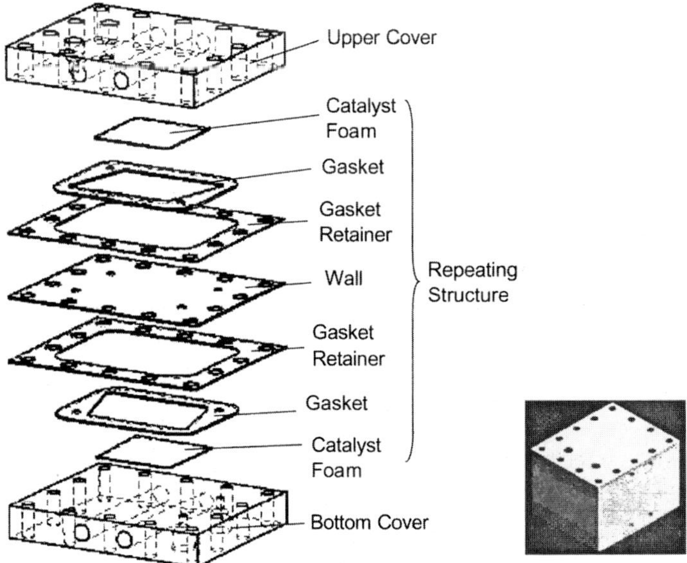

Figure 2.26 Exploded view of the micro structured catalytic test reactor [57] (by courtesy of AIDIC).

Catalyst testing was performed both in a fixed bed and over the coatings at a high GHSV of 100000 and 200000 h^{-1}, respectively. The S/C ratio was two and the O_2/C ratio 0.5. In fixed-bed testing, the support alone showed minimal activity and above 600 °C homogeneous reactions occurred, producing primarily CO and some lower hydrocarbons such as methane and propane. At a reaction temperature of 525 °C, the product composition was 25.7% hydrogen, 9.7% carbon monoxide, 0.2% methane and 8.7% carbon dioxide.

No coking was observed after 8 h of operation even with simulated gasoline (43% isooctane, 35% toluene, 22% *n*-heptane). A hot-spot as high as 800 °C was detected for the packed bed.

On testing of the metal foams at a catalyst loading of 200 mg, 72% conversion of the gasoline was achieved. Further increases in catalyst loading led to coke formation, which was attributed to lower reaction temperatures towards the reactor outlet. The better dilution of the initial hot-spot was assumed to lead to the lower reaction temperatures observed. Additionally, suspicion arose that some of the feed was channelling through the metal foam.

TeGrotenhuis et al. [58] performed a 1 000 h stability test in a micro structured reactor for the steam reforming reaction with a catalyst not specified. A mixture of 74% isooctane, 20% xylene and 5% methylcyclohexane as simulated gasoline was fed to the reactor at a S/C ratio of three and a 650 °C reaction temperature. A regeneration step was performed after 500 h and finally the catalyst converted 97% of the feed.

2.5 Combustion in Micro Channels as Energy Source for Fuel Processors

The catalytic combustion of various fuels is frequently applied to supply energy for liquid fuel evaporation and steam reforming reactions. Hydrogen not converted by the fuel cell anode is one possible energy source that is frequently used. Residual carbon monoxide stemming either from the effluent of CO-tolerant fuel cells, from membrane separation processes or even from the cathode effluent of molten carbonate fuel cells (MCFCs) is another option. Finally, part of the fuel may be combusted separately, which is frequently done in many systems running on hydrocarbon steam reforming as the energy demand of these reactions is relatively high compared with alcohol steam reforming.

2.5.1 Catalytic Hydrogen Combustion

2.5.1.1 Mechanistic Investigations of Hydrogen Combustion

The explosion limits of the reaction range from 4 to 75% for hydrogen in air, hence the safety issue is most crucial for this oxidation reaction:

$$H_2 + \tfrac{1}{2} O_2 \rightarrow H_2O \qquad \Delta H_R = -251.8 \text{ kJ mol}^{-1} \qquad (2.15)$$

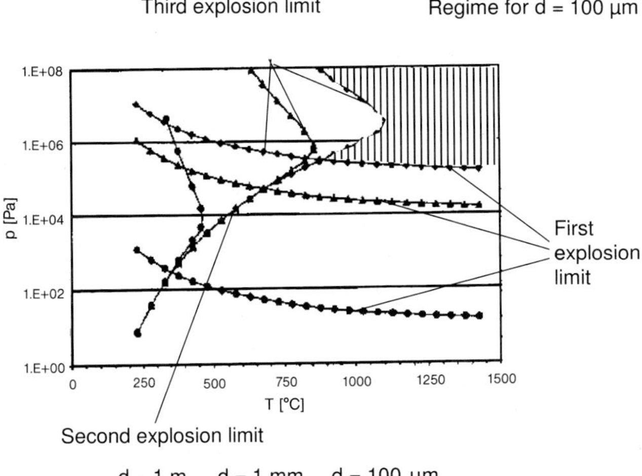

Figure 2.27 Explosion limits for a stochiometric H_2/O_2 mixture and different reactor diameters as determined by Veser [59] (by courtesy of Springer-Verlag).

A detailed mechanistic investigation of the explosion limits revealed [59] that the first and the third explosion limits of the reaction are dependent on the reactor dimensions (see Figure 2.27). The first explosion limit is reached when the mean free path of the molecules becomes smaller than the reactor dimensions, which was already the case at very low pressure (5 mbar). For the third explosion limit, which is responsible for explosions at ambient temperature, both the kinetic and the thermal explosion limit are increased. Veser states that this is not achieved by thermal quenching but rather by 'radical quenching', which means by kinetic effects. A core statement of the investigations is that none of the three explosion limits can be crossed at ambient pressure in a 100 µm micro channel, i.e. the reactor is 'inherently' safe.

2.5.1.2 **Catalytic Hydrogen Combustion 1 [CHC 1]: Single-channel Micro Reactor for Catalytic Hydrogen Combustion**

Veser et al. [60] developed a single-channel reactor and applied it to hydrogen oxidation in the scope of a thorough analysis of this reaction and the related topic of analysis of its explosive regime. The reactor was designed as a modular and flexible tool for various high-temperature reactions. A stainless-steel housing took up the silicon chips which were carrying the micro channels.

The wafers were coated with silicon dioxide (400 nm thickness) and silicon nitride by low pressure chemical vapor deposition (LPCVD) alternately. The chips were fabricated by photolithography and etching. The catalyst (for the application Pt) was introduced as a wire (150 µm thickness), which was heated resistively for igniting the reaction. The ignition of the reaction occurred at 100 °C and complete conversion was achieved at a stochiometric ratio of the reacting species generating a thermal power of 72 W (Figure 2.28).

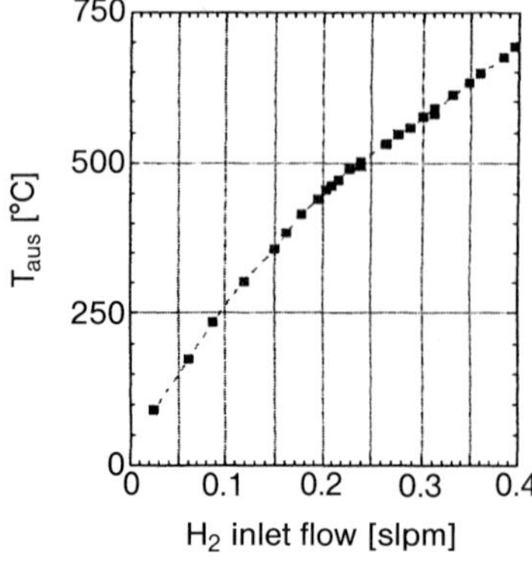

Figure 2.28 Temperature rise of the gas exit temperature with increasing hydrogen content in air [60].

Figure 2.29 Photograph of the ceramic reactor housing and the quartz-glass tubulare microreactor (visible through the center hole) [59].

2.5.1.3 Catalytic Hydrogen Combustion 2 [CHC 2]: Quartz-glass Micro Reactor for Catalytic Hydrogen Combustion

Later, Veser [59] performed the reaction in a quartz-glass micro reactor with a diameter of 600 μm and 20 mm length (Figure 2.29). The ceramic housing of the reactor and the reactor itself were stable for temperatures exceeding 1100 °C. Again, a Pt wire of 150 μm diameter was used as a catalyst and electrically heated for start-up. Residence times down to 50 μs could be achieved.

The fact that no homogeneous reactions, which are explosive, could be detected, demonstrates the possibility of separating homogeneous and heterogeneous reactions in micro reactors owing to the higher surface area to volume ratio of this reactor type. Increasing the flow rate leads to increased temperature of the gas exiting the reactor. This was attributed to reduced heat losses by conduction through the reactor walls.

2.5.1.4 Catalytic Hydrogen Combustion 3 [CHC 3]: Combined Mixer/Cross-flow Combustor/Heat Exchanger for Determination of the Kinetics of Hydrogen Oxidation

Görke et al. [61] investigated the catalytic oxidation of hydrogen in an integrated cross-flow heat exchanger/reactor which was combined with a mixer for feed homogenization. The reactor was built as a measurement tool for the determination of the kinetics of fast and highly exothermic heterogeneous gas-phase reactions. It had a height of 1.6 mm and a width and length of 14 mm each. The channels were 200 μm wide and 70 μm deep and each of the 16 stainless-steel foils of the stack carried 38 channels. Twelve foils were used for cooling the reactor with water and four foils were used as the reaction zone and were impregnated with Pt directly on the stainless-steel surface.

Using 324 measuring points taken at temperatures between 35 and 75 °C, hydrogen concentrations between $1.6 \cdot 10^{-3}$ and $11.0 \cdot 10^{-3}$ mol Ndm^{-3} and oxygen concentrations between $1.7 \cdot 10^{-3}$ and $7.3\ 10^{-3}$ mol Ndm^{-3}, a kinetic expression for the reaction was determined on the basis of a Langmuir–Hinshelwood model (Figure 2.30). The Mears criterion was applied to verify that no mass transfer limitation was to be expected for the system from the gas phase to the non-porous catalyst:

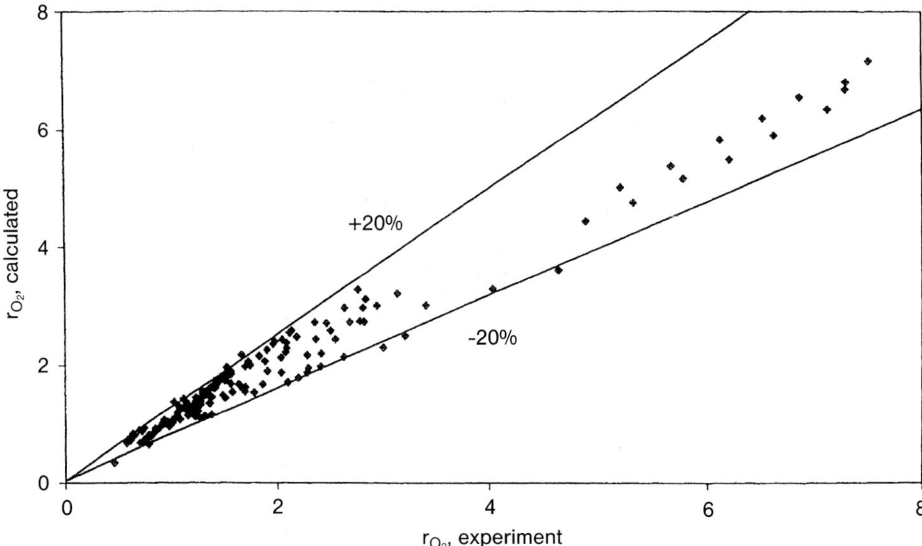

Figure 2.30 Parity plot based on the Langmuir–Hinshelwood kinetics of the O_2 reaction rate r_{O2} for hydrogen oxidation as determined over microstructured Pt-impregnated stainless-steel foils [61] (by courtesy of O. Görke).

$$\frac{r_{\mathrm{eff}}\; d_{\mathrm{h}}}{2\, k_{\mathrm{g}}\; c} < 0.15 \qquad (2.16)$$

where d_h is the hydraulic diameter of the micro channel, r_{eff} the determined (effective) rate of reaction, k_g the mass transfer coefficient and c the concentration of the species under investigation. No heat transfer limitation in the boundary layer was found according to the Anderson criterion:

$$\frac{-\Delta H_{\mathrm{r}}\; r_{\mathrm{eff}}\; d_{\mathrm{h}}\; E_{\mathrm{a}}}{\alpha\; R\; T_{\mathrm{r}}^{2}} < 1 \qquad (2.17)$$

where ΔH_r is the reaction enthalpy, r_{eff} the measured rate of reaction, d_h the hydraulic diameter, E_a the activation energy, α the heat transfer coefficient, R the gas constant and T_r the temperature in the gaseous phase. If the criterion is fulfilled, heat transfer limitation in the gas phase of the micro channel is negligible. The Nusselt number was calculated as 5.7.

For a cooling flow of 7.5 Ndm^3 min^{-1} the temperature increase of the water was 1.8 K at a power generation of 30 W of the reaction. The overall heat transfer coefficient was calculated to 15 kW m^{-1} K^{-1}. No increase in the gas outlet temperature was observed, proving that the heat was removed completely. From the kinetic experiments an activation energy of 17.3 kJ mol^{-1} and an adsorption enthalpy of 252 kJ mol^{-1} for oxygen were found for the highly endothermic reaction. The frequency factor was also high as well (1.16 · 10^9 mol dm^{-3} s^{-1}), which points to the high reaction rate.

2.5.1.5 Catalytic Hydrogen Combustion 4 [CHC 4]: Cross-flow Combustor/Heat Exchanger for Hydrogen Oxidation

Hagendorf et al. [62] applied a cross-flow heat exchanger developed at Karlsruhe Research Center (FZK) fabricated of 100 μm thick copper foils of 14 mm × 14 mm size for the H_2/O_2 reaction in the explosive regime. Alumina was used as catalyst carrier introduced by in situ CVD and Pt as active component was in situ wet impregnated. The channel dimensions were 100 μm × 200 μm for the reactor side and 70 μm × 100 μm for the cooling gas side (nitrogen). In this early work, in a 1 cm^3 reactor volume complete conversion could be achieved at temperatures below 220 °C generating 150 W of thermal energy, which was removed by the cooling fluid.

Janicke et al. [63] applied the reactor for the same reaction in a far more detailed study, stressing the possibility of applying this system as a catalytic burner for automotive applications. Here channel dimensions of 140 μm × 200 μm were used for the reaction area and again 70 μm × 100 μm for the cooling channels. Alumina layers were generated by a sol–gel method and Pt was introduced via up to threefold wet impregnation to improve the loading level (see Section 2.10). Catalyst characterization was performed by SEM and EDX and even coatings of 10 μm thickness were found. Krypton adsorption was performed, revealing a low surface area of 0.17 $m^2\,g^{-1}$ as the coated carriers, and not the coating alone, were analyzed. A Pt particle size of 15 nm was determined by XRD, which did not change during the reaction.

The feed flow rates ranged between 0.2 and 0.4 $Ndm^3\,min^{-1}$ for hydrogen and between 0.1 and 0.2 $Ndm^3\,min^{-1}$ for oxygen. Nitrogen was used as a diluent for the feed at a flow rate of 1.0 $Ndm^3\,min^{-1}$. It also served as a cooling medium at a flow rate between 3.0 and 7.0 $Ndm^3\,min^{-1}$ [63].

At low Pt loadings, which correspond to a single impregnation procedure, the reactor was externally heated to 80 °C, leading to complete conversion and a power generation of 72 W. About 70% of the heat generated was lost (Figure 2.31). At a flow rate of 1.6 $dm^3\,min^{-1}$ a temperature increase of 300 K was observed. Occasionally, owing to experimental errors, which occurred during the post-mounting catalyst deposition technique, in some cases hot-spot formation and subsequent homogeneous reactions were observed in the reactor in- or outlet [63].

At high Pt loadings, which correspond to a threefold impregnation procedure, ignition took place after an induction period with a gradual temperature increase from room temperature. At a reaction temperature of 219 °C, 0.2 and 0.4 $Ndm^3\,min^{-1}$ flow rates for oxygen and hydrogen, respectively, and a 1.0 $Ndm^3\,min^{-1}$ nitrogen flow at 7.0 $Ndm^3\,min^{-1}$ cooling gas flow rate , full conversion of the hydrogen was achieved. The temperature could be easily controlled by the cooling gas flow rate. Interestingly, this induction period became shorter after several experiments, which was attributed to the formation of a thin and highly active oxide layer at the surface of the platinum initially in the reduced state (Figure 2.32) [63].

The phenomenon of increasing combustion catalyst activity was frequently observed in literature before [64, 65]. At both levels of Pt loading, the coolant temperature at the exit was higher than the temperature of the products which was explained by hot-spot formation.

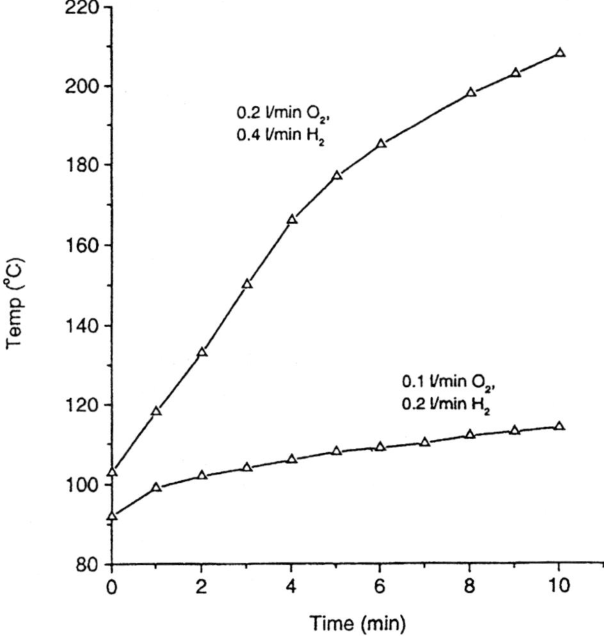

Figure 2.31 Temperature of the exiting reactant gas as a function of flow rates for hydrogen and oxygen at low Pt catalyst loading [63] (by courtesy of Elsevier Ltd.).

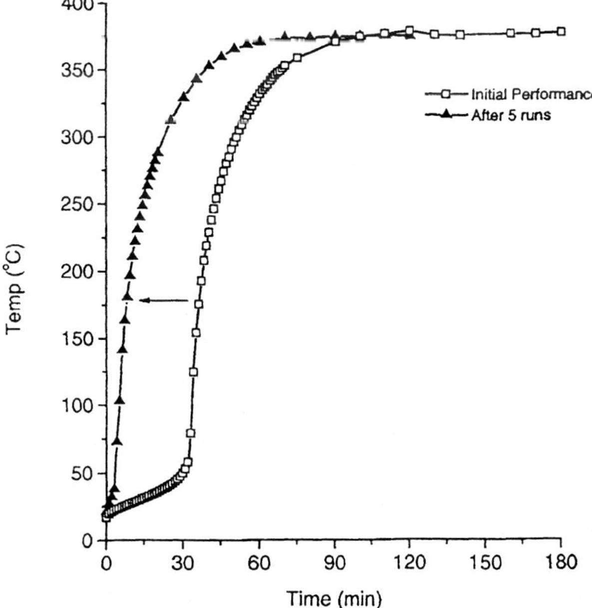

Figure 2.32 Initial performance (□) and temperature response of the micro reactor after five runs (▲) for high Pt loading [63] (by courtesy of Elsevier Ltd.).

Further experiments were carried out by Janicke et al. [66] applying cooling oil. Surprisingly for a cross-flow heat exchanger, an exit temperature of 207 °C was determined for the heat carrier, whereas the product gas exited at 70 °C. This phenomenon was attributed to the fact, that most of the energy was transferred to the oil at the feed inlet. Subsequently the product was then cooled by the cold oil, which had heated up not yet.

2.5.1.6 Catalytic Hydrogen Combustion 5 [CHC 5]: Combination of a Mixer, a Cross-flow Combustor/Heat Exchanger and a Heat Exchanger for Product Quenching for Hydrogen Oxidation

Haas-Santo et al. [67] made use of the [CHC 3] cross-flow heat exchanger/reactor for the oxidation of excess hydrogen stemming from electrolytic cells of future space vehicles. The equipment consisted not only of the micro structured reactor (1 cm^3 volume) with integrated cooling channels, but also of a micro mixer for feed mixing and a micro structured heat exchanger (1 cm^3 volume) located downstream for product cooling (see Figure 2.33). All channel systems had dimensions of 200 × 70 μm.

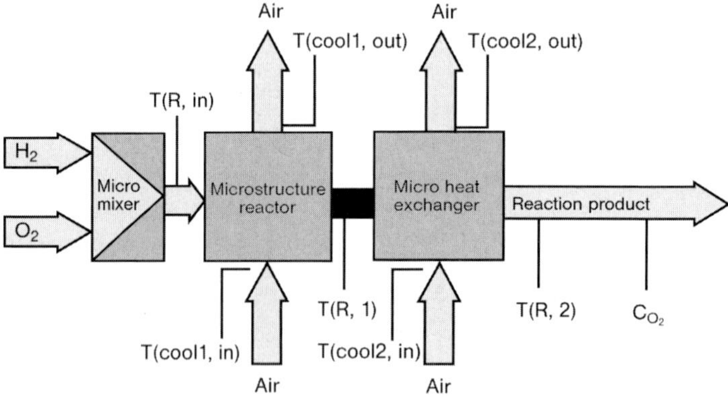

Figure 2.33 Schematic of the micro structured reaction system consisting of a micro mixer, a micro reactor, a micro heat exchanger and temperature sensing points [67] (by courtesy of Springer-Verlag).

The reaction channels were coated by the sol–gel technique, revealing a 2–3 μm thick alumina layer with pore diameters ranging between 1 and 5 nm having a maximum at 4 nm. Owing to the preparation technique, no surface area data could be obtained but a surface enhancement factor of 430 $m^2\ m^{-3}$ was determined. Pd was introduced as active component by wet impregnation. Subsequently, the catalyst was activated at 80 °C and reduced in H_2.

Experiments performed with the miniaturized process engineering device at Forschungszentrum Karlsruhe revealed complete conversion after 120–150 s TOS at a flow rate of 1 $Ndm^3\ min^{-1}$ H_2 and a stochiometric amount of O_2. The cooling air flow rate was 20 $Ndm^3\ min^{-1}$. Direct ignition of the reaction occurred at 25 °C and outlet temperatures of the cooling air up to 220 °C were observed. The catalyst was stable for 200 h TOS (Figure 2.34) [67].

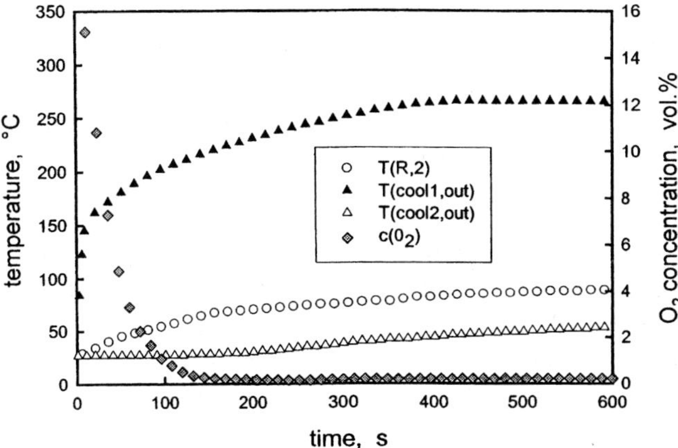

Figure 2.34 Experimentally measured temperature at sensing points indicated in Figure 2.35 and oxygen concentration as a function of time. Hydrogen flow 1 $Ndm^3\ min^{-1}$; cooling air flows each 20 $Ndm^3\ min^{-1}$; cooling inlet temperature 30 °C [67] (by courtesy of Springer-Verlag).

At Astrium Laboratories, the project partner of Forschungszentrum Karlsruhe, 3–5 electrolysis cells were used to generate a pure H_2 flow of up to 1.16 $Ndm^3\ min^{-1}$ revealing a liquid product (water) temperature of 40 °C at 30 $Ndm^3\ min^{-1}$ cooling gas flow rate. Steady state of the system was achieved after 20 min when changing the number of cells feeding the system [67].

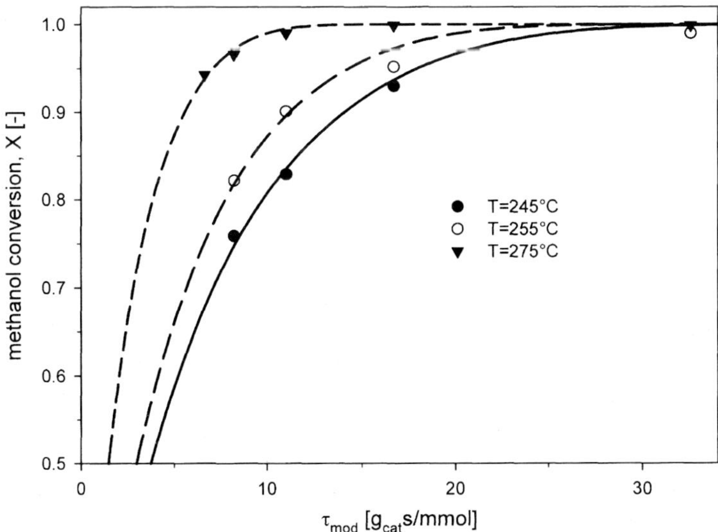

Figure 2.35 Methanol conversion for the total oxidation reaction vs modified space time as a function of reaction temperature [68] (by courtesy of ACS).

2.5.2 Catalytic Combustion of Alcohol Fuels

No work exclusively dedicated to alcohol combustion in micro channels is known to the authors. However, in some integrated devices running on methanol discussed in Section 2.7.2, methanol combustion is applied to feed either the reformer alone ([ISMol 6], [ISMol 7] and [ISMol 8]) or additionally evaporators (ISMol 5). Platinum (ISMol 8) and cobalt oxide catalysts (ISMol 7) were applied, the latter by Reuse et al. [68]. For the cobalt oxide catalyst, at reaction temperatures exceeding 250 °C, > 99% conversion was achieved for the combustion reaction, which was fed with an oxygen surplus at an O_2/CH_3OH molar ratio of 1.9 at a modified space time of more than 26 g_{cat} s $mmol^{-1}$ as indicated in Figure 2.35. An activation energy of 130 kJ mol^{-1} was determined for the reaction.

2.5.3 Catalytic Hydrocarbon Combustion (CHCC)

2.5.3.1 Catalytic Hydrocarbon Combustion 1 [CHCC 1]: Ceramic Micro Reactor for Butane Combustion

Wang et al. [69] developed a ceramic micro reactor for butane combustion. The reactor was made of co-fired ceramic tapes (LTCC DuPont 951 AX), which are inexpensive and also allow for the integration of resistors and conductors. The minimum thickness of the tapes available starts at 111 μm. They consist of oxide particles, glass frit and organic binder and may be machined by laser, milling, chemically or even by photolithography [70]. Upon firing, the organic binder burns out and the oxide particles are joined by sintering. The tapes may be stacked and co-fired to form monoliths. A single-channel micro reactor carrying an integrated heating device was realized. The channel was either 47 mm long with a cross-section of 500 × 800 μm or 17 mm long with a cross-section of 1 000 μm × 800 μm (Figure 2.36).

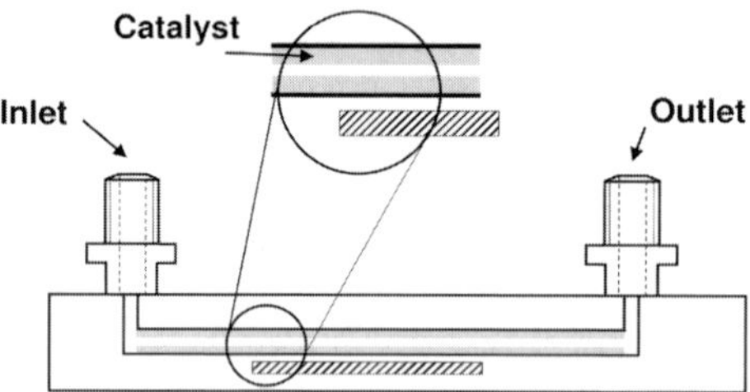

Figure 2.36 Cross-sectional view of the reactor [69] (by courtesy of Kluwer Academic Publishers).

Various platinum and palladium catalysts were generated by both wet impregnation with a 4 wt.% palladium solution and filling the channel with a catalyst (1 wt.% Pd on Al_2O_3) slurry. Both procedures were performed at the readily mounted reactors.

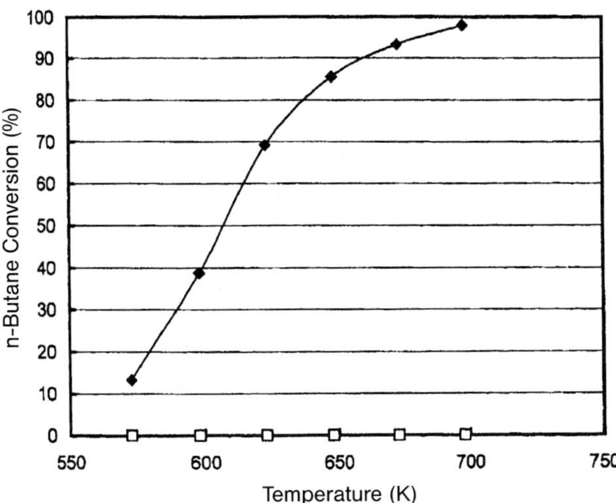

Figure 2.37 Light-off curves for *n*-butane oxidation with (◆) and without (□) palladium generated in a micro-reactor made of ceramic tapes [69] (by courtesy of Kluwer Academic Publishers).

The feed was composed of 11% butane and 89% oxygen at a flow rate of 10 $Ncm^3\ min^{-1}$. At a reaction temperature of 430 °C, about 98% conversion was achieved, carbon dioxide and water being the only products detectable (see Figure 2.37). A significantly higher rate of reaction was determined for the catalyst, which was attributed to the better dispersion of the active material. The channel dimensions had only a minor effect on activation energy, showing that diffusion limitations were not important.

The activation energy of the reaction was determined as 90 kJ mol^{-1} regardless of the catalyst preparation technique and channel size. When varying the partial pressure of the reactants, a zero reaction order was determined for oxygen, whereas two regimes were identified for butane. For butane contents below 8%, a reaction order of 0.7 was determined, which was explained by rate limitations due to the dissociative adsorption of butane. Above 8% butane content, zero reaction order was found and the rate of oxidation of carbonaceous species was assumed to be limiting.

2.5.3.2 Catalytic Hydrocarbon Combustion 2 [CHCC 2]: MEMS System for Butane Combustion

Arana et al. [71] designed a small-scale catalytic combustor focusing on minimized heat losses (Figure 2.38). They pointed out that for very small systems, heat conduction in heat exchangers is dominated by heat transfer along the walls, which therefore needs to be minimized. Additionally, at device diameters below a few millimeters, conventional insulation material rather increases the heat losses. Hence new approaches to system design are required for small-scale systems. Their so-called suspended tube reactor consisted of four rectangular ducts (called tubes) made from silicon nitride with a very low wall thickness of 2 μm in order to minimize axial heat transfer. The ducts had a width of 200 μm and a height of 480 μm. At

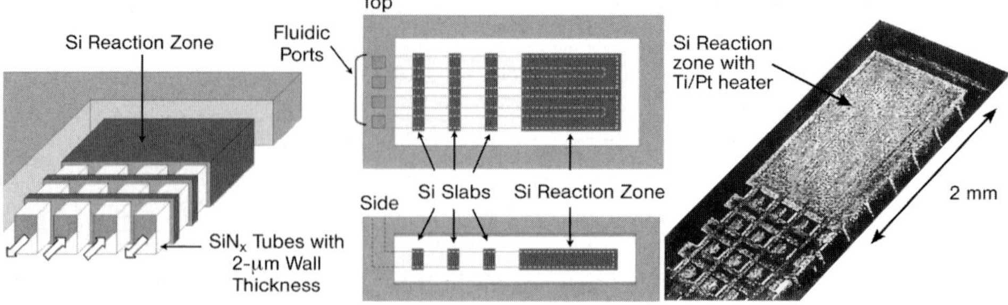

Figure 2.38 Schematic of the suspended-tube reactor (left and middle) and SEM (right) showing four suspended SiN_x tubes connected to the Si reaction zone, Si slabs thermally linking the four tubes and a meandering Ti/Pt resistive heater/temperature sensor [72] (by courtesy of L. R. Arana).

each end, these ducts were embedded in a silicon substrate, which carried the connection to external tubing at the front and the reaction zone at the end. Heat was transferred between the tubes in the reaction zone by the relatively high heat conductivity of the silicon.

The first and second pair of tubes were connected to a U-shaped reactor by the silicon reaction zone. In the center of the tubes, silicon slabs served as a recuperation zone for the fluids entering and leaving the system. Their impressive efficiency was demonstrated by CFD simulations revealing the highest heat transfer resistance in the fluid itself and thus temperature profiles across the width of the ducts (see Figure 2.39).

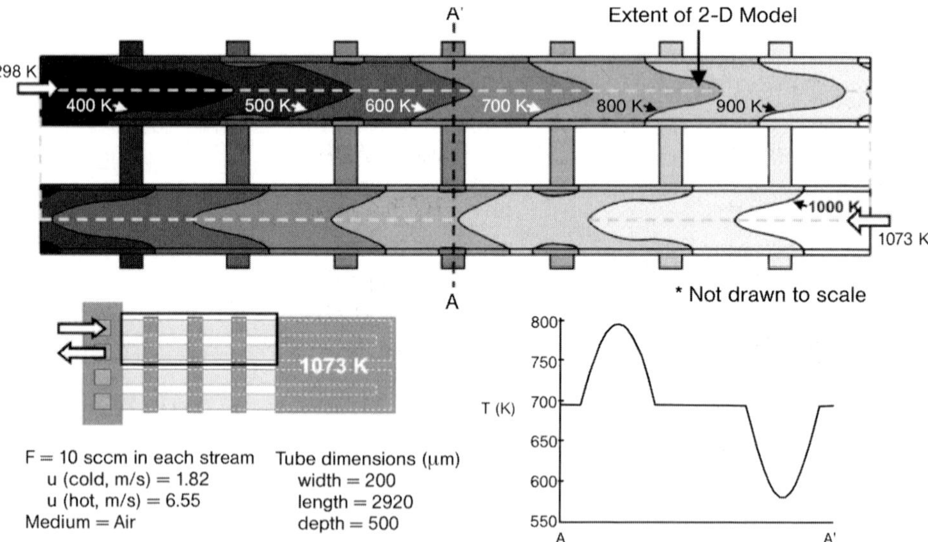

Figure 2.39 Temperature profile from 2-D CFD simulation for the recuperation ducts of the suspended tube reactor [71] (by courtesy of SAGE Publications).

Posts made of silicon coated with silicon nitride were introduced into the silicon reaction zone. They improved the heat conductivity and served as supports for the wash-coat catalyst. To minimize heat losses to the environment, vacuum packaging of the system and application of reflective shields, to overcome radiation losses especially for high-temperature applications, were proposed. The vacuum required for complete elimination of convective heat losses was determined as 0.013 mbar or lower. By applying vacuum packaging, reduction of steady-state heat losses by 60% was determined. The remaining losses were mainly attributed to radiation at temperatures exceeding 500 °C. Meandering thin-film platinum lines were integrated on the outer surface of the silicon reaction zone and on the supporting chip. They could be used either for resistance heating or temperature measurements. The reactor could be heated to 900 °C by applying these elements. The fabrication of the device was performed by deep reactive ion etching (DRIE) or KOH etching combined with silicon fusion bonding.

In this way, a mold was formed on to which silicon nitride was deposited. After removing the mold by etching, the desired free-standing ducts were formed. Further details of the fabrication process are described in Section 2.9. The pressure drop of a single duct was determined as 60 mbar at a flow rate of 60 $Ncm^3\ min^{-1}$ with posts integrated (Figure 2.40). As the combination of an endothermic reaction producing hydrogen with an exothermic reaction supplying energy was the application for which the reactor was designed, ammonia cracking and butane combustion were performed independently.

Pt/γ-alumina wash-coat catalyst introduced by co-impregnation was used for butane combustion. After brief heating for 10 s, the reaction ignited and proceeded

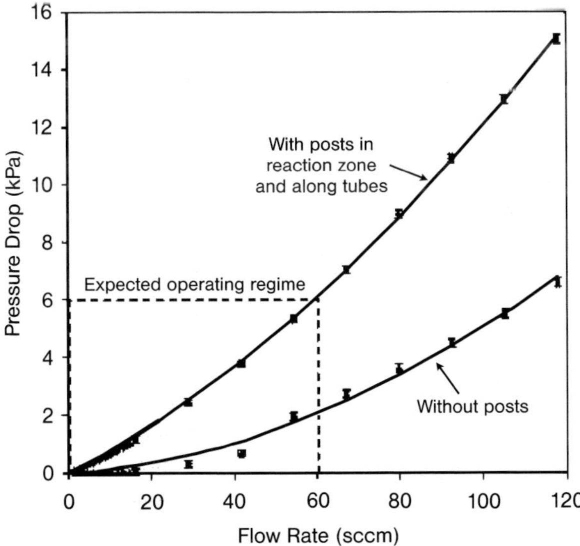

Figure 2.40 Pressure drop vs. flow rate in one channel of the suspended-tube reactor with and without posts in the reaction zone (nitrogen used at 25 °C) [71] (by courtesy of SAGE Publications).

without external heating. Full conversion of a stoichiometric butane/air mixture was achieved for a butane feed rate of 0.8 $Ncm^3 min^{-1}$ [72]. In this way, 0.8 W of energy was produced by the reactor. As the reactor was not insulated by vacuum packaging and reflective shields, 1.8 W was necessary to achieve 97% conversion of the ammonia during cracking experiments. This translates to an LHV of the hydrogen produced of 1.6 W or a power density of 13 kW dm^{-3}.

2.5.3.3 Catalytic Hydrocarbon Combustion 3 [CHCC 3]: Silicon Micro Reactor for Butane Combustion

Tanaka et al. [73] developed another MEMS system for the catalytic combustion of butane. It is composed of a combustion chamber 8 mm wide, 14 mm long and 150 μm deep which was prepared by anisotropic wet etching of a silicon substrate. The substrate was then covered with Pyrex™ glass applying anodic bonding. Combustion was performed on a platinum/titania catalyst.

2.5.4 Homogeneous Combustion in Micro Channels

2.5.4.1 Modeling of Homogeneous Methane Combustion in Micro Channels

Raimondeau et al. [74] modeled the homogeneous high-temperature combustion of a preheated stoichiometric mixture of methane and air at a flow rate of 2 m s^{-1} and a reference temperature of 1 000 °C, which corresponds to the methane ignition temperature under these conditions. For a channel of 100 μm diameter, no gradients of species concentration and temperature were determined by the calculations. It was demonstrated that temperature losses through the wall lead to flame extinction, which was more pronounced with decreasing channel diameter.

In addition, flame propagation is not possible in stainless-steel channels owing to the high heat conductivity and affinity to radicals. But even for materials which do not adsorb radicals, a minimum channel width exists below which no homogeneous combustion is possible any longer. Insulating materials such as silicon nitride and inert layers such as alumina are required to maintain the homogeneous reaction.

For a preheated wall, the conditions required for self-sustaining operation were investigated with CFD modeling by Norton and Vlachos. [75]. Wall materials having very low heat conductivity appeared not only to generate severe hot-spots but also limited the energy transfer from the combustion zone to the preheating zone of the feed, which moved the location of the combustion zone downstream the channel. An optimum value of the wall thermal heat conductivity was determined in the range of 3–5 W m^{-1} K^{-1} (see Figure 2.41), which corresponded to ceramics such as silica and alumina. In the low wall thermal conductivity regime, limited heat transfer within the wall retarded preheating, whereas for high wall conductivity, heat loss from the entire burner surface dominated.

A minimum flow velocity was required to maintain a flame. When the flow velocity of the feed was increased, the location of the flame first moved upstream and then downstream again, while the width of the reaction zone decreased and the maximum

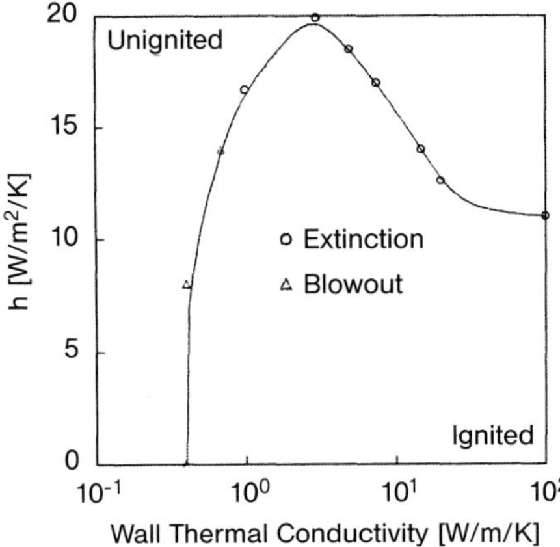

Figure 2.41 Stability diagram for methane combustion in a micro-burner conditions. flow velocity 0.5 m s^{-1}, channel length 600 μm and channel width 200 μm [75] (by courtesy of Elsevier Ltd.).

temperature achieved increased. Owing to the spread reaction zone at low flow velocities, the external temperature losses caused the extinction of the flame at a certain minimum value.

2.5.4.2 Homogeneous Combustion in Micro Channels 1 [HCC 1]: Homogeneous Hydrogen Combustion in a Micro Combustor

Mehra and Ayón [76] developed the first micro structured hydrogen combustor made of silicon. The device was developed to power a micro gas turbine. However, it could also be applied for fuel processing and is therefore included in this chapter.

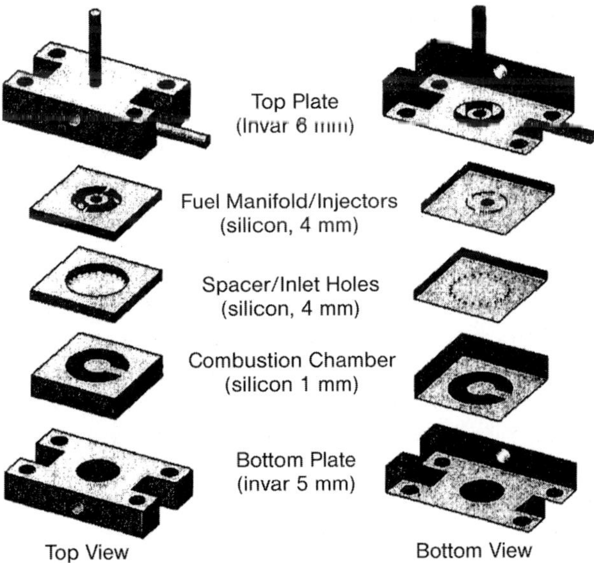

Figure 2.42 Exploded view of the micro combustor test rig. The silicon parts are fitted between two Invar plates [76] (by courtesy of SAGE Publications).

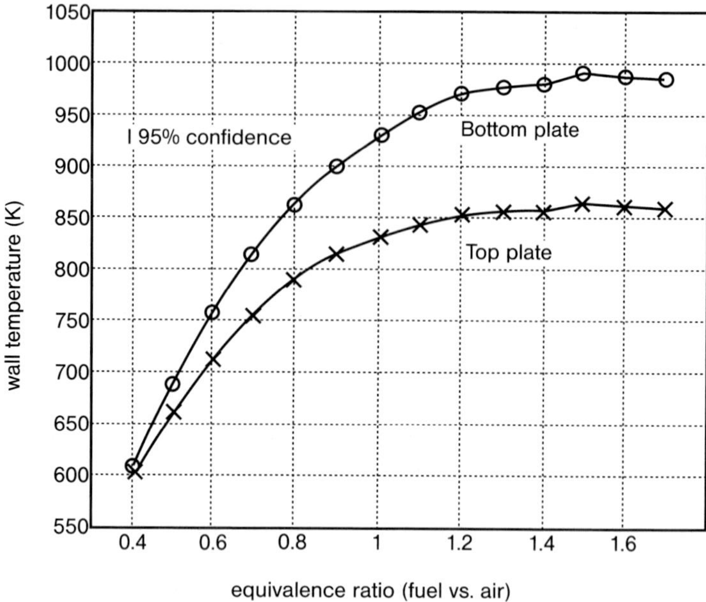

Figure 2.43 Wall temperatures of the micro combustor increase with the equivalence ratio of fuel to air up to 1000 K [76] (by courtesy of SAGE Publications).

The combustor was composed of three fusion-bonded silicon wafers. Hydrogen was added to the air flow by 76 injector holes of 30 μm diameter. The mixture then entered the annular-shaped combustion chamber through 24 combustor inlet ports of 340 μm diameter and left the device through a circular exhaust. The dimensions of the combustion chamber were 5 and 10 mm diameter at a height of 1 mm, which corresponds to a volume of 66 mm^3. The fabrication of the device was performed by dry isotropic and anisotropic etching (Figure 2.42).

A temperature of about 1450 °C was achieved at the gas outlet for an H_2/O_2 ratio of 1.2. Owing to significant heat losses, the silicon walls could be kept below their melting-point (Figure 2.43). The device was run for 15 h at 150 W power generation.

2.5.4.3 Homogeneous Combustion in Micro Channels 2 [HCC 2]: Homogeneous Hydrogen Combustion in a 2-D Micro Combustor

A two-dimensional micro combustor was developed by Tanaka et al. [77] for hydrogen combustion. It comprised of two combustion chambers, two inlets for air and one inlet for hydrogen feed (see Figure 2.44). As the quenching distance of the reaction was assumed to be 640 μm, the chamber depth was chosen higher and varied between 1 and 3 mm. The total volume of both combustion chambers was fixed at 181 mm^3. The device was fabricated as a silicon green compact by spark plasma sintering (SPS) followed by reaction sintering in a nitrogen atmosphere to achieve the silicon nitride ceramic. A minimum depth of the combustion chamber of 2 mm was necessary to maintain the flame.

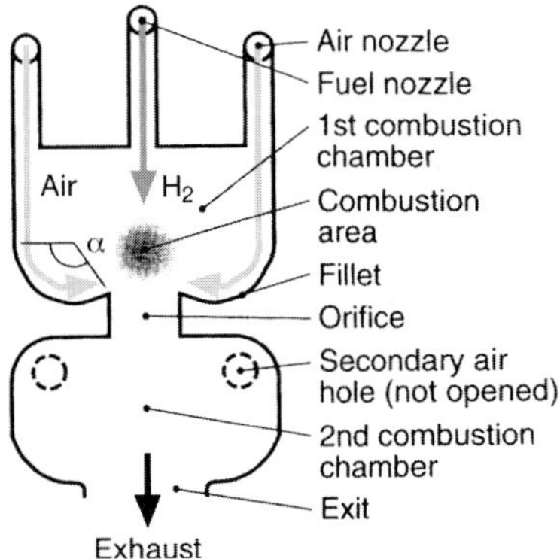

Figure 2.44 Design principle of a two-dimensional micro combustor [77] (by courtesy of IOP Publishing Ltd.).

The feed rates of the device were set to 1.7 $Ncm^3\ min^{-1}$ hydrogen and 4 $Ncm^3\ min^{-1}$ air. The flow rate in the hydrogen inlet was 11.5 $m\ s^{-1}$, being higher than the maximum propagation speed of the hydrogen flame (3 $m\ s^{-1}$). This flow rate corresponded to a heating power of 310 W. The exit temperature of the gas increased with increasing total flow rate. At a stoichiometric feed ratio, hydrogen conversion increased from 90% to more than 99.75% when the depth of the combustion chamber was increased from 2 to 3 mm.

2.6 Micro Structured Reactors for Gas Purification (CO Clean-up)

2.6.1 Water-gas Shift

Catalytic gas purification of reformate containing significantly more than 1% of carbon monoxide is performed by the water-gas shift reaction. Frequently, especially in the case of partial oxidation reactions, water is added to the reformate to shift the equilibrium of the reaction in the desired direction:

$$H_2O + CO \rightleftharpoons H_2 + CO_2 \qquad \Delta H_R = -41\ \text{kJ mol}^{-1} \tag{2.18}$$

The equilibrium constant of the reaction may be easily calculated according to the following equation [78]:

$$K(T) = e^{-4.33+\frac{4577.8}{T}} \tag{2.19}$$

The reaction is slightly exothermic, which is not favorable for the carbon monoxide equilibrium conversion when running shift reactors in the adiabatic mode.

Conventionally the reaction is performed in two stages, the so-called high- and low-temperature water-gas shift. In large-scale industrial processes, Fe_2O_3/Cr_2O_3 catalysts are applied for high-temperature shift (which is then performed between

350 and 450 °C [79]) and Cu/ZnO catalysts for low temperature shift (200–300 °C). However, these catalysts are rather optimized for long-term stability and resistance against the harsh industrial environment than for activity and are therefore not suited for mobile fuel processors without modification. Thus, especially for the high-temperature shift reaction, noble metal catalysts are also under investigation nowadays. However, their stability against coke formation is limited to a temperature of about 400 °C and consequently high-temperature shift is performed at lower temperatures compared with the industrial process using these catalysts.

2.6.1.1 Simulation of the Effect of Integrating Heat-exchange Capabilities into Water-gas Shift Reactors

TeGrotenhuis et al. [80] simulated the the water-gas shift reaction at a noble metal catalyst. Diffusion effects were taken into consideration both for the gas phase and for the catalyst. For a given feed composition of the reformate (9% CO, 9% CO_2, 36% H_2O, 45% H_2, which corresponds to the product of an octane steam reformer at a S/C ratio of three), an optimum temperature profile was calculated for the reaction, at which the maximum reaction rate occurs (Figure 2.45). Running a reactor at this profile, 82% conversion occur in the first third of it, whereas the remaining 66% of the reaction volume is required to achieve a further 8% conversion. As the reaction temperature needs to be limited to about 400 °C owing to activity considerations (see above), the optimum profile should be limited accordingly. By doing so, the overall size required for the reactor is increased merely by 12%, which points to the importance of following the optimum profile especially in the lower temperature (and conversion) range.

A comparison was made between a reactor running at the optimum temperature profile and a two-stage adiabatic reactor; the amount of catalyst required was calculated to be 2.3 times higher for the adiabatic reactor. Hence much less catalyst,

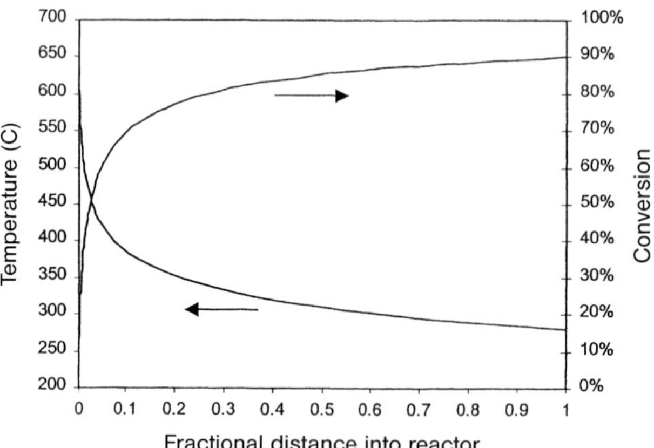

Figure 2.45 Optimum temperature profile and corresponding conversion profile for a WGS reactor with a steam reformate fee at an initial composition of 9% CO, 9% CO_2, 36% H_2O and 45% H_2 [80] (by courtesy of W. E. TeGrotenhuis).

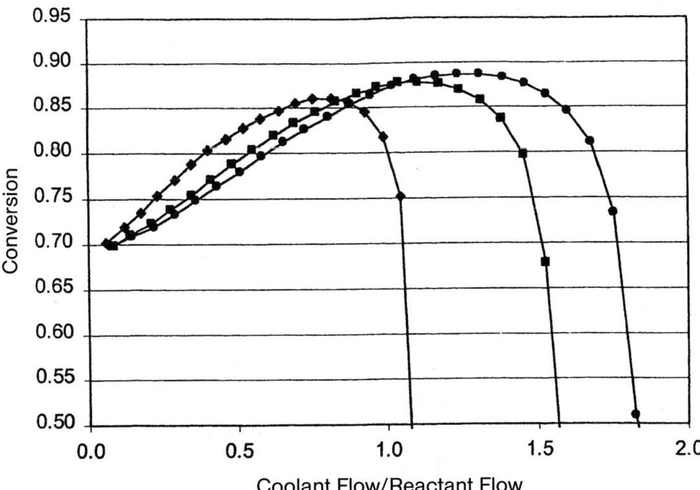

Figure 2.46 Effect of coolant temperature on CO conversion for the water-gas shift reaction in a micro channel reactor at constant reformate feed inlet temperature of 350 °C. Coolant temperature: (◆) 125; (■) 200; (●) 225 °C. Initial gas composition 9% CO, 9% CO_2, 36% H_2O, 45% H_2 [80] (by courtesy of W. E. TeGrotenhuis).

merely a single reactor, and no intermediate heat exchanger or water injection are required when applying an integrated heat exchanger for water-gas shift. Further investigations revealed that there exists an optimum feed flow ratio of reformate/cooling gas for given inlet temperatures of both gases. Finally, it could be demonstrated that the inlet reformate temperature has a very minor effect on the maximum conversion achievable compared with the inlet temperature of the coolant gas (see Figure 2.46).

2.6.1.2 **Catalyst Testing for the Water-gas Shift Reaction in Micro Structures**

Tonkovich et al. [81] compared the performance of a commercial ruthenium/zirconia powder catalyst from Degussa with a laboratory-made ruthenium/zirconia catalyst prepared on a nickel foam monolith for the water-gas shift reaction. Methane formation occurred for the powder catalyst, which was much less pronounced for the monolith. The selectivity towards methane could be reduced at shorter residence times. However, the activity of the laboratory-made catalyst was lower, which was partially attributed to the lower catalyst mass (modified residence time).

2.6.1.3 **Water-gas Shift 1 [WGS 1]: Stack-like Reactor Applied to Water-gas Shift Testing**

Germani et al. [82] wash-coated Cu/ZnO catalyst on to micro channels and compared their performance with that of conventional monoliths for the low-temperature water-gas shift. Up to six plates could be put into a stack-like reactor heated by cartridges, which had a maximum operation temperature of 600 °C (see Figure 2.47). The reactor had capabilities for measuring the inlet and outlet temperature of the gases via thermocouples.

Figure 2.47 Stack-like reactor developed by IMM for catalyst testing applied to water-gas shift (source: IMM).

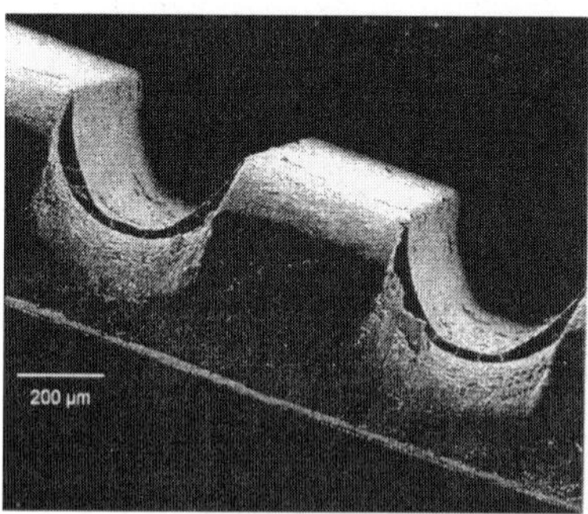

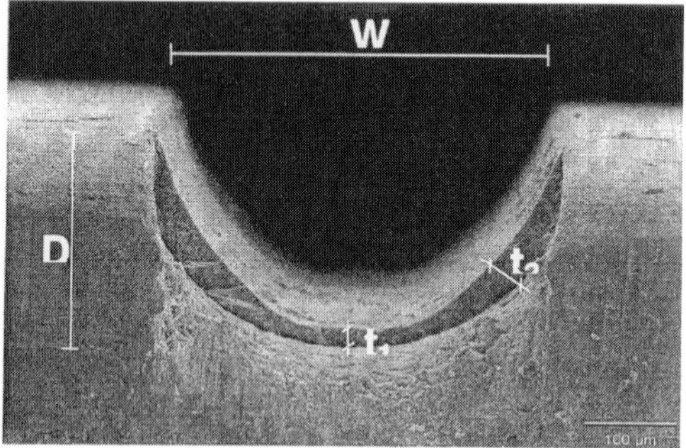

Figure 2.48 SEM of a micro channel coated with γ-alumina. $W = 430$ µm, $D = 250$ µm, $t_1 = 20$ µm and $t_2 = 38$ µm [82] (by courtesy of G. Germany).

The plates were 50 mm × 50 mm in size, carried 49 channels each and the channels were 600 µm wide and 400 µm deep. γ-Alumina was used as catalyst carrier (see Figure 2.48) and the active species were introduced by impregnation, resulting in a copper content of 20.5% and a zinc content of 10%. Good distribution of the zinc was found, whereas some inhomogeneities were identified for the copper by SEM–EDX, which was attributed to the higher mobility of Cu^{2+} during the thermal treatment of the coatings. Four plates were introduced into the reactor, resulting in an overall catalyst loading of 240 mg.

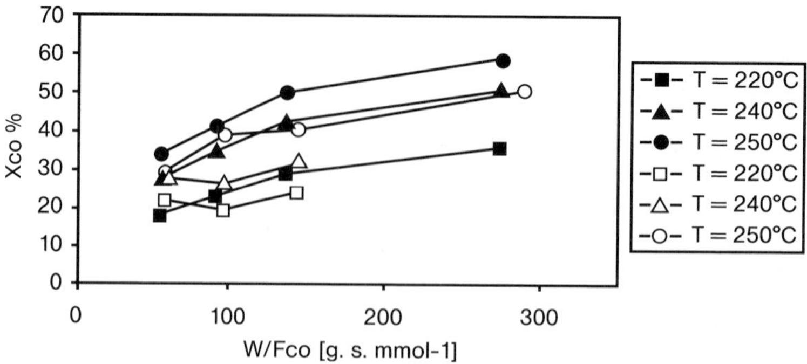

Figure 2.49 CO conversion on monolith (open symbols) and catalyst deposited in micro channels (closed symbols). Feed: 10% steam and 2–10% carbon monoxide, balance argon [82] (by courtesy of G. Germany).

The feed was composed of 10% steam and 2–10% carbon monoxide, balance argon (Figure 2.49). Up to 60% conversion was achieved at a 280 °C reaction temperature and a modified residence time of 280 g_{cat} (s mmol)$^{-1}$.

2.6.1.4 Water-gas Shift 2 [WGS 2]: Stack-like Reactor Applied to Water-gas Shift

Görke et al. [83] investigated the catalytic performance of Au/CeO_2 and Ru/ZrO_2 catalysts for the water-gas shift reaction. The externally heated testing reactor took up 25 micro structured foils each carrying 38 channels, which were 20 mm long, 200 μm wide and 70 μm deep (Figure 2.50).

The Au/CeO_2 catalysts were prepared by coating of micro structured FeCrAlloy (1.4767) foils with ceria nanoparticles, subsequent annealing in air at 950 °C and finally impregnating with gold solution; 3 wt.% gold was deposited on the ceria support (Table 2.5). The adhesion of the coating to the foils was not very strong, as documented by cracks found with electron microscopy.

The Ru/ZrO_2 catalysts were prepared by first annealing the FeCrAlloy and thus generating an α-alumina layer on the alloy surface (see Section 2.10.7). Then both coating with zirconia and impregnation with a ruthenium solution were performed without an intermediate calcination step.

Another sample was prepared on stainless-steel foils (1.4301) omitting the annealing step. Both samples contained 5 wt.% Ru on zirconia (Table 2.5). The coatings containing zirconia adhered much better to the surface. XRD analysis revealed elementary ruthenium and zirconia (baddeleyite) for both samples. However, apparently chromium migrated into the coating from the stainless-steel foil (and not from the FeCrAlloy), and was detected by XRD. The thickness of the catalyst coatings was determined as 5–20 μm for the Au/CeO_2 and 1–10 μm for the Ru/ZrO_2 catalysts by SEM.

The feed composition for catalyst testing was within the ranges 1–25% H_2O, 1–25% CO and 50–98% N_2. GHSV was in the range from 133 000 to 360 000 h^{-1}

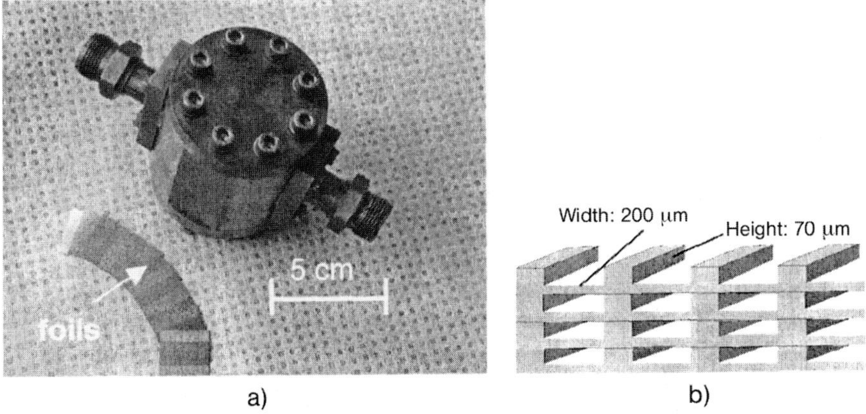

Figure 2.50 (a) Mounted stack-like reactor used for the water-gas shift reaction and (b) schematic of the stacked foils [83] (by courtesy of Elsevier Ltd.).

Table 2.5 Substrate, specific mass, composition and thickness of the catalyst layer of micro structured foils used for the water-gas shift reaction [83].

	Catalyst		
	WGS Au/CeO$_2$-I	***WGS Ru/ZrO$_2$-I***	***WGS Ru/ZrO$_2$-II***
Substrate	FeCrAlloy	FeCrAlloy	Stainless steel
Catalyst mass/foil surface area (mg cm^{-2})	7.4	0.44	0.44
Composition	3 wt.% Au	5 wt.% Ru	5 wt.% Ru
Layer thickness (μm)	5–20	1–10	1–10

and WHSV from 0.5 to 25 Ndm^3 (min g_{cat})$^{-1}$. At a feed composition of 25% H_2O and 25% CO (balance nitrogen), reaction temperature 310 °C and a GHSV of 180 000 h^{-1}, 96% conversion was determined using a ruthenium sample prepared on FeCrAlloy, whereas it was only 22% using a ruthenium sample prepared on a stainless-steel surface.

The gold catalyst showed only negligible conversion at this reaction temperature. The value of 96% conversion found by the authors exceeds the thermodynamic equilibrium (Figure 2.51), which might be due to the formation of methane disturbing the reaction system. However, the authors claim to have detected no

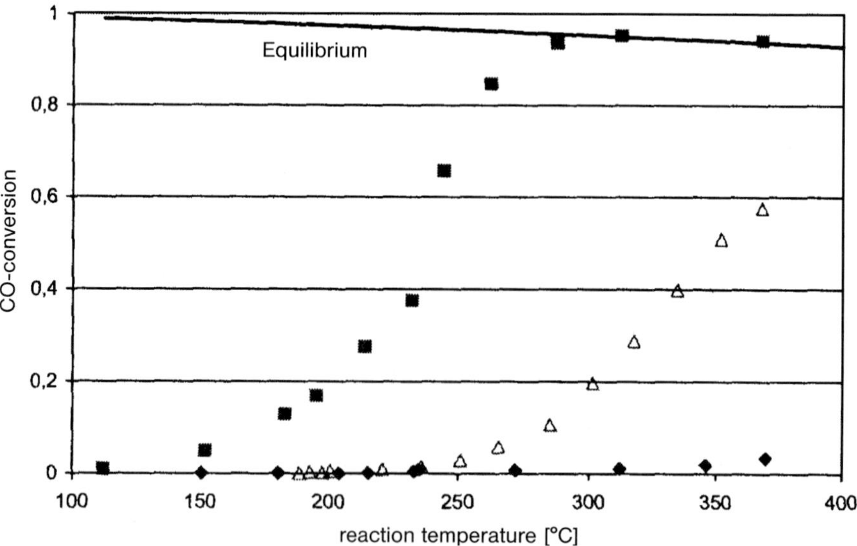

Figure 2.51 Experimental results for CO conversion of the water-gas shift reaction vs. reaction temperature for an average residence time of 20 ms. (◆) Au/CeO$_2$-I; (■) Ru/ZrO$_2$-I; (△) Ru/ZrO$_2$-II. 25% CO, 25% H_2O, balance nitrogen [83] (by courtesy of Elsevier Ltd.).

methane in the product. Another explanation is an error in the dosing equipment leading to an elevated concentration of carbon monoxide, thus shifting the equilibrium. The lower activity of the ruthenium catalyst was regarded as sufficient to run a low-temperature water-gas shift reactor at 1–3% carbon monoxide concentration in the feed.

2.6.1.5 Water-gas Shift 3 [WGS 3]: Sandwich-type Reactor ([HCR 4]) Applied to Water-gas Shift Catalyst Testing

Kolb et al. [84] applied the [HCR 4] reactors (see Section 2.4.3) to catalyst screening on the water-gas shift reaction. Testing was performed at a total flow rate of 30 $Ncm^3\ min^{-1}$, which corresponds to a GHSV of 15 000 h^{-1} and a WHSV of 1.5 $Ndm^3\ (min\ g_{cat})^{-1}$. Tests were performed for both low- and high-temperature water-gas shift. The feed was composed of simulated high-temperature shift product for low-temperature shift (3% CO, 14% CO_2, 25% H_2O and 55% H_2) and of a simulated steam reforming product for high-temperature water-gas shift (9% CO, 8% CO_2, 34% H_2O and 49% H_2).

Temperature ramps were applied for testing, which were set to 300, 325 and 350 °C and held for 1 h each for low-temperature shift. For high-temperature shift testing, the temperature ramps were set to 350, 375 and 400 °C for the same duration. These low reaction temperatures compared with industrial conditions for high-temperature shift (up to 450 °C) were applied because mostly precious metal catalysts were tested in the screening protocol, which are subject to coke formation at higher reaction temperatures.

Table 2.6 gives an overview of some of the catalysts tested and their activity for low- and high-temperature shift. Under the conditions of low-temperature shift testing, the highest activity was found for the $Pt/Ru/CeO_2$ sample, which even exceeded the thermodynamic equilibrium of the water-gas shift reaction. However, this catalyst showed little selectivity towards the water-gas shift reaction, as it generated between 5 and 15% methane in the off-gas (at 300 and 350 °C reaction temperature, respectively). These high amounts of methane were not only formed by methanation of the carbon monoxide but also of the carbon dioxide. The methanation functionality of the sample degraded rapidly. After 45 min, the total CO conversion had decreased from 70 to 50% at a 300 °C reaction temperature.

The $Pt/Rh/CeO_2$ sample also generated methane, but to a minor extent (2% in the off-gas at 300 °C). Both Pt/CeO_2 and $Pt/Pd/CeO_2$ catalysts were less active but no methane was detected in the off-gas and deactivation within 1 h test duration was moderate compared with the other samples. A proprietary laboratory-made catalyst achieved equilibrium conversion of 60% under the same conditions. The catalyst showed no selectivity towards methanation.

For high-temperature shift at 350 °C reaction temperature, the $Pt/Rh/CeO_2$ catalyst was most active but again very selective towards methane. The $Pt/Ru/CeO_2$ catalyst also produced methane, which was, again, not the case for both Pt/CeO_2 and $Pt/Pd/CeO_2$ catalysts. The last three samples showed moderate deactivation within 1 h test duration even at 400 °C. All catalysts could be regenerated by treatment in air at 500 °C for 30 min.

Table 2.6 Catalysts tested and conversions achieved for high- and low-temperature water-gas shift [84].

Catalyst	*Base material*	*Active component 1 (wt.%)*	*Active component 2 (wt.%)*	*Active component 3 (wt.%)*	*CO conversion: low-temperature shift (300 °C)*	*CO conversion: high-temperature shift (350 °C)*
Prop	–	–	–	–	60	–
$Pt/CeO_2/Al_2O_3$	Commercial Pt catalyst	4.1	17.0	–	20	27
$Pt/Rh/CeO_2/Al_2O_3$	Commercial Pt catalyst	4.0	4.0	17.0	23	39
$Pt/Ru/CeO_2/Al_2O_3$	Commercial Pt catalyst	4.1	4.1	13.1	68	24
$Pt/Pd/CeO_2/Al_2O_3$	Commercial Pt catalyst	4.1	4.1	12.3	6	9
$Pt/CeO_2/ZrO_2/Al_2O_3$	Commercial Pt catalyst	4.3	11.0	3.6	22	24
Commercial $Fe_2O_3/Cr_2O_3/CuO$	Blend	88.0	9.4	2.2	0	–

2.6.2 Preferential Carbon Monoxide Oxidation

The removal of low concentrations of carbon monoxide from the pre-cleaned reformate of hydrocarbon and ethanol reformers is commonly performed by oxidation with air. Owing to the lower carbon monoxide concentration achieved by the low temperatures of methanol reforming, in this case the reformate goes directly to the preferential CO oxidation (PrOx):

$$CO + 0.5\ O_2 \rightarrow CO_2 \qquad \Delta H_R = -283\ \text{kJ mol}^{-1} \tag{2.20}$$

The reaction is accompanied by the undesired side reaction of part of the hydrogen present in the reaction mixture:

$$H_2 + 0.5\ O_2 \rightarrow H_2O \qquad \Delta H_R = -242\ \text{kJ mol}^{-1} \tag{2.21}$$

Another problem that occurrs in PrOx reactors at low load levels is the re-production of carbon monoxide at the catalyst by the reverse water-gas shift reaction (RWGS) in an oxygen-deficient atmosphere [85]:

$$H_2 + CO_2 \rightarrow H_2O + CO \qquad \Delta H_R = +41\ \text{kJ mol}^{-1} \tag{2.22}$$

Generally, the corresponding catalysts tend to have a narrow operation window and heat removal and temperature management of the reaction are crucial.

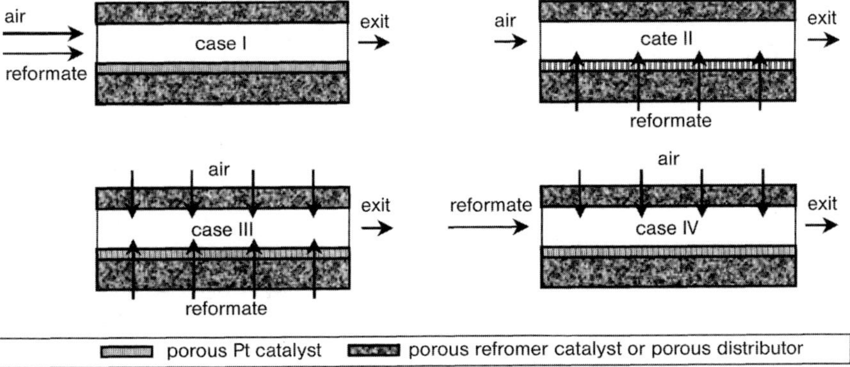

Figure 2.52 Strategies for feeding air and reformate in a PrOx reactor [85] (by courtesy of Elsevier Ltd.).

A description of various feed addition strategies in a porous reactor concept was given by Schuessler et al. [85] (see Figure 2.52). In case I, both reformate and air are added at the front of the channel. This is the scenario valid for most common micro channel applications. Case II involves gradual addition of reformate to the air feed and case IV the opposite. Case III involves the gradual addition of both feed components through porous channel walls, the reformate entering through the porous PrOx catalyst itself.

Kinetics are provided which were determined in the relevant parameter space [85]:

$$r_{PrOx} - \left[\frac{k_4\ p_{CO}}{1 + K_{CO}\ p_{CO}} - k_{inh}\ p_{CO} \right] p_{O_2}{}^{0.5} \tag{2.23}$$

$$r_{H_2Ox} = k_5\ p_{CO}{}^{-0.33}\ p_{O_2}{}^{0.5} \tag{2.24}$$

$$r_{RWGS} = k_5\ p_{CO}{}^{-0.33}\ p_{O_2}{}^{0.5} \tag{2.25}$$

These expressions are valid for carbon monoxide and oxygen in the range between 10 and 50 000 ppm. For hydrogen oxidation, the rate equation is valid for mole fractions exceeding 40%. Both RWGS and hydrogen oxidation are inhibited by carbon monoxide and therefore have a negative reaction order.

Simulation of the four cases described above was performed taking into consideration mass transfer of the two species limited in mass transfer (oxygen and carbon monoxide). The calculations were carried out for a pressure of 1.5 bar and a high reaction temperature of 300 °C; 1% carbon monoxide and 2% oxygen were assumed for the feed composition. Case II did not reduce the carbon monoxide content below 4 000 ppm at all load levels owing to elevated hydrogen oxidation at the reactor inlet. As indicated in Figure 2.53, case I had the narrowest operation window. Case III had a wider operation window than case I, but the carbon monoxide conversion achieved was not lower than 400 ppm. Case IV offered the widest dynamic range for load changes.

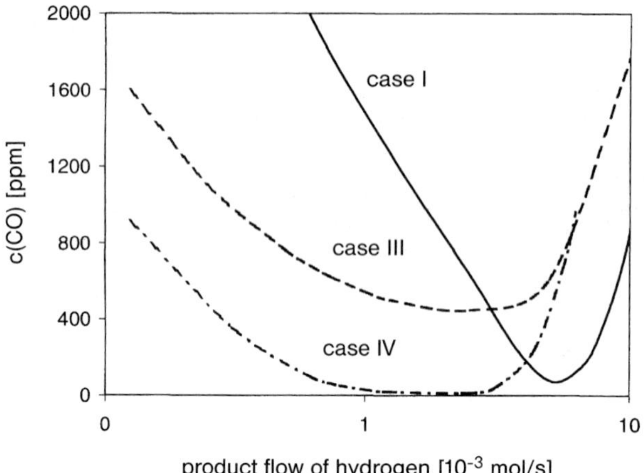

Figure 2.53 Carbon monoxide concentration as a function of reactor load for the various cases shown in Figure 2.54 [85] (by courtesy of Elsevier Ltd.).

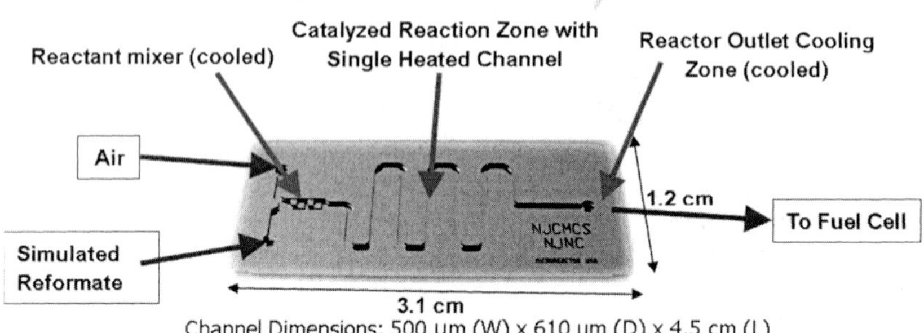

Figure 2.54 Single-channel silicon reactor with pre-mixer and outlet cooler [86] (by courtesy of R. Besser).

2.6.2.1 Preferential Carbon Monoxide Oxidation 1 [PrOx 1]: MEMS-like Reactor Applied to Studies of the PrOx Reaction in Micro Channels

Besser et al. [86] studied the reaction in a silicon reactor fabricated by applying MEMS technology, namely photolithography and DRIE by inductively coupled plasma. Each reactor incorporated dual gas inlets, a pre-mixer, a single reaction channel and an outlet zone where the product flow was cooled (see Figure 2.54). The single channel was 500 μm wide, 470 μm deep and 45 mm long.

Pt/Al_2O_3 was deposited in the channel prior to covering it by anodic bonding with a Pyrex® plate. The catalyst suspension contained the alumina sol and the platinum metal salt solution and was dried and calcined in the micro channel. The surface area of the catalyst was very high (480 $m^2 g^{-1}$). The final content of platinum in the coating was 2 wt.%. The coating thickness ranged between 2 and 15 μm. Platinum dispersion was 20% and the platinum particle size was 2.5–8 nm as determined by SEM (Figure 2.55).

The reactor was reduced in pure hydrogen at 400 °C for 4 h prior to introducing the feed, then 5 $Ncm^3 min^{-1}$ simulated reformate and 0.5 $Ncm^3 min^{-1}$ air were fed

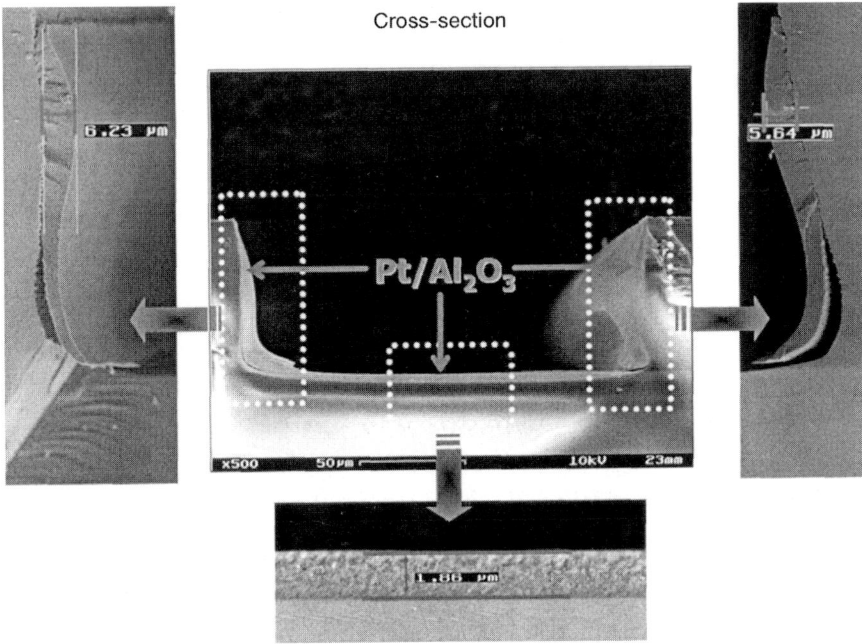

Figure 2.55 SEM cross-sectional view of micro channel with thin-film coating. The thickness of the coatings on the sidewall is ~6 µm, and on the bottom a thickness of ~2 µm was observed. The delamination of the catalyst coating is an artefact caused by sample preparation [86].

to the reactor, which corresponds to an O_2/CO ratio of 1.3. The former was composed of 1.7% CO, 68% H_2, 21% CO_2 and balance N_2. The experiments were performed by applying a temperature ramp program starting at ambient and finishing at 300 °C.

Full conversion of the feed was achieved at a 170 °C reaction temperature. Numerical simulations of the reaction system were performed applying CHEMKIN software and a network of eight species in the gas phase, eight surface species and 28 reactions not provided here. The simulation described the experimental performance of the reactor very well. It revealed that oxidation of carbon monoxide occurs by the reaction between adsorbed CO and OH species and not by the reaction between adsorbed CO and O species, as the rate of the latter reaction was 10 orders of magnitude lower. Thus a simplified mechanism of the reaction according to Besser [86] could be formulated as follows ((S) standing for adsorbed species):

$$
\begin{aligned}
&O_2 && + 2\,Pt\,(S) && \rightarrow 2\,O\,(S) \\
&H_2 && + 2\,Pt\,(S) && \rightarrow 2\,H\,(S) \\
&O\,(S) && + H\,(S) && \rightarrow OH\,(S) \\
&CO\,(S) && + OH\,(S) && \rightarrow CO_2\,(S) + H\,(S) \\
&CO_2\,(S) && && \rightarrow CO_2
\end{aligned}
\tag{2.26}
$$

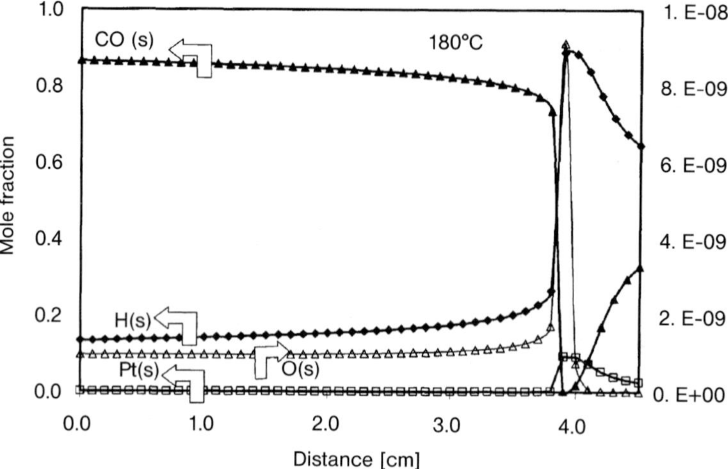

Figure 2.56 CHEMKINsimulation for the mole fractions of surface species as a function of residence time (here expressed as channel length) [86] (by courtesy of R. Besser).

The simulation revealed that the catalyst surface was almost completely covered by carbon monoxide up to certain channel length (residence time). Then the surface concentration of carbon monoxide abruptly decreased, when no carbon monoxide was left in the gas phase. At the same time, the concentration of adsorbed oxygen and hydrogen increased by almost an order of magnitude, which led to consumption of hydrogen due to water formation. The degree of water formation was merely limited by the amount of oxygen present in the gas phase. When all carbon monoxide and water had been consumed, carbon monoxide was formed again via the reverse water-gas shift reaction (Figure 2.56) [86].

2.6.2.2 Preferential Carbon Monoxide Oxidation 2 [PrOx 2]: Single-plate Reactor Based on MEMS Technology

Bednarova et al. [87] designed a single-plate PrOx reactor for a 1 W fuel cell based on MEMS technology, similar to [PrOx 1]. The results of the simulation work described above were applied to optimize the reactor performance. The plate carried a pre-mixer and 29 channels each 12 mm long, 500 μm wide and 400 μm deep. Flow uniformity was verified by simulations. Full conversion of carbon monoxide was achieved at much lower temperature with [PrOx 1] and the selectivity towards carbon dioxide was 40%.

2.6.2.3 Preferential Carbon Monoxide Oxidation 3 [PrOx 3]: Integrated Micro Structure Heat Exchanger for PrOx Applied in a 20 kW Fuel Processor

Dudfield et al. [88] presented results generated in the scope of the Mercatox program funded by the European Community aimed at a combined methanol steam reformer/combustor with consecutive CO clean-up by PrOx. First, various catalysts were tested for the reaction as micro spheres in a test reactor which was similar to a macroscopic shell-and-tube heat exchanger (Figure 2.57).

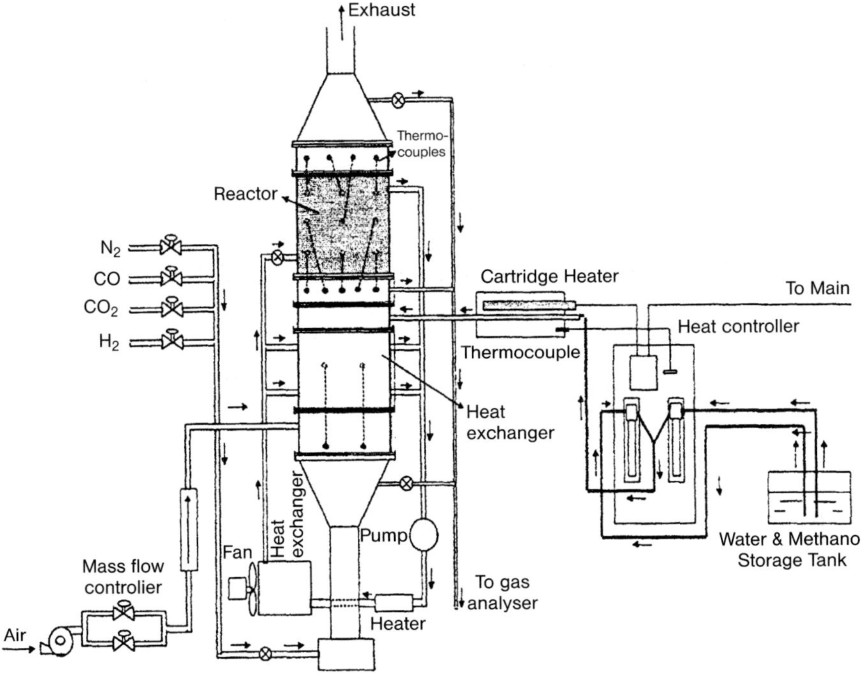

Figure 2.57 Schematic of the catalyst test assembly [88] (by courtesy of Elsevier Ltd.).

A 10 Ndm^3 min^{-1} feed composed of 75.0% H_2, 0.7% CO and balance CO_2 with air for oxidation was fed into the relatively large test reactor carrying 230 cm^3 catalyst microsperes of 1 mm diameter. Of the non-precious metal catalysts, that composed of hopcalite showed the highest activity, achieving almost full conversion in the temperature range 130–160 °C. The minimum CO output achieved was 40 ppm. A minimum O_2/CO ratio of 2.5 was determined for this catalyst to achieve a carbon monoxide conversion exceeding 90%. The catalysts tested are summarized in Table 2.7.

Table 2.7 Catalysts evaluated for CO PrOx application [88].

Catalyst carrier	*Catalyst code*	*Active component 1 (wt.%)*	*Active component 2 (wt.%)*
Hopcalite	LU-1	–	–
Aluminum stannate	LU-2	Cu 5	La 3
Ferric oxide	LU-3	Cu 3	Ce 2
Silica + 4 wt.% Ce	LU-4	Mn 8	Cu 2
Silica	LU-5	Pd 2.5	–
Silica	LU-6	Ru 2	–
Silica	LU-7	Ru/Pt 2.25	–
Alumina	LU-8	Ru/Pt 3.5	–
Hopcalite	LU-9	Ru/Pt 2.5	–

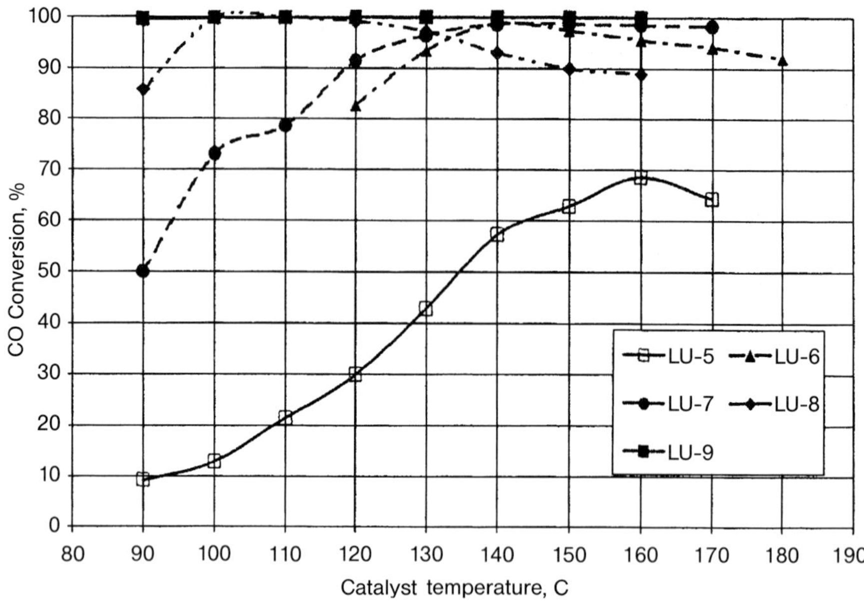

Figure 2.58 Experimental results for the conversion of CO on precious metal catalysts [88] (by courtesy of Elsevier Ltd.).

At an O_2/CO ratio of 2.0, this catalyst converted 98% of the carbon monoxide and as much as 4.6% of the hydrogen, which corresponds to 3.5% absolute loss of hydrogen. Of the precious metal catalysts, the Pt/Ru samples showed the highest activity, corresponding to conversions exceeding 99.8%. Especially the sample based on hopcalite carrier operated at this high level of conversion in the wide temperature range between 90 and 160 °C and achieved a low CO output of 7 ppm (Figure 2.58).

2.6.2.4 Preferential Carbon Monoxide Oxidation 4 [PrOx 4]: Stack-like Reactor Applied to PrOx

Görke et al. [83] applied the [WGS 1] reactor for the PrOx reaction using Au/CeO_2, CuO/CeO_2 and Au/α-Fe_2O_3 catalysts. The CeO_2-supported catalysts were prepared by coating of micro structured stainless steel (1.4301) foils with ceria nanoparticles, subsequent calcination in air at 450 °C and impregnation with gold or copper solutions. In this way, 3 wt.% gold or 1.9 wt.% copper were introduced into the ceria support. The Au/α-Fe_2O_3 catalysts were prepared by impregnating the stainless-steel foils with an iron solution and impregnation of the iron salt crystals generated with gold solution, which was performed without an intermediate calcination step. Finally, the sample was calcined at 400 °C. The adhesion of the coatings containing iron oxide to the foils was not very strong, as shown by cracks found with electron microscopy. The thickness of the catalyst coatings was determined as 0.3–3 μm for the Au/CeO_2,CuO/CeO_2 samples and to 1–10 μm for the Au/α-Fe_2O_3 catalysts by SEM (Table 2.8). The size of the gold nanoparticles was determined to be between 50 and 200 nm by SEM [83].

Table 2.8 Substrate, specific mass, composition and thickness of the catalyst layer of micro structured foils used for PrOx reaction [83].

	Catalyst		
	PrOx Au/CeO$_2$-II	***PrOx Au/α-Fe$_2$O$_3$***	***PrOx Cu/CeO$_2$***
Substrate	Stainless steel	Stainless steel	Stainless steel
Catalyst mass/foil surface area (mg cm^{-2})	0.61	9.8	0.61
Composition	3 wt.% Au	5 wt.% Ru	5 wt.% Ru
Layer thickness (μm)	0.3–3	1–10	0.3–3

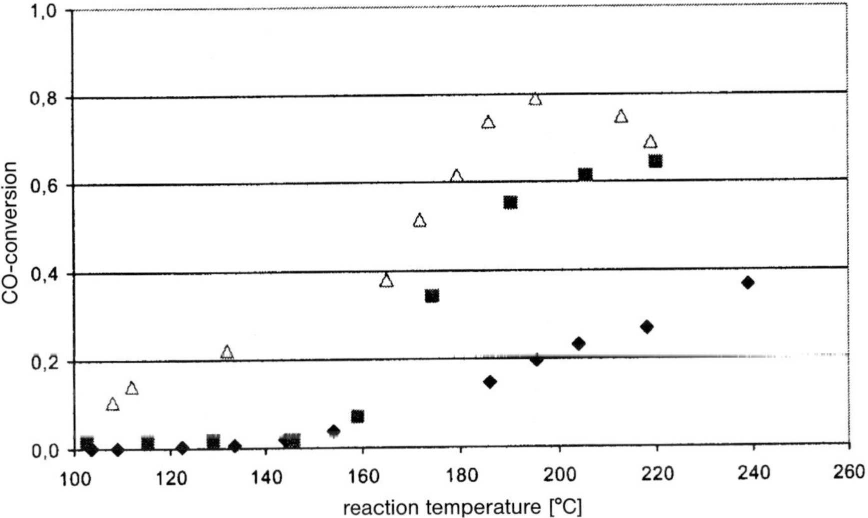

Figure 2.59 Experimental results for CO conversion of the PrOx reaction vs. reaction temperature for an average residence time of 14 ms. (◆) Au/α-Fe$_2$O$_3$; (■) Au/CeO$_2$-II; (△) Cu/CeO$_2$. 8% CO, 40% H$_2$, 6% O$_2$, balance nitrogen [83] (by courtesy of Elsevier Ltd.).

The feed composition for catalyst testing was within the ranges 46–53% N_2, 6% O_2, 1–8% CO and 40% H_2. The GHSV was 257 000 h^{-1}. At a feed composition of 40% H_2O, 8% CO and 6% O_2 (balance nitrogen), which corresponds to a stoichiometric O/CO ratio of 1.5, and a 150 °C reaction temperature, the highest conversion (79%) was achieved with the copper catalyst, followed by the Au/CeO_2 sample, which converted 60% of the carbon monoxide (Figure 2.59) [83].

The Au/α-Fe_2O_3 catalyst showed a much lower conversion of 20%. Satisfactory values of CO selectivity exceeding 50% were determined for the copper catalyst in the temperature range 180–210 °C.

The performance of the copper catalyst was investigated under more realistic conditions of 1% carbon monoxide in the feed. More than 99% CO conversion was achieved in a lower temperature range of 140–180 °C. However, the selectivity of the catalyst towards carbon monoxide was only 20% at the very high O/C ratio of 12 set for the experiments [83].

2.6.2.5 Preferential Carbon Monoxide Oxidation 5 [PrOx 5]: Integrated Heat Exchanger/Reactor for PrOx

Cominos et al. [89] developed a micro structured test reactor with integrated heat-exchanging capabilities for the PrOx reaction. The reactor (see Figure 2.60) had outer dimensions of 52 mm × 53 mm × 66 mm. It was split into three parts, namely two heat exchangers composed of six plates and the reactor, which were thermally decoupled by insulation material. This set of components was part of an integrated methanol fuel processor design described in Section 2.7.2.5. Four thermocouples were integrated into the arrangement to determine the temperature at the outlet of each component. The reactor itself incorporated 19 plates each with 75 micro channels coated with catalyst and sealed by graphite gaskets. The dimensions of the channels were 250 μm width, 125 μm depth and 30 mm length.

Various catalysts (see Figure 2.61) on ceria and alumina wash coats were tested in the reactor for their performance. The feed was composed of 58% hydrogen, 21% carbon dioxide, 1.12% carbon monoxide, 4.6% oxygen and 15% nitrogen. Ru/Pt, Rh and Pt/Rh on alumina were found to be the most active, achieving a carbon monoxide reduction from 12 000 ppm to below 4 ppm at 126, 140 and 144 °C reaction temperature, respectively. The Pt/Rh catalyst containing 2.5 wt.% of each noble metal was found to be the most stable with time, maintaining its high activity at a WHSV of 2.0 Ndm^3 (min $g_{cat})^{-1}$ for 20 h. The optimum reaction temperature for maximum selectivity towards carbon monoxide of 35% was 140 °C. As a reaction temperature of 144 °C was necessary to achieve full conversion of CO, 4% hydrogen was lost from the feed.

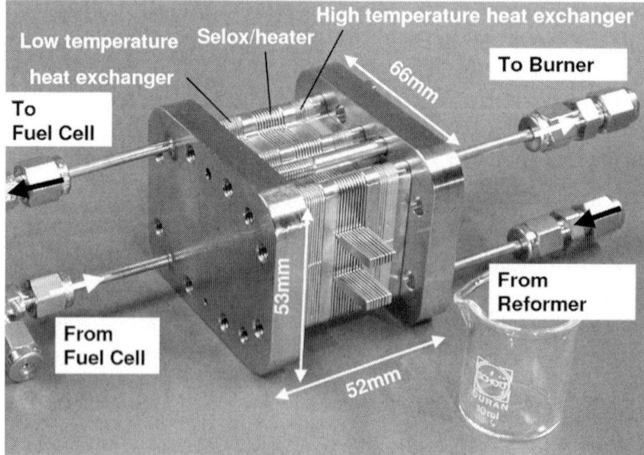

Figure 2.60 Integrated reactor/heat exchanger for the preferential oxidation of carbon monoxide developed by Eindhoven University and IMM [89] (source IMM).

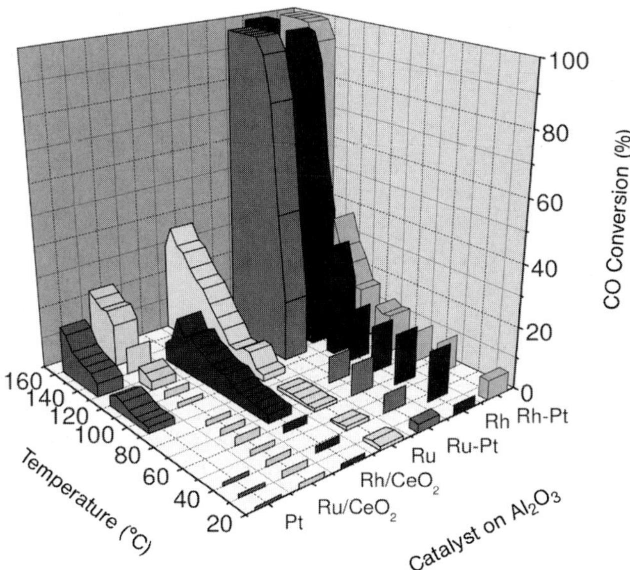

Figure 2.61 Carbon monoxide conversion over various catalysts with respect to reaction temperature at an O_2/CO ratio of 4 and 200 ms residence time [89] (source IMM).

Delsman et al. [90] performed numerical simulations for the [PrOx 5] reactor. They found that a two-dimensional model is necessary to describe the reactor performance accurately. Additionally, it was demonstrated that it was not possible to achieve the temperature gradients for which the device was designed experimentally, which was attributed to the low length/width ratio of 1. Further, the performance of a Pt/Co/α-alumina catalyst was investigated in the [PrOx 5] reactor. Here 0.5% CO, 1.6% O_2, 56% H_2 and 18% CO_2, balance helium, were fed to the integrated device at a temperature of 300 °C. Nitrogen was used as coolant. At a WHSV of 0.8 Ndm^3 $(min\ g_{cat})^{-1}$, the carbon monoxide could be reduced to 7 ppm, but the catalyst was deactivated and 300 ppm CO were found after 4 h. The CO selectivity was 17% regardless of the conversion level.

To minimize the device volume and reduce axial heat transfer through the walls, the length/width ratio of a redesigned reactor was reduced from 1 to 3, thus improving the heat-exchanging mechanism. A laser-welded version of the reactor/heat exchanger assembly, which removes the space demand for gaskets and screws, was designed and fabricated. It was considerably smaller (45 mm × 17 mm × 50 mm) than the [PrOx 5] reactor.

2.6.2.6 Preferential Carbon Monoxide Oxidation 6 [PrOx 6]: Stack-like Reactor Applied to PrOx

A stack-like reactor was applied for catalyst testing for the reaction by Chen et al. [38]. Four micro structured foils made of stainless steel formed the stack, which were 340 μm thick, carrying 48 channels 170 μm deep and 500 μm wide (see Figure 2.14).

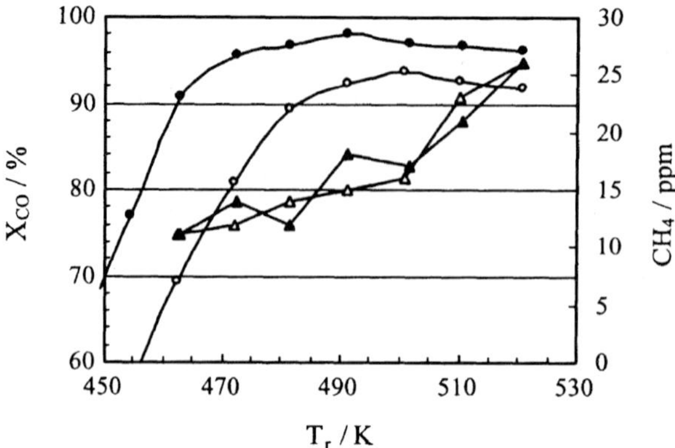

Figure 2.62 CO conversion vs. reaction temperature in a micro channel reactor. CO = 5000 ppm; GHSV = 500 000 h^{-1}; O_2/CO ratio = 1.0 (○, △), 1.25 (●, ▲). (○, ●) CO conversion; (△, ▲) outlet CH_4 concentration [38] (by courtesy of ACS).

Rh/Al_2O_3 and $Rh/K/Al_2O_3$ catalysts were prepared by wash coating the alumina, dip coating the potassium and incipient wetness impregnation of the rhodium. The catalyst was reduced in mixture of 10% hydrogen in nitrogen prior to testing.

The feed was composed of 40% H_2, 20%CO_2, 0.2–1.0% CO and 0.2–1.5% O_2 on a dry basis; 10% water was added to the feed and the GHSV of the standard experiments was 20 000 h^{-1}. The $Rh/K/Al_2O_3$ catalyst outperformed its potassium-free counterpart when tested in ceramic monoliths.

At a 300 °C reaction temperature, 1 200 ppm methane was found under standard conditions. Testing in the micro structured reactor at a very high GHSV of 500 000 h^{-1}, a 230 °C reaction temperature and an O_2/CO ratio of 1.0 revealed a very low CO concentration of < 20 ppm at the reactor outlet regardless of the feed concentration. Less than 30 ppm methane was detected under these conditions. Increasing the O_2/CO ratio above 1.0 did not further improve this performance (Figure 2.62).

A carbon monoxide conversion of 93.6% was found for the Rh/Al_2O_3 compared with > 99% at a reaction temperature of 200 °C and values of the O_2/CO ratio of 1.0 or higher. The highest conversion achieved was 99.82% for a CO concentration of 5 000 ppm, which corresponds to < 10 ppm at the reactor outlet. At reaction temperatures exceeding 250 °C, significant amounts of methane were formed owing to the methanation reaction, which is in line with findings of Kolb et al. [84], showing a selectivity of rhodium towards methanation under the conditions of the water-gas shift reaction.

2.6.3
Micro Structured Membranes for CO Clean-up

Membrane purification is an elegant alternative to catalytic CO clean-up systems because it has the capability of producing highly purified hydrogen [91]. Additionally, the fuel cell anode can be designed to run dead-end, which decreases pressure losses significantly [91]. However, a relatively high pressure (about 10 bar) and temperature are required to deliver the driver of the membrane separation process. Hence there is a need for membranes which allow operation at lower pressures. Here micro technology offers unique possibilities by applying ultra-thin membranes on micro structured supports, which also have a low thickness. The following section gives a few examples dealing with membrane separation processes on porous or micro structured supports, not at all claiming completeness. The separation factor s is defined as

$$s = \frac{\dot{n}_1 \, \Delta p_1}{\dot{n}_2 \, \Delta p_2} \tag{2.27}$$

where $\dot{n}_i$ represents moles of species i transferred through the membrane and Δp_i is the partial pressure difference of species i through the membrane.

2.6.3.1 Micro Structured Membranes for CO Clean-up 1 [MMem 1]: Palladium-based Reactor for Membrane-supported Water-gas Shift

Barbieri et al. [92] realized a Pd77Ag23 membrane reactor to increase the hydrogen production of the water-gas shift reaction. By applying membrane technology, hydrogen was removed along the reaction pathway and therefore the equilibrium of the reaction was shifted, leading to higher theoretically possible conversion. A palladium alloy film about 1 μm thick was first produced by sputtering, then this film was placed on a porous stainless-steel support. The patented method applied allowed for much higher pore size/film thickness values than conventional methods. Tubular membranes of 13 mm outer diameter, 10–20 mm length and 1.1 and 1.5 μm thickness were prepared.

A commercial Cu based catalyst supplied by Haldor-Topsoe was applied to the water-gas shift reaction. At 210 °C, a permeating flux of 4.5 Ndm^3 m^{-2} s^{-1} was determined for pure hydrogen at a very low pressure drop of 0.2 bar. Then the membrane reactor was coupled with a conventional water-gas shift reactor. At 260–300 °C reaction temperature and a GHSV of 2 085 h^{-1}, the maximum conversion achievable due to the thermodynamic equilibrium could be exceeded by this new technology by 5–10%.

2.6.3.2 Micro Structured Membranes for CO Clean-up 2 [MMem 2]: Palladium Membrane Micro Reactor

Franz et al. [93] developed a palladium membrane micro reactor for hydrogen separation based on MEMS technology, which incorporated integrated devices for heating and temperature measurement. The reactor consisted of two channels separated by the membrane, which was composed of three layers. Two of them, which were made of silicon nitride introduced by low-pressure chemical vapor deposition (0.3 μm thick) and silicon oxide by temperature treatment (0.2 μm thick), served as perforated supports for the palladium membrane. Both layers were deposited on a silicon wafer and subsequently removed from one side completely

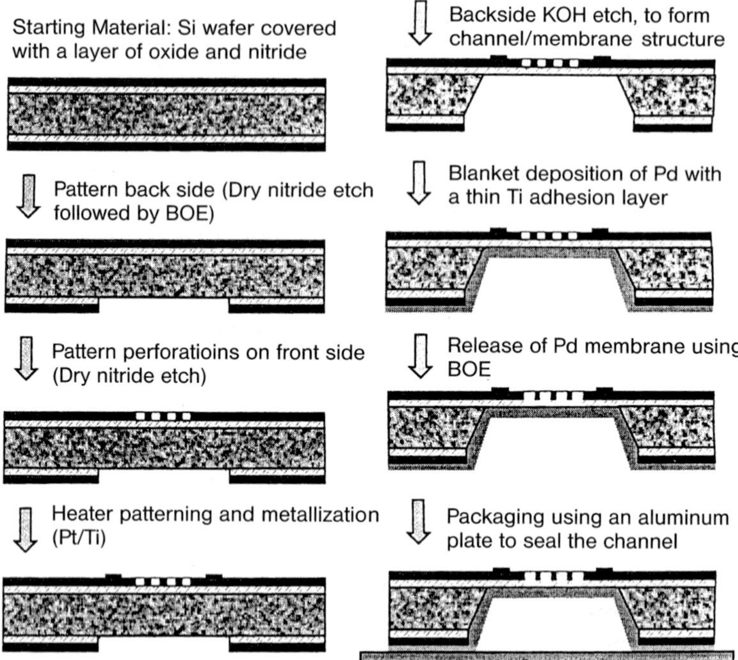

Figure 2.63 Micro fabrication sequence for the silicon component of the palladium membrane reactor [93] (by courtesy of Springer-Verlag).

and from the other side partially to achieve the support function. The removal was performed by dry SF_6 etching and wet buffered oxide etching (BOE). Then one channel was etched out of the wafer with KOH.

An electron beam was applied to deposit the platinum/titanium films on the device, which served as heaters and temperature sensors. Then a 'blanket' of palladium was deposited on one side of the support using a titanium film as adhesion layer. In this way, a 700 μm wide and 17 mm long membrane of 0.2 μm thickness was realized. An aluminum plate sealed the first channel. The second channel was fabricated from polydimethylsiloxane by applying a molding process (see Figure 2.63). The membrane showed high mechanical stability. When pressurized below the support structure, rupture occurred at more than 5 bar pressure. Surprisingly, the pressure tolerated was much lower (1.4 bar), when the membrane was pressurized from the opposite direction. A pressure drop of 1 bar was even tolerated at a temperature of 500 °C when the higher pressure was put below the support structure.

The performance of the membrane was tested with a mixture of 90% nitrogen and 10% hydrogen at atmospheric pressure and 500 °C. The permeate side of the membrane was set under vacuum. A hydrogen flux of 100 $Ndm^3\ m^{-2}\ s^{-1}$ was determined at the pressure drop of 1 bar.

2.6.3.3 Micro Structured Membranes for CO Clean-up 3 [MMem 3]: Palladium Membranes in Micro Slits

Kusakabe et al. [94] prepared palladium membranes in micro slits ethched into a copper foil. The foil had dimensions of 1 cm × 1 cm at 50 µm thickness. The palladium membrane was formed by electrodeposition in a bath contaning 20 g Ndm^{-3} $PdCl_2$, citric acid and ammonium sulfate with the pH adjusted to 7.0. A current density between 20 and 30 mA cm^{-2} was found to be optimal for avoiding metallic crystal growth. The electrolysis temperature affected the permeability of the membrane considerably and optimum separation factors were achieved at a temperature between 35 and 45 °C. The thickness of the membrane was determined as 4 µm. The micro slits were introduced from the reverse side of the copper foil by photolithographic ethching. They were about 100 µm wide. The total area of free-standing membrane over the slits was 5.4 mm^2. Permeation experiments were performed at ambient pressure using a mixture of nitrogen and hydrogen at the retenate side and argon as sweep gas on the permeate side of the membrane. As shown in Figure 2.64, hydrogen permeance increased with increasing permeation temperature, whereas nitrogen permeance was not affected over the whole temperature range under investigation (200–400 °C). A separation factor of 2000 was determined at 300 °C. The hydrogen flux was 1.3 Ndm^3 m^{-2} s^{-1} under these conditions.

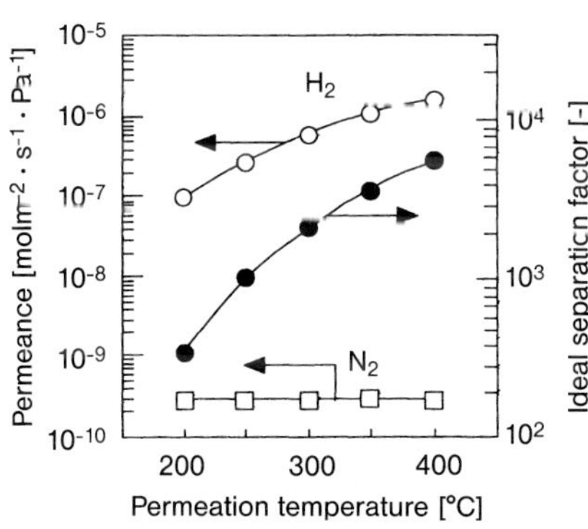

Figure 2.64 Effect of permeation temperature on permeance and separation factor for a membrane fabricated at 40 mA cm^{-2} current density and 45 °C electrolysis temperature [94] (by courtesy of Springer-Verlag).

2.6.3.4 Micro Structured Membranes for CO Clean-up 4 [MMem 4]: Supported Palladium Membrane

Lee et al. [95] fabricated a palladium membrane on a 400 mm thick silicon wafer. The process applied was very similar to the method applied by Franz et al. [93]. Silicon nitride and silicon oxide were deposited on the wafer and then the silicon was removed from the reverse side by DRIE. A 100 µm thick palladium layer forming the membrane was deposited on the support by an evaporation method.

2.6.3.5 Micro Structured Membranes for CO Clean-up 5 [MMem 5]: Sputtered Tantalum Membrane

Mukherjee et al. [96] used sputtered tantalum, as tantalum is reported [97] to be a cheap alternative with better permeation properties when coated with palladium compared with membranes made completely of palladium.

2.6.3.6 Micro Structured Membranes for CO Clean-up 6 [MMem 6]: Pd and Pd77Ag23 Membranes

Tong et al. [98] prepared Pd and Pd77Ag23 membranes by applying a similar procedure to that of Franz et al. [93]. The membranes introduced on the support by sputtering had a thickness between 500 and 1 000 nm. They were stable against a differential pressure up to 4 bar and were operated at 450 °C for more than 1000 h. A hydrogen flux up to 90 Ndm^3 m^{-2} s^{-1} and a separation factor of 1 500 for H_2/He were determined for these membranes.

2.6.3.7 Micro Structured Membranes for CO Clean-up 7 [MMem 7]: Free-standing Pd, Pd/Cu and Pd/Ag Membranes

Gielens et al. [99] fabricated free-standing membranes of Pd, Pd/Cu and Pd/Ag of 500–1 000 nm thickness by sputtering on to a silicon wafer. Micro channels were introduced on two glass plates by power blasting, and were bonded to the silicon wafer for separation experiments.

A hydrogen flux between 50 and 90 Ndm^3 m^{-2} s^{-1} was determined for these membranes on increasing the operating temperature from 350 to 450 °C. The H_2/He separation factor was > 2 750. An increased hydrogen flux compared with membranes deposited on porous supports was demonstrated. Addition of steam to the retenate flow (20% H_2, 60% He, 20% H_2O) reduced the hydrogen flux from 90 to 15 Ndm^3 m^{-2} s^{-1}.

The original permeability could be regained when the steam addition was stopped. However, this was not the case when adding carbon dioxide to the feed instead of steam. This decreased the hydrogen flux from 90 to 13 Ndm^3 m^{-2} s^{-1} and the permeability was lower than before, when carbon dioxide addition was stopped. This effect was attributed to carbon formation on the membrane surface.

2.7 Integrated Micro Structured Reactor Fuel Processing Concepts

One of the benefits of applying micro structured heat exchanger/reactors is the potential of combining an endothermic reaction such as steam reforming with an exothermic reaction such as catalytic combustion. This idea was proposed for the macro scale as a so-called catalytic plate reactor (CPR) by Reay [100].

An early application of a combined steam reformer/catalytic combustor on the meso scale was realized by Polman et al. [101]. They fabricated a reactor similar to an automotive metallic monolith with channel dimensions in the millimeter range (Figure 2.65). The plates were connected by diffusion bonding and the catalyst was introduced by wash coating. The reactor was operated at temperatures between 550 and 700 °C; 99.98% conversion was achieved for the combustion reaction and 97% for the steam reforming side. A volume of < 1.5 dm^3 per kW electrical power output of the reformer alone was regarded as feasible at that time, but not yet realized.

Frauhammer et al. [102] developed a ceramic monolith which may be used as a counter- or co-current heat exchanger. This is performed by partially closing ducts

Figure 2.65 Concept of reforming with heat supply through heat-conductive substrates by catalytic combustion (left) and photograph of the realized catalytic test reactor (right) [101] (by courtesy of Elsevier Ltd.).

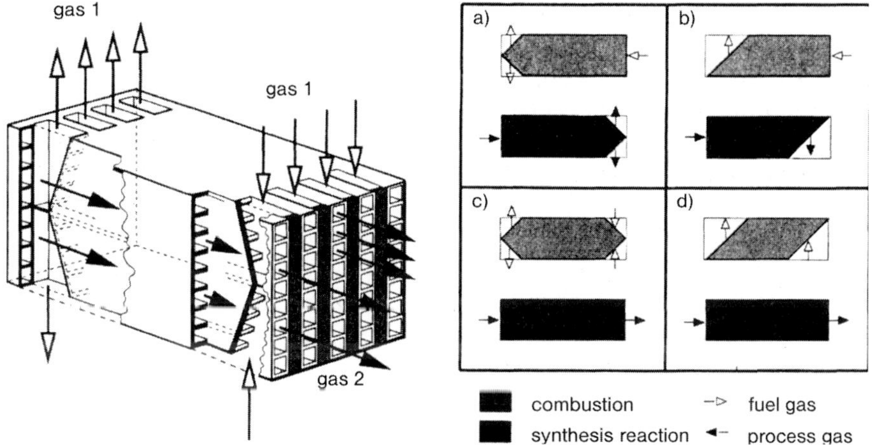

Figure 2.66 Preparation of a ceramic monolith to obtain a cross-flow heat exchanger (left) and different options of monolith preparation (right) [102].

and opening walls of the monolith as shown Figure 2.66, according to a patent by Minjolle [103]. However, this method was regarded as costly and time consuming by the authors.

Von Hippel et al. [104] patented a special reactor head, which allows for a distribution of the two gas flows through each individual channel. Even at a 1 200 °C monolith temperature the heads did not heat up to more than 200 °C, hence silicone rubber was applied for sealing the heads. This concept was applied for coupling methane combustion and steam reforming in separate flow paths [105].

2.7.1.1 **Parametric Study for Coupling Highly Exothermic and Endothermic Reactions**

Kolios et al. [106] performed an extensive study revealing that coupling of exothermic and endothermic reactions is possible under safe and stable operation conditions only in catalytic wall reactors and not in coupled packed beds. In the latter case instability and thermal runaway of the reactor may occur. Additionally, the two reactions taking place at the two sides of the same reactor wall need to be coupled by highly heat-conductive material. This is the case for micro structured heat exchanger/reactors. However, the authors proposed a folded wall reactor operated in the counter-current mode as the optimum solution for the problem. This design was successfully applied to methanol steam reforming. Folded metal foils, which are similar to the foils used for automotive exhaust gas systems, were incorporated into the reactor and formed the flow ducts. The feed for the catalytic combustion reaction entered the ducts from the side through a spacer over the whole length of the reactor, which allows for optimum control of the process.

2.7.1.2 Co-current Operation of Combined Meso-scale Heat Exchangers and Reactors for Methanol Steam Reforming

Gritsch et al. [107] presented a new concept for coupling endothermic and exothermic reactions (Figure 2.67). The distributed feed addition along the reactor channel was regarded as too complicated. Therefore, a co-current flow arrangement was proposed for the two reactions, including two heat exchangers heating the feed of each flow by the hot product gases. The feasibility of this concept was investigated in a single combustion channel combined with two neighboring reforming channels.

The test reactor was fabricated from structured metal foils and the channel dimensions were on the millimeter scale. The reactor had a length of 250 mm and a width of 50 mm and the combustion and reforming channels were 2.3 and 1.1 mm wide, respectively. More than 99% conversion was achieved when the methane feed concentration exceeded 5%. The distributed hot-spot of the reactor was limited to 150 K. Homogeneous combustion was observed for feed containing both hydrogen and methane. To achieve distributed energy generation, the feed had to be dosed at various positions in the reactor.

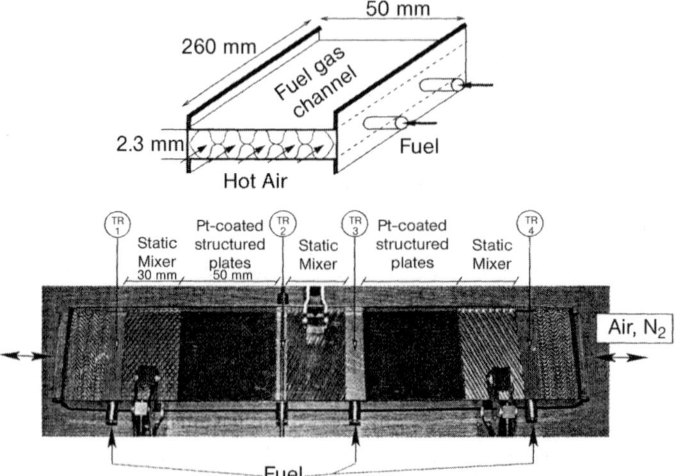

Figure 2.67 Schematic and photograph of new developed reactor system for methanol steam reforming [107] (by courtesy of Motorola).

Alternatively, carbon dioxide was added as an inhibitor to avoid the homogeneous reactions. This problem could likely be obviated when applying micro structures, as they act as flame arrestors (see Section 2.5.1).

2.7.1.3 Feasibility Study for Combined Methane Oxidation/Steam Reforming in an Integrated Heat Exchanger

Zanfir and Gavriilidis [108] studied the feasibility of this concept in detail by numerical simulations based on a two-dimensional model for methane as fuel. Literature kinetic data determined for nickel catalysts applied in industrial large-scale steam reforming were used. First-order kinetics for typical precious metal combustion catalysts were chosen to describe the combustion process. Co-current flow of reformer and combustor gases was assumed for the reactor model. Pressure close to ambient, an S/C ratio of 3.4 and a 520 °C feed temperature were assumed for the process, the last two being close to the conditions of the industrial process. An almost 260 K temperature rise was calculated for the reactor wall temperature along the feed flow path owing to the slower kinetics of the steam reforming reaction. The gradient of the axial temperature increased when the half-height of the channels was increased from 0.5 to 2 mm at constant WHSV. Larger channel dimensions resulted in less efficient heat transfer and higher temperature gradients in the gas phase. Very low temperature gradients below 0.5 K were found through the reactor wall, whereas temperature gradients from 20 to 70 K were observed from the wall to the gas phase depending on the channel half-height, which was varied from 0.5 to 2 mm.

The methane conversion increased with increasing channel height in the first half of the reactor owing to the higher temperature achieved for both the reforming and the combustion side. Decreasing the channel half-height from 2 to 0.5 mm increased the methane conversion from 92 to 95%. This effect was much more pronounced when keeping the inlet velocities rather than the inlet flow rates constant. In this case, the outlet conversion dropped from 100% at a 0.5 mm channel half-height to 70% at 2 mm half-height (see Figure 2.68).

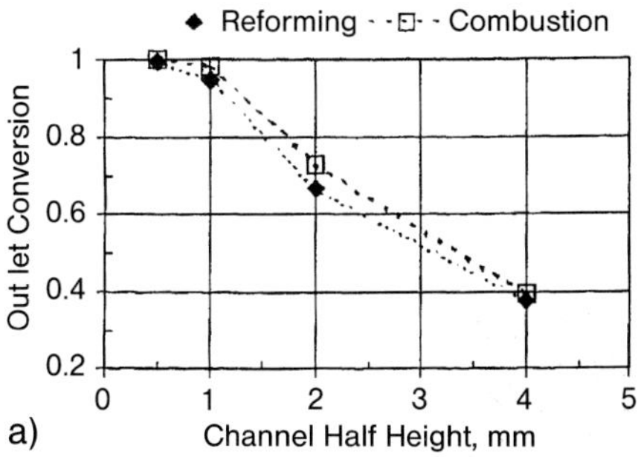

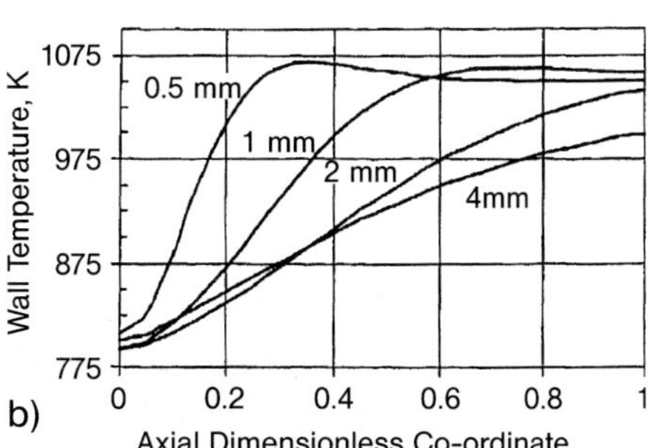

Figure 2.68 Results from numerical calculations for combustion-assisted methane steam reforming. (a) Outlet conversion dependence on channel half-height; (b) wall temperature as a function of dimensionless reactor length. Calculation results determined at constant inlet velocity [108] (by courtesy of Elsevier Ltd.).

Actually, full conversion was achieved after 45% of the reactor total length. The temperature rise was fast for the small channels, approaching a maximum after 30% of the reactor length, thus leading to higher conversion. In the large channels both mass transfer became limiting and the amount of catalyst became insufficient for the residence time applied.

In another case study, the thickness of the catalyst layer was increased at constant WHSV by increasing the inlet flow rate. Increasing the catalyst layer thickness from 10 to 60 μm led to a decrease in conversion from 100 to < 70% for both reactions. A faster increase in the axial temperature was achieved for the thin coatings.

2.7.2
Integrated Systems Fuelled by Methanol

2.7.2.1 Integrated Systems Fuelled by Methanol 1 [ISMol 1]: Integrated Methanol Fuel Processor (Casio)

Little information is publicly available on the status of fuel processor development in industry. Figure 2.69 shows the fuel cell system developed by Casio [2]. It contains a silicon wafer methanol reformer, which was developed by Casio's research division [2]. The catalyst was developed under the guidance of the University of Japan. It achieved 98% methanol conversion and allowed the operation of a hand-held computer for 20 h. Commercialization is aimed for in 2004–2005.

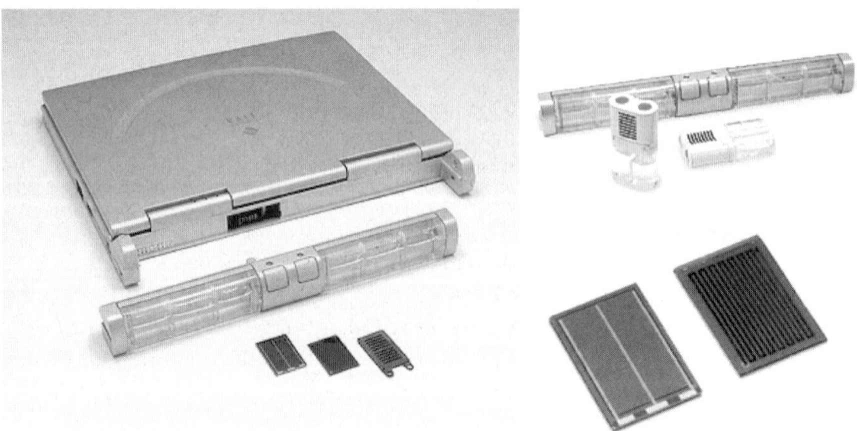

Figure 2.69 Casio prototype laptop (by courtesy of Casio).

2.7.2.2 Integrated Systems Fuelled by Methanol 2 [ISMol 2]: Integrated Methanol Fuel Processor (Motorola)

Motorola is cooperating with Engelhard and Michigan University to develop a micro structured steam reformer in a project funded by the US Commerce's Department of Technology Administration [109]. Figure 2.70 shows four slides presented by Motorola [110]. The integrated design of the fuel processor/fuel cell is shown on

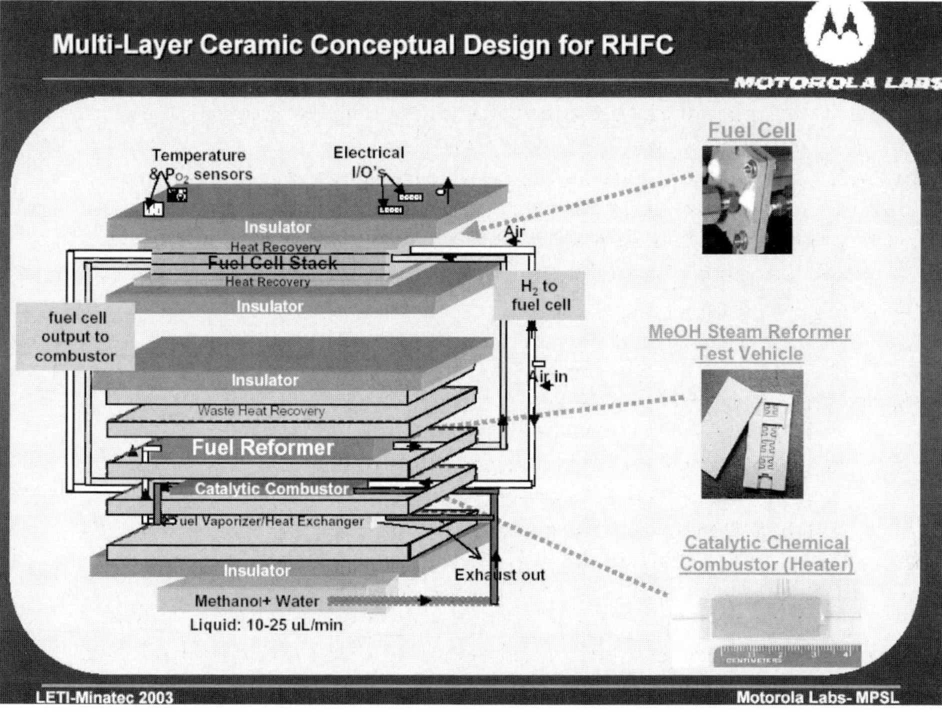

Figure 2.70 Motorola prototype methanol steam reformer/fuel cell system (by courtesy of Motorola).

the left. It consists of an evaporator, a combustor, a reformer, heat exchangers, insulation layers and a fuel cell utilizing ceramic technology. The maximum power output of this device, which is patented [111], is 1 W.

2.7.2.3 Integrated Systems Fuelled by Methanol 3 [ISMol 3]: Integrated Autothermal Methanol Fuel Processor (Ballard)

Schuessler et al. [85] of XCELLSiS (later BALLARD) presented an integrated methanol fuel processor system based on autothermal reforming, which coupled fuel/water evaporation with exothermic preferential oxidation (PrOx) of carbon monoxide. The reactor technology was based, in contrast to most other approaches, on a sintering technique.

The reforming reactor was built of copper powder, which could be sintered at temperatures between 500 and 700 °C, being low enough to avoid damage to the catalyst. In the same fabrication stage, the Cu/ZnO catalyst with a particle size between 300 and 500 μm was incorporated into the device. Copper and aluminum powder were used as inert materials for parts such as channels and diffusion layers.

Platinum catalyst was introduced into an alumina-based suspension and then spray-coated. The evaporation was performed using porous cylinders of 7 mm diameter and 1 mm height introduced into each individual plate before performing

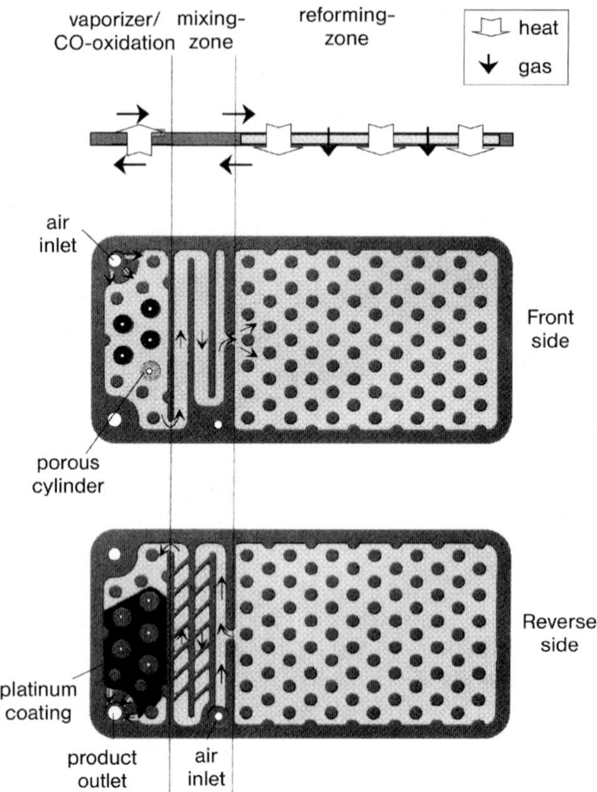

Figure 2.71 Single plate of the integrated fuel processor (IFP) developed by Ballard [85] (by courtesy of Elsevier Ltd.).

the final sintering step which sealed the reactor. The evaporation zone was closely coupled with the PrOx reaction taking place at the reverse side of the same plate (see Figure 2.71).

Autothermal reforming was performed, again unlike other approaches, not along the length of the plate, but rather perpendicular through the porous catalyst/sintered copper particle mixture [112] (pore diameter 2 μm, 60% porosity). As the whole reaction zone was only a few millimeters long, hot-spot formation could be avoided by thermal conduction through the wall material. The porous structure allowed for a high GHSV of 15 000 h^{-1} at a low flow rate of 0.1 m s^{-1}, which corresponded to a very low Reynolds number of 0.01 [112]. Temperature gradients of < 10 K were determined. Load changes did not lead to changing temperature profiles in the reactor. The methane concentration was kept below 100 ppm.

After the reformer outlet, air was added again before the reformate entered the PrOx zone. The pressure drop of the whole coupled evaporator/reformer/PrOx reactor was limited to 200 mbar at full load. The whole IFP had a mass of 1.8 kg for a volume of 0.5 dm^3. At a reaction temperature of 280 °C, a methanol feed of 15 mol h^{-1} and a water/methanol ratio of 1, almost complete (> 99%) conversion of methanol could be achieved. The carbon monoxide concentration was reduced from 1.3 vol.% at the reformer outlet to about 0.1 vol.% after the PrOx reactor. The selectivity of the PrOx stage was around 50%. The system was started up by a surplus of air, leading to total oxidation of the methanol especially in the PrOx region, thus providing energy. Owing to this indirect heating, a 10 min start-up time was required for the system.

2.7.2.4 Integrated Systems Fuelled by Methanol 4 [ISMol 4]: Integrated Methanol Steam Reforming Fuel Processor for 20 kW Power Output

The [PrOx 3] reactor (see Section 2.6.2) and an improved second version of it carrying also a different ratio of platinum and ruthenium on the catalyst were tested separately and switched in series by Dudfield et al. [88] prior to combining it with a 20 kW methanol steam reformer. The reactors had dimensions of 46 mm height, 56 mm width and 170 mm length, which corresponds to a volume of 0.44 dm^3 and a weight of 590 g. They contained 2 g of catalyst each.

A reformate flow rate of 25–175 $Ndm^3\ min^{-1}$, simulating methanol steam reformer product gases, and 2.5–17.5 $Ndm^3\ min^{-1}$ air were fed to the reactors. The simulated reformate was composed of 68.9% H_2, 0.6% CO, 22.4% CO_2, 6.9% H_2O and 0.4% CH_3OH, the last to simulate incomplete conversion. The carbon monoxide output of the single reactors and of both switched in series is shown in Figure 2.72. The CO output of the two reactors switched in series was < 10 ppm and the optimum air volume split between the first and second reactors was determined as 70/30.

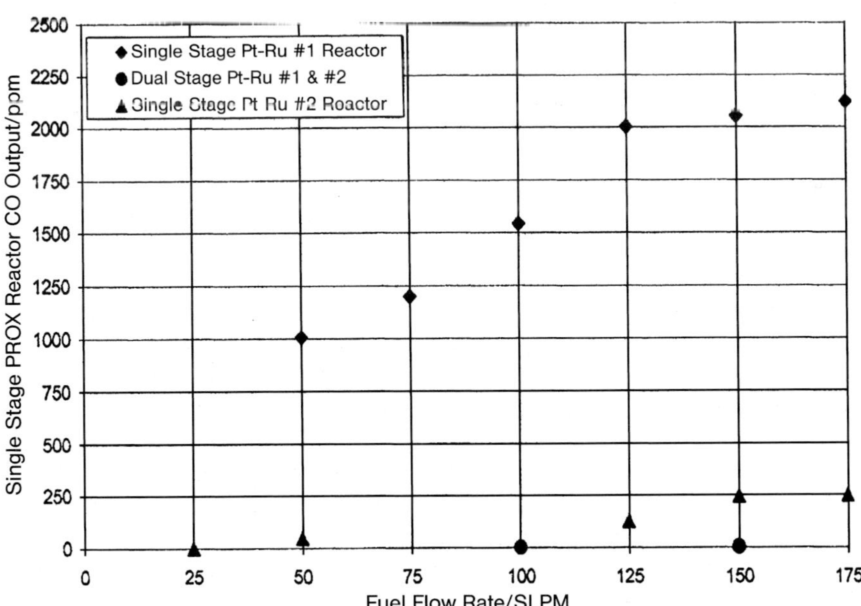

Figure 2.72 CO output vs. reformate feed flow rate of two generations of [PrOx 3] reactors separately and switched in series [88] (by courtesy of Elsevier Ltd.).

To meet the requirements of the 20 kW fuel processor, the [PrOx 3] reactor was scaled up to a still dual-stage design of 4 dm^3 volume. Each reactor was now 108 mm high, 108 mm wide and 171 mm long and had a weight of 2.5 kg. Each carried 8.5 g of catalyst.

For reformate flow rates up to 400 $Ndm^3\,min^{-1}$, the CO output was determined as < 12 ppm for simulated methanol. The reactors were operated at full load (20 kW equivalent power output) for ~100 h without deactivation. In connection with the 20 kW methanol reformer, the CO output of the two final reactors was < 10 ppm for more than 2 h at a feed concentration of 1.6% carbon monoxide. Because the reformer was realized as a combination of steam reformer and catalytic burner in the plate and fin design as well, this may be regarded as an impressive demonstration of the capabilities of the integrated heat exchanger design for fuel processors in the kilowatt range.

2.7.2.5 Integrated Systems Fuelled by Methanol 5 [ISMol 5]: Integrated Methanol Fuel Processor for 100 W Power Output

Schouten et al. [113] presented the design of a methanol fuel processor for an electrical power output of 100 W, which was done in the scope of a project funded by the European Union (MiRTH-e). In contrast to [ISMol 3], the processor was composed of three separate devices.

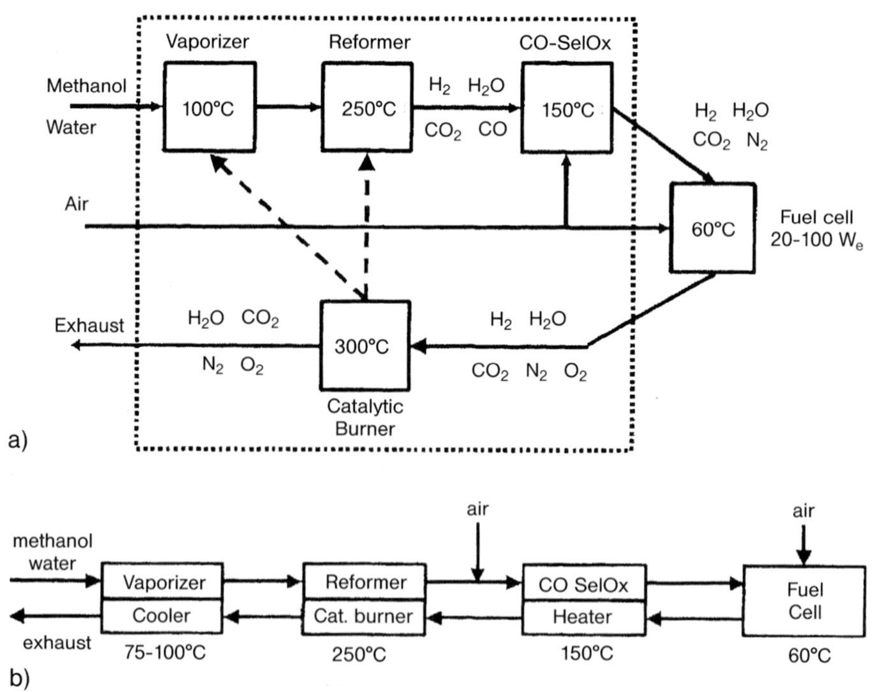

Figure 2.73 Overall flow scheme of the 100 W fuel processor [ISMol 4] [113] (by courtesy of SCS).

Figure 2.73 shows the flow scheme developed by Delsman et al. [114]. The methanol/water mixture was foreseen to be evaporated by the hot off-gases of the catalytic burner and fed to the integrated steam reformer/burner reactor. A Cu/ZnO catalyst supported by alumina was applied for the steam reforming reaction. The reformate was then fed to the PrOx reactor, which was cooled by the fuel cell anode off-gas. Two heat exchangers integrated into the PrOx unit were switched before and after the PrOx reactor to cool the reformate to the operating temperature of the PrOx (about 150 °C) and to cool further the cleaned reformate to the fuel cell operating temperature (about 65 °C). The residual hydrogen contained in the same off-gas was combusted in the burner over a platinum/alumina catalyst.

Start-up was effected by feeding methanol directly to the burner. The integrated reformer/burner reactor and the integrated PrOx reactor/cooler fabricated at IMM are shown in Figure 2.74. Characterization of the single devices and their assembly is still pending.

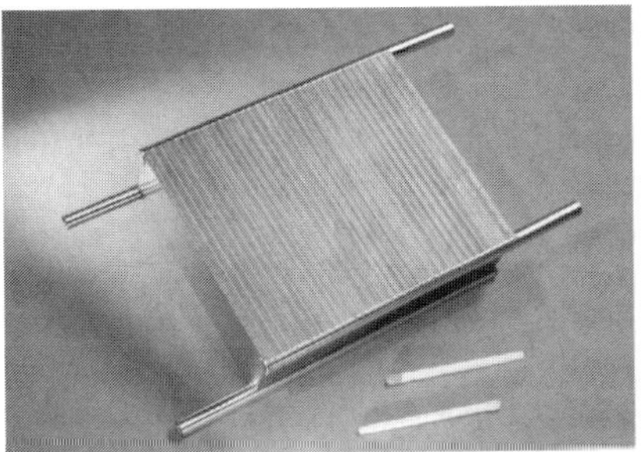

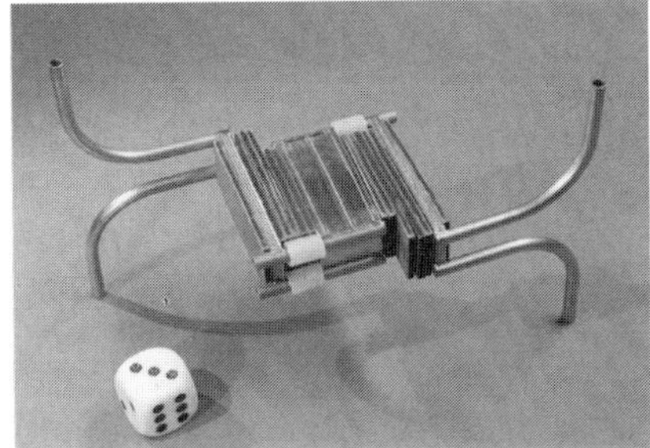

Figure 2.74 (Left) reformer/burner and (right) PrOx reactor/heat exchanger of the 100 W fuel processor [ISMol 4] (source: IMM).

2.7.2.6 Integrated Systems Fuelled by Methanol 6 [ISMol 6]: Integrated Methanol Fuel Processor for 15 W Power Output

Palo et al. [18] presented the concept of an integrated fuel processor for portable military applications. System specifications were a weight of < 1 kg and a volume of < 100 cm^3, which transfer to a required power density of > 0.15 kW dm^{-3}.

The concept of the system was to vaporize/preheat a methanol/air mixture, combust it in a separate combustor and feed the methanol steam reforming reaction with the energy of the hot combustion gases (see Figure 2.75). Light-off of the combustion gases occurred at 70 °C [115]. The combustion gases were further used subsequently to supply the fuel pre-heater/evaporator of the combustor and finally the fuel pre-heater/evaporator of the reformer. Unconverted methanol and water were removed in a separator from the reformate prior to GC analysis.

Full conversion was achieved in the steam reformer separately tested at a 300 ms contact time and 300 °C reaction temperature. At a very high reaction temperature of 375 °C, even a 50 ms contact time was sufficient to achieve full conversion. Methane formation was below the detection limit of 100 ppm [115]. At higher reaction temperatures its formation was to be expected owing to the methanation reaction of the carbon monoxide (see Section 2.4).

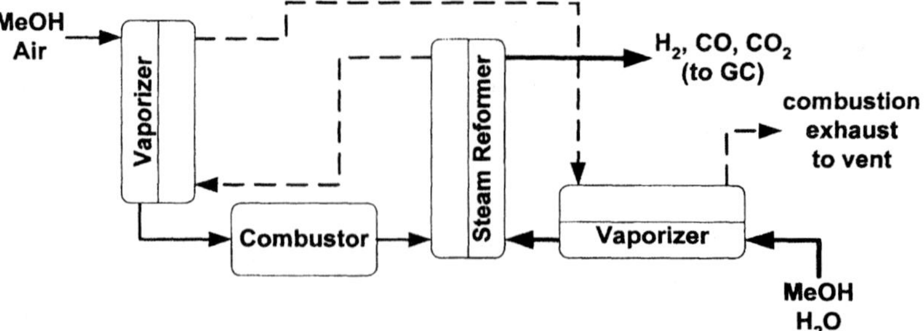

Figure 2.75 Methanol fuel processor concept for 15 W electrical power output [18] (by courtesy of Springer-Verlag).

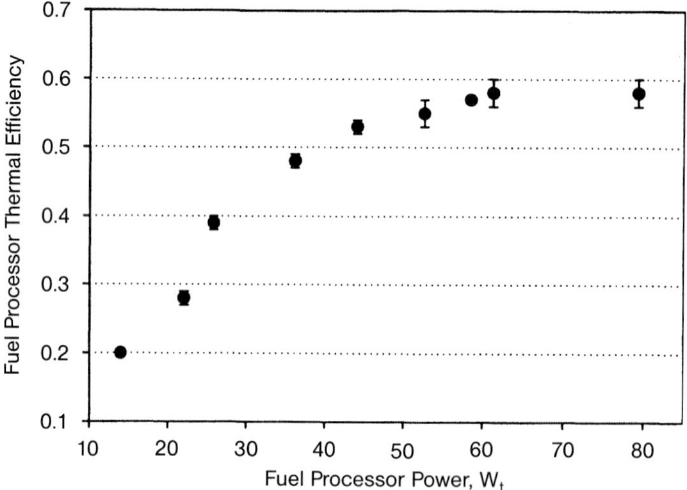

Figure 2.76 Thermal efficiency vs. system power of the methanol bread-board system [18] (by courtesy of Springer-Verlag).

Results generated at a partially integrated demonstrator, which was still heated electrically for start-up, were presented (Figure 2.76). At a 350 °C reaction temperature, a 140 ms contact time and a S/C ratio of 1.8, full conversion of the methanol was achieved. The carbon monoxide concentration of 0.8% detected at this high reaction temperature was surprisingly low [18]. Operating at 13 W_e, the fuel processor efficiency was calculated as 45% and the estimated overall efficiency, including the fuel cell, as 22%. An energy density of about 720 W h kg^{-1} is achievable, which exeeds the energy density of lithium ion batteries several-fold [18].

2.7.2.7 Integrated Systems Fuelled by Methanol 7 [ISMol 7]: Integrated Methanol Fuel Processor for the Sub-watt Power Range

Jones et al. [116] presented an integrated and miniaturized device for methanol steam reforming consisting of two evaporators/pre-heaters, a reformer and a combustor with a total volume of < 0.2 cm^3 for a power range between 50 and

500 mW, which corresponds to flow rates of 0.03–1 $Ncm^3 h^{-1}$ of methanol/water mixture.

The energy for the steam reforming reaction was transferred from the combustor device having 3 W power capacity which was fed by an H_2/O_2 mixture for start-up and later by up to 0.2 $Ncm^3 h^{-1}$ methanol and up to 14 $Ncm^3 h^{-1}$ air. The combustor was said to have a volume of < 1 mm^3, the reformer volume being higher (5 mm^3).

Holladay et al. [19] calculated, that even a fuel processor efficiency as low as 5% outperforms a lithium ion battery. In this later paper, more detailed data on the device were given. The flow rate of methanol for the reforming reaction was said to be 0.02–0.1 $Ncm^3 h^{-1}$ and > 99% conversion were achieved at a S/C ratio of 1.8 and a 325 °C reaction temperature. The methanol flow rate fed to the burner ranged between 0.1 and 0.4 $Ncm^3 h^{-1}$. Combustion air flow was 8–20 $Ncm^3 min^{-1}$. The burner temperature exceeded 400 °C. The thermal power of the device was 200 mW at 9% efficiency. Assuming a 60% fuel cell efficiency and 80% hydrogen conversion, the net efficiency of the system amounted to 4.5% and the power output was 100 mW.

In a second prototype, the reaction temperature was reduced to 250 °C, which reduced the carbon monoxide concentration from 1.2 to < 1%. Later, the first fuel processor prototype was linked to a meso-scale high-temperature fuel cell developed at Case Western University by Holladay et al. [117], which was tolerant to carbon monoxide concentrations up to 10%. Hence no CO clean-up was necessary to run the fuel processor. A 23 mW power output was demonstrated according to Holladay et al. [118]. This value was lower than expected, which was attributed to several factors. First, the hydrogen supply was lower in the reformate. Second, the presence of carbon monoxide (2%) lowered the cell voltage. Third, the presence of carbon dioxide (25%) generated a magnified dilution effect at the gas diffusion layer material of the fuel cell, which was considerably less porous than conventional materials.

2.7.2.8 Integrated Systems Fuelled by Methanol 8 [ISMol 8]: Integrated Reformer/Combustor Reactor

Reuse et al. [68] combined endothermic methanol steam reforming with exothermic methanol combustion. The reactor consisted of a stack of 40 foils, 20 dedicated to each reaction (see Figure 2.77). The total length of the foils was 78 mm and their thickness was 200 μm. The foils carried 34 S-shaped channels each with a length of 30 mm, a depth of 100 μm and a width of 310 μm. A special plate in the center of the stack allowed for temperature measurements. The plates were made of FeCrAlloy and an α-alumina film 5 μm thick was generated on their surface by temperature treatment at 1000 °C for 5 h to improve the adherence of the catalyst coatings (see Section 2.10.7).

A total of 206 mg [119] of commercial Cu/Zn catalyst from SüdChemie (G-66MR) ground to the nanometer range was coated into the channel system at 5 μm thickness to promote the steam reforming reaction. A cobalt oxide catalyst was prepared by impregnating the corundum layer (see above) with cobalt nitrate and calcining at 350 °C for 2 h; 434 mg [119] of the CoO catalyst were applied for the combustion reaction (see Section 2.5).

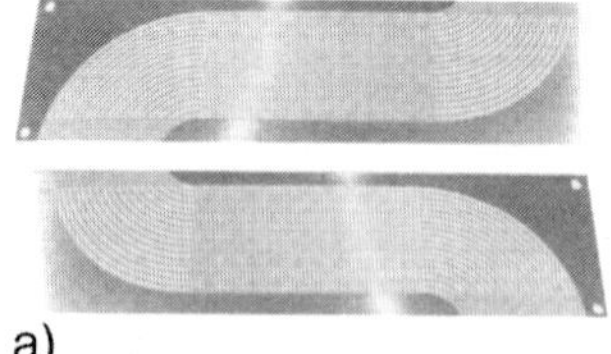

a)

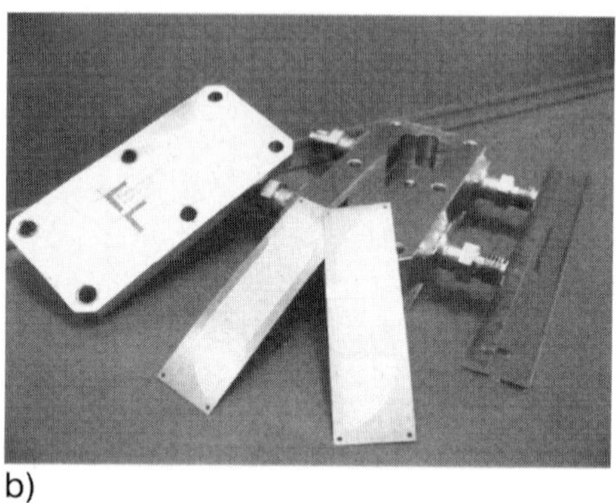

b)

Figure 2.77 Integrated reformer/combustor for methanol steam reforming [68] (by courtesy of ACS).

The reactor was operated in the co-current mode. The feed for methanol steam reforming was composed of 37.7% methanol, 45.3% water and balance argon. The feed for methanol combustion was composed of 10% methanol, 18.9% oxygen and 71.1% nitrogen. For the steam reforming reaction side, a H_2O/CH_3OH molar ratio of 1.2 was fed to the reactor.

Under stationary conditions, at reaction temperatures between 250 and 260 °C, > 95% conversion and a carbon dioxide selectivity of > 95% were achieved [119].

2.7.2.9 Integrated Systems Fuelled by Methanol 9 [ISMol 9]: Chip-like Methanol Reformer/Combustor

A 200 mW methanol fuel processor was presented by Hu et al. [32]. A 9% efficiency was determined for the device running at 1 vol.% carbon monoxide in the reformate stream.

2.7.2.10 Integrated Systems Fuelled by Methanol 10 [ISMol 10]: Micro Integrated Heat Exchanger/Reactor for Methanol Steam Reforming

The aim of the work was to develop a methanol fuel processor for a 50 kW automobile engine. An integrated heat exchanger/reactor was fabricated and results were presented by Hermann et al. [120] of GM/OPEL. The specifications for the system were ambitious, amongst others:

- volumetric power density > 5 kW dm^3
- gravimetric power density < 2.5 kW kg
- transient response to load changes from 10 to 90% in milliseconds.

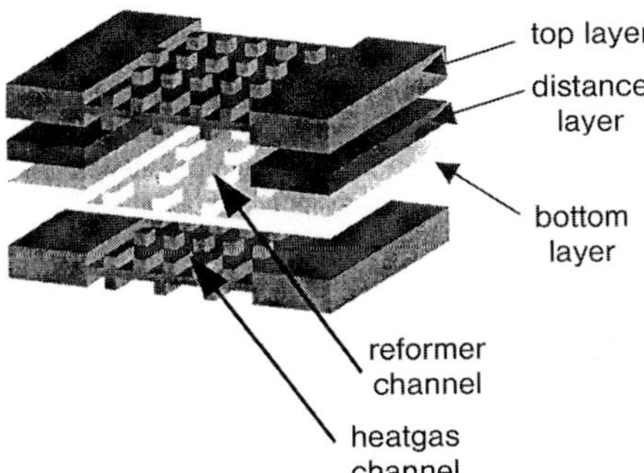

Figure 2.78 Layer assembly of the 5 kW micro structured methanol reformer [120] (by courtesy of VDE-Verlag).

The device described in detail was a 5 kW combined methanol steam reformer/catalytic combustor. The reactor was composed of three types of plates forming a stack. Instead of micro channels, a fin-like structure was chosen (see Figure 2.78). The fins served as mechanical support and improved heat transfer. A total of 225 plates were incorporated into the reactor. The coating density of the reaction walls was 250 $m^2\ g_{cat}^{-1}$, which corresponded to a thickness of 200 µm. The theoretical power output of the device was 4.97 kW with a corresponding power density of 5 kW dm^{-3}. The reactor was designed for a maximum operating pressure of 4 bar and a maximum reaction temperature of 350 °C. At a flow rate of 30 $m^3\ h^{-1}$ nitrogen, the pressure drop of the device was 600 mbar.

Results presented were determined at a partial load of the device (1–2 kW for the LHV of the hydrogen produced). At a burner off-gas (heating gas) inlet temperature of 350 °C, a S/C ratio of 1.5 and a pressure of 3 bar, full conversion of the methanol was achieved and 0.9 $m^3\ h^{-1}$ hydrogen were produced. The hydrogen production rate was regarded as competitive with literature data.

Later on, a meso-scale combined reformer/catalytic combustor with 10 kW power output was realized by GM/OPEL but was not presented in detail. With this device, the carbon monoxide of the reformate content increased as expected with increase in reformer outlet temperature from 0.5% at 250 °C to 2% at 300 °C. Increasing the residence time led to a higher CO output owing to the reverse water-gas shift reaction. Increasing the S/C ratio from 1.2 to 1.8 at a reaction temperature of 300 °C increased the hydrogen concentration in the reformate slightly from 72 to 73% and decreased the carbon monoxide content from 1.5 to 1.0% owing to the beneficial effect of steam addition on the water-gas shift reaction. Comparable results were generated concerning conversion and selectivity at lower power density but reduced pressure drop.

2.7.2.11 Integrated Systems Fuelled by Methanol 11 [ISMol 11]: Micro Integrated Heat Exchanger/Fixed-bed Reactor for Methanol Steam Reforming

Catalysts from Süd Chemie were applied for methanol steam reforming by Stimming et al. [121]. G66-MR, a general-purpose catalyst containing 11% Al, 37% Cu and 52% Zn with 121 $m^2\ g^{-1}$ surface area, and C18-HA, a catalyst optimized for methanol synthesis containing 2–3% Al, 50–60% Cu and 25–35% Zn with

104 m^2 g^{-1} surface area, were introduced as wash coats into micro channels. The C18-HA catalyst, which was designed for methanol synthesis, showed inferior performance to G66-MR, which converted the methanol completely at a reaction temperature of 275 °C and a WHSV of 0.13 Ndm^3 $(g_{cat}\,min)^{-1}$, which was only slightly inferior to fixed-bed testing. As the catalyst activity was too low compared with the design criterion, a micro fixed-bed reactor was built with integrated heat-exchanging capabilities.

The reactor contained 60 micro fixed-bed passages taking up 15.9 g of catalyst and 62 heating passages and was designed to supply a fuel cell of 500 W electrical power. However, the heating could not be performed directly by the combustion of methanol, which had to be carried out at higher temperatures than the steam reforming process [122]. Therefore, an evaporator was switched between the two devices and the power supply of the steam reformer was performed by a heat transfer fluid (oil).

More than 90% conversion was achieved at the design point at a space velocity up to 12.5 h^{-1}, an oil temperature of 270 °C and a reaction temperature of 250 °C. During the first 4 h of operation, 15% of the initial activity of the catalyst was lost, but then the activity remained stable for another 4 h. The catalyst could be regenerated by oxidation and subsequent reduction [122]. The micro structured reactor had a start-up time demand of 18 s after being heated to the operating temperature, which was considered an improvement over conventional fixed-bed technology. Load changes were performed in the 1 : 5 range without significant changes in product composition.

2.7.2.12 Integrated Systems Fuelled by Methanol 12 [ISMol 12]: Integrated Methanol Evaporator and Hydrogen Combustor

Tonkovich et al. [123] claimed a 90% size reduction due to the introduction of micro channel systems into their device, which made use of the hydrogen off-gas of the fuel cell anode burnt in monoliths at palladium catalyst to deliver the energy for the fuel evaporation. A metallic nickel foam 0.63 cm high was etched and impregnated with palladium to act as a reactor for the anode effluent. It was attached to a micro structured device consisting of liquid feed supply channels and outlet channels for the vapor, the latter flowing counter-flow to the anode effluent.

A bench-scale evaporator was built first, consisting of the nickel foam monolith and heat exchanger plates 5.7 cm wide and 7 cm long. The stainless-steel channels fabricated by EDM were 254 µm deep and the vapor channel depth was varied from aspect ratios of 4 to 18, the latter being the optimum value determined by experiments.

At Pd loadings of > 5 wt.%, 80–90% hydrogen conversion was found for a synthetic anode effluent containing 6.7% hydrogen and 3.35% oxygen. The heat exchanging efficiency of the bench-scale evaporator increased linearly with the aspect ratio of the vapor channels and reached 92% at an aspect ratio of 18.

The full-scale reactor/evaporator had a total size of 7.6 cm × 10.2 cm × 5.1 cm and was composed of four monoliths of 5 cm^2 cross-sectional area and four heat exchangers with 7.2 cm^3 cross-sectional area. This device was claimed to evaporate

208 Ncm^3 min^{-1} methanol by converting 25 Ndm^3 min^{-1} hydrogen at a level of 100%, an efficiency of 93.2% and a heat flux of 145 W cm^{-2}. The methanol vapor exited the device at 203 °C and the combustion gases at 300 °C. Cleaning of the evaporation channels was necessary on a regular basis by burning off the coke deposits formed during operation.

2.7.2.13 Integrated Systems Fuelled by Methanol 13 [ISMol 13]: Integrated Methanol Evaporator and Methanol Reformer

A combined evaporator and methanol reformer was developed by Park et al. [124] to power a 5 W fuel cell. However, the device was still electrically heated by heating cartridges. Both the evaporator and the reformer channels, which were identical in size, were prepared on metal sheets 200 μm thick by wet chemical etching. The channel dimensions were length 33 mm, width 500 μm and depth 200 μm. Therefore, the channels were completely etched through the sheets and the channel depth could be varied by introducing several of these sheets into the reactor. The flow distribution between the 20 channels of the device was performed by triangular inlet and outlet fields. Both devices had outer dimensions of 70 mm × 40 mm × 30 mm.

The evaporator was fed a mixture of methanol and water and operated at a temperature of 120 °C. Prior to coating the channels with a commercial CuO/ZnO/Al_2O_3 catalyst (Synetix 33–5 from ICI), an alumina sol was coated as interface to the channel surface. The catalyst was reduced in 10% hydrogen in nitrogen at 280 °C prior to exposing it to the reaction mixture. Methanol conversion increased at S/C 1.1 from 55 to 90% on increasing the reaction temperature from 200 to 260 °C at 6 Ncm^3 h^{-1} liquid feed flow rate.

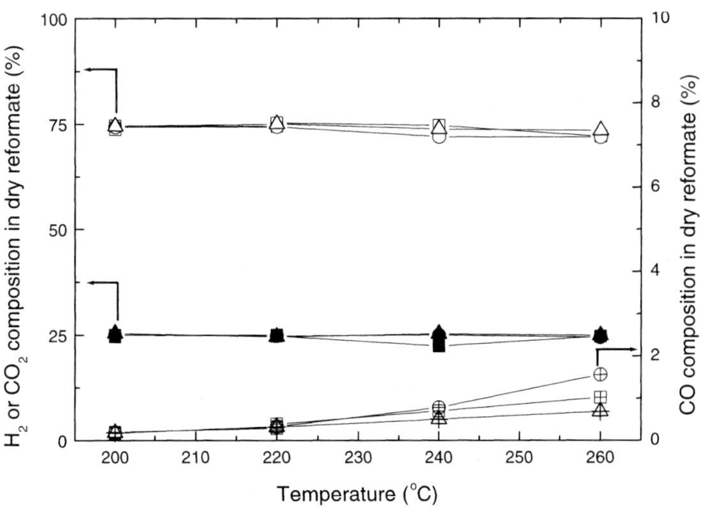

Figure 2.79 Composition of dry reformate as a function of reaction temperature. Liquid feed flow rate 6 Ncm^3 h^{-1}. Circles, S/C 1.1; squares, S/C 1.5; triangles, S/C 2.0 [124] (by courtesy of ACS).

At the latter temperature, a carbon monoxide concentration of < 2% was found in the product. However, the product composition did not change significantly between 200 and 260 °C reaction temperature (75% hydrogen and 25% carbon dioxide) except for the carbon monoxide, which increased from 0.17 to 1.6% at S/C 1.1 (see Figure 2.79). Increasing the S/C ratio was beneficial in terms of reducing the carbon monoxide in the effluent from 1.6 to 0.8%.

2.7.3 Integrated Systems Fuelled by Methane

2.7.3.1 Integrated Systems Fuelled by Methane 1 [ISM 1]: Integrated Reformer/Combustor Reactor

Cremers et al. [125] presented a reactor combining endothermic methane steam reforming with the exothermic combustion of hydrogen stemming from the fuel cell anode off-gas (see Figure 2.80). NiCroFer 3220H® was applied as reactor material. The reactor was designed to power a fuel cell with 500 W electrical power output. The steam reforming side of the reactor was operated at a S/C ratio of 3 and temperatures exceeding 750 °C. A nickel spinel catalyst developed earlier [121] was applied for promoting the steam reforming reaction.

Figure 2.80 Combined methane reformer/combustor designed for 500 W electrical power output [126].

2.7.3.2 Integrated Systems Fuelled by Methane 2 [ISM 2]: Integrated Reformer/Combustor Reactor

Mazanec et al. [127] presented another device with the same dedication as [ISM 1]. At an 865 °C reaction temperature, 12 bar pressure and an S/C ratio of 2, 90% conversion and 70% carbon monoxide selectivity were determined. The device had passed a total test time of more than 2 000 h.

2.7.3.3 Design Study for the Multi-stage Adiabatic Mode

For small- to medium-scale stationary applications, Johnston and Haynes [128] proposed a series of adiabatic reactors with heat exchangers in between rather than directly removing the heat in a single reactor. Thus a saw-tooth-like temperature

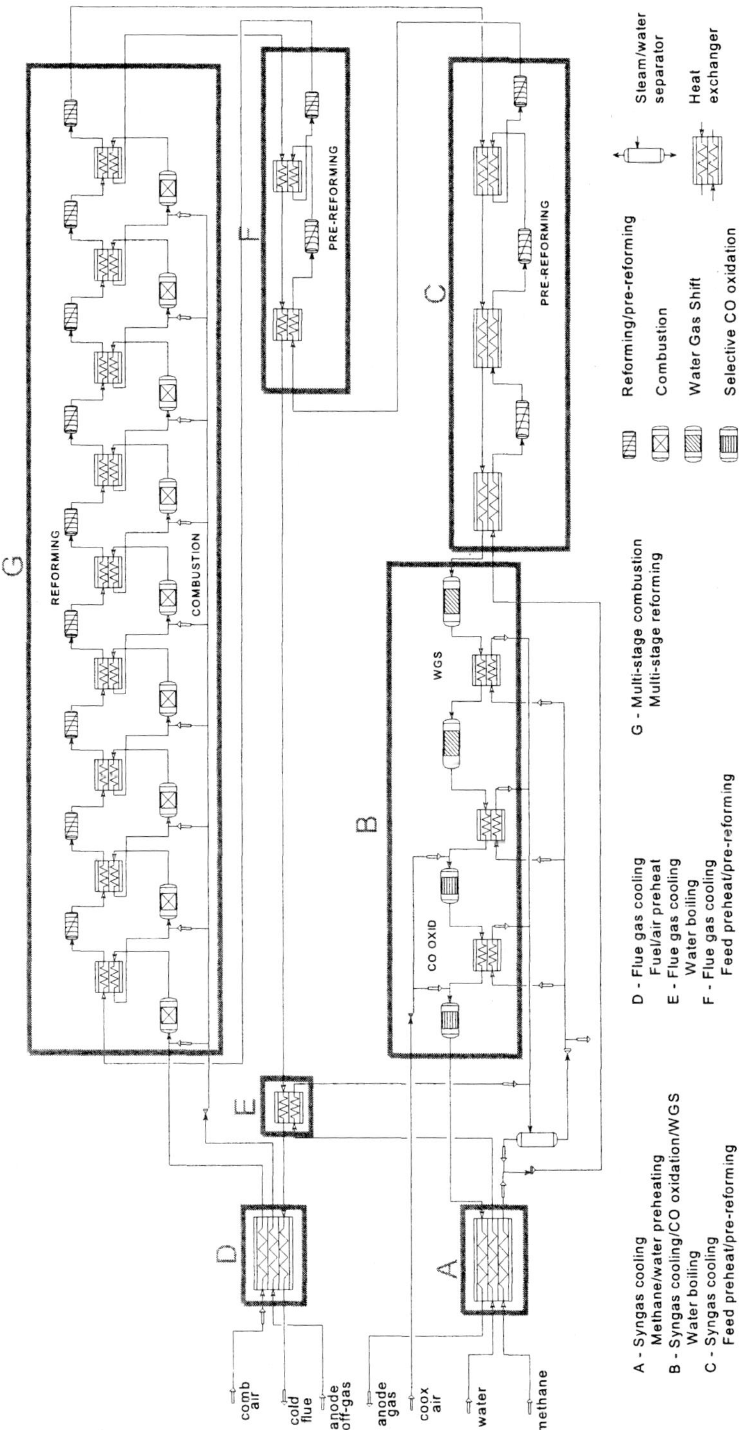

Figure 2.81 Network of 27 catalyst beds (five for pre-reforming, nine for reforming, nine for catalytic combustion to supply the heat, two for water-gas shift and two for preferential oxidation) and of 20 heat exchangers [128] (by courtesy of A. M. Johnston).

profile would result, as known from conventional large-scale industrial processes. The proposed flow paths are more on the meso-scale and conventional, although smaller catalyst particles instead of coatings are proposed. The study shows a network of 27 catalyst beds (five for pre-reforming, nine for reforming, 9 for catalytic combustion to supply the heat, two for water-gas shift and two for preferential oxidation) and of 20 heat exchangers (Figure 2.81). Power supply was designed to stem from the anode off-gas. However, many of the devices are expected to be integrated by printed circuit heat exchanger (PCHE) technology. A reforming temperature below 800 °C and an S/C ratio of 2.6 were regarded as sufficient to run the system. A theoretical efficiency of 85.2% was calculated for the system.

2.7.4 Integrated Systems Running on Various Fuels

2.7.4.1 Integrated Systems Running on Various Fuels 1 [ISV 1]: Integrated Evaporator/Burner Device for Automotive Applications

Drost et al. [129] developed an evaporator combined with a micro scale combustion chamber for homogeneous combustion of hydrocarbons (Figure 2.82). The main focus of the work was to maintain a stable combustion of the fuel avoiding NO_x formation. Evaporation tests were carried out under isothermal conditions. Fifty-four parallel channels, 270 μm wide and 1 000 μm deep with a length of 20.52 mm, were cut into a copper substrate with a diamond saw.

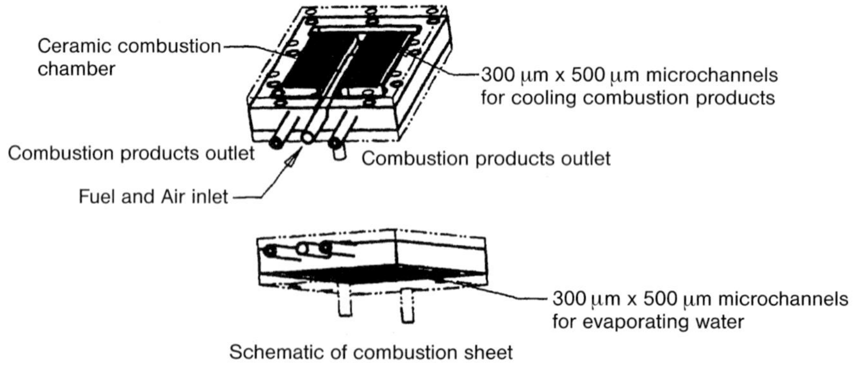

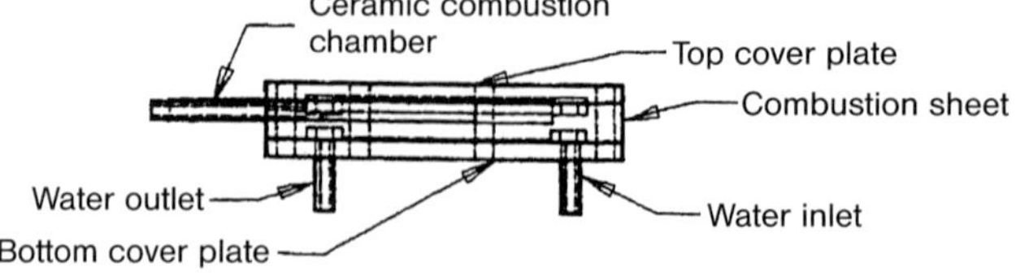

Figure 2.82 Schematic of the integrated microchannel combustor/evaporator [129] (by courtesy of Taylor & Francis Ltd.).

Basically an unstable two-phase flow was found at the exit of the evaporator, which was indicated by high Nusselt numbers in the range 20–30. Then a combined combustor/evaporator was designed which consisted of a ceramic tube as combustion chamber placed in a stainless-steel housing (dimensions 41 mm × 60 mm × 20 mm). Ignition wires were used to initiate the reaction. The combustion gases then entered channels which were combined with evaporation channels for the water cooling the combustion gases.

Both channel systems fabricated by EDM were 35 mm long, 300 μm wide and 500 μm deep. An alternative design was developed for methane combustion, where a sintered metal plate served as flame arrestor. The device had a size of 50 mm × 50 mm × 10 mm. For a cooling water flow rate of 1.32 g s^{-1}, heat fluxes up to 28 W cm^{-2} were achieved at a combustion efficiency of 82%. The best combustion efficiency of 91% was achieved at a heat flux of 6 W cm^{-2}. NO_x emissions of the system were determined to range between 5 and 15 ppm.

2.7.4.2 Integrated Systems Running on Various Fuels 2 [ISV 2]: Combined System of Integrated Reformer/Heat Exchanger and Evaporator/Heat Exchanger Devices for Automotive Applications

Whyatt et al. [130] presented results generated at a combined system of independent evaporators, heat exchangers and reformers for isooctane steam reforming. The machining of micro structured components was done by photochemical etching and diffusion bonding. Four integrated reformers/cross-flow heat exchangers switched in series were fed by four independent water vaporizers which acquired their energy by combustion of anode off-gas performed in an independent burner. These devices also preheated the combustion air. The combustion gases supplied first the cross-flow heat exchanger/reformers and then the water vaporizer units with energy. The fuel was evaporated in a second group of four units which also superheated the steam/isooctane vapor in a second stage and preheated the water in a third stage. This unit was heated by the hot reformate.

The reformers achieved up to 98.6% conversion at a reaction temperature of 750 °C, an S/C ratio of three and a product composition of 70.6% H_2, 14.6% CO, 13.7% CO_2 and 0.9% CH_4. Doubling the reformate output was claimed to be feasible in 20 s. The reformate generated was sufficient to feed a 13.7 kW PEM fuel cell. An overall efficiency of 44% of the system was calculated assuming 90% conversion and 100% selectivity for the water-gas shift reaction and 54% overall efficiency for the fuel cell.

2.7.4.3 Integrated Systems Running on Various Fuels 3 [ISV 3]: Combined System of Integrated Reformer/Heat Exchanger and Evaporator/Heat Exchanger Devices for Automotive Applications

Whyatt et al. [131] designed two reformer reactors of 107 and 68 cm^3 total volume for reforming of methane, propane, butane, ethanol, methanol, isooctane and a mixture of hydrocarbons simulating sulfur-free gasoline. The two reactors were tested in a test bench in combination with micro structured heat exchangers for feed preheating using product (hot reformate) and evaporators for water and liquid feed preheating (Figure 2.83).

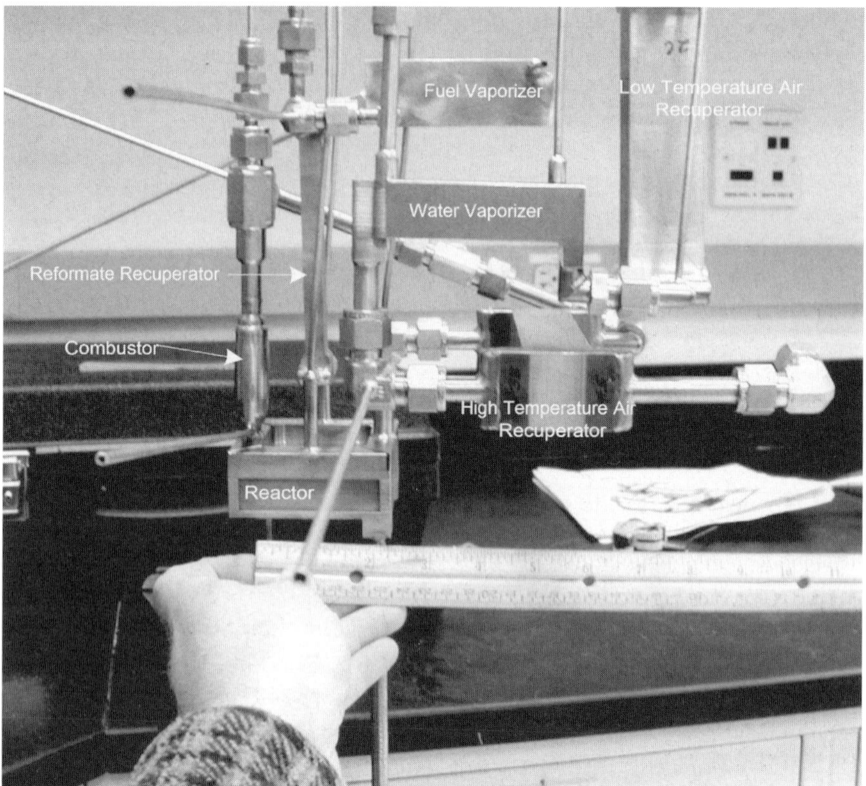

Figure 2.83 Photograph of the assembled reforming system [131] (by courtesy of G. A. Whyatt).

Table 2.9 Conversion, power density, reaction temperature and reformate composition for various fuels [131].

Parameter	*Fuel*						
	Methane	*Propane*	*Butane*	*Methanol*	*Ethanol*	*Iso-octane*	*Sulfur-free gasoline*
C_1 Feed-rate (10^{-4} mol s^{-1})	8.13	8.34	11.62	8.07	10.4	7.51	10.8
Conversion to C_1 (%)	95	99.8	99	100	98.9	100	99.6
Power density (kW dm^{-3})	2.26	4.06	5.98	1.69	2.25	1.70	2.27
Reaction temperature (°C)	725	775	751	725	724	725	725
Dry reformate composition							
H_2 (%)	75.7	72.3	71.6	70.4	70.0	70.9	70.1
CO (%)	12.7	14.8	14.6	14.1	13.7	14.7	14.9
CO_2 (%)	10.4	12.0	12.9	14.9	15.2	13.3	14.1
CH_4 (%)	2.0	0.7	0.8	0.6	0.9	1.1	0.9

The experiments were carried out at ambient pressure. All hydrocarbons were tested at a S/C ratio of three and all alcohols at a corresponding oxygen to carbon ratio. Decreasing conversion was found for the various fuels with increasing feed rates except for methanol owing to the very high reaction temperature of 725 °C. Table 2.9 summarizes some of the results presented for the various fuels. The proprietary catalyst showed only minor deactivation after 70 h TOS. It was deactivated reversibly by sulfur. Load changes of the liquid input from 100 to 10% resulted in a system response after 5–10 s.

2.7.4.4 Integrated Systems Running on Various Fuels 4 [ISV 4]: Integrated Evaporator/Reformer/Burner Device for Automotive Applications

To decrease the start-up time and the electrical power demand requested for the air supply system, Whyatt et al. [10] redesigned the [ISV 3] system completely. The aim was to meet the US Department of Energy ambient temperature start-up time demand targets, which are < 1 min by 2005 and < 30 s by 2010. Figure 2.84a shows the flow schematics of the device and the prototype is shown in Figure 2.84b.

The energy for start-up and continuous operation was provided by homogeneous fuel combustion rather than catalytic combustion to decrease the time demand of start-up. The combustion was ignited by a spark-plug. Hydrogen was used as the energy source, but the option of combusting atomised gasoline is under development. The power was supplied to the reformer and the evaporator, which was placed behind the reformer in a duct. After mixing the fuel with the combustion air, the combusted gases were passed through this duct and heated the devices by a cross-flow arrangement.

The pressure drop of the heating gas passing through the reformer and evaporator was reduced by the small width of these devices to less than 3 and 0.2 mbar, respectively. The combustion gases finally preheated the combustion air in a low-temperature heat exchanger designed to provide 85% efficiency, which was placed downstream in the duct. The pressure drop of this heat exchanger was 2.5 mbar. It was bypassed during start-up to minimize further the combustion air supply pressure drop. To compensate for the lack of air preheat during start-up, it was designed to burn more fuel. Water was vaporized by the heat of the combustion gases.

In contrast to the previous [ISV 2], fuel vaporization was now performed by injecting the liquid fuel into the superheated steam. This avoided the start-up time demand of the fuel vaporizer. The feed mixture was then, after further preheating by the hot reformate in a small separate heat exchanger, fed into the reformer. The steam reformer was designed to operate at an outlet temperature of 650 °C and an S/C ratio of three. It was composed of reforming reaction channels of low height, which were as wide as the whole device. The combustion gases passed around the single channels. The overall pressure drop on the reformate side through the reformer and the evaporator was 56 mbar.

During start-up, it was designed to produce a larger amount of steam by the evaporator, which would be used to heat the gas purification units downstream. The very high S/C ratio during start-up was also intended to release the load for

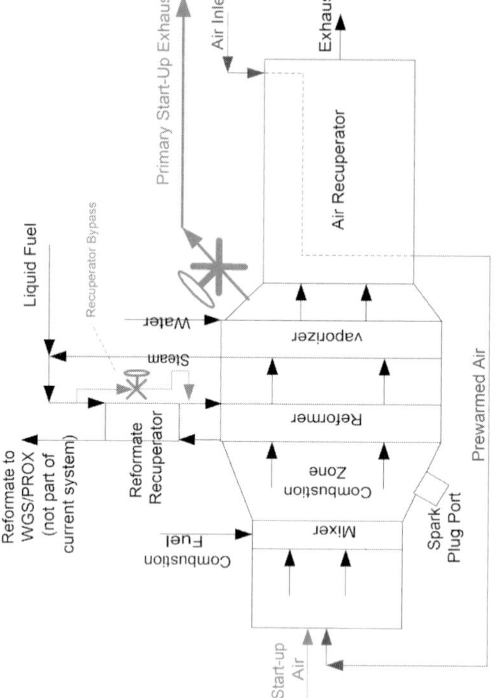

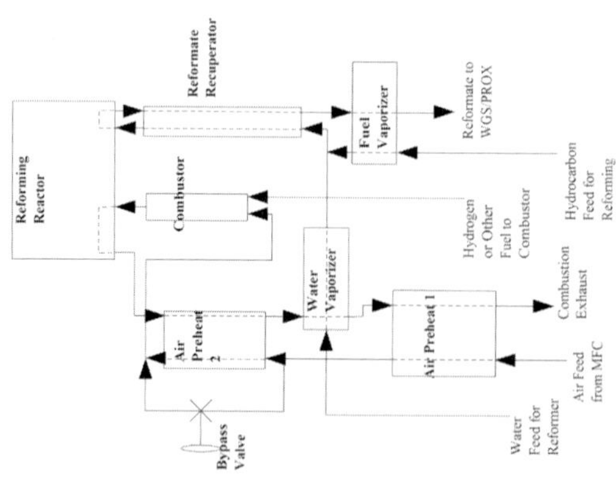

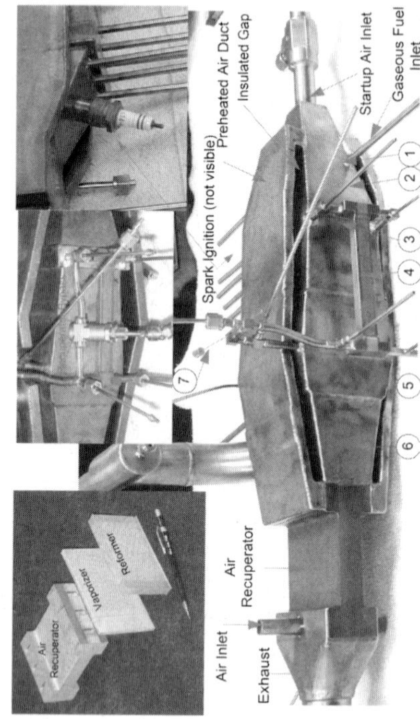

Figure 2.84 (a) Flow schematics of the [ISV 4] device; (b) the prototype and some details [10] (by courtesy of G. A. Whyatt).

Table 2.10 Estimated mass and thermal energy for the components of a 50 kW_e fuel processor [10].

Component	*Mass (kg)*	*Average operating temperature (°C)*	*Stored thermal energy (MJ)*
Inconel 600 steam reformer	5.5	875	2.09
Water vaporizer	7.4	300	1.04
Air recuperator	24.7	100	0.97
Reformate recuperator	1.8	554	0.48
WGS (differential)	9.0	343	1.5
PrOx	6.6	200	0.90
Total	55.0		6.97

the water-gas shift reactor, which would be beyond the operating temperature at this time. Rapid cold-start testing of the device with and without recuperator bypass was performed. The duration of the whole test (here only results without bypass are presented) was as short as 60 s. After 5 s of heating and dosing water, the fuel pump was switched on and after 15 s the excess water flow was reduced to the level of stationary operation. At this point, the temperature of the reformer feed was about 110 °C and the reformate left the reactor at a temperature of 220 °C. The full reformate flow was achieved after 30 s and the temperature of the reformate was then 500 °C. After 60 s, the steam left the evaporator at a temperature of 400 °C and the reformate outlet temperature had reached a temperature of 700 °C.

The total mass of a future 50 kW fuel processor was estimated to be 55 kg (see Table 2.10), which corresponds to a total energy demand of 7 MJ for start-up heating. From this, a power demand for the air blower, which has to provide some 22 m^3 min^{-1} of air to the system during start-up, was calculated to be about 1 kW. As this power is only required during the rapid start-up of the system (60 s), a normal battery could provide the power. An efficiency of 78% was calculated for the entire future fuel processor.

2.7.4.5 Integrated Systems Running on Various Fuels 5 [ISV 5]: Combined Evaporator/Reformer/Burner Device

A device capable of processing various fuels such as methanol, octane and diesel was presented by Hu et al. (Figure 2.85) [132]. It was composed of a catalytic combustor feeding a separate steam reformer. Both devices were supplied by separate evaporators.

For hydrocarbon steam reforming, 95% conversion was achieved at GHSV values between 36 000 and 144 000 h^{-1} at very low reforming temperatures between 450 and 600 °C. Synthetic, sulfur free diesel was processed at a GHSV of 144 000 h^{-1} and an S/C ratio of 3 at a 530 °C reaction temperature. More than 90% conversion was achieved for almost 10 h duration. On the other hand, methanol reforming was performed at very high reaction temperature between 350 and 425 °C.

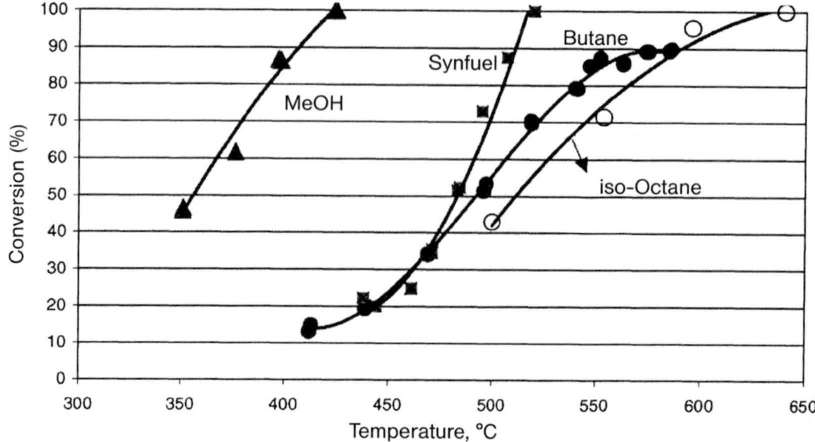

Figure 2.85 Steam reforming of different hydrocarbon fuels. GHSV = 144 000 h^{-1}, S/C = 3, 0.1 MPa [132] (by courtesy of ACS).

2.7.4.6 Integrated Systems Running on Various Fuels 6 [ISV 6]: Integrated Reformer/Burner Device for Various Fuels

Irving et al. [133] presented a micro reactor filled with catalyst particles which was capable of reforming gasoline, diesel, methanol and natural gas by steam reforming at temperatures up to 800 °C. Not only this reactor, but also fuel mixing, heat exchange, evaporation and preheating were done in an integrated device made both of stainless steel and ceramics.

Gasoline steam reforming at a flow rate of 0.1 g min^{-1} gasoline feed was performed at high S/C ratios between 5 and 8, revealing 70% hydrogen in the off-gas without methane formation and catalyst deactivation. Sulfur in the gasoline led to H_2S formation. Applying lower S/C ratios for isooctane steam reforming led to methane concentrations as high as 5% in the reformate. As much as 3.75 g of catalyst was used to achieve 100% conversion and methane concentrations well below 5% for a mixture of 60% isooctane, 20% toluene and 20% dodecane and 476 ppm sulfur at a feed rate of 0.3 g min^{-1} and an S/C ratio of 4.

Irving et al. [134] presented some results from a micro structured device. The proprietary catalyst was said to have passed 1 000 h stability tests for natural gas and results were presented indicating 60 h stability for diesel processing.

2.7.4.7 Integrated Systems Running on Various Fuels 7 [ISV 7]: Integrated Steam Reformer/Heat Exchanger for Isooctane

Fitzgerald et al. [135] presented a micro structured isooctane heat exchanger/steam reformer heated by combustion gas with a total volume of 30 cm^3 which produced enough hydrogen for a 0.5 kW PEM fuel cell. At ambient pressure, a temperature of 650 °C, a residence time of 2.3 ms and a high S/C ratio of 6, up to 95% conversion was achieved at 90% hydrogen selectivity. Decreasing the S/C ratio at a constant residence time decreased the isooctane conversion, but not the hydrogen selectivity (Figure 2.86).

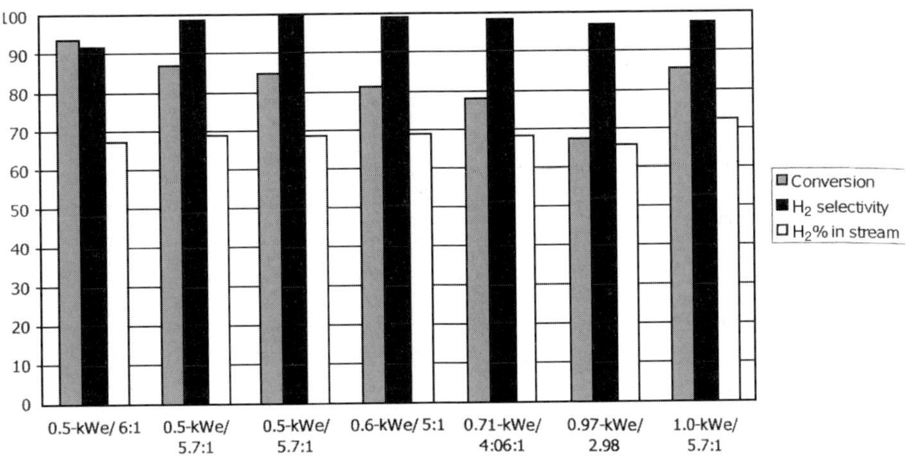

Figure 2.86 Isooctane steam reformer performance. At constant residence time the hydrogen selectivity is not affecteded by decreasing the S/C ratio while the isooctane conversion is lowered [135] (by courtesy of S. P. Fitzgerald).

2.7.4.8 Integrated Systems Running on Various Fuels 8 [ISV 8]: Design of an Integrated MEMS Reformer/Burner Device for Butane

Tanaka et al. [73] presented the design of a combined steam reformer/combustor for butane based on MEMS technology. A suspended membrane design was applied to avoid heat losses of the system. A silica membrane 3 μm thick was deposited on silicon by CVD. The channel system was part of the Pyrex® glass covers, which was connected to the silicon by anodic bonding. Copper and platinum/titania were introduced as catalysts by sputtering. However, methanol steam reforming and hydrogen combustion were applied for separate testing using integrated electric heaters and revealed low conversion levels under the experimental conditions applied.

2.8 Comparison of Micro Structured Fuel Processor Systems with Conventional Technologies

Conventional technology derived from automotive exhaust gas purification systems works on either ceramic or metallic monoliths. Both the former and the latter, generated from prestructured and rolled metallic foils, achieve very low equivalent channel diameters in the sub-millimeter range and are therefore considered as 'microtechnology'. This technology allows for running the processes in an adiabatic mode, which is an option for both autothermal reforming and gas purification, but hot-spot formation remains an issue. Therefore, Groppi and Tronconi [136] proposed the application of new, highly heat-conductive materials to overcome the problems of hot-spot formation in monolithic reactors, especially for selective oxidation reactions. Steam reforming reactors also require heat-exchanging capabilities to supply energy to the process.

Micro structured heat exchanger reactors offer unique capabilities of improving fuel processor performance as described in the sections above. However, manufacturing costs are currently considerably higher than for conventional mature technology. This will be overcome in the future with the application of cheap manufacturing and coating techniques suitable for mass production (see Sections 2.9 and 2.10), a procedure which is currently under way.

However, especially smaller systems in the power range well below 1 kW have an even stronger driver towards compactness and system integration. Therefore, systems benefit additionally from the higher integration potential of the plate heat exchanger design with decreasing size.

Additionally, highly efficient heat exchangers are required in fuel processors and the micro structured plate heat exchanger design seems to be the best solution so far to maintain the crucial system efficiency competitive.

2.8.1.1 Comparison on a Larger Scale Between a Shell and Tube Heat Exchanger, a Porous Metal Structure and a Plate and Fin Heat Exchanger Applied to Preferential CO Oxidation

The performance of a micro structured plate and fin heat exchanger (0.25 dm^3 reactor volume), wash coated with 2.0 g of catalyst, was assessed in comparison with similar alternative technologies by Dudfield et al. [88]. The first was a shell and tube heat exchanger filled with 4.66 g of catalyst microspheres (0.25 dm^3 reactor

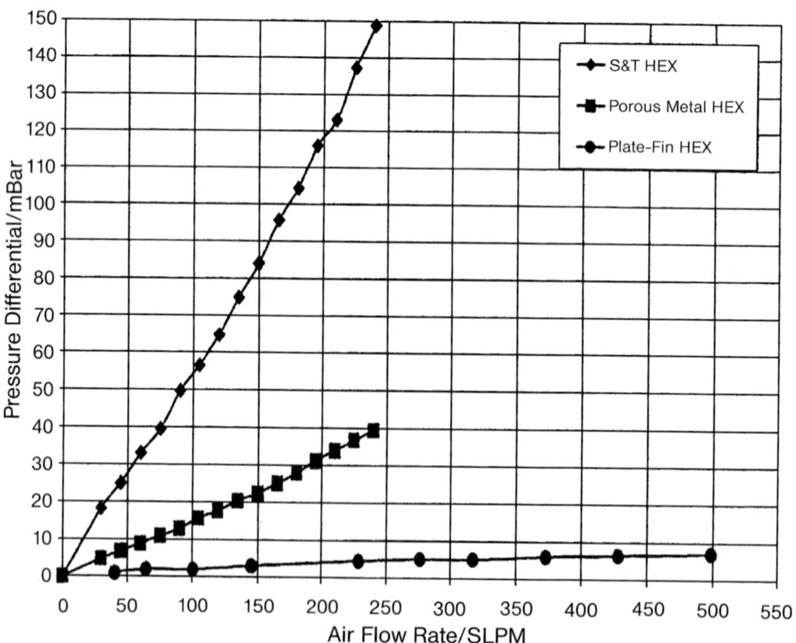

Figure 2.87 PrOx reactor CO output vs. reformate flow rate for three reactor heat exchanger (HEX) designs [88] (by courtesy of Elsevier Ltd.).

volume). The second was a heat exchanger in which steel granules were sintered to generate a porous structure, which was then wash coated with 3.4 g of catalyst (0.25 dm^3 reactor volume). The plate and fin design was realized as a sandwich and made of aluminum; the other devices were made of stainless steel. All devices were cooled by oil. The pressure drop was compared for the reactors and revealed lower values for the plate and fin reactor. It was 30 times higher for the sintered structure and eight times higher for the shell and tube heat exchanger.

The performance of the reactors was compared at a temperature of 150 °C, 10–175 $Ndm^3\ min^{-1}$ flow rate of reformate, simulating methanol steam reformer product gases, and 1–17.5 $Ndm^3\ min^{-1}$ air. The simulated reformate was composed of 69.7% H_2, 0.6% CO, 22.4% CO_2, 6.9% H_2O and 0.4% CH_3OH, the last to simulate incomplete conversion. The thermal management of the sintered porous structure was worst, leading to hot-spot formation up to 150 K, despite the integrated heat-exchange capabilities. The two remaining devices showed comparable performance, which was slightly in favor of the plate and fin design. Hot-spots were limited to 20 K here. Thus the plate and fin heat exchanger outperformed the other devices considerably leading to a maximum CO output of 250 ppm at 175 $Ndm^3\ min^{-1}$ feed rate (see Figure 2.87).

2.8.1.2 Comparison Between Packed Bed and Coating in Micro Tubes Applied to Methanol Steam Reforming

Bravo et al. [29] compared the performances of commercial catalysts deposited on a glass capillary with that of a packed bed of the same catalyst. Under identical experimental conditions, the coating showed 10–20% higher conversion.

2.8.1.3 Comparison Between Coated Micro Structures and a Conventional Monolith Applied to Autothermal Methanol Reforming

Chen et al. [36] performed a comparison of micro structured steel and aluminum plates with a conventional monolith by varying the GHSV. Full conversion could be maintained for autothermal methanol reforming in the micro structures up to a GHSV of 40 000 h^{-1}, whereas conversion dropped to 80% at 20 000 h^{-1} at the monolith. Even at 186 000 h^{-1}, still 95% conversion could be achieved in the stainless-steel micro reactor. No significant performance differences were observed between the steel and aluminum plates.

2.8.1.4 Comparison Between a Micro Structured Monolith and Conventional Monoliths Applied to Partial Oxidation of Methane

Mayer et al. [43] compared their results generated at a micro structured monolith (see Section 2.4.3) with literature data [137]. The degree of conversion and the hydrogen selectivity of the rhodium monolith outperformed both metal-coated foam monoliths, Pt–Rh gauzes and extruded monoliths. This was partially attributed to the higher activity of the rhodium monolith, but also to the lower cross-sectional channel area of the metallic monolith, which reduced mass transfer limitations, and to the improved heat conductivity.

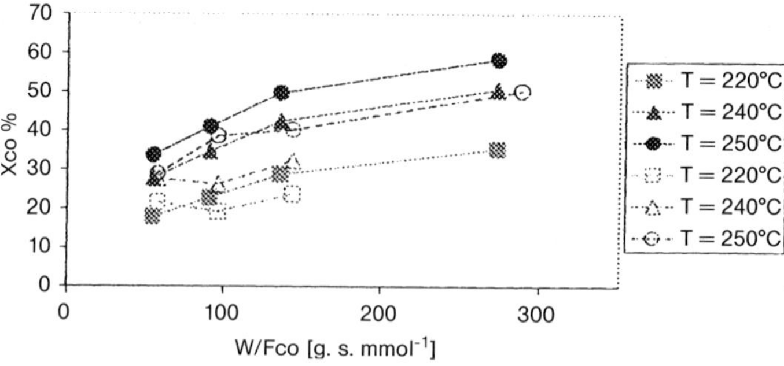

Figure 2.88 CO conversion found for low-temperature water-gas shift at various reaction temperature vs. modified residence time (catalyst weight/carbon monoxide flow). Results from a micro channel stack reactor (closed symbols) are compared with conventional cordierite monoliths (open symbols) [82].

2.8.1.5 Comparison Between Coated Micro Structures and a Conventional Monolith Applied to Water-gas Shift

Germani et al. [82] compared the performance of their catalyst coating developed for water-gas shift in a micro structured reactor with that of the same catalyst coated on a cordierite monolith under identical reaction conditions. Higher conversion was achieved in the micro channels at same modified residence time under all experimental conditions applied. Figure 2.88 shows the CO conversion vs. a modified residence time (catalyst weight/flow of carbon monoxide) measured at various reaction temperatures.

2.8.1.6 Comparison Between Coated Micro Structures and a Conventional Monolith Applied to Preferential Oxidation of Carbon Monoxide

Another comparison of micro structures with monoliths was carried out by Chen et al. [38] for the preferential oxidation of carbon monoxide (see Section 2.6.2).

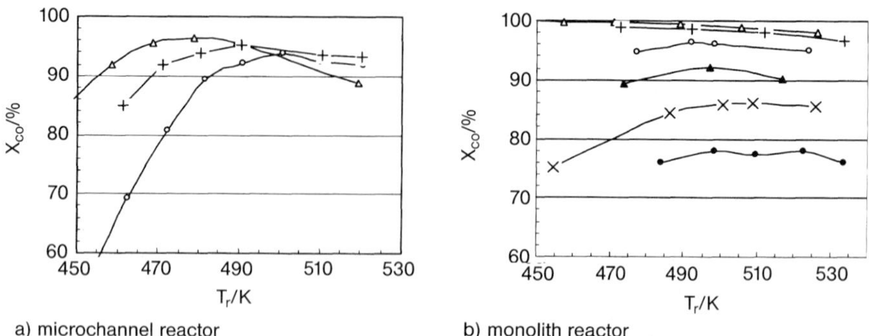

Figure 2.89 CO conversion vs. reaction temperature in (a) micro channel and (b) monolith reactors. CO feed concentration 5000 ppm; O_2/CO ratio 1.0. GHSV (micro channels): △, 100 000h^{-1}; +, 300,000h^{-1}; ○, 500 000h^{-1}. GHSV (monolith): △, 20 000h^{-1}; +, 50 000h^{-1}; ○, 100 000h^{-1}; ▲, 200 000h^{-1}; ×, 300 000h^{-1}; ●, 400 000h^{-1} [38] (by courtesy of ACS).

Comparing the performance of the micro structured reactor with a ceramic monolith at 230 °C reaction temperature and a GHSV of 300 000 h^{-1}, the conversion in the micro reactor was 94%, whereas 86% was found for the monolith, which was attributed to the improved heat and mass transfer in the metallic micro-structures (see Figure 2.89). The GHSV value of 500 000 h^{-1} corresponds to a dry gas flow rate of 440 $Ndm^3\ h^{-1}$. However, the stability of the catalyst coated on the monolith was superior to that of the catalyst coated on the micro structured stainless-steel plates.

2.9 Fabrication Techniques for Micro Structured Energy Generation Systems

Unlike chemical reactors, which are frequently fabricated as a few or even single devices, reactors and fuel processors for future distributed energy generation are heading for mass production. Consequently, fabrication costs will become the most crucial issue when the technology is mature concerning technical feasibility. This section discusses various fabrication techniques applied for micro structured fuel processors and energy generation systems taking the cost issue into consideration. However, it does not at all claim completeness, especially as far as the applications cited are concerned, but rather gives a survey of the techniques frequently found in the literature.

2.9.1 Materials Applied

The general focus of material choice for medium- to large-scale micro structured devices is on metals. Mostly stainless steels of various compositions, depending on the operation temperature, are used.

To meeet the demands of high-temperature applications such as hydrocarbon steam reforming and partial oxidation reactions, nickel-based alloys are frequently taken into consideration [10].

Another option are iron–chromium–aluminum (FeCrAl) alloys [24, 55, 57], which are mechanically stable up to very high temperatures exceeding 1 200 °C. In addition, this alloy type offers the opportunity of a unique way of generating catalyst coatings, which will be discussed below (see Section 2.10).

In the high-temperature range, ceramics also offer advantages. Interesting work has been performed by Wang et al. [69] on a ceramic micro reactor made of ceramic tapes, which might well be a future option depending on the application.

Copper and aluminum are alternative metals for low-temperature processes such as alcohol reforming [22, 85] and gas purification. The higher heat conductivity of these metals (401 and 236 $W\ m^{-1}\ K^{-1}$), respectively, compared with stainless steel (ca. 15 $W\ m^{-1}\ K^{-1}$) makes them attractive, in case isothermal conditions are required, which may well be the case for evaporators or reactors with heat-exchanging capabilities. On the other hand, the efficiency of small-scale counter-flow heat exchangers suffers from heat transfer through the wall material. Corrosion needs to be addressed in the case of aluminum in low-temperature applications [13]. Comparing the energy demand for fuel processor start-up, which results from the product of specific heat capacity and density, aluminum is the most favorable option compared with copper and stainless steel (at identical geometry).

Polymers are a viable material option for the fabrication of highly efficient low-temperature heat exchangers, designed to withdraw the last portion of energy out of the fuel processor off-gases before releasing them to the environment.

For small-scale MEMS-like systems, the materials commonly applied in this field are silicon [20, 71] (as a material with high heat conductivity) and silicon nitride [71] (as an insulation material).

2.9.2 Micro Structuring Techniques

The following techniques for the generation of micro structured channel systems are discussed below, focusing on their applicability for future mass production rather than providing detailed information about the techniques:

- micro milling
- electrodischarge machining (EDM)
- wet chemical etching
- punching
- embossing
- laser micro machining (ablation)
- sintering.

The special issue of micro injection molding applied for polymer heat exchangers will not be considered here. Additionally, for MEMS-like systems discussed above, other techniques, mostly based on etching, e.g. deep reactive ion etching (DRIE), are applied [20, 71, 86]. These techniques allow for future mass production of the small-scale (sub-watt fuel processors) devices to which they are applied.

2.9.2.1 Micro Milling

Micro milling was applied frequently in many early applications of micro structuring. Microfabrication of the parallel channels was performed by micro milling of metal tapes at the Karlsruhe Research Center (Forschungszentrum Karlsruhe) [138]. In the case of aluminum alloys, ground-in monocrystalline diamonds were used [139]. In the case of iron alloys, ceramic micro tools have to be used owing to the incompatibility of diamonds with that material [44]. It is certainly a useful tool for experimental work and rapid prototyping, but not a choice with respect to future mass production.

2.9.2.2 Electrodischarge Machining

Two types of EDM, which is controlled spark micro machining under a dielectric fluid, exist. They are namely EDM die-sinking, where an electrode is moved into the workpiece creating reverse-shaped ablation, and EDM wire-cutting, where the discharge process is achieved by a wire (minimum diameter about 20 μm) moving through the workpiece. Both techniques are excellent tools for rapid prototyping and fabrication of mold inserts and frequently applied by IMM for these purposes [25, 27]. However, they are not suitable for the mass production of micro structured devices.

2.9.2.3 Wet Chemical Etching

Wet chemical etching, which applies photo-resist for masking and usually iron chloride solution for etching, is a mature and automated technique industrially available for many applications. It is competitive for mass production and allows for a relatively wide range of channel depths from about 100 up to 600 μm and more, which covers the channel size usually applied in micro structured reactors for processing gases.

2.9.2.4 Punching

Punching is a cheap technique suitable for mass production. To achieve a sealed micro channel system, the punched plates need to be separated from each other by unstructured plates. It appears that only cross-flow heat exchangers may be fabricated out of punched plates, as straight channel systems are required. Figure 2.90 shows a cross-flow heat exchanger design fabricated at IMM out of punched plates, which were subsequently diffusion bonded afterwards at Heatric facilities. Soldering would be an alternative bonding technique for this kind of design.

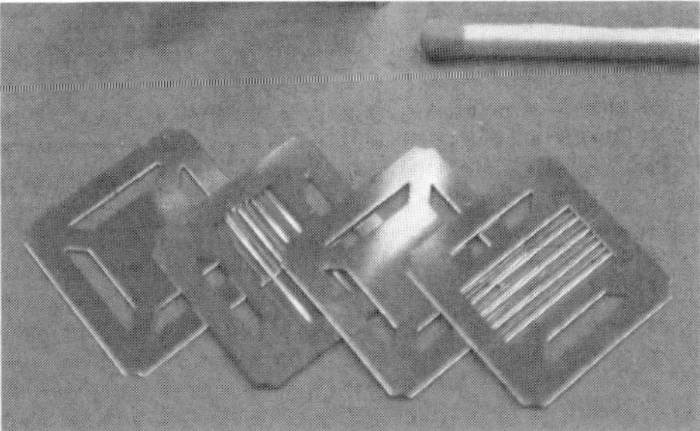

Figure 2.90 Cross-flow heat exchanger design applying punching technology developed by IMM (source: IMM).

2.9.2.5 Embossing

Embossing is a cheap technique for micro structuring highly suitable for mass production [12] similar to punching. Rolled foils of structured metal are currently applied in the field of automotive exhaust gas systems. The metallic monoliths fabricated in this way achieve channel dimensions in the sub-millimeter range and are actually 'microtechnology'. However, the introduction of integrated heat-exchanging capabilities seems to be a challenging task (for applications on the meso-scale, see Section 2.7).

2.9.2.6 Laser Micro Machining (Ablation)

Laser ablation is a technique frequently applied in many industrial applications. However, for generating micro structures of several hundred micrometers depth the method appears not to be cost competitive. However, for smaller channel dimensions in the range well below 100 μm, laser ablation might well be a cost competitive option especially for small-scale applications.

2.9.2.7 Sintering

Schuessler et al. [85] applied sintering of copper and aluminum powder to form micro structured plates for an integrated autothermal methanol reformer (see Section 2.7.2). The powders were compressed before sintering at a pressure of 1000 bar. Sintering of copper is performed at temperatures between 500 and 700 °C, which allows for the bonding of the plates in a second sintering step (see the next section).

2.9.3 Bonding Techniques

Numerous sealing/bonding techniques for micro structured metal plates exist and will be discussed below:

- gaskets (polymer, graphite or metal)
- conventional welding
- laser welding
- electron beam welding
- diffusion bonding
- soldering
- sintering.

The special issue of bonding polymer heat exchangers will not be considered here.

MEMS-like systems are frequently sealed by anodic bonding of Pyrex® glass covers [20, 73, 86], or melting of low melting-point glass frits [71], which may well be suitable for future mass production of small devices. Further details will not be discussed here.

When considering the various bonding techniques, the application needs to be carefully taken into consideration.

2.9.3.1 Gaskets

Gaskets will always be applied for testing reactors, which need to allow for the exchange of the micro structured foils e.g. in reactors, where various catalysts are screened. To seal this kind of devices, screws are therefore needed to compress the gaskets. In the case of applications in the medium temperature range (200–700 °C), gaskets prepared from graphite foils allow the easiest operation. At IMM, these gaskets are cut out of the graphite plate material by laser ablation on a routine

Figure 2.91 Hot gas-driven 5 kW evaporator sealed by graphite gaskets developed at IMM (source: IMM).

basis. However, compression of the graphite material requires considerable mechanical pressure, hence the devices are rather bulky. A typical example of the medium scale, a 5 kW hot gas-driven evaporator with removable plates, is shown in Figure 2.91. The same device would have approximately 20% of this size if welding techniques were applied rather than gaskets. For temperatures exceeding 700 °C, metallic gaskets and sealing techniques should be applied.

All remaining bonding techniques apply elevated temperatures and are irreversible. For process engineering devices such as micro structured heat exchangers and evaporators, the operating temperature of the bonding technique is not critical. The same applies for chemical reactors, as long as the catalyst or the catalyst coating is introduced after the reactor bonding, which will be denoted 'post-coating' in the following. However, when the catalyst coating needs to be done before the bonding ('precoating'), the temperature applied during the bonding step needs to be well below the maximum operating temperature of the specific catalyst, which may reach from about 450 °C to more than 800 °C depending on the application. For a survey of coating techniques, see Section 2.10.

2.9.3.2 **Conventional Welding**

Conventional welding may well be applied for prototypes and small series production especially for bonding of reactor/device periphery such as inlet diffusers and fluidic connections. In the case of chemical reactors, overheating needs to be avoided if precoating techniques are applied to avoid damage to the catalyst coating.

2.9.3.3 **Laser Welding**

Laser welding is a viable option for the fabrication of micro structured devices [43]. For large-scale applications the length of the weld needs to be minimized to achieve competitive pricing [12]. However, for small-scale devices this is generally valid. An advantage of laser welding is the low and locally limited amount of energy brought into the device. This allows for precoating of reactors. Another advantage is the applicability in the high-temperature range. Figure 2.92 shows a 10 kW counter-flow heat exchanger developed by IMM sealed by laser welding.

Figure 2.92 10 kW counter-flow heat exchanger developed by IMM, which was bonded by laser-welding (source: IMM).

2.9.3.4 Electron Beam Welding

This technique, nowadays increasingly applied in the automotive industry, combines the advantages of laser welding with competitive pricing. This makes it attractive for larger devices in the kilowatt range. It has been applied for bonding of micro structured plate heat exchangers [43, 140].

2.9.3.5 Diffusion Bonding

Takeda et al. [141] described the conditions applied for the diffusion bonding of nickel-based alloys (Hastelloy). A test piece of a plate and fin heat exchanger (dimensions 40 mm × 40 mm × 3 mm, channel dimensions 1000 μm × 1000 μm) was first cleaned with a 1% solution of nitric acid and hydrogen fluoride to remove the oxidation layer. A pressure as low as 6 mPa was necessary to run the process. For NiCrFe and other materials, other authors claim that an even lower pressure of 1 μPa is required. A welding temperature of 1 150 °C and a contact pressure of

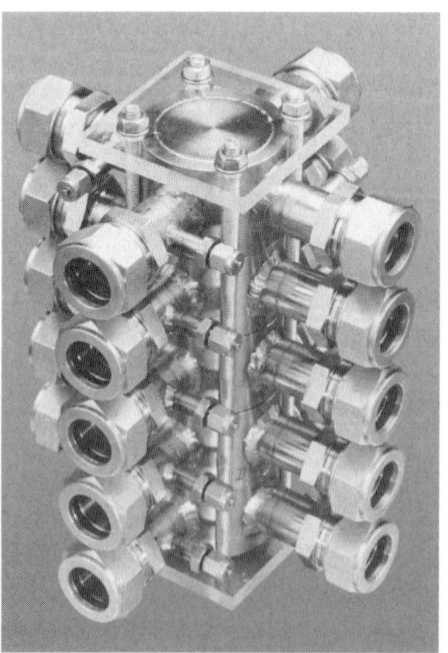

Figure 2.93 Stack of cross-flow heat exchangers developed by Karlsruhe Research Center [144].

440 bar for a duration of 30 min were identified as optimized conditions. Significant deformation of the test piece was required according to the authors. However, other authors claim 3% deformation to be sufficient [44]. Leak tightness was verified at ambient temperature up to 630 bar.

Heatric, a member of the Meggit group, is one of the few companies capable of a proprietary diffusion bonding process. The company applies diffusion bonding for the fabrication of large meso-scale heat exchangers with exchange capabilities in the megawatt range mainly for off-shore applications. IMM and Heatric have a strategic alliance aimed at the common development of processes up to industrial scale.

Karlsruhe Research Center frequently applies diffusion bonding for their cross-flow heat exchangers (see Figure 2.93) [142, 143]. The technique applies high temperature as mentioned above, hence only catalyst coating techniques which are performed after the bonding procedure are suitable for diffusion bonded reactors (see Section 2.10). However, the diffusion bonding of FeCrAlloy materials seems to be creating significant problems [83].

2.9.3.6 Brazing

Soldering is a low-cost bonding technique highly suitable for mass production. It allows for precoating techniques but is, however, limited to the melting-point of the solder applied. However, solders are available up to a melting-point exceeding 1 200 °C. A 40 kW evaporator was fabricated by Xcellsis (now Ballard) applying high-vacuum brazing [12]. Stainless-steel and nickel-based alloys may be bonded by brazing using Ni and other solders. Heat exchangers fabricated by Atotech [121] were also bonded by brazing. According to Takeda et al. [141], brazing is limited to operating temperatures of 650 °C. However, contact of the solder with catalysts in chemical reactors needs to be carefully avoided.

2.9.3.7 Sintering

Sintering as a micro structuring (see the section above) and bonding technique was applied by Schuessler et al. [85] of Ballard for their compact methanol fuel processor (see Figure 2.94). The stack of plates and the endplate are connected in a single bonding step.

Figure 2.94 Single plate and stack of plates with endplate of the integrated fuel processor fabricated at Ballard [85] (by courtesy of Elsevier Ltd.).

2.10
Catalyst Coating Techniques for Micro Structures and Their Application in Fuel Processing

This section deals with coating techniques for micro structured reactors. The various techniques available are briefly discussed and their applicability for the different bonding techniques as discussed in the previous section is taken into consideration.

In addition to the coating techniques discussed below, one may also consider making the whole micro reactor out of the catalytically active material [44]. However, as precious metals are frequently used, this is not a cost-competitive option for mass production, of course.

Additionally, thin-film layers of the catalyst may be deposited on to the surface of the micro channels by sputtering [73] or CVD [63]. In the latter case, aluminum isopropoxide, $Al[(CH_3)_2CHO]_3$, was used as alumina precursor, which was passed through a ready-stacked reactor at 300 °C for 1 h in a flow containing nitrogen and oxygen in addition. The flow direction was subsequently changed and the procedure repeated.

These methods do not, however, generate the surface area that is required in most cases to achieve sufficient reactor productivity.

2.10.1
Coating of Ready-made Catalyst

Coating of commercial catalyst [24, 124, 145] is a viable option which allows for applying mature catalyst technology in micro structures. However, care has to be taken that the catalyst properties are not affected by the coating procedure and that the conventional catalyst is suited for application under the conditions in micro structures.

Bravo et al. [29] dealt with the coating of a commercial CuO/ZnO catalyst on quartz and fused-silica capillaries for future application in micro channels. The catalyst was mixed with boehmite as binder and water at a mass ratio of 44 : 11 : 100. The boehmite was treated with hydrochloric or nitric acid before. The capillaries were pretreated with a hot sulfuric acid/solid oxidation step before coating. The capillaries were filled with the catalyst/binder suspension and then cleared with air. In this way, catalyst coatings up to 25 μm thick were obtained. The coatings were applied to methanol steam reforming (see Section 2.4.1).

2.10.2
Wash Coating

Exemplarily, Zapf et al. [145] described a manual wash coating/impregnation method, a typical procedure for the preparation of supported catalysts on micro-structures. Micro structured stainless-steel plates (X2CrNiMo17 12 2 and MoTi17 12 2) structured by photochemical etching were used. The micro channels had a semi-circular cross-section and were fabricate with the dimensions 500 μm × 300 μm, 750 μm × 300 μm and 500 μm × 70 μm (width × depth). The first step of the coating procedure was the preparation of a slurry, which consisted of 20 g of γ-alumina (average particle size 3 μm), 75 g of de-ionized water, 5 g of binder poly(vinyl alcohol) and 1 g of acetic acid. The binder was dissolved in water and the suspension stirred at 60 °C for 2 h and left overnight in a vessel for the release of air bubbles. Alumina and acetic acid were added and stirred overnight. Micro structured substrates were first cleaned in an ultrasonic bath with 2-propanol (step 1, see Figure 2.95).

After positioning the substrates in the pocket of a holder, the area which should not be coated outside the micro structures could be covered by a mask if necessary (step 2).

The prepared wash coat slurry was deposited in the micro structured channels with a coating knife (step 3).

The channels were filled completely and surplus slurry was removed by a razor blade (step 4).

After drying the deposited slurry, the layer thickness in the channels shrunk and an open cross-section for the fluid developed (step 5).

The final coating thickness was adjusted by the amount of alumina in the aqueous suspension. The wash coats were calcined at 600 °C (step 6), which resulted in a porous coating with an internal surface area of approximately 70 $m^2 g^{-1}$ measured by BET surface analysis.

Scanning electron-microscopic (SEM) studies revealed the porous surface of these wash-coats. A falling test was developed and applied to the coatings. At an impingement velocity of 3 m s^{-1} losses of the wash coat did not exceed 1%. The deviation of the coating mass over a set of parallel micro channels at a certain axial position was determined as less than 5%. To achieve coating profiles of equal thickness in the radial direction, optimization of the slurry viscosity was required

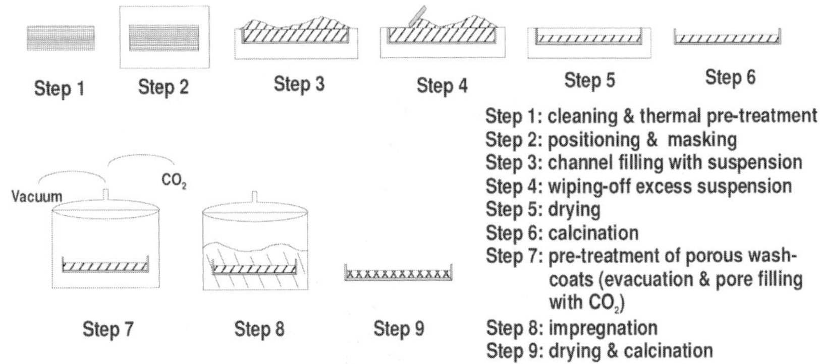

Figure 2.95 Principle of manual wash coating of micro structures (source: IMM).

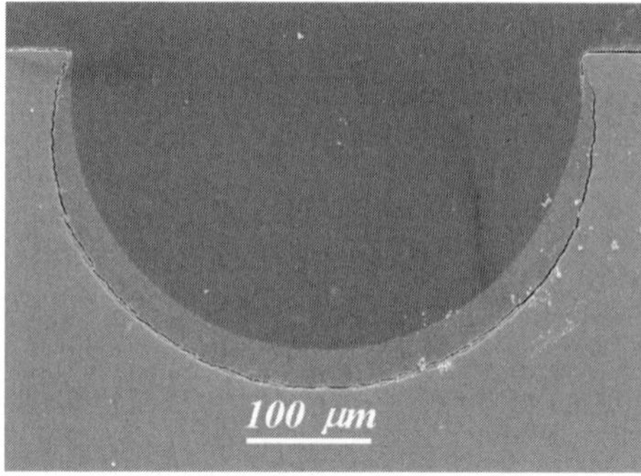

Figure 2.96 SEM cross-sectional view of micro channels coated with boehmit/ceralox [146].

[146]. Wash coats made of various source aluminas were prepared by applying this procedure (Figure 2.96) [147]. The catalysts obtained after subsequent impregnation were applied to methanol steam reforming [25, 28], propane steam reforming [52], water-gas shift [84] and preferential oxidation [89], to name but a few reaction systems.

Pfeifer et al. [45] deposited CuO/ZnO and PdO/ZnO nanoparticles on micro channels by wash coating. A stable dispersion of the nanoparticles was achieved when polymers such as hydroxylpropylcellulose were added to the solvent 2-propanol. The wash coating was performed on both single micro structured foils and stacked foils (post-coating, see Figure 2.97). These coatings were applied to methanol steam reforming [22].

Wash coating techniques for metal foils exist from automotive exhaust system technology [148, 149]. Adomaitis et al. [148] presented the Metreon process, which was developed for coating of unstructured FeCrAlloy metal foils. The foils wrapped up in coils were run through rolls, where they were formed and heat treated first to generate an alumina layer. Then a single or up to four wash coating steps followed, each comprising drying and calcination steps. Finally, the metal foil was recoiled. The coatings generated this way were highly resistant to thermal shock.

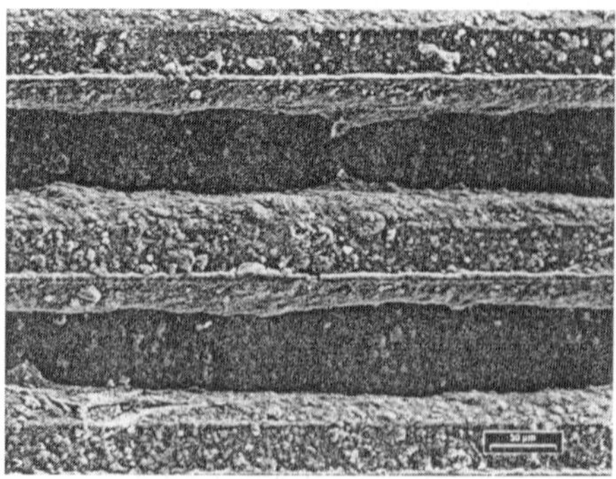

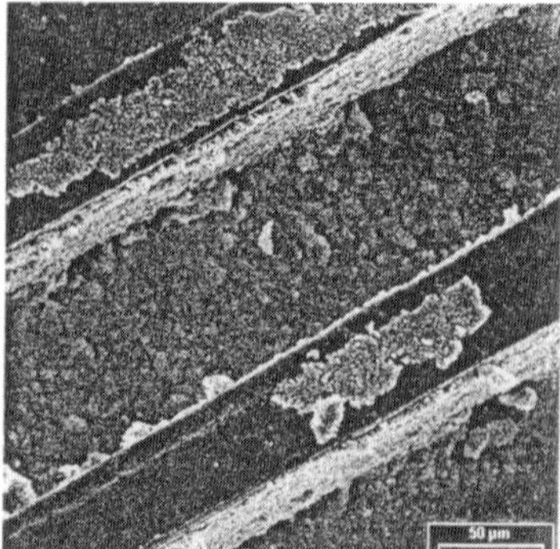

Figure 2.97 SEM picture of a post-coated aluminum foil with channel dimensions 100 × 100 µm [45] (by courtesy of P. Pfeifer).

As wash coating of micro structured metal foils has to be cost efficient, it must be performed in an automated continuous procedure. Continuous wash coating (or flow coating) of micro structured metal foils, contrary to continuous impregnation (see Chapter 4), demands a considerably larger machinery set-up. The method similar to the procedure described by Adomaitis et al. [148] (see above) was transferred by IMM [150] to perform the continuous coating of micro structured stainless-steel foils using a special coating device fabricated by Coatema (Dormagen, Germany), a leading supplier of coating machinery for the textile industry.

A continuous micro structured foil fabricated by wet-chemical etching was unwound from a roll and continuously coated with slurry. If required, the metal foil could be pre-cut to any plate size by laser cutting at defined positions. This would ease later separation of the plates after coating. These etched and laser-cut foils are products already commercially available. First tests were performed with micro structured foils and described in detail in Chapter 4.

2.10.3
Spray Coating

Spray coating of commercial catalysts was successfully applied by Schuessler et al. [85].

2.10.4
Sol–Gel Coating

The sol–gel coating technique offers the opportunity of a future automated coating process for micro structured reactors. After reactor bonding, the sol, which has a much lower viscosity than wash coat slurries, may simply be pumped through the reactor.

Janicke et al. [63] applied a sol–gel-type deposition technique by filling micro channels with aluminum hydroxide solution, which was dried and calcined at 550 °C thereafter. Platinum was introduced as catalyst by impregnation on the coating up to three times, resulting in different loadings of the precious metal.

Haas-Santo et al. [151] performed an extensive study on optimizing the operation conditions of sol–gel coating on micro structured metal foils. Tetraethylorthosilane (TEOS), aluminum *sec*-butyrate (AISB), aluminum triisopropylate and titanium tetraisopropylate served as alkoxides to generate titania, alumina and silica coatings. Acetylacetone was used as stabilizer (the optimum molar ratio of acetylacetone to alkoxide = 1 for alumina sol) and nitric acid as catalyst, whereas an alcohol served as solvent to generate the sols. FeCrAlloy foils were used, which were treated for 15 h at 1 000 °C before coating. An alumina layer of about 1 µm thickness was created on the steel surface, which enhanced the adhesion of the sol–gel coating considerably. Dip-coating was applied as the coating method for single metal foils; stacks of foils were coated by pipetting the sol into the channel system put into a vertical position. To remove the carbonaceous species from the coatings, heat treatment at 500 °C was required. Higher temperatures decreased the surface area of the coatings, which was attributed to sintering and phase transition to corundum, especially at 800 °C.

Sols made of AISB with acetylacetone in ethanol were found to be stable for months, generating coatings of similar surface area and pore volume distribution. Silica sol viscosity was found to be very sensitive to pH and aging time. The surface area of the silica sol coatings increased with increasing pH from 250 $m^2 g^{-1}$ at pH 6 to 800 $m^2 g^{-1}$ at pH 8. However, above pH 8 immediate gelation took place. Hence a trade-off was necessary between time demand for coating, as gelation took place earlier at higher pH, and high surface areas which were achieved at high pH.

The highest surface area enhancement factors were determined for the alumina coatings prepared with AISB, followed by alumina from aluminum triisopropylate, silica and titania. Alumina coatings were deposited in a ready-mounted micro structured reactor. The coatings had a thickness of 2–3 µm. They were impregnated with palladium and successfully applied to hydrogen oxidation [67] (see Section 2.5.1).

Chen et al. [152] prepared a Pt/Al_2O_3 catalysts by different sol–gel procedures, either by depositing the carrier sol and the catalyst solution simultaneously or subsequently on to micro channels made of silicon wafers.

2.10.5 Anodic Oxidation

The layer generated by anodic oxidation of aluminum and aluminum alloys was found to be amorphous hydrated alumina [153]. It has the morphology of a packed array of hexagonal cells, each containing a pore in the center. Hence the layer has a highly ordered porous structure and even distribution of the layer thickness, in case an equally sized anode is positioned close to the surface which is oxidized. However, to apply anodic oxidation as a cheap method suitable for mass production in ready-made micro reactors, which are mounted and sealed, e.g. by diffusion bonding, the oxidation procedure needed to be modified. Therefore, Wunsch et al. [154] demonstrated anodic oxidation of aluminum or aluminum alloys ($AlMg_3$) in 1.5% oxalic acid at a stack of micro structured foils. Aluminum wires positioned at the inlet and outlet of the stack served as cathodes at 50 V. After 3–4 h, a satisfactorily even film distribution in the range 10–12 µm was achieved up to 14 mm deep in the axial direction of the stack of channels. The film thickness increased almost linearly with time. Deeper inside the channel system, a decrease of the layer thickness from 7 to 3 µm at 15 (see Figure 2.98) and 40 mm, respectively, was determined for a stack of greater dimensions. The catalyst carriers were impregnated with Pt. However, owing to the inaccessibility of the in situ-generated catalyst, surface area data for the layers were only available as surface magnification factors (400 $m^2 m^{-3}$).

Gorges et al. [155] introduced anodic spark deposition (ASD), a modification of anodic oxidation, for the formation of polycrystalline ceramic oxide layers on passivating metals (Figure 2.99). The method is therefore limited to Ti, Al and Zr

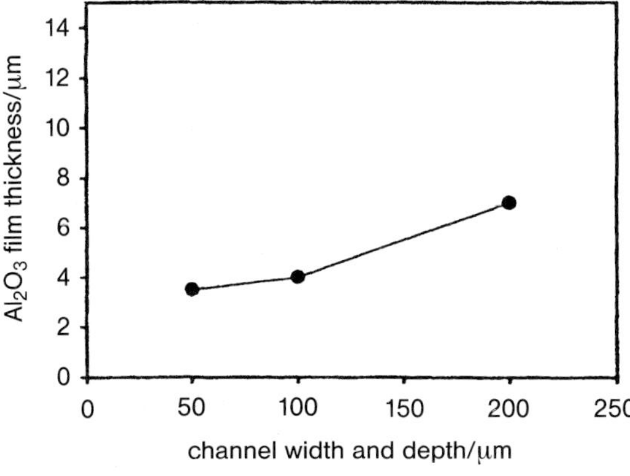

Figure 2.98 Influence of channel dimensions on alumina layer thickness generated by anodic oxidation [154] (by courtesy of R. Wunsch).

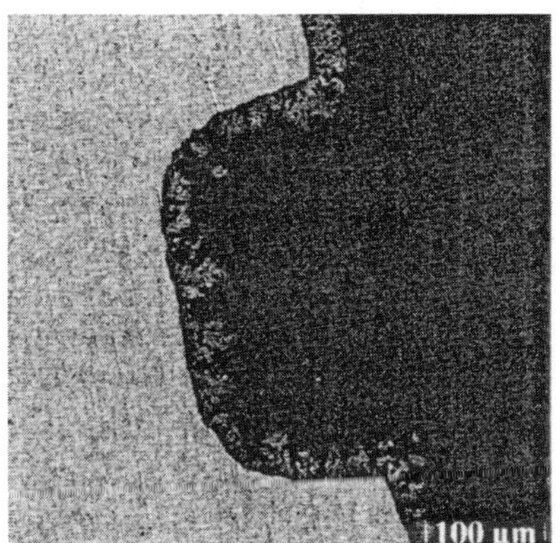

Figure 2.99 SEM images of TiO_2 coatings on solid Ti microstructures (left) and on ceramic microstructures covered with a titanium PVD layer (right) [155] (by courtesy of G. Gorges).

surfaces. Voltages exceeding 100 V were applied. By local melting of the layer into the substrate, strongly adhered coatings of up to 40 μm thickness could be generated and the incorporation of dopant elements was possible. A titania coating generated by ASD was found to be composed of rutile and anatase by XRD and had a surface area of 51 $m^2 g^{-1}$. By coating a ceramic green body with titanium applying PVD, ASD could even be carried out on a non-conductive base material. However, the applicability of this method to ready-made stacks of metal foils has not yet been investigated.

2.10.6
Electrophoretic Deposition

Födisch et al. [156] applied electrophoretic precipitation of industrial catalyst powders at 100 V (DC). After 2 min, uniform deposition of the catalyst powder on the surface was achieved. As an alternative to impregnation methods, palladium was deposited by electrochemical deposition [7.5 V (DC) for 3 min] and reductive deposition (Pd^{2+} was reduced to metallic palladium by formalin) on alumina layers previously generated by anodic oxidation.

Pfeifer et al. [45] performed electrophoretic deposition of alumina nanoparticles on a stack of micro structured metal foils. Various dispersions of alumina nanoparticles were pumped through the stack for 20 h and a voltage of 50 V at a current density of 0.5 A cm^{-2} was set to the stack. The best composition of the dispersion to achieve a well-adhering, homogeneous coating was found to be water and glycerol. A surface enhancement factor of 102 m^2 m^{-3} and a layer thickness of about 3 μm were determined by SEM for the coating generated.

2.10.7
Oxidation of FeCrAlloys

By thermal treatment of FeCrAlloys, which contain up to 5 wt.% aluminum, an alumina layer is formed on the surface of the alloy, which might be used as catalyst carrier by itself [55] or serve as an intermediate layer improving the adhesion of the coating [57, 68, 83, 151]. This procedure corresponds to the coating technique applied for metallic monoliths applied in the automotive industry, which are in fact structured FeCrAlloy plates, which are rolled up to form a monolith. The mature technology allows not only for coating the monolith reactors after the bonding step, but also before the bonding using, e.g., pin connections [157] of the foil stack with the can body. The introduction of the coating on the plane metal foils improves the homogeneity of the coating considerably.

2.10.8
Introduction of ZSM-5 Zeolite into Micro Channels

Rebrov et al. [158] synthesized ZSM-5 zeolite in micro channels. The main focus of this work was to assess the performance benefits of zeolitic coatings in micro channels compared with conventional zeolite-based pellets and powders. The coatings were performed in a sandwich of two plates of 1 cm length and width at a thickness of 2 mm. The plates carried seven channels each, which were 1 cm long and had a diameter of 500 μm. The plates were positioned at a distance of 280 μm from each other in the housing of the reactor. A zeolitic film of one crystal thickness was formed under the optimum synthesis conditions, which were determined as a water/silicon ratio of 130 and a template/aluminum ratio of 2 at a temperature of 130 °C after 35 h on a flat plate. The Si/Al ratio of the zeolite, which also lowers the crystal size when decreased, was optimized to a value of 28.

To obtain a uniform distribution of the zeolite in the micro channels, a two-step procedure was developed, including nucleation growth at high temperature at the horizontally oriented plates followed by a growth period at the vertically oriented plates, which was performed at lower temperature and water/silicon ratio. The crystals were oriented parallel to the surface of the carrier. Nitrogen adsorption revealed the typical micropore distribution of ZSM-5 for the coating. Thermal cycling of the coated plates revealed good adhesion of the coating and no migration of Cr from the passive metal layer into the zeolite.

By XRD analysis, high crystallinity of the coatings exceeding 90% was determined. The crystallization took place on the whole platelet and the crystal layer had to be removed mechanically from the flat surface.

Input from the earlier work summarized above [158] was used by Rebrov et al. [159] for further optimization on 1 cm^2 sized, 2 mm thick plates carrying channels of 200 μm diameter. The optimum template/aluminum ratio of 2 could be maintained, but the water/silicon ratio had to be lowered to 108 at the same silicon/aluminum ratio of 28 and a slightly lower temperature of 135 °C. Larger crystals (6 μm × 4 μm × 3 μm) were formed in this way compared with the former method. For the synthesis, diffusion limitation occurred rather than crystal formation limitation. Stirring resulted in more uniform, but thinner coatings. Pretreatment of the stainless-steel support with dilute template solution improved the crystal growth in the upper part of the channels.

Finally, large 1.5 μm × 1 μm × 2 μm crystals with a very narrow size distribution (within 0.2 μm) [113] were achieved by first immersing the platelets in a concentrated synthesis mixture at a water/silicon ratio of 24 for 2 h at 160 °C. This resulted in 0.3 μm long crystals which were further treated by applying the initial method for flat plates for 35 h at 130 °C. The coatings were tested for their performance in the selective catalytic reduction of NO after exchanging cerium into the zeolite as catalyst. Similar, but always higher, reaction rates were found for the zeolite coatings compared with pelletized zeolites, which is remarkable, as the conventional measurements were carried out under an extremely turbulent flow regime in a Berty-type recycle reactor. Therefore, diffusion limitations in the micropores of the pelletized zeolite obviously created transport limitations.

Another approach to gain the advantages of micro structures in order to avoid high pressure drops, heat and mass transfer limitations and unequal flow distributions was presented by Louis et al. [160]. ZSM-5 zeolites were coated on stainless-steel grids as an alternative to granules or pellets usually applied in zeolitic catalysis. The H-ZSM-5 coatings were tested for the one-step oxidation of benzene by nitrous oxide to phenol. The grids had a total area of 9 cm^2, a wire diameter of 250 μm and a mesh size of 800 μm. Fifteen grids formed a stack separated by steel rings. By acid pretreatment of the grids, defects were generated which are known to become crystallization centers during the synthesis of the zeolite.

The synthesis gel was prepared and after 2–3 h of aging it was poured into the support packing. After three synthesis runs of 40 h duration at 171 °C, the coating was calcined and exchanged with ammonium chloride to the protonated form. An Si/Al ratio of 65 and a BET surface area of 302 $m^2\ g^{-1}$ were found for the zeolite. Loadings of about 10, 55 and 95 $m^2\ g^{-1}$ zeolite, the last corresponding to a layer thickness of 38 μm, were found after the individual synthesis steps. Three loadings were regarded as an upper limit, as the grids tended to lose their void fraction at higher loadings.

References

1 *www.fuelcelltoday.com/FuelCellToday/IndustryInformation/IndustryInformationE/External/IndustryInformationDisplayArticle/0, 631,00.html*, 1 November **2003**.

2 Cominos, V., Hofmann, C., *State-of-the-art fuel cell systems for portable devices and the applicability of the MiRTH-e system*, Control No. ENK6-CT-2000-00110, funded by the European Comission, IMM Institut für Mikrotechnik Mainz, Mainz, **2002**.

3 Lemons, R. A., Fuel cells for transportation, in *Proceedings of the 1st Grove Fuel Cell Symposium*, London, **1989**, 251.

4 Teagan, W. P., Bentley, J., Barnett, B., Cost reductions of fuel cells for transport applications: fuel processing options, *J. Power Sources* **1998**, *71*, 80–85.

5 Jochem, E., Schirrmeister, E., Potential economic impact of fuel cell technologies, in Vielstich, W., Lamm, A., Gasteiger, H. A. (Eds.), *Handbook of Fuel Cells*, Vol. 4, Wiley, Chichester, **2003**, 1130.

6 Garch, J., Jörissen, L., PEMFC fuel cell systems, in Vielstich, W., Lamm, A., Gasteiger, H. A. (Eds.), *Handbook of Fuel Cells*, Vol. 4, Wiley, Chichester, **2003**, 1244.

7 Amphlett, J. C., Baumert, R. M., Mann, R. F., Peppley, B. A., System analysis of an integrated methanol steam reformer/PEM fuel cell power generating system, in *Proceedings of the 27th Intersociety Energy Conversion Engineering Conference* (3–7 Aug. 1992), San Diego, CA, **1992**, 3343–3348.

8 Emonts, B., Hansen, J. B., Grube, T., Höhlein, B., Peters, R., Schmidt, H., Stolten, D., Tschauder, A., Operational experience with the fuel processing system for fuel cell drives, *J. Power Sources* **2002**, *106*, 333–337.

9 Edwards, N., Ellis, S. R., Frost, J. C., Golunski, S. E., van Keulen, A. N. J., Lindewald, N. G., Reinkingh, N. G., On-board hydrogen generation for transport applications: the HotSpot™ methanol processor, *J. Power Sources* **1998**, *71*, 123–128.

10 Whyatt, A., Fischer, C. M., Davis, J. M., Development of a rapid-start on-board automotive steam reformer, in *Proceedings of the AIChE Spring Meeting* (25–29 April 2004), New Orleans, **2004**, 120C, published on CD.

11 Ahmed, S., Krumpelt, M., Hydrogen from hydrocarbon fuels for fuel cells, *Int. J. Hydrogen Energy* **2001**, *26* (4), 291–301.

12 Weisser, M., *Effiziente Energiewandlung mit Brennstoffzellen für mobile Anwendungen, http://www.xcellsis.com* (Forum Mikrotechnologie, Hannover Messe), 27 April **2001**.

13 Pfeifer, P., Schubert, K., Liauw, M. A., Emig, G., Electrically heated microreactors for methanol steam reforming, *Trans. IChemE* **2003**, *81A*, 711–720.

14 Ramshaw, C., *Chem. Eng.* **1985**, *415*, 30.

15 Ratnasamy, C., Wagner, J. P., Tackett, D., Production of hydrogen by autothermal reforming of methane and simulated natural gas, in *Proceedings of the AIChE Spring Meeting* (25–29 April 2004), New Orleans, **2004**, 12e, published on CD.

16 Takahashi, K., Kobayashi, H., Takezawa, N., *Chem. Lett.* **1985**, 759.

17 Larson, A. T., *US 2 425 625*, **1947**.

18 Palo, D., Rozmiarek, R., Steven, P., Holladay, J., Guzman, C., Wang, Y., Hu, J., Dagle, R., Baker, E., Fuel processor development for a soldier-portable fuel cell system, in Matlosz, M., Ehrfeld, W., Baselt, J. P. (Eds.), *Microreaction Technology – IMRET 5: Proc. of the 5th International Conference on Microreaction Technology*, Springer-Verlag, Berlin, **2001**, 359–367.

19 Holladay, J. D., Jones, E. O., Phelps, M., Hu, J., Microfuel processors for use in a miniature power supply, *J. Power Sources* **2002**, *108*, 21–27.

20 Pattekar, A. V., Kothare, M. V., Karnik, S. V., Hatalis, M. K., A microreactor for in-situ hydrogen production by catalytic methanol reforming, in Matlosz, M., Ehrfeld, W., Baselt, J. P. (Eds.), *Microreaction Technology – IMRET 5: Proc. of the 5th International Conference on Microreaction Technology*, Springer-Verlag, Berlin, **2001**, 332–342.

21 Pattekar, A. V., Kothare, M. V., A microreactor for hydrogen production in micro fuel cells, *IEEE J. Microelectromech. Syst.* **2004**, *13*, 7–18.

22 Pfeifer, P., Fichtner, M., Schubert, K., Liauw, M. A., Emig, G., Microstructured catalysts for methanol-steam reforming, in Ehrfeld, W. (Ed.), *Microreaction Technology: 3rd International Conference on Microreaction Technology, Proc. of IMRET 3*, Springer-Verlag, Berlin, **2000**, 372–382.

23 Pfeifer, P., Schubert, K., Fichtner, M., Liauw, M. A., Emig, G., Methanol-steam reforming in microstructures: difference between palladium and copper catalysts and testing of reactors for 200W fuel cell power, in *Proceedings of the 6th International Conference on Microreaction Technology, IMRET 6* (11–14 March 2002), AIChE Pub. No. 164, New Orleans, **2002**, 125–130.

24 Reuse, P., Tribolet, P., Kiwi-Minsker, L., Renken, A., Catalyst coating in microreactors for methanol steam reforming: kinetics, in Matlosz, M., Ehrfeld, W., Baselt, J. P. (Eds.), *Microreaction Technology – IMRET 5: Proc. of the 5th International Conference on Microreaction Technology*, Springer-Verlag, Berlin, **2001**, 322–331.

25 Cominos, V., Hardt, S., Hessel, V., Kolb, G., Löwe, H., Wichert, M., Zapf, R., A methanol steam micro-reformer for low power fuel cell applications, in *Proceedings of the 6th International Conference on Microreaction Technology, IMRET 6* (11–14 March 2002), AIChE Pub. No. 164, New Orleans, **2002**, 113–124.

26 Cominos, V., Hardt, S., Hessel, V., Kolb, G., Löwe, H., Wichert, M., Zapf, R., A methanol steam micro-reformer for low power fuel cell applications, *Chem. Eng. Commun.* **2005**, in press.

27 Kolb, G., Cominos, V., Drese, K., Hessel, V., Hofmann, C., Löwe, H., Wörz, O., Zapf, R., A novel catalyst testing microreactor for heterogeneous gas phase reactions, in *Proceedings of the 6th International Conference on Microreaction Technology, IMRET 6* (11–14 March 2002), AIChE Pub. No. 164, New Orleans, **2002**, 61–72.

28 Ziogas, A., Hessel, V., Kolb, G., Löwe, H., Zapf, R., Becker-Willinger, C., Bolz, H., Pannwitt, A.-K., Schmidt, H., Catalyst screening in a modular catalyst testing microreactor for heterogeneous gas phase reactions, in *Proceedings of the European Conference on Chemical Engineering, ECCE-4* (21–25 Sept. 2003), European Federation of Chemical Engineering, Granada, **2003**, published on CD.

29 Bravo, J., Karim, A., Conant, T., Lopez, G. P., Datye, A., Wall coating of a CuO/ZnO/Al_2O_3 methanol steam reforming catalyst for micro channel reformers, *Chem. Eng. J.* **2004**, *101*, 113–121.

30 Pfeifer, P., Schubert, K., Liauw, M. A., Emig, G., PdZn catalysts prepared by washcoating microstructured reactors, *Appl. Catal. A* **2004**, *270*, 165–175.

31 Chin, Y.-H., Dagle, R., Hu, J., Dohnalkova, A. C., Wang, Y., Steam reforming of methanol over highly active Pd/ZnO catalyst, *Catal. Today* **2002**, *77*, 79–88.

32 Hu, J., Wang, Y., Chin, C., Palo, D., Rozmiarek, R., Dagle, R., Cao, J., Holladay, J., Baker, E., Catalytic hydrogen production for small portable power applications, *Am. Chem. Soc. Div. Fuel Chem. Prepr.* **2002**, *47*, 725–726.

33 Chin, Y.-H., Wang, Y., Dagle, R. A., Li, X. S., Methanol steam reforming over Pd/ZnO: catalyst preparation and pretreatment studies, *Fuel Process. Technol.* **2003**, *83*, 193–201.

34 Men, Y., Gnaser, H., Zapf, R., Kolb, G., Hessel, V., Ziegler, C., Parallel screening of $Cu/CeO_2/\gamma\text{-}Al_2O_3$ for steam reforming of methanol in a 10 channel micro-reactor, *Catal. Commun.* **2004**, submitted for publication.

35 Men, Y., Gnaser, H., Zapf, R., Kolb, G., Hessel, V., Ziegler, C., Parallel screening of $Cu/CeO_2/\gamma\text{-}AI_2O_3$ for steam reforming of methanol in a 10 channel micro reactor, *Appl. Catal. A* **2004**, submitted for publication.

36 Chen, G., Yuan, Q., Li, S., Pillot, C., Li, H., Hydrogen generation by methanol autothermal reforming in microchannel reactors, in *Proceedings of the AIChE Spring Meeting* (30 March – 3 April 2003), AIChE, New Orleans, **2003**.

37 Chen, G., Yuan, Q., Li, S., Microchannel reactor for methanol autothermal reforming, *Chin. J. Catal.* **2002**, *23*, 22–23.

38 Chen, G., Yuan, Q., Li, H., Li, S., CO selective oxidation in a microchannel reactor for PEM fuel cell, *Chem. Eng. J.* **2004**, *101*, 101–106.

39 Horny, C., Kiwi-Minsker, L., Renken, A., Micro structured string-reactor for autothermal production of hydrogen, *Chem. Eng. J.* **2004**, *101*, 3–9.

40 Maki, T., Ueyama, T., Mae, K., in *Proceedings of the 7th International Conference on Microreaction Technology, IMRET 7* (7–10 Sept. 2003), Lausanne, **2003**, 185, Book of abstracts.

41 Xu, J., Froment, G. F., *AIChE J.* **1989**, *35*, 88.

42 Find, J., Lercher, J. A., Cremers, C., Stimming, U., Kurtz, O., Crämer, K., Characterization of supported methane steam reforming catalyst for micro-reactor systems, in *Proceedings of the 6th International Conference on Microreaction Technology, IMRET 6* (11–14 March 2002), AIChE Pub. No. 164, New Orleans, **2002**, 99–104.

43 Mayer, J., Fichtner, M., Wolf, D., Schubert, K., A microstructured reactor for the catalytic partial oxidation of methane to syngas, in Ehrfeld, W. (Ed.), *Microreaction Technology: 3rd International Conference on Microreaction Technology, Proc. of IMRET 3*, Springer-Verlag, Berlin, **2000**, 187–196.

44 Fichtner, M., Mayer, J., Wolf, D., Schubert, K., Microstructured rhodium catalysts for the partial oxidation of methane to syngas under pressure, *Ind. Eng. Chem. Res.* **2001**, *40*, 3475–3483.

45 Pfeifer, P., Görke, O., Schubert, K., Washcoats and electrophoresis with coated and uncoated nanoparticles on microstructured metal foils and micro-structured reactors, in *Proceedings of the 6th International Conference on Microreaction Technology, IMRET 6* (11–14 March 2002), AIChE Pub. No. 164, New Orleans, **2002**, 281–285.

46 Tonkovich, A. L. Y., Zilka, J. L., Powell, M. R., Call, C. J., The catalytic partial oxidation of methane in a microchannel chemical reactor, in Ehrfeld, W., Rinard, I. H., Wegeng, R. S. (Eds.), *Process Miniaturization: 2nd International Conference on Microreaction Technology, IMRET 2, Topical Conf. Preprints*, AIChE, New Orleans, **1998**, 45–53.

47 Kikas, T., Bardenshteyn, I., Williamson, C., Ejimofor, C., Puri, P., Fedorov, A. G., Hydrogen production in the reverse-flow autothermal catalytic microreactor, *Ind. Eng. Chem. Res.* **2003**, *42*, 6273–6279.

48 Kurungot, S., Yamaguchi, T., Nakao, S.-I., Rh/γ-Al_2O_3 catalytic layer integrated with sol–gel synthesized microporous silica membrane for combact membrane reactor applications, *Catal. Lett.* **2003**, *86*, 273–278.

49 Oklany, J. S., Hou, K., Hughes, R., A simulative comparison of dense and microporous membrane reactors for the steam reforming of methane, *Appl. Catal. A* **1998**, *170* (*1*), 13–22.

50 de Vos, R. M., Verweij, H., Highselectivity, high-flux silica membranes, *Science* **1998**, *279*, 1710–1711.

51 Armor, J. N., Applications of catalytic inorganic membrane reactors to refinery products, *J. Membr. Sci.* **1998**, *147* (*2*), 217–233.

52 Kolb, G., Zapf, R., Hessel, V., Löwe, H., Propane steam reforming in micro channels, *Appl. Catal. A* **2004**, submitted for publication.

53 Guimaraes, A. L., Dieguez, L. C., Schmal, M., *Stud. Surf. Sci. Catal.* **2001**, *138*, 395.

54 Tornianien, P. M., Chu, X., Schmidt, L. D., Comparison of monolith-supported metals for the direct oxidation of methane to syngas, *J. Catal.* **1994**, *146*, 1–10.

55 Aartun, I., Gjervan, T., Venvik, H., Görke, O., Pfeifer, P., Fathi, M., Holmen, A., Schubert, K., Catalytic conversion of propane to hydrogenin microstructured reactors, *Chem. Eng. J.* **2004**, *101*, 93–99.

56 Heitnes Hofstad, K., Sperle, T., Rokstad, O. A., Holen, A., Partial oxidation of methane to syngas over a Pt/10% Rh gauze, *Catal. Lett.* **1997**, *45* (*1–2*), 1–2, 97–105.

57 Schwank, J., Tadd, A., Gould, B., Autothermal reforming of simulated gasoline in compact fuel processor, *Chem. Eng. Trans.* **2004**, *4*, 75–80.

58 TeGrotenhuis, W. E., Brooks, K. P., Davis, J. M., Fischer, C. M., King, D. L., Pederson, L. R., Stenkamp, V. S., Wegeng, R. S., Whyatt, G. A., in *Proceedings of the 7th International Conference on Microreaction Technology, IMRET 7* (7–10 Sept. 2003), Lausanne, **2003**, 19, Book of abstracts.

59 Veser, G., Experimental and theoretical investigation of H_2 oxidation in a high-temperature catalytic microreactor, *Chem. Eng. Sci.* **2001**, *56*, 1265–1273.

60 Veser, G., Friedrich, G., Freygang, M., Zengerle, R., A modular microreactor design for high-temperature catalytic oxidation reactions, in Ehrfeld, W. (Ed.), *Microreaction Technology: 3rd International Conference on Microreaction Technology, Proc. of IMRET 3*, Springer-Verlag, Berlin, **2000**, 674–686.

61 Görke, O., Pfeifer, P., Schubert, K., Determination of kinetic data in the isothermal microstructure reactor based on the example of catalyzed oxidation of hydrogen, in *Proceedings of the 6th International Conference on Microreaction Technology, IMRET 6* (11–14 March 2002), AIChE Pub. No. 164, New Orleans, **2002**, 262–274.

62 Hagendorf, U., Janicke, M., Schüth, F., Schubert, K., Fichtner, M., A Pt/Al_2O_3 coated microstructured reactor/heat exchanger for the controlled H_2/O_2-reaction in the explosion regime, in Ehrfeld, W., Rinard, I. H., Wegeng, R. S. (Eds.), *Process Miniaturization: 2nd International Conference on Microreaction Technology, IMRET 2, Topical Conf. Preprints*, AIChE, New Orleans, **1998**, 81–87.

63 Janicke, M. T., Kestenbaum, H., Hagendorf, U., Schuth, F., Fichtner, M., Schubert, K., The controlled oxidation of hydrogen from an explosive mixture of gases using a microstructured reactor/heat exchanger and Pt/Al_2O_3 catalyst, *J. Catal.* **2000**, *191*, 282–293.

64 Baldwin, T. R., Burch, R., *Appl. Catal.* **1990**, *66*, 337.

65 Briot, P., Primet, M., *Appl. Catal.* **1991**, *68*, 301.

66 Janicke, M. T., Holzwarth, A., Fichtner, M., Schubert, K., Schüth, F., *Stud. Surf. Sci. Catal.* **2000**, *130*, 437.

67 Haas-Santo, K., Görke, O., Schubert, K., Fiedler, J., Funke, H., A microstructure reactor system for the controlled oxidation of hydrogen for possible application in space, in Matlosz, M., Ehrfeld, W., Baselt, J. P. (Eds.), *Microreaction Technology – IMRET 5: Proc. of the 5th International Conference on Microreaction Technology*, Springer-Verlag, Berlin, **2001**, 313–320.

68 Reuse, P., Renken, A., Haas-Santo, K., Görke, O., Schubert, K., Hydrogen production for fuel cell application in an autothermal microchannel reactor, *Chem. Eng. J.* **2004**, *101*, 133–141.

69 Wang, X., Zhu, J., Bau, H., Gorte, R. J., Fabrication of micro-reactors using tape-casting methods, *Catal. Lett.* **2001**, *77*, 173–177.

70 Bau, H. H., Anatasuresh, S. G. K., Santiago-Aviles, J. J., Zhong, J., Kim, M., Yi, M., Esponosa-Vallejos, P., Sola-Laguna, L., *Proceedings of International Mechanical Engineering Conference and Exposition* **1998**, 491.

71 Arana, L. R., Schaevitz, S. B., Franz, A. J., Schmidt, M. A., Jensen, K. F., A microfabricated suspended-tube chemical reactor for thermally efficient fuel processing, *IEEE J. Microelectromech. Syst.* **2002**, *12*, 147–158.

72 Arana, L. R., Schaevitz, S. B., Franz, A. J., Schmidt, M. A., Jensen, K. F., A suspended-tube microreactor for thermally-efficient fuel processing, in *Proceedings of the 6th International Conference on Microreaction Technology, IMRET 6* (11–14 March 2002), AIChE Pub. No 164, New Orleans, **2002**, 147–158.

73 Tanaka, S., Chang, K.-S., Min, K.-B., Satoh, D., Yoshida, K., Esashi, M., MEMS-based components of a miniature fuel cell/fuel reformer system, *Chem. Eng. J.* **2004**, *101*, 143–149.

74 Raimondeau, S., Norton, D., Vlachos, D. G., Masel, R. I., Modeling of high-temperature microburners, *Proc. Combust. Inst.* **2002**, *29*, 901–907.

75 Norton, D. G., Vlachos, D. G., Combustion characteristics and flame stability at the microscale: a CFD study of premixed methane/air mixtures, *Chem. Eng. Sci.* **2003**, *58* (*21*), 4871–4882.

76 Mehra, A., Ayón, A. A., Microfabrication of high-temperature silicon devices using wafer bonding and deep reactive ion etching, *IEEE J. Microelectromech. Syst.* **1999**, *8*, 152–160.

77 Tanaka, S., Yamada, T., Sugimoto, S., Li, J.-F., Esashi, M., Silicon nitride ceramic-based two-dimensional micro-combustor, *J. Micromech. Microeng.* **2003**, *13*, 502–508.

78 Moe, J., *Chem. Eng. Process* **1962**, *58*, 33.

79 Petterson, L. J., Westholm, R., *Int. J. Hydrogen Energy* **2001**, *26*, 243.

80 TeGrotenhuis, W. E., King, D. L., Brooks, K. P., Holladay, B. J., Wegeng, R. S., Optimizing microchannel reactors by trading-off equilibrium and reaction kinetics through temperature management, in *Proceedings of the 6th International Conference on Microreaction Technology, IMRET 6* (11–14 March 2002), AIChE Pub. No. 164, New Orleans, **2002**, 18–28.

81 Tonkovich, A. L., Zilka, J. L., LaMont, M. J., Wang, Y., Wegeng, R., Microchannel chemical reactor for fuel processing applications. – I. Water gas shift reactor, *Chem. Eng. Sci.* **1999**, *54*, 2947–2951.

82 Germani, G., Quiney, A., van Veen, A., Schuurman, Y., Mirodatos, C., Kolb, G., The water gas shift reaction for automotive applications: preparation and testing of microstructured catalysts, in *Proceedings of the 7th International Conference on Microreaction Technology, IMRET 7* (7–10 Sept. 2003), Lausanne, **2003**, 173–175.

83 Görke, O., Pfeifer, P., Schubert, K., Water gas shift reaction and selective oxidation of CO in microreactors, *Appl. Catal. A* **2004**, *263*, 11–18.

84 Kolb, G., Zapf, R., Pennemann, H., Hessel, V., Löwe, H., Wash-coat catalysts applied for the water-gas shift reaction in micro channels, in *Proceedings of the AIChE Spring Meeting* (26–29 April 2004), New Orleans, **2004**, 2d, published on CD.

85 Schuessler, M., Portscher, M., Limbeck, U., Monolithic integrated fuel processor for the conversion of liquid methanol, *Catal. Today* **2003**, *79–80*, 511–520.

86 Besser, R. S., Ouyang, X., Chen, H., Shin, W. C., Bednarova, L., Lee, W., Pau, S., Pai, C. S., Taylor, J. A., Mansfield, W. M., Ho, P., Preferential oxidation in microchannel reactors for scalable fuel processing, in *Proceedings of the AIChE Spring Meeting* (26–29 April 2004), **2004**, 171a, published on CD.

87 Bednarova, L., Ouyang, X., Ho, P., Chen, H., Laval, A., Qian, D., Shin, W. C., Lee, W., Besser, R. S., Multi-channel microreactor for preferential oxidation of CO for 1 W fuel cell, in *Proceedings of the AIChE Spring Meeting* (26–29 April 2004), **2004**, 16d, published on CD.

88 Dudfield, C. D., Chen, R., Adcock, P. L., A carbon monoxide PROX reactor for PEM fuel cell automotive application, *Int. J. Hydrogen Energy* **2001**, *26*, 763–775.

89 Cominos, V., Hessel, V., Hofmann, C., Kolb, G., Zapf, R., Delsmann, E., de Croon, M. H. J. M., Schouten, J. C., Micro reactor for selective oxidation of CO in low power fuel cell applications, in *Proceedings of the 16th Intern. Congress of Chemical and Process Engineering* (22–26 Aug. 2004), Prague, **2004**, submitted for publication.

90 Delsman, E. R., de Croon, M. H. J. M., Kramer, G. J., Cobden, P. D., Hofmann, C., Cominos, V., Schouten, J. C., Experiments and modeling of an integrated preferential oxidation-heat exchanger microdevice, *Chem. Eng. J.* **2004**, *101*, 123–131.

91 Han, J., Lee, S.-M., Chang, H., Direct methanol fuel-cell combined with a small back-up battery, *J. Power Sources* **2002**, *112* (*2*), 484–496.

92 Barbieri, G., Bernardo, P., Mattia, R., Drioli, E., Bredesen, R., Klette, H., Pd-based membrane reactor for water-gas shift reaction, *Chem. Eng. Trans.* **2004**, *4*, 55–60.

93 Franz, A. J., Ajmera, S. K., Firebaugh, S. L., Jensen, K. F., Schmidt, M. A., Expansion of microreactor capabilities trough improved thermal management and catalyst deposition, in Ehrfeld, W. (Ed.), *Microreaction Technology: 3rd International Conference on Microreaction Technology, Proc. of IMRET 3*, Springer-Verlag, Berlin, **2000**, 197–206.

94 Kusakabe, K., Miyagawa, D., Gu, Y., Maeda, H., Morooka, S., Development of a self-heating catalytic microreactor, in Matlosz, M., Ehrfeld, W., Baselt, J. P. (Eds.), *Microreaction Technology – IMRET 5: Proc. of the 5th International Conference on Microreaction Technology*, Springer-Verlag, Berlin, **2001**, 70–77.

95 Lee, S. H., Maeda, R., Mizukami, F., Hanaoka, T., in *Proceedings of the 7th International Conference on Microreaction Technology, IMRET 7* (7–10 Sept. 2003), Lausanne, **2003**, 141, Book of abstracts.

96 Mukherjee, S., Hatalis, M. K., Kothare, M. V., in *Proceedings of the 7th International Conference on Microreaction Technology, IMRET 7* (7–10 Sept. 2003), Lausanne, **2003**, 176, Book of abstracts.

97 Buxbaum, R. E., Kinney, A. B., *Ind. Eng. Chem. Res.* **1996**, *35*, 530.

98 Tong, H. D., Gielens, F. C., Gardeniers, J. G. E., Hansen, H. V., Nijdam, W., van Rijn, C. J. M., Elwenspoek, M. C., in *Proceedings of the 7th International Conference on Microreaction Technology, IMRET 7* (7–10 Sept. 2003), Lausanne, **2003**, 343, Book of abstracts.

99 Gielens, F. C., Tong, H. D., Vorstman, M. A. G., Keurentjes, J. T. F., in *Proceedings of the 7th International Conference on Microreaction Technology, IMRET 7* (7–10 Sept. 2003), Lausanne, **2003**, 170, Book of abstracts.

100 Reay, A. D., *Heat Recovery Syst. CHP* **1993**, *13*, 383.

101 Polman, E. A., Der Kinderen, J. M., Thuis, F. M. A., Novel compact steam reformer for fuel cells with heat generation by catalytic combustion augmented by induction heating, *Catal. Today* **1999**, *47*, 347–351.

102 Frauhammer, J., Friedrich, G., Kolios, G., Klingel, T., Eigenberger, G., von Hippel, L., Arntz, D., Flow distribution concepts for new type monolithic co- or countercurrent reactors, *Chem. Eng. Technol.* **1999**, *22*, 1012–1016.

103 Minjolle, L., *US 4 271 110.*

104 von Hippel, L., Arntz, D., Frauhammer, J., Eigenberger, G., Friedrich, G., *DE 19 653 989.*

105 Frauhammer, J., Eigenberger, G., von Hippel, L., Arntz, D., A new reactor concept for endothermic high-temperature reactions, *Chem. Eng. Sci.* **1999**, *54 (16)*, 3661–3670.

106 Kolios, G., Frauhammer, J., Eigenberger, G., Efficient reactor concepts for coupling of endothermic and exothermic reactions, *Chem. Eng. Sci.* **2002**, *57*, 1505–1510.

107 Gritsch, A., Kolios, G., Eigenberger, G., Reaktorkonzepte zur autothermen Führung endothermer Hochtemperaturreaktionen, *Chem. Ing. Tech.* **2004**, *76*, 722–725.

108 Zanfir, M., Gavriilidis, A., Catalytic combustion assisted methane steam reforming in a catalytic plate reactor, *Chem. Eng. Sci.* **2003**, *58*, 3947–3960.

109 *http://www.motorola.com/mediacenter/news/detail/0, 3099_2539_23,00.html*, Motorola, Engelhard, University of Michigan win ATP award to develop miniature fuel-cell power source, Press Release (5 Aug. 2003), 29 October **2003**.

110 *www.minatec.com/minatec2003/* (International Meeting on Micro and Nanotechnologies, Grenoble, France, 22–26 September), 29 October **2003**.

111 Koripella, R. C., Dyer, C. K., Gervasio, F. D., Rogers, S. P., Wilconx, D., Ooms, W. J., Fuel processor with integrated fuel cell utilizing ceramic technology, *US 6 569 553*, **2003**.

112 Schüssler, M., Lamla, O., Stefanovski, T., zur Megede, D., Vorstellung eines kaltstartfähigen Reaktors zur autothermen Reformierung von Methanol, *Chem. Ing. Tech.* **2000**, *72*, 982.

113 Schouten, J. C., Rebrov, E. V., de Croon, M. H. J. M., Miniaturization of heterogeneous catalytic reactors: prospects for new developments in catalysis and process engineering, *Chimia* **2002**, *56*, 627–635.

114 Delsman, E., Rebrov, J., de Croon, M. H. J. M., Schouten, J., Kramer, G. J., Cominos, V., Richter, T., Veenstra, T. T., van den Berg, A., Cobden, P., de Bruijn, F. A., Ferret, C., d'Ortona, U., Falk, L., MiRTH-e: micro reactor technology for hydrogen and electricity, in Matlosz, M., Ehrfeld, W., Baselt, J. P. (Eds.), *Microreaction Technology – IMRET 5: Proc. of the 5th International Conference on Microreaction Technology*, Springer-Verlag, Berlin, **2001**, 368–375.

115 Palo, D. R., Holladay, J. D., Rozmiarek, R. T., Guzman-leong, C. E., Wang, Y., Hu, J., Chin, Y.-H., Dagle, R. A., Baker, E. G., Development of a soldier-portable fuel cell power system. Part I: a bread-board methanol fuel processor, *J. Power Sources* **2002**, *108*, 28–34.

116 Jones, E., Holladay, J., Perry, S., Orth, R., Rozmiarel, B., Hu, J., Phelps, M., Guzman, C., Sub-watt power using an integrated fuel processor and fuel cell, in Matlosz, M., Ehrfeld, W., Baselt, J. P.

(Eds.), *Microreaction Technology – IMRET 5: Proc. of the 5th International Conference on Microreaction Technology*, Springer-Verlag, Berlin, **2001**, 277–285.

117 Holladay, J. D., Wainright, J. S., Jones, E. O., Gano, S. R., Power generation using a mesoscale fuel cell integrated with a microscale fuel processor, *J. Power Sources* **2004**, *130*, 111–118.

118 Holladay, J., Jones, E., Palo, D. R., Phelps, M., Chin, Y.-H., Dagle, R., Hu, J., Wang, Y., Baker, E., Miniature fuel processors for portable fuel cell power supplies, *Mater. Res. Soc. Symp. Proc.* **2003**, *756*, FF9.2.1–FF9.2.6.

119 Renken, A., Microstructured reactors: novel approaches for chemical reaction engineering, in *Proceedings of the 27th International Exhibition-Congress on Chemical Engineering, Environmental Protection and Biotechnology, ACHEMA* (19–24 May 2003), DECHEMA, Frankfurt, **2003**, 13.

120 Hermann, I., Lindner, M., Winkelmann, H., Düsterwald, H. G., Microreaction technology in fuel processing for fuel cell vehicles, in *Proceedings of the VDE World Microtechnologies Congress, MICRO.tec 2000* (25–27 Sept. 2000), VDE Verlag, Berlin, EXPO Hannover, **2000**, 447–453.

121 Stimming, U., Cremers, C., Find, J., Lercher, J. A., Reuse, P., Renken, A., Kurtz, O., Crämer, K., Haas-Santo, K., Görke, O., Schubert, K., Wurtziger, O., Reinhardt, G., Schütz, W., in *ACHEMA Congress*, Frankfurt/Main, **2003**.

122 Cremers, C., Dehlsen, J., Stimming, U., Reuse, P., Renken, A., Haas-Santo, K., Görke, O., Schubert, Micro structured-reactor-system for the steam reforming of methanol, in *Proceedings of the 7th International Conference on Microreaction Technology, IMRET 7* (7–10 Sept. 2003), Lausanne, **2003**, 56.

123 Tonkovich, A. L. Y., Jimenez, D. M., Zilka, J. L., LaMont, M. J., Wang, J., Wegeng, R. S., Microchannel chemical reactors for fuel processing, in Ehrfeld, W., Rinard, I. H., Wegeng, R. S. (Eds.), *Process Miniaturization: 2nd International Conference on Microreaction Technology, IMRET 2, Topical Conf. Preprints*, AIChE, New Orleans, **1998**, 186–195.

124 Park, G.-G., Seo, D. J., Park, S.-H., Yoon, Y.-G., Kim, C.-S., Yoon, W.-L., Development of microchannel steam reformer, *Chem. Eng. J.* **2004**, *101*, 87–92.

125 Cremers, C., Stummer, M., Stimming, U., Find, J., Lercher, J. A., Kurtz, O., Crämer, K., Haas-Santo, K., Görke, O., Schubert, K., Micro structured reactors for coupled steam-reforming and catalytic combustion of methane, in *Proceedings of the 7th International Conference on Microreaction Technology, IMRET 7* (7–10 Sept. 2003), Lausanne, **2003**, 100.

126 Pfeifer, P., Bohn, L., Görke, O., Haas-Santo, K., Schubert, K., Mikrostrukturkomponenten für die Wasserstofferzeugung aus unterschiedlichen Kohlenwasserstoffen, *Chem. Ing. Tech.* **2004**, *76*, 618–620.

127 Mazanec, T. J., Perry, S. T., Tonkovich, A. Y., Wang, Y., Micro channel gas-to-liquids conversion: shrinking the problem, in *Proceedings of the Cat Con* (6 May **2003**).

128 Johnston, A. M., Haynes, B. S., Heatric steam reforming technology, in *Proceedings of the AIChE Spring National Meeting – 2nd Topical Conference on Natural Gas Utilization* (10–12 March 2001), AIChE, New Orleans, **2001**, 112.

129 Drost, M. K., Call, C., Cuta, J., Wegeng, R. S., Microchannel combustor/evaporator thermal process, *Microscale Therm. Eng.* **1997**, *1*, 321–332.

130 Whyatt, G. A., TeGrotenhuis, W. E., Wegeng, R. S., Pederson, L. R., Demonstration of energy efficient steam reforming in microchannels for automotive fuel processing, in Matlosz, M., Ehrfeld, W., Baselt, J. P. (Eds.), *Microreaction Technology – IMRET 5: Proc. of the 5th International Conference on Microreaction Technology*, Springer-Verlag, Berlin, **2001**, 303–312.

131 Whyatt, G. A., Fischer, C. M., Davis, J. M., Progress on the development of a microchannel steam reformer for automotive applications, in *Proceedings of the 6th International Conference on Microreaction Technology, IMRET 6* (11–14 March 2002), AIChE Pub. No. 164, New Orleans, **2002**, 85–98.

132 Hu, J., Wang, W., VanderWiel, D., Chin, C., Palo, D., Rozmiarek, R.,

DAGLE, R., CAO, J., HOLLADAY, J., BAKER, E., Fuel processing for portable power applications, *Chem. Eng. J.* **2003**, *93*, 55–60.

133 IRVING, P. M., LLOYD ALLEN, W., HEALEY, T., THOMSON, W. J., Catalytic micro reactor systems for hydrogen generation, in MATLOSZ, M., EHRFELD, W., BASELT, J. P. (Eds.), *Microreaction Technology – IMRET 5: Proc. of the 5th International Conference on Microreaction Technology*, Springer-Verlag, Berlin, **2001**, 286–294.

134 IRVING, P. M., MOELLER, T. M., MING, Q., Fuel processing using microstructured devices for hydrogen production, in *Proceedings of the 7th International Conference on Microreaction Technology, IMRET 7* (7–10 Sept. 2003), Lausanne, **2003**, 188.

135 FITZGERALD, S. P., WEGENG, R. S., TONKOVICH, A. L. Y., WANG, Y., FREEMAN, H. D., MARCO, J. L., ROBERTS, G. L., VANDERWIEL, D. P., A compact steam reforming reactor for use in an automotive fuel processor, in *Proceedings of the 4th International Conference on Microreaction Technology, IMRET 4* (5–9 March 2000), AIChE Topical Conf. Proc., Atlanta, GA, **2000**, 358–363.

136 GROPPI, G., TRONCONI, E., Design of novel monolith catalyst supports for gas/solid reactions with heat exchange, *Chem. Eng. Sci.* **2000**, *55*, 2161–2171.

137 HICKMANN, D. A., SCHMIDT, L. D., *J. Catal.* **1992**, *136*, 300.

138 BIER, W., KELLER, W., LINDER, G., SEIDEL, D., SCHUBERT, K., Manufacturing and testing of compact micro heat exchangers with high volumetric heat transfer coefficients, ASME, DSC-Microstructures, *Sensors, and Actuators* **1990**, *19*, 189–197.

139 SCHUBERT, K., BIER, W., LINDER, G., SEIDEL, D., Profiled microdiamonds for producing microstructures, *Ind. Diamond Rev.* **1990**, *50*, 235–239.

140 BRANDNER, J., FICHTNER, M., SCHUBERT, K., LIAUW, M., EMIG, G., A new microstructured device for fast temperature cycling for chemical reactions, in MATLOSZ, M., EHRFELD, W., BASELT, J. P. (Eds.), *Microreaction Technology – IMRET 5: Proc. of the 5th International Conference on Microreaction Technology*, Springer-Verlag, Berlin, **2001**, 164–174.

141 TAKEDA, T., KUNITOMI, K., HORIE, T., IWATA, K., Feasibility study on the applicability of a diffusion-welded compact intermediate heat exchanger to next-generation high temperature gas-cooled reactor, *Nucl. Eng. Des.* **1997**, *168*, 11–21.

142 BIER, W., KELLER, W., LINDER, G., SEIDEL, D., SCHUBERT, K., MARTIN, H., Gas-to-gas heat transfer in micro heat exchangers, *Chem. Eng. Process.* **1993**, *32*, 33–43.

143 SCHUBERT, K., BRANDNER, J., FICHTNER, M., LINDER, G., SCHYGULLA, U., WENKA, A., Microstructure devices for applications in thermal and chemical process engineering, *Microscale Therm. Eng.* **2001**, *5*, 17–39.

144 *www.fzk.de*, Forschungszentrum Karlsruhe, 17 July **2004**.

145 ZAPF, R., BECKER-WILLINGER, C., BERRESHEIM, K., HOLZ, H., GNASER, H., HESSEL, V., KOLB, G., LÖB, P., PANNWITT, A.-K., ZIOGAS, A., Detailed characterization of various porous alumina based catalyst coatings within microchannels and their testing for methanol steam reforming, *Trans IChemE* **2003**, *A81*, 721–729.

146 ZAPF, R., HESSEL, V., Nanoporöse Katalysatorschichten für Anwendungen im Mikroreaktor, *Chem. Ing. Tech.* **2004**, *76*, 513–514.

147 ZAPF, R., HESSEL, V., KOLB, G., ZIOGAS, A., BERRESHEIM, K., GNASER, H., Porous alumina-supported catalyst coatings in steel microchannels, in *Proceedings of the 7th International Conference on Microreaction Technology, IMRET 7* (7–10 Sept. 2003), Lausanne, **2003**, 276–278.

148 ADOMAITIS, J. R., GALLIGAN, M. P., KUBSH, J. E., WHITTENBERGER, W. A., *SAE Paper 962080* **1996**, 185.

149 LYLYKANAS, R., LAPPI, P., *SAE Paper 910614*, **1991**.

150 MÜLLER, A., unpublished results, IMM Institut für Mikrotechnik Mainz, Mainz.

151 HAAS-SANTO, K., FICHTNER, M., SCHUBERT, K., Preparation of microstructure compatible porous supports by sol–gel synthesis for catalyst coatings, *Appl. Catal. A* **2001**, *220*, 79–92.

152 CHEN, H., OUYANG, X., SHIN, W. C., PARK, S. M., BEDNAROVA, L., BESSER, R. S., LEE, W. Y., Sol–gel Pt/Al_2O_3 catalyst films for microchannel reactors, in *Proceedings of the 7th International*

Conference on Microreaction Technology, IMRET 7 (7–10 Sept. 2003), Lausanne, **2003**, 132.

153 Diggle, J., Downie, T., Goulding, C., *Chem. Rev.* **1969**, *69*, 365.

154 Wunsch, R., Fichtner, M., Schubert, K., Anodic oxidation inside completely manufactured microchannel reactors made of aluminum, in *Proceedings of the 4th International Conference on Microreaction Technology, IMRET 4* (5–9 March 2000), AIChE Topical Conf. Proc., Atlanta, GA, **2000**, 481–487.

155 Gorges, R., Kässbohrer, J., Kreisel, G., Meyer, S., Surface-functionalization of microstructures by anodic spark deposition, in *Proceedings of the 6th International Conference on Microreaction Technology, IMRET 6* (11–14 March 2002), AIChE Pub. No. 164, New Orleans, **2002**, 186–191.

156 Födisch, R., Kursawe, A., Hönicke, D., Immobilizing heterogeneous catalysts in microchannel reactors, in *Proceedings ofthe 6th International Conference on Microreaction Technology, IMRET 6* (11–14 March 2002), AIChE Pub. No. 164, New Orleans, **2002**, 140–146.

157 Määttanen, M., Lylykangas, R., Mechanical strength of a metallic catalytic converter made of precoated foil, *SAE Paper 900505* **1990**, 71–79.

158 Rebrov, E. V., Seijger, G. B. F., Calis, H. P. A., de Croon, M. H. J. M., van den Bleek, C. M., Schouten, J. C., The preparation of highly ordered single layer ZSM-5 coating on prefabricated stainless steel microchannels, *Appl. Catal. A* **2001**, *206*, 125–143.

159 Rebrov, E. V., Seijger, G. B. G., Calis, H. P. A., de Croon, M. H. J. M., van den Bleek, C. M., Schouten, J. C., Synthesis and characterization of ZSM-5 zeolites on prefabricated stainless steel microchannels, in *Proceedings of the 4th International Conference on Microreaction Technology, IMRET 4* (5–9 March 2000), AIChE Topical Conf. Proc., Atlanta, GA, **2000**, 250–255.

160 Louis, B., Kiwi-Minsker, L., Renken, A., Synthesis of ZSM-5 coatings on stainless steel grids and their catalytic performance for partial oxidation of benzene by N_2O, *Appl. Catal. A* **2001**, *210*, 103–109.

3
Catalyst Screening

3.1
Introduction

3.1.1
Catalyst Screening During the Last Decade

Over the last decade, screening seemed to have gone through a climax if one considers the number of published articles and research activities and has now settled at a stable level with established technologies. Patent production reached a maximum in 1996–1997. Since 2002, also the number of publications decreased [1]. These facts could be due to a normal development process of a new booming technology, but they could also be an indication that new input from other technologies is necessary. Some indications indicate that screening could profit from the principles of micro technology and from new computational evaluation methods (see Section 3.6, *Computational Evaluation Methods*). On the other hand, parallel miniaturized processes will certainly increase and supplement the micro scene by new methods and approaches.

The first uses of microtechnology for screening applications were presented recently. For instance, Watts and Haswell [2] presented first work on microfluidic combinatorial organic chemistry. Most of the examples described apply to glass, polymer or silicon reactors, which restricts their usage to low-pressure operation similar to pharmaceutical applications. They concluded that micro reactors could be a tool for rapid reaction development and process optimization.

In addition to showing the few examples of screening in microstructured reactors, it is one aim of this chapter to point out future applications in this field, since many such advantages are obvious today, albeit were not realized so far. In this context, this chapter will place a strong focus on the challenges of performing screening in the laminar-flow regime. As this normally implies the use of small dimensions, micro reactors or better reactors equipped with micro structures are presented. A survey of this kind certainly cannot be complete and comprehensive (as this is the aim of other chapters in this book series) and comprises to some extent the author's personal view because some detailed information about processes was available only for the author's work and not for other works. Nevertheless, applications at the periphery of micro structures such as ceramic monoliths and small fixed-bed reactors are also described here.

Chemical Micro Process Engineering. V. Hessel, H. Löwe, A. Müller, G. Kolb.

ISBN: 3-527-30998-5

The next section describes the state of the art of screening technology by briefly referring to its major parts of the workflow. The unsolved issues are derived from there. A further section relates the features of processing in microstructured reactors to these future issues of screening. It will be shown how chemical micro process engineering may contribute to future screening. Later in this chapter, more detailed information on each of these subjects will be given.

3.1.2
Current Situation and Future Challenges for

In the following, the major aspects of current screening in the sequence of its workflow will be discussed and unsolved issues are pointed out.

3.1.2.1 Library Size and Design

It is now widely accepted that no longer just the size of a catalyst library guarantees success [3]. Large numbers of solid compounds can easily be produced by split/mix strategies, but the quality of these arbitrarily selected catalyst libraries remains low. Tools for library design and library generation are becoming essential for efficient sample generation [1].

3.1.2.2 Sample Handling and Characterization

Solid-phase chemistry, a major topic in combinatorial screening, is influenced by problems such as sample handling, method development and characterization, which still remain unsolved and hinder its further widespread usage. Miniaturization and automation of solid handling devices became a rather complicated task, much more challenging than the liquid dispensing by laboratory robots used in homogeneous catalysis. Research activities now concentrate on apparent simple tasks such as shaping, drying and calcination of catalysts [4].

The amount of catalyst mass increases from a few micro- or milligrams used in primary screening up to several grams typically used in the fixed-bed reactor tubes used for secondary screening. Consequently, preparation steps such as grinding, pressing and sieving have to be automated to assure reproducible catalyst particles capable of producing the desired quantities [5]. Manual procedures executed in this step reduce the quality and fail if a large number of particles or coats have to be produced. Especially the transfer of thin films to bulk catalysts constitutes one of the main key limitations during scale-up.

3.1.2.3 Automated Measurement and Analysis

Flego [1] recommends the use of micro devices for automated measurement and microanalysis of high-throughput in situ characterization of catalyst properties. Murphy et al. [5] stress the importance of the development of new reactor designs. Micro reactors at Dow were described for rapid serial screening of polyolefin catalysts. De Bellefon ete al. used a similar approach in combination with a micro mixer [6]. Bergh et al. [7] presented a micro fluidic 256-fold flow reactor manufactured from a silicon wafer for the ethane partial oxidation and propane ammoxidation.

3.1.2.4 Data Handling

In the early days of catalyst screening, speed was the only important matter. This meant collecting as much information as possible on a certain catalyst under defined process parameters. This approach produces a large number of non-interrelated single data points with a low degree of information. As soon as correlations between these data can be found, the 'information density' increases. This is the case if reaction kinetics are derived from single data points or if a supervised artificial neural network has learned to predict relations between data points.

Efficient testing requires efficient data handling and test protocols, which permit the automated change of reaction conditions. Nevertheless, the number of leads transferred to production scale remains still fairly low, with some exceptions, such as the development of the new family of propylene–ethylene copolymers recently announced by Dow and Symyx [8].

3.1.2.5 In Situ Surface Science Studies to Provide Micro Kinetics

The four sections above followed the typical workflow of screening approaches of the early days. In recent years, it became evident that additional steps have to be added on top of the workflow. One such step is the analysis of the true kinetics and the interplay of its elemental reactions by modern surface-science techniques.

Combinatorial screening at first glance might appear superficial as a number of identical experiments are executed under the same conditions in a parallel fashion. This is certainly too simple an interpretation. As soon as fluid dynamic information is added to the pure experimental data, more complex aspects of catalysis are derivable from overall conversion data. This is a different approach to established screening methods, which normally concentrate purely on the experimental collection of data about the conversion or the selectivity of a certain catalyst.

Such a very basic approach utilizes the in situ characterization of a catalyst during the reaction to design more efficient versions of a catalyst. These in situ methods give a basic understanding of surface phenomena between single molecules. A surface reaction always follows adsorption. Adsorption is described by advanced simulation techniques such as the *Monte Carlo method* or the *methods of molecular dynamics*, which calculate the so-called configurational diffusion of molecules adsorbed to surfaces [9]. These statistical methods increase the knowledge of surface interactions but they are not widely used in catalyst development. This is partly because these surface studies on monocrystalline surfaces are executed under low vacuum conditions and are not easily transferable to realistic reaction conditions. Another approach is the so-called micro kinetics of heterogeneous catalysis [10]. The authors divided the reaction into its elementary steps adsorption, reaction and desorption and studied the selectivity of the intermediate products using parameters estimated from theories of chemical bonding. Despite these promising approaches, empirical strategies are still the most commonly applied tool in industrial catalyst development. This is certainly due to the rough industrial environment which also requires robust equipment for screening. Such a more empirical approach was enabled by the reactors of Adler et al. [11] (Figure 3.1). They reported the application of a unit construction system based upon modularized single-channel reactors

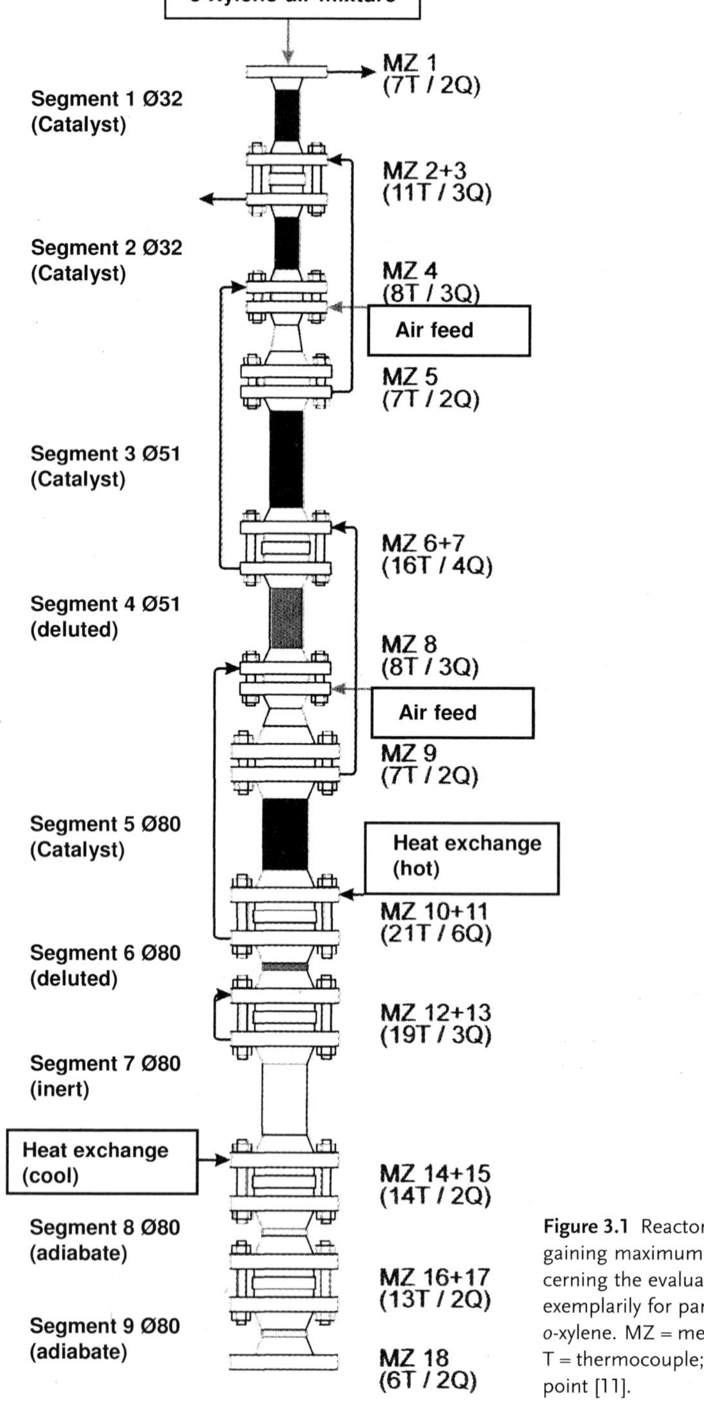

Figure 3.1 Reactor combination for gaining maximum information concerning the evaluation of parameters exemplarily for partial oxidation of *o*-xylene. MZ = measuring cell; T = thermocouple; Q = sampling point [11].

applicable to integral or differential reactor operations. The dimensions were close to those of conventional tube reactors.

3.1.2.6 Multidisciplinary Knowledge Beyond Chemistry and Chemical Engineering Needed for Future Catalyst Screening

The addition of further steps to the workflow of catalyst screening demands novel skills not typical of past catalyst development.

In chemical process engineering, chemists usually work closely together with process engineers. Apparatus design will seldom be executed specially for certain chemistries. More often standard vessels or heat exchangers will be assembled at a plant. This approach is difficult to apply in combinatorial screening because the great complexity of parallel reactors will often exceed the complexity of standard chemical devices. Engineers from other disciplines such as specialists in computational fluid dynamics and software programming and mechanical engineers with a profound knowledge of 3-D CAD tools will implement the scientific scene. In such a way, combinatorial acquires an additional background meaning here the combination of technical disciplines. This interdisciplinary approach is an issue which in screening certainly will be of increasing importance. Consequently, some interdisciplinary design approaches, which were available to the authors, will be described in detail in this chapter.

3.1.3 Features of Chemical Micro Process Engineering to Impact on Catalyst Screening

The trend of decreasing reactor scale in high-throughput technologies seems to proceed from tube reactors towards micro structured reactors. Thus, a thorough analysis of their possible contribution to the field of catalyst screening is needed. Such know-how, in turn, will open the design of new specially micro structured reactors for certain types of screening workflow.

3.1.3.1 Flow Conditions in Small-sized Reactors

Micro structured reactors will not only increase reactor compactness and reduce the size of sometimes expensive samples but will also allow a thorough fluidic description of the flow in the reactor. The mechanism, which assists here, is the laminar-flow regime, which develops owing to the small reactor dimensions.

Ideal plug-flow conditions can also be established in so-called nano-flow reactors with catalyst particle sizes from 50 to 200 μm. These reactors were operated in 16- and 64-barrel mode at Avantium for the regression of intrinsic kinetics [4].

3.1.3.2 Analytical Expressions of Laminar Flow for Consolidation of Screening Experiments

Laminar flow reactors are equipped with micro structured reaction chambers which account for the desired low Reynolds numbers owing to their small dimension. Mass transport perpendicular to the laminar channel flow is dominated by diffusion, a phenomenon known as dispersion. Without the influence of diffusion, laminar flow reactors could hardly be used in heterogeneous catalysis. There would be no mass transport from the bulk flow to the coated walls, as laminar flow, in contrast to turbulent flow, cannot mix the flow macroscopically.

To consolidate the experimental screening data in a quantitative fashion, it is desirable to obtain information on the fluid mechanics of the reactant flow in the reactor. Experimental data are difficult to evaluate if the experimental conditions and especially the fluid dynamic behavior of the reactants flow are not known. This is the case, for example, in a typical tubular reactor filled with a packed bed of porous beads. The porosity of the beads in combination with the unknown flow of the reactants around the beads makes it difficult to describe the flow close to the catalyst surface. A way to achieve a well-described flow in the reactor is to reduce its dimensions. This reduces the Reynolds number to a region of laminar-flow conditions, which can be described analytically.

3.1.3.3 Impact of Laminar-flow Descriptions on Computational Evaluation Methods

The use of laminar flow has some direct impact on the computational evaluation as, e.g., first-principle order simulation approaches may be used [12].

It is assumed that in the near future catalyst screening will make more intensive use of evaluation methods, which are introduced in Section 3.6. Future aspects such as data management problems will intensify these endeavors by stressing more the importance of evaluation instead of further intensifying the speed of experiments. Genetic algorithms and neural nets will extract information which is buried in the tangled mass of data. By reducing the reactor dimensions, the fluid dynamics in the reactor will also be better understood, and the exact definition of the flow regime will open the way to analytic reactor descriptions and model discrimination.

3.1.3.4 Heat Transport and Thermal Overshooting

The reaction is also influenced by the heat of reaction that develops during the conversion of the reactants, a problem in tubular screening reactors. In micro structures, the heat transport through the walls of the channels is facilitated owing to their small dimensions. The catalysts are deposited on the walls of these micro structures and will thus have the appropriate environment for exothermic reactions by enabling fast quenching of the reaction with near isothermal conditions. Hence also the heat and mass balance in the reactor will be decoupled, which permits the analytical description of the flow in the screening reactor.

3.1.3.5 Exploration of Novel Reaction Regimes by Micro-space Operation

Micro-space operation as an enabling technology allows one to perform operations which are not common or even not feasible when using conventional equipment. This refers first of all to safe processing in the explosive regime or under otherwise hazardous conditions. By choosing the width of micro structures below the range of the explosion quench diameter, the reaction can also be executed in the explosion regime. In this way, processing with undiluted gases, at exceptionally high temperatures or pressures, is enabled. As a second aspect, processing at unusually short and defined residence times can be applied. The residence time is adjustable from a few milliseconds to seconds. Thus, processing similar to short-time processing reactors such as monoliths and catalyst gauzes is achievable. Owing to the uniquely defined flow path, the residence time distribution may even be improved, at least under some preferred circumstances.

3.1.3.6 Up-scaling

Up-scaling from very small dimensions to industrial tubular reactors still remains an unsolved problem. A solution to this problem could be the increase in throughput by a combination of a large number of identical channels with known fluid dynamics to a multi-channel reactor. This approach is known as the up-numbering strategy and is already applied in industrial practice. This new approach tries to avoid up-scaling completely by using large area micro structures with high volume flow. These structures are already in use in heat exchangers in heavy industrial environments [13], but for its application in reaction technology, new techniques such as sputter-coating and automated washcoating for large-area micro structures are demanded. First attempts in this direction are presented in Section 3.2 and in more detail in Section 4.12.4, *Online Reactor Manufacturing*.

Reactor up-scaling during the screening process, for example from primary leads to secondary process parameter screening, is facilitated if a flexible combination of meso- and micro-scale reactors could be used in the same experimental set-up.

3.1.4
Structure of the Contents of the Chapter

According to the analysis given below, the technologies used for synthesis, testing and evaluation of catalysts are described in the sequence of their appearance in the process flow. Emphasis is placed on machine construction and process design.

Then, a survey of micro reactors for heterogeneous catalyst screening introduces the technological methods used for screening. The description of microstructured reactors will be supplemented by other, conventional small-scale equipment such as mini-batch and fixed-bed reactors and small monoliths. For each of these reactors, exemplary applications will be given in order to demonstrate the properties of small-scale operation. Among a number of examples, methane oxidation as a sample reaction will be considered in detail. In a detailed case study, some intrinsic theoretical aspects of micro devices are discussed with respect to reactor design and experimental evaluation under the transient mode of reactor operation. It will be shown that, as soon as fluid dynamic information is added to the pure experimental data, more complex aspects of catalysis are derivable from overall conversion data, such as the intrinsic reaction kinetics.

3.2
Catalyst Preparation Methodology

Microreaction technology as a new branch in chemical processing became an efficient tool with significant advantages over conventional reactors [14–20]. One important area is heterogeneously catalyzed gas-phase reactions, which are very common in reaction engineering on an industrial scale [21–30]. Catalyzed processes often provide important compounds used in a large variety of different processes and products. In addition, the catalysts and their carriers themselves have become the focus of scientific investigations – on their preparation, morphology, porosity, composition, etc. This new technology offers new routes in chemistry by utilizing laminar flow reactors [12, 31–34]. New ways of reactor operation became available such as the possibility of executing a reaction in the explosive regime as shown by the synthesis of ethylene oxide [22] or the H_2/O_2 reaction [35] and also the facility of reaching higher product yields [36].

Coating on micro structures is still dominated by manual coating techniques. For future automated production of catalysts and even complete reactors, automated coating procedures will become crucial. Some new approaches in coating technology especially suited for micro structures are described in Section 4.12.4, *Online Reactor Manufacturing*.

3.2.1
Catalyst Deposition

3.2.1.1 Manual Impregnation Procedure

In order to improve the filling of the pores of the alumina layer (for manufacturing details, see Section 2.10) with catalytically active substances, the samples were evacuated in an desiccator and conditioned with carbon dioxide, which is water soluble in contrast to air. If this procedure was successful, then the so pre-conditioned γ-alumina layers were dip-coated with a solution of salts of transition metals. Another possibility is to impregnate the catalyst carrier by applying solutions of catalytically active substance subsequently drop by drop with a dispenser.

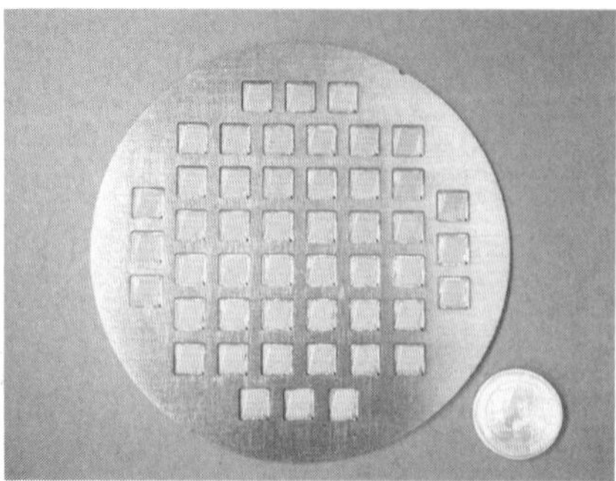
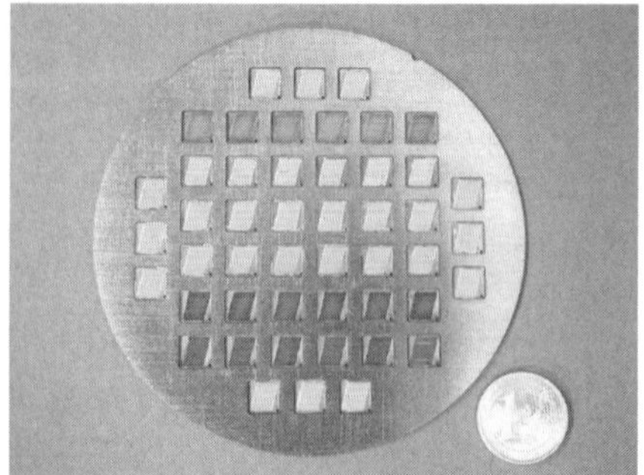

Figure 3.2 A 48-fold titer-plate coated with ã-alumina after calcination (left) and after impregnation (right) [38] (by courtesy of VDI-Verlag GmbH).

If necessary, this procedure can be repeated either to increase the amount of catalyst or to apply different substances which cannot be mixed in the same solution. After the impregnation of the alumina layer with metal salts, the samples were dried and calcined again. A 48-fold titer-plate manually coated with γ-alumina after calcination and after impregnation is shown in Figure 3.2.

3.2.1.2 Semi-automated Impregnation Method

Continuous flow reactors allow a new way of catalyst preparation. The fluidic pathways can be used for the transport of impregnation liquid or solid particles if means are supplied to localize the catalysts at defined positions inside the reactor. One possibility is the flow impregnation of wash-coated micro structures with catalyst solutions [38].

The set-up for this semi-automated impregnation method consisted of a number of plates which were supplied for the feed of the impregnation liquid, the sealing of the wells and the removal of the excess liquid. The contact time for the impregnation was adjusted by closing the exit channel for a defined time. The flow manifold was designed to fit into a typical laboratory sample robot and was operated in a semi-continuous mode. When the titer-plate was inserted into the manifold, the cover was closed and thus the titer-plate was sealed. The exits of the wells were connected to a continuously operated water jet suction pump via capillary tubes. The flow was adjusted by a manual valve. The cover of the manifold was equipped with feeding funnels which were filled with the individual impregnation liquids. In order to fill the wells with the catalyst solutions, the valve was opened for less than 1 s and then closed again for the desired duration of the impregnation. Excess impregnation liquid was removed by opening the valve again.

The time for the preparation of the 48 single mixtures could thus be reduced from several days to about 24 h, of which 20 h can be fully automated. Hence the time for the sample preparation was in the same range as the estimated time for the forthcoming sample testing and analysis with the parallel reactor (see Table 3.1).

Table 3.1 Procedure for semi-automated sample preparation on a titer-plate.

Parameter	*Process step*				
	Pretreatment of titer-plate	*Wash-coating*	*Calcination*	*Flow impregnation*	*Calcination*
Time expended (h)	0.5	3	10	Variable*	10
Temperature (K)	298	298	873	298	Variable†
Amount of substance (ml)	–	48	–	48	–

* Depends on the desired amount of catalyst in the alumina layer; 1 h is often sufficient. Note: only the time for the penetration of the liquid into the pores is considered. The time for the preparation of the impregnants is not included because this can be done by a laboratory sample robot.

† Depends on the catalytic reaction. In order to hold the layer stable during the reaction, the calcination temperature should be higher than the reaction temperature [38].

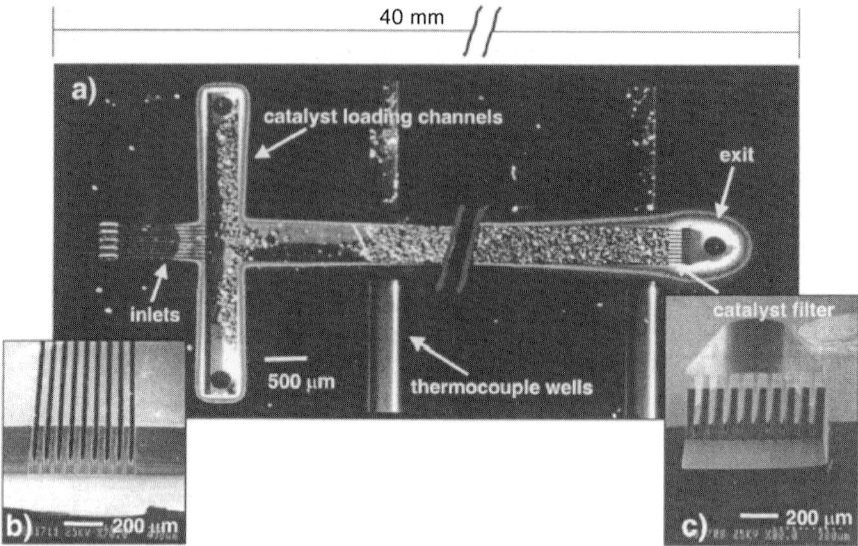

Figure 3.3 Fast serial gas-phase screening reactor [39] (by courtesy of AIChE).

3.2.1.3 **Catalyst Powder Injection**

A similar flow approach but in the area of solid transport was reported by Ajmera et al. [39]. They also used suction for the fast exchange of catalyst powder in a single-channel reactor applied for catalyst testing (Figure 3.3). For this purpose, they applied below atmospheric pressure at the reactor exit to draw in a defined amount of catalyst powder in a flat fixed bed. Over-pressure was applied in order to reverse this procedure after testing was finished.

Both approaches were actually using the characteristics of the laminar flow in micro structures, which enable a defined deposition of catalysts as the characteristics of the laminar flow remain unchanged. The deposition is not affected by unforeseeable turbulent effects. The reactors were used for catalyst preparation and for catalyst testing simultaneously.

3.2.1.4 **Catalyst Pellet Preparation**

Sample preparation on pellets was reported by Miyazaki et al. [40]. They screened catalysts for the direct oxidation of propylene to propylene oxide. The reactor was explained above. For the catalyst preparation, pellets were impregnated by dispensing defined volumes of metal salt solutions in small vials. The washer-shaped pellets had a diameter of 4 mm and a height of 1 mm. The gases passed over the surface of the pellet, where they reacted. Prior to the impregnation, the pellets were rinsed in deionized water. After impregnation, the pellets were heated to 110 °C and dried overnight. The library was then calcined at 500 °C for 2 hours. Experiments were executed at a maximum temperature of 350 °C, clearly below the calcination temperature, in order to prevent sintering of the catalysts. The propylene conversion during the reaction was limited to 2% maximum in order to prevent hot-spots. The most active catalyst component was found to be rhodium, either elementary or in binary combinations. A special matrix inversion method had to be developed because co-products developed during the reaction with overlapping spectrometer patterns.

3.2.1.5 Parallel Sputter Coating

Sputtering technologies offer an even faster procedure at the cost of the versatility of the catalysts produced. Sputtering produces catalysts with a dense surface layer. These catalysts are different from industrially used catalysts, which usually have a larger surface area. The BET surface area of sputtered catalysts is below 1 $m^2 g^{-1}$ and thus much lower than the surface area of the wash-coated catalysts (usually above 60 $m^2 g^{-1}$). However, if catalysts for a fast reaction have to be screened, the gas components will mainly react at the surface of the catalyst and the porosity of the catalyst is not important.

Danielson et al. introduced sputtering as a new tool for catalyst preparation in 1997 [41]. The group of Weinberg then further developed this technology in 1999 [42, 43]. They used a mask technology, which allowed the production of multi-component catalysts in a sequence of steps producing stacked layers of elementary catalysts. Mixed catalysts were obtained by molecular diffusion during a sintering process.

Their methodology was further modified in order to permit the manufacture of stacked layers with thickness gradients [5]. These thickness gradients were combined in such a way that catalysts with individual compositions, consisting of up to three components, could be manufactured in a one-step process. Existing thin-film technologies for elementary catalyst coating were modified and a new method, the so-called Simultaneous Gradient Sputtering process (SGS), was developed [37, 38]. Details of this process exemplary for other combinatorial thin-film technologies are described in the following.

During this process, the layer thickness of three catalyst components on a titer-plate was simultaneously varied in such a way that 48 single catalyst mixtures were obtained in a one-step process. Thus, the amount of each catalyst component in the mixture could either be decreased or increased, resulting in a homogeneous multi-component layer.

The catalysts to be deposited are inserted into the sputter plant as the so-called targets. The shuttle carries the titer-plate and revolves below the metal targets (Figure 3.4). In every rotation up to three sublayers are deposited. For a binary system, the sublayer thickness was 10 nm in each revolution. The thickness gradients were realized by aperture orifices which shaped the particle beam (Figure 3.5).

Every orifice is supplied with an individual geometry. The rotation frequency of the revolving shuttle influenced the homogeneity of the layer. A faster rotational speed will result in thinner sublayers and thus in an increased layer homogeneity. However, the rotational speed is restricted by mechanical properties of the turn table (Table 3.2).

For the single component zirconium, the gradient of the layer thickness was measured on a silicon monocrystalline wafer. The resulting thickness gradient is

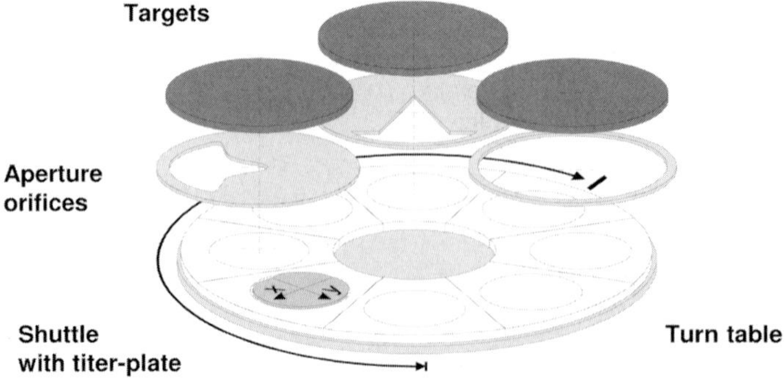

Figure 3.4 Principle of the SGSP with three targets [37].

Figure 3.5 Details of sputtering plant for gradient sputtering (a single aperture orifice is shown in the lower right photograph) (source: IMM).

Table 3.2 Typical process parameters of the SGS process [38].

Process properties						
Pressure (mbar)	*Gas*	*Power (W)*	*Targets (ϕ_{max}/mm)*	*Rotational speed (rpm)*	*Deposit rate (nm min^{-1})*	*Substrates (ϕ_{max}/mm)*
$5 \cdot 10^{-3}$	Ar	1000	200	10	2	125

shown in Figure 3.6 as a function of the wafer diameter and can be approximated by a linear fit.

For binary catalysts, two of these profiles have been combined, with the second profile being mirrored to the first one. Platinum was chosen for the second component. By adjusting the target powers (150–2500 W), individual gradients were defined.

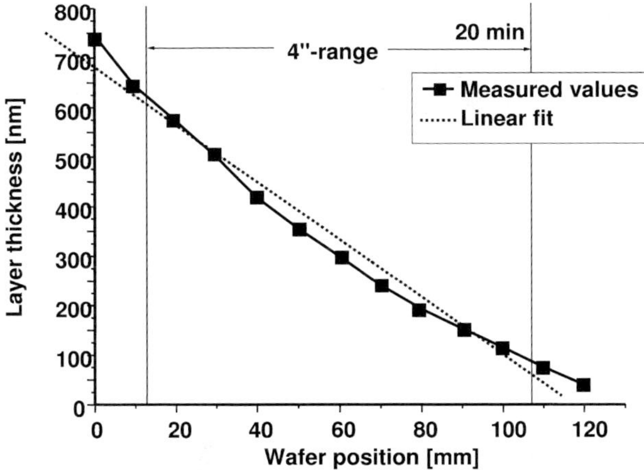

Figure 3.6 Distribution of the layer thickness for a single component (zirconium) [37] (by courtesy of Elsevier Ltd.).

The homogeneity of this binary layer composition was examined using secondary neutral particles mass spectrometry (SNMS) [37]. This procedure removes the catalyst coat stepwise and delivers the amount of the catalyst components in the respective layer depending on the sputtering time. One second of sputtering time corresponds here to a layer removal of approximately 1 nm. Hence the sputtering time is also a measure of the layer thickness. The maximum layer thickness was 500 nm.

Layers consisting of two components were realized by employing two targets. This is demonstrated in Figure 3.7 with the components zirconium and platinum on a stainless-steel substrate. Zirconium and platinum were evenly distributed within the layer. The surface roughness of the steel substrate was responsible for the steady incline of the iron concentration curve.

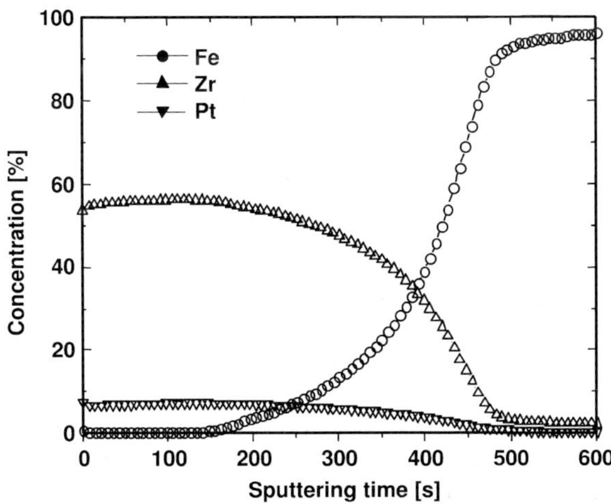

Figure 3.7 SNMS analysis of a binary catalyst on a steel substrate [37] (by courtesy of Elsevier Ltd.).

To increase the number of catalysts, the binary catalyst system zirconium/platinum was extended to a ternary system by adding a third component such as molybdenum. The expected quality of such a composition is given in Figure 3.8. First results of such a ternary layer are presented in Figure 3.9 (left). The layer was deposited on a silicon monocrystalline wafer and the layer composition again examined with SNMS.

By oscillating the titer-plate, a third gradient can be sputtered by moving the substrate half way underneath the target, thus also introducing a temporal effect into the sputter coating. The effective speed of this oscillating table is lower than the rotating speed of the table as in the case with only two targets. This leads to a deposition of less homogeneous films. Nevertheless, the average composition in the layer is constant and widely stable. This is not the case on the catalyst surface. The composition here is not well defined and a method to stabilize it was searched for. Since the vertical depth of the sputtered sublayers was only 2 Å, it seemed possible to achieve homogenization by solid-state diffusion. A number of sputtered ternary catalysts were sintered at 850 °C under vacuum conditions for 4 h. The result of one composition is shown in Figure 3.9 (right) as concentration of the components versus layer depth. The unit of the layer depth is again given in seconds, and 1 s corresponds to approximately 1–2 nm. This is a result of the consecutive removal of the layer by the SNMS process which uses the sample as a target and removes the layer by (de-)sputtering at a rate of 1–2 nm s^{-1}. Comparing the untreated catalysts on the left with the sintered catalysts on the right in Figure 3.9 proves that heat-induced molecular diffusion reduces the inhomogeneity. During the sintering process, the plates were laid on a heated plate on the uncoated side of the wafer. Thus, the coated upper surface of the samples was not as well heated as the internal sublayers.

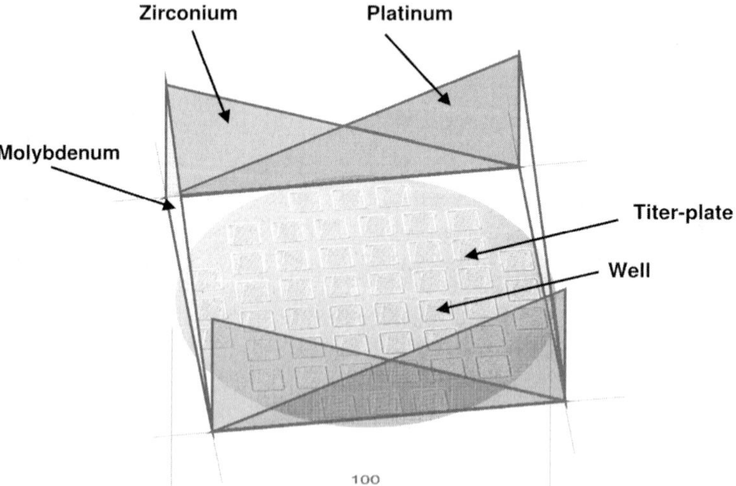

Figure 3.8 Expected qualitative ternary layer composition represented by three wedge-shaped profiles [37] (by courtesy of Elsevier Ltd.).

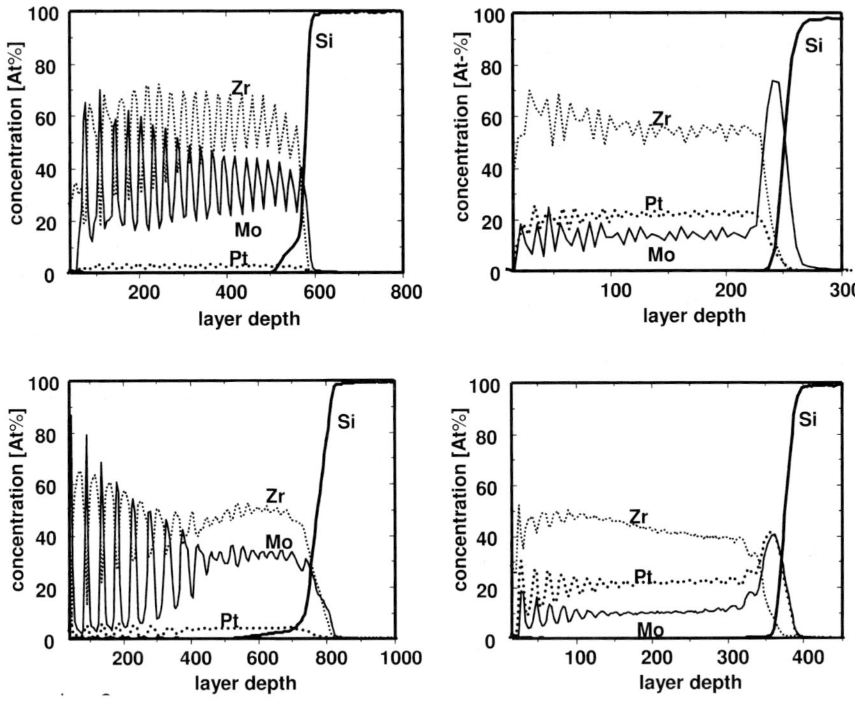

Figure 3.9 Layer composition of two ternary catalyst mixtures examined with SNMS before tempering (left) and after tempering at 850 °C for four hours (right) [37] (by courtesy of Elsevier Ltd.).

This is the reason why the composition of the internal sublayers is already balanced whereas the sublayers close to the surface are still inhomogeneous (Figure 3.9, right).

Up to 48 ternary catalyst mixtures were prepared simultaneously in less than 1 h. Hence the sputtering procedure is much faster than the wet chemical route and in fact one of the fastest syntheses available. This advantage is gained at the expense of low layer porosity. Thus, sputtered catalysts are new artificial catalysts and not directly comparable to catalysts prepared by wet-chemical procedures. These catalysts offer the advantage of quick preparation and characterization compared with alumina-based catalysts. They can also be used for obtaining so-called intrinsic kinetics because there is no influence of diffusion.

3.3
Parallel Batch Screening Reactors

3.3.1
Reactor 1 [R 1]: Agitated Mini-autoclaves

High-pressure reactions are often executed discontinuously. This batch-type operation facilitates sealing of the reactor and allows active mixing with adjustable residence times. Lucas and Claus [44] used agitated mini-autoclaves (Figure 3.10) for the heterogeneous gas/liquid selective hydrogenation of citral and acrolein. The screening set-up consists of five mini reactors having a reactor volume of 45 ml. The mini reactors are integrated in an insulated heating block and are suitable for a maximum pressure of 115 bar and a maximum temperature of 300 °C.

The activity of catalysts can be characterized by the pressure decrease observed during the reaction (Figure 3.11). Such discrimination demands the knowledge of the stoichiometry of the reaction as a volume increase during the reaction would otherwise falsify the results.

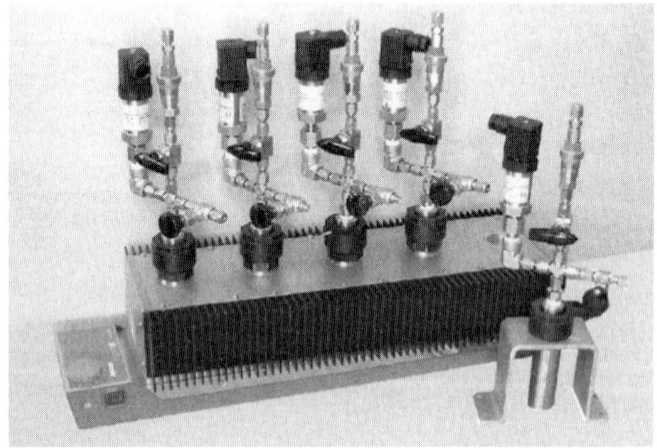

Figure 3.10 Photograph of a 5-fold multibatch reactor [44].

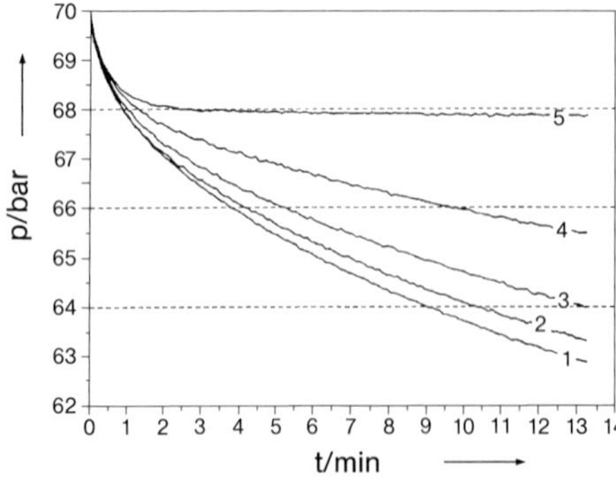

Figure 3.11 Parallelization of activity testing of Rh/Al_2O_3 catalyst with and without addition of $SnCl_2$ for the hydrogenation of citral at 298 K. $SnCl_2$ addition (mg) 1, 0; 2, 1.2; 3, 2.8; 4, 8.3; 5, 22.0 [44].

For the hydrogenation of citral different reaction products, e.g. citronellal, citronellol, geraniol and nerol, can be expected:

Citral

Rh/Al_2O_3 ($SnCl_2$)

Citronellol — Citronellal — Nerol — Geraniol

The set-up of Lucas et al. allows the fast investigation of various parameters, e.g. temperature, pressure and mixing speed, affecting the outcome of the reaction. Exemplarily, the results of the hydrogenation of citral are shown (Table 3.3) using different amounts of the catalyst additive $SnCl_2$.

Table 3.3 Conversion and selectivities for the hydrogenation of citral in the liquid phase using Rh/Al_2O_3 catalysts with and without added $SnCl_2$ [44].

	Batch reactor				
	1	***2***	***3***	***4***	***5***
$SnCl_2$ added (mg)	0	1.2	2.8	8.3	22.0
Conversion of citral (%)	84.0	76.0	65.9	41.1	1.6
Selectivities (%)					
Citronellal	98.2	97.5	96.4	91.6	50.0
Citronellol	1.4	1.6	1.8	2.8	0
Geraniol + nerol	0.4	0.9	1.8	5.6	50.0

Starting material, 1.83 ml citral; solvent, 7.7 ml *n*-hexane; internal standard, 0.67 ml dodecane. Reaction conditions: $p_{start} = 70$ bar; $T = 298$ K; $t = 15$ min; stirring at 840 rpm.

3.3.2 Reactor 2 [R 2]: Agitated Mini-autoclaves

The mini multi-well batch reactors presented by Desrosiers et al. [45] (Figure 3.12) were applied to the direct amination of benzene to aniline. The screening reactor

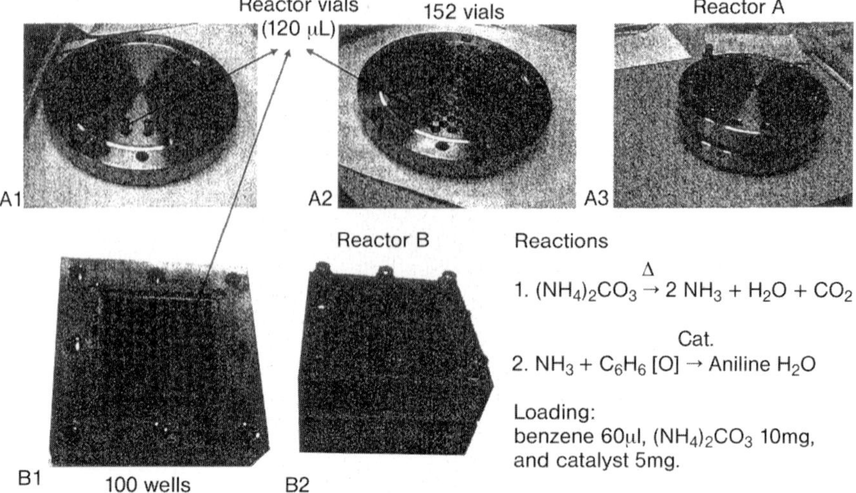

Figure 3.12 High-throughput batch reactor used as primary screen for the discovery of aniline cataloreactants. The reactor consists of a circular block with an array of 15 × 10 catalysts [45] (by courtesy of Elsevier Ltd.).

concept used was either optimized towards high-throughput experiments (primary screening: up to 152 wells per plate, each with a volume of 0.12 ml) or high-pressure experiments (secondary screening: 24 wells per plate, each with a volume of 0.5 ml).

In order to increase conversion, hydrogen, which was produced during the reaction, was converted into water by so-called cataloreactants, solid reducible oxides. The primary screening was targeted to identify the best oxidant, benzene/ammonia activator and cataloreactant support. The latter is a reducible metal oxide which acts as hydrogen scavenger.

$$C_6H_6 + NH_3 \rightleftharpoons C_6H_5\text{–}NH_2 + H_2$$

$$NiO + H_2 \longrightarrow Ni + H_2O$$

The catalysts, based on nickel/nickel oxide, were prepared in wells in situ by a freeze-drying method of precursors (Figure 3.12); the mass of catalyst amounted to 5 mg. The sealed wells were heated until a high pressure of approximately 200–300 bar was generated. The aniline yield was detected by an adsorption/optical method using spray impregnation and a CCD camera together with a UV lamp. In the case of the secondary screening, various cataloreactants were screened at different temperatures. Typical results are shown in Figure 3.13 indicating that nickel was the best cataloreactant with 4% maximum aniline yield at 325 °C.

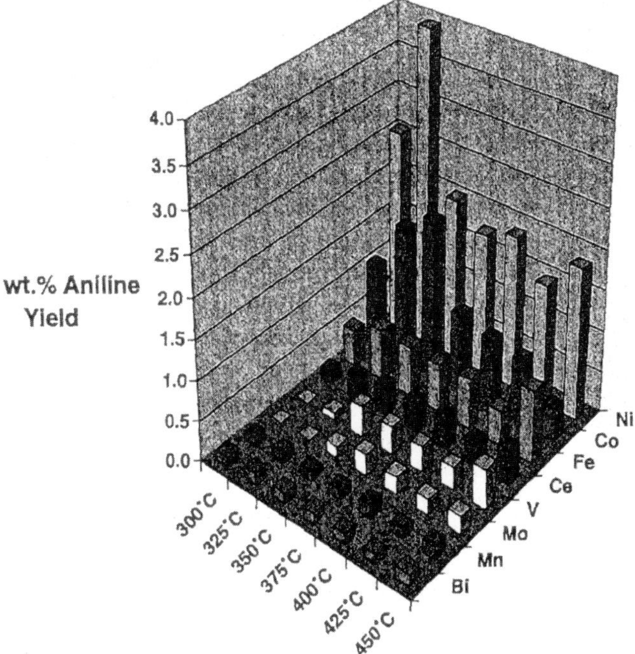

Figure 3.13 Activity of metal oxide oxidants in the temperature range 300–450 °C. The composition of the catalyst is 2 wt.% Rh and 20 wt.% of a metal mixture based on the weight of the ZrO_2 support [45] (by courtesy of Elsevier Ltd.).

3.3.3
Reactor 3 [R 3]: Agitated Mini-autoclaves

Further miniaturization of mini-autoclaves would result in well-type batch reactors. These are conveniently arranged as two-dimensional array reactors, the so-called titer-plate known from pharmaceutical applications. In pharmacy, titer-plates are operated under atmospheric pressure, usually at room temperature. That titer-plates could also be used as process equipment for batch screening under elevated pressure was demonstrated by the Stanford Genome Technology Center [46], where such a titer-plate was integrated in a so-called automated multiplex oligonucleotide synthesizer (AMOS). This synthesizer consists of a 96-well titer-plate integrated in a liquid reagent handling system equipped with valves, manifolds and tubing (Figure 3.14). Reagents were delivered to the wells via jets activated by pressurized argon in order to avoid pump devices.

The individual wells can be operated under a positive atmospheric pressure of inert argon. Each well is divided into two chambers separated by a filter. The actual synthesis is executed in the upper part of the well, which is filled with small glass beads, and the filter acts as a barrier for the reagent liquid inside the well. The liquid is discharged by an increase in the argon pressure inside the well, which then transports the liquid through the filter to the waste collection chamber.

Figure 3.14 Multiplex titer-plate synthesizer [46] (by courtesy of Stanford Genome Technology Center).

3.3.4 Lawn-format Assays

The expense of screening depends very much on the number of samples tested. Consequently, the density format of titer-plates has increased in recent years from 96-well up to the 9600-well format. The next big step towards miniaturization would be the complete avoidance of any container, which then results in the smallest well possible and a well-less, so-called lawn-format assay develops. This is exactly what is proposed by a number of authors (see review [47]). Screening in a lawn format does not mean avoiding any structure or arrangements. Samples are still prepared on beads, which are produced by split-mix synthesis, but the beads are arrayed directly on the well-less assay. A typical matrix applied for such biological screening is the agarose lawn. Active beads are then picked from the assay matrix and decoded for compound identification.

3.3.5 Catalyst Screening by Multistep Synthesis

Even the lawn format still does not represent the final limit of miniaturization. The one well/one catalyst or one bead/one catalyst strategy, where catalyst identity is spatially coded, can be replaced by in situ synthesis combined with mass spectrometry [48]. The advantage of this strategy is the use of a mass spectrometer for the synthesis, reaction and analysis. The described electrospray ionization procedure helps to avoid the cleavage of chemical bonds, which would falsify the results. The synthesis step does not have to deliver clean and isolated products. Instead, after synthesis, the reactants are first separated by a quadrupole. In a second step, they are further reacted in an octapole and the reaction products are finally isolated in a second quadrupole and analyzed. Figure 3.15 describes the screening process in detail [49].

With this experimental set-up, highly active, cationic ruthenium–carbene catalysts are used in ring-opening metathesis polymerization (ROMP). Four different structural features of the catalyst $[\{R_2P(CH_2)_nPR_2\text{-}\kappa^2P\}XRu{=}CHR]^+$ (the halogen

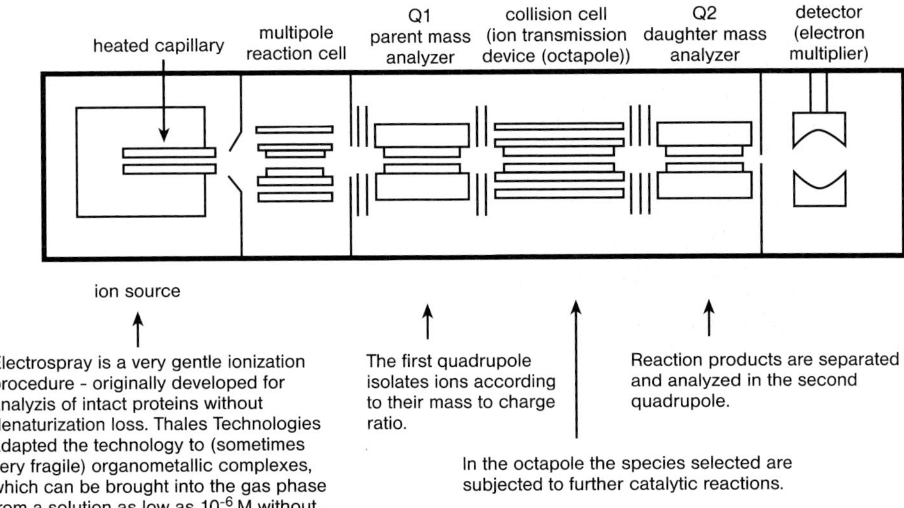

Figure 3.15 Schematic drawing of electrospray mass spectrometry for in situ synthesis and screening [49] (by courtesy of Thales Technologies).

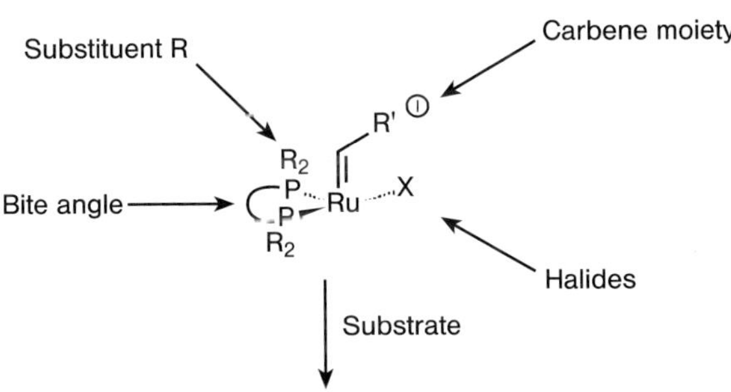

Figure 3.16 Structural variants for the screening of $[\{R_2P(CH_2)_nPR_2\text{-}\kappa^2P\}XRu{=}CHR]^+$ cationic ROMP catalysts [48].

ligand, the diphosphane bite-angle, the steric bulk of the phosphane and the carbine ligand) were investigated (Figure 3.16) [48]:

As a result, it could be shown that both in the solid state and in solution the $[\{R_2P(CH_2)_nPR_2\text{-}\kappa^2P\}XRu{=}CHR]^+$ cation dimerizes to form dicationic, dinuclear complexes $[\{\{R_2P(CH_2)_nPR_2\text{-}\kappa^2P\}XRu{=}CHR\}_2]^{2+}$. In different crossover and trapping experiments, monomeric species could be found in solution. During the next pre-equilibrium step these monomeric complexes should form an olefin π-complex. The product formation presumably proceeds via a metallacyclobutane according to the Chauvin metathesis mechanism (Figure 3.17) [48].

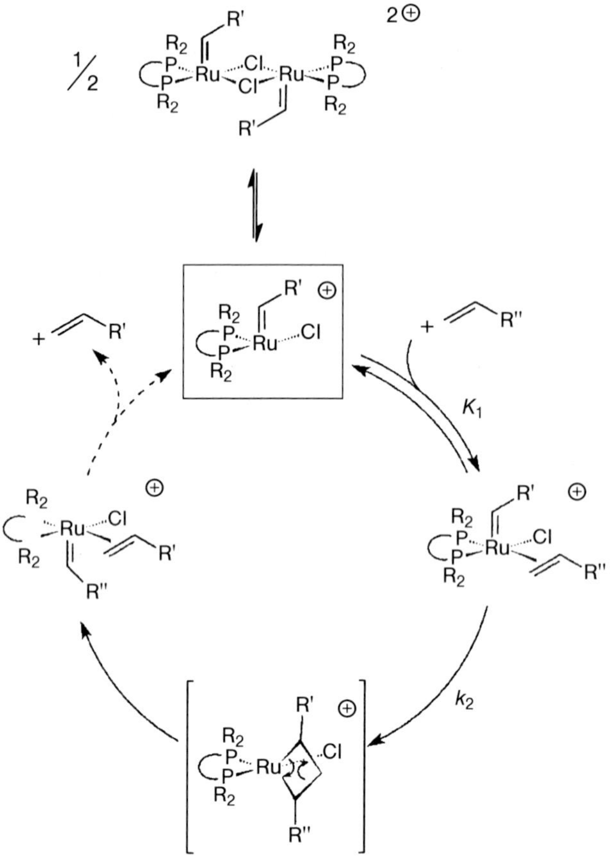

Figure 3.17 Proposed mechanism for olefin metathesis by dicationic dinuclear complexes $[\{\{R_2P(CH_2)_nPR_2\text{-}\kappa^2P\}XRu{=}CHR\}_2]^{2+}$ [48].

Usually, the optimization of a given catalyst for a given chemical reaction requires independent synthesis of each individual complex. This means that for typical organometallic complexes in homogeneous catalysis the synthesis is the most time-consuming reaction.

The new proposed screening method (screening before synthesis), the rapid assay of the variety of structural effects on metathesis rate mentioned above, combined with mechanistic analysis, gives the opportunity to find optimized catalysts for ROMP with great savings in time and effort.

3.4 Screening Reactors for Steady Continuous Operation

3.4.1 Multiple Micro Channel Array Reactors

A characteristic of this type of reactor is the steel substrate which is preferably used as the reaction chamber (but also titanium or aluminum). This allows the use of micro structures under high temperature and robust experimental conditions.

3.4.1.1 Reactor 4 [R 4]: Stacked Platelet Screening System

The reactor system of Zech and co-workers [50, 51] is a good example of an integrated approach as it combines devices from different suppliers witha complex screening system. The reactor was manufactured at IMM and the sampling device was provided by AMTEC, Chemnitz. The catalyst and their preparation method were supplied by the TU Chemnitz. The housing of the reactor module consists of 35 stacked frames which can incorporate the same number of catalyst wafers (Figure 3.18). The modular concept of the reactor allows the use of micro structured catalyst wafers made of different materials such as metals, ceramics, silicon and glass.

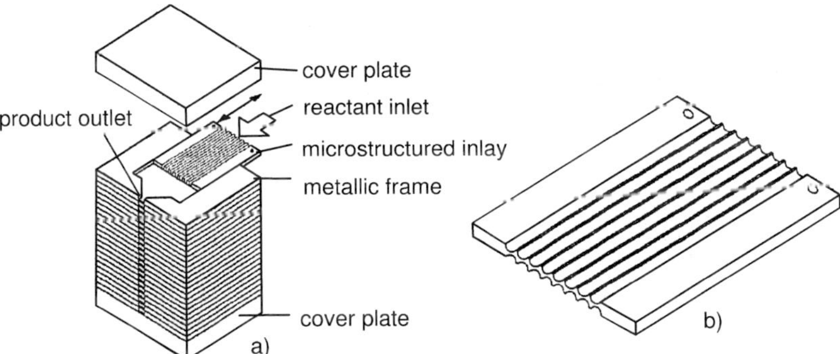

Figure 3.18 Schematic view of the reactor module consisting of a stack of metallic frames (a). Single catalyst carrier platelet (b) [51] (by courtesy of SCS).

Catalytic methane combustion

In the case of catalytic methane combustion, aluminum was chosen as an appropriate material for the catalyst wafers since anodic oxidation of aluminum can be used to obtain porous surfaces. Such micro structured aluminum platelets were coated by wet impregnation with Pt, V and Zr precursors [50].

The analytical system consisted of an x/y-positioning robot equipped with a sampling capillary was supervised by a CCD camera to address each micro channel individually. This sampling capillary was connected to a quadrupole mass spectrometer to analyze the different reaction mixtures subsequently. With this set-up it was possible to screen up to 35 catalysts per day (Figure 3.19).

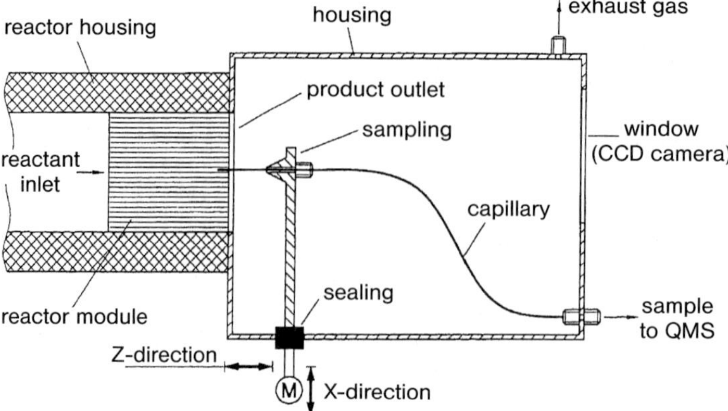

Figure 3.19 Schematic drawing of the sampling device used for sequential sampling of reaction products at the outlets of the parallel micro channel reactor [51] (by courtesy of SCS).

A library of 35 different catalysts fixed on electrochemically oxidized aluminum either in oxalic acid (Lib 1) or sulfuric acid (Lib 2) was tested at 450 °C and 1.1 bar. The methane-to-oxygen ratio was set to 1 in order to establish the potential of the catalyst to form intermediates. Figure 3.20 shows experimental results for a residence time of 550 ms and a screening time of 60 s. The conversion rate followed directly the platinum content in the catalysts. The higher the platinum content, the higher is the degree of conversion. Catalyst carrier formed by anodization of

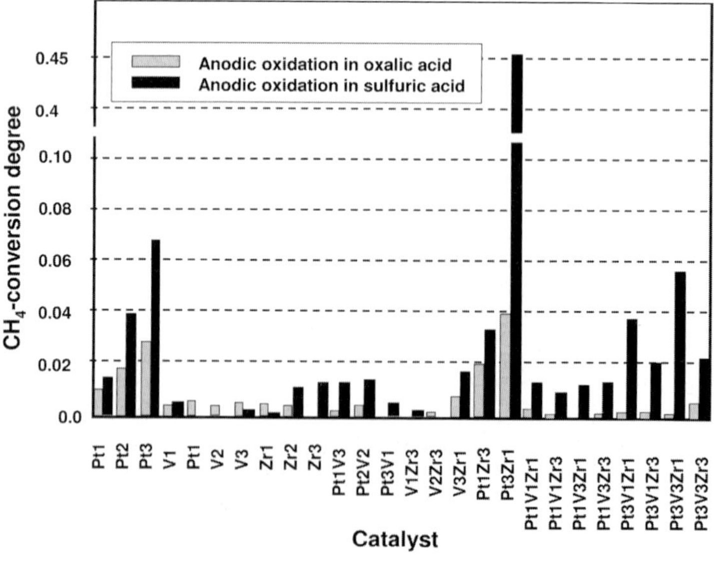

Figure 3.20 Conversion of methane for 35 different catalyst samples on a micro structured and oxidized aluminum substrate [52] (by courtesy of AIChE).

aluminum in sulfuric acid always shows higher activity. In that case the achieved pore diameter (~15 nm diameter) is much smaller and, therefore, the surface area much higher than for Al_2O_3 layers formed by anodic oxidation in oxalic acid (~30 nm diameter) [52].

These data showed good reproducibility and the variations are within the range of measurement error.

Partial hydrogenation of 1,3-butadiene

For the partial hydrogenation of 1,3-butadiene, anodically oxidized aluminum catalyst carriers were impregnated with Pd, Co, and Cu. The concentrations of the catalytically active component were varied to form a ternary catalyst library with at least 132 different catalysts (Figure 3.21) [51].

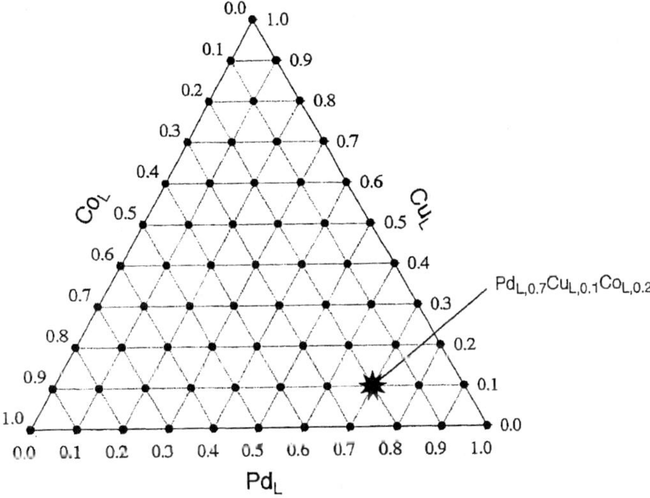

Figure 3.21 Ternary catalyst library prepared by wet impregnation of oxidized aluminum carriers with Pd, Co and Cu salts [51] (by courtesy of SCS).

The catalysts were screened under seven different conditions by varying the reaction temperature and residence time. In each case the time on stream was 4 h, and each of the 35 catalyst platelets was under investigation for 75 min.

The results for the selected reaction conditions are shown in Figure 3.22. The ternary diagrams show the sum of *n*-butene yield, the reaction products of the hydrogenation of 1,3-butadiene, at the corresponding catalyst composition. It can be shown that the binary Cu and Co catalyst composition is inactive. An increase in Pd increases the yield of *n*-butene. However, binary catalyst compositions with Pd/Cu or Pd/Co show higher performance than single Pd. Furthermore, it is demonstrated that high catalytic activity can be observed also at isolated catalyst compositons within the field of ternary catalysts [51].

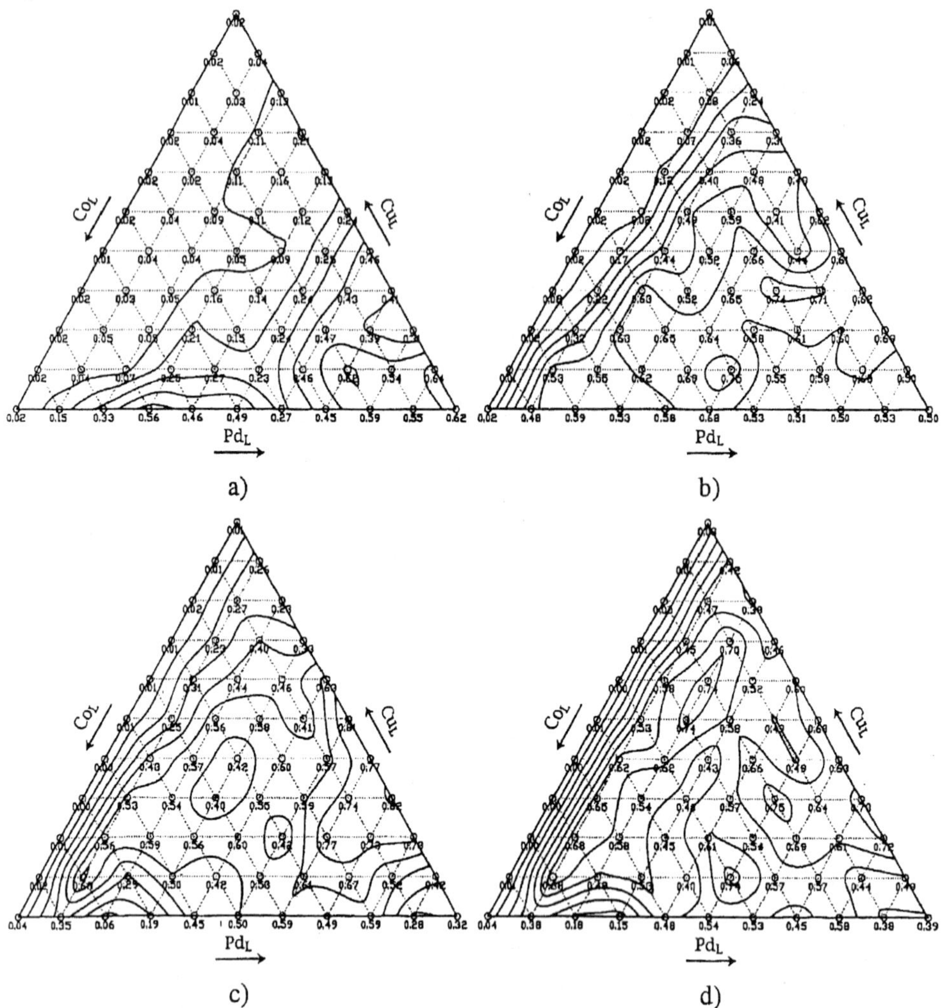

Figure 3.22 Selected results for the screening of 132 different Pd/Co/Cu catalysts used for the partial hydrogenation of 1,3-butadiene. Reaction conditions: $C_{butadiene} = 1.6$ vol.%, Ar balance. (a) Lib 1, $T = 50$ °C, $\tau = 60$ ms; (b) Lib 2, $T = 50$ °C, $\tau = 60$ ms; (c) Lib 2, $T = 50$ °C, $\tau = 300$ ms; (d) Lib 2, $T = 130$ °C, $\tau = 60$ ms [51] (by courtesy of SCS).

3.4.1.2 **Reactor 5 [R 5]: 10-fold Parallel Reactor with Exchangeable Flow Distribution Section**

Kolb et al. [53] developed a modular 10-fold parallel reactor made of titanium which included an exchangeable distribution section for the reactant delivery to the single platelets (Figure 3.23). The externally heatable reactor is specially designed for the determination of catalyst activity in parallel (primary screening) and serial screening (catalyst optimization). The reactor contains 10 drawers which carry the micro-structured and catalyst coated platelets. Underneath each platelet a graphite foil is put to compensate for material roughness. The drawers are introduced from the

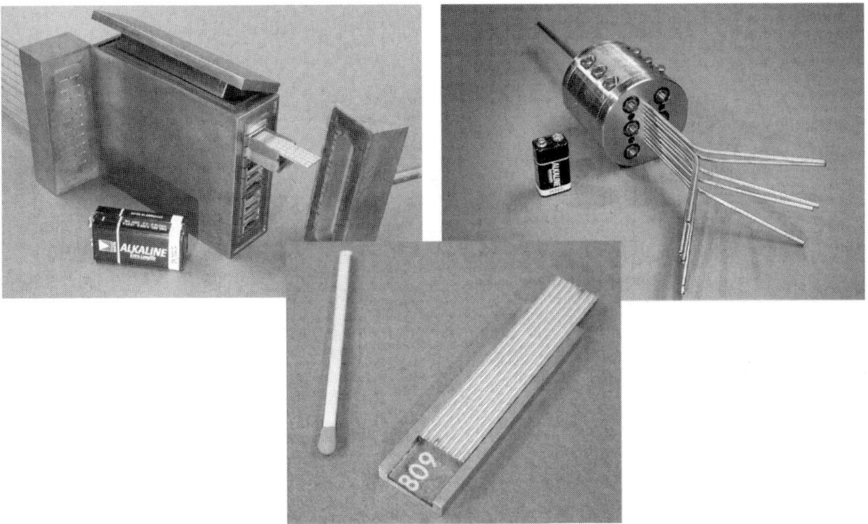

Figure 3.23 Screening reactor fully made of titanium (left) and with stainless steel (right) with catalysts in separate drawers (middle) (gas distribution section removed) [53] (by courtesy of AIChE).

front of the reactor. Putting each catalyst platelet into a separate drawer makes the equipment handling easier for the operator and generates space for the gaskets which separate the platelets from each other. These gaskets are an integral part of the main gasket, which seals the reactor against the environment. Graphite was used as the material for the gaskets. The 10 drawers are covered by a top plate and compressed from the top. This reactor could be operated as a 10-fold parallel reactor as well as a single-channel reactor. In the latter case, the 10 channels were combined to a single channel by changing the distribution end-caps. This allowed the adjustment of the residence time without changing the mass flow and thus enabled it to maintain the flow conditions.

Parallel screening of up to 10 plates and secondly serial operation of up to 10 identically coated catalyst plates were investigated. The latter gave the opportunity of modifying the residence time via the catalyst mass under identical flow conditions. Owing to the low stability of Ti, at the reaction temperature of 500 °C a steel housing was put around the titanium reactor. This additional housing increased the dimensions of the reactor (160 mm × 70 mm × 120 mm). Heating of the reactor had to be executed externally. The micro structured plates had dimensions of 300 μm depth and 500 μm width with a length of 100 mm.

Experiments performed with the reactor, in that case made of titanium, by Wörz at BASF in a proprietary reaction revealed 60% yield for the desired product at residence times as low as 40 ms in the micro reactor. This performance was superior to that of the experiments performed in an aluminum capillary, which corresponds well with the reactor design of the industrial process (Figure 3.24). A 2000% gain in space-time yield was found for the porous coated micro structures compared with the aluminum capillaries.

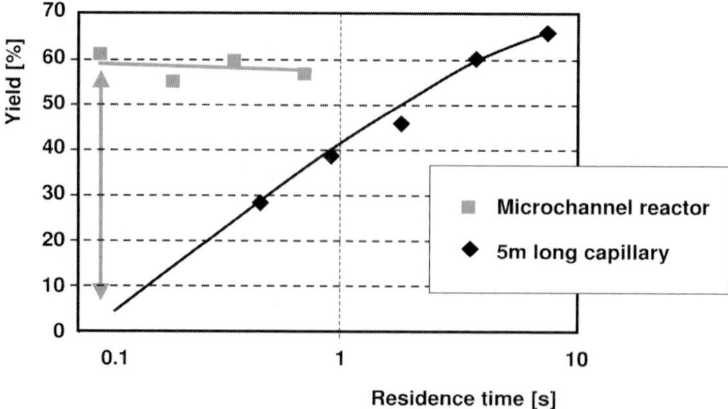

Figure 3.24 Experimental results for comparison a proprietary reaction performed in a conventional 5 m long reactor and in a microstructured reactor [53] (by courtesy of AIChE).

Methanol steam reforming

To assess the performance of the reactor, methanol steam reforming over CuO/Cr_2O_3/Al_2O_3 and CuO/Mn_2O_3/Al_2O_3 catalysts was chosen as a test reaction. Alumina washcoats were introduced into the micro structures and the active components were deposited via impregnation and subsequent heat treatment. Various alumina carriers such as γ-alumina (specific surface area 70 $m^2\ g^{-1}$), boehmite (specific surface area 160 $m^2\ g^{-1}$), corundum (specific surface area 7 $m^2\ g^{-1}$) and two mixed systems of boehmite and corundum were tested in parallel using the same loadings of active component (1.5 mg CuO/Cr_2O_3 and 1.75 mg CuO/Mn_2O_3) at flow rates of 120 and 240 ml min^{-1}, a temperature of 250 °C and a methanol-to-steam molar ratio (C/S) of 1 : 2.

The highest activity for the Cu/Cr system was achieved by the mixed system of boehmite/corundum 1 : 1 and the lowest by the pure corundum carrier. However, by using the same loadings of active component, it was found that the magnitude of the existing surface area is a major but not the only factor affecting the performance of the catalyst. For example, the mixed system of boehmite/corundum 1 : 1 (0.55 m^2) has a lower actual surface area than γ-alumina (0.94 m^2) but the system activity proved to be three times higher than that using γ-alumina. The reason may be attributed to the fact that steric, morphological and electrostatic effects also play an important role. This assumption also holds for the Cu/Mn system. There the highest activity was achieved by the system with pure boehmite carrier (0.75 m^2) and the lowest by a mixed system of boehmite/corundum 1 : 1.

The Cu/Cr system was also investigated in the serial operation mode. Owing to initial deactivation of the catalyst at the first 10–12 hours, the measurements were carried out once a steady state has been reached. It could be demonstrated that a subsequent increase in the residence time by introducing one to eight catalyst platelets always led to the same relative increase in conversion and hydrogen yield (100%), because no other products were detected. For all experiments the maximum conversion achieved was 18%. Thus, mass transport limitations are absent from the reaction system under the experimental conditions applied.

3.4.1.3 Reactor 6 [R 6]: Micro Reactor for Steam Reforming Catalyst Testing

Cominos and co-workers [30, 54] developed a stack-like micro reactor for testing catalysts for methanol steam reforming for PEM fuel cells. The stainless steel micro reformer test device was built with dimensions 75 mm × 45 mm × 110 mm and incorporates 5–15 stainless-steel plates with micro channels of 500 μm width and 350 μm depth (Figure 3.25). The inlet region of the test device was designed such that it was large enough to allow for uniform flow conditions at the inlet of the channels while the outlet region was designed to be smaller such that product gases can leave the device fast for subsequent analysis to take place. The holes seen on the side of the top cover and bottom part of the device were for inserting eight heating cartridges. Graphite seals were used to keep the stack of micro structured plates leak tight [30].

As carrier material, Cu/Zn catalysts were prepared by introducing γ-alumina wash-coats of an average thickness of 10 μm into the micro channels. The average pore diameter was 45 nm and the BET surface area of the catalysts was determined as 72 $m^2\ g^{-1}$. The residence times were adjusted from 100 to 200 ms at total flow rates between 500 and 900 ml min^{-1} using only five coated plates. Increasing the temperature from 200 to 275 °C increased the conversion from 37 to 65%. Carbon monoxide formation started at temperatures exceeding 250 °C. When 15 plates were introduced into the reactor, 80% conversion was achieved at 290 °C, a residence time of 600 ms and a steam to carbon ratio of 2. The product stream contained more than 50 vol.% hydrogen and 0.25 vol.% CO. From these results, a power output of the testing device of 40 W and a power density of 1.7 kW l^{-1} were calculated.

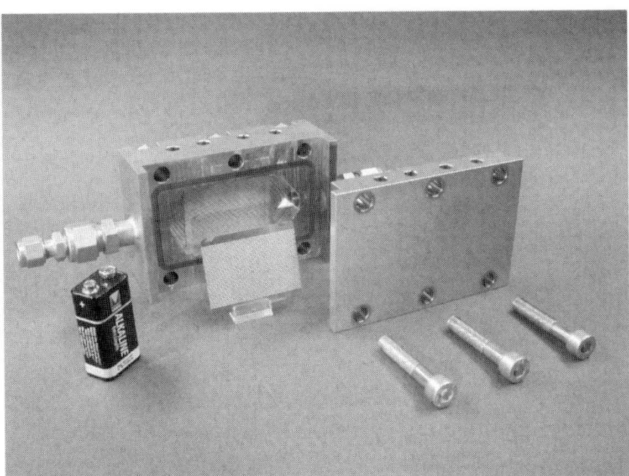

Figure 3.25 Unmounted micro reactor for steam reforming catalyst testing with exchangeable catalyst carrier platelets (source: IMM).

3.4.1.4 Reactor 7 [R 7]: High-throughput Micro Reactor with Parallel Micro Compartments

A high-throughput micro reactor with parallel flow passages, so-called micro compartments, with typical dimensions in the micro flow range was fabricated as part of a broad design study by Mies et al. [55]. The reactor was designated for operation in the temperature range 100–800 °C for the study of gas-phase reactions with large heat effects ($\Delta H_{298} = \pm 500$ kJ mol^{-1}).

The reactor comprises, in addition to the compartment–reactor part, a diffuser and a quench/gas-sampling section as well as temperature-sensing elements. The reactor was molybdenum based owing to its excellent thermal conductivity, approaching that of copper and notably exceeding that of stainless steel. The reactor part contains eight microstructured compartments, each with a cross-section of 2.28 mm × 10.18 mm and a length of 40 mm. Each compartment carries eight plates which are inserted into small recesses in the wall of the compartments. The plates do not contain further micro channels, but act as a plate stack itself as a channel passage, since small open gaps remain after stacking of the eight plates within a compartment. The eight plates are made of molybdenum and have 100 μm thicknesses. As they are coated with a 10 μm catalyst layer and the height of the

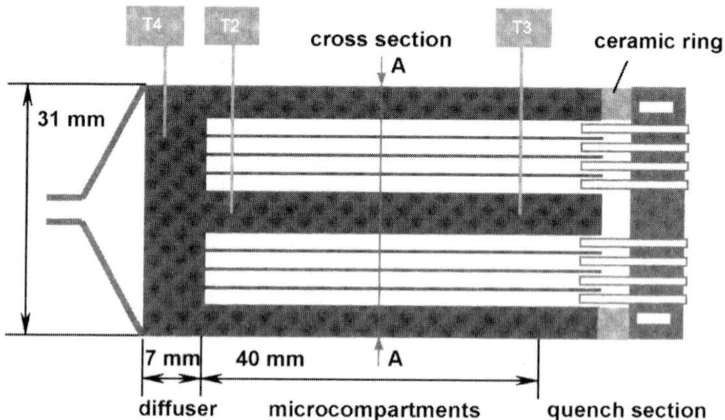

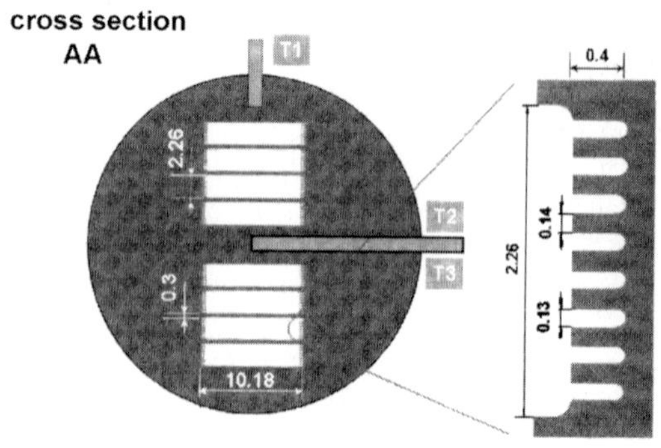

Figure 3.26 Schematic of the high-throughput micro reactor with eight parallel flow passages, the micro compartments (right). Cross-section of the reactor showing details of the micro compartments (left). Dimensions are given in mm [55] (by courtesy of ACS).

micro compartments is 130 μm, a small gap of 10 μm remains for the fluid passage after insertion of the coated plates. Micro electro discharge machining was applied for manufacture of the small compartments into a solid reactor block (Figure 3.26).

Special care during the design of the high-throughput micro reactor was taken to ensure flow equilibration and excellent thermal management. It was shown by CFD simulation that an even flow distribution throughout all compartments can be reached for total flows ranging from 50 to 1000 $cm^3\ min^{-1}$ (STP). The diffuser also contained structured compartments which guarantee improved pressure recovery due to negligible pressure drop and virtually clog-free behavior. The direction of the reactor compartments was shifted by 90° relative to the diffuser compartments (Figure 3.27). In a design study, involving two parameters of the diffuser and reactor compartments, it could be shown that the flow distribution is very sensitive for such smart structural details (Figure 3.28).

In the quench section, the effluent gases are cooled within milliseconds to prevent consecutive reactions of the product mixture. Even at the highest flow velocities assumed, corresponding to 1000 $cm^3\ min^{-1}$ (STP), and a temperature of the outlet gas of 500 °C, a temperature drop of about 300 °C can be reached within only a 3 mm passage, which is equivalent to a quenching time of 0.5 ms.

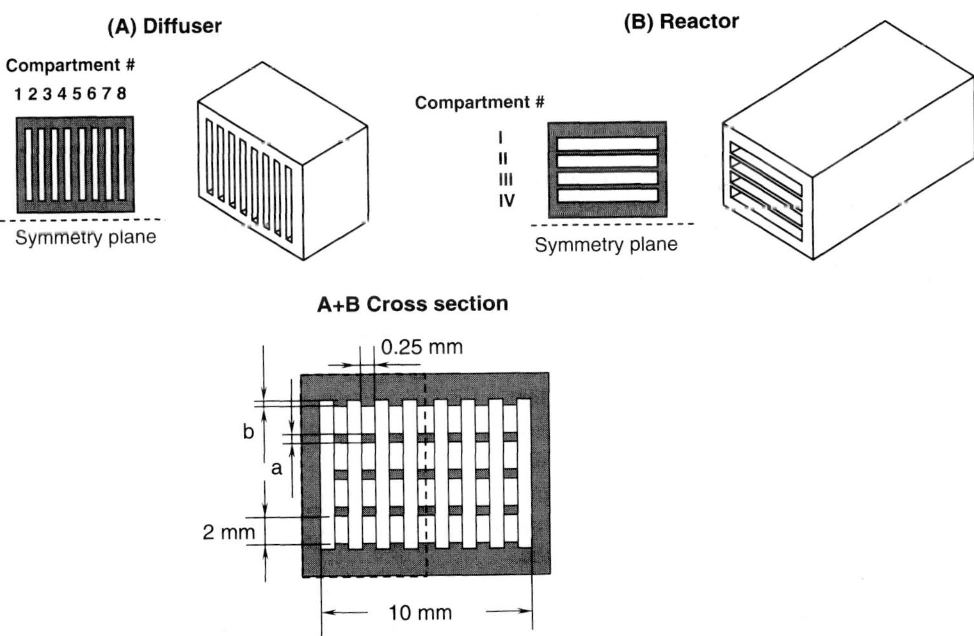

Figure 3.27 Schematic of the flow diffuser and the reactor (only half of both designs is shown for reasons of symmetry). The two parameters *a* and *b* taken for a design study to reflect the relationship between structural details of the diffuser and reactor compartments and the flow distribution are indicated [55] (by courtesy of ACS).

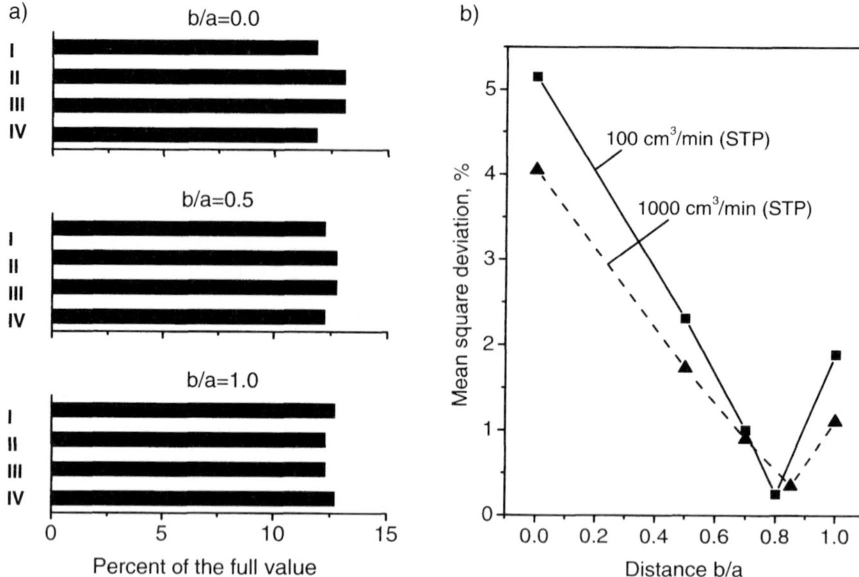

Figure 3.28 Flow distributions in the reactor compartments as a function of the ratio of the structural parameters *b/a* (a). Mean-square deviation from the average value at various *b/a* ratios and various flow velocities (b) [55] (by courtesy of ACS).

In the gas-sampling section, the reaction products from a selected compartment are analyzed without any interference to the other compartments. To judge the amount of cross-over, simulations were made using a 10% oxygen in helium mixture and 10% nitrogen in helium mixture, each flowing in adjacent compartments. The aim of the simulation was to quantify the amount of oxygen cross-over into the nitrogen flow dependent on the penetration depth of the sampling capillary into the nitrogen-flow compartment. It was found that a minimum penetration of 4 mm is required assuming favorable conditions, i.e. low flow rates (Figure 3.29).

The corrosion resistance of the reactor can be improved by deposition of an ultra-thin α-alumina layer (200 nm).

3.4.1.5 Reactor 8 [R 8]: Modular Screening Reactor Unit

A modular reactor similar to the approach of Adler et al. [11] was introduced by Müller and co-workers [37, 38, 56]. The screening procedure was separated into a number of process operations. In chemical process engineering, these so-called unit operations are essential components of every complex plant. As catalyst screening involves many different processes such as heat exchange, flow distribution, sampling, analysis and reaction, such a subdivision into unit operations is justified. The flexibility of such a system was demonstrated with two exemplified configurations later, one of which was used for transient studies and one for steady-state experiments (Figure 3.30).

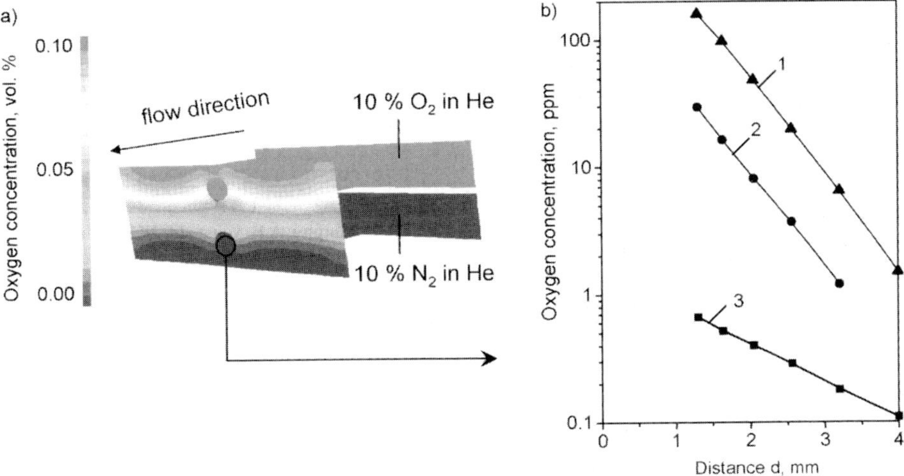

Figure 3.29 (a) Oxygen concentration profile at the inlet and outlet of the compartments of the high-throughput micro reactor. The inlets of the sampling tubes have to penetrate into the compartments to minimize flow cross-over. (b) Area averaged oxygen concentration at one capillary outlet. Total flow velocity: 50 (1), 75 (2) and 100 $cm^3\ min^{-1}$ (3) [55] (by courtesy of ACS).

Figure 3.30 Two versions of the unit construction system built at IMM for transient (left) and for steady-state screening (right) [37] (source IMM).

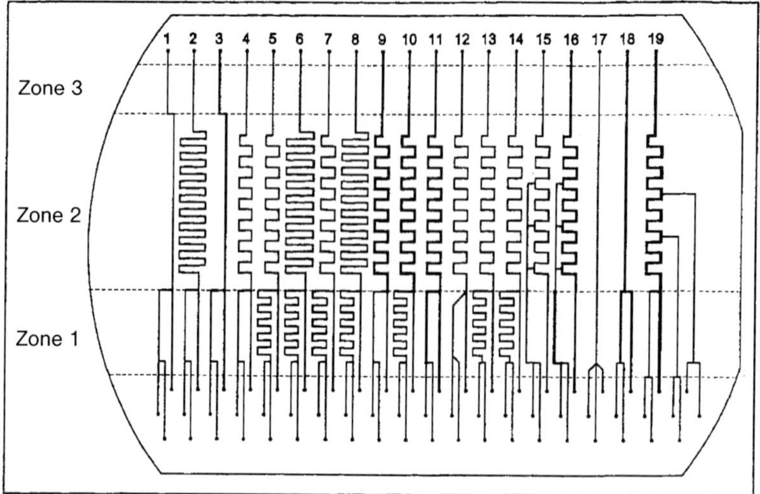

Figure 3.31 Different channel configurations on a silicon chip reactor for the screening of process conditions at the TU Chemnitz [51] (by courtesy of SCS).

3.4.2 Chip-type Screening Reactors

3.4.2.1 Reactor 9 [R 9]: Laboratory Automaton Integrated Chip-like Microsystem

Sample integrations close to pharmaceutical approaches were examined in 1997 [57]. Here, a chip-like microsystem was integrated into a laboratory robot which was equipped with a miniaturized micro titer-plate. Micro structures were introduced later [52] for catalytic gas-phase reactions. The authors also demonstrated [58] the rapid screening of reaction conditions on a chip-like reactor for two immiscible liquids on a silicon wafer (Figure 3.31). Process conditions such as residence time and temperature profile were adjustable. A third reactant could be added to permit a two-step reaction and also a heat transfer fluid which was used to quench the products.

Experimental results concerning performing chemical reactions in the proposed chip-like reactor are not given in [58].

3.4.2.2 Reactor 10 [R 10]: Chip-based Catalytic Reactor

The single-channel reactor of Jensen and co-workers [19, 59, 60] is mentioned because these authors opened the path to a completely new field in catalysis by combining the MEMS (micro-electro-mechanical systems) technology with a chip-based catalytic reactor (Figure 3.32). The reactor consist of a 20 mm long, 625 μm wide, 300 μm deep channel which covers an inner volume of about 3.75 μl. The flow is split at the inlet among several interleaved channels with a width of 25 μm to enhance mixing (compare Figure 3.3). Perpendicular to this reaction channel are 400 μm wide loading channels to place or remove catalyst particles quickly.

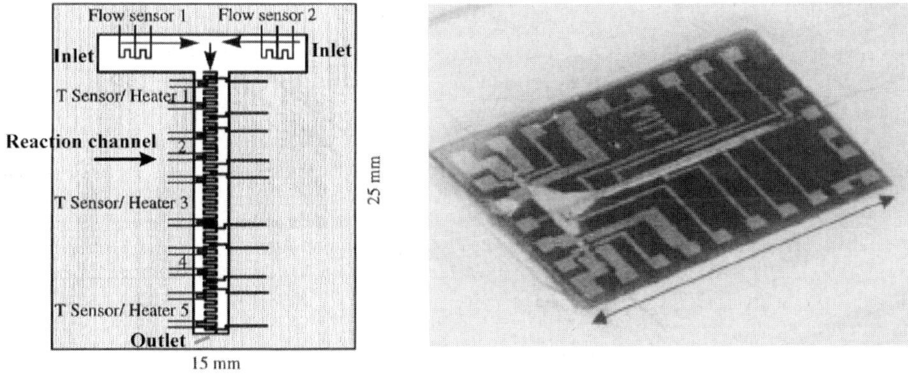

Figure 3.32 Chip-type micro reactor equipped with flow and temperature sensors [19, 62] (by courtesy of Springer Verlag).

By placing vacuum at the exit of the reactor the catalyst particle could be unloaded, and using the inert gas stream at the inlet of the reactor, new catalyst can be placed. At the outlet of the reaction channel two posts 25 μm apart act as a filter to retain the catalyst particles. In addition, some channels were used to hold thermocouples or optical fibers to monitor the experimental conditions. To prevent corrosion of silicon by gaseous chlorine, the channels are covered with a silicon oxide layer of about 5 nm thickness. The reactor is capped with a Pyrex wafer [39]. A more detailed description of the reactor is given in [61].

The operation of the catalytic reactor was demonstrated by the synthesis of phosgene starting from carbon monoxide and chlorine. The experimental set-up is shown in Figure 3.33.

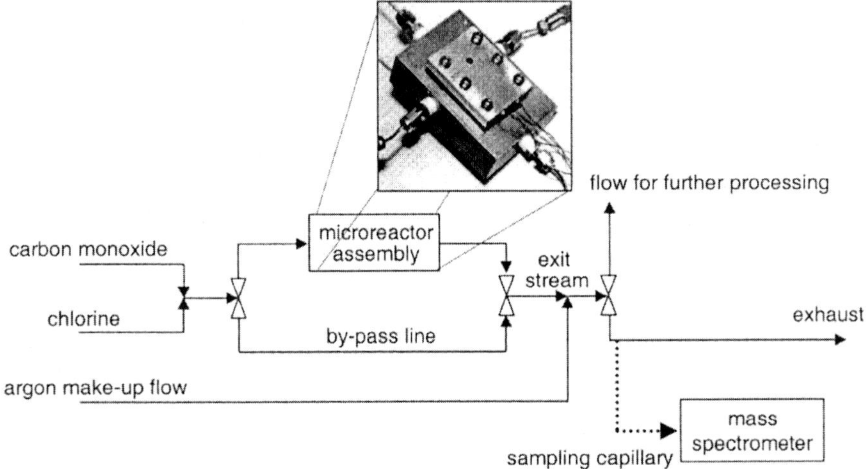

Figure 3.33 Experimental set-up for the synthesis of phosgene [39] (by courtesy of AIChE).

The carbon monoxide and chlorine gas feeds are mixed in the T-junction of the reactor and then guided through the catalyst bed of the reactor. The catalyst, carbon particles with a diameter of 53–73 μm, is preconditioned by heating the reactor at 150 °C for 2 h under a constant argon flow. A mixture of 2/3 CO and 1/3 chlorine (4.5 sccm min^{-1}) is fed into the reactor. The reactor was incrementally heated to 220 °C; the pressure at the inlet was ~132 kPa and nominally atmospheric at the outlet [39].

As a result, no temperature increase due to the exothermal chemical process was observed. With this single-channel reactor, a phosgene productivity of 3.5 kg per year is projected for continuous operation.

3.4.2.3 **Reactor 11 [R 11]: Chemical Processing Microsystem**

Symyx, one of the pioneers in combinatorial screening, presented a chemical processing micro system in 2000 for the screening of 256 catalysts on a silicon or quartz glass wafer [63]. The reactant flow was distributed by a micro structured manifold etched into a silicon wafer (Figure 3.34).

Micro structured wells (2 mm × 2 mm × 0.2 mm) on the catalyst quartz wafer were manufactured by sandblasting with alumina powder through steel masks [7]. Each well was filled with ~1 mg catalyst. This 16 × 16 array of micro reactors was supplied with reagents by a micro fabricated gas distribution wafer, which also acted as a pressure restriction. The products were trapped on an absorbent plate by chemical reaction, condensation or absorption. The absorbent array was removed from the reactor and sprayed with dye solution to obtain a color reaction, which was then used for the detection of active catalysts by a CCD camera. Alternatively, the analysis was also carried out with a scanning mass spectrometer. The above-described reactor configuration was used for the primary screening of the oxidative dehydrogenation of ethane to ethylene, the selective oxidation of ethane to acetic acid, and the selective ammonoxidation of propane to acrylonitrile.

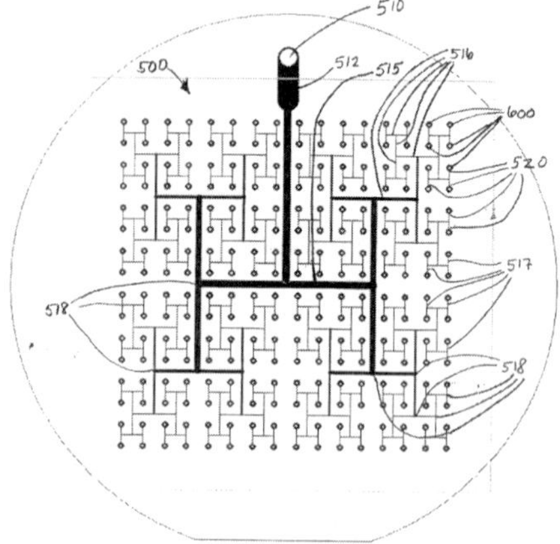

Figure 3.34 Parallel reactor for catalyst screening of wafer-type substrates and silicon wafer supply manifold for the distribution of the reactants to the 256 wells [7] (by courtesy of Kluwer Academic Publishers).

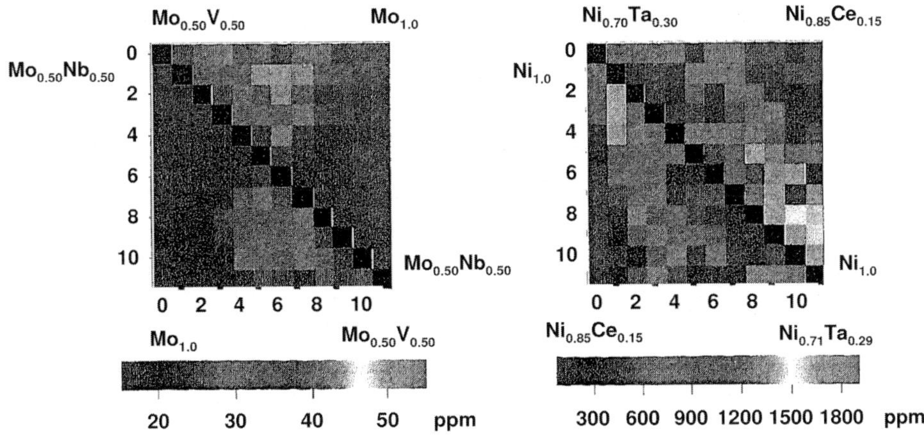

Figure 3.35 Ethylene produced by oxidative dehydrogenation of ethane over Mo–V–Nb and Ni–Ce–Ta oxide catalyst libraries. The detection of ethylene was performed in a scanning mass spectrometer using a photothermal deflection method. Inactive Mo–V–Nb oxide catalyst (a); active Ni–Ce–Ta oxide library (b) [7] (by courtesy of Kluwer Academic Publishers).

Oxidative dehydrogenation of ethane

The oxidative dehydrogenation of ethane is an interesting alternative route to produce ethylene [7]. As catalysts multimetal oxides were used and tested with high-throughput combinatorial methods. A combination of all possible ternary (M1–M2–M3–Ox) with 10% gradient steps gives millions of unique compositions. Keeping in mind that this number can easily be increased by testing all catalyst compositions with a set of reaction conditions, e.g. temperature, pressure, flow rate, reactant concentrations, the screening procedure cannot be done in a limited time. The authors reduced the number of catalyst mixtures to ~100 000 by preparing of a ternary mixture two metal oxides and one metal which can be oxidized or reduced by hydrocarbons or oxygen [7].

Primary screening can be done on wafer-based ternary mixed metal oxide libraries. For the oxidative dehydrogenation of ethane, two interesting libraries consist of Ni–Ce–Nb and Ni–Ce–Ta oxides. The maximum amount of ethylene produced is 1800 ppm at 400 °C in nickel-rich regions of the catalyst mixture (Figure 3.35b) compared with inactive Mo–V–Nb oxide catalysts (Figure 3.35a) [7].

Ni–Ta–Nb oxide catalysts also show high activity for the conversion of ethane to ethylene in primary screening tests. For further optimization bulk catalysts were prepared to perform secondary screening in an 48-channel fixed-bed reactor at 300 °C (see Table 3.4).The highest selectivity (86%) for ethylene was achieved with an $Ni_{0.62}Ta_{0.10}Nb_{0.28}O_x$ catalyst [7].

Promising catalyst mixtures were tested over a 400 h run on-stream. Under the given reaction condition, no significant loss in activity or selectivity was observed [7].

Table 3.4 Secondary sceening of different $Ni_xTa_yNb_z$ oxide catalysts [7].

Catalyst composition	*Ethane conversion (%)*	*Ethylene selectivity (%)*
Ni	11	54
$Ni_{0.88}Ta_{0.12}$	12	55
$Ni_{0.62}Ta_{0.38}$	19	84
$Ni_{0.89}Nb_{0.11}$	18	82
$Ni_{0.63}Nb_{0.37}$	19	85
$Ni_{0.70}Ta_{0.10}Nb_{0.20}$	20	85
$Ni_{0.62}Ta_{0.10}Nb_{0.28}$	21	86

Reaction conditions: $O_2 : C_2H_6 : N_2 = 0.088 : 0.42 : 0.54$ sccm at 300 °C; 50 mg catalyst.

3.4.3 Pellet-type and Ceramic Reactors

3.4.3.1 Reactor 12 [R 12]: Alumina Tablets Equipped Parallel Gas-phase Reactor

Some authors recommend the use of pellets for fast catalyst screening. Among the pioneers was Senkan [64–66, 134], who developed a parallel gas-phase reactor for the screening of up to 80 catalysts (Figure 3.36). The reactor is composed of four micro structured ceramic plates each having 20 wells. The wells can be equipped with standardized γ-alumina pellets (23 mg γ-Al_2O_3, 4 mm diameter, 1 mm height), which are impregnated with metal salt solutions. Product gases exiting the micro reactor channels were sequentially withdrawn by a capillary sampling line and analyzed with a quadrupole mass spectrometer.

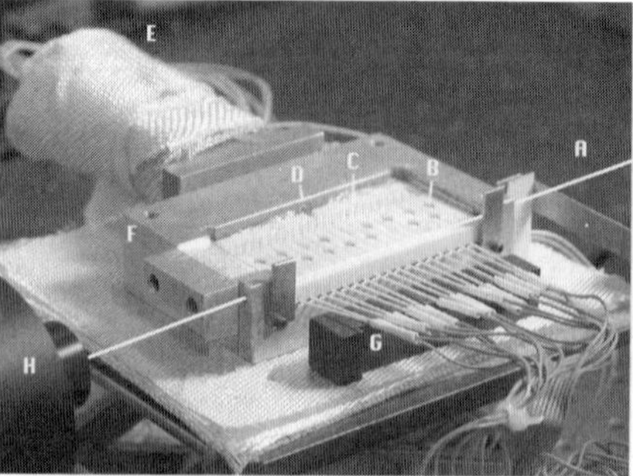

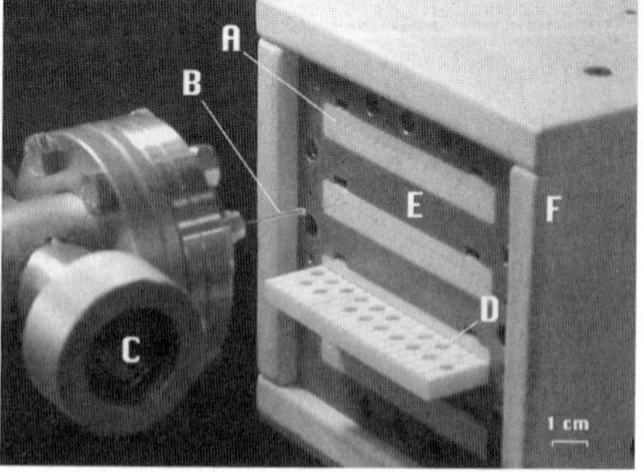

Figure 3.36 Reactor set-up of the multiple-chamber reactor system at the University of California. Left, upper three chambers removed; right, detail of sample gas delivery via capillary B to mass spectrometer C, lower drawer moved forward to show the catalyst pellets D [134].

Dehydrogenation of cyclohexane

The above-described set-up has been used for the primary screening of various reactions, e.g. the dehydrogenation of cyclohexane using different catalyst compositions of the ternary system Pt/Sn/In [64]. Concerning this first example, a collection of 66 catalyst combinations was operated for 24 h to check the long-term behavior. All combinations show deactivation, especially the most active catalysts. This clearly demonstrated that screening also depends very much on the selection of the appropriate time scale for a reaction (in this study, at least 24 h).

Selective oxidation of propylene to propylene oxide

In the second example, the selective oxidation of propylene to propylene oxide (PO) was investigated with a huge library of γ-alumina pellets impregnated with various single metals, binary metal combinations and catalyst loadings [40].

$$\text{propylene} + O_2 \xrightarrow{\text{Rh-catalyst}} \text{propylene oxide} + \text{acrolein} + \text{acetone} + CO_2$$

To focus only on the binary compositions, the catalysts were evaluated under the following experimental conditions: feed 40% propylene and 10% oxygen; temperature, 250 °C; and 20 000 h^{-1} of GHSV (gas hourly space velocity). Three different catalyst preparation procedures were used [40]:

Set A. Selection from periodic groups. Two elements, one PO active and the other PO inactive, were selected from each periodic group, resulting in 153 different catalysts in total.

Set B. Selection from top 20. Twenty of the most active elements were selected and combined to form at least 190 different catalysts.

Set C. Rh-based catalysts. As the single metal with the highest PO activity, rhodium was identified. Rhodium was combined with other elements at three different loading levels to form 102 catalysts. The rhodium level was kept constant at 1% in each catalyst composition.

The study revealed that rhodium is a particularly PO-active metal, regardless of whether it is used as a single component or in binary combinations. Exemplary results for the catalyst Set A are shown in Figure 3.37. For the experimental results for the catalyst Set B, see [40].

As a consequence of the experimental results for catalyst Sets A and B, appropriate rhodium-containing catalysts were tested as Set C. Figure 3.38 shows the reactor outlet concentration for propylene oxide, acrolein and acetone. A large number of the catalysts tested produce high concentrations of propylene oxide of up to 2000 ppm at 1% conversion of propylene. The combinations Rh–Sn and Rh–In are very effective for propylene oxide formation. In most cases the binary catalysts have higher activity at lower propylene loading. In Figure 3.38, it can also be seen that with the exception of Rh–Sn, all other Rh-based catalysts were not effective with regard to acetone and acrolein formation. Some of these catalysts exhibited significant conversion to carbon dioxide [40]. The best catalyst, of course, must have a high selectivity for propylene oxide and a low selectivity for the total oxidation to carbon dioxide. Rh–Ag, Rh–Zn, and Rh–Cr binary catalysts have both and they should be further evaluated [40].

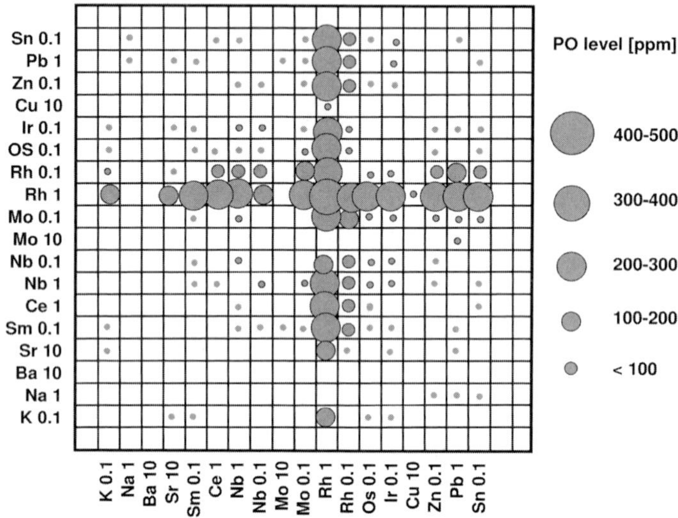

Figure 3.37 Reactor exit concentrations of propylene oxide for the binary catalyst Set A. Feed gas, 20 000 h^{-1} GHSV; propylene : oxygen ratio, 4 : 1; temperature, 250 °C; 40% propylene [40] (by courtesy of Elsevier Ltd.).

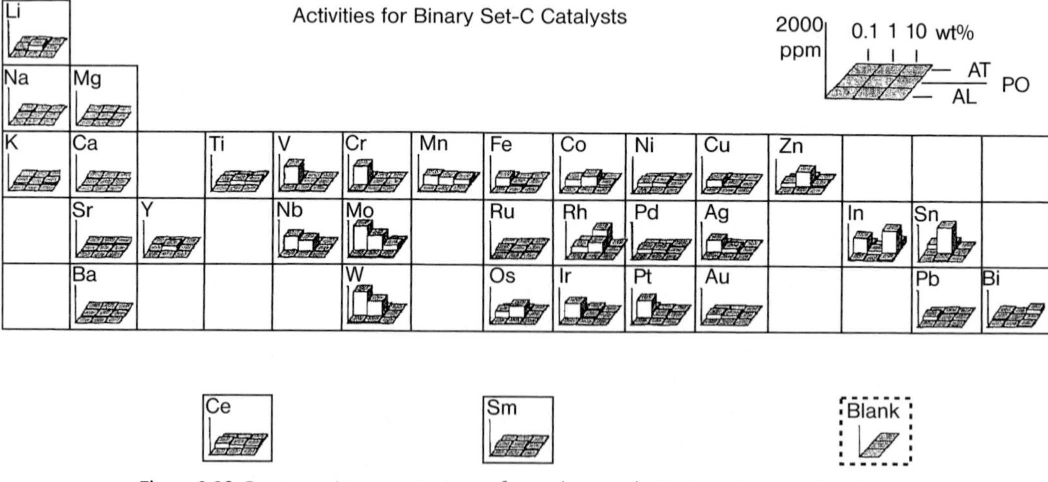

Figure 3.38 Reactor exit concentrations of propylene oxide (PO), acetone (AT) and acrolein (AL) for the binary catalyst Set C. Feed gas, 20 000 h^{-1} GHSV; propylene : oxygen ratio 4 : 1; temperature, 250 °C; 40% propylene [40] (by courtesy of Elsevier Ltd.).

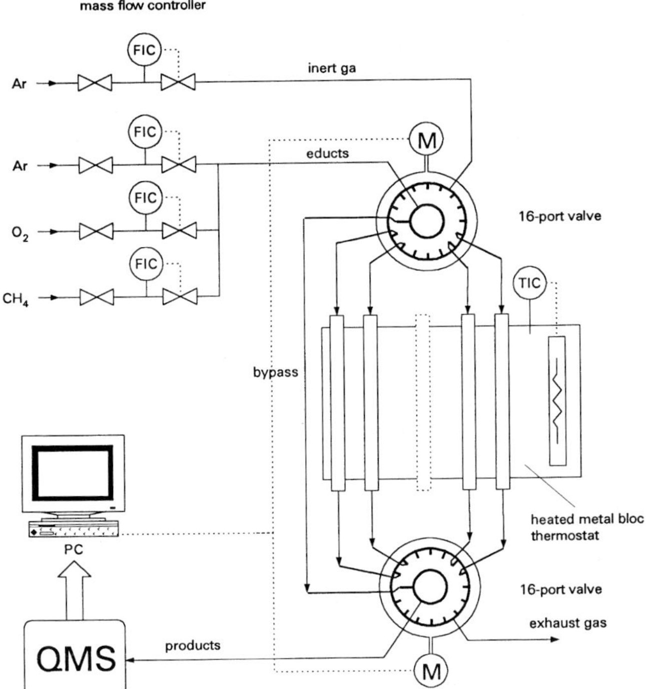

Figure 3.39 Schematic of the 15-fold automatic testing device [67].

3.4.3.2 Reactor 13 [R 13]: Ceramic Monolith Reactor

A 15-fold glass tube parallel packed-bed reactor was introduced [67, 68], which is similar to conventional catalyst testing equipment (Figure 3.39). The premixed reactant gas is supplied by a 16-port valve either to the bypass or to one of the 15 reactors. By a second 16-port valve, the product gas stream of a selected reactor can be channeled to the quadrupole mass spectrometer. The full automation of the screening set-up allows the investigation of 15 catalysts per day [67].

Oxidative coupling of methane

As a first example, the authors investigated the oxidative coupling of methane to higher hydrocarbons on various $Mn/Na_2WO_4/SiO_2$ catalysts. It is known from the literature that the activity and selectivity of the catalyst increases if the Mn amount is low with Na necessary to disperse Mn on the catalyst carrier surface. Each of the 15 catalyst compositions shown in Table 3.5 were filled into the quartz reactor. After calcination 6 h in oxygen at 850 °C, the catalysts were used for the oxidative coupling of methane reaction. It turned out that the highest activities for the oxidative coupling of methane were obtained with 2.5–4 wt.% Na_2WO_4, whereas the Mn content had no significant effect. Nevertheless, the highest proportion of ethylene, the target product, was achieved for catalysts which contained 0.1–1.0 wt.% Mn and 2.5–5 wt.% Na_2WO_4 (see Table 3.5) [67].

Table 3.5 Experimental results for testing of an $Mn/Na_2WO_4/SiO_2$ catalyst array for the oxidative coupling of methane [67].

Na_2WO_4 (%)	Conversion of methane (%)					Ethylene selectivity (%)					Ratio ethylene/ethane				
	Mn 0	Mn 0.1	Mn 1.0	Mn 1.9	Mn 5.0	Mn 0	Mn 0.1	Mn 1.0	Mn 1.9	Mn 5.0	Mn 0	Mn 0.1	Mn 1.0	Mn 1.9	Mn 5.0
0	9	14	12	10	11	35	32	31	31	29	0.5	0.6	0.6	0.5	0.5
0.1	13	15	15	15	16	31	35	37	36	37	0.6	0.7	1.0	0.9	1.0
1.0	4	17	17	17	17	34	39	40	39	45	0.1	1.2	1.2	1.1	1.2
2.5	5	19	11	16	17	41	51	79	47	46	0.1	1.2	0.8	0.8	0.9
4.0	3	17	19	11	9	75	79	77	78	81	0.1	1.5	1.7	0.8	0.6
5.0	2	15	12	10	9	79	82	82	80	82	0.1	1.3	1.0	0.7	0.5

T = 800 °C; 36 000 h^{-1} GHSV; feed 76% CH_4, 14% O_2, 10% Ar.

Oxidation of carbon monoxide

For the oxidation of carbon monoxide, Au nanoparticles as catalyst on different carriers were tested. It was shown recently that nanodispersed Au particles on metal oxide catalyst carriers allow the oxidation of carbon monoxide even at temperatures below 0 °C. Different catalyst carriers, e.g. ZrO_2, TiO_2, SiO_2, MgO, ZnO_2, Nd_2O_3, Y_2O_3, CeO_2 and Mn_2O_3, were investigated. The catalysts were prepared individually by drying, calcination in an air flow or reduction in hydrogen at different temperatures [67].

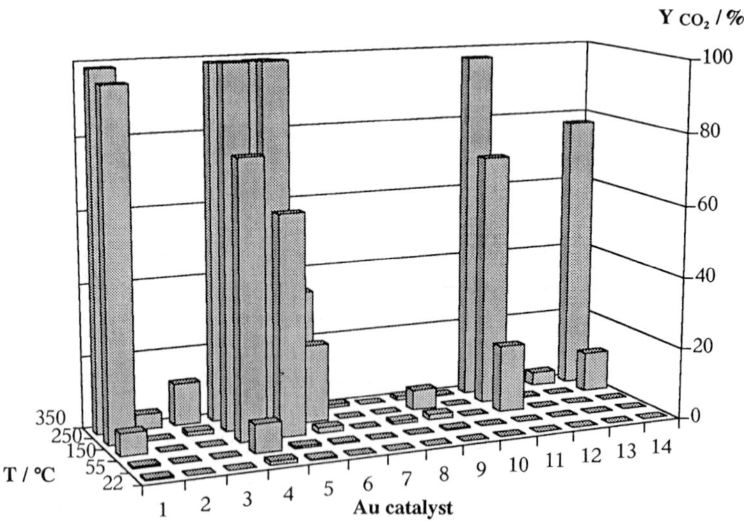

Figure 3.40 Oxidation of Co on nanodispersed Au particles on metal oxide catalyst carriers as a function of temperature: 20 000 h^{-1} GHSV; feed, 1% CO, 20% O_2, 4% Ar, 75% He [67].

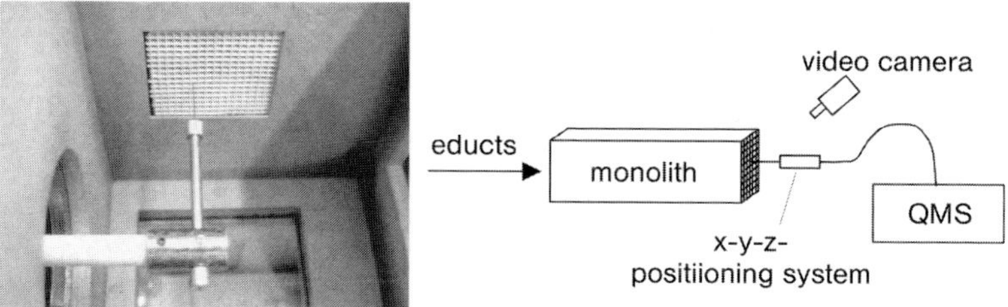

Figure 3.41 *X–Y–Z* positioning device for scanning the individual catalyst-coated micro channels of a ceramic monolith (left) and plant set-up for high-throughput experiments at the ACA Berlin-Adlershof (right) [69] (by courtesy of Springer Verlag).

The results, summarized in Figure 3.40, indicate that in particular catalysts based on ZrO_2 and TiO_2 were active. Also, similar activities were found for Au/Nd_2O_3 catalysts. Surprisingly, all catalysts show remarkable activity at temperatures above 150 °C, which is contrary to the results found in the literature. Obviously the preparation conditions of the catalyst play a major role in the development of the catalyst activity and further investigations should be made [67].

For the screening of 14 catalysts at five different temperatures, only 16 h were needed.

The same groups also reported on a 64-fold ceramic block reactor and a ceramic monolithic reactor for the screening of up to 250 catalysts in parallel (Figure 3.41). The catalyst array was prepared via an incipient wetness method by combination of different amounts of Pt, Zr and V on the alumina walls of the monolith by means of an automatic liquid handler. Gas samples from each channel of the monolith were analyzed sequentially by a quadrupole mass spectrometer by moving a capillary sampling line into the channels with the help of a three-dimensional positioning system [69].

3.4.3.3 **Reactor 14 [R 14]: High-pressure Fixed-bed Reactor**

Schüth's group developed in the past a number of reactors similar to conventional testing methods with different degrees of sample integration. For multiphase reactions a 25-fold stirrer vessel reactor was developed [70] and for heterogeneous gas-phase reactions a 16-fold fixed-bed reactor was presented [71], which was later followed by a 49-fold parallel reactor [135]. The reactor in Figure 3.42 was used for methanol production from Syngas at up to 50 bar and was essentially an improved version of the 49-fold reactor described in [135].

Some of their methods have in the meantime been commercialized by hte AG, a German supplier of services in combinatorial catalysis. High-pressure applications have not been reported very often owing to their huge expenditure. A 14-fold stirrer autoclave was presented [73] for liquid/gas reactions.

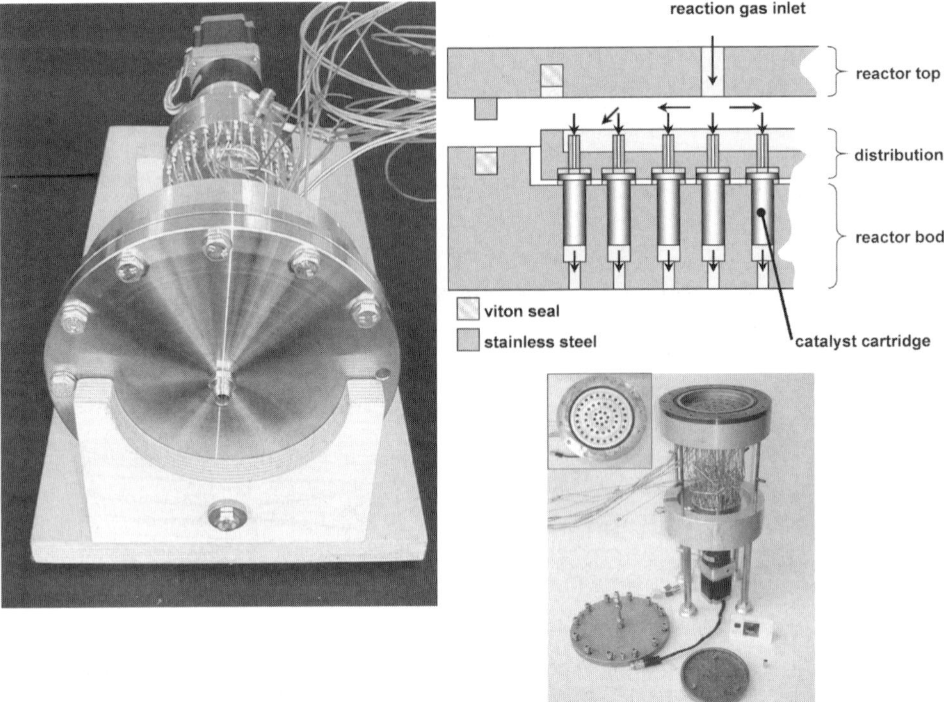

Figure 3.42 Parallel 49-fold fixed-bed reactor for high-pressure use at the MPI-Mühlheim. Mounted reactor (left), schematic sectional view (right) and unmouted reactor parts (bottom) [135] (by courtesy of Elsevier Ltd.)

3.4.3.4 Reactor 15 [R 15]: Multiple-bead Pellet-type Catalyst Carrier Reactor

Being in a way the link between pellet-type reactors and well-type reactors, Klein et al. presented a multiple-bead reactor [74] in combination with a split and pool synthesis. The reactor shown in Figure 3.43 consists of pellet-type catalyst carriers, so-called beads, which are positioned in square containers.

Figure 3.43 Multiple-bead reactor (by courtesy of hte AG).

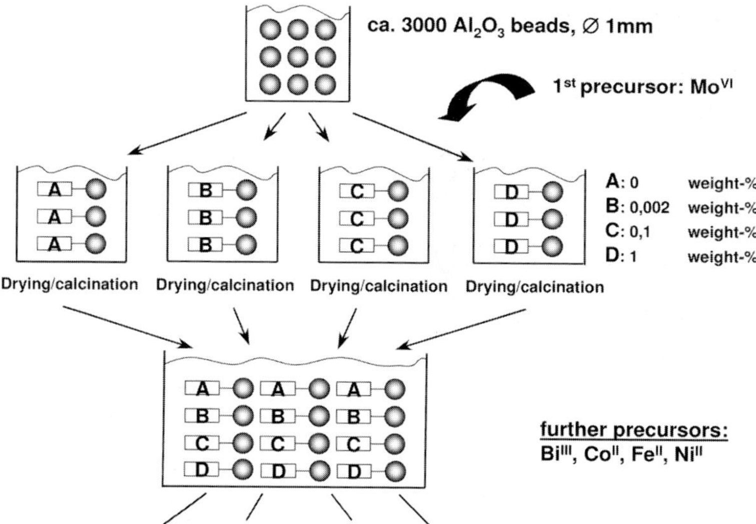

Figure 3.44 Schematic of the split and pool procedure [74] (by courtesy of Elsevier Ltd.).

The pool consisted of γ-alumina beads with a diameter of 1 mm. Gases passed these beads through micro structured pore membranes in the cover and the base plate of the containers. The synthetic parameter space of the split–pool library had to be extended by synthetic parameters such as drying, calcination conditions and even mechanical parameters such as the stirrer speed. This is an indication that inorganic solid-state catalysts occupy a much larger parameter space than organic catalysts. On the other hand, the large amount of sample mass compared with organic synthesis allows the detection of the elemental composition of the bead. Thus, a simple and direct encoding method could be used for the identification of a single bead in the pool. In addition to the MS and GC measurements executed, the silicon-based array reactor permits also infrared detection of catalyst activity. A real example of the split and pool procedure is given in Figure 3.44.

3.4.4
Well-type Screening Reactors

3.4.4.1 Infrared/Thermography Monitored Screening Reactor

Reactions that introduce a temperature difference either by the emission of heat or by consuming heat are accessible to infrared–thermographic methods. Reetz et al. [75] reported the screening of enantioselective reactions on a modified micro titer-plate. A time-resolved picture of a sector of this titer-plate showed increased activity of the catalysts in three individual wells (Figure 3.45).

To validate IR imaging for screening reactions, the activity and enantioselectivity of metal catalysts (*S,S*)-**4a–c** for the hydrolysis of epichlorohydrin was tested. Time-resolved changes in temperature indicated that the (*S,S*)-**4c** catalyst as the most active and an *S*-configured epoxide (**1c**) is preferred for hydrolysis (Figure 3.45).

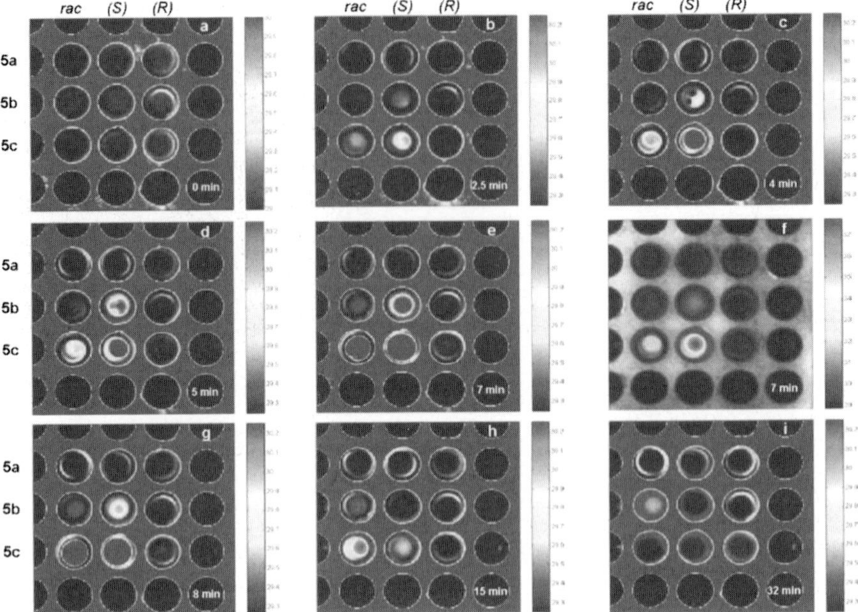

Figure 3.45 Infrared–thermographic picture of the time-dependent activity of selected catalysts for the hydrolysis of epichlorohydrin after 0, 2.5, 4, 5, 7, 8, 15 and 32 min from (a) to (i), with (e) and (f) recorded at the same time but different scales [75].

H_2O, (S,S)- **4**

1 → **2** + **3**

1 a) R = CH_2OCH_2Ph
b) M = Ph
c) M = CH_2Cl

4 a) M = Mn (Cl)
b) M = Cr (Cl)
c) M = Co

3.4.4.2 Reactor 16 [R 16]: Catalyst Filled Borings Reactor

Klein and co-workers [76, 77] investigated catalysts made by combinatorial hydrothermal synthesis directly in the reactor vials. The reactor for the catalyst syntheses shown Figure 3.46 consists of a silicon base-plate fixed in a circular housing. To generate the 37 vials, a PTFE disk of 4 mm thickness with drilled holes (diameter ~1.4 mm) is placed directly on the silicon base-plate and press-fitted by an addition stainless-steel plate [76].

To generate a catalyst 2 μl of solution of ~1 $\mu mol\ l^{-1}$ were pipetd into each vial to achieve 50–150 μg of solid matter. After closing the reactor, the whole system was

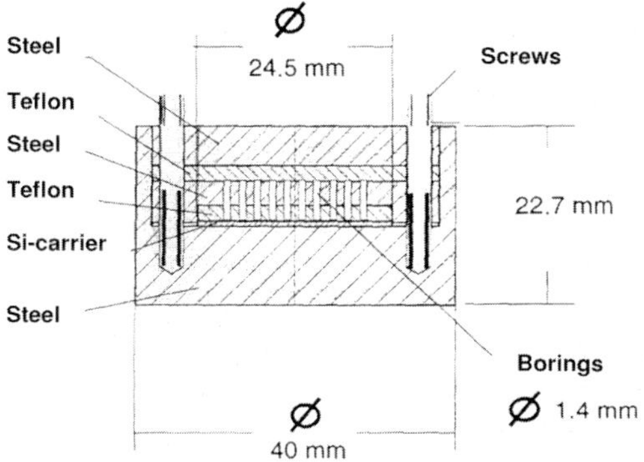

Figure 3.46 Schematic of the reactor used for hydrothermal catalyst synthesis (left) and photograph of the unmounted reactor parts (right) [76].

heated to temperatures which normally were used for hydrothermal synthesis. After cooling, the reactor is opened and the silicon wafer with the formed catalyst spots is calcined at the necessary temperatures. With this method, combinatorial syntheses were performed to achieve modifications of TS-1 silicalites [76].

Oxidation of propene

To investigate the activities of the synthesized catalysts the silicon wafer with the catalyst spots is mounted in an experimental setup (Figure 3.47). Propene and air are mixed prior to the reactor. Via a bundle of glass capillaries (two glass-capillaries are mounted in a stainless-steel tube), the reactant gas mixture is delivered to the catalyst spot surface while the second capillary, connected to a mass spectrometer, guides the reaction mixture to the analyzer. Each of the catalyst spots can be addressed by using a commercial pipetting robot. A detailed cross-sectional view of the reactor is given in Figure 3.47 [77].

The oxidation of propene was chosen as a test reaction to validate the catalyst activities. Propene and synthetic air with a ratio of 0.4 and an overall flow rate up to 7 ml min^{-1} were used to screen 33 different catalysts (Table 3.6) [77].

Table 3.6 Synthesized and screened amorphous, micro porous mixed oxide catalysts [77].

1	*2*	*3*	*4*	*5*	*6*
Sc_2Si	Y_2Si	V_5Si	Ta_2Si	Cr_2Si	Fe_2Si
Co_2Si	Rh_2Si	Ni_2Si	Cu_6Si	Ag_2Si	Au_2Si
In_2Si	Si	Sn_5Si	$Bi_{10}Si$	Te_2Si	W_3Ti
W_3Zr	Mo_3Ti	Mo_3Zr	Sb_3Ti	Sb_3Zn	Cu_3Ti
Cu_3Zr	In_3Ti	In_2Zr	Re_3Ti	Re_3Zr	Cr_3Ti
Cr_3Zr	Fe_3Ti	Fe_3Zr			

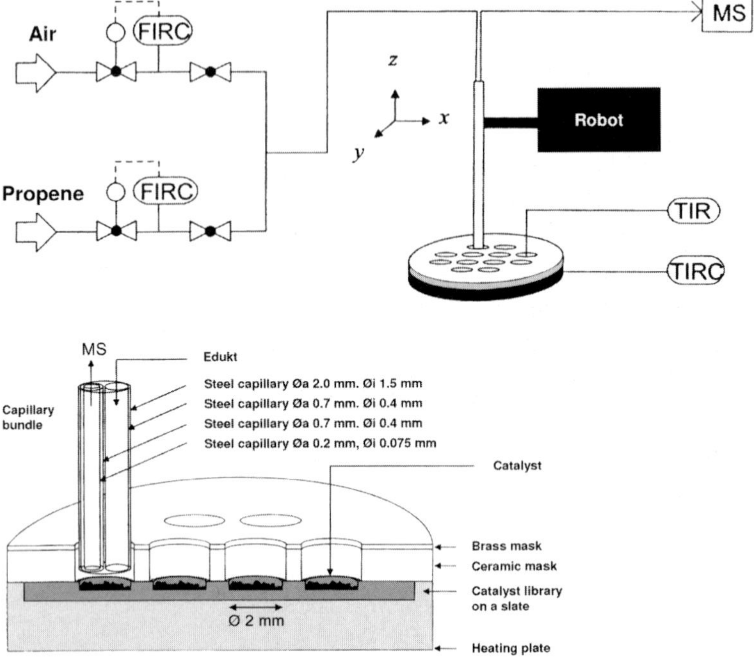

Figure 3.47 Schematic of the experimental set-up for catalyst testing with high-resolution mass spectrometry (left) and cross-sectional view of the reactor (right) [77].

For the oxidation of propene, different reaction products are expected, e.g. carbon dioxide, acrolein, benzene, propylene oxide, 1,5-hexadiene, allyl alcohol [77]:

As an example, one experiment will be described. The catalysts were heated to 500 °C and the flow rate and ratio of the gases adjusted to values given above. Each catalyst was measured for a period of 1 min, resulting in an overall time of 33 min for the investigation of one catalyst-spotted silicon wafer. It could be shown that an automated system is be able to generate many experimental data, e.g. the mass spectra of the whole reaction product composition. When the mass peaks are separated, e.g. mass peaks with m/z 67 typical for 1,5-hexadiene, an increase is observable with increasing temperature (Figure 3.48). In this case an amorphous In_3Sn catalyst was used [77].

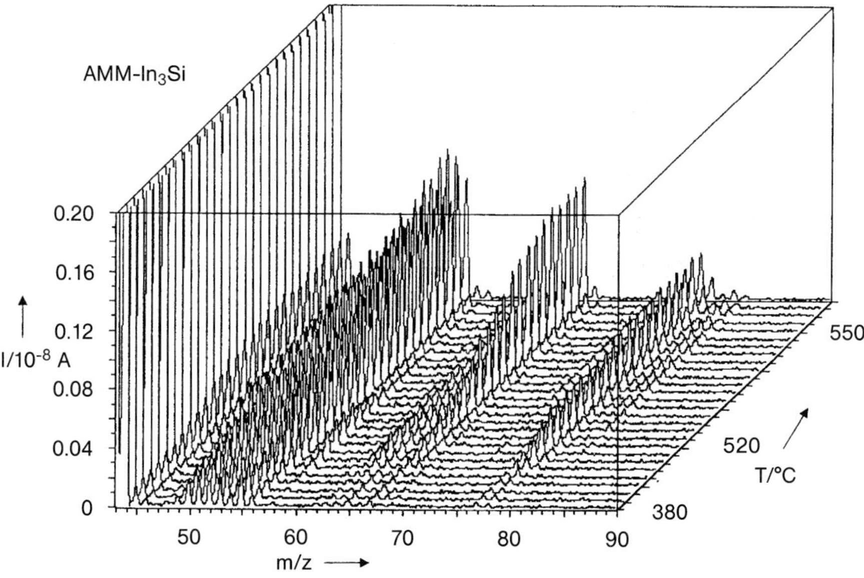

Figure 3.48 Increase of mass peak *m/z* 67 with temperature as a result of the oxidation reaction of propene using amorphous In_3Sn catalysts [77].

3.4.4.3 Reactor 17 [R 17]: Sputtered Catalyst Spots on Quartz Wafer Reactor

To investigate CO oxidation reactions, a ternary library of transition metals was prepared by Cong and co-workers [78, 79]. They prepared catalyst samples by co-sputtering of the individual targets and by sol–gel techniques on a quartz wafer. The time for the preparation of a sputtered library was reported to be only 1 h.

To scan the large number of catalysts, a scanning mass spectrometer was used. By applying this analytical technique, measurement, of both the activity and selectivity of more than 100 catalysts in a library within 3 h is possible.

The experimental set-up consists of two parts, the reaction chamber where the catalytic reaction occurs, and the analysis chamber where the reaction products are analyzed (Figure 3.49). Some ancillary equipment is necessary to position the library for analysis and, in addition, some control units for monitoring temperature, gas flow rate and pressure are incorporated. The feed gas is delivered through a cylindrical glass tube directly to the catalyst spot placed on the glass wafer and the product gas is sucked through a second glass tube, which is placed inside the gas delivery tube, to the ionization zone of the mass spectrometer. To minimize cross-talk between neighboring catalytic spots on the wafer, both glass tubes are positioned in close proximity directly over the catalytic spot to be investigated. The temperature of the catalytic spot can be adjusted individually by laser irradiation and monitored with an IR sensor. If the reaction conditions are stable, the measurement can be performed [80].

A ternary library of noble metals is formed by deposition of solutions containing Pt, Pd and Rh by sol–gel techniques on a 3 in quartz wafer to form catalyst spots.

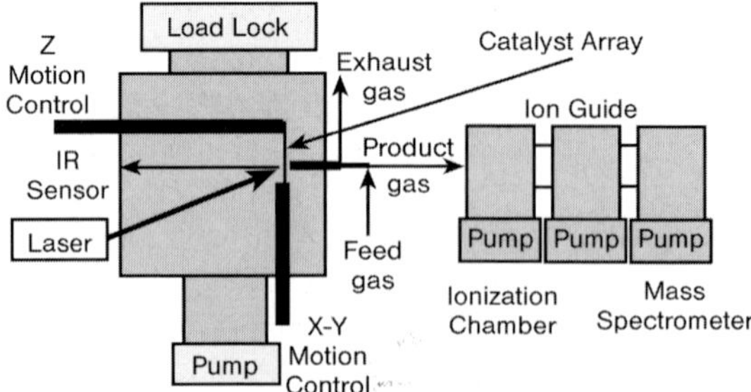

Figure 3.49 Schematic of experimental set-up for a scanning mass spectrometer-assisted catalyst screening device [80] (by courtesy of AIChE).

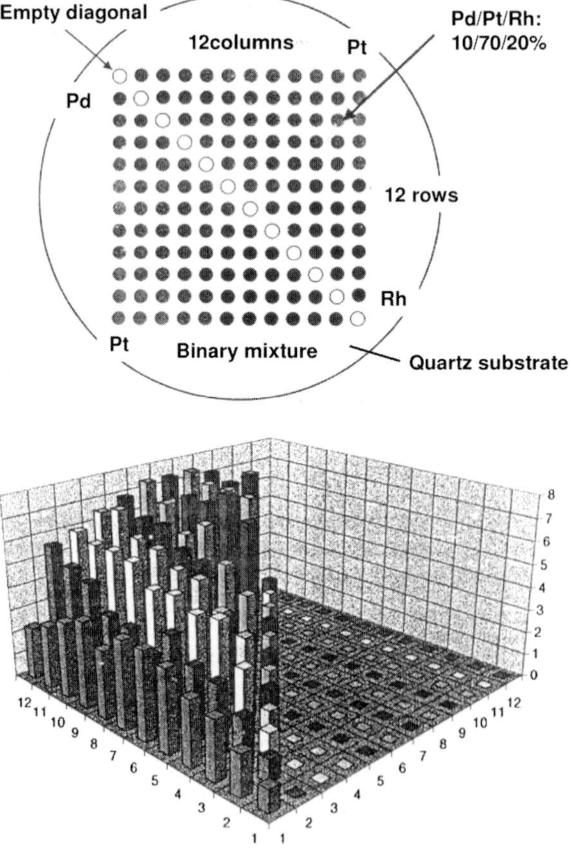

Figure 3.50 Intensity of the CO_2 signal measured by a scanning mass spectrometer. Reaction conditions: reaction temperature, 350 °C; stochiometric Co and O_2 ratio; Ar as carrier gas. The library is redundant and mirrored at the diagonal while the diagonal itself does not contain any catalytically active material to allow baseline calibration. (2,1) element is pure Rh, (12,1) element is pure Pt and (12,11) element is pure Pd. The numbers in the x,y-plane indicate the catalyst composition and the numbers in z-direction indicate CO signal intensity [43] (by courtesy of Elsevier Ltd.).

All the solutions which are necessary to form an equilateral triangular 11 × 11 × 11, 61-element ternary phase diagram were prepared automatically by a liquid dispensing machine. Mixtures with compositions of each component ranged from 0 to 100% in 10% increments. On the 3 in quartz wafer used, 144 wells form a 12 × 12 square grid for the catalyst spots. Each of these elements has a diameter of 3 mm. The whole array is calcined at 500 °C for 12 h and subsequently for an additional 12 h at 800 °C under a 6% H_2 in argon atmosphere [80].

The performance of a ternary library consisting of various mixtures of Pd, Pt and Rh was used in the conversion of carbon monoxide to carbon dioxide. At a temperature of 350 °C the conversion of carbon oxides increases with increase in Pd concentration. Along the binary line of Pd/Pt, the conversion goes through a maximum with increasing Pt content (Figure 3.50). In general, these trends are also observed when the temperature is increased to 400 °C but the overall conversion is higher. A further increase of the temperature up to 450 °C leads to complete conversion with nearly all catalysts, except for Pt–Rh binary compositions.

Since a scanning mass spectrometer was used for measuring the conversion, 144 experiments were carried out within 2 h [80].

3.4.4.4 Reactor 18 [R 18]: Polymerization Reactions Screening Reactor

Fluorescence spectroscopy as a means for judging process conditions in a polymerization reaction was reported [81, 82]. The authors optimized parameters such as the reactant ratio and the catalyst amount to reach the smallest variability of material properties. The samples were deposited on a 96-well micro reactor array and examined with a spectrofluorimeter during the reaction (Figure 3.51).

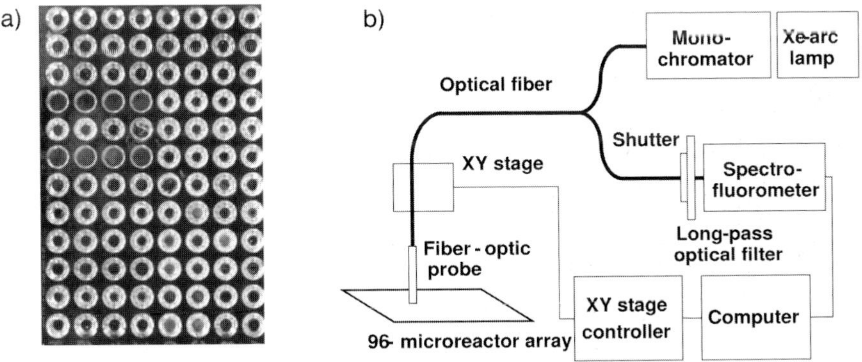

Figure 3.51 96-well reactor array for the parameter screening of polymerization conditions [82] (by courtesy of ACS).

3.4.4.5 Reactor 19 [R 19]: Photochemical Active Catalyst Parallel Screening Reactor

Improvement of the efficiency for the photochemical splitting of water was the aim of Morris and Mallouk [83]. A parallel optical screening method was developed to select photocatalytically active catalysts by their absorption spectra with UV and visible light.

3.4.4.6 Reactor 20 [R 20]: Microstructured Chips with Catalyst-coated Channels

Besser et al. [84] studied the hydrogenation of cyclohexane over a platinum catalyst. They used micro structured chips with channel widths of 100 and 5 μm and demonstrated that the conversion in the smaller channels is larger than expected, owing to a larger surface/volume ratio (Figure 3.52). The conversion data found were consistent with data from macro-scale reactors.

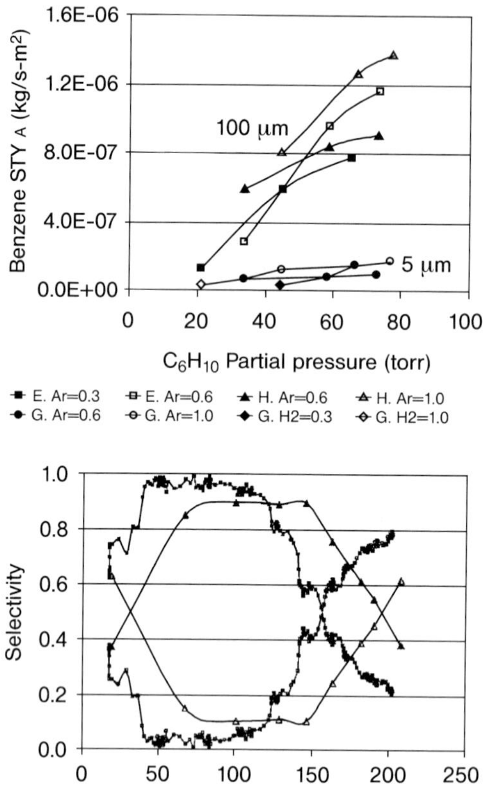

Figure 3.52 Conversion and selectivity data for the hydrogenation of cyclohexane on a platinum catalyst in micro structured channels with widths of 100 and 5 μm [133] (by courtesy of ACS).

3.4.4.7 Reactor 21 [R 21]: 64-Channel Tubular Disk Fixed-bed Reactor

The ceramic 64-channel reactor of Rodemerck and co-workers [67, 68] bridges the gap between micro- and macroscopic fixed-bed reactors (Figure 3.53). The catalyst containers consist of a massive tubular disc of alumina with bores for the catalyst powder clamped between two massive cordierite discs. This set-up can be operated up to 550 °C at pressures up to 3 bar.

In this set-up, two new catalysts were found for the oxidative dehydrogenation of ethane to ethylene. With catalysts Cr/Mo–O_x and Co/Cr/Sn/W–O_x an industrially relevant product yield of more than 60% was reached, as shown in Figure 3.54.

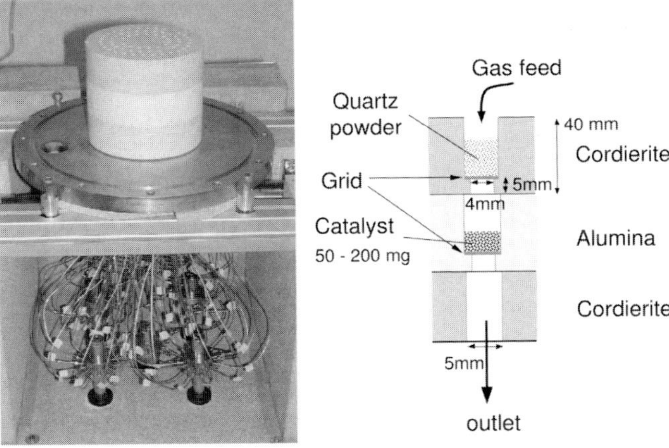

Figure 3.53 Ceramic 64-fold tubular reactor consisting of three ceramic discs (left) and internal cross-sectional view of a single channel including pressure restriction (right) [67].

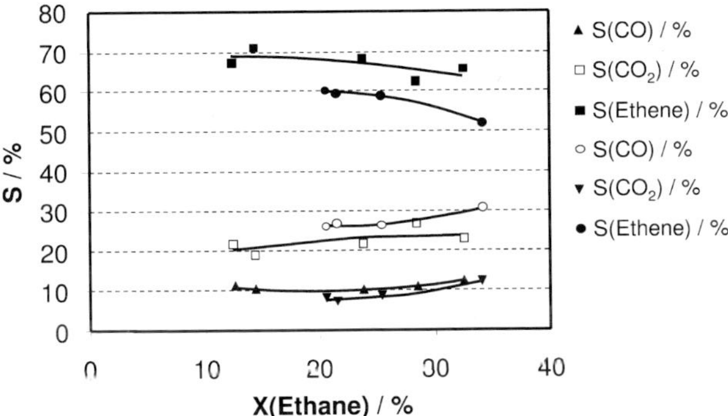

Figure 3.54 Selectivity for two catalysts for the dehydrogenation of ethane to ethene (upper curves, Cr/Mo–O_x; lower curves, Co/Cr/Sn/W–O_x) [67].

3.4.4.8 **Reactor 22 [R 22]: The Microstructured Titer Plate Reactor Concept**

Well-type flow reactors offer some advantages over stacked-plate reactors. The most prominent advantage is the possible thermal isolation of wells on a two-dimensional array by the supply of appropriate heat sinks underneath the titer-plate. Highly active catalysts neighboring wells with less active catalysts could introduce so-called thermal cross-talk. Thermal cross-talk means the apparent activity increase of a catalyst introduced by heating from a neighboring catalyst and not by the real reaction rate itself. The thermal balance of a titer-plate for chemical studies thus differs from the demands on titer-plates operating under operation conditions without heat development as, for example, in the pharmaceutical industry. An example of the design of a titer-plate in catalysis is described in detail below.

The screening method described here was originally developed in the pharmaceutical industry to examine samples positioned on so-called titer-plates with up to several thousand species nowadays. This approach was transferred to high-temperature catalyst screening [37, 38]. Owing to the danger of thermal cross-talk, this huge number of species cannot be expected in the field of heterogeneous catalyst screening. Nevertheless, an increase in efficiency and flexibility was desired. The standardized format of these plates enabled the build-up of a library of catalysts with a very dense format for later reference. The standard format of the titer-plate is the 48-well format with the respective number of individual catalysts. Plates with other configurations and different numbers of wells are possible.

The parallel reactor for the steady screening of the titer-plates consisted of several modules. Each of them was responsible for just a single operation (Figure 3.55). The gas flow, for example, was preheated and evenly distributed within the distribution module and delivered to the wells on the titer-plate. The latter was clamped between the distribution module and the insulation module. The insulation module separated the heated section of the parallel reactor from the unheated section and was further cooled by the heat exchanger module on top of it. The last module just above the heat exchanger module was a multi-port valve, which delivered the product gas to the gas chromatograph.

The concept of using micro structured titer-plates in a screening device helps to prevent some of these difficulties when using conventional screening devices. One advantage of micro structured devices in general is the possibility of realizing isothermal reaction conditions. Tubular reactors filled with catalysts as powder or as small beads will have a radial temperature profile whose gradient depends on the reaction rate. Without the possibility of cooling (or heating) the catalyst particles,

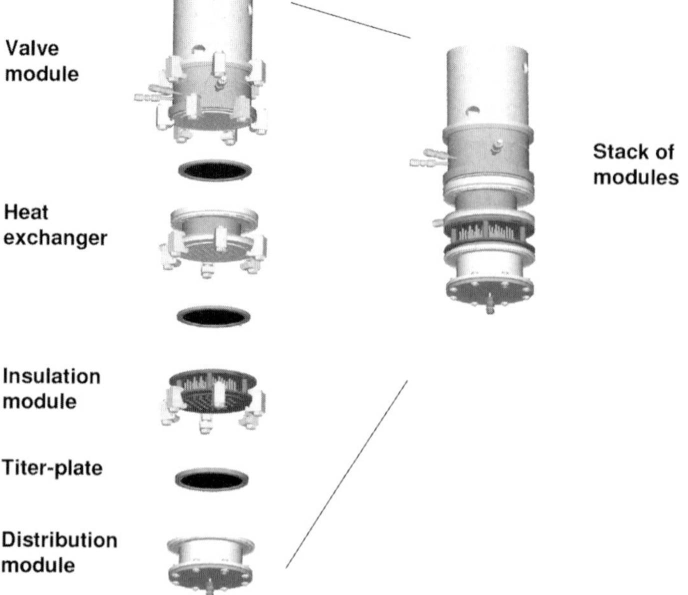

Figure 3.55 Modular set-up of 48-fold parallel gas-phase screening reactor [38] (by courtesy of VDI-Verlag GmbH).

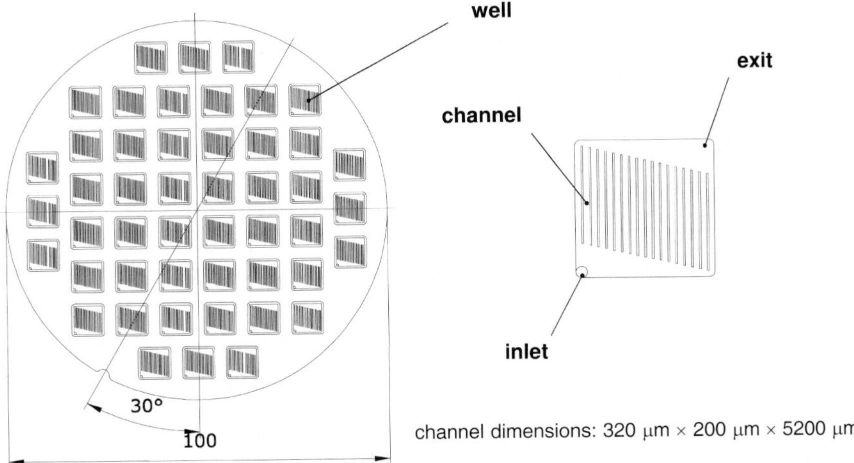

Figure 3.56 CAD drawing of a 48-fold titer-plate [38] (by courtesy of VDI-Verlag GmbH).

the additional heat produced inside such a reactor will distort the apparent reaction rate obtained. This problem can be accounted for by planar reactors with the possibility of conditioning the reactor walls. A general problem occurs, of course, with such a planar reactor. In contrast to tubular reactors, planar reactors will not necessarily have a symmetric flow, which could result in internal regions not adequately supplied with feed flow. To consider this, one has to supply guiding structures inside such a flat planar reactor. Owing to their small dimensions, a number of these planar reactors can be arranged on one titer-plate following similar approaches as in the pharmaceutical industry, and consequently such a reactor can also be called a well (Figure 3.56). Different from conventional titer-plates, these wells are flow reactors, which are electrically heated, and the walls of the wells are held at a constant temperature by the use of heat sinks below the plate.

To check this assumed isothermal behavior, one first has to examine the temperature rise in a single well due to the chemical reaction [38]. As test reaction, the catalytic partial oxidation of methane was selected:

$$CH_4 + \tfrac{1}{2}\,O_2 \rightleftharpoons CO + 2\,H_2 \qquad \Delta H_r = -35.7\ \text{kJ mol}^{-1}$$

The space velocity in the well was chosen as 9000 h^{-1}, which delivered an average velocity in the channels of 0.0133 m s^{-1} and a residence time τ of 0.4 s. The temperature distribution was calculated numerically with the well geometry given above. The numerical method applied was the finite-volume method of the software package CFX4 with an upwind/central differential scheme in a three-dimensional flow. Owing to the larger pressure loss in micro structures compared with macro-scale devices, compressible flow was assumed for the calculations. The well was assumed to be coated only at the two vertical walls and at the bottom wall of the channel. To account for the exothermic heat developing during the chemical reaction at the walls, appropriate heat sources were allocated to these walls (Figure 3.57).

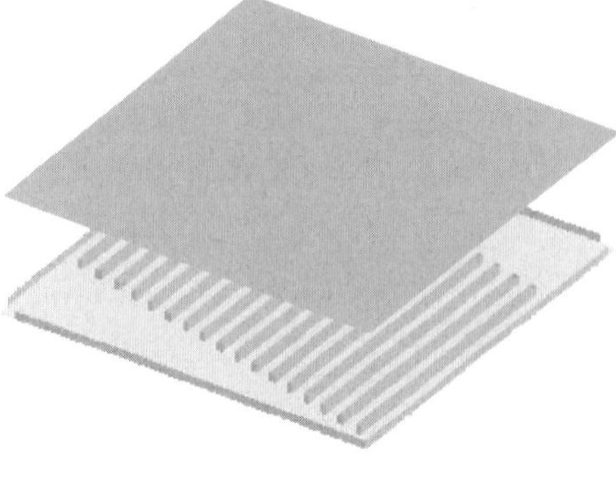

Figure 3.57 Micro structured well with channel structure and cover plate (all channel walls were allocated heat sources, all other faces outside the channels and the cover plate were defined as isothermal) [37] (by courtesy of Elsevier Ltd.).

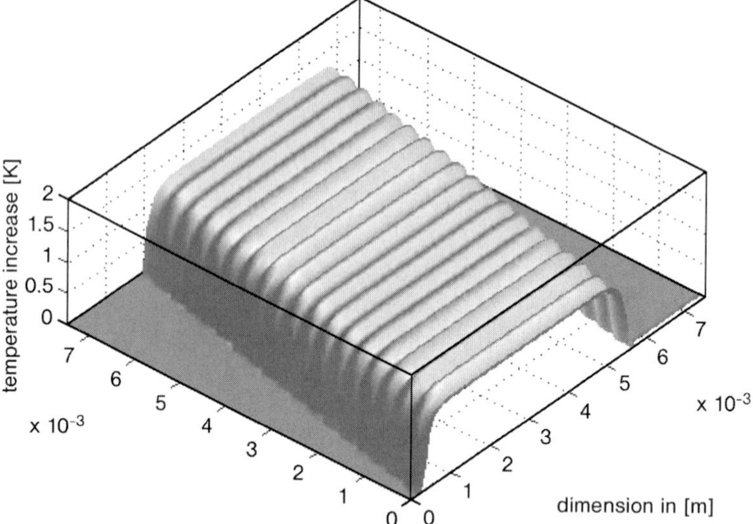

Figure 3.58 Internal well-surface temperature at the coated bottom wall [38] (by courtesy of VDI-Verlag GmbH).

These heat sources were derived from the reaction enthalpy and related to the wall surface area. This delivered a specific heat of 259.15 W m^{-2} developed at the walls. All other surfaces were set to be isothermal with a surface temperature of 500 °C. The justification for this boundary condition will be given by executing a temperature calculation inside the titer-plate. As fluidic boundary conditions, the flow velocity at the inlet to the well was fixed and the outlet pressure was set equal to the ambient pressure.

Figure 3.58 shows the results of the numerical calculation of the temperature distribution for this well in an isothermal reaction regime. Isothermal conditions were realized by applying appropriate heat sinks close to the well.

Two temperature surfaces were given: the temperature increase of the coated bottom wall of the channels (Figure 3.58), and the temperature increase at the uncoated cover plate (Figure 3.59). The difference between the two surface

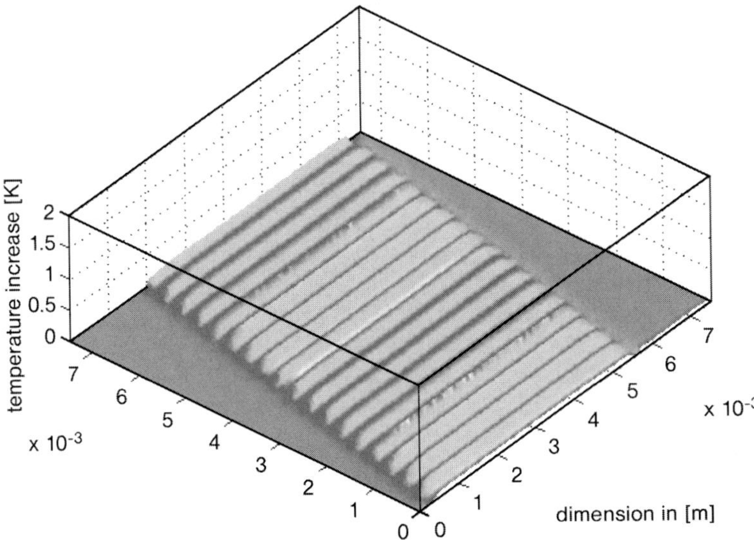

Figure 3.59 Internal well-surface temperature at the uncoated cover plate [38] (by courtesy of VDI-Verlag GmbH).

temperatures is due to the reaction heat developing at the coated walls. The ripples in the temperature surface of the bottom plate indicate the slightly higher temperatures in the corners of the channel, which is also demonstrated in Figure 3.60.

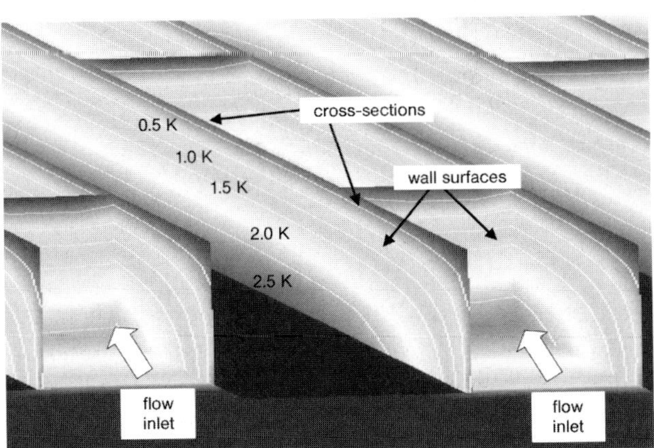

Figure 3.60 Distribution of gas temperature increase in the channel of a well with coated walls close to the flow inlet (without cover plate). The gas temperature increases asymptotically owing to the reaction and the lowest temperature is reached at the border to the uncoated (not reactive) cover plate. The channel walls are removed to show the temperature inside the channel. The temperature in the cross-section is shown at two different positions [38] (by courtesy of VDI-Verlag GmbH).

In Figure 3.60, the temperature increase of the gas flow at the entrance to the channels is given. Obviously the gas is heated by the wall reaction during its path through the channel indicated by the asymptotic profile of the isotherms along the channel (see line 0.5 K). The gas is slightly less heated in the center of the channel and reaches the temperature of the isothermal cover plate at the top of the channel. It is obvious that also the gas itself shows close to isothermal behavior.

For comparison to the isothermal conditions in the well, an adiabatic calculation with the same well geometry was executed. This time, all uncoated faces were set to be adiabatic while the coated channel faces obeyed the above-given boundary conditions. Figure 3.61 shows that there is a tremendous temperature increase if the exothermal reaction heat is not allowed to leave the control volume.

These results indicate that only by supplying appropriate isothermal boundary conditions will the temperature of the reactants not exceed the wall temperature.

These assumed boundary conditions have to be derived from a heat balance of the titer-plate. The reaction heat was assumed to be released equally distributed in the wells, hence the heterogeneous conditions were ignored. The titer-plate was accompanied by two steel carrier plates (plate depth 5 and 8 mm) accounting for the heat sinks and the equilibration of the reaction heat. Both plates were in good thermal contact with the titer-plate by graphite foils, which also effected sealing of the reactants. Due to the symmetry, the problem was reduced to one quadrant of the titer-plate. The reaction heat of every single well was set to 0.041 W (equivalent to a space velocity of 9000 h^{-1}). It was assumed that there is no heat flux through the cylindrical wall. All other faces were isothermal with a wall temperature of 500 °C. This was assumed to be a reasonable boundary condition considering the huge metallic mass of the carrier plates with respect to the titer-plate. Inside the

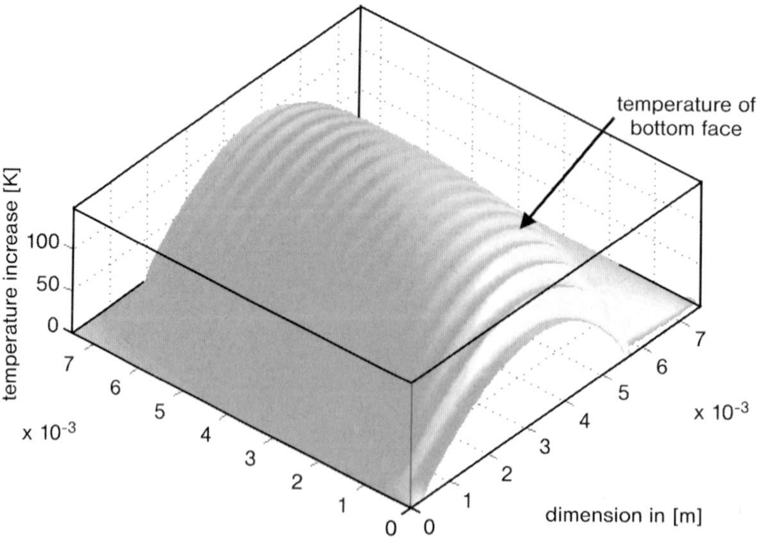

Figure 3.61 Surface temperature at the bottom wall of a coated well assuming all uncoated walls to be adiabatic [38] (by courtesy of VDI-Verlag GmbH).

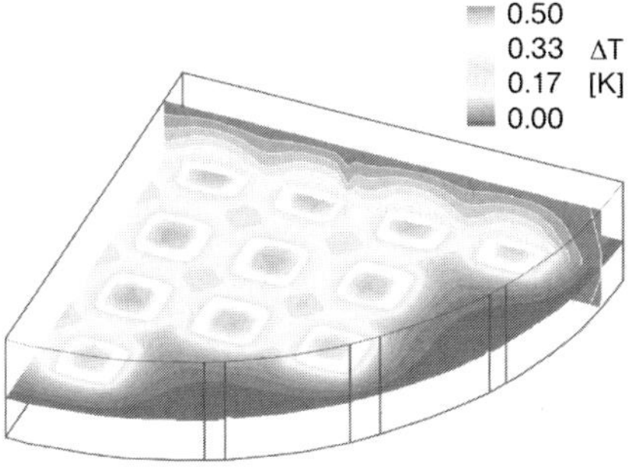

Figure 3.62 Horizontal and vertical temperature distribution in the titer-plate (11 wells shown) [38] (by courtesy of VDI-Verlag GmbH).

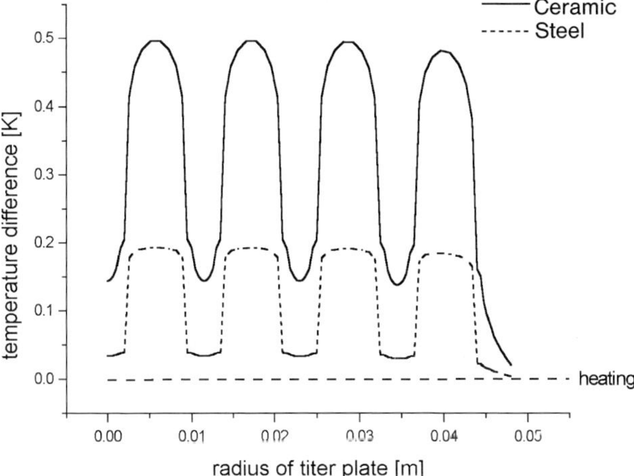

Figure 3.63 Gas temperatures in a ceramic and a steel titer-plate [38] (by courtesy of VDI-Verlag GmbH).

titer-plate the temperature increased by a mean value of 0.5 K (Figure 3.62). This comes close to the isothermal numerical calculation in a single well from above, which delivered a maximum temperature increase of 1.3 K. The difference is due to the simplified assumption of a homogeneous heat distribution inside the single wells on the titer-plate.

The temperature increase in the single wells is clearly visible in the horizontal cross-section and in the vertical cut. The gas temperature in the wells also depends on the material of the titer-plate. In Figure 3.63, the gas temperature obtained in a steel titer-plate (heat conductivity 30 W mK^{-1}) was compared with that in a titer-plate made of a less conductive material such as an aluminum oxide ceramic (heat conductivity 3 W mK^{-1}). In either case, the gas temperature does not exceed the temperature of the solid material by more than 0.5 K and in the solid material between the wells this temperature drops towards the heater temperature.

Very active catalysts produce a huge amount of heat, which can influence other catalysts in neighboring positions on the titer-plate. To account for such a situation, a single well was selected to represent such a hot-spot and given heat generated by a four times higher conversion than for all the other wells. The results are given in Figure 3.64.

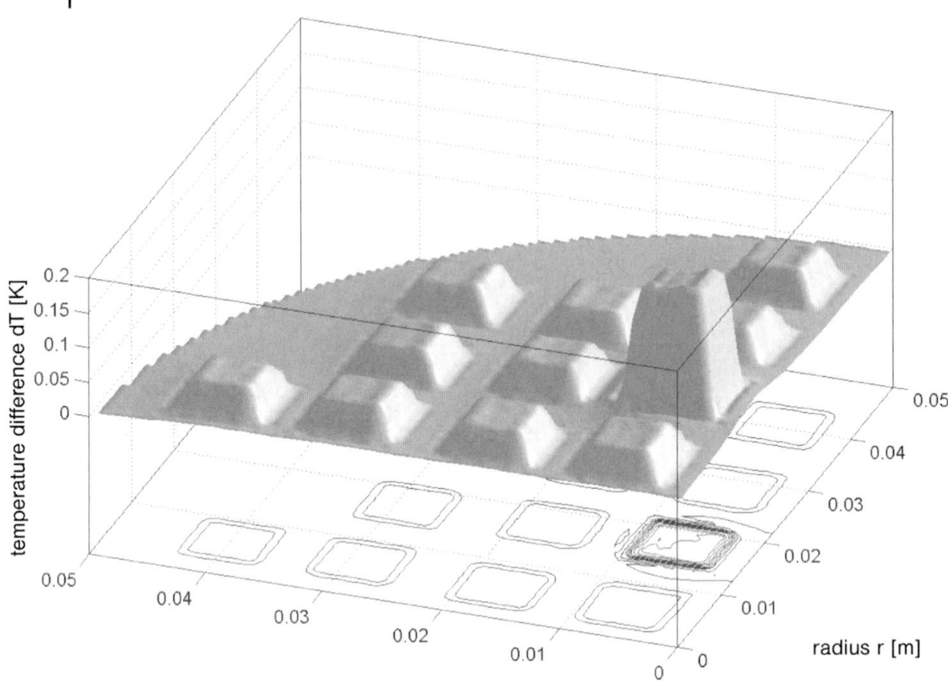

Figure 3.64 Temperature distribution in the titer-plate close to a very active catalyst [38] (by courtesy of VDI-Verlag GmbH).

The isotherms around the most active well (see parentheses in Figure 3.64) are a measure of the cross-talk to a neighboring position. The shallow curvature of the isotherms indicates that the heat flow due to the local temperature gradient can be neglected.

Catalyst Screening for Combustion of Methane

The reactor set-up was applied to the catalytic combustion of methane at low temperatures [56]. For the catalytic combustion of methane, a number of parallel and consecutive reactions are known to occur [85]. Cullis et al. [86] also found that under low-temperature conditions the conversion of methane is nearly completely due to the formation of these total oxidation products, a result which was confirmed in the experiments.

Catalyst activity tests

After the coating with catalysts, the titer-plates were inserted into the reactor to test the activity of the catalysts. Steady reaction conditions were adjusted in all of the following experiments. The reactor was heated to 475 °C and held under a pressure of 0.2 bar(g). The heat exchanger was operated at 50 °C and the throughput for a single well was adjusted to 1 ml min^{-1}, which resulted in a space velocity of 9000 h^{-1}. The residence time in the wells was 0.4 s.

An under-stoichiometric mixture of methane and oxygen (2 : 1) was diluted with 70% nitrogen, preheated and distributed within the distribution module and continuously delivered to the single wells.

Catalysts prepared by the wash-coating method were first used in order to check the reproducibility of the measured values. For this reason, six elementary metal salts (platinum, zirconium, molybdenum, nickel, silver and rhodium) were dissolved and impregnated on a titer-plate. The catalysts were pre-reduced inside the reactor with 5% hydrogen in 95% nitrogen at 250 °C. The results were recorded before and after the pre-reduction. The repeated measurements indicated good reproducibility in both cases. The conversion of methane with the rhodium catalyst is better after the pre-reduction.

Thermal or fluidic cross-talk

In order to check if fluidic or thermal cross-talk exists, the same six metal catalysts were coated in six rows on one titer-plate. Active catalysts were positioned close to wells with less active catalysts (Figure 3.65). The result was that the two active rows (platinum and rhodium) did not influence the neighboring less active rows (zirconium and silver). Thus, cross-talk was not observed. In the next test, elementary, binary and ternary mixtures of the same metal catalysts were examined. Each mixture was deposited on four different wells to check the reproducibility (Figure 3.66). Except for some of the ternary mixtures, the results were well reproduced and were consistent with the known well activity of rhodium (row 1) and the mixture of rhodium and platinum (row 5), which had the highest activity.

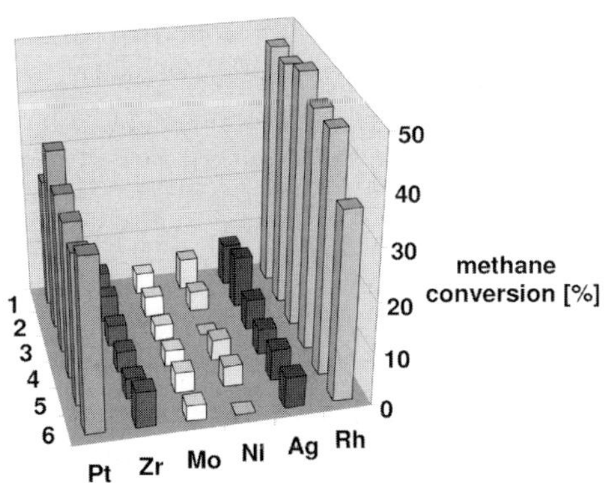

Figure 3.65 Six metallic catalysts each deposited on six different wells on a titer-plate (conversion is correlated with the correct geometric position of the wells) [37] (by courtesy of Elsevier Ltd.).

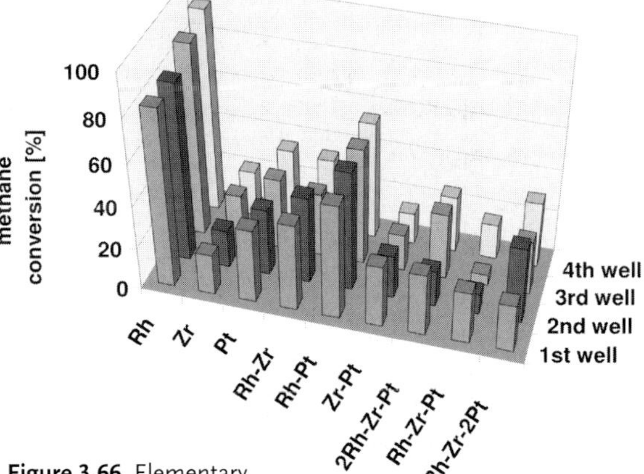

Figure 3.66 Elementary (rows 1–3), binary (rows 4–6) and ternary (rows 7–9) catalysts deposited on four different wells [37] (by courtesy of Elsevier Ltd.).

3.4.4.9 **Physical Parameter Screening Reactor**

The majority of studies have concentrated on the screening of chemical properties. Screening of physical parameters such as the condensation behavior of a liquid mixture, the wettability of a solution or crystal morphology has not often been reported. Polymorphism, the ability of a liquid mixture to crystallize in different crystal structures, was the objective of a study executed at Solvias (Switzerland) [87]. A 96-well quartz micro titer-plate was filled with solvent mixtures. Nitrogen was delivered to the individual wells in order to dry the solvents. A number of solvent mixtures were prepared and the crystal growth was observed. The different crystal structures could be characterized by Raman microscopy. The crystal structure of the tested drugs influences the performance of the final product.

3.5 Reactors for Transient/Dynamic Operation

3.5.1 Transient Operations in Microstructured Gas-phase Reactors

With the introduction of micro reactors, transient reactor operations also became of interest for production owing to their low internal reactor volume and thus fast dynamic behavior. In 1999, Liauw et al. presented a periodically changing flow to prevent coke development on the catalyst and to remove inhibitory reactants in a micro channel reactor [88]. This work was preceded in 1997 by Emig and Seiler, of the same group, who presented a fixed-bed reactor with periodically reversed flow [89]. In 2001, Rouge et al. [27] reported the catalytic dehydration of isopropanol in a micro reactor.

The use of transient reaction conditions for the characterization of catalysts began in 1950 with the single-pellet reactor [90]. The single-pellet reactor is basically a Wicke–Kallenbach cell operated under reactive environments. It was extended in 1974 into the single-pellet tube reactor, which is closer to industrial environments. A one-dimensional dispersion model was used to obtain the effective dispersion coefficient from pulsed input signals. Chromatographic techniques for the evaluation of kinetics in a micro pulse reactor were introduced in 1955 [91]. The model of Schneider and Smith [92] is often used for the evaluation of adsorption rates or effective diffusion coefficients from experimental data. From 1961 to 1969, Polanski and Naphthali [93] developed frequency response experiments where the reactor volume is periodically changed and the resulting pressure fluctuation is used to describe adsorption rates on a catalyst surface. In 1998, Colin et al. [94] applied the frequency response method to rectangular micro channels close to and below 1 μm. They studied the slip effect at the channel wall with pulsed sinusoidal signals. In 1999, a highly miniaturized flow-through calorimeter on a silicon chip was presented [95] for the evaluation of kinetics using an integrated thermochemical detector.

Despite recent promising strategies, the principle of micro process engineering is still not widely used in combinatorial catalysis. One drawback certainly is the increasing distance to industrial applications with decreasing dimensions. On the other hand, the small structures possess laminar flow conditions, which are fully accessible by analytical and not only numerical macroscopic descriptions. This offers the chance to describe thoroughly the fluidic, diffusive and reactive phenomena in catalysis in order to find intrinsic kinetics on using, for example, non-porous sputtered catalysts.

Screening in the stationary mode will only give information about the activity of a single catalyst or a catalyst mixture. When a proper catalyst for a certain reaction is found, the next important information is the reaction kinetics. To obtain this information, a number of methods and reactors are recommended in the recent literature [10, 91, 96–101]. Most of them apply transient reactor operations to find detailed kinetic information. Micro reactors are particularly suited for such an operation since their low internal reaction volumes permit a fast response to process parameter changes, for example, concentration or temperature changes. This feature has been applied by some authors to increase the product yield in micro reactors [34, 98, 102].

Micro reactors operated in the pulsed mode were introduced by Kokes et al. in 1955 [91], but have been intensively used only in the last 10 years. Such transient studies to obtain insight into reaction mechanisms were undertaken by Gleaves et al. with the *temporal analysis of products* (TAP) reactor 1997 [100]. They observed rate coefficients of elementary reaction steps such as adsorption and desorption by applying pulses of reactants to a catalytic micro reactor combined with a quadrupole mass spectrometer.

Not only concentration pulses have been used as input signals. Wojciechowski used temperature ramps with his *temperature scanning reactor* [99, 103] and Kobayashi and Kobayashi [104] applied concentration step functions. Typical process parameters, which can be changed, are the pressure, the temperature or the composition of the gas mixture. Fast mixture or pressure pulses can be realized by the injection of reaction gas into the system by a micro-dispense valve. An appropriate flow sensor will then record the transition into the next stationary mode.

3.5.1.1 **Reactor 23 [R 23]: Microstructured Titer Plate Transient Reactor Concept**

To accommodate such measurements, the reactor set introduced in Section 3.4.4.8, *The Microstructured Titer Plate Reactor Concept*, was equipped with a second insulation module and a second rotary valve. The latter replaced the gas distribution module of the steady reactor configuration (Figure 3.67). The input signal was injected by an injection valve into the reactor inlet valve, which mainly consisted of a short wound 1/16 in Teflon tube, which sent the signal through one of the tubes of the insulation module and a bore in the foil heater to the reaction plate. From here, the product gas was delivered via one of the tubes of the second insulation module, the Teflon tube of the second (exit) valve directly to the sensor. The diameter of the internal tubes ($\phi = 1$ mm) was the same either in the Teflon tubes or in the steel tubes of the insulation modules. The gas also passed etched channels in the distribution plate of the inlet and the exit valve, which is a concession to the application of these valves also for steady screening. Owing to the etching process, the cross section of these channels in the distribution plate was U-shaped with a width of 0.5 mm, a depth of 0.3 mm and a length of 50 mm for each channel. The

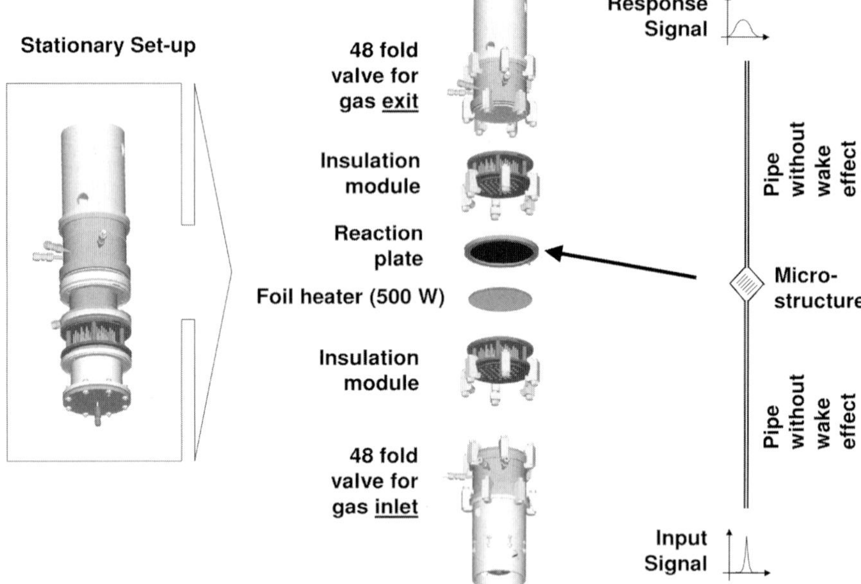

Figure 3.67 Modular set-up of configuration for transient screening (change from the stationary configuration to the transient configuration by the exchange of modules) [38] (by courtesy of VDI-Verlag GmbH).

reactor core of this set-up consisted of an exchangeable reaction plate, which could be structured with various micro geometries. This feature results from the fact that gas inlet and outlet positions could be selected among 48 bores.

For the following tests, a wound spiral was manufactured by fine mechanical means. The reaction plate was heated by a 500 W foil heater, which had a number of bores to allow the gas to pass through the heater.

For the transient measurements, a special foil heater of 0.7 mm thickness was designed (and manufactured by Watlow). The internal wiring scheme of the electrical resistance wire is shown in Figure 3.68. The 0.5 mm bores were located in the center of the white spots of the wiring scheme. The wiring is the remaining part of the metal foil after the etching procedure. A bottom and a top layer consisting of electrically insulating ceramic fiber material cover the resistance wire. The foil heater thus possessed some flexibility and could withstand the sealing force of the hydraulic press.

For the dispersion measurements and for the coupled dispersion/reaction measurements, the same reaction plate was used. A 1.5 m long channel with a rectangular cross-section of 1 mm width and 0.5 mm depth was manufactured out of a 1.4571 steel plate (Figure 3.69). The plate was coated in a sputtering plant with a 300 nm layer of platinum. The reactants entered the channel through a 0.5 mm bore from below the plate. The reactants exited the channel through a similar bore in the cover plate. To allow long channels on the reaction plate, the channel was

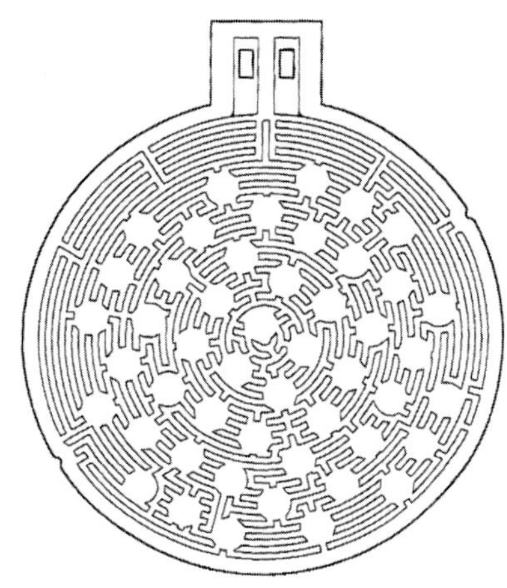

Figure 3.68 Electrical foil heater and internal scheme of resistance wire (by courtesy of Watlow Electric, Kronau, Germany).

Figure 3.69 Reaction plate coated with a dense layer of platinum [38] (by courtesy of VDI-Verlag GmbH).

wound into a spiral. The curvature of the spiral was small because only the peripheral part of the reaction plate was structured, hence the influence of secondary flow was negligible. The gas channel was sealed by a graphite foil. The available reactor temperatures were thus limited to ~600 °C.

The mean velocity u_0/c of the carrier gas was determined by Taylor's restriction $D << k_r$ and resulted in $u_0/c = 3.24$ m s^{-1}. This value delivered a mean residence time of $\tau = cL/u_0 = 0.46$ s.

Before the pulses were injected into the reactor, the reactor was supplied with the carrier gas nitrogen until steady-state conditions were reached. For the high-temperature experiments, the reactor was heated for at least 1 h and the steady state was observed with a temperature sensor on the reaction plate. The reactor temperature was adjusted to 450 °C at a pressure close to atmospheric pressure.

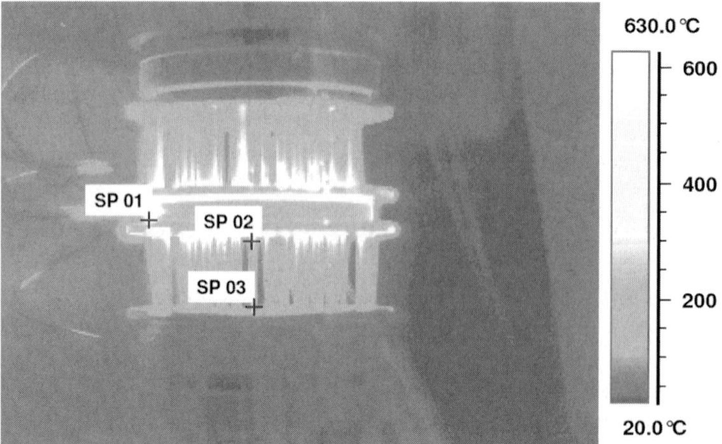

Figure 3.70 Infrared image of the central part of the reactor consisting of the reaction plate and two insulation modules (the temperature decrease in the pipes is only due to passive cooling by radiation and convection) [38] (by courtesy of VDI-Verlag GmbH).

The increase in the reactor temperature was observed with an infrared thermography camera to detect possible inhomogeneously heated zones (Figure 3.70). In Figure 3.71, the temperature increase during the start-up of the reactor was recorded at three different positions in the reaction module. Position SP01 represented the internal reactor temperature measured at the lead-through of the heater, SP02 the temperature of the gas tube close to the reaction zone and SP03 the temperature of the gas tube at its outermost end. After ~30 min, SP03 reached 100 °C and it could be expected that this temperature woul not be exceeded very much. After ~60 min, a steady temperature distribution was reached and the tube end temperature did not reach a critical value for the O-ring seals in the rotary valves. This means that the desired thermal insulation was realized just by passive cooling (active cooling with a liquid cooling medium is excluded here because of the high temperatures).

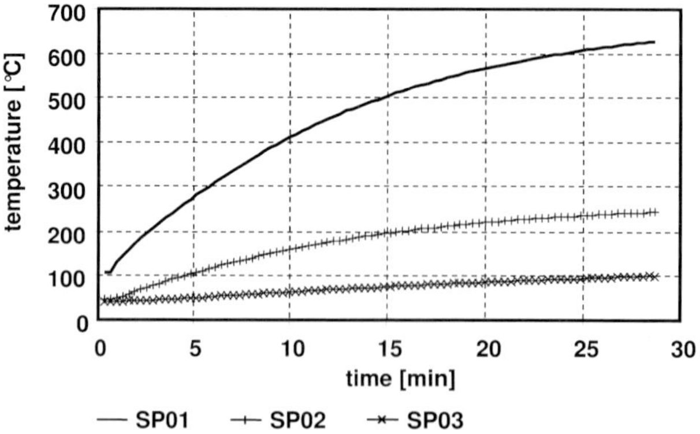

Figure 3.71 Transient temperature course at three different reactor positions as indicated [38] (by courtesy of VDI-Verlag GmbH).

The temperature of the tube ends of the upper insulation module was higher than that of the tube ends in the lower insulation module, which is a result of additional heat transfer from the reaction zone by natural convection in the upper module. In both cases, the heat loss is sufficient to prevent the rotary valves next to the heat exchanger modules from being damaged by overheating.

Transient catalytic gas–solid phase reaction

The set-up described in Section 3.4.1.1, *Stacked Platelet Screening System*, for catalytic steady methane combustion was also used for transient studies [38]. Here, the influence of the reactor temperature on the conversion of methane was examined during pulsed operation. The heat performance of the foil heater was slowly increased while the reactor was continuously supplied with gas pulses consisting of pure oxygen at a volume flow of 129.5 ml min^{-1} together with a flow of 0.5 ml min^{-1} of methane. The volume flow of the carrier gas nitrogen was adjusted to 130 ml min^{-1} at atmospheric pressure, which delivered a residence time in the coated spiral of 0.4 s. The catalyst was not reduced. The result is given in Figure 3.72 for gas pulses of 2 s followed by breaks of 2 s. No significant conversion was observed below 200 °C whereas at 325 °C the methane conversion reached ~20%, which further increased to more than 50% at 440 °C. This was consistent with the conversion values obtained under steady operation with a platinum catalyst ranging from 18 to 80% depending on the pretreatment of the catalyst under similar residence times and temperatures.

Figure 3.72 also indicates a slight decrease of the signal plateau which, at first glance, was an unexpected result. A reactive dispersion model given in [38] was applied to deduce rate constants for different reaction temperatures. As the model signal an analytical trapezoidal response function was adjusted to the real valve input signal. The temperature-dependent diffusion coefficient was calculated according to a prescription by Hirschfelder, given, for example, in [105] or [106], derived from the Chapman–Enskog theory. For the dimensionless formulation,

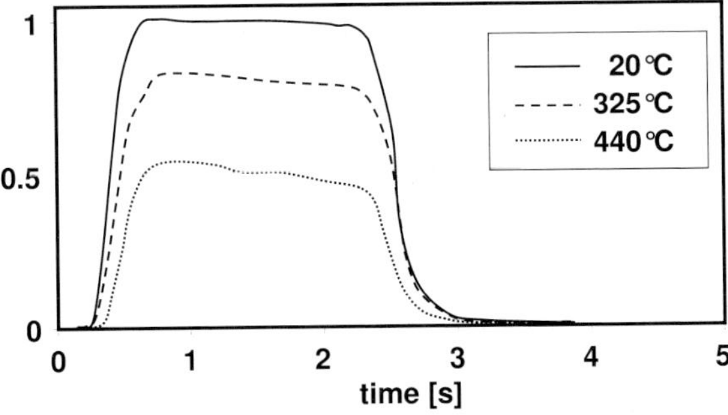

Figure 3.72 Methane concentration of the measured response of a trapezoidal input signal for different reactor temperatures (INKA-type two-way injection valve from Lee Hydraulische Miniaturkomponenten, 2 s injection time, improved FID sensor) [38] (by courtesy of VDI-Verlag GmbH).

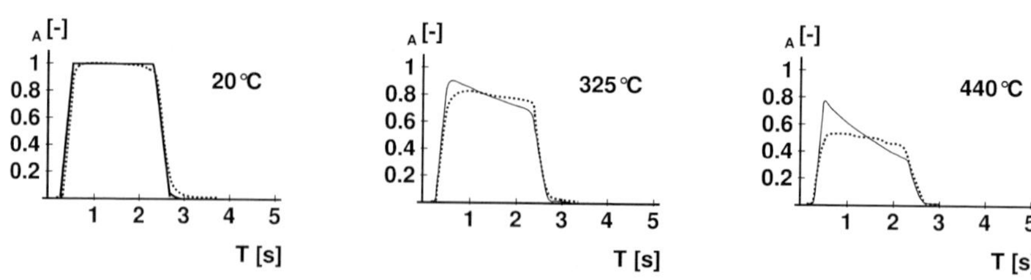

Figure 3.73 Comparison of model-calculated and measured response of a trapezoidal input signal for three reactor temperatures (solid line = model, broken line = experimental data, two-way INKA-type injection valve, 2 s injection time, FID sensor) [38] (by courtesy of VDI-Verlag GmbH).

the equation is divided by M/A (with M the injected mass and A the cross-sectional area). This analytical function is compared in Figure 3.73 with the experimental values for three temperatures. The qualitative behavior of the measured pulses is well reproduced; especially the observed decrease of the plateau is also obtained. The overall fit is less accurate than for the non-reactive case but sufficient for the evaluation of the rate constant.

To obtain the real rate constant, the assumed reaction rate constants in the model were adjusted until the integral values of the calculated pulses were identical with the measured integral values and then drawn as a function of the inverse temperature, $1/T$. This Arrhenius-type plot is given in Figure 3.74 together with a first-order approximation. The straight line indicated a reaction mechanism of first order. This reaction mechanism is confirmed by the results of Mazaki and Inoue [107] and Spencer and Pereira [108]. The slope of the function is, according to the following equation (where k is the rate constant, k_0 the pre-exponential term and E the activation energy):

$$\ln\frac{k}{k_0} = -\frac{E}{\overline{R}}\frac{1}{T} \tag{3.1}$$

$$\text{slope} = \frac{E}{\overline{R}} = 2505\,\frac{1}{K}, \qquad \overline{R} = 8.314\,\frac{\text{J}}{\text{mol K}}, \qquad E = 20.8\,\frac{\text{kJ}}{\text{mol}} \tag{3.2}$$

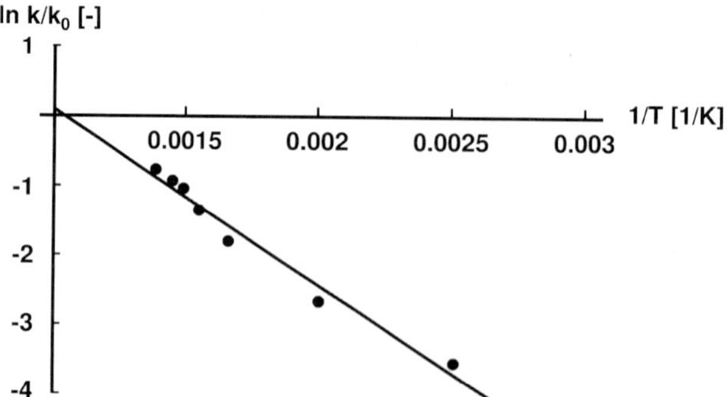

Figure 3.74 Arrhenius plot of the methane oxidation at a platinum surface [38] (by courtesy of VDI-Verlag GmbH).

a function of the activation energy E. This activation energy was close to the value 24 kJ mol^{-1} reported by Mazaki and Inoue for methane oxidation at a platinum catalyst on an alumina substrate at temperatures above 350 °C.

In the literature, higher values for the activation energy can also be found [86, 108]. One reason for this could be the neglect of a pre-reduction of the platinum catalyst and also the low porosity of the sputtered catalyst. Another important factor might be the fact that here intrinsic kinetic data were measured rather than the diffusion-affected kinetic data in [86, 108].

3.5.2 Dynamic Sequential Screening in Liquid/Liquid and Gas/Liquid Reactors

Rapid liquid-phase screening in a single-channel reactor is restricted by the large dispersion effects, which demand extremely long distances between two samples and hence long testing times. A possibility for handling this problem is the use of a multiphase system in which a barrier liquid – immiscible with the sample liquid – separates the fluids. A further approach to overcoming the large dispersion effects of single-phase systems is the application of finely dispersed two-phase systems.

3.5.2.1 Reactor 24 [R 24]: High-throughput Gas/Liquid and Liquid/Liquid Dynamic Sequential Screening Reactor

In this context, a new concept for high-throughput screening was developed [109]. De Bellefon et al. [109] reported a dynamic sequential method to screen liquid/liquid- and liquid/gas-phase catalytic reactions by applying the method of the injection of different samples followed by barrier liquid. Although the pulsed input to establish spatially separated samples, this method might also be applicable to the study of the dynamic behavior in gas-phase reactions.

The authors applied this concept to both gas/liquid (see Figure 3.75) and liquid/liquid systems (see Figure 3.76). This set-up consisted in the core of a tubular reactor with an interdigital micro mixer as dispersion unit (compare Figure 3.77). The peripheral equipment consisted of an automated pipetting robot, a fraction collector and a gas-chromatograph equipped with an automatic injector.

The micro mixer was supplied continuously with two carrier fluids, which were either two immiscible liquids or a gas and a liquid. During screening, pulses of the dissolved catalyst and the substrate were injected simultaneously by a pipetting robot. The pulses were subsequently mixed in a micro mixer and formed short defined reacting segments, which moved along the tubular reactor. The reactor itself consisted of a tubular glass reactor with a length of 156 cm and an inner diameter of 2.85 mm in the case of gas/liquid applications. For liquid/liquid applications, a glass tube with a smaller inner diameter of about 4 mm a length of 80 cm was chosen.

A more detailed description of these reactors can be found in [110].

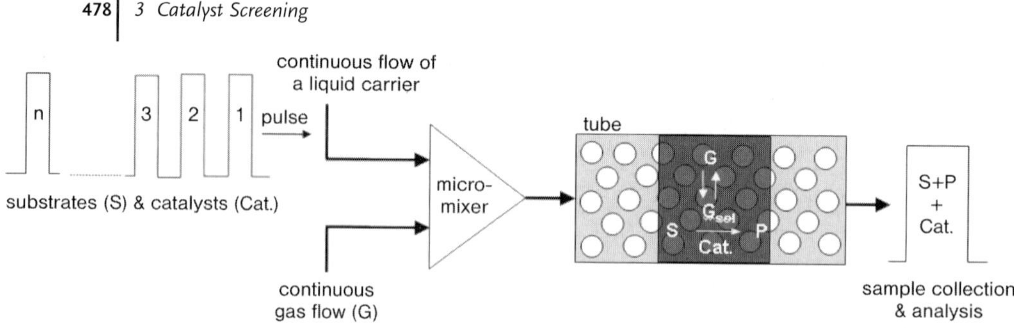

Figure 3.75 Scheme of the screening device in gas/liquid systems [109].

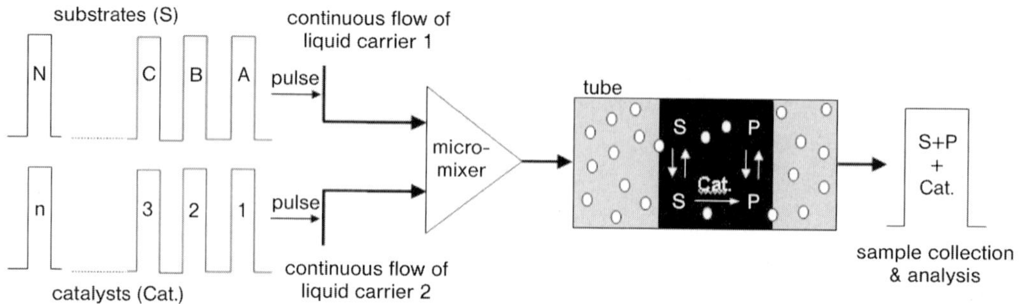

Figure 3.76 Scheme of the screening device in liquid/liquid systems [109].

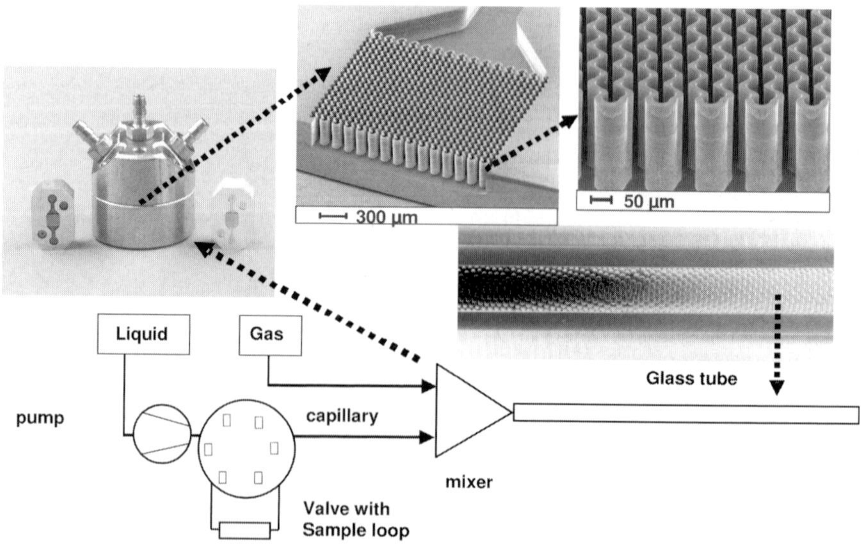

Figure 3.77 Apparatus for the gas/liquid screening set-up, consisting of a micro mixer–tube reactor combination [111] (by courtesy of AIChE).

Gas/liquid multiphase catalyst screening

As a reaction in a gas/liquid multiphase system, the enantioselective hydrogenation of methyl (*Z*)-α-acetamidocinnamate was investigated [109]:

Rh-diphosphane, H_2

water/glycol/SDS

A continuous foam flow was generated by the micro mixer, which was composed of small bubbles of hydrogen with a diameter of about 200 μm in the liquid [ethylene glycol/water, 60 : 40 (w/w) sodium dodecyl sulfate as surfactant] (see Figure 3.77). The reaction rate was proportional to the catalyst concentration and decreased with increasing surfactant concentration [6].

For serial screening, minimization of the dispersion is essential to obtain a higher number of experiments per unit time. For this reason, the individual parts of the screening set-up were investigated for their contribution the overall signal broadening (Figure 3.78) [111].

Especially if the residence time of an individual set-up component is taken into account, it is obvious that the equipment for pulse injection (HPLC valve) and for dispersion (micro mixer) together have a larger impact on pulse broadening than the tubular reactor. In addition to different gas/liquid ratios, also various inner diameters of the tubular reactor were investigated (unpublished results). In Figure 3.79 it can be seen that tubes with a diameter in the range 2–4 mm have the smallest impact on dispersion ([112]). Smaller diameters lead to increased dispersion whereas larger diameters result in drainage effects (Figure 3.79) [111].

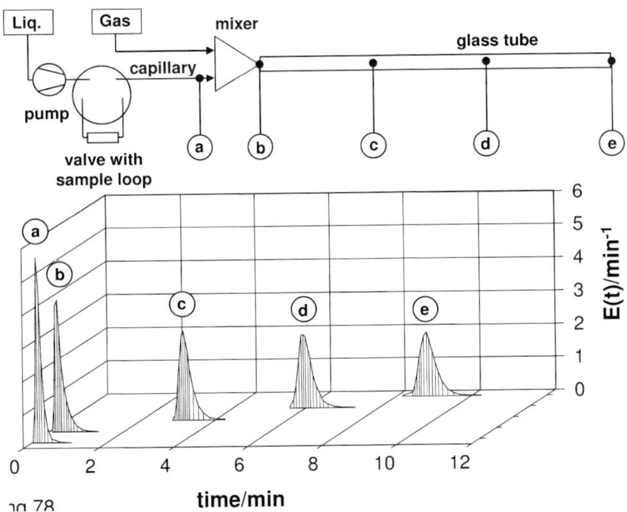

Figure 3.78 Exit age distribution function at different measuring points of the screening set-up [111] (by courtesy of AIChE).

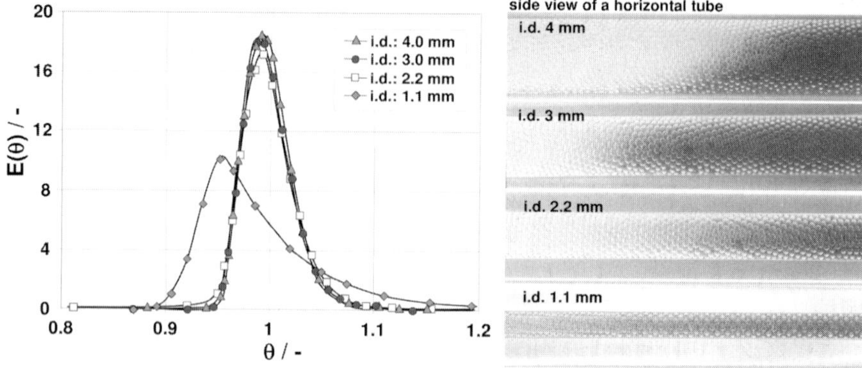

Figure 3.79 Dimensionless exit age distribution function using tubular reactors with different inner diameters [111] (by courtesy of AIChE).

Liquid/liquid phase catalyst screening

This screening concept was also applied to liquid/liquid systems. As a test reaction, the isomerization of allylic alcohols to carbonyls with water-soluble catalysts in a biphasic heptane/water system was chosen [109, 113]. The catalysts (metal precursor, Rh, Ru, Pd, Ni; ligands. sulfonated phosphane or disphosphane ligands) were injected the liquid carrier 2 (water). The substrates (different allylic alcohols) were injected into liquid carrier 1 (heptane):

OH, R, R_1 —(Rh-cat., water/heptane)→ O, R, R_1

The residence time achieved with the given set-up was 100 s. The highest conversion of 53% was found with the rhodium catalyst based on the ligand tris(*m*-sulfophenyl)phosphane. Results gained with the micro mixer-based set-up were in good agreement with results found for a mini-batch reactor.

3.5.2.2 Multi-port Valves, Injection Valves and Sensors

For the realization of gas pulses, a number of methods can be applied. In gas chromatography, a gas sample is usually collected in a sample loop and injected by an injection valve. However, in order to have some more flexibility concerning the length and the frequency of the pulse, fast switching electromagnetic valves, for example three-way valves of type LFY from Lee Hydraulische Miniaturkomponenten (Figure 3.80) could be used instead.

An alternative is a fast-switching electromagnetic two-way valve connected to a pipe T-fitting (Figure 3.81). These valves are used commercially for ink-jet printing and biomedical applications and allow a shortest injection time of less than 1 ms. The valve itself is free from wake effects and possesses a small internal volume.

The switching time for the three-way valve is $\Delta t = 20$ ms and the pulse length is adjustable by a commercial electronic pulse generator. The pulse length can thus be varied from 50 ms to 10 s. A sequence of pulses can also be delivered with

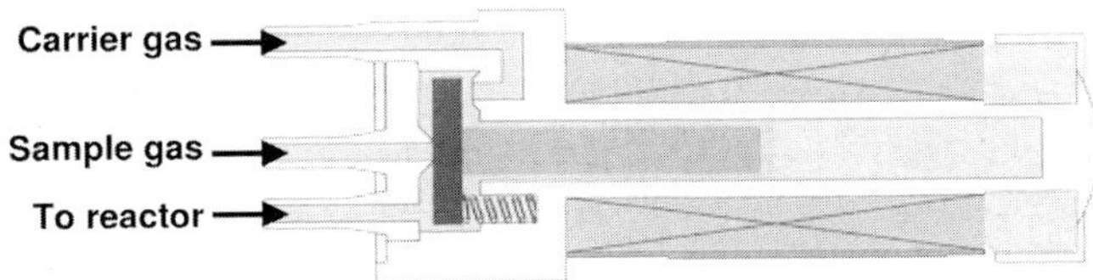

Figure 3.80 Three-way injection valve (LFY type) (by courtesy of Lee Hydraulische Miniaturkomponenten, Frankfurt, Germany).

adjustable break lengths also from 50 ms to 10 s. The time for the activation of the valve must be added to the overall pulse length; this is especially important for short pulses. The wake effect of the three-way valve is slightly larger than that of the two-way valve as indicated in Figure 3.80. The internal volume of the three-way valve is only 18 μl.

These valves were applied together with a flame ionization detector (FID) to measure transient signals using a conventional gas chromatograph [38]. The FID detects the change in conductivity caused by organic radicals, which develop during the combustion of the sample gas probe in an electric field. For these measurements, the built-in separation column in the gas chromatograph had to be bypassed, because separation of the gas pulse into its individual components was not intended. The carrier gas flow normally supplied to the FID was replaced by the sample gas flow, which consisted of nitrogen in excess and the sample gas pulse with a steady volume flow of 130 ml min^{-1}. These detectors are normally fed with hydrogen and air and are therefore not sensitive to oxygen but they are sensitive to hydrocarbons.

Multi-port valves belong to the key components in screening. Most equipment for analysis exists only in single-channel versions. Exceptions are, for example, radiation beam measurements of Atkins and Senkan [114] and tools for parallel analysis at Symyx. As long as equipment for parallel analysis does not exist for every type of desired measurement, multi-port valves will always be an essential part of every screening device.

The complex internal set-up of a typical rotary valve is shown in Figure 3.82 [38]. The complexity of a rotary valve causes a number of problems, of which cross-talk is the most critical item. Nevertheless, a rotary valve is certainly the easiest to realize type of valve if a large number of individual gas streams have to be switched. Especially the transformation of the array-type arrangement in the reaction section to a rotary arrangement in the valve is a critical item. Here, a micro structured plate was applied to allow such a transformation. As an alternative, the individual

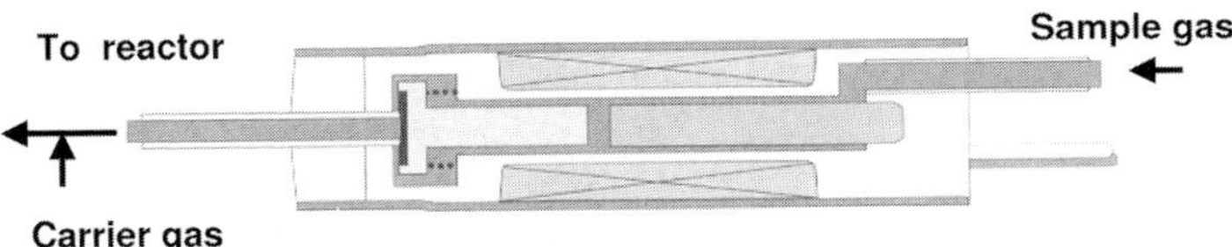

Figure 3.81 Two-way injection valve (INKA type) combined with a pipe T-fitting (by courtesy of Lee Hydraulische Miniaturkomponenten, Frankfurt, Germany).

sample gas streams leaving the hot reactor section might be connected to the valve by pipe connections. Such an approach was applied in combination with a rotary valve [72]. Other solutions for switching gas streams were realized by solutions similar to conventional laboratory sample robots. This means that the gas recipient has no direct connection to the reactor. The sample gas position, which has to be allocated, was controlled by a computer program [115]. However, a rotary valve is considered to be a more robust (and inexpensive) solution.

In order to avoid an expensive array of tube fittings from the reaction module to the rotary valve, a stack of plates was used for the diversion of the sample gas from the chequered arrangement on the titer-plate to the circumferential arrangement in the rotary valve. The distribution of all 48 gas streams was executed on a single plate, the so-called distribution plate. The structured surface of the distribution plate was manufactured by wet chemical etching. In a first version of this plate, the channels were just deepened structures in the surface of a metal plate. This version was a source of cross-talk because sealing between the etched structures was difficult to achieve owing to the huge surface area of the graphite seal lying on the unstructured surface area of the plate. In an improved version of this distribution plate, this problem was solved. This time the channels were surrounded by sealing structures parallel to the channels, which consisted of a double wall labyrinth seal around every channel (Figure 3.82, photograph). The sealing surface area was drastically reduced by the removal of most of the sealing surface (except the sealing walls), thus increasing the sealing pressure. These walls had a thickness of only 200 μm, which allowed a very high sealing force. The structure of the walls was

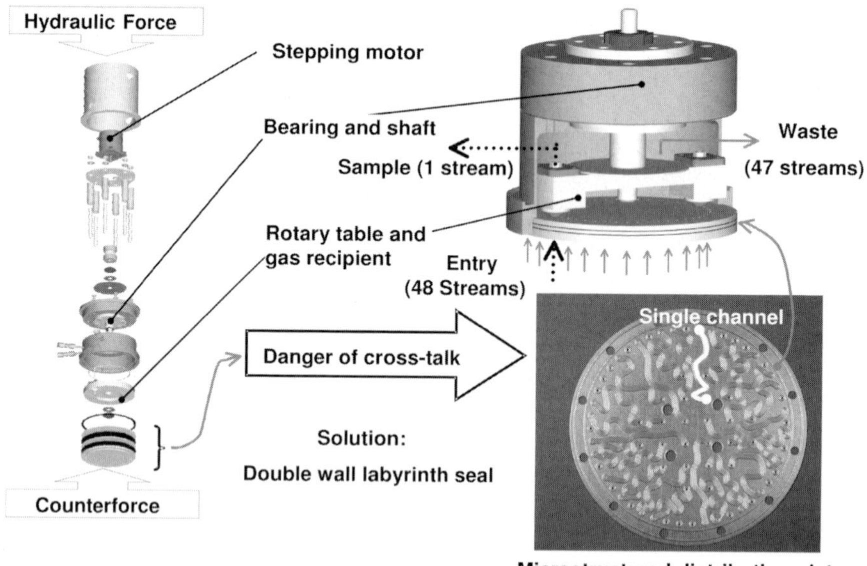

Figure 3.82 Internal set-up of a 48-fold rotary valve for the delivery of product gas to a sequential online analysis such as gas chromatography [38] (by courtesy of VDI-Verlag GmbH).

clearly marked in the surface of the graphite seals used, which is an indication of a proper sealing. The sealing force was separately adjusted by internal screws, and the sealing force was not affected by the closing force of the hydraulic press used for sealing the reactor. This set-up also allowed quick maintenance in case of local blockage because the distribution plate was accessible. After passing the distribution plate, the product gas was delivered to a gas recipient, which consisted of a spring-loaded Teflon gas recipient, and delivered the gas further to the analysis. The part of the product gas which was not delivered to the analysis was collected in a chamber before it left the reactor via a pressure controller. In such a way, the flow of the product gas through the wells was never interrupted except for the very short valve switching time. The valve was driven by a programmable stepping motor and was synchronized with a gas chromatograph for automated measurement.

3.6 Computational Evaluation Methods

In current screening procedures, data evaluation is becoming more and more the bottleneck. The number of samples increases and new sampling strategies have to be developed. A review about computational design strategies in the medicinal industry describes possible future trends in computational evaluation methods [116]. The authors recommend to concentrate screening on 'focused' libraries instead of trying to obtain a maximum diversity of samples. Natural products could be a major source of inspiration for focused libraries.

Combinatorial libraries are often synthesized using simple molecules with only a few synthetic steps, whereas natural products are produced in complex multi-step reactions with complex molecules. Also, the parameter space between natural products and chemical libraries seems to be different. Synthesized drugs occupy an intermediate space between natural and chemical libraries, also a hint that the space of the drug libraries should be adjusted.

Potential scaffolds for drug design were analyzed by Kohonen mapping. Natural products and drugs share similar pharmacophores, which indicates that natural products should be an interesting source for new scaffold architectures. Despite the large diversity of protein surfaces, only a few of these sites are biologically relevant. This is also an indication that combinatorial libraries should concentrate on privileged scaffolds. The authors concluded that about 10% of all pharmaceutical expenditure will be on computational design strategies in 2006. They also stress that computational design is less efficient for vast libraries, another reason why many companies applying computational methods now concentrate on focused libraries.

Rothenberg et al. emphasized the relevance of a statistically accurate sample distribution in the parameter space [117]. They studied the effect of the distribution of samples along the time axis and found that using an equidistant distribution along the time axis is much less efficient than the use of an equidistant distribution along the ordinate.

Another modern approach is the strategy of condensing information derived from test data. This leads directly to transient measurements, which 'condense' stochastic conversion data by either numerical or analytical methods to a function (usually the reaction kinetics).

3.6.1
Evaluations Following Biological Means

A separate class of experimental evaluation methods uses biological mechanisms. An artificial neural network (ANN) imitates the processes proceeding in the human brain, especially its layered structure and its network of synapses. On a very basic level, such a network is capable of learning rules, for example, the relations between activity and component ratio or process parameters. An evolutionary strategy was proposed by Mirodatos et al. [118]. They combined a genetic algorithm with a knowledge-based system and added descriptors, for example the catalyst pore size, the atomic or crystal ionic radius and the electronegativity. This strategy allowed a reduction in the number of materials necessary for a study.

Holena and Baerns [119] stressed the problem of overtraining, which means that an ANN may well reproduce the screening data of the training-data set but would deliver bad results for an unknown set of data. A method to prevent this effect is, for example, the early stopping of training. The perceptron-type network was equipped with one hidden layer and network optimization was executed with the Levenberg–Marquardt approximation, a back-propagation method to minimize the error caused by the Gauss-Newton quadratic minimization method. With this method they could predict the yield of propene for the oxidative dehydrogenation of propane with an error of 5.4%.

Two types of ANNs can be distinguished, compositional nets for the prediction of performance from composition and synthesis of a catalyst and kinetic nets for the prediction of performance from reaction conditions [120]. The sensitivity of a net depends very much on the experimental errors. The prediction is not very much influenced as long as the experimental deviation error is less than 2%. Also, the training environment affects the model quality. A net was trained for an isomerization reaction network with long-chain hydrocarbons involved. The net was then applied to a reaction network with hydrocarbons of smaller chain length. This resulted in good performance prediction. It was stated that the greater knowledge of the net trained originally with a more complex reaction scheme allowed prediction with a small mean square error.

Corma et al. [121] also used a neural net, which they adapted to the desired results in the training phase by supervised learning. During training, the network adapts the weights between the neurons, thus acting as a kind of complex function approximator between training data as the input data and desired results as the output data. They also found a multilayer perceptron to possess better performance than a radial-basis network or a self-organizing network using the Kohonen learning rule. As input data for the learning process, again the catalyst composition of 13 elements was fed into the net and adapted to the desired output data, the yield and the selectivity. Predicted and experimentally observed data for the oxidative dehydrogenation of ethane, used as the test reaction, agreed sufficiently. They also reported problems with data singularities, a known disadvantage of neural nets. Neural nets are better suited to predict smooth data correlations.

Serra et al. [122] used an evolutionary strategy for the design of catalyst libraries to evaluate a synthesis route for styrene from toluene.

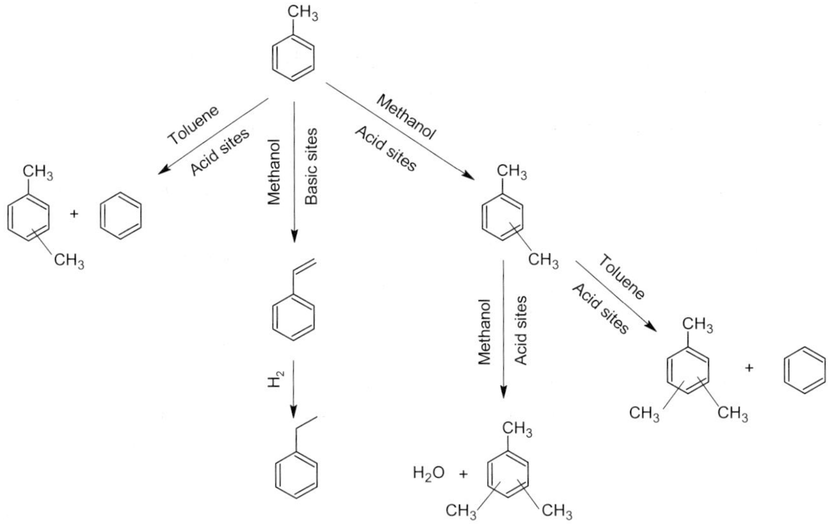

The following parameters were selected for screening: five zeolites, nine basic cations, three different types of precursors, three different preparation methods, four cation loadings and two binary mixtures of zeolites. The number of combinations of these parameters was larger than 250 000, and it was not practicable to screen all within a reasonable time. The selected parameters for synthesis were discrete and a genetic algorithm was developed (Figure 3.83). In addition, this generic algorithm was hybridized with a knowledge-based system which enables one to extract, update and validate knowledge from all the catalysts [122].

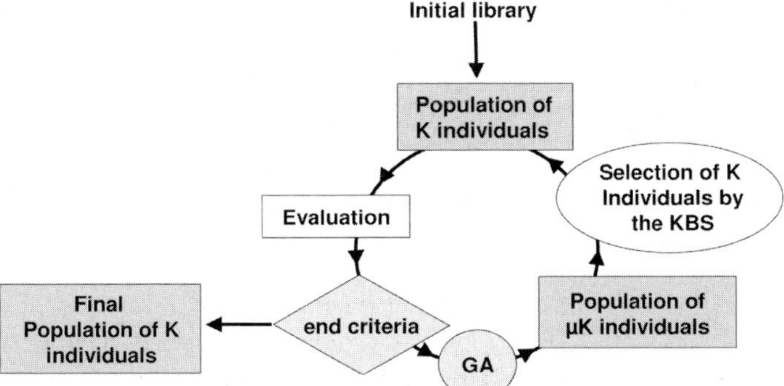

Figure 3.83 Schematic of the evolutionary strategy. GA = generic algorithm; KBS = knowledge-based system [122] (by courtesy of Elsevier Ltd.).

During the investigations, the learning proceeded on two classes of catalysts defined as 'bad' and 'good' according an objective function, which gives more importance to the production of the target molecule styrene than to the consecutive reaction product ethylbenzene.

96 catalysts are fabricated and tested randomly. From this pool of catalysts the seven 'good' ones were selected and used as a first generation on which the learning proceeded. In order to obtain a better tendency overview, the algorithm parameters were tuned. A next generation of catalysts was designed, thus promoting the browsing of the parameter space. For this second generation, the learning process was retrained and a further selection of 'good' catalysts was made. This procedure can be repeated several times; in this case three generations were made. In contrast to the literature data, the best results for the reaction were obtained with zeolite-based catalysts. This example shows the possibility of high-throughput screening

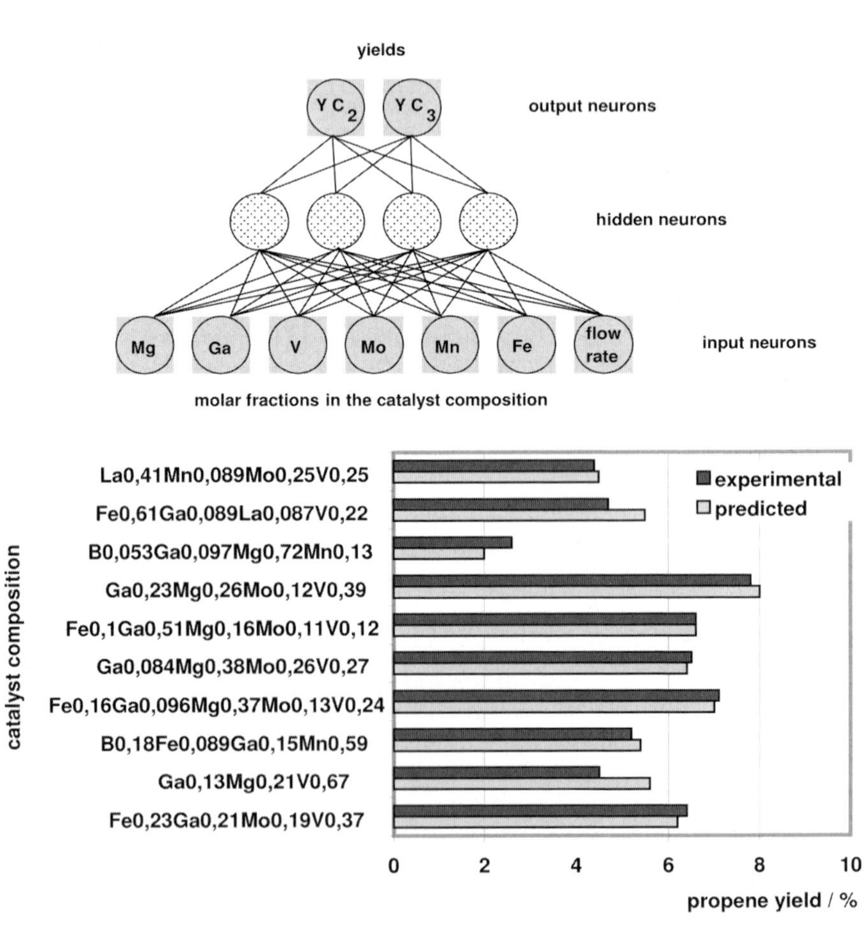

Figure 3.84 Set-up of a neural network and comparison between predicted and experimentally obtained values of catalyst activity for the dehydrogenation of propane to propene [123] (by courtesy of ACA).

even in the presence of complex reactions such as the side-chain alkylation of toluene to styrene and ethylbenzene [122].

Rodemerck [123] and Corma et al. [121] prepared catalyst compositions by genetic algorithms and combined this approach with an ANN. This seems to be one of the most promising strategies for lean data mining. Rodemerck et al. trained a neural network to predict the composition of new catalysts (Figure 3.84). Together with a genetic algorithm, new generations of catalysts were produced by mutation and recombination of catalysts from the old generation following an iterative process.

3.6.2 Numerical Evaluation Methods

Numerical simulations were conveniently used to describe complex fluid dynamic behavior in micro structures [36, 101]. Van der Linde et al. [101] solved the coupled diffusion equations for reacting species and compared the results with data from the oxidation of CO on alumina-supported Cr using the step-response method. Transient periodical concentration changes in micro channels were numerically calculated by various authors [34, 88, 124].

In 2001, Holzwarth et al. [125, 136] stressed the importance of transient studies as an alternative to steady continuous reactor operations. A combination of microkinetic analysis together with transient experiments allowed the determination of the global catalytic conversion from elementary reaction steps. A prerequisite for such an analysis is the correlation of experimental data with the data of a model. In Figure 3.85, experimental and model responses of an impulse of reactants were correlated. Agreement between the data helped to derive the reaction mechanism.

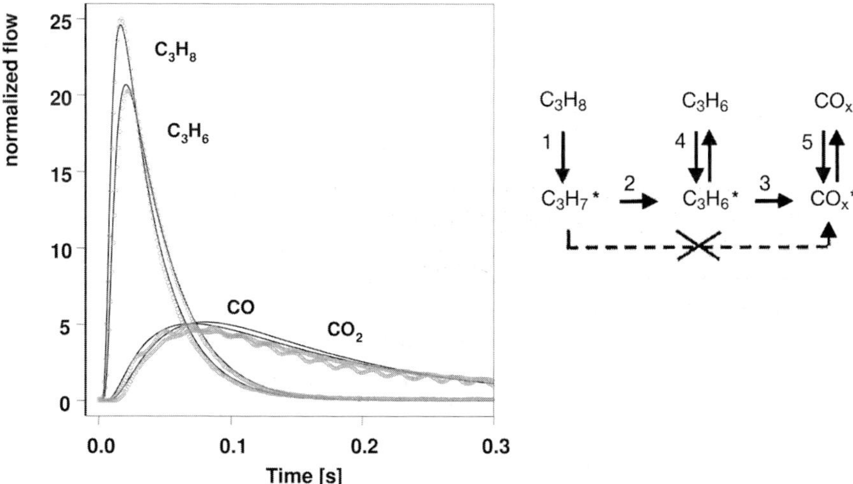

Figure 3.85 Reactor response to an impulse of the reactants propane and oxygen compared with the model response (right: proposed reaction mechanism derived from model data) [136] (by courtesy of Elsevier Ltd.).

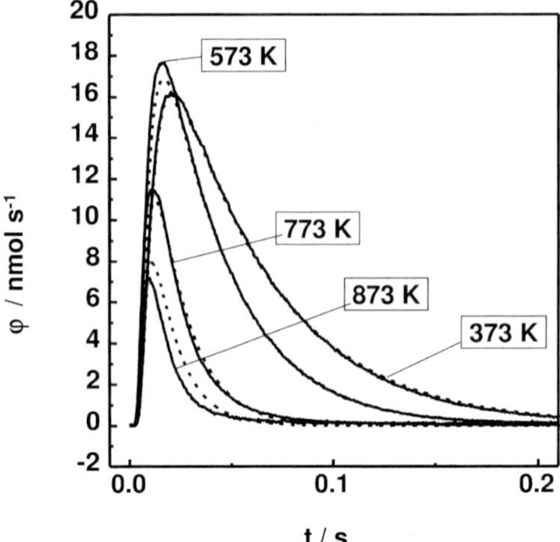

Figure 3.86 Comparison of model and experimental data of the oxygen flux leaving the reactor in a single-channel TAP fixed-bed reactor [126] (by courtesy of Elsevier Ltd.).

In another recent publication [126], experimental impulse responses were compared with model data, which were obtained from a numerical solution of the non-linear differential equations for an impulse of O_2 and Ne (Figure 3.86).

In the same publication, a method for the parallelization of TAP (temporal analysis of product) experiments (for reactor, see Figure 3.87) was also indicated. It was stated that "... high-throughput transient kinetics carried out in addition to high-throughput catalyst synthesis and testing both accelerate the search for new catalytic materials and bring fundamental insights into reaction mechanisms." One reason for this statement is certainly the fact that transient screening, in contrast to steady screening experiments, gives a denser information content. This is due to the simple fact that, during transient tests, a complete function (the kinetics) is recorded instead of single stochastic data.

Figure 3.87 TAP reactor equipped with a multi-sample holder for fast parallel transient screening [126] (by courtesy of Elsevier Ltd.).

3.6.3
Kinetics Derived from Signal Dispersion

For the fast evaluation of such response signals, a diffusion/dispersion model is necessary, which must also include information about the heterogeneous wall reaction. A simplified dispersion model assuming a first-order reaction is presented below, which allows the prediction of the concentration distribution in rectangular micro channels [38]. These results prepare the way for an extended type of secondary screening in which case screening means the evaluation of catalysts applying various reactor geometries and transient operations for the collection of kinetic information. Some specific aspects of dispersion in laminar flow typical for micro structures will be discussed in detail in the following.

The limitation of such a model to first-order reaction rates is not as restricting as it seems. In fact, many reactions might at least be considered as of 'pseudo'-first order, which means that they behave macroscopically like first-order reactions. This is the case for diluted fluids and for non-catalytic gas/solid reactions such as the so-called shrinking core or shrinking particle model. Other examples are electrochemical reactions [106].

For the above applied oxidation of methane to carbon dioxide on some metal oxide catalysts, also a first-order reaction was assumed [10, pp. 182 and 193]. However, in combinatorial catalysis it may be sufficient to have a first rough idea about the underlying kinetics. Without having prior information about the kinetics, the performance of a reactor is provided with a huge uncertainty. This is obvious if one considers the wide variation of reaction rates. Pre-exponential factors of reaction rate constants derived by the transition-state theory vary widely from approximately 10 to 10^{16} s^{-1} [10]. This first information might then be used to develop a pilot plant for the up-scaling and for further detailed kinetic examinations.

A characteristic of micro channel reactors is their narrow residence-time distribution. This is important, for example, to obtain clean products. This property is not imaginable without the influence of dispersion. Just considering the laminar flow would deliver an extremely wide residence-time distribution. The near wall flow is close to stagnation because a fluid element at the wall of the channel is, by definition, fixed to the wall for an endlessly long time, in contrast to the fast core flow. The phenomenon that prevents such a behavior is the known dispersion effect and is demonstrated in Figure 3.88.

The velocity profile in the channel is assumed to be fully developed and steady. Any kind of disturbance, as for example a concentration pulse or a local pressure change, is primarily distorted by the so-called hydrodynamic dispersion. Without considering diffusion, such a signal would be spread extremely long in an axial direction. Molecular diffusion counter balances this effect and equilibrates the signal strength in the cross-section very fast. This equilibration is due to the large radial concentration gradients originating from the laminar parabolic velocity profile. The disturbance reaches the center of the channel before the signal has arrived at the walls. After a sufficiently long distance, the laminar flow profile develops into a plug flow profile. The average length of signal t_{disp} and the averaged signal height

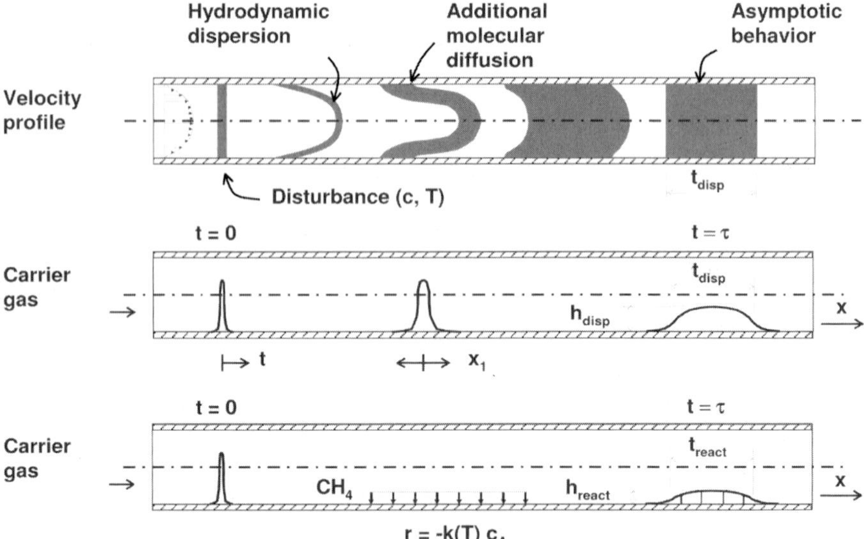

Figure 3.88 Dispersion in a laminar flow channel reactor with inert walls and catalytically active walls [38] (by courtesy of VDI-Verlag GmbH).

h_{disp}, after the signal has traveled a time τ in the channel, is then a measure of the dispersion inside the channel characterized by the dispersion coefficient k_r. This signal is further reduced in height by the wall reaction h_{react}.

Similar to the sample delivery in a reactor operated in differential mode, the sample is injected into the steady laminar carrier flow in the channel which moves at a mean speed u_0/c.

The concentration of methane at room temperature at any channel position is described by the impulse response without reaction [Eq. (3.4)] [the dispersion coefficient for a rectangular channel with channel depth b and aspect ratio ε is given in Eq. (3.3)]:

$$k_r = \frac{2272}{8505} \frac{b^2\, u_0^2}{D} \frac{(1+\varepsilon^4)^2}{(2+\varepsilon+3\,\varepsilon^2)^2\,(1+\varepsilon^6)} \tag{3.3}$$

$$c_A(x,t,T = 293\text{ K}) = \frac{\frac{M}{4\,a\,b}}{\sqrt{\pi}\,\sqrt{4\,k_r\,t}} \exp\left[-\frac{x_1^2}{4\,k_r\,t}\right] \tag{3.4}$$

or by the extended impulse response as follows:

$$s(x_1,t) = \frac{M}{2\,A}\left[\operatorname{erf}\left(\frac{x_1+\frac{X}{2}}{\sqrt{4\,k_r\,t}}\right) - \operatorname{erf}\left(\frac{x_1-\frac{X}{2}}{\sqrt{4\,k_r\,t}}\right)\right] \tag{3.5}$$

At the reaction temperature, the concentration of methane obeys the following equation (apparent velocity increase k_v given in [38]ss):

$$c_A(x_1, t, T = 723\ \mathrm{K}) =$$

$$\frac{M}{2A}\left[\operatorname{erf}\left(\frac{x_1 + t\,k_v + \frac{X}{2}}{\sqrt{4\,k_r\,t}}\right) - \operatorname{erf}\left(\frac{x_1 + t\,k_v - \frac{X}{2}}{\sqrt{4\,k_r\,t}}\right)\right]\exp\left[-D\,\phi^2\left(\frac{a+b}{a\,b}\right)^2 t\right] \qquad (3.6)$$

Mass transport in laminar flow is dominated by diffusion and by the laminar velocity profile. This combined effect is known as dispersion and the underlying model for the theoretical derivation of a kinetic study had to be derived from the dispersion model, which Taylor [127] and Aris [128] developed. Taylor concluded that in laminar flow the speed of an inert tracer impulse initially given to a channel will have the same speed as the steady laminar carrier gas flow originally prevailing in this channel.

A reactive dispersion model based on the Taylor dispersion model was proposed, which predicts a change of speed if the tracer impulse consists of reactants which react at the walls of the channel (see Figures 3.89 and 3.90).

Brenner found a clever and convincing explanation for this phenomenon [129]. He regarded the stochastic Brownian motion of a tracer particle through a channel with reactive walls (Figure 3.91). Such a molecule will irreversibly react at the channel walls as soon as it touches the wall. An interesting side effect is that the speed of the particle will also be reduced because it spends some time in the slow near-wall region, in contrast to a particle, which does not approach the wall. On the other hand, the latter particle will be accelerated compared with the mean flow velocity. Brenner stated: “This phenomenon stems from the fact that only those solute molecules ‘smart enough’ to stay away from the wall region survive their trip

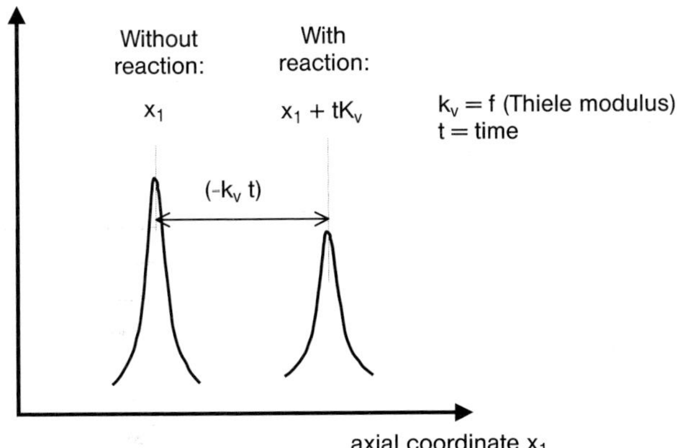

Figure 3.89 A chemical reaction changes the speed of an impulse of reactants in a single-channel reactor [38] (by courtesy of VDI-Verlag GmbH).

Taylor dispersion model:

$$\bar{u}(tracer) = \bar{u}(carrier\ gas)$$

$$c_A(x,t) = \frac{M/4ab}{\sqrt{\pi}\sqrt{4k_r t}} \exp\left[-\frac{(x_1)^2}{4k_r t}\right]$$

Reactive dispersion model:

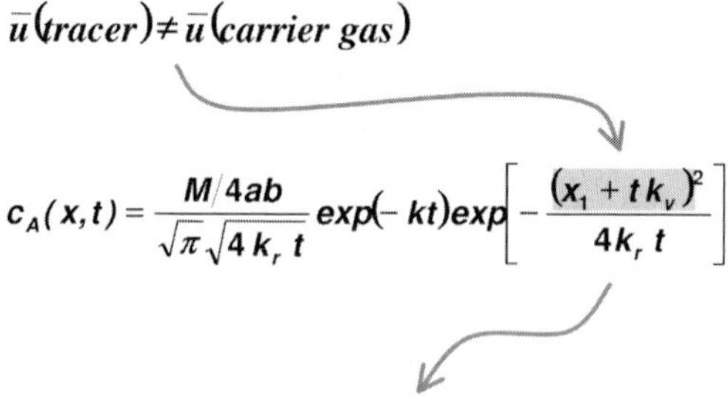

Figure 3.90 In contrast to the (non-reactive) Taylor model, the tracer gas speed for a reacting gas is unequal to the carrier gas speed. This fact can be derived from the exponential term in the solution of the concentration of the tracer gas c_A at the channel exit [38].

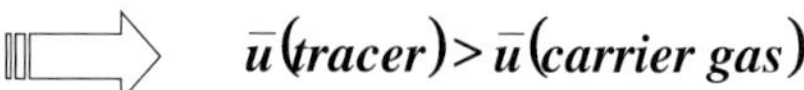

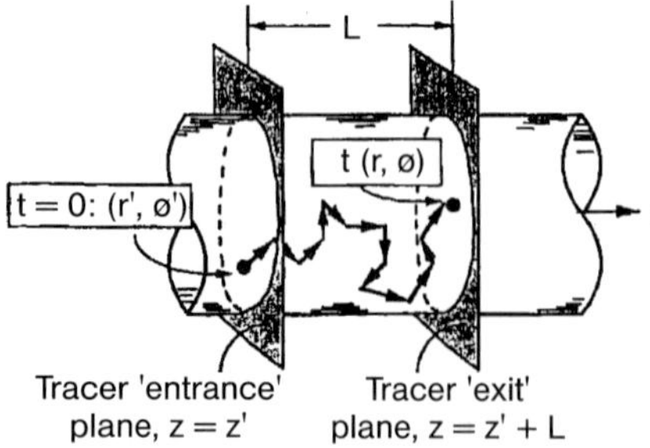

Figure 3.91 Stochastic motion of a Brownian tracer particle in a laminar flow profile [129] (by courtesy of ACS).

downstream, ..., those molecules 'foolish' enough to meander (by lateral motion) over to the tube wall are destroyed by the reaction. As such, those solute molecules that exit the tube (...) have not sampled the slower moving streamlines of the Poiseuille flow existing near to the wall."

The theoretical foundation for this kind of analysis was, as mentioned, originally laid by Taylor and Aris with their dispersion theory in circular tubes. Recent contributions in this area transferred their approach to micro reaction technology. In 1999, Angeli et al. [130] studied a reaction in a catalytic wall micro reactor applying the so-called eigenvalue method for a vertically averaged one-dimensional solution under isothermal and non-isothermal conditions. Dispersion in etched micro channels was examined [131], and a comparison of electro-osmotic flow with pressure-driven flow in micro channels was given by Locascio et al. in 2001 [132].

A method was proposed to obtain the kinetic rate constant at a fixed temperature with a one-point measurement. This method is comparable to gas chromatographic concentration measurements and can in principle be executed with a convenient gas chromatograph equipped with a flame ionization detector (FID) [38]. The background of this method is introduced in the following.

In macroscopic reactors, the knowledge of the velocity profile in the channel cross section is a necessary and sufficient prerequisite for the description of the material transport. In microscopic dimensions down to a few micrometers, diffusion also has to be considered. In fact, without the influence of diffusion extremely broad residence time distributions would be found because of the laminar flow conditions in the latter. Superposition of convection and diffusion is called dispersion. Taylor [127] was among the first who noticed this strong dominating effect in laminar flow. It is possible to transfer his deduction to rectangular channels. A complete fluid dynamic description of the flow including effects such as the influence of the wall, the aspect ratio and a chemical wall reaction on the concentration field in the cross section was given in [38].

Here, the response functions of the diffusion equation for a number of discrete input signals were calculated based on the solution for a Dirac impulse input signal. A set of transformation relations for the injected mass M and the axial coordinate x_1 was used to obtain the solutions for a reacting gas directly from the solutions of a non-reacting gas:

$$M \Rightarrow M \exp\left[-D\,\phi^2 \left(\frac{a+b}{a\,b}\right)^2 t\right] \qquad (3.7)$$

$$x_1 \Rightarrow x_1 + t\,k_v$$

The response function for a reacting gas (the dispersion coefficients k_r and the coefficient k_v were given in [38]):

$$c_A(x,t) = \frac{\frac{M}{4\,a\,b}}{\sqrt{\pi}\sqrt{4\,k_r\,t}} \exp\left[-D\,\phi^2 \left(\frac{a+b}{a\,b}\right)^2 t - \frac{(x_1 + t\,k_v)^2}{4\,k_r\,t}\right] \qquad (3.8)$$

was derived from the response function obtained for a Dirac impulse for a non-reacting gas:

$$c_A(x,t) = \frac{\frac{M}{4\,a\,b}}{\sqrt{\pi}\sqrt{4\,k_r\,t}} \exp\left(-\frac{x_1^{\,2}}{4\,k_r\,t}\right) \qquad (3.9)$$

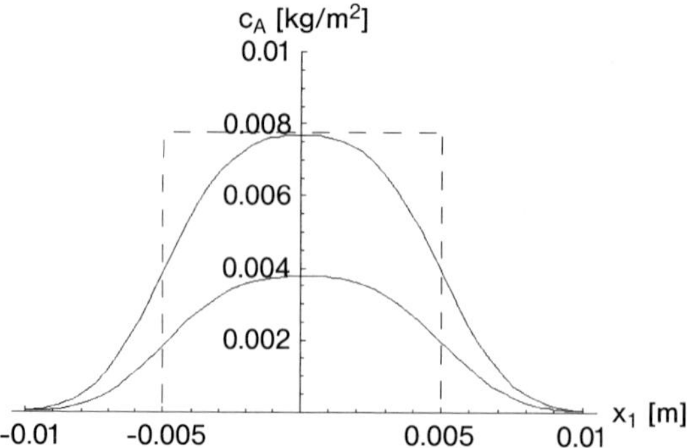

Figure 3.92 Response to the extended pulse function for $t = 0$ s (broken line) and $t = 0.01$ s for a non-reacting gas (upper solid line) and a reacting gas ($\phi = 0.3$, lower solid line) [38] (by courtesy of VDI-Verlag GmbH).

The response to the step function for a non-reacting gas was given by integrating the response of the pulse function as

$$h(x_1, t) = \frac{M}{4\,a\,b}\frac{1}{\sqrt{\pi}\sqrt{4\,k_r\,t}}\int_{x_1}^{\infty}\exp\left(-\frac{x_1^{\,2}}{4\,k_r\,t}\right)\mathrm{d}x_1 = \frac{M}{8\,a\,b}\left[1-\mathrm{erf}\left(\frac{x_1}{\sqrt{4\,k_r\,t}}\right)\right] \quad (3.10)$$

The solution for the reactive case was then obtained without calculation directly from Eq. (3.10) applying the transformations [(Eq. 3.7)]:

$$h(x_1, t) = \frac{M}{8\,a\,b}\left[1-\mathrm{erf}\left(\frac{x_1 + t\,k_v}{\sqrt{4\,k_r\,t}}\right)\right]\exp\left[-D\,\phi^2\left(\frac{a+b}{a\,b}\right)^2 t\right] \quad (3.11)$$

A graphical representation is shown in Figure 3.92. For the extended pulse function of width X, the following solution is obtained in the same manner:

$$s(x_1, t) = \frac{M}{8\,a\,b}\left[\mathrm{erf}\left(\frac{x_1 + t\,k_v + \frac{X}{2}}{\sqrt{4\,k_r\,t}}\right) - \mathrm{erf}\left(\frac{x_1 + t\,k_v - \frac{X}{2}}{\sqrt{4\,k_r\,t}}\right)\right]\exp\left[-D\,\phi^2\left(\frac{a+b}{a\,b}\right)^2 t\right] \quad (3.12)$$

The reaction introduces a spatial shift into the solution ($t \cdot k_v$) which leads to an asymmetric behavior of the response function in the moving frame of reference. The slight asymmetry of the lower solid line in Figure 3.92 is not visible here. This effect is more distinct in a very flat rectangular channel and even more in a liquid system as shown in Figure 3.93 for the system water in acetone at 25 °C. Diffusion coefficients of liquids are close to 10^{-5} cm^2 s^{-1}, a value four orders of magnitude smaller than the diffusion coefficients of gases. In consequence, the dispersion and thus the pulse shift are also more distinct. Figure 3.93 shows the concentration in a flat channel with such a small aspect ratio of 0.1 for different Thiele moduli ϕ. Obviously, the maximum of the pulses has moved towards the channel exit (towards the right side in Figure 3.93).

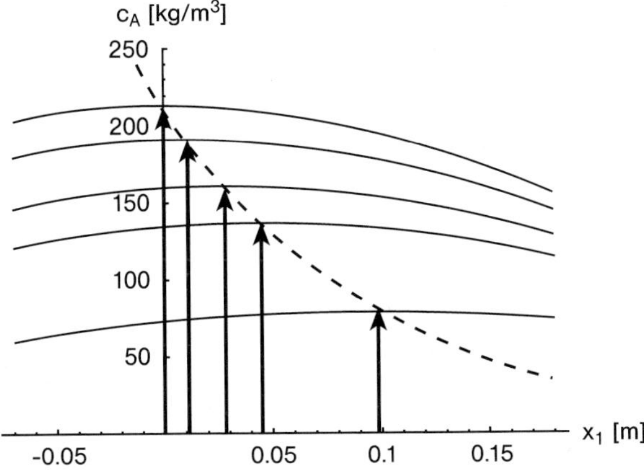

Figure 3.93 Shift of concentration pulses (Dirac impulses) for an aspect ratio of 0.1 in the system water/acetone (the arrows indicate the individual peak maximum; the hatched line connects the peak maxima) [38] (by courtesy of VDI-Verlag GmbH).

This ability of the heterogeneous wall reaction could be a means to measure the kinetic rate coefficient k directly from the arrival time of the pulse at the channel exit with a one-point measurement (see above). An experimental procedure for the fast determination of reaction kinetics can profit from the apparent velocity increase in the pulse movement due to the wall reaction in order to obtain the kinetics. Information about the intrinsic kinetics is obtainable by measuring the arrival time of the peak maximum of a pulse function at the reactor outlet. Only the position of the peak maximum has to be measured and not the reactor outlet concentration itself. As the effect is not strong, especially for gases, very flat channels and high-resolution equipment are certainly necessary. The examples above correspond to slow reactions such as the hydrogenation of benzene ($\phi > 0.05$) or the oxidation of ethylene ($\phi = 0.08$). For fast reactions such as the oxidation of methanol to formaldehyde ($\phi = 1.1$), the reaction time will reduce and hence the difficulties of observing the pulse shift will increase but also the observed asymmetry will be more distinct. Micro reactors seem to be especially suited for use as a means for measuring such pulse shifts because they exhibit short residence times and thus high time resolutions.

The above analytical solution was expanded to three dimensions. In such a way, the reactor geometry or the channel can be designed. An appropriate simplified model, given in [38], can be derived from the diffusion equation. Appropriate boundary conditions at the channel walls account for the heterogeneous wall reaction. The concentration of a species A which reacts at the channel wall irreversibly to a species B was given as a function of the lateral channel dimensions y and z and the axial channel dimension x_1. For an inert gas and for y and z equal to zero (coordinate center indicated in Figure 3.94), Eq. (3.13) reduces to the solution of a non-reactive fluid given above:

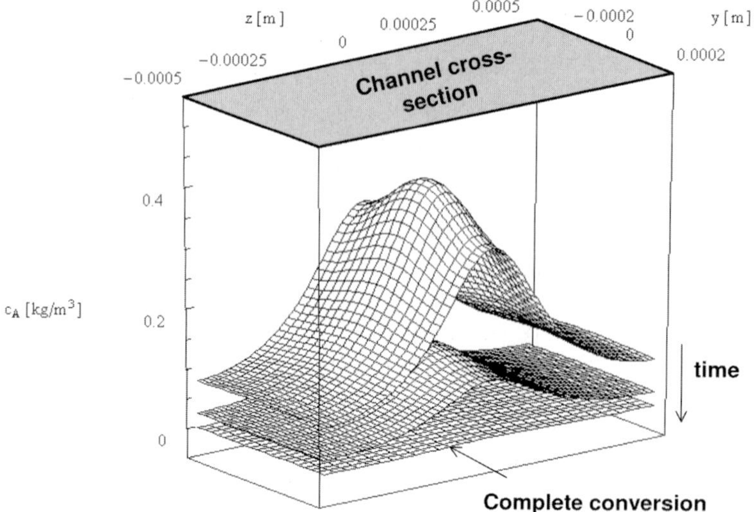

Figure 3.94 Concentration in the cross-section of a rectangular channel for three different times [38] (by courtesy of VDI-Verlag GmbH).

$$c_A(x_1,y,z,t) = \frac{c\,M}{4\,a\,b\,\sqrt{\pi}\,\sqrt{4\,k_r\,t}} \left[1 - \underbrace{\frac{(a+b)^2\,(p\,y^2 + q\,z^2)\,\phi^2}{2\,a^2\,b^2}}_{\text{het}}\right]$$
$$\cdot \left[1 - \frac{2835\,(147 - 44\,c)}{4096\,c^2\,f}\,\frac{y^4 - 2\,b^2\,y^2 + z^4 - 2\,a^2\,z^2}{a^4 + b^4}\right]^{-1} \tag{3.13}$$
$$\underbrace{\cdot \exp\left[-D\,\phi^2\left(\frac{a+b}{a\,b}\right)^2 t}_{\text{hom}} - \frac{(x_1 + t\,k_v)^2}{4\,k_r\,t}\right]$$

In Eq. (3.13), the Thiele modulus ϕ was introduced instead of the reaction rate constant k. A preliminary choice was made of the fit parameters $p(d)$ and $q(d)$, given in Eqs. (3.14) as functions of the maximum channel concentration and the aspect ratio ε:

$$c_{A,w}(z) = c_{A,0}\,p \Leftrightarrow p = d\,\frac{1}{1+\varepsilon}$$
$$c_{A,w}(y) = c_{A,0}\,q \Leftrightarrow q = d\,\frac{\varepsilon}{1+\varepsilon} \tag{3.14}$$

A further improvement of Eq. (3.13) might be obtained by the following case study. An iterative procedure allows the calculation of the fit parameters p and q.

A comparison of the marked terms het and hom in Eq. (3.13) shows that for a small Thiele modulus the term het is also small whereas the term hom is small only if the product of ϕ^2 and the time t is small. This means that for small t the concentration profile is dominated by the heterogeneous character. On the other hand, for times, which are not too short, the influence of the exponential term hom predominates. Then, the dependence of the dimensionless parameters p and q on d is weak and d is assumed to be well enough fitted by the value 0.2.

For larger ϕ or very small times, the heterogeneous character of Eq. (3.13) increases, in which case p and q might be obtained iteratively from the set of equations shown in Eqs. (3.14).

The procedure is the following: Eq. (3.13) is calculated for given p and q. Then the resulting concentration at the respective walls $c_{A,w}$ is compared with the assumptions in Eqs. (3.14). If the result is not satisfactory, either p or q can be changed individually or the global fit parameter d can be changed to execute another iteration.

In the following, a Thiele modulus of $\phi_D = 0.1$ was chosen, in which case d is well approximated by $d = 0.2$. In Figure 3.94, Eq. (3.13) is presented in the moving frame of reference $x_1 = 0$ for three different times. At $t = 0.5$ s nearly the entire original present component A has reacted irreversibly to component B and the concentration of A is close to zero everywhere.

If the term hom in Eq. (3.13) is small, the concentration distribution has a more heterogeneous character. In Figure 3.95, the gradient at the wall is much steeper for small times t than for longer times. In the latter case, the distribution thus exhibits a more homogeneous character as a result of vertical and horizontal concentration equilibration by dispersion.

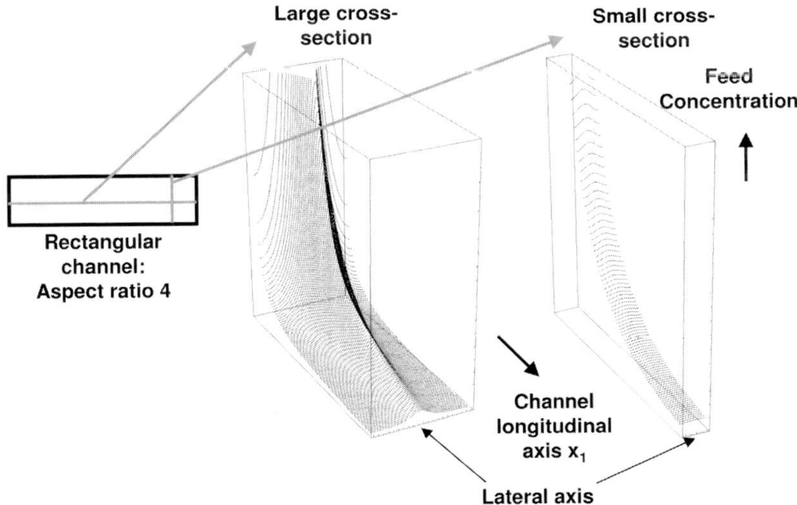

Figure 3.95 Concentration distribution (aspect ratio 4, $\phi = 0.05$, $d = 0.2$) in the plane $y = 0$ (right) and $z = 0$ (left) for different times (longest axis: time $t = 0$–0.5 s, the width of the boxes is proportional to the channel dimensions) [38] (by courtesy of VDI-Verlag GmbH).

References

1 Flego, C., Combinatorial catalysis: a help for the chemical industry of the near future, *Chim. Oggi* **2003**, *9*, 69–74.

2 Watts, P., Haswell, S. J., Microfluidic combinatorial chemistry, *Curr. Opin. Chem. Biol.* **2003**, *7*, 380–387.

3 Merrit, A. T., Gerritz, S. W., Combinatorial chemistry – rumors of the demise of combinatorial chemistry are greatly exaggerated, *Curr. Opin. Chem. Biol.* **2003**, *7*, 305–307.

4 Maxwell, I. E., van den Brink, P., Downing, R. S., Sijpkes, A. H., Gomez, S., Maschmeyer, T., High-throughput technologies to enhance innovation in catalysis, *Top. Catal.* **2003**, *24*, 1–4, 125–135.

5 Murphy, W., Volpe, A. F., Weinberg, H., *High-throughput approaches to catalyst discovery, http://www.cmbi.edu.cn/view/0307/301.pdf*, 5 March **2004**.

6 de Bellefon, C., Pestre, N., Lamouille, T., Grenouillet, P., High-throughput kinetic investigations of asymmetric hydrogenations with microdevices, *Adv. Synth. Catal.* **2003**, *345*, 190–193.

7 Bergh, S., Cong, P., Ehnebuske, B., Guan, S., Hagemeyer, A., Lin, H., Liu, Y., Lugmair, C. G., Turner, H. W., Volpe, A. F., Weinberg, W. H., Woo, L., Zysk, J., Combinatorial heterogeneous catalysis: oxidative dehydrogenation of ethane to ethylene, selective oxidation of ethane to acetic acid, and selective ammonoxidation of propane to acrylonitrile, *Top. Catal.* **2003**, *23*, 65–79.

8 *www.highthroughputexperimentation.com*, **2004**.

9 Keil, F., *Diffusion und Chemische Reaktion in der Gas/Feststoffkatalyse*, Springer-Verlag, Berlin, **1999**.

10 Dumesic, J., Rudd, D., Aparicio, L., Rekoske, J., Trevino, A., *The microkinetics of heterogeneous catalysis*, University press, Oxford, **1993**.

11 Adler, R., Henning, T., Bauer, M., Schröder, F., Schreier, M., Neue modulare Integralversuchreaktoren zur Informationsgewinnung über festbett-katalytische Gasphasenreaktionen, *Chem. Ing. Tech.* **2002**, *74*, 783–787.

12 Hessel, V., Hardt, S., Löwe, H., Chemical processing with microdevices: device/plant concepts, selected applications and state of scientific/commercial mplementation, chemical engineering and technology, *Proceedings of the 6th Italian Conference on Chemical and Process Engineering, ICheaP-6* **2003**, *3*, 479–484.

13 *www.heatric.com*, Heatric (Poole, UK), 30 May **2004**.

14 Ehrfeld, W., Hessel, V., Haverkamp, V., Microreactors, in *Ullmann's Encyclopedia of Industrial Chemistry*, Wiley-VCH, Weinheim, **1999**.

15 van den Berg, A., Lammerink, T. S. J., Micro Total Analysis Systems: Microfluidic Aspects, Integration Concepts and Applications, *Topics Curr. Chem.* **1997**, *194*, 21–50.

16 Gavriilidis, A., Angeli, P., Cao, E., Yeong, K. K., Wan, Y. S. S., Technology and application of microengineered reactors, *Trans. IChemE.* **2002**, *80/A*, 3–30.

17 Fletcher, P. D. I., Haswell, S. J., Pombo-Villar, E., Warrington, B. H., Watts, P., Wong, S. Y. F., Zhang, X., Micro reactors: principle and applications in organic synthesis, *Tetrahedron* **2002**, *58*, 4735–4757.

18 Wegeng, R. S., Drost, M. K., Brenchley, D. L., Process intensification through miniaturization of chemical and thermal systems in the 21st century, in Ehrfeld, W. (Ed.), *Microreaction Technology: 3rd International Conference on Microreaction Technology, Proc. of IMRET 3*, Springer-Verlag, Berlin, **2000**, 2–13.

19 Jensen, K. F., Microreaction engineering – is small better? *Chem. Eng. Sci.* **2001**, *56*, 293–303.

20 de Mello, A., Wootton, R., But what is it good for? Applications of microreactor technology for the fine chemical industry, *Lab Chip* **2002**, *2*, 7N–13N.

21 Besser, R. S., Ouyang, X., Surangalikar, H., Fundamental characterization studies of a microreactor for gas-solid heterogeneous catalysis, in *Proceedings of the 6th International Conference on Microreaction Technology, IMRET 6* (11–14 March 2002), AIChE Pub. No. 164, New Orleans, **2002**, 254–261.

22 Kestenbaum, H., Lange de Olivera, A., Schmidt, W., Schüth, H., Ehrfeld, W.,

Gebauer, K., Löwe, H., Richter, T., Synthesis of ethylene oxide in a catalytic microreactor system, *Stud. Surf. Sci. Catal.* **2000**, *130*, 2741–2746.

23 Wiessmeier, G., Hönicke, D., Heterogeneously catalyzed gas-phase hydrogenation of cis, trans, trans-1,5,9-cyclododecatriene on palladium catalysts having regular pore systems, *Ind. Eng. Chem. Res.* **1996**, *35*, 4412–4416.

24 Srinivasan, R., Hsing, I.-M., Berger, P. E., Jensen, K. F., Firebaugh, S. L., Schmidt, M. A., Harold, M. P., Lerou, J. J., Ryley, J. F., Micromachined reactors for catalytic partial oxidation reactions, *AIChE J.* **1997**, *43*, 3059–3069.

25 Fichtner, M., Mayer, J., Wolf, D., Schubert, K., Microstructured rhodium catalysts for the partial oxidation of methane to syngas under pressure, *Ind. Eng. Chem. Res.* **2001**, *40*, 3475–3483.

26 Rebrov, E. V., de Croon, M. H. J. M., Schouten, J. C., Design of a microstructured reactor with integrated heat exchanger for optimum performance of highly exothermic reaction, *Catal. Today* **2001**, *69*, 183–192.

27 Rouge, A., Spoetzl, B., Gebauer, K., Schenk, R., Renken, A., Microchannel reactors for fast periodic operation: the catalytic dehydration of isopropanol, *Chem. Eng. Sci.* **2001**, *56*, 1419–1427.

28 Wörz, O., Jäckel, K.-P., Richter, T., Wolf, A., Microreactors – a new efficient tool for reactor development, *Chem. Eng. Technol.* **2001**, *24*, 138–143.

29 Wan, Y. S. S., Chau, J. L. H., Gavriilidis, A., Yeung, K. L., TS-1 zeolite microengineered reactors for 1-pentene epoxidation, *Chem. Commun.* **2002**, 878–879.

30 Cominos, V., Hardt, S., Hessel, V., Kolb, G., Löwe, H., Wichert, M., Zapf, R., A methanol steam micro-reformer for low power fuel cell applications, in *Proceedings of the 6th International Conference on Microreaction Technology, IMRET* 6 (11–14 March 2002), AIChE Pub. 164, New Orleans, **2002**, 113–124.

31 Hessel, V., Löwe, H., Micro chemical engineering: components – plant concepts – user acceptance: Part I, *Chem. Eng. Technol.* **2003**, *26*, 13–24.

32 Hessel, V., Löwe, H., Micro chemical engineering: components – plant concepts – user acceptance: Part II, *Chem. Eng. Technol.* **2003**, *26*, 391–408.

33 Hessel, V., Löwe, H., Micro chemical engineering: components – plant concepts – user acceptance: Part III, *Chem. Eng. Technol.* **2003**, *26*, 531–544.

34 Commenge, J. M., Falk, L., Corriou, J. P., Matlosz, M., Optimal design for flow uniformity in microchannel reactors, *AIChE J.* **2000**, *48*, 345–358.

35 Hagendorf, U., Janicke, M., Schüth, F., Schubert, K., Fichtner, M., A Pt/Al_2O_3 coated microstructured reactor/heat exchanger for the controlled H_2/O_2-reaction in the explosion regime, in Ehrfeld, W., Rinard, I. H., Wegeng, R. S. (Eds.), *Process Miniaturization: 2nd International Conference on Microreaction Technology, IMRET 2, Topical Conf. Preprints*, AIChE, New Orleans, **1998**, 81–87.

36 Makhijani, V. B., Raghavan, J., Przekwas, A., Przekwas, A. J., Simulation of biochemical reaction kinetics in microfluidic systems, in Ehrfeld, W. (Ed.), *Microreaction Technology: 3rd International Conference on Microreaction Technology, Proc. of IMRET 3*, Springer-Verlag, Berlin, **2000**, 441–450.

37 Müller, A., Drese, K., Gnaser, H., Hampe, M., Hessel, V., Löwe, H., Schmitt, S., Zapf, R., Fast preparation and testing methods using a microstructured modular reactor for parallel gas phase catalyst screening, *Catal. Today* **2003**, *81*, 377–391.

38 Müller, A., A *Modular Approach to Heterogeneous Catalyst Screening in the Laminar Flow Regime*, VDI Verlag, Düsseldorf **2004**.

39 Ajmera, S. K., Losey, M. W., Jensen, K. F., Schmidt, M. A., Microfabricated packed-bed reactor for phosgene synthesis, *AIChE J.* **2001**, *47*, 1639–1647.

40 Miyazaki, T., Ozturk, S., Onal, I., Senkan, S., Selective oxidation of propylene to propylene oxide using combinatorial methodologies, *Catal. Today* **2003**, *81*, 473–484.

41 Danielson, E., Golden, J. H., McFarland, E. W., Reaves, C. M., Weinberg, W. H., Wu, X. D., *Nature* **1997**, *389*, 944–948.

42 Gong, P., Doolen, R. D., Fan, Q., Giaquinta, D. M., Guan, S.,

McFarland, E. W., Poojary, D. M., Self, K., Turner, H. W., Weinberg, W. H., Kombinatorische Parallelsynthese von Hochgeschwindigkeitsrasterung von Heterogenkatalysator-Bibliotheken, *Angew. Chem.* **1999**, *111* (4), 508–511.

43 Jandeleit, B., Schaefer, D. J., Powers, T. S., Turner, H. W., Weinberg, W. H., Kombinatorische Materialforschung und Katalyse, *Angew. Chem.* **1999**, *111*, 2649.

44 Lucas, M., Claus, P., Parallelisiertes Screening von heterogen katalysierten Gas/Flüssigphasenreaktionen in einem Multibatch-Reaktor, *Chem. Ing. Tech.* **2001**, *73*, 252–257.

45 Desrosiers, P., Guan, S., Hagemeyer, A., Lowe, D. M., Lugmair, C., Poojary, D. M., Turner, H., Weinberg, H., Zhou, X., Armbrust, R., Fengler, W., Notheis, U., Application of combinatorial catalysis for the direct amination of benzene to aniline, *Catal. Today* **2003**, *81*, 319–328.

46 *www.sequence.stanford.edu/group/techdev/oligosyn.html*, Oligonucleotide synthesizer (Stanford), 5 March **2004**.

47 Marron, B. E., Jayawickreme, C. K., Going to the well no more: lawn format assays for ultra-high-throughput screening, *Curr. Opin. Chem. Biol.* **2003**, *7*, 395–401.

48 Volland, M. A. O., Adlhart, C., Kiener, C. A., Chen, P., Hofmann, P., Catalyst screening by electrospray ionization tandem mass spectrometry: hofmann carbenes for olefin metathesis, *Chem. Eur. J.* **2001**, *7*, 4621–4632.

49 *www.thalestech.com*, **2004**.

50 Zech, T., Hönicke, D., Lohf, A., Golbig, K., Richter, T., Simultaneous screening of catalysts in microchannels: methodology and experimental setup, in Ehrfeld, W. (Ed.), *Microreaction Technology: 3rd International Conference on Microreaction Technology, Proc. of IMRET 3*, Springer-Verlag, Berlin, **2000**, 260–266.

51 Zech, T., Claus, P., Hönicke, D., Miniaturized reactors in combinatorial catalysis and high-throughput experimentation, *Chimia* **2002**, *56*, 611–620.

52 Zech, T., Hönicke, D., Efficient and reliable screening of catalysts for microchannel reactors by combinatorial methods, in *Proceedings of the 4th International Conference on Microreaction Technology, IMRET 4* (5–9 March 2000), AIChE Topical Conf. Proc., Atlanta, GA, **2000**, 379–389.

53 Kolb, G., Cominos, V., Drese, K., Hessel, V., Hofmann, C., Löwe, H., Wörz, O., Zapf, R., A novel catalyst testing microreactor for heterogeneous gas phase reactions, in *Proceedings of the 6th International Conference on Microreaction Technology, IMRET 6* (11–14 March 2002), AIChE Pub. No. 164, New Orleans, **2002**, 61–72.

54 Cominos, V., Hessel, V., Hofmann, C., Kolb, G., Löwe, H., Zapf, R., Delsman, E. R., de Croon, M. H. J. M., Schouten, J. C., Fuel processing in micro-reactors for low power fuel cells, *Chem. Eng. Commun.* **2005**, accepted for publication.

55 Mies, M. J. M., Rebrov, E. V., de Croon, M. H. J. M., Schouten, J. C., Design of a molybdenum high throughput micro-reactor for high temperature screening of catalyst coatings, *Chem. Eng. J.* **2004**, *101*, 225–235.

56 Müller, A., Drese, K., Gnaser, H., Hampe, M., Hessel, V., Löwe, H., Schmitt, S., Zapf, R., A 'combinatorial' approach to the design of a screening reactor for parallel gas phase catalyst screening, *Chim. Oggi* **2003**, *21*, 60–68.

57 Schober, A., Schlingloff, G., Thamm, A., Kiel, H. J., Tomandl, D., Gebinoga, M., Döring, M., Köhler, J. M., Mayer, G., System integration of microsystems/chip elemens in miniaturized automata for high-throughput synthesis and screening in biology, biochemistry, and chemistry, *Microsyst. Technol.* **1997**, *4*, 35–39.

58 Zech, T., *Miniaturisierte Screening-Systeme für die kombinatorische heterogene Katalyse*, Vol. 3, Nr. 732, VDI-Verlag, Düsseldorf, **2002**.

59 Jensen, K. F., Firebaugh, S. L., Franz, A. J., Quiram, D., Srinivasan, R., Schmidt, M. A., Integrated gas phase microreactors, in Harrison, J., van den Berg, A. (Eds.), *Micro Total Analysis Systems*, Kluwer, Dordrecht, **1998**, 463–468.

60 Franz, A. J., Quiram, D. J., Srinivasan, R., Hsing, I.-M., Firebaugh, S. L., Jensen, K. F., Schmidt, M. A., New operating regimes and applications feasible with microreactors, in Ehrfeld,

W., RINARD, I. H., WEGENG, R. S. (Eds.), *Process Miniaturization: 2nd International Conference on Microreaction Technology, IMRET 2, Topical Conf. Preprints,* AIChE, New Orleans, **1998**, 33–38.

61 LOSEY, M. W., SCHMIDT, M. A., JENSEN, K. F., Microfabricated multiphase packed-bed reactors: characterization of mass transfer and reactions, *Ind. Eng. Chem. Res.* **2001**, *40,* 2555–2562.

62 JENSEN, K. F., HSING, I.-M., SRINIVASAN, R., SCHMIDT, M. A., HAROLD, M. P., LEROU, J. J., RYLEY, J. F., Reaction engineering for microreactor systems, in EHRFELD, W. (Ed.), *Microreaction Technology – Proc. of the 1st International Conference on Microreaction Technology, IMRET 1,* Springer-Verlag, Berlin, **1997**, 2–9.

63 BERGH, S., GUAN, S., *WO 00/51720,* SYMYX, **2000**.

64 SENKAN, S., KRANTZ, K., OZTURK, S., ZENGIN, V., ONAL, I., Hochdurchsatz-Screening von Heterogenkatalysator-Bibliotheken unter Verwendung eines Mehrkammerreaktorsystems und der Massenspektrometrie, *Angew. Chem.* **1999**, *111,* 2965–2971.

65 SENKAN, S. M., *Nature* **1998**, 350.

66 SENKAN, S. M., MIYAZAKI, T., KRANTZ, K., OZTURK, S., LEIDHOLM, C., Discovery and optimization of heterogeneous catalytic materials using combinatorial metho dologies, in *Proceedings of the CombiCat 2000 North America,* The Catalyst Group Resources, Philadelphia, **2002**.

67 RODEMERCK, U., IGNASZEWSKI, P., LUCAS, A., CLAUS, P., Parallel synthesis and fast catalytic testing of catalyst libraries for oxidation reactions, *Chem. Eng. Technol.* **2000**, *23,* 413–416.

68 RODEMERCK, U., IGNASZEWSKI, P., LUCAS, M., CLAUS, P., BAERNS, M., Parallel synthesis and testing of heterogeneous catalysts, in EHRFELD, W. (Ed.), *Microreaction Technology: 3rd International Conference on Microreaction Technology, Proc. of IMRET 3,* Springer-Verlag, Berlin, **2000**, 287–293.

69 RODEMERCK, U., IGNASZEWSKI, P., LUCAS, M., CLAUS, P., BAERNS, M., Parallel synthesis and fast screening of heterogeneous catalysts, *Top. Catal.* **2001**, *13,* 249–252.

70 HOFFMANN, C., WOLF, A., SCHÜTH, F., Parallele Synthese und Prüfung von Katalysatoren nah an konventionellen Testbedingungen, *Angew. Chem.* **1999**, *111,* 2971.

71 SCHÜTH, F., HOFFMANN, C., WOLF, A., Parallel synthesis and testing of catalysts under nearly conventional testing conditions, *Angew. Chem. Int. Ed.* **1999**, *38* (*18*), 2800–2803.

72 HOFFMANN, C., THOMPSON, S., BUSCH, O., WOLF, A., KIENER, C., SCHMIDT, H.-W., SCHÜTH, F., in *Proceedings of the Microreaction Technology: Parallel Preparation and Testing of Catalysts* (21 June 2001), DECHEMA, Frankfurt/Main, **2001**.

73 RAMPF, F. A., HERRMANN, W. A., High-Throughput-Katalysatorscreening unter Hochdruckbedingungen, *Chem. Ing. Tech.* **2001**, *73,* 1–2, 97–99.

74 KLEIN, J., ZECH, T., NEWSAM, J. M., SCHUNK, S. A., Application of a novel split & pool principle for the fully combinatorial synthesis of functional inorganic materials, *Appl. Catal. A* **2003**, *254,* 121–131.

75 REETZ, M. T., BECKER, M. H., KÜHLING, K. M., HOLZWARTH, A., Zeitaufgelöste IR-thermographische Detektion und Screening von enantioselektiven katalytischen Reaktionen, *Angew. Chem.* **1998**, *110,* 2792–2794.

76 KLEIN, J., LEHMANN, C. W., SCHMIDT, H.-W., MAIER, F. W., Kombinatorische Materialbibliothek im µg-Maßstab am Beispiel der Hydrothermalsynthese, *Angew. Chem.* **1998**, *110,* 3557–3561.

77 ORSCHEL, M., KLEIN, J., SCHMIDT, H.-W., MAIER, W. F., Erkennung der Selektivität von Oxidationsreaktionen auf Katalysatorbibliotheken durch ortsaufgelöste Massenspektrometrie, *Angew. Chem.* **1999**, *111,* 2961–2965.

78 CONG, P., DOOLEN, R. D., FAN, Q., GIAQUINTA, D. M., GUAN, S., MCFARLAND, E. W., POOJARY, D. M., SELF, K., TURNER, H. W., WEINBERG, W. H., Kombinatorische Parallelsynthese und Hochgeschwindigkeitsrasterung von Heterogenkatalysator-Bibliotheken, *Angew. Chem.* **1999**, *111,* 508.

79 CONG, P., DOOLEN, R. D., FAN, Q., GIAQUINTA, D. M., GUAN, S., MCFARLAND,

E. W., Poojary, D. M., Self, K., Turner, H. W., Weinberg, W. H., High-throughput synthesis and screening of combinatorial heterogeneous catalyst libraries, *Angew. Chem. Int. Ed.* **1999**, *38*, 484–487.

80 Cong, P., Giaquinta, D., Guan, S., McFarland, E., Self, K., Turner, H., Weinberg, H., A combinatorial chemistry approach to oxidation catalyst discovery and optimization, in Ehrfeld, W., Rinard, I. H., Wegeng, R. S. (Eds.), *Process Miniaturization: 2nd International Conference on Microreaction Technology, IMRET 2, Topical Conf. Preprints*, AIChE, New Orleans, **1998**, 118–123.

81 Potyrailo, R. A., Olson, D. R., Medford, G., Brennan, M. J., Development of combinatorial chemistry methods for coatings: high-throughput optimization of curing parameters of coating libraries, *Anal. Chem.* **2002**, *74*, 5676–5680.

82 Potyrailo, R., Wroczynski, R., Lemmon, J., Flanagan, W., Siclovan, O., Fluorescence spectroscopy and multivariate spectral descriptor analysis for high-throughput multiparameter optimization of polymerization conditions of combinatorial 96-microreactor arrays, *J. Comb. Chem.* **2003**, *5*, 8–17.

83 Morris, N., Mallouk, T., A high-throughput optical screening method for the optimization of coloidal water oxidation catalysts, *J. Am. Chem. Soc.* **2002**, *124*, 11114–11121.

84 Besser, R. S., Ouyang, X., Surangalikar, H., Hydrocarbon hydrogenation and dehydrogenation reactions in microfabricated catalytic reactors, *Chem. Eng. Sci.* **2003**, *58*, 19–26.

85 Fichtner, M., Mayer, J., *Mikroreaktorsystem für die partielle Oxidation von Methan: Teilprojekt 5: Entwicklung und Herstellung von Mikroreaktoren aus Metallen mit spanabhebender Materialbearbeitung für die Beispielreaktion Methanoxidation*, Internal Annual Report, Forschungszentrum, Karlsruhe, **1997**.

86 Cullis, C. F., Keene, D. E., Trimm, D. L., Studies of the partial oxidation of methane over heterogeneous catalysts, *J. Catal.* **1970**, *19*, 378–385.

87 Hilfiker, R., berghausen, J., Blatter, F., De Paul, S. M., Szelagiewicz, M., von Raumer, M., High-throughput screening for polymorphism, *Chim. Oggi* **2003**, *9*, 75–79.

88 Liauw, M., Baerns, M., Broucek, R., Buyevskaya, O. V., Commenge, J.-M., Corriou, J.-P., Falk, L., Gebauer, K., Hefter, H. J., Langer, O.-U., Löwe, H., Matlosz, M., Renken, A., Rouge, A., Schenk, R., Steinfeld, N., Walter, S., Periodic operation in microchannel reactors, in Ehrfeld, W. (Ed.), *Microreaction Technology: 3rd International Conference on Microreaction Technology, Proc. of IMRET 3*, Springer-Verlag, Berlin, **2000**, 224–234.

89 Emig, G., Seiler, H., Reduction-oxidation cycles in a fixed-bed reactor with periodic flow reversal, *Chem. Eng. Technol.* **1999**, *22* (6), 479–484.

90 Roiter, V. A., Korneichuk, G. P., Leperson, M. G., Stukanowskaja, N. A., Tolchina, B. I., *Zhr. Fiz. Khim.* **1950**, *24*, 459.

91 Kokes, R., Tobin, H., Emmett, P., New microcatalytic–chromatographic technique for studying catalytic reactions, *J. Am. Chem. Soc.* **1955**, *77*, 5860–5862.

92 Schneider, P., Smith, J. M., *AIChE J.* **1968**, *14*, 763.

93 Polanski, L., Naphthali, L., *AIChE J.* **1969**, 1541.

94 Colin, S., Aubert, C., Caen, R., Unsteady gaseous flow in rectangular microchannels: frequency response of one or two pneumatic lines connected in series, *Eur. J. Mech. B/Fluids* **1998**, *17*, 79–104.

95 Zieren, M., Willnauer, R., Köhler, J. M., Schwesinger, N., Detection of catalytic reactions in a miniaturized FIA set-up using a static micromixer and a flow-through chip calorimeter, in *Proceedings of the 13. Ulm–Freiberger Kalorimetrietage*, Freiberg, **1999**, 110.

96 Bennett, C., *Catal. Rev.-Sci. Eng.* **1976**, *13*, 121.

97 Mirodatos, C., Use of isotopic transient kinetics in heterogeneous catalysis, *Catal. Today* **1991**, *9* (*1–2*), 83–95.

98 Walter, S., Liauw, M., Fast concentration cycling in microchannel reactors, in *Proceedings of the 4th International Conference on Microreaction Technology, IMRET 4* (5–9 March 2000), AIChE

Topical Conf. Proc., Atlanta, GA, **2000**, 209–214.

99 Wojciechowski, B., The temperature scanning reactor I: reactor types and modes of operation, *Catal. Today* **1997**, *36* (*2*), 167–190.

100 Gleaves, J., Yablonski, G., Phanawadee, P., Schuurmann, Y., TAP-2: an interrogative kinetics approach, *Appl. Catal. A* **1997**, *160* (*1*), 55–58.

101 van der Linde, S., Nijhuis, T., Dekker, F., Kapteijn, F., Moulijn, J., Mathematical treatment of transient kinetic data: combination of parameter estimation with solving the related partial differential equation, *Appl. Catal. A* **1997**, *151*, 27–57.

102 Jahn, R., Snita, D., Kubicek, M., Marek, M., Evolution of spatiotemporal temperature patterns in monolithic catalytic reactors, *Catal. Today* **2001**, *70*, 393–409.

103 Rice, N. M., Wojciechowski, B., The temperature scanning reactor II: theory of operation, *Catal. Today* **1997**, *36* (*2*), 191–207.

104 Kobayashi, H., Kobayashi, M., *Catal. Rev.-Sci. Eng.* **1974**, *10*, 139.

105 Baerns, M., Hofmann, H., Renken, A., *Chemische Reaktionstechnik, Lehrbuch der Technischen Chemie*, Georg Thieme, Stuttgart **1987**, 68.

106 Cussler, E. L., *Diffusion Mass Transfer in Fluid Systems*, Cambridge University Press, Cambridge, **1998**, 104.

107 Mazaki, R., Inoue, H., *Rate equations of solid-catalyzed reactions*, University of Tokyo Press, Tokyo, **1991**.

108 Spencer, N. D., Pereira, C. J., Partial oxidation of CH_4 to HCHO over a MoO_3–SiO_3 catalyst: a kinetic study, *AIChE J.* **1987**, *33*, 1808–1812.

109 de Bellefon, Tanchoux, N., Caravieilhes, S., Grenouillet, P., Hessel, V., Microreactors for dynamic high throughput screening of fluid–liquid molecular catalysis, *Angew. Chem.* **2000**, *112*, 3584–3587.

110 Hessel, V., Hardt, S., Löwe, H., *Chemical Micro Process Engineering – Fundamentals, Modeling and Reactions*, Wiley-VCH, Weinheim **2004**.

111 Pennemann, H., Hessel, V., Kost, H.-J., Löwe, H., de Bellefon, C., Investigations on pulse broadening for transient catalyst screening in gas/liquid systems, *AIChE J.* **2003**, *50* (*8*), 1814–1823.

112 IMM Institut für Mikrotechnik Mainz GmbH, unpublished.

113 de Bellefon, C., Abdallah, R., Lamouille, T., Pestre, N., Caravieilhes, S., Grenouillet, P., High-throughput screening of molecular catalysts using automated liquid handling, injection and microdevices, *Chimia* **2002**, *56*, 621–626.

114 Atkins, M. P., Senkan, S. M., *Radiation activation and screening of catalyst libraries for catalyst evaluation and reactors therefor, WO 99/19724*, BP Chemicals Ltd. and LASER Catalyst Systems Inc., **1997**.

115 Zech, T., Hönicke, D., in *Proceedings of the Microreaction Technology: Parallel preparation and Testing of Catalysts* (21 June 2001), DECHEMA, Frankfurt/Main, **2001**.

116 Rose, S., Stevens, A., Computational design strategies for combinatorial libraries, *Curr. Opin. Chem. Biol.* **2003**, *7*, 331–339.

117 Rothenberg, G., Boelens, H. F. M., Iron, D., Westerhuis, J. A., A high-throughput approach to kinetic studies, in *Proceedings of the CombiCat 2002 North America*, The Catalyst Group Resources, Philadelphia, **2002**.

118 Mirodatos, C., Serra, J. M., Corma, A., Farrusseng, D., Baumes, L., Flego, C., Perego, C., Styrene from toluene by combinatorial catalysis, *Catal. Today* **2003**, *81*, 425–436.

119 Holena, M., Baerns, M., Feedforward neural networks in catalysis, a tool for the approximation of the dependency of yield on catalyst composition, and for knowledge extraction, *Catal. Today* **2003**, *81*, 485–494.

120 Serra, J. M., Corma, A., Argente, E., Valero, S., Botti, V., Neuronal networks for modeling of kinetic reaction data applicable to catalyst scale up and process control and optimization in the frame of combinatorial catalysis, *Appl. Catal. A* **2003**, *254*, 133–145.

121 Corma, A., Serra, J. M., Argente, E., Botti, V., Valero, S., Application of artificial neuronal networks to combinatorial catalysis: modeling and predicting ODHE catalysts, *ChemPhysChem* **2002**, *3*, 939–945.

122 Serra, J. M., Corma, A., Farrusseng, D., Baumes, L., Mirodatos, C., Flego, C., Perego, C., Styrene from toluene by combinatorial catalysis, *Catal. Today* **2003**, *81*, 425–436.

123 Roedemerck, U., *Hochdurchsatz-Experimentation, Jahresbericht*, Institut für Angewandte Chemie Berlin-Adlershof, e.V., Berlin, **2003**.

124 Commenge, J. M., Falk, L., Corriou, J. P., Matlosz, M., Optimal design for flow uniformity in microchannel reactors, in *Proceedings of the 4th International Conference on Microreaction Technology, IMRET 4* (5–9 March 2000), AIChE Topical Conf. Proc., Atlanta, GA, **2000**, 24–30.

125 Holzwarth, A., Schmidt, H.-W., Maier, W. F., IR-thermographische Erkennung katalytischer Aktivität in kombinatorischen Bibliotheken heterogener Katalysatoren, *Angew. Chem.* **1998**, *110*, 2788.

126 Van Veen, A. C., Farruseng, D., Rebeilleau, M., Decamp, T., Holzwarth, A., Schuurman, Y., Mirodatos, C., Acceleration in catalyst development by fast transient kinetic investigation, *J. Catal.* **2003**, *216*, 135–143.

127 Taylor, G. I., Dispersion of soluble matter in solvent flowing slowly through a tube, *Proc. Roy. Soc. London A* **1953**, *219*, 186–203.

128 Aris, R., *Proc. Roy. Soc. London A* **1955**, *223*, 67.

129 Brenner, H., *ACS J. Surf. Colloids* **1990**, 6, 12.

130 Angeli, P., Gobby, D., Gavriilidis, A., Modeling of gas–liquid catalytic reactions in microchannels, in Ehrfeld, W. (Ed.), *Microreaction Technology: 3rd International Conference on Microreaction Technology, Proc. of IMRET 3*, Springer-Verlag, Berlin, **2000**, 253–259.

131 Dutta, D., Leighton, D. T., Dispersion reduction in pressure-driven flow through microetched channels, *Anal. Chem.* **2001**, *75*, 57–70.

132 Ross, D., Johnston, T. J.,, Locascio, L. E., Imaging of electroosmotic flow in plastic microchannels, *Anal. Chem.* **2001**, *73* (*24*), 2509.

133 Surangalikar, H., Ouyang, X., Besser, R. S., Experimental study of hydrocarbon hydrogenation and dehydrogenation reactions in silicon microfabricated reactors of two different geometries, *Chem. Eng. J.* **2002**, *90*, 1–8.

134 Senkan, S., Combinatorial heterogeneous catalysis – a new path in an old field, *Angew. Chem. Int. Ed.* **2001**, *40*, 312–329.

135 Kiener, C., Kurtz, M., Wilmer, H., Hoffmann, C., Schmidt, H.-W., Grunwaldt, J.-D., Muhler, M., Schüth, F., High-throughput screening under demanding conditions: Cu/ZnO catalysts in high pressure methanol synthesis as an example, *J. Catal.* **2003**, *216*, 110–119.

136 Holzwarth, A., Denton, P., Zanthoff, H., Mirodatos, C., Combinatorial approaches to heterogenous catalysis: strategies and perspectives for academic research, *Catal. Today* **2001**, *67* (4), 309–318.

4
Micro Structured Reactor Plant Concepts

4.1
Micro Reactor or Micro Structured Reactor Plant (MRP)

The use of micro structured components for process engineering has gained increasing importance in chemical, pharmaceutical and life science applications in recent years. Small devices – reactors, heat exchangers, static mixers and other process components – can be fabricated in configurations measured in millimeters and embedded with micrometer-sized pores or channels. Owing to large specific surface areas, devices with these small dimensions provide more efficient mass and heat transfer. This results in greater selectivity and higher yield for chemical reactions. Micro process engineering and the application of apparatus such as micro structured reactors, heat exchangers and mixers developed into a self-dependent discipline. Especially in branches such as the chemical industry, the automotive industry and environmental technology, an increasing number of chemical reactions and physical transformations in micro structures were executed under highly selective and inherently safe conditions.

The use of micro reactors for chemical production was proposed since the idea of such small scaled devices became reality in the late 1980s and early 1990s. More process flexibility, capacity, variability and inherent safety were predicted, and more or less realistic micro reactors should allow one to supplement existing large-scale plants [1–6]. One major concept relies on the capability of distributed production or production on-site to minimize transport issues [7]. Sometimes the advantage of a distributed production may also be the fact that performing the process on-site shifts the side products of the chosen reaction, their selling or environmentally friendly disposal directly to the costumer. Also a distributed performance of reactions with hazardous reactants, intermediates and products, one of the very early arguments for the usage of micro reactors, became questionable against the backdrop of terrorist attacks after September 11, 2001 [8]. It should also be addressed that sometimes the raw materials for performing a reaction are no less hazardous than the desired product, e.g. the synthesis of phosgene requires chlorine and carbon monoxide, both pressurized toxic gases supplied in cylinders, which are as hazardous as the reaction product phosgene.

Nevertheless, some theoretical and practical work has been done in the last few years to bring micro reactors from the early stage of exotic devices to the application

Chemical Micro Process Engineering. V. Hessel, H. Löwe, A. Müller, G. Kolb.

ISBN: 3-527-30998-5

level. Concepts were developed to transfer the proposed advantages of micro reactors to pilot-scale or even to production-scale plant. To understand the advantages and disadvantages of micro reactors, and to find paths to incorporate them into the knowledge base of chemists and chemical engineers, a new scientific discipline, 'Chemical Micro Process Engineering' has become mature [9–12].

A huge number of publications meanwhile have included the term 'micro reactors' and their applications [9] and some attempts were made to maintain order in the field, e.g. by sorting from the point of design and manufacturing issues [5, 13], by a classification of unit operations [14] or by sorting of the chemical reactions performed [15]. The latter becomes more common than the others, but a generic look seems to be necessary.

Considering the fluid involved in a micro reactor and not the micro reactor and its internal structure, the fluid itself is structured. Fluid structuring is among others one of the most important issues. This leads to the assumption that the advantage (and also the disadvantage) of chemical processing is at first sight independent of the reactor used. If this prediction is true then the outer size of the micro reactor becomes irrelevant; credit card-sized, laboratory-sized or production scale, the general driving forces running a chemical reaction are always the same. Then, the term 'micro reactor' is no longer valid and should be replaced by the term 'micro structured reactor', which covers all devices which are able to structure fluids down to the micro scale. In the smallest reactor comprising only one micro structured fluid stream, the reaction conditions are as in a bigger device with hundreds or even thousands of fluid streams in parallel.

Nevertheless, it is becoming the done thing to name single-channel or multi-channel credit card-sized devices as 'chip reactors', and single reactors such as mixers essentially made of steel as 'micro reactors'; this holds also for a set of such reactors. Apart from these findings, the concept of 'numbering-up' instead of scale-up becomes realistic. To draw the logical conclusion from these considerations, chemical plants consisting of micro structured items should be named 'micro structured reactor plants' (MRPs) instead of the often used term 'micro plants'. The latter is mostly used to characterize small-scale miniplants which provide no micro structured devices in the narrow sense.

Of course, it makes no sense to mount huge numbers of small micro structured reactors together to achieve higher throughput; the micro structured device has to be adapted and enlarged with the restriction of not increasing the micro structured fluid flow behavior remarkably. With higher throughput a secondary problem arises: the environment of the micro structured reactor must be able to provide the required flow rates, pressure and temperature, etc., of the fluids. From the engineering point of view it makes no sense to shrink pumps, tubes and other devices down to the micro-scale and to replace them. It seems to be much better to incorporate a microstructured reactor in a given environment of a chemical plant. This approach, named the 'multiscale concept', where micro structured devices are equal to the macroscopic environment, gives the best opportunity to transfer 'micro' into a 'macro' world.

4.2
Applicable Principles for Micro Structured Reactor Plant (MRP) Design

A literature survey for *microstructured reactor plants* (MRPs) produced only a very limited number of hits compared with approximately 1500 for *micro reactors*, a surprising result. It could mean either plant technology is not possible from a technological or economical point of view or not necessary as there are no applications for this technology. Yet another aspect is much more evident. The chemical industry developed standards over the last two centuries but micro reaction technology was busy enough with reactor development and proof of the concept without bothering about interfaces between the reactors and the macroscopic environment. This now handicaps the setting up of complex plants.

There is also a kind of division of labor between industrial suppliers of apparatus and plant engineering companies. As installation of a plant amounts to about 10% of all costs [16], pressure from the engineering companies was great enough to introduce common qualification standards and also standardized interfaces between industrial apparatus.

Owing to safety aspects, apparatus is much more restricted than mechanical machines in plants. Construction and manufacturing of machines are widely unrestricted if basic principles, fixed, for example, in national regulation collections such as the German 'VDMA-Einheitsblätter' are considered.

The apparatus is essentially classified by the formula $D \times p$ (product of internal diameter D and pressure p), which defines the wall thickness and the prescribed material. Regulations concerning the manufacturing, assembly and repair of apparatus and vessels are collected in the American ASME Boiler and Pressure Vessel Code. This code was originally developed for the refinery industry and is now widely accepted. In Europe, a CE certification also is required if a certain value of the product of internal volume and pressure is exceeded. Owing to their small internal volume, this is normally not applicable to micro devices. Flanges, the fluidic interfaces between vessels, are classified according to their internal nominal pressure (PN 1–6000) and nominal diameter (DN 6–3600). The nominal diameter corresponds approximately to the internal flange diameter. The same classification is applied for pipes.

Micro technology developed rapidly from the early 1990s. As with every new technology, standards are only set after a certain degree of maturity of a technology has been reached. This seems to be the case at present, as a number of initiatives and commercial suppliers are currently starting to create their own standards for plants [17–23]. Micro reaction technology is currently located in a transition state from applications with single isolated devices to complex operations applying recycle loops and separation units. Consequently, it must be seen as a competitor or supplement to the established miniplant technology. This certainly requires in the near future the setting of common standards as, for example, proposed by the German *μChemTec* initiative [24].

Some of the principles in process design now will be reviewed in order to define the current working procedures in industry. Process development is divided into a sequence of test phases and evaluation steps. Vogel [16, p. 293] recommends a cyclic procedure of process development (Figure 4.1). Typically, a new synthesis will be optimized in a discontinuous laboratory batch process and defined in a first process concept. Transfer from the discontinuous laboratory process to a continuous process operation is first realized in the micro plant. The complexity of test phases then increases from micro plant to miniplant and further to pilot plant operations.

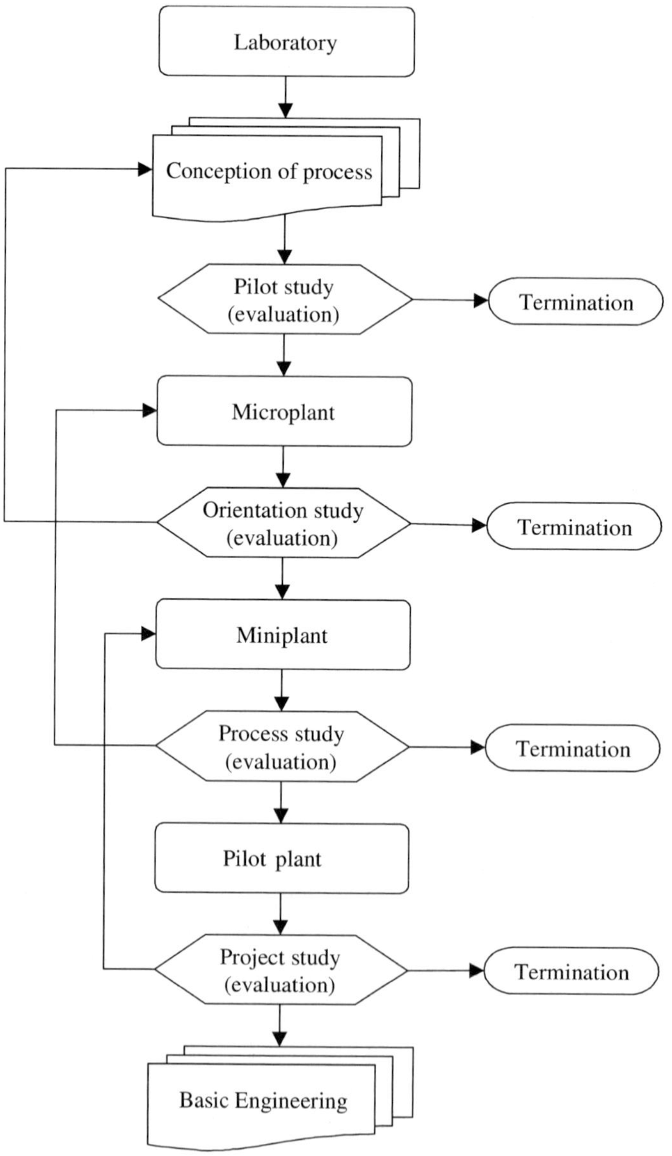

Figure 4.1 Steps in process development [16].

Whereas in a micro plant only the most important process steps and some of the recycle loops are tested, in the miniplant the complete process with all recycle loops is executed and processed continuously for several weeks. For environmental and economic reasons, solvents and unconverted reactants are recycled. Here, it is especially important to find out if small impurities below the detection limit will possibly accumulate or form deposits. To minimize the up-scaling risk for the final production plant, often a pilot plant is the next scale-up step. Now the final product is produced and often first samples are sent to customers. Every step in this scale-up procedure is followed by an evaluation step defined in a number of evaluation studies, which assist in the decision as to whether the project has to be terminated, repeated or further processed.

Integrated process development

With the introduction of advanced process simulation tools [25], the so-called *integrated process development* established. Here the most time-consuming pilot plant phase is bypassed by a combined approach of experimental miniplant testing and numerical process modeling.

The principle of *integral process development* [26] covers much more than just the optimization of a process. This approach begins with computer-aided decision procedures in the conception phase. Tools are available in which the process structure is suggested, for example: should the process be a batch or a continuous operation? The software tool for process synthesis PROSYN uses databases which include knowledge of experts, material data and calculation models for unit operations. Interfaces to process simulation tools such as ASPENPLUS and material databases are also supplied. PROSYN also delivers an economic evaluation of the future production process.

Process management systems

Every production plant will be controlled by a Process Control System (PCS). A PCS not only controls set-points but also delivers data which define the product quality. *Process management systems* allow the direct access of the production management to the production line.

Supply chain management

Information such as yield, energy consumption or personal costs helps to overview a process continuously during operation and to deal with spare parts and raw materials. This so-called *supply chain management* is now widely accepted. The predominant supplier of business software SAP claims to have given more 21 000 licenses to companies which use their management tools for business and production management [27].

In the following, the procedure of process development is described, including typical aspects which need special consideration in micro chemical engineering. The sections follow the sequence of their appearance during process development.

4.2.1
Miniplant Technology – A Model for the Micro Structured Reactor Plant Concept

As miniplant technology must be considered as a model for micro structured reactor plant technology, a closer look into its working principles is very illustrative. All characteristics of a miniplant were described by the inventors of this technology, Dow, in 1979 [28] and BASF [29], which also introduced the combined simulation/experimentation concept (Figure 4.2).

One of the main characteristics of a miniplant is certainly its continuous operation day and night for several weeks, which requires a fully automated mode of operation controlled by a process control system. A miniplant should be equipped with all recycle loops and separation units. For safe and uninterrupted operation, pumps and critical devices must be installed redundantly. As the measurement and instrumentation will have to be changed frequently during the experimentation and therefore free access becomes necessary, the operation of the plant should be executed in a non-hazardous area. Also, fast access to the plant for the operators is necessary, which is often enabled by the installation of the plant below a ventilated hood with free entry to the plant from all sides. A miniplant is very close to the final production plant if it is possible to test a 'slice' of the production plant. In the case of a rectification column this means, for example, keeping the tray spacing but not the diameter of the column the same. There are unit operations which can be up-scaled with the help of simulation tools such as single- and two-phase fluidic process steps. Others, for example filtration, solid-phase processes and heterogeneous catalytic reactions, are difficult to up-scale from a miniplant. In contrast, there is also a problem with down-scaling from the intended production plant to a miniplant as the boiler volume in the stripping section of a rectification column

Figure 4.2 A miniplant running at BASF [29].

cannot fall below a certain limit. This is defined by the tolerable temperature difference in the boiler and the danger of foam development [16, p. 309]. The specific hold-up volume (related to the volume flow) can be much larger than those found in a production plant, which makes it very difficult to obtain steady-state operation in a miniplant.

An important aspect in miniplant technology is the possibility of developing a modular set-up of the plant. This means essentially introducing modularized equipment as discussed, for example, by Rinard and others [30–32]. A standardized design of a reactor enables flexible and cost-efficient manufacturing and also gradual optimization of the reactor design. Even more critical than the well-known problems with scale-up is the problem with recycle loops [33]. Substances below the detection limit can accumulate in a process after several hundred recycle periods. Substances developing during a reaction in an amount of 10–100 ppm will reach a concentration of 1–10% after 1000 recycle periods. This clearly indicates that the whole process can be influenced by trace elements.

All these aspects also have to be taken into account in the conception and design phase of micro structured reactor plant technology, introduced in the following.

4.2.2
The Micro Unit Operations Concept

There are a few micro structured reactor plants of table-top size on the market [21, 22, 34–42] and even one of larger scale for production purposes [43–45]. Nevertheless, engineers do not have much experience in designing micro structured reactor plants and, of course, systematic tools for designing and control are also lacking[13].

In scouting work, Hasebe [13] described a concept for the application and design rules for a micro structured reactor plant. First the reasons for performing a chemical reaction in a micro device must be clear. Either the target product cannot be produced conventionally, or the production efficiency of the target product can be drastically improved. Currently many chemical reactions are performed in micro reactors (see, e.g., [9]), but it is not clear how many of them fit with the above-mentioned rules.

The economic benefit is one of the dominant problems if a micro structured reactor plant is used for chemical production. Without any doubt, an overall flow rate through a micro structured device can be achieved that is comparable to that with a conventional batch process. However, the residence time is very short because of the dimensions of a microstructured device. If the reaction kinetics are slow, an additional device is necessary to increase a dwell time. Hence, much effort should be devoted to increasing the reaction rate instead of transferring the standard protocol to a micro structured reactor [13].

4.2.3
Design Problems of Chemical Micro Structured Reactor Plants

Owing to the small size of micro devices, new design methods are required. Similar to conventional unit operations, micro-scale pendants exist, named 'micro unit operations' [13], which have characteristic design problems. Some of them are described in the following.

Design margins

In a micro structured device, the size strongly affects the function itself. With this prerequisite, the functional and physical design cannot be executed separately. For example, in a simple micro channeled device such as shown in Figure 4.3, the cross-sectional area and the residence time are assumed to be dominant factors to affect the functionality. It is obvious that with increase in the cross-sectional area the advantages of a micro channel will be lost. For a given flow rate and keeping the cross-sectional area constant, an extension of the device or an increase in the number of channels will increase the residence time. As a result, the uncertainties of the model and parameters may not be compensated by the design margins [13].

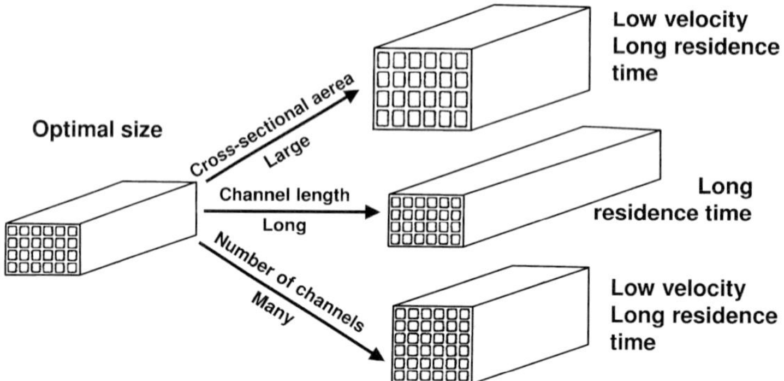

Figure 4.3 Effect of design margins [13] (by courtesy of S. Hasebe, Kyoto University).

Shape of the device

In a conventional design, each unit operation is modeled using terms and values which do not depend on the location inside the device. On the one hand, convection and diffusion in a micro structured device strongly influence the functioning of the device, and on the other hand the convection and diffusion conditions are affected by the shape of the device. To obtain an optimized fluidic micro device, some constraints on the shape of the device are necessary. These are constraints, e.g., on the average residence time, the residence time distribution and the temperature distribution [13].

Reevaluation of the neclected terms

Internal structures of micro devices (sometimes the micro device itself) are very small compared with conventional chemical equipment. However, micro fluidic devices are large enough that the physical laws established for the macroscopic world can be used to describe the behavior in the device. However, some additional terms must be taken into account, which are neglected in designing conventional reactors. An efficiency analysis of micro heat exchangers [46] is used to explain this briefly: CFD simulations of micro heat exchangers of different materials, e.g. copper, stainless steel and glass, were performed to analyze the influence of the material properties on the heat transfer performance. The conditions are given in Table 4.1.

Table 4.1 Conditions for CFD simulations of heat transfer performance [13].

Geometic data		***Operating conditions***		
Channel length (mm)	10	Hot streams	Fluid	Water
Channel width (μm)	100		Inlet average velocity (m s^{-1})	0.2
Channel height (μm)	100		Inlet temperature (K)	373
Plate length (mm)	20	Cold streams	Fluid	Water
Plate width (mm)	3.9		Inlet average velocity (m s^{-1})	0.2
Plate height (μm)	150		Inlet temperature (K)	293

By evaluating the performance of micro heat exchangers, the temperature changes of heat transfer fluids were used, and the heat transfer efficiency of devices made of stainless steel or glass were found to be higher than that of devices made of copper (see Table 4.2).

Table 4.2 Heat transfer performance for copper, stainless-steel and glass micro heat exchangers [13].

Material	***Thermal conductivity (W m^{-1} K^{-1})***	***Temperature change (K)***
Copper	388	59.4
Stainless steel	16.3	72.7
Glass	0.78	64.8

Because of the high thermal conductivity of copper, the temperature profile becomes averaged in the longitudinal direction. In contrast, for materials with lower thermal conductivity, such as stainless steel or glass, the temperature gradient remains steep and higher heat transfer efficiency is observed. This leads to the assumption that heat transfer efficiency depends largely on the longitudinal heat conduction inside the channel walls [13]. Similar results were also presented by other authors [47–49]. However, longitudinal heat conductivity plays a minor role in designing macroscopic heat exchangers and therefore it can be neglected.

Numbering-up

The possibility of numbering-up is one of the greatest advantages of using micro reactors for the construction of a micro structured reactor plant. It is an easy way to adapt throughput to economic or technical demands. The question is how to do numbering-up of micro devices having the same function so as to end up with an aggregated device. Four major possibilities are shown in Figure 4.4.

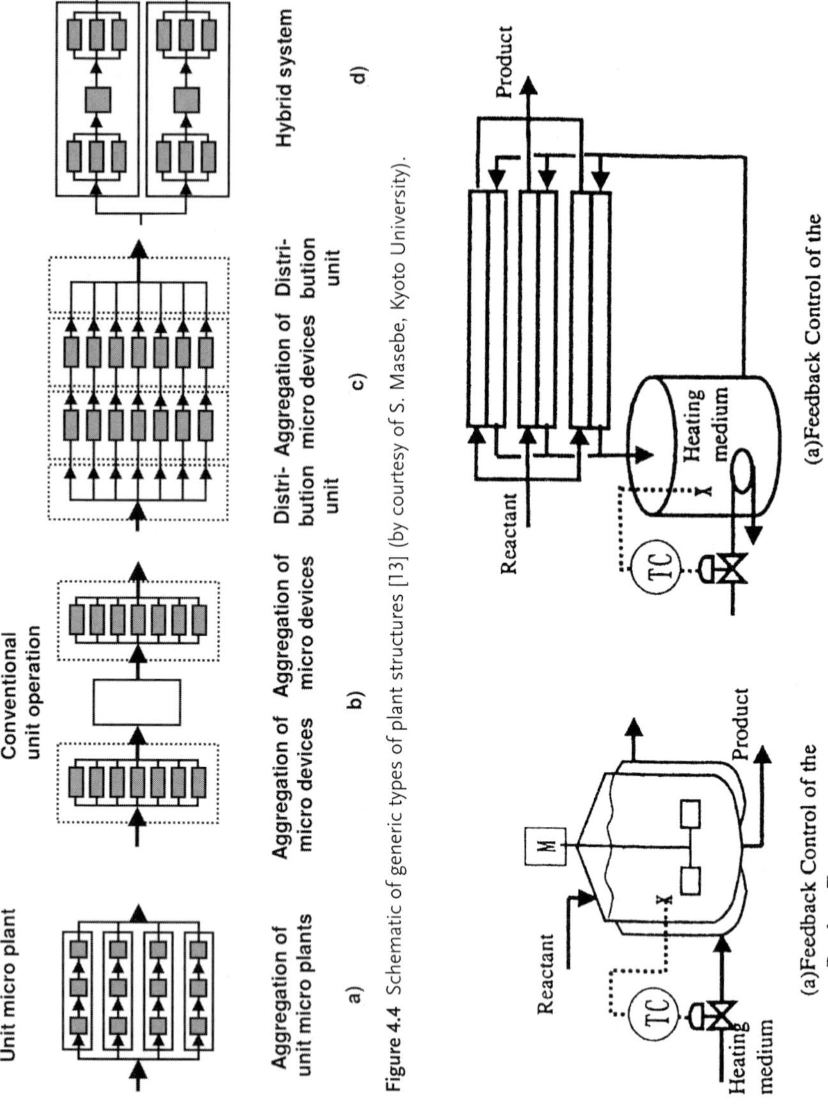

Figure 4.4 Schematic of generic types of plant structures [13] (by courtesy of S. Masebe, Kyoto University).

Figure 4.5 Schematic of two possibilities of monitoring and adjusting temperature in a conventional batch reactor and a micro fluidic device [13] (by courtesy of S. Masebe, Kyoto University).

In general, to increase the production rate it is possible to perform the chemical process in an micro structured reactor plant which can be formed by increasing the number of unit micro structured reactor plants operating in parallel. Each micro structured reactor plant consists of several micro unit operations [Figure 4.4(a)]. In some cases the residence time of a reaction is too long or intermediate storage is necessary, and a conventional unit operation must be used as a part of the process [Figure 4.4(b)]. For shorter residence times a structure similar to Figure 4.4(c) seems to be better. If two micro unit operations are working at different temperatures they must be decoupled thermally. The structures shown in Figure 4.4(b) and (c) are much better than that in Figure 4.4(a). For a production process it is desirable to use a conventional environment, i.e. conventional pumps instead of micro pumps. In this case a hybrid system is formed [Figure 4.4(d)] [13].

Instrumentation and control of micro structured reactor plants

For a conventional chemical plant, it is easy to add or replace measurement devices after plant construction. It is obvious that this cannot be done in a micro reactor. Hence the design of the micro structured reactor plant and the design of control elements and instrumentation must be executed simultaneously. It must be taken into account that disturbances caused, e.g., by the measurement, must be avoided and a counter measure against each disturbance of the process must be developed. The observed, controlled or manipulated variables are selected to implement counter measures derived above. As a consequence, the process itself is changed if there is no instrumentation or manipulation device to perform necessary counter measures [13].

In a conventional chemical plant, the feedback control method is dominantly used to keep the operating conditions constant. The temperature of a batch reactor is controlled by measuring the temperature of the reaction mixture inside the vessel. Normally, the temperature of a heating or cooling medium is constant, and by adjusting the flow rate of that medium the temperature of the reaction mixture can be easily changed and controlled. The product temperature gives a feedback control itself (Figure 4.5). To control the temperature in a micro fluidic device, an indirect measurement is another issue. The temperature of a fluid flowing through a micro device can be controlled by keeping the temperature of the heating or cooling medium constant. Because the thermal efficiency of a micro device is high, the temperature of the reactants can be kept in a constant range (Figure 4.5). From this point of view, the temperature of the heating or cooling medium gives feedback control [13].

4.3 Process Conception and Economics

The above-mentioned *integrated process development* combines simulation tools with miniplant equipment to bypass the set-up of a pilot-scale plant. Similar to this combined approach, a micro structured reactor plant can bridge laboratory-scale and micro-scale development. One could think of a future *micro-integrated process development* which consists of the three-step sequence microplant–miniplant–production plant. Even the transition from micro- to mini-scale is no longer very distinct as in micro structured devices flow rates up to 1 $m^3 h^{-1}$ were realized, thus already reaching or exceeding the flow rate of a miniplant [50]. In the following, the expression micro structured reactor plant represents an individual class of plants which is intended for small-scale production and also process development but not necessarily limited to the latter.

4.3.1 Market Study and Availability of Micro Structured Reactors

The designer of a plant needs information about the availability of devices and unit operations. A market study executed by the Institut für Mikrotechnik Mainz and YOLE Développment [51] helps to make a first survey of commercially available devices. It also estimates future needs and objectives of the chemical industry and delivers a comparison between offered and required components (Figure 4.6). The providers of micro structured devices can deliver most of the components required by the chemical industry. On the other hand, there is a lack of separation devices but this is not fully transparent in Figure 4.6. Extraction devices are under represented and the important rectification units were not asked for by the interviewers, possibly because they hardly exist.

The study also reveals differences between providers of large-scale components and providers of micro-scale components (Figure 4.7). Macro-scale manufacturers, as expected, seem to prefer stainless steel whereas the application range of micro-scale suppliers is spread over a wider variety of materials.

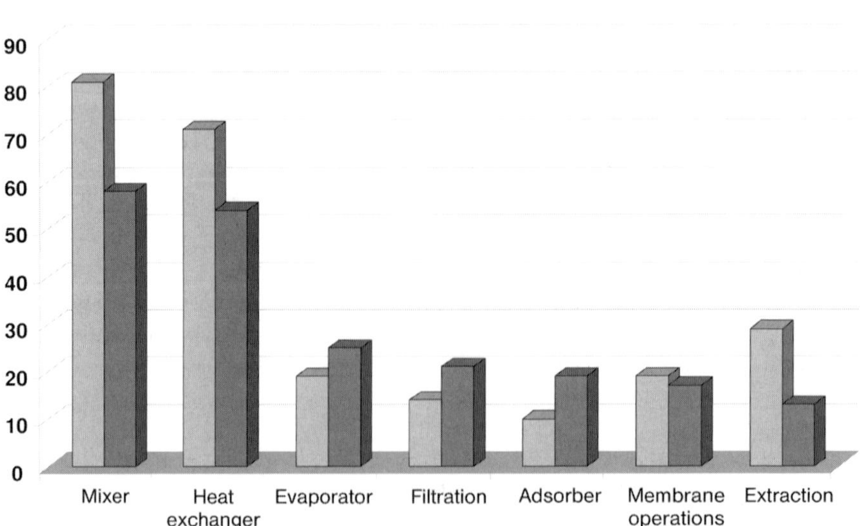

Figure 4.6 Comparison between offered and required process units [51].

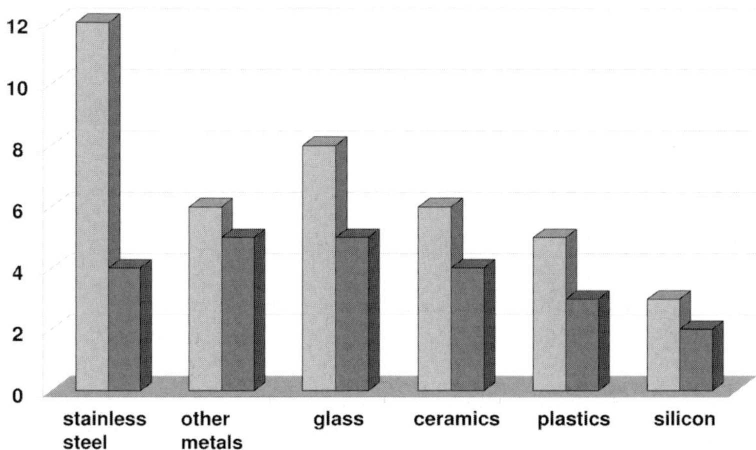

Figure 4.7 Materials offered at present (left column, all suppliers – 22 answers; right column, micro-scale suppliers – 11 answers) [51].

More than 100 micro structured devices are listed on the homepage of the µChemTec consortium [24]. The devices cover physical applications such as flow distribution, mixing, heat transfer, phase transfer, emulsification and suspension, as well as chemical applications such as chemical and biochemical processing. Some separation units such as membrane separation and capillary electrophoresis are also offered. Control devices such as valves, micro pumps for product analysis and mass flow controllers supplement the catalog.

4.3.2 Pilot Study

A pilot study should summarize the overall cost estimation considering manufacturing, investment and production. As in every commercial plant, the desired yield will be correlated with this value to judge the profitability of a microstructured reactor plant compared with the conventional macro-scale process. Cost estimates for the industrial process (costs for raw materials, waste management, etc., are usually given by the client). In front of the technological state of the art also a risk estimate is given. The following aspects will also have to be considered in micro structured reactor plant design.

Basic flow chart

The basic flow chart contains all the information necessary to obtain a quick survey concerning reactants, utilities and product.

Process flow chart and process description

The process will be described with all apparatus, machines and product streams. The latter are identified by specification numbers and listed in a table together with properties such as mass flow, concentration, liquid/gas phase, density. If already known, also process parameters such as pressure and temperature are listed. A list of utilities covering heating fluids, cooling agents, pressurized air and nitrogen and the electrical power supply will be delivered.

Waste disposal

Substances which cannot be recycled or used economically in any other way will be disposed of. Toxic or other difficult to dispose of substances are clearly marked. The individual disposal in a sewage plant, by waste combustion or by neutralization will be recommended.

Estimation of investment

The investment cost will be given, indicated as ISBL (inside battery limit: plant, laboratories, process control system, etc.), OSBL (outside battery limit) or infrastructure costs. It is expected that the OSBL costs for a micro structured reactor plant will be a minor issue as here mainly costs for housing, cooling facilities, etc., are included.

Calculation of manufacturing costs

Based upon the process flow diagram and the investment costs, the product manufacturing costs or the necessary yield in the micro structured reactor plant to reach an overall profit of, for example at least 20%, will be calculated.

Evaluation of technology

It will be indicated if the technology used for this process will be based upon existing knowledge or if a completely new process will have to be developed. On a smaller scale, the apparatus will also be evaluated in the same way.

Process safety and reserve machinery capacity

As in a continuous flow process downstream apparatus is directly linked to upstream devices, the complete production line is interrupted in case of a failure upstream. It will be examined if this failure behavior can be corrected either by the design of several parallel lines or by supplying buffer tanks, additional spare pumps, machines, etc.

4.4 Early Concepts for Micro Structured Reactor Plant Design

Plants used in micro reaction technology span a wide range of geometrical scales and operational limits. Also a variety of application modes such as commercial, research or training purposes are found. Supplier-specific set-ups exist which intend to cover the whole range of reactors and devices as well as open architecture systems such as so-called hybrid plants. Hybrid plants involve micro structured reactors and commercial macro-scale devices such as pressure controllers, valves and mass flow sensors. These systems enable the supplier to put the focus on the reactor development as peripheral equipment already exists. However, this approach will not always work as micro scale features such as a narrow residence time distribution may deteriorate in a macro-scale device. A lot of macro scale measuring devices will not work owing to flow ranges, wake effects or dead volume influences because they were originally developed for production plants [52]. One important application is certainly the mobility of small process units either for analytical purposes and of course for the automotive industry. Finally, production plants and the problems they are faced with during scale-up, such as flow distribution and interfaces to the macroscopic world, are also an issue.

Ten years ago, when the first investigation of processing chemical reactions in micro structured reactors became successful [53–57], ideas were generated to transfer this technology to chemical engineering and to replace partially or even completely existing plants with micro structured ones. Some advantages of microstructured reactors were put forward: the potential for chemical production in terms of process flexibility and capacity, the predicted inherent safety and the possibility of distributed production [1–4, 7]. The first conceptual approaches concerning micro reactor implantation were driven by similarities of the well-known miniplant technologies [58–61].

It must be pointed out again that even today confusion of terms can be observed when chemical engineers discuss 'miniplants' or 'microplants'. In most of these cases they identify with the terms mentioned above chemical plants made of glassware with volumes in the range of up to a few liters. To summarize, at that early stage no specialized micro structured reactors for production purposes were available. Most of the fabricated micro structured devices were made in terms of micro fabrication capabilities and not adapted to the chosen chemical process. It is no wonder that at first visionary theoretical work either had to be based on conventionally fabricated chemical reactors or did not outline reactor design in detail [30].

4.4.1
Paradigm Change Drives Miniplant Design Methodology

4.4.1.1 Reduction of Process Complexity for Distributed Chemical Manufacture

The miniplant concept was proposed first by Ponton at the University of Edinburgh (UK) in 1993 [1, 58–60]. The potential benefits of distributed processing, as outlined above and reported elsewhere [2, 4], were identified, in particular focusing on an increase in process safety. However, the critical evaluation of the miniplant concept also revealed a number of potential drawbacks. For instance, the specific production costs of a small production units not only depend on reactor miniaturization, but are also determined by control instruments and other peripheral equipment.

A particularly important feature of the work of Ponton is the stress on the importance of reducing process complexity of distributed small-scale devices, e.g.

for reasons of maintenance and process reliability. Hence miniplants should not be just a miniaturized reflection of industrial reactors, but resemble more a derivative of this benchmark. This, however, may require the finding of new or at least adapted process parameters when operating micro reactors.

As major unit operations contributing to system complexity of large-scale plants, separation and recycling were identified, whereas usually less technical expenditure is needed for carrying out a reaction. Hence processes with near exclusive formation of products or relatively simple separation thereof have to be identified. As the type of processing, semi-batch operation with solid feed or product is recommended. The demand for process simplification holds, in addition, for energy supply, which has to be guaranteed even at remote locations. For this purpose, the use of electrical heating and cooling is advised. The construction of the equipment should be robust since maintenance may not be possible for longer periods at remote sites. To ensure safety and to minimize pollution, encapsulation of the whole plant is suggested.

On the basis of ideas similar to the concept of Ponton [1, 60], a systematic approach, referred to as design methodology, was developed by Rinard at the University College New York (USA) [30]. Starting from a historical analysis of the main drivers governing plant development, it is suggested that one should change the relative weighting of these drivers. The design of modern plants should be much more oriented according to standards of environment, safety and process control, despite relying solely on productivity. These tasks are ideally met by miniplants being modular assembled from small components. Despite utilizing small existing processing components, miniplant construction using micro structured devices is additionally taken into account.

4.4.1.2 Historical Analysis of Chemical Plant Development

Rinard starts his analysis by reporting the rapid growth of plants, e.g. in the petrochemical industry, in recent decades and the consequent impact on the environment. This increase in plant size was mainly motivated by economy of scale, i.e. a decrease in specific capital costs. In addition, the efficiency of the plants increased using new technologies, e.g. to save energy and to achieve advances in instrumentation and process control. Within this history of technological improvements, disadvantages of using large-scale reactors were figured out, such as potential risks of transport and storage of hazardous chemicals and toxic emissions during production. Another drawback of existing equipment, being custom-designed and custom-built, is the low flexibility regarding process variation. Hence times for process adaptation or development are long and costly.

These disadvantages should be overcome by miniplants, generally defined similar as in [1, 60], strictly based on a modular design. In order to adapt to varying needs, several modules may be operated in parallel. Typical module capacities range from 100,000 to 1,000,000 lb/yr. Operation of the miniplants should be so reliable and simple that the majority of these plants can be operated by personnel not specially skilled in process technology. Start-up and shutdown have to be performed fast to allow just-in-time production. The entire plant should be transportable including footprint and containment volume.

Safety and Controllability as Major Design Issues

Rinard stresses, in particular, that prior attempts mainly focused on improving the process flow sheet rather than giving priority to safety (see also [62]), controllability (see also [63, 64]) or other operational issues. Existing technology suffers further from the accuracy of simulations which could enable a real conceptual design strategy. In addition, process control is limited, e.g., by the inability to obtain comprehensive information owing to restrictions on the number of control units and for method-related reasons, completing basic design before operability considerations. Environmental aspects are usually addressed thereafter ('end-of-the-pipe' approach). Safety reflections are also performed at a late stage in design.

As a solution for a different type of processing, stressing operability and environmental aspects, a simplicification of chemical processes is suggested, referred to as the KISS (keep it simple) principle. This, in particular, can be achieved by suitable choice of process chemistry and of plant components. As an example of a conceptually simple process, the Degussa route [65] for producing hydrogen cyanide is mentioned (see Figure 4.8). For this process, a simple miniplant operation is sketched using a microwave oven for energy supply [66], benefiting from simple separation. As suitable process equipment, plant components based on combined operations are introduced, i.e. membrane reactors. For product separation, only sufficiently small components should be employed, hence favoring gas membrane and adsorptive separators over other separation equipment such as distillation.

$$CH_4 + NH_3 \xrightarrow{Pt} HCN + 3\ H_2$$

Another issue with the miniplant concept relies on standardization, rendering a different strategy for increasing throughput. Whilst current scale-up processes are based on an incremental progression of know how, this should be achieved in miniplants in one stage. The small standardized modules should, at best, be fabricated using mass production techniques.

Encapsulation in containment vessels allows near zero emission rates to be reached in miniplants, e.g. by purging the reactor with an inert gas sent to a scrubber. This embedding of the miniplant should also dramatically reduce the risk of explosion or environmental contamination in case of an accident. Even if modules of the miniplant are damaged or break, the robust encasement will be mechanically and chemically stable enough to prevent pollution.

4.4.1.4 Supply-chain Systems

Considering supply-chain systems, the use of relatively non-hazardous raw materials is envisaged. Hazardous products should be used only locally and intermediately. If both products and raw materials are hazardous for a given process, it is recommended to include further processes until returning to non-hazardous raw materials. In this case, an entire sequence of plants is needed. In this context, Rinard outlines a supply-chain system for the manufacture of toluene diisocyanate. Although chlorine is used in the system, it is not present in either the starting raw materials or the product.

4.4.2
Reactor 1 [R 1]: Concept for an HCN Miniplant

Ponton [60] discusses exemplarily a miniplant concept for performing the Andrussov process, yielding hydrogen cyanide from methane, oxygen and ammonia with a platinum catalyst. HCN is a widely used but highly toxic chemical which requires extreme safety issues, in particular when it is transported or shipped. A miniplant should allow one to produce this toxic material from comparably low toxic ammonia and methane directly on-site at the customer and on-demand in small or even bigger quantities.

$$CH_4 + NH_3 + 1.5\ O_2 \xrightarrow{Pt} HCN + 3\ H_2O$$

An alternative route for HCN formation, the Degussa process, is also discussed. Only methane and ammonia are used to form HCN.

$$CH_4 + NH_3 \xrightarrow{Pt} HCN + 3\ H_2$$

The process looks simple at the first sight and, of course, a flowsheet is given (see Figure 4.8) [60]. For performing chemical reaction, some unusual technologies for heating of the reaction system were proposed: a microwave generator for preheating of the reaction mixture and RF heating to keep the temperature constant over the catalytic bed. To use the byproduct hydrogen practically it can be guided through a fuel cell to produce electrical DC power.

Following a similar motivation, a schematic flow scheme of a miniplant for hydrogen cyanide formation via the Degussa route was proposed in [30].

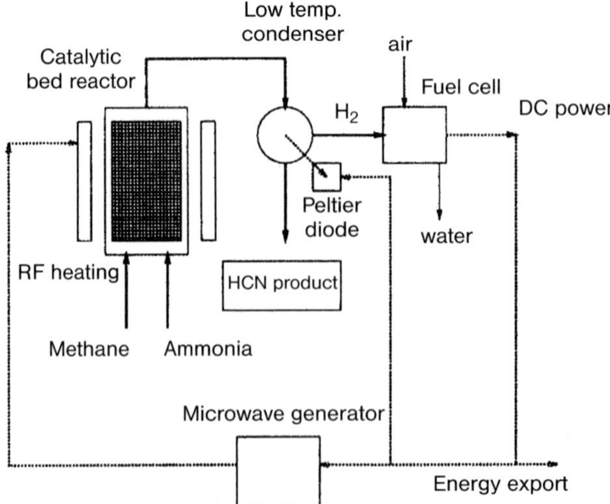

Figure 4.8 Flowsheet of an HCN miniplant [60] (by courtesy of Springer Verlag).

4.4.3
Reactor 2 [R 2]: Concept for a Disposable HF Miniplant

4.4.3.1 Use of Polymers as Disposable Construction Material

With the example of an HF plant, Ponton aimed at developing guidelines for inexpensive plant construction. The idea of using reactors of limited lifetime and made of disposable and recyclable materials, referred to as disposable batch plant [58, 60], was oriented on the highly sophisticated chemical manufacture of living organisms, animals and plants. Ideally, such systems would require no internal cleaning, repair or maintenance.

As a cheap, recyclable construction material for micro reactors with sufficient short-term chemical resistance, polymers were explicitly mentioned. A further argument for the use of polymers is that for this material flexible computer-aided rapid prototyping methods are available in order to produce reactor components of complex shapes at moderate cost. The low thermal stability of polymers, however, demands advanced heating concepts when carrying out high-temperature reactions. For example, this may be accomplished by means of microwave heating, avoiding the generation of hot surfaces.

In order to reduce the technical expenditure of processing, the use of pumps for the transport of process fluids should be avoided, and such transport should be simply based on peristalsis or gravity feed, hence eliminating moving parts. For product analysis, sophisticated non-invasive techniques are proposed.

4.4.3.2 Capacity of a Disposable Plant for HF Production

The disposable batch plant concept was exemplarily verified by a theoretical study concerning distributed HF production. Assuming microwave heating as energy source for the reaction zone, calculations show that a power of 1 kW would be required for producing 20 kg of aqueous HF at 0.6 kg h^{-1}. The reactor volume and the total plant area thereby derived amounted to 0.04 m^3 and 0.5 m^2, respectively. The total weight of the plant referred to is about 120 kg including chemicals.

$$NaF \xrightarrow{H_2SO_4} HF$$

The concept of the disposable batch plant was further extended by using robotic systems, moving plant items from station to station, referred to as the table-top pipeless plant. As an example of the use of this concept, an ethanol production unit was outlined consisting of a fermenter/stillpot, topped by a packed column section, and a partial reflux condenser.

4.5
Fluidic and Electrical Interconnects – Device-to-device and Device-to-world

Fluidic and electrical interfacing of micro fluidic devices to themselves and to the environment ('world') is a topic where investigations should have been started.

This is needed to operate at a system/plant level and to exert process control in a way similar to analytical techniques for chips on the one hand and to chemical plants on the other.

So far, only a few reports have been dedicated solely to this topic. The available ones most often were developed for chip devices based on control concepts known from micro electronics. Although the concepts are probably not directly transferable to micro structured reactors of larger size, they may nonetheless serve as describing generic paths for how to approach the problem.

The following concepts are some selected examples of interconnects and do not give a comprehensive overview of the subject.

4.5.1
Reactor 3 [R 3]: Fluidic Manifold Concept – Micro Structured Reactor-to-micro Structured Reactor

A plate-like set-up consisting of functionalized layers as base for a plant set-up is introduced for the design and construction of micro structured reactor plants [67]. The authors recommend the so-called manifold for examinations during the basic engineering of a new process. As the single plates of a manifold need to be manufactured every time a new plant set-up is designed, a CAD tool with a direct connection to a CNC machining facility must be available for the user. The fluidic pathways are milled channels in different layers and sealed by intermediate sealing layers. Channels for heating agents are supplied in the top and bottom layers of the manifold. Valves are integrated; pumps and a mass flow controller are connected externally by clamp-type fittings (Figure 4.9).

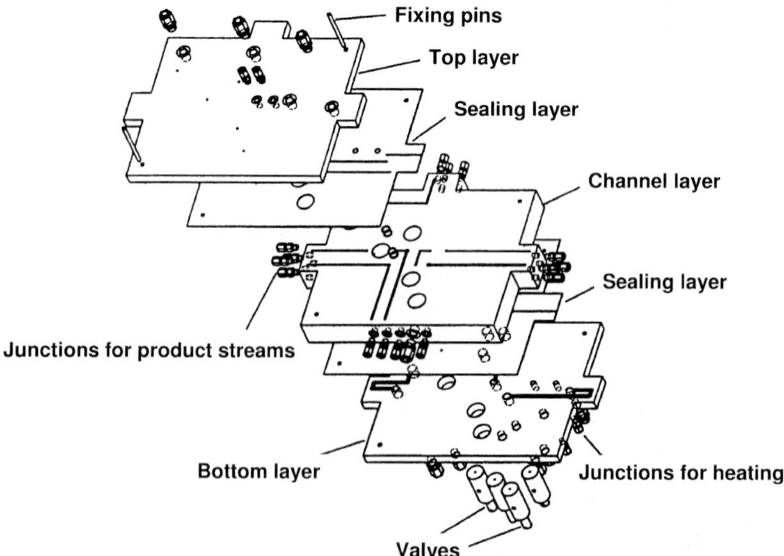

Figure 4.9 Layered manifold for an integrated plant set-up [67] (by courtesy of Springer Verlag).

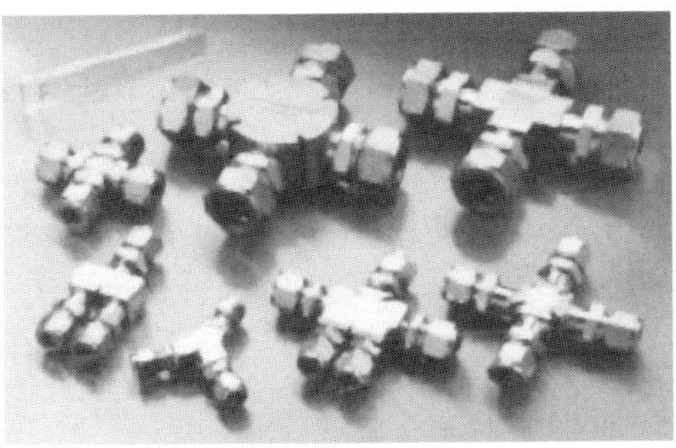

Figure 4.10 Modular unit construction set at the Forschungszentrum Karlsruhe [68] (by courtesy of Forschungszentrum Karlsruhe.

4.5.2
Reactor 4 [R 4]: Commercially Available Fluidic Interconnects – Micro Structured Reactor-to-micro Structured Reactor

In the laboratory, Swagelok® pipe fittings are often used to set up plants. The Forschungszentrum Karlsruhe used these interfaces to define a modular platform which consists of modules for a number of standard unit operations. The modules are all equipped with these pipe fittings (Figure 4.10) and can thus be combined using the Swagelok® interface standard ([68]; see also [69]).

4.5.3
Reactor 5 [R 5]: Specially High-pressure Fluidic Interconnect – Chip-to-chip

An interface or housing for micro fluidic applications was designed and fabricated. The device, made of PEEK, allows facile non-permanent coupling of standard capillary tubing to silicon/glass micro mixer chips. No additional adhesive material was used to for a secure and tight interconnect [70].

A special type of a glass/silicon sandwich micro mixer was used for the experiments, operating according to the principle of distributive mixing. Two inlet flows for different liquids are split into a series of separate multichannel flows by a repetition of flow splitting. Nozzles in the glass or silicon sheets allow the fluid streams to converge and mix. Subsequently, the formed channels are merged in a broad outlet channel again (Figure 4.11).

Figure 4.11 Image of visualized flow streams in a distributive mixer. The fluorescence of a buffered solution of quinidine hydrochloride (20 μM) in 2 M KCl (pH 9.4) quenched with KI solution was investigated [70] (by courtesy of RSC).

Table 4.3 Characteristic dimensions of the micro mixer chip and the housing [70].

Micro mixer chip		*Housing*	
Material	Glass/silicon	Material	PEEK
Outer dimensions (mm)	1.3 × 5 × 10	Outer dimensions (mm)	38 × 35 × 30
Internal volume (nl)	600	Diameter of holes for inserting capillaries (μm)	400
Number of channels	16–64	Flatness tolerances of the cavity for chip insertion (μm)	< ±5
Pitch (inlet to outlet holes) (μm)	1400	Thickness of silicone elastomer seal (μm)	100
Tolerances of inlet and outlet holes (μm)	±5	Diameter of the through-holes in the seal (μm)	300
Tolerances of holes to external edges of the chip (μm)	±30	Capillaries	Fused silica (375 μm)

The thin and sheet-like design of the micro mixer makes an interconnect with a fluidic environment difficult. Also the small external dimensions of the micro mixer (see Table 4.3) cause some problems, e.g. the access ports are smaller than most off-the shelf parts, handling of the chip itself is difficult and tedious, the distances between inlet and outlet holes are very small, the chip is made from a brittle material, and consequently, high pressures cannot be applied by clamping [70].

The interface main body is made of two PEEK parts, the upper and the lower housing, by precision micro machining (see Table 4.3). The upper part consists of two halves which after machining are connected together permanently. This was necessary owing to the restrictions imposed by the small hole-to-hole pitch from the micro mixer chip. Straight-through holes could not be machined into a solid upper housing so each half has curved semi-circular cross-sectional paths. By mounting the two halves together, circular holes are formed which guide the externally inserted capillaries directly to the holes in the micro mixer chip (Figure 4.12). The lower part has a precision-machined cavity which allows the

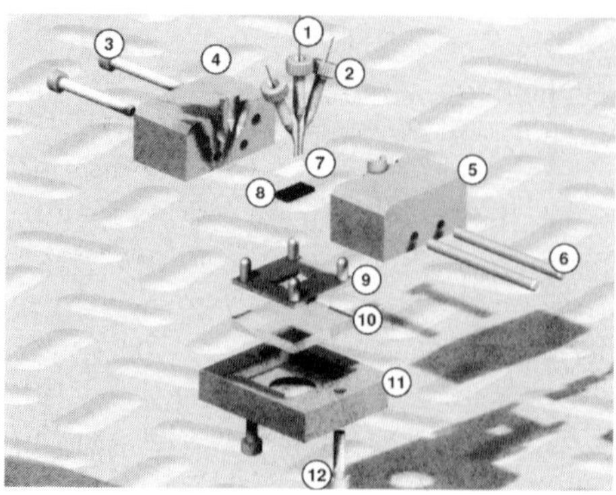

Figure 4.12 Schematic exploded view of the chip interface. The upper part of the housing is split into two halves and the lower part shows the cavity for the micro mixer chip. The actual chip (10) is small compared with the housing (4, 5, 11) which seals the device and connects it to the three capillary fittings (2) [70] (by courtesy of RSC).

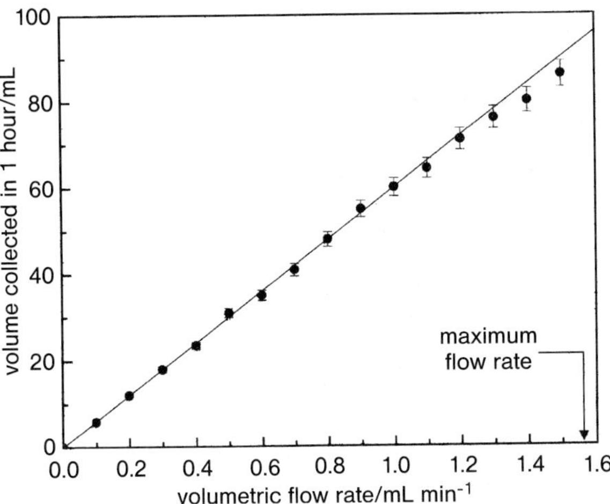

Figure 4.13 Variation of the volumetric flow rate compared with the collected fluid volume after passing the device [70] (by courtesy of RSC).

micro mixer chip to be fitted. Before assembling the pre-mounted housing parts, a silicone elastomer seal is inserted between them to ensure the tightness of the device [70].

To monitor the tightness of the device, the volume flow at the inlet of the device and the liquid volume coming from the outlet of the device are compared. As a result, the variation between volume flow and collected fluid volume after passing the device is essentially linear for flow rates between 0.001 and 1.50 ml min^{-1}, while at flow rates above 1.57 ml min^{-1} appreciable leakage was observed owing to the large back-pressures generated by higher flow rates (Figure 4.13) [70]. A value for pressure tightness is not given [70].

4.5.4
Reactor 6 [R 6]: Specially Fluidic Sequencing Interconnect – Chip-to-world

The structurally programmable micro fluidic system consists of passive valves and flow conduits, which provide different pressure drops. These pressure drops depend on the structure and surface properties of the fluidic paths. The pressure drops can be controlled by tailoring the length of the flow paths and the relative restriction ratio of the passive valves [71].

The micro channel structure of the device is fabricated in a glass wafer by common procedures (Figure 4.14). To allow sealing of the channels, the whole surface is coated with CYTOP, a Teflon™-like polymer. On the one hand it forms a bondable layer and on the other it makes the micro channel surface strongly hydrophopic. Bonding with a CYTOP-coated cover glass plate occurs under moderate pressure at 180 °C. Because sometimes the CYTOP layer peels off and disturbs the fluid flow behavior, the whole device is fabricated in polydimethylsiloxane (PDMS) [71].

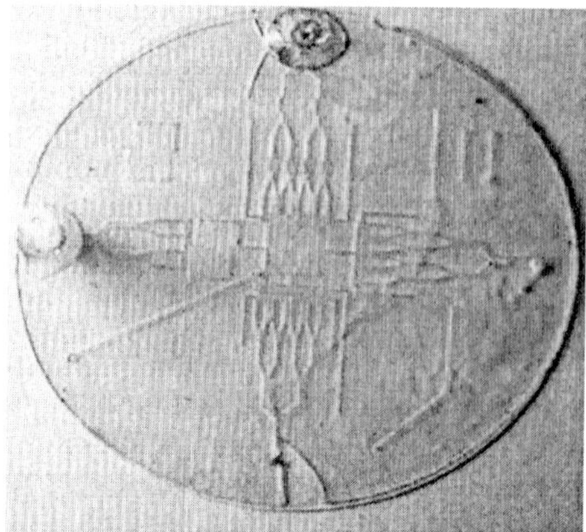

Figure 4.14 Photograph of a wafer with four programmable test devices [71] (by courtesy of Kluwer Academic Publishers.

From the Hagen–Poiseuille equation, the pressure drop in a micro fluidic system with laminar flow can be derived (compare [9]):

$$\Delta P = \frac{12\,L\,\mu\,Q}{w\,h^3}$$

where L is the length of the micro channel, μ is the dynamic viscosity of the fluid, Q is the flow rate and w and h are the width and the height of the micro channel, respectively. If the hydraulic diameter of a micro channel is constant, the pressure drop can be controlled by varying the length of the channel or the flow rate. An abrupt decrease in the hydraulic diameter of the channel causes a pressure drop at the point of restriction [71].

Typical results

[R 6] In Figure 4.15 a schematic of the programmable micro fluidic system used is shown. All channel pairs are adjusted to have the same volume. At the first split-off point, a passive micro valve, R1, ensures the fluid filling in the opposite channel system 1 (see upper branch in Figure 4.15) first. The micro valve R1 is designed so that channel 2 can be filled if the pressure exceeds the pressure drop of R1 caused

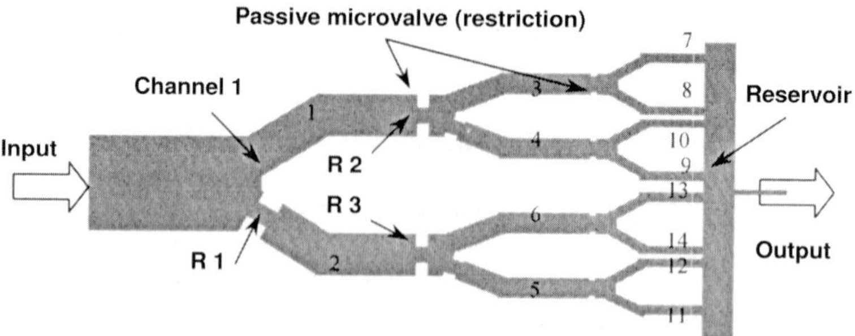

Figure 4.15 Schematic of a micro fluidic array with programmable passive micro valves [71] (by courtesy of Kluwer Academic Publishers).

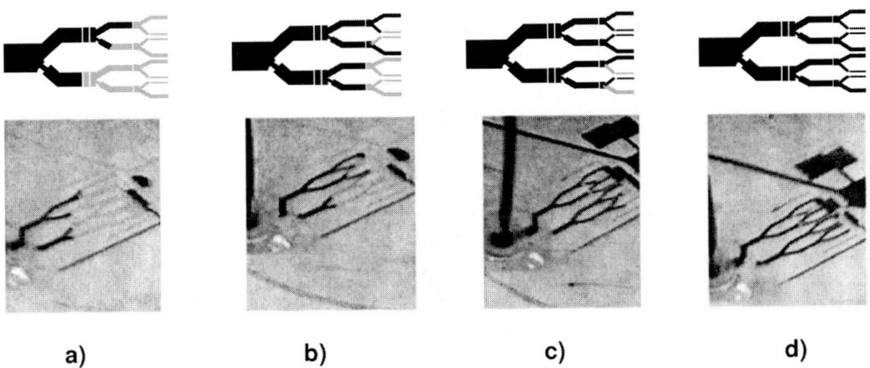

Figure 4.16 Series of photographs showing the programmed filling sequence of micro channels [71] (by courtesy of Kluwer Academic Publishers).

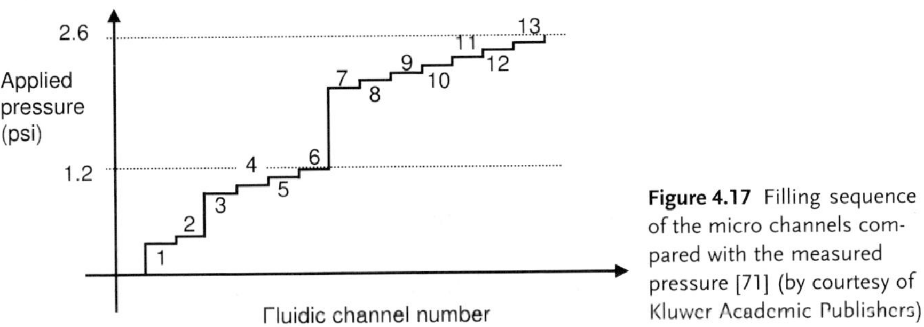

Figure 4.17 Filling sequence of the micro channels compared with the measured pressure [71] (by courtesy of Kluwer Academic Publishers).

by the passive micro valve R2. At the end of each channel pair, passive valves further regulate the fluid flow. By extending this arrangement, fluid can be manipulated to an exact location, i.e. for the example given in Figure 4.15 the filling sequence of the channel is programmed from 1 to 14 [71].

[R 1] The programmable micro fluidic system is characterized by filling in a dyed fluid at a flow rate of 1 μl min^{-1} (Figure 4.16). At the input of the system the pressure is measured while the fluid is subsequently entering the micro channel. For each filled channel a discrete pressure step can be observed (Figure 4.17). A maximum pressure of at least 2.6 psi is necessary for complete filling of the channels [71].

4.5.5
Reactor 7 [R 7]: Electrical Interconnect for Fluid Driving – Chip-to-world

Similar to computer technologies, so-called plug and play micro fluidic devices were developed. These devices are composed of a fluid driving unit and a polymer chip containing micro fluidic channels and reservoirs. The one and only connection is an electrical bus system which connects the chip with the external control unit. By filling the reservoirs with reagents, the chip can be used for performing chemical reactions or biochemical analysis [72].

The micro fluidic device is made from PMDS by silicone rubber molding on a glass base-plate and some prior lithographic steps. Also the fluid driving unit, a silicon-based micro pump, is bonded on the opposite site of the glass plate and connected to the micro channels by drilled holes. Owing to this set-up, problems with tubing and fluid interconnects can be avoided. The micro pump is a valveless diaphragm pump driven by piezoelectric actuator. A wide range of flow behavior, e.g. bi-directional flow and pulsed flow, can be realized. The flow rate ranges from 20 to 200 nl s^{-1} and the pressure achievable is about 5 kPa [72].

Laminar flow formation can be generated by using two micro pumps, each feeding different micro channels with dimensions of 150 μm × 150 μm which are connected in the shape of a Y- or T-piece. Additional capillary valves allow one to synchronize the fluid flow and to generate a laminar flow regime in the main channel (Figure 4.18). By changing the pressure of one pump and keeping the pressure of the other constant, the flow ratios could be adjusted [72].

Since the reaction efficiency in micro channels depends on the achieved mixing performance, specific structures or surface modifications are widely used to enhance mixing. Because the developed micro pumps can be controlled very easily, dynamic or alternated pumping of the fluids is possible. To monitor the capability of the system in terms of mixing speed and quality, a bioluminescence reaction is used:

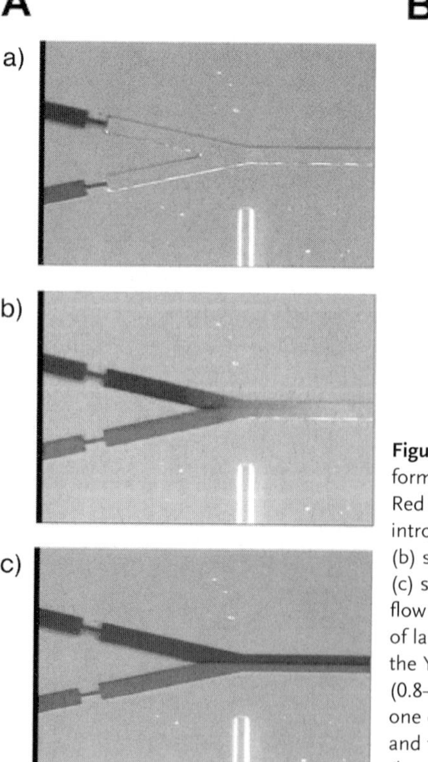

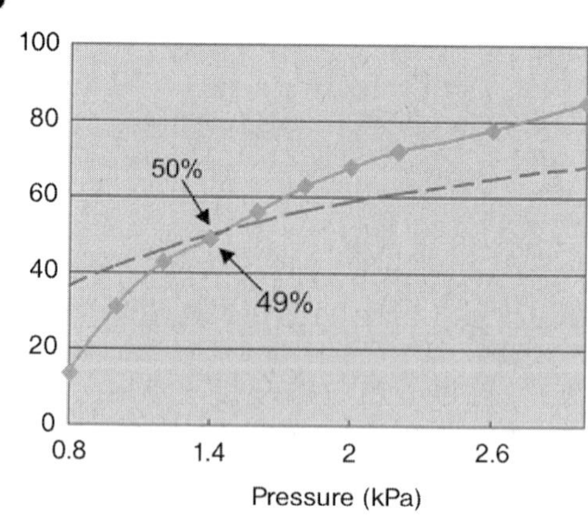

Figure 4.18 Photographs of the sequence of laminar flow formation with the assistance of capillary valves. (A) (a) Red and blue aqueous solutions ($1.7 \cdot 10^{-3}$ Pa s) are introduced into the channels and stopped at the valves; (b) solutions pressurized to break through the valves; (c) stable laminar flow pattern achieved. (B) Ratio of the flow width controlled by the pumping pressure. The width of laminar flow is measured at 300 μm downstream of the Y-shaped junction. At various pumping pressure (0.8–3.0 kPa) of one micropump and keeping the other one constant (1.4 kPa), the solid line shows the flow ratio and the dashed line shows the calculated values [72] (by courtesy of RSC).

solution A contains firefly luciferase, an enzyme, and luciferin; solution B contains ATP (adenosine triphosphate). For steady flow operations, both solutions are pumped into the channels at constant pressure (2 kPa), and the flow is stopped when it reaches the steady state. Then, the time course of the luminescence is measured. For alternated pulsed flow, high-speed switching of the pumps is applied. To achieve highly efficient mixing, narrow channels, 30 μm wide and 150 μm high, are made at the Y-junction. Alternate switching makes thin plugs of each solution. This leads to an increase in the interface between the two liquids and to enhanced diffusion-based mixing. The results with steady flow have implications for diffusion-based mixing and the reaction kinetics of the luciferase-catalyzed bioluminescence reaction. The maximum peak of luminescence appears at 46 mm downstream of the Y-junction, i.e. both solutions are completely mixed by diffusion after a short period of time. In the case of alternate pulsed flow in the range 40–100 Hz, the maximum peak of luminescence is observed immediately downstream the Y-junction, owing to the fast diffusive mixing at the interface between the two solutions. A further increase in pulse frequency leads to a decrease in mixing efficiency because a regular liquid plug is no longer formed (Figure 4.19) [72].

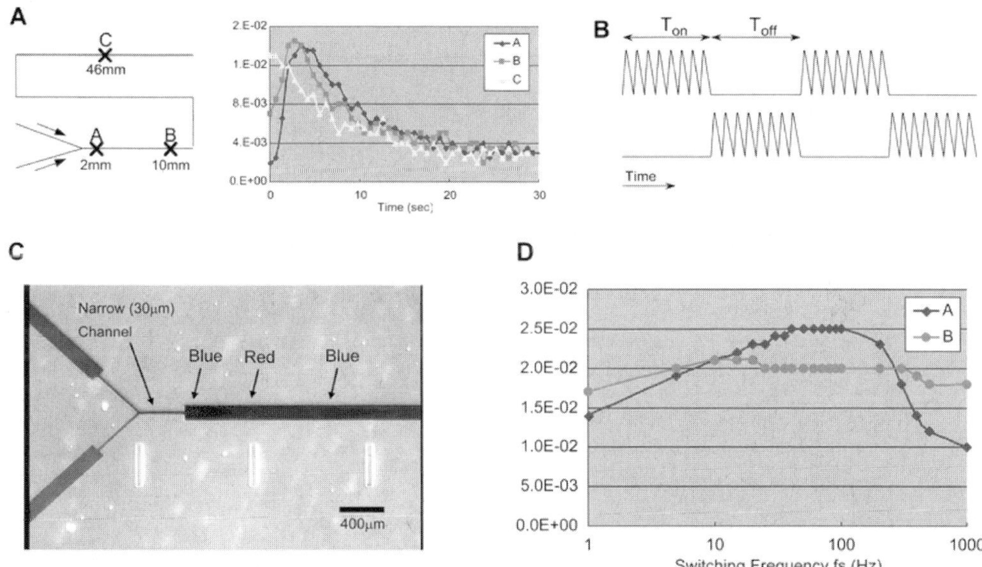

Figure 4.19 (A) Time courses of the bioluminescence at three measurement points (2, 10, 46 mm) downstream of the Y-shaped junction. The measurement was started immediately after flow stopping. (B) Typical pattern of the driving signal for alternate pulse flow. (C) Photograph of the alternate pulse flow generated by switching at 3 Hz. This frequency is selected for the visualization of thin skins, which enhances the diffusion-based mixing along the flow axis. (D) Mixing performance of the alternate pulsed flow at a frequency ranging from 1 Hz to 1 kHz. The intensity of bioluminescence is measured at points A (◆) and B (●) [72] (by courtesy of RSC).

4.5.6
Reactor 8 [R 8]: Electrical Integrated Circuit Interconnect (ASIC) – Chip-to-world

A micro fluidic system is deposited directly on top of an ASIC, a semiconductor-based Application Specific Integrated Circuit. The ASIC contains the custom specific circuitry for the control of the micro fluidic system, e.g. for liquid transport, process compartments, fluidic channels and on-chip analysis tools. Together, the ASIC and the micro fluidic system form a complete lab-on-a-chip, shown as a cross-section in Figure 4.20. The micro fluidic structures are made similarly to common semiconductor technologies in a polymer layer on top of the ASIC with a thickness of 10 μm. The polymer layer is patterned using an aluminum hard mask and reactive ion etching (RIE) in an SF_6/O_2 plasma. The smallest achieved lateral dimension, the width of micro channels is also 10 μm. These pattern [73]. To seal the micro fluidic system, a Pyrex™ glass coated with a polymer glue is bonded on top by thermally induced polymerization [73].

To incorporate an ASIC in a micro fluidic system, the common electrical top layers made of AlCuSi or AlCuGe cannot be used as electrode material owing to their low corrosion resistance. These metal layers are replaced by etching and the internal tungsten plugs of the ASIC become accessible. Electrode materials with high corrosion resistance suited for chemical processing are then deposited by physical vapor deposition (PVD) processes. For the first experiments chromium electrodes with an area of 10 μm × 10 μm each are deposited in an array of 10 000 contacts (Figure 4.21) [73].

To form a highly resistive interface between the ASIC surface and the fluidic system, a polymer layer of Cyclotene™ is deposited on the ASIC surface before fabrication of the polymeric micro fluidic system. Cyclotene™ has high electrical resistance and good chemical stability. The excellent transmission properties ranging

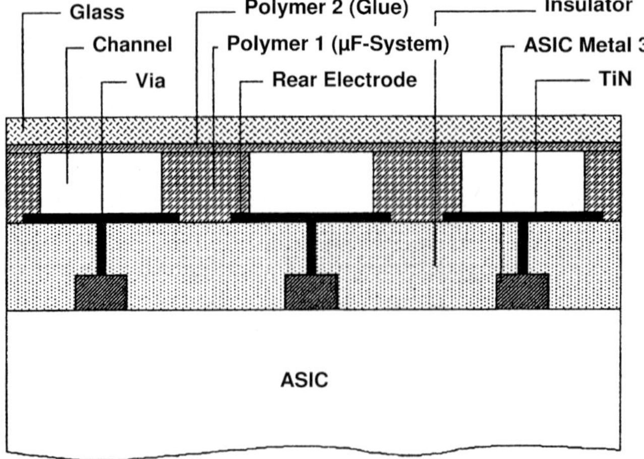

Figure 4.20 Schematic cross-section of a lab-on-a-chip [73] (by courtesy of AIChE).

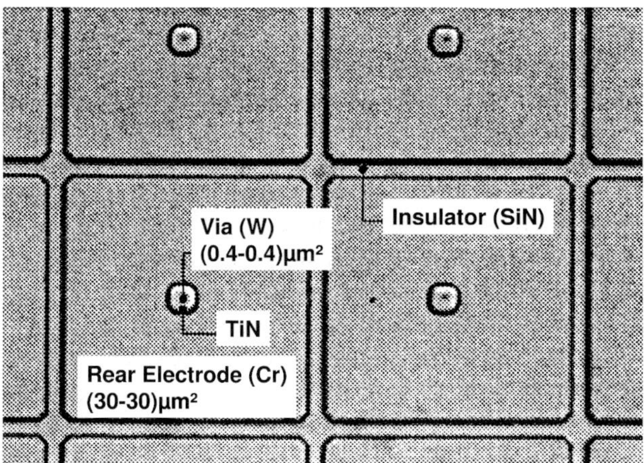

Figure 4.21 Detail of the ASIC interface. Tungsten plugs surrounded by TiN and chromium forming the electrodes patterned by an insulating SiN mesh [73] (by courtesy of AIChE).

from the UV to near IR recommends Cyclotene™ as an appropriate material for use as a transparent material for optical spectroscopy. A necessary thermal post-processing with temperatures up to 210 °C did not affect the functionality of the ASIC [73].

4.6 Table-top Laboratory-scale Plants

There are central embracing characteristics for the engineering approach deployed during the development of the following basic plant concepts. Examples are presented of plants with a uniform appearance, standardized pitch dimensions, interfaces, etc., either as closed systems or as open systems with interfaces to other suppliers. This strategy is completely different from the hybrid plant concepts.

Many manufacturers, e.g. Cellular Process Chemistry, CPC GmbH, Mainz, Microglas Chemtec, Mainz, Ehrfeld Mikrotechnik and the Fraunhofer initiative FAMOS [17–22], are currently working on miniaturized reaction systems. Most of these systems are available on the market or can be purchased on request. Common to all is the endeavor to provide complete systems equipped with a chemical environment which allows one to perform certain chemical reactions without any additional equipment. Changeable parts and optional deliverable equipment open up a way to have a flexible chemical laboratory on the desk. Each manufacturer addresses different customers, i.e. chemical research, pharmaceutical drug synthesis or education.

One major disadvantage of these systems is that it is nearly impossible to replace parts from one manufacturer with those from another. Hence all of these systems are isolated applications.

4.6.1
Reactor 9 [R 9]: CPC Table-top Reactors

CPC developed a series of table-top micro reaction systems called CYTOS™ (Figure 4.22), SEQUOS™ and OPTIMOS™ based on a standardized platform. CYTOS™ is the basic laboratory system with internal and external modularity for high flexibility by running different chemical reactions. The internal modularity offers the realization of variable reaction times up to 45 min by using several residence time units. The external modularity provides a system configuration for a multi-step synthesis. CPC Systems' individual connecting principle minimizes the dead volume in the system and provides the reaction with isothermal conditions through the whole system [74, 75].

The CYTOS™ system is designed for:

- liquid–liquid reactions
- residence or dwell times up to 90 min
- temperature range from –70 to 230 °C
- flow rates from 0.1 up to 20.0 ml min^{-1}
- targeted throughput from g to kg.

SEQUOS™ is an automated synthesis tool for drug discovery and process development; it is an automated synthesis system which maintains precise control of reaction time, temperature and reactant concentration ratios, which allows syntheses to be conducted on a continuous basis from 50 mg up to the gram scale. By modification of the SEQUOS™ system it is also possible to produce quickly 100 g or kilogram quantities of target compounds utilizing the same instrument and reaction conditions. This eliminates the time and effort currently required for scale-up [37].

The SEQUOS™ system is designed for:

- synthesis of compound libraries
- screening and optimization of reaction components
- screening of homogeneous catalysis reaction conditions
- direct scale-up to 100 g or kg quantities of compounds on same instrument.

OPTIMOS™ is a system designed for the process development especially to determine quickly the optimal reaction condition parameters for a chemical operation without having to run many batch syntheses. This system is highly integrated, equipped with analytical tools, e.g. FTIR technology, and can be used for continuous process development with real-time monitoring and analysis of chemical reactions to its micro reactor technology-based product offerings [37].

A new concept for combinatorial experimentation has been completed. Whilst it aims at the synthesis of focused compound libraries and at selecting the right reagent choices to conduct synthesis, it organizes around the same scaleable process used throughout the chemical synthesis process. Within this concept, variations of

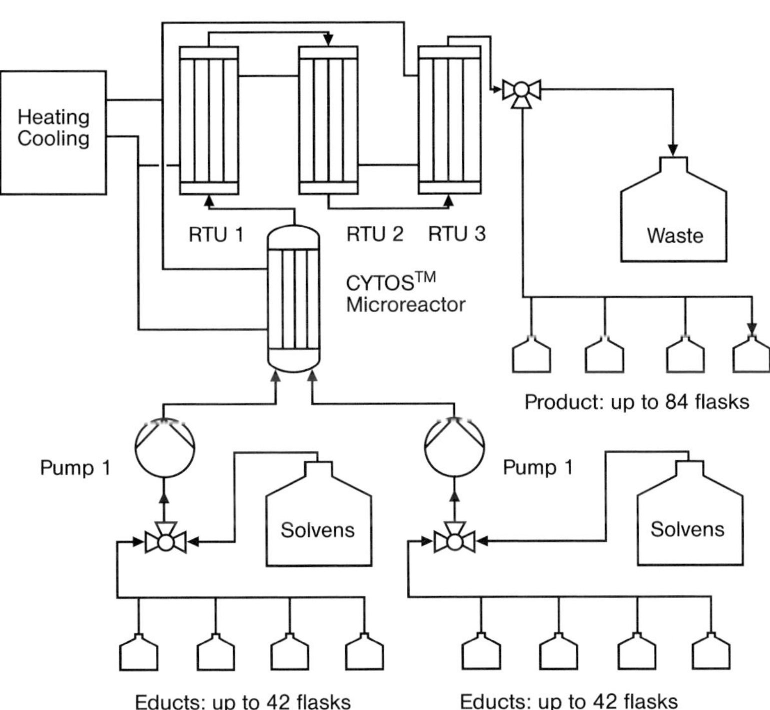

Figure 4.22 Photograph of the CPC CYTOS™ system [74] (top) and working scheme [76] (bottom) (by courtesy of CPC).

starting materials A and B_x are sequentially reacted in a micro reactor process plant to yield a variety of C_x (Figure 4.23).

In one example sequence the technology has been used to synthesize a variety of amides from amines and acid anhydrides. It is currently also being used to optimize reactions through the variations of auxiliaries and solvents and might become an interesting and more direct alternative to solid-phase synthesis.

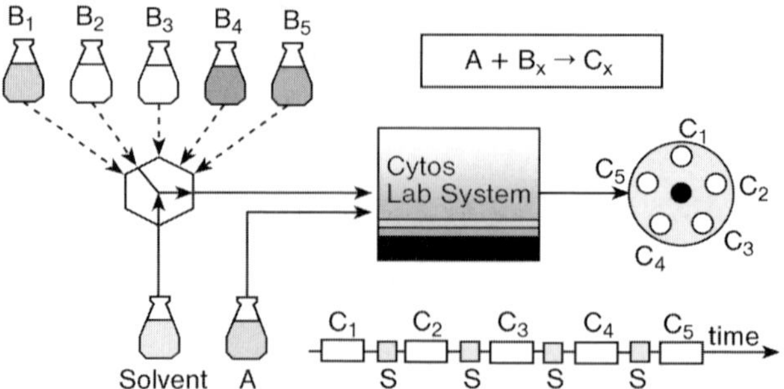

Figure 4.23 CPC concept for combinatorial chemistry [74] (by courtesy of CPC).

Selected examples of a 'better' chemistry performed in CPC table-top reactors

Beyond the removal of the generalized doubt of applicability that confronted the technology, its potential for a better chemistry was and remains a worthwhile field of examination. Important examples (Table 4.4) of improved reactions so far are nitrations (entry 1), lithiations leading to organometallic intermediates and the reaction of these organometallic intermediates in general with electrophiles (2). Other examples are reduction (3) and oxidation reactions (4), several highly exothermic cyclocondensations (5) and even some tricky, entropically driven *trans*-acetalisations (6) and special applications (7 and 8). The manufacturing-scale nitration of Viagra^{TM} intermediate (entry 1), for instance, required several years of chemical development. Running the nitration under conventional conditions initially resulted in a significant amount of losses to side reactions, such as the decarboxylated product as a result of a lack of temperature control. Such sensitive chemistry is an obvious candidate for synthesis in micro reactors. Laboratory yields of the Viagra™ intermediate were immediately achieved in a CPC micro reactor and maintained throughout the production run. This reaction protocol demonstrates how chemical R&D processes can be significantly accelerated.

In another example, the lithiation of starting material yields an unstable intermediate that is prone to β-elimination. Lithiation on a laboratory scale was possible under cryogenic conditions. The subsequent addition to a ketone was effected at –60 °C to yield an overall 83% of the alcohol. In a micro reactor, not only could the lithiation be conducted at –15 °C and the addition at 0 °C, a change towards significantly lower investment and operating costs when performed in production, but also the yield could be increased to 93%. Conventionally it was not possible to scale the process. Again, the example shows how the tighter reaction control in micro reactor permits new syntheses of functional molecules in a straightforward manner [37].

Comparison of effort between conventional and micro reactor synthesis

Economic comparisons of conventional technologies with micro reactors have also been discussed [37]. An exemplary comparison for an investment decision in a chemical development pilot unit is given in Table 4.5. In this calculation a micro reactor array is used in place of a 50 l batch vessel in a pilot-plant environment.

Table 4.4 Examples of efficient reactions in a CPC system.
Throughput rates in g h^{-1} are given in parentheses after the yields [37].

	Reaction	Yield (%) micro reactor	Yield (%) from literature
1	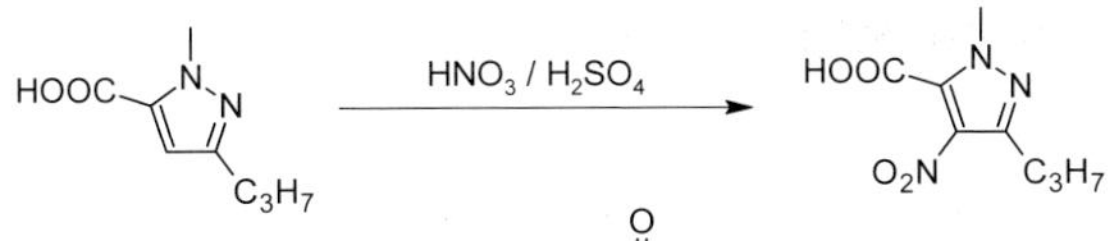 HOOC, N, N, C_3H_7 — HNO_3 / H_2SO_4 → HOOC, N, N, O_2N, C_3H_7	73 (5.5)	
2	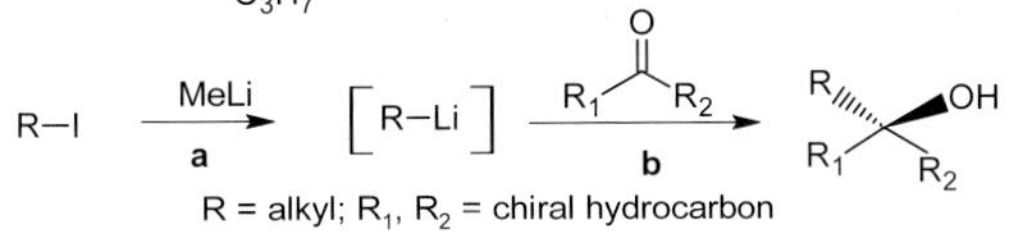 R–I — MeLi, a → [R–Li] — O, R_1, R_2, b → R, OH, R_1, R_2; R = alkyl; R_1, R_2 = chiral hydrocarbon	93 a: –15 °C b: 0 °C	83 a: –78 °C b: –60 °C
3	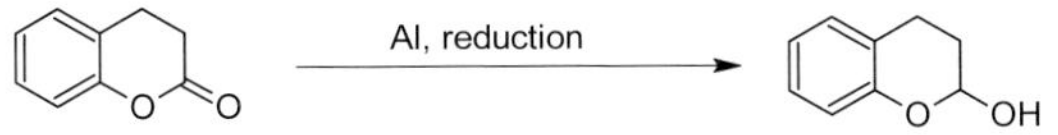 O, O — Al, reduction → O, OH	61 (4.0)	43
4	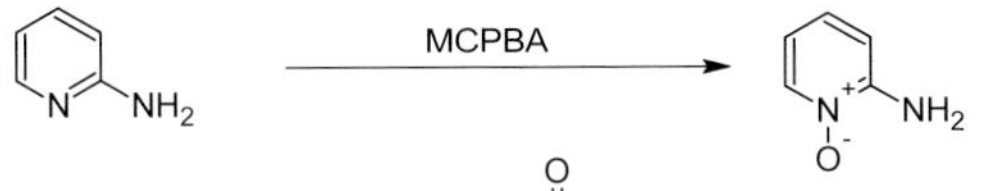 N, NH_2 — MCPBA → N+, NH_2, O-	82 (19)	71
5a	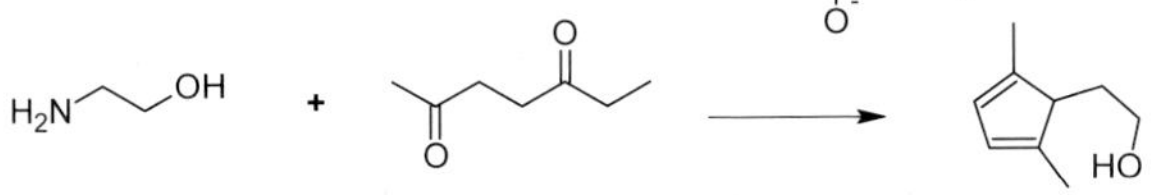 H_2N, OH + O, O → HO	95 (136)	74
5b	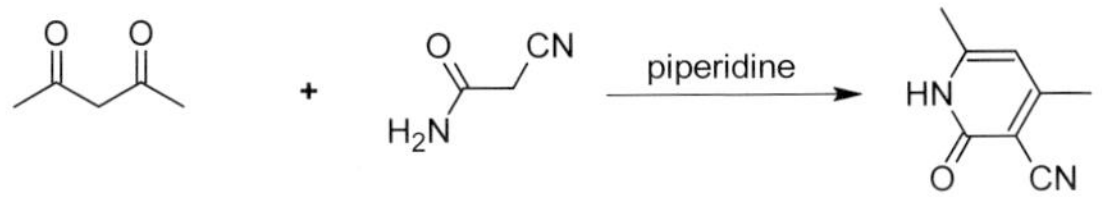 O, O + O, CN, H_2N — piperidine → HN, O, CN	89 (6)	68
6	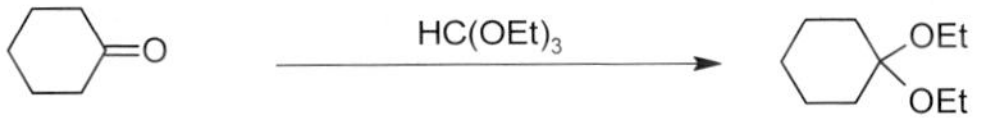 O — $HC(OEt)_3$ → OEt, OEt	97 (50)	85
7a	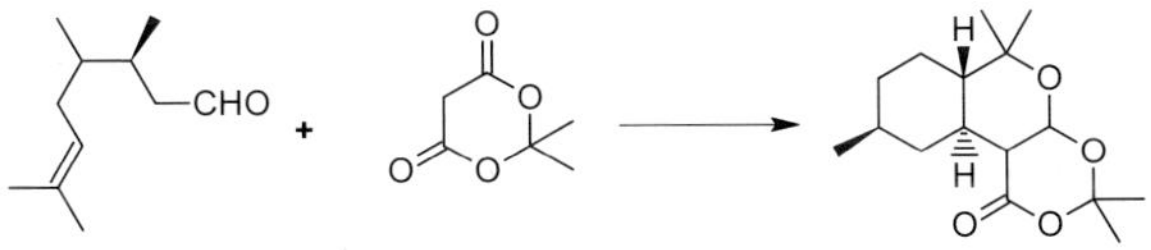 CHO + O, O, O, O → H, O, H, O, O, O	85 (4)	78
7b	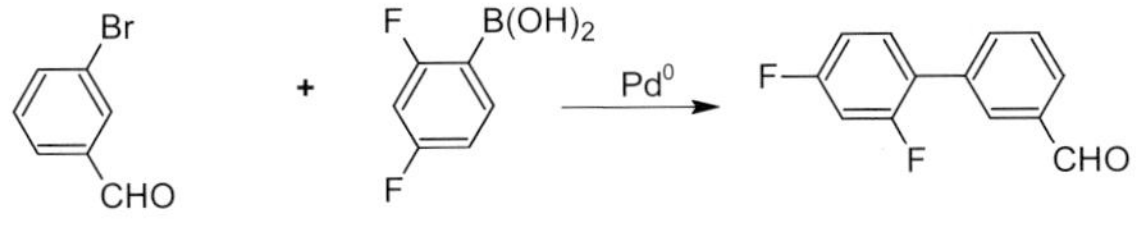 Br, CHO + F, $B(OH)_2$, F — Pd^0 → F, F, CHO	88 (3)	63
7c	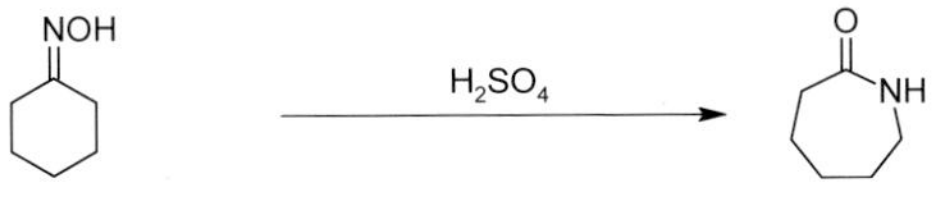 NOH — H_2SO_4 → O, NH	98 (19)	88
8a	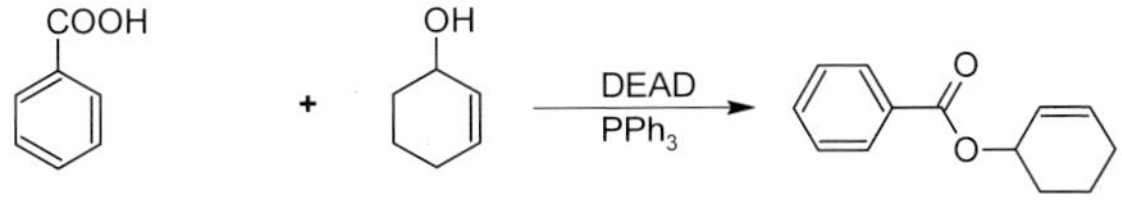 COOH + OH — DEAD, PPh_3 → O, O	98 (4)	82
8b	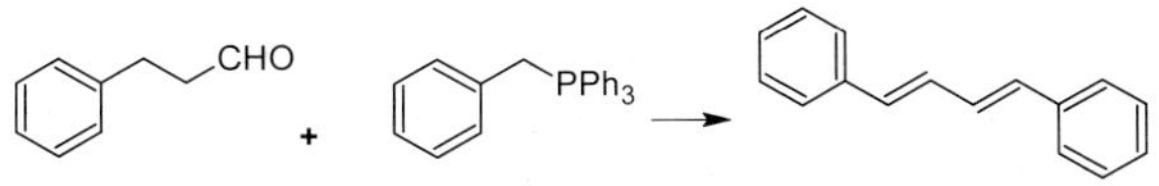 CHO + PPh_3 →	67 (9)	61

Table 4.5 Comparison of costs of production in a batch vessel and in a micro reactor [37].

	50 l batch vessel	*Micro reactor array*
Investment (€)	96 632	430 782
Scale-up effort (man days)	10	0
Mean yield (%)	90	93
Specific solvent consumption (l kg^{-1})	10.0	8.3
Required personnel per facility (men)	2	1
Production rate (kg yr^{-1})	427	536
Specific production cost (€ yr^{-1})	7 227	2 917
Cost advantage of micro reactor array (€ yr^{-1})		2 308 529
Return on investment (yr)		0.14

Each new product introduced to this batch environment goes through a scale-up effort of 10 man days. The resources required to operate the batch environment are twofold. When running the same process in a micro reactor array the scaling effort is saved and fewer personnel are required to operate the continuous process. In this cost comparison calculation it is assumed that yields are increased moderately and further the solvent consumption is reduced marginally owing to lower thermal buffering requirements. Whilst the total installed cost of a micro reactor unit in this environment exceeds the cost of a conventional unit, the internal rate of return due to operating cost savings boils is a fraction of a year.

4.6.2
Reactor 10 [R 10]: Microinnova 'Chemical Production Anywhere' Concept

Microinnova (Graz, Austria) has introduced a new plant concept named 'Chemical Production Anywhere' for on-site-production of chemicals (Figure 4.24). The aim was to construct a small portable unit to produce hazardous or explosive chemicals, whereever they are needed. This avoids dangerous transportation or storage of these chemicals. Based on these considerations, a complete ready-to-use plant is fixed into a suitcase. The suitcase is equipped with two storage bottles filled with starting materials, pumps, reactor, heat exchangers and typical sensors such as pressure, temperature and throughput. All this is operated by a built-in state-of-

Figure 4.24 Photograph of a 'Chemical Production Anywhere' plant set-up [77].

the-art process control unit based on Compact-Fieldpoint™ technology. The user interface was designed by LabView™. The user interface can be run on the Compact-Fieldpoint™ integrated web engine. Since the web engine has its own IP address, experiments or production runs can be controlled or monitored by a normal web browser. Only power supply is needed for operation. Even this can be provided by a mobile generator.

The 'Chemical Production Anywhere' is a suitcase pilot-style set-up equipped with, e.g., two HNP Microgear Pumps [78] or other pumps on request. The system can be equipped for different types of micro structured reactors, e.g. Micromixer SIMM V2, Caterpillar or Starlaminator (all provided by IMM). The throughput is sensed by using two Weber differential heating sensors. Temperature is measured by three Pt100 (3 mm diameter) and pressure is monitored by two SS316 pressure sensors (0–3 V; 24 V; 0–10 bar). Dwell time can be easily adjusted by different tube lengths. To heat the fluidic system a special heater, the Microinnova Quick Heater (1000 W), and two cross-flow micro heat exchangers are used. For process control a commercially available device, a National Instruments Field Point plug-and-play system (8 × analog In, 8 × analog Out, 8 × Pt100, 4 × digital Out, 2 × RS232) is used. The control of the system is possible via LabView™ 7.0 or the self-written Microinnova Process Control software in connection with a PC.

4.6.3
Reactor 11 [R 11]: Microinnova 'Lab Experiment Toolbox' Concept

The second set up concept from Microinnova, the 'Lab Experiment Toolbox', has some minor differences from the first concept. It was designed to speed up process development and reduce the time to market of chemicals or compounding products (Figure 4.25) [77]. A modular toolbox has been developed for laboratory-scale experiments. Small tube connections are used to connect various unit operation devices to each other. Process control and user interface are based on Fieldpoint™ and LabView™ or Microinnova Process Control. The integration of the sensors into the system is different to the previous concept. The Fieldpoint™ unit has a front panel, where sensors can be added on a plug-and-play basis. All experimental data are stored in spreadsheets, which can be transferred to common spreadsheet

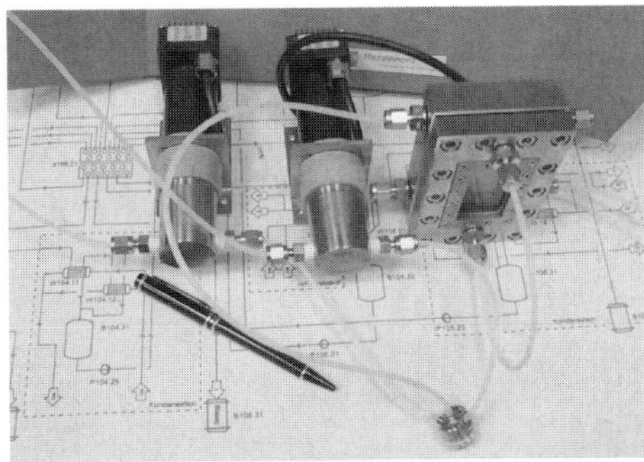

Figure 4.25 Photograph of parts of the Lab Experiment Toolbox [77].

programs for further data processing. Online analytics such as UV spectroscopy, IR spectroscopy or micro gas chromatography can be integrated easily into the system. Owing to short residence times, many experiments can be carried out in a very short time. Multistep syntheses can be carried out and even automated reporting for standard experiments are possible.

The 'Lab Experiment Toolbox' comes with fixings, e.g. for two HNP microgear pumps [78] or other pumps on request. The system can be equipped different types of micro structured reactors, e.g. Micromixer SIMM V2, Caterpillar or Starlaminator (all provided by IMM). The throughput is sensed by using two balances with an RS232 interface. Temperature is measured by a Pt100 (3 mm diameter) and pressure is monitored by SS316 pressure sensor (0–3 V;24 V; 0–10 bar). Dwell time can be easily adjusted by different tube lengths. For process control a commercial available device, a National Instruments Field Point plug-and-play system (8 × analog In, 8 × analog Out, 8 × Pt100, 4 × digital Out, 2 × RS232) is used. The control of the system is possible via LabView™ 7.0 or the self-written Microinnova Process Control software in connection with a PC.

Both plant concepts, [R 10] and [R 11], can be equipped with different types of online analytics, e.g. pH value, conductivity, UV, IR, and different process control units. Incorporation of additional reactor device allows one to perform even complex chemical reactions, including gas/liquid reactions. The pressure is limited to 8 bar, in special cases up to 30 bar, and the maximum temperature is 200 °C.

4.6.4
Reactor 12 [R 12]: Mikroglas Chemtech Micro Reaction System 'MikroSyn'

A family of table-top microreaction systems named mikroSyn is provided by MiKroglas Chemtech (Figure 4.26) [77]. The basic system has a size of 700 mm × 70 mm × 330 mm and a weight of ~30 kg. It can be equipped with two rotary pumps, one gear pump, valves and pressure and temperature sensors. The reactor itself is interchangeable with all reactor modules provided by Mikroglas Chemtec.

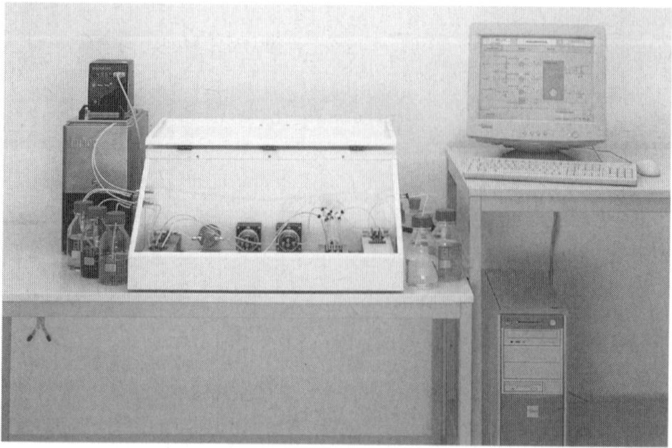

Figure 4.26 Photograph of the basic mikroSyn microreaction system [77].

To feed the system with starting materials, rotary pumps with ceramic pump heads are used. The flow rate can be adjusted from 0.1 to 45 ml min^{-1} and a maximum pressure of 7 bar. Depending on the demands of the chemical reaction, the system can also be equipped with three-way, pressure relief and check valves. The heating/cooling circulation is controlled by a thermostat and the heating or chilling fluid is pumped by a gear pump with Ryton™ gear wheels and Teflon™ seals. Flow rates from 6.0 to 560 ml min^{-1} are adjustable with a maximum pressure of 5.2 bar and the temperature of the heat-transfer medium ranges from –20 to 120 °C. Glass-encapsulated Pt100 thermocouples and piezoceramic pressure sensors up to 10 bar complete the microSyn system.

The system can be controlled manually or by a SIMATIC S7–300 (Siemens) control system. The latter configuration allows one to adjust all parameters by a user interface. To generate different programs all data can be stored, externally edited and loaded again to execute. All measured data are available online and they can also be stored, exported to an editable file or printed directly (Figure 4.27).

For some chemical reactions it is necessary to quench the product formed immediately or to add a further reaction step to intercept unstable intermediates. An upgraded version of the micro reaction system, the mikroSyn II, is equipped with a third rotary pump. Additional, different micro fluidic components provided by the manufacturer can be incorporated in the system.

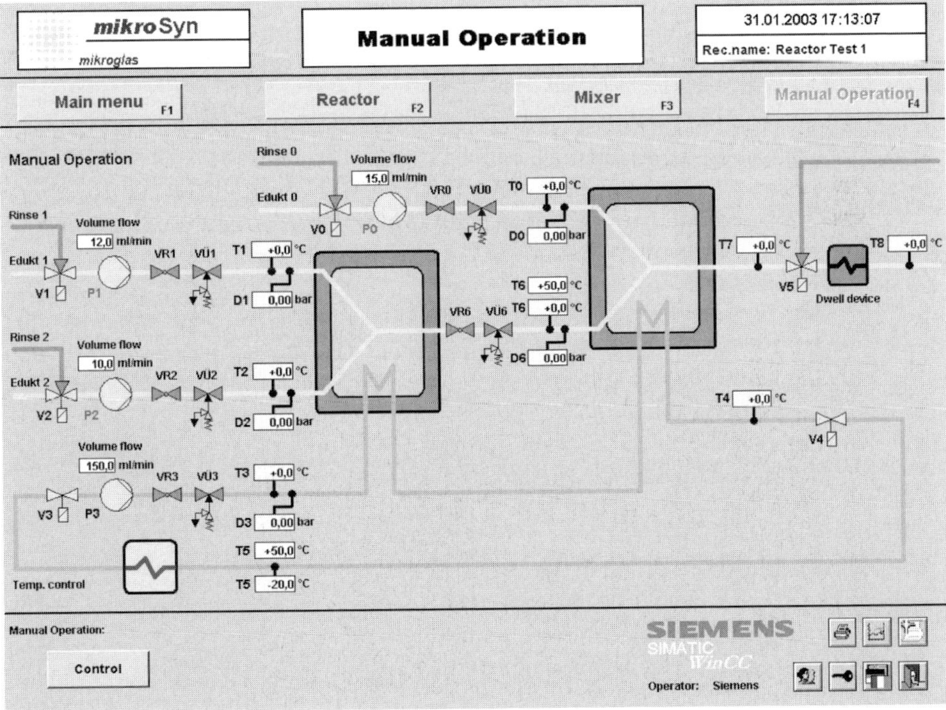

Figure 4.27 Screen print of the control system [77].

4.6.5
Reactor 13 [R 13]: Modular Micro Reaction System FAMOS (Fraunhofer-Allianz Modulares Mikroreaktionssystem)

Six Fraunhofer institutions in Germany developed a 'box of bricks' for performing chemical reactions in micro-scale table-top systems [17, 18] (Figure 4.28). On a hexagonal base plate, up to seven different reaction modules can be placed and interconnected via tubing. The base plate carries all important fluidic, electric and electronic supplies. The last two are connected via a bus system with a PC. This allows one to run desired process steps virtually and, of course, to monitor process parameters such as temperature, pressure, flow rate and analytical tools, when chosen chemical process is running. All parameters can even be changed during a running process. The major advantage of this toolkit concept is the standardized design, which allows easy exchange of individual modules and upgrading with additional modules. All modules can be heated or cooled individually and, a very important issue, the modules are thermally isolated and decoupled.

The modules have also a hexagonal shape and they fit exactly in symmetrically arranged sockets provided for them in the base plate. Each module socket is equipped with a clamping mechanism and symmetrically arranged fluidic interfaces. An important degree of freedom is the ability to rotate each module in 60° steps. This makes it possible to arrange the modules in a way which shortens fluidic interconnects and provides equidistant flow paths between the hexagonal sockets to generate equal hold-ups.

Some of these modules were fabricated in solid ceramic materials allowing chemical reactions with highly corrosive reactants or products or even at very high temperatures. Also special ceramic foams were developed which can be used as a catalyst carrier.

Some major requirements for the FAMOS toolkit were identified:

1. The toolkit does not artificially fragment unit operations such as mixing and heat exchange across different modules, but integrates them into individual function-oriented micro reaction modules. A controlled thermal environment and continuous monitoring of the micro reaction process with sensors and analyzers are other crucial elements of the concept.
2. The toolkit itself allows one to perform chemical reactions and processes which require widely different conditions or processing operations in the micro reaction environment. A variety of available materials and application-customized modules are important.
3. Equal weight is given to considerations of ease-of-operation and access to special 'MRT know-how' (for example with software), as to the compatibility of the toolkit with the laboratory or production facility equipment to which the user is used. Micro structures are used only where they make sense for the reaction in question. Frivolous miniaturization is avoided, so as to preserve the macroscopic operability of the toolkit [79].

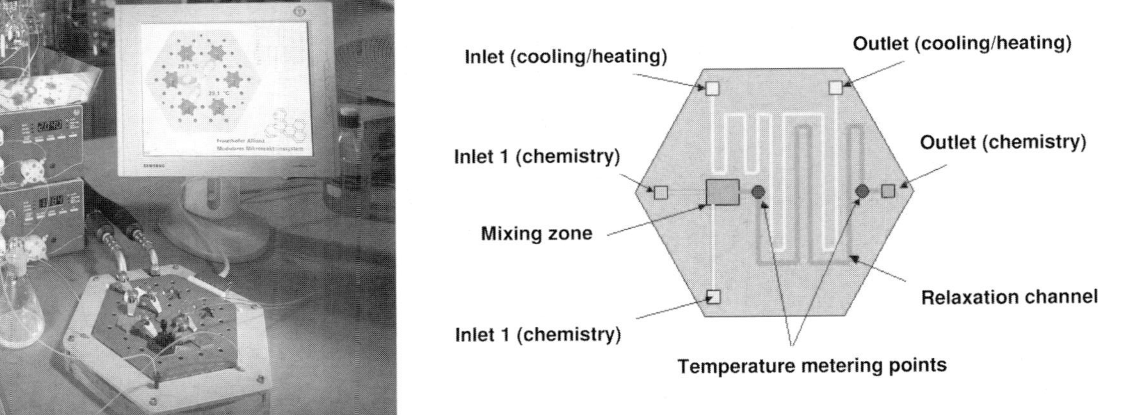

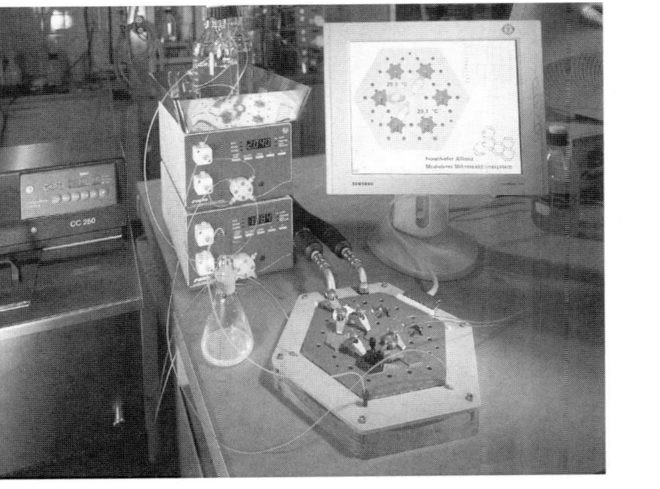

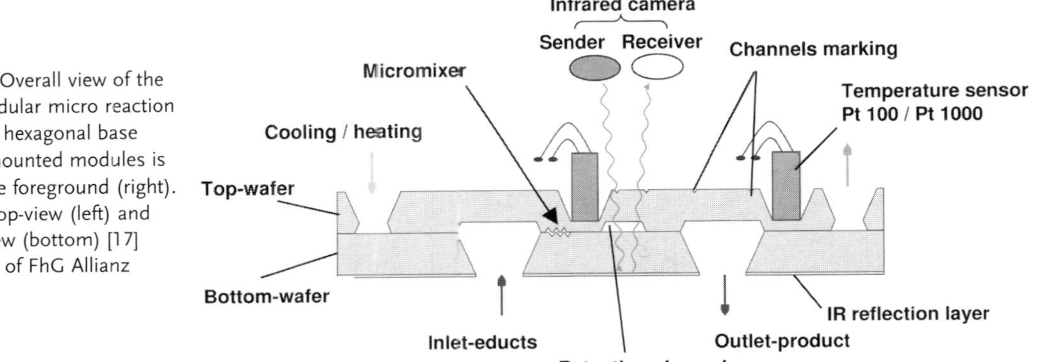

Figure 4.28 Overall view of the FAMOS modular micro reaction system. The hexagonal base plate with mounted modules is shown in the foreground (right). Schematic top-view (left) and sectional view (bottom) [17] (by courtesy of FhG Allianz FAMOS).

The system allows the setting up of a micro-reaction process by a well-equipped toolkit with a few function-oriented modules.

Modularity is only needed where micro reaction technology is really demanded: i.e. for micro fluidic modules. The motto of the developers is thus in analogy with the credo of the authors of this book, not to make microstructured devices as small as possible: 'as many tools as necessary, not as many as possible'.

A special typed software (MicroSim) was designed to allow the user to analyze, calculate, predict and optimize experimental runs in the toolkit. That means that the user is not to be required to do any modeling in the sense of formulating mathematical equations or representing geometric configurations. In contrast to CFD simulations, the time required for simulation of a micro reaction process with its peripherals is in the range of a few minutes. Furthermore, MicroSim runs on standard PCs.

The FAMOS toolkit is the first one which has the potential to be incorporated into academic education at universities and institutes. The system is small enough to put it on a laboratory bench and flexible enough to run numerous chemical reactions. By changing the process parameters easily via a PC, kinetic investigations are possible within a couple minutes depending on the chemical reaction and the connected analytical device. This should also allow one to gather much information for an in-depth chemical reaction investigation.

4.6.6 Reactor 14 [R 14]: EM Modular Microreaction System (Ehrfeld Mikrotechnik)

The Modular Microreaction System offered by Ehrfeld Mikrotechnik (Wendelsheim, Germany) permits many essential basic chemical engineering operations, such as mixing, dispersing, extracting, separating, heat transfer and others [19, 80, 81]. All these unit operations can be performed in separate modules which can be mounted very flexibly on a base plate. The modules are equipped with micro structures guiding the fluid flow. Interconnects are realized either by direct mounting of the module cubes together or by additional fluid distribution via tubing [82]. Also micro-scale analytical devices to maintain a chemical reaction can be mounted [83] (Figure 4.29).

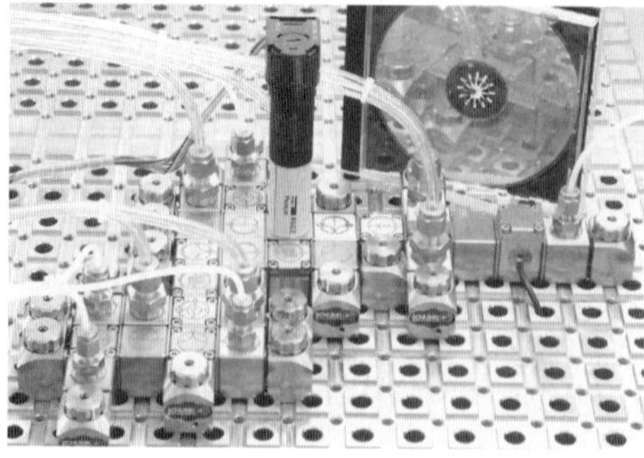

Figure 4.29 Photograph of an example of the EM Modular Microreaction System [41] (by courtesy of Ehrfeld Mikrotechnik BTS GmbH).

A throughput of about 10 l h^{-1} is achieved by operating the system in a continuous mode. For higher throughputs a special type of mixer, LH 1000, was developed which also fits on the base plate. This mixer is based on the principle of multi-lamination caused by dividing the main fluid into 50 μm wide flow layers. With a possible flow volume of about 1000 l h^{-1} for aqueous fluids the throughput is increased of a factor of 10 compared with the standard modules [83]. So far no information has been given on the increase in pressure drop on incorporation of the LH 1000 mixer in the existing standard system.

In addition, the modular system can be equipped with a cryoreactor which allows one to perform a chemical reaction within a temperature range from −50 to 200 °C. For both the LH 1000 high-throughput micro mixer and the cryoreactor no experimental results and characterization could be found in the literature.

Single-stage syntheses

A single-stage reaction is exemplarily described where two reactants, a vinyl compound and an aryl halide, are mixed and brought to reaction temperature by passing through a micro heat exchanger [83]. Then the reaction is carried out in a cartridge filled with catalyst and the reaction mixture is subsequently cooled in an additional heat exchanger. The cartridge-type module should allow a fast change of catalyst by exchanging with another previously prepared one. Sensor modules can determine the reaction mixture composition and monitoring pressure, temperature and flow rate. All parameters can be monitored and regulated via software (Figure 4.30).

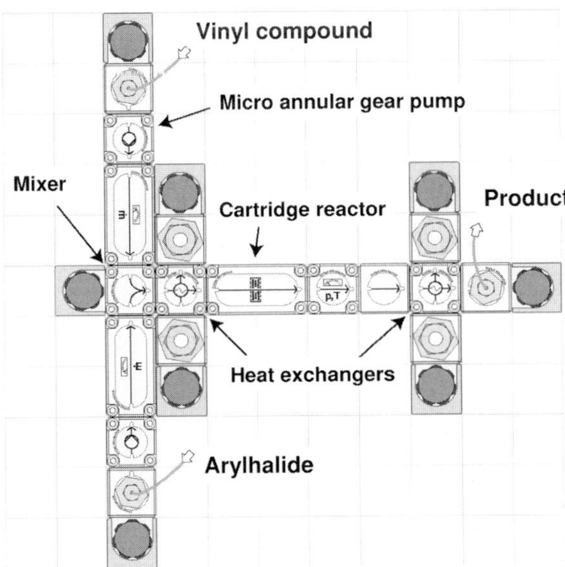

Figure 4.30 Scheme of EM Modular Microreaction System for performing a reaction between aryl halides and vinyl compounds [83] (by courtesy of Ehrfeld Mikrotechnik BTS GmbH).

For such a modular configuration, it is claimed that a large number of heterogeneously catalyzed reactions can be performed, e.g. Heck and Suzuki coupling, acid-catalyzed esterifications and reductions of functional groups [83]. However, for the mentioned process, the catalyst and reactants are not named and no reaction conditions or experimental results, e.g. yield, selectivity or nature of side products, are given.

Multi-stage syntheses

The EM Modular Reaction System can also be used to perform multi-step syntheses [83]. For the production of pharmaceuticals, in this case for the synthesis of vitamin A, an ylid is formed from a phosphonium salt and a base in the first stage at 2 °C. In a second stage, the ylid reacts with an aldehyde at 60 °C in a flow-through capillary reactor. In a third stage the crude product is hydrolyzed at 20 °C in an additional micro mixer to form the target product vitamin A acetate, as illustrated. For the claimed reaction, no further experimental details were given.

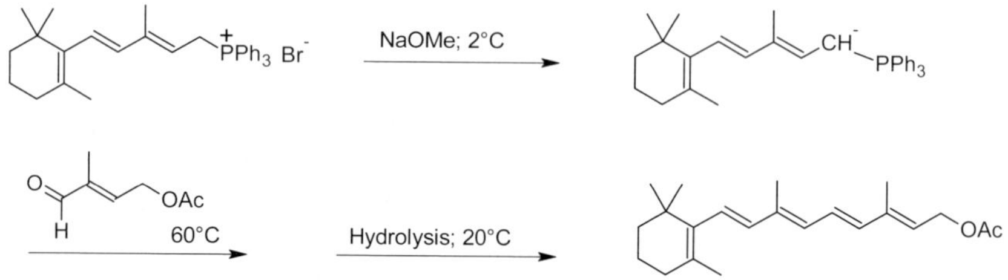

4.6.7
Reactor 15 [R 15]: Integrated Chemical Synthesizer

A modular micro structured chemical reaction system (ICS) similar to [R 14] is claimed by Bard [84]. The system can use various replaceable and interchangeable cylindrical or rectangular reactors. Generally, the ICS system can include fluid flow handling and control components, mixers, reaction 'chip-type' units, separator devices, process variable detectors and controllers and a computer interface for communicating with the master control center (see Figure 4.31) [84].

Some support structures are also included for detachably retaining the various components of the system. Preferably the support structure can be of the 'assembly board type', which provides prearranged flow channels and connector ports. The desired components of the system can be fastened into these connectors by pins. The flow control system that makes up the ICS system can include pumps, flow channels, manifolds, flow restrictors, valves, etc. These components are equipped with the necessary fittings that allow them to be sealed with the prearranged or selectively located flow channels or connectors. The flow system can also include detachable mixing devices, e.g., static or ultrasonic, or with a 'chip-like' design. The reaction units, whether 'chip-like' or not, can be of thermal, electrochemical, photochemical or pressure type [84].

The separation components can provide for membrane separation, co-current or countercurrent flow extraction, chromatographic separation, electrophoretic

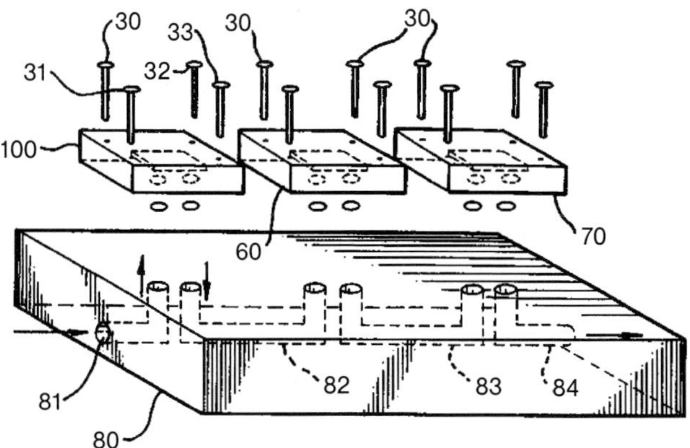

Figure 4.31 Scheme of the Integrated Chemical Synthesizer (ICS): 30–33, pins; 60 and 100, reaction units; 70, analyzer unit; 80, assembly board; 81–84, channels [84].

separation or distillation. Detectors can include electrochemical, spectroscopic or fluorescence based detectors to monitor the reactants, intermediates or final products [84].

The ICS system, e.g. a kit, provides a reaction system capable of handling a variety of reactions by using a reactor unit having a reaction chamber with internal diameters from about 1 μm up to about 1 mm, and more preferably 1–100 μm. In accordance with the internal dimensions, the reaction volumes range from about 1 nl up to about 10 μl [84].

A schematic of a 'chip-type' processing unit is given in Figure 4.32. The assembly board includes a reactor (R) formed in a manner similar to unit 100 (see Figure 4.32), but includes a heat transfer system. The reactor (R) communicates with a chip-type mixer (M_X) at the upstream end and a chip-type detector (D_1), e.g. unit 100 (see Figure 4.32), at the downstream end. The detector (D_1) communicates with a chip-type separator, e.g. unit 60 (see Figure 4.32), which in turn is in fluid communication with a second chip-type detector unit (D_2), e.g. unit 70 (see Figure 4.32). The system operates as follows: reagents A and B via pressure-actuated pumps P_A and P_B, and valves V_A and V_B sequentially or simultaneously flow to the mixer M_X. If isolation of a reagent is necessary, after reagent A is fed to the mixer M_X and discharged to the reactor R_1, a wash fluid W is conveyed via pump P_W and valve V_W to the mixer MX and discharged. Signals from the detectors D1, D2, thermocouple TC and flow meter FM are transmitted to the computer through an electrical interface to control the flow of reagents A and B and temperature, or any additional reagents according to the process to be performed [84].

With the schematic processing unit given, the thermal conversion of *tert*-butanol to *tert*-butyl chloride, can be performed [84].

$$(CH_3)_3C\text{-}OH + HCl \longrightarrow (CH_3)_3C\text{-}Cl + H_2O$$

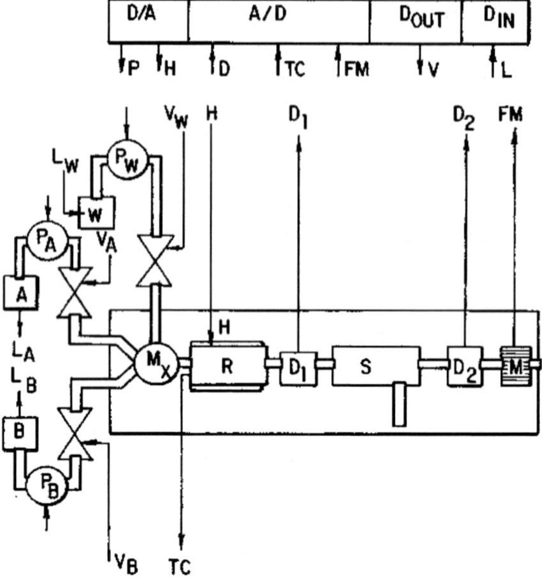

Figure 4.32 Exemplarily workflow of an Integrated Chemical Synthesizer (ICS) set-up. For a detailed description, see [84].

By incorporation of transparent process cell, performing photochemical reactions is possible. The example shows the conversion of dibenzyl ketone to bibenzyl by irradiation of the reaction mixture with a 450 W xenon lamp. The CO produced by the reaction is vented and the bibenzyl product is purified, if desired, through a micro structured chromatographic separator unit and withdrawn off-line [84]:

$$\xrightarrow{h\nu} CO + $$

An electrochemical reaction, the reduction of benzoquinone, is exemplarily described. An electrochemical micro structured reactor is divided into a cathode and anode chamber by a Nafion™ hollow-fiber tube. The anode chamber is equipped with a platinum electrode and the cathode chamber contains the analyte and carbon or zinc electrodes. Current density and flow rate are controlled to maximize current efficiency as determined by analysis of the formed hydroquinone by an electrochemical detector. Hydroquinone is extracted subsequently in a micro extractor from the resulting product stream [84].

$$\xrightarrow{+\ 2\ H^{+}\ +\ 3e^{-}}$$

In a multiphase membrane reactor, the conversion of benzylpenicillin to 6-aminopenicillinic acid is performed. The type of microstructured reactor used is a fermentation reactor which contains the enzyme penicillin acylase immobilized on the wall of a hollow-fiber tube. The hollow-fiber tube extracts 6-aminopenicillinic acid at the same time selectively. Benzylpenicillin is converted at the outer wall of the hollow fiber into the desired product, which passes into the sweep stream inside the fiber where it can be purified, e.g. by ion exchange. The non-converted benzylpenicillin is recycled back through the reactor [84].

penicillin acylase, H_2O, pH 8.5

For all these examples, no further experimental details are given. Additional information concerning yield, selectivity or conversion is missing.

4.6.8
Reactor 16 [R 16]: Integrated Micro Laboratory Disk Synthesizer

This type of integrated chemical synthesizer comprises a device array of micron-sized wells and connecting channels in a substrate that interfaces with a station for dispensing fluids to and collecting fluids from the array and for performing analytical measurements of material in the well. The station is also connected to control apparatus to control the fluid flow to the channels and wells and to collect measurement data from the substrate. All the mentioned elements are independent and together they can perform a variety of tasks in parallel [85].

On a so-called micro laboratory disk (see Figure 4.33) the individual wells of the array and their sequence can be varied depending on the synthesis or analysis to be performed. Thus the function of the arrays can be readily changed with only the additional need to choose a suitable interface for monitoring and controlling the flow of fluids to the particular array being used and the test or synthesis to be performed [85].

In one embodiment, the integrated chemical synthesizer can be used to perform various clinical diagnostics, such as assays for DNA in parallel, using the well-known protocols of the polymerase chain reaction (PCR), primers and probe technology for DNA assay. In another embodiment the synthesizer system can be used for immunoassays for antibodies or antigens in parallel for screening purposes. In still other embodiments, the synthesis of a series of chemical compounds, or a series of peptides or oligonucleotides, can be performed in parallel. Each well in the array is designed to accomplish a selected task in appropriate modules on a substrate, each module containing the number of wells required to complete the task. The wells are connected to each other to a sample source and to a source of reagent fluids by means of connecting micro channels. This capability permits broad-based clinical assays for disease not possible by sequential assay, permits improvement in statistics of broad clinical assays such as screening of antibodies

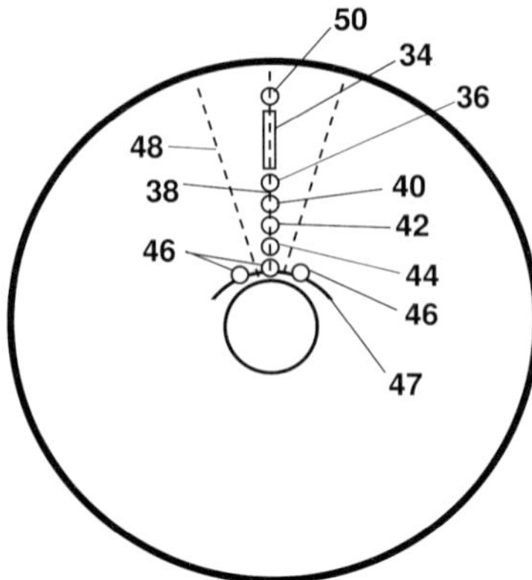

Figure 4.33 Schematic of a single module of the main part of the Integrated Chemical Synthesizer (ICS). 14, Micro laboratory disk; 34, loading capillary channels; 36, 38, 50, fluid micro channels; 40, 42, 44, wells; 47, exit channel [85].

because of the parallelism, and permits a reduction in costs and an improvement in the speed of testing [85]. In addition, so-called cross-over channels permit a high-density arrangement of modules and wells in an array that can be serviced from more than one channel either to transmit fluids into the well or out of the well or to transmit the material being treated to a succeeding well.

The array is formed in a suitable dielectric substrate and the channels and wells are formed therein using maskless semiconductor patterning techniques. The station and control means, such as computer, use existing technology that includes commercially available apparatus [85].

In addition to clinical and biochemical reactions, e.g. DNA analysis, immunological assays or syntheses of large number of oligomers and peptides, comparably small molecules can be synthesized. The synthesis of an array of different substituted 1,4-benzodiazepine compounds is claimed [85].

Figure 4.34 Surface-mounted components such as valves, gauges and pressure regulators of the NeSSI consortium [23] (by courtesy of CPAC).

4.6.9
Reactor 17 [R 17]: The NeSSI Modular Micro Plant Concept

The American consortium NeSSI [23] (New Sampling Sensor Initiative) is a CPAC-sponsored (Center for Process Analytical Chemistry) open initiative formed in 2000 to create a standard for process analysis. The object is to implement modular and smart process analytics. It also aims at the future integration of micro analytical devices and micro structured reactors. The concept is derived from the semiconductor industry and follows the ISA SP76 standard. Companies such as Parker-Hannifin, Kinetics and Swagelok are involved. Further information is also given on the web site [23].

A selection of components surface-mounted to carrier plates is shown in Figure 4.34. The carrier plates support the internal fluidic pipes and can also be equipped with electrical heaters to permit trace heating of the fluid. The surface-mounted components of this system are compatible with the backbone concept introduced in the following.

4.6.10
Reactor 18 [R 18]: The Micro Structured Reactor Backbone Interface Concept

Within the framework of a BMBF-funded project, five research institutes are developing a standardized system for the combination of micro structured devices and laboratory equipment [86–88]. The idea is to integrate devices from many different suppliers to build up a complex chemical plant. This is actually a normal procedure in industrial plant engineering as the large variety of chemical apparatus cannot be delivered from just one supplier. However, this approach is not widely used in micro structured reactor plant technology. Here usually one supplier tries to cover the whole range of devices.

'Plug and produce' plant assembly

Although a number of micro structured devices and process equipment such as valves and pumps are currently on the market, only a limited number of applications have been reported so far which combine micro structured reactors from the same or even different suppliers. Aside from problems with capacity differences, one key limitation of these devices is that they use different kinds of interfaces, which hampers their direct connection. Standardized interfaces would allow the execution of complex chemical processes. Starting by connecting a few micro structured devices, entire set-ups are built based on a 'plug and produce' plant-assembly systematic.

The concept presented here in much detail as an example of cooperative project work in micro structured reactor plant development is based on the bus system and simultaneously handles a number of tasks such as mechanical stability, fluidic flow and signal transmission. A key feature of the so-called backbone interface is its open architecture. It does not rely on standardized reactors or devices, thus allowing the combination of devices from various suppliers. A robust interface was developed which withstands high pressures and temperatures. Thermal cross-talk was minimized through the use of different heat-conducting materials.

Micro structured devices are often specialized products designed for a single purpose. Costs for individually developed and manufactured devices are certainly higher than those for products from the shelf. The non-existent compatibility of devices from different suppliers represents another restraint, especially because many suppliers offer only a limited number of devices and thus cannot provide the complete range of unit operations.

In 2000, the industrial platform 'modular micro process engineering' (μPVT) was founded with so far 45 research institutes, suppliers and industrial users [89]. The target of this platform was to create a concept for the establishment of a uniform standard and a modular approach to process technology which should avoid the above-mentioned disadvantages and offer a cost-efficient solution. A manufacturer-spanning building set which consists of compatible micro process components had to be developed.

For the transaction of the voluminous development, a BMBF research project was launched in October 2001 [42, 87] with the partners *Dechema* in Frankfurt/main, the *Forschungszentrum Karlsruhe*, the *Fraunhofer Institut für Zuverlässigkeit und Mikrointegration* (IZM) in Berlin, the *Institut für Angewandte Chemie Berlin-Adlershof* (ACA), the *Institut für Mikrotechnik Mainz* (IMM) and the *Technische Universität Darmstadt* (TUD).

4.6.10.1 The Backbone Interface Concept

In the following, the concept of micro modular process engineering is introduced together with the backbone interface developed in order to realize this modular approach. The integration of sensors and an electronic bus system is also described, and the physical characterization of the backbone is discussed within a case study of the enantioselective synthesis of organoboranes. Within the second case study, the sulfonation of toluene with gaseous sulfur trioxide, the backbone system together with the micro structured devices used is finally assessed based on its application to chemical synthesis.

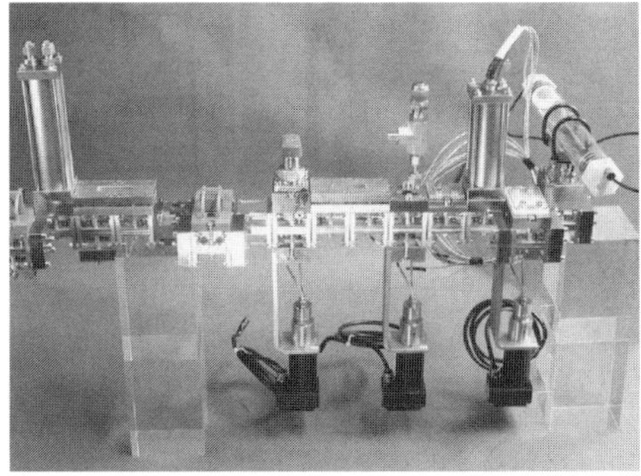

Figure 4.35 Example of a microstructured reactor plant based upon the modular fluidic backbone: 1, heat exchanger; 2, mixer; 3, valve; 4, safety valve; 5, pump; 6, heated residence-time module; 7, mixer–settler extractor; 8, heated mixer-tube reactor; 9, thermal decoupler [86].

Through the collaboration of the above-mentioned five project partners, a so-called *backbone interface* has been developed based on the bus concept where the flow is passed through a central spine (Figure 4.35). Such a 'microplant' is seen as an alternative or a supplement to the miniplant approach in plant engineering. Similar to this technology, a closed-loop set-up with refeeding of side products is realized if unit operations for product separation can be integrated. With the present state of development this demands the integration of conventional miniplant glassware, for example, by the installation of a distillation column into the micro structured reactor plant set-up. Such a combination is feasible because the flow rates of a micro structured reactor plant are similar to typical flow rates of miniplant separation units.

Micro structured reactor plant components can also be looked at as a supplement to miniplant components if one has to deal with highly exothermic reactions or high pressure environments. Owing to their low internal volume, micro structured devices withstand large internal pressures. In such a way, a pressurized subregion of a plant can be kept at high pressure using micro structured devices (appropriate control equipment exists), whereas the rest of the plant is kept under low pressure using miniplant equipment. Examples [90, 91] exist of very fast and/or highly exothermic reactions or reactions in the explosive regime (unthinkable with standard glassware).

Backbone design

The modular backbone introduced here allows both commercial and demonstration-type micro structured devices to be coupled in all three dimensions in a flexible and easy manner. Micro structured heat exchangers, reactors and mixers of different manufacturers are surface mounted on this backbone. Owing to the standardized interfaces, devices can easily be exchanged, for example, to evaluate different types of mixers. The backbone itself consists of elements which can be combined individually and flexibly in all directions, according to the demands of the plant to be built. The backbone provides the flow paths for fluids and electrical conduits for power supply and signal transmission of sensors and actuators.

The construction permits the setting up of compact systems with low internal volumes of 35.3–950.0 mm^3 (per backbone element) which can be operated up to

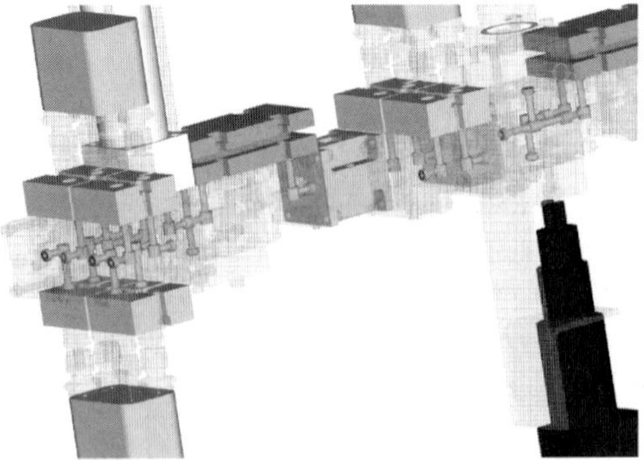

Figure 4.36 Internal view of backbone structure [86].

300 °C and 100 bar, being limited only by the physical properties of the gaskets used (e.g. steel, graphite, Viton®, Kalrez®). By exchanging tubes and bores of various diameters, a large range of volume flows is accessible. For liquid-phase processes, volume flows of up to $32\,l\,h^{-1}$ at a pressure drop of 150 mbar (per spine unit, $d_i = 1.5$ mm) can be realized and extended to $130\,l\,h^{-1}$ (70 mbar, $d_i = 3.0$ mm) by the exchange of the pipe inlets, thereby enabling pilot-plant operation.

Parallel bus concept

The principle of both the fluidic and the electric paths are based on the bus concept, as shown in Figure 4.37.

During the experiments, the micro structured reactor plant concept was also assessed by measuring properties such as seal reliability and thermal cross-talk between parallel fluidic channels. The performance of the micro structured reactor plant with respect to parameters such as product yield, selectivity and compactness, flexibility and robustness were examined using exemplary chemical reactions. These are advanced multi-step reactions in organometallic boron chemistry (liquid/liquid),

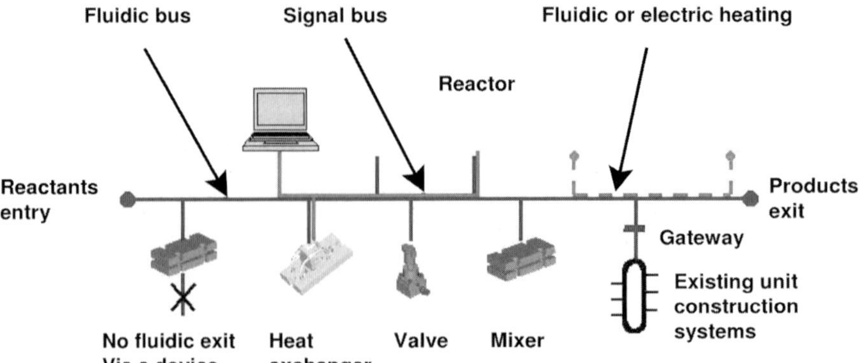

Figure 4.37 Parallel bus concept [86].

fast and exothermic sulfonation of alkyl-substituted aromatics (gas/liquid and liquid/liquid) and Darzens glycide ester synthesis (liquid/liquid).

4.6.10.2 Case Study 1 [C 1]: Physical Characterization of the Set-up for an Enantioselective Synthesis

Driven by the rapidly growing demand for highly enantiomeric pure substances in the pharmaceutical industry in order to provide safer and cheaper drugs, the development of a process which can achieve enantioselectivities approaching 100% has become desirable and challenging.

One case study within the framework of this project is thus to test the concept of a micro structured reactor plant by applying the fast reaction of the enantioselective synthesis via organoboranes yielding chiral-substituted alcohols. This is typically a batch process carried out in the laboratory using conventional glassware and in the present case has been converted into a continuous process carried out by micro structured devices. This set-up has been used to characterize the physical properties of the backbone system.

Leak test

[R 18] [C 1] As a first test, a number of elements were assembled and the spacing of the housings was measured. It turned out that the spacing is independent of the number of screwed elements. The small errors of the individual housing dimensions compensate each other. This is a prerequisite for the build-up of complex three-dimensional plants. The latter require a continuity of spacing also over long distances. The leak tightness was characterized using gaseous nitrogen under a pressure of 76 bar. Three backbone elements were combined consisting of two straight pipes, two elbows and six PTFE seals ($d = 6$ mm, $t = 1$ mm). The pressure was recorded as a function of time (see Table 4.6).

Table 4.6 Pressure of three combined backbone elements recorded as a function of time [86].

Time (min)	0	10	24	58	81	1008
Pressure (bar)	75.9	75.9	75.8	75.8	75.8	74.8

The leak rate was calculated according to the German DIN standard. This leak rate is related to the lengths of all seals (90 mm) and the internal volume (837 mm^3). The specific leak rate obtained was 0.00076 mg s^{-1} m^{-1}, which corresponds to a leak rate class of $L_{0,001}$ comparable to commercial valves. In a plant it is desired that the pressure drop along pipes and fittings is at least one order of magnitude smaller than the pressure drop of the devices. A typical micro structured device will possess a pressure drop of approximately 1 bar when operated with a volume flow (water assumed) of approximately 1 l h^{-1}. The calculated pressure drop in a straight pipe inlay under laminar conditions was measured to approximately 1 mbar, which is three orders of magnitude smaller than the typical pressure loss of a micro structured device. It is obvious that the pipes will not dominate the pressure loss of the plant as in fact their contribution to the pressure loss is nearly negligible.

Determination of thermal cross-talk

[R 18] [C 1] Figure 4.38 shows a single-backbone element as the smallest unit of the backbone. The element is equipped with three different pipe inlays, a straight pipe, an elbow and a T-piece. All three inlays are inserted in the front plate, in which case they also have a thermal contact via this plate. This could be a desired effect if, for example, trace heating is intended. However, in most cases this is an unwanted situation.

In order to study the influence of thermal cross-talk, liquids with different temperatures were conveyed through the backbone and their temperatures measured with thermocouples and with thermographic infrared measurements. To achieve realistic measurements, elbows were equipped with thermocouples with direct contact to the fluids (see Figure 4.39).

Figure 4.39 shows the complete set-up used for the temperature tests. A realistic set-up was chosen by combining three heat exchangers, a glass mixer and a mixer-tube reactor with six backbone elements. Initially the micro structured reactor plant was at room temperature.

The test set-up chosen for the infrared measurements showed regions with clearly distinct temperature values. The heat exchangers 1 and 2 were supplied with the cooling agent 2-propanol cooled in a cryostat to –11 °C. The surface temperature, as indicated in the thermographic image (Figure 4.40), was much lower than the temperature of the neighboring backbone element, an indication of thermal decoupling between the backbone structure and the pipe inlays which deliver the cooling agent to the heat exchangers. Thermocouples inserted in the pipe inlays to the heat exchanger showed a temperature of –6 °C at the heat exchanger exit.

Nearly complete thermal decoupling is enabled by the use of a special backbone element used for insulation (see Figure 4.39). This element is made of fiber-reinforced PTFE (Teflon®), which is chemically fairly inert and possesses an extremely small heat conductivity coefficient of 0.25 W mK^{-1} compared with the standard backbone materials steel (16 W mK^{-1} and aluminum (204 W mK^{-1}).

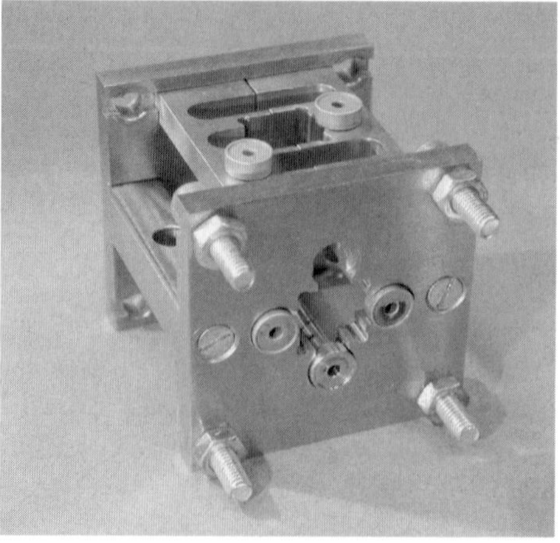

Figure 4.38 Standardized single-backbone element consisting of two face plates, two spacer bars and fork-type support plates equipped here with three pipe inlays (elbow, straight pipe and T-piece, from left to right, see Figure 4.89) [86].

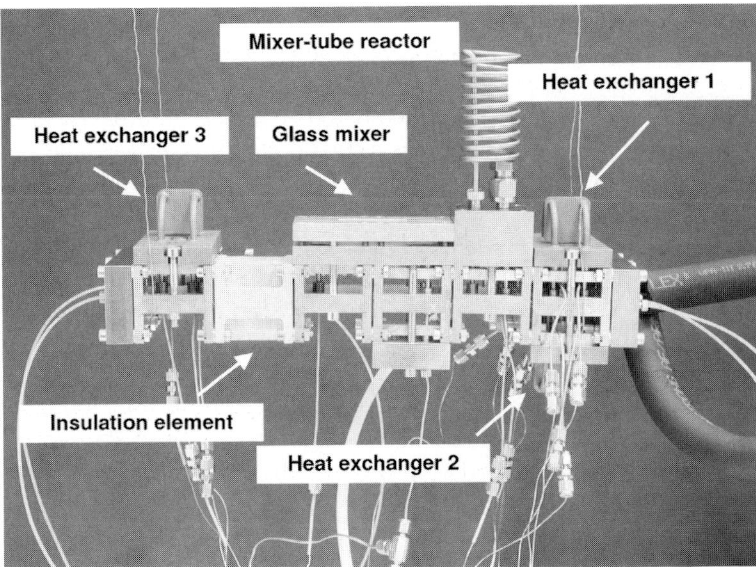

Figure 4.39 Micro structured reactor plant set-up for studying cross-talk behavior equipped with additional thermocouples (the horizontal backbone consists here of six elements without branching) [86].

Such an insulation element permits the definition of distinct temperature regions. For example, in Figure 4.38, a region on the left side (consisting of one element) with ambient temperature conditions is thermally separated from a region cooled to –10 °C on the right side (consisting of four elements). The temperatures, indicated in color code in the infrared image, are only qualitative temperatures owing to different surface emissivities. Hence these thermographic images were complemented by temperatures measured directly in the fluid with thermocouples inside the pipe inlays. All the devices exhibit individual temperatures which were different from the backbone temperature. This desired effect can also be increased by the use of steel or PTFE as housing material for the backbone elements. Here aluminum was used as housing material. The largest temperature difference of approximately 60 K exists between the mixer-tube reactor and heat exchanger 3. Remarkable also is that the glass mixer (see arrow in Figure 4.39) is well suited to insulate the fluid owing to the low heat transfer coefficient of glass. This result was confirmed by measuring the temperature difference between product inlet and outlet of the glass mixer, which amounted to only 3 K.

The insulation effect of the PTFE element is obvious if the temperature course is given as a function of the element position (see Figure 4.41). For this reason, the surface temperature of the six horizontal spacer bars was recorded. The surface temperatures of the spacer bars depend on the emissivity coefficient of the materials and on the surface characteristics. To eliminate these effects, the spacer bars were painted.

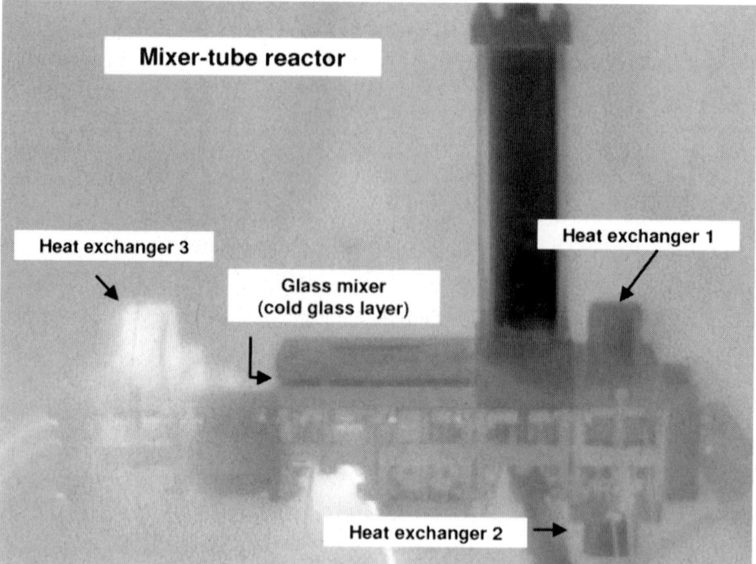

Figure 4.40 Thermographic image of the set-up shown in Figure 4.39. The temperatures range from approximately −10 °C (surface temperature of the perpendicular tube of the mixer-tube reactor) to 50 °C (surface temperature of heat exchanger 3) [86].

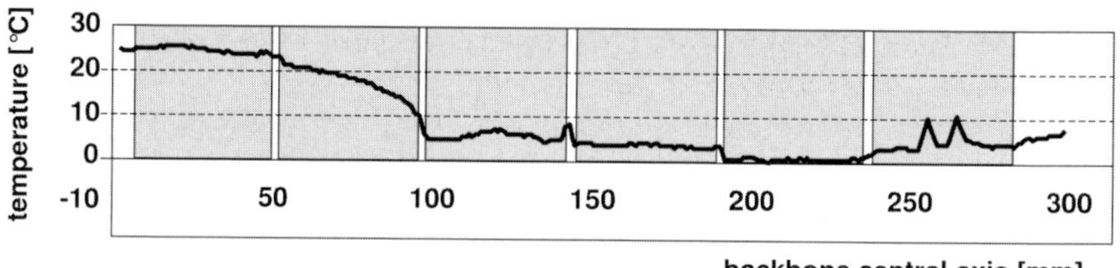

Figure 4.41 Temperature course along the central axis of the backbone. The gray boxes indicate the positions of the individual backbone elements; the peaks in the right box are due to cables lying in front of the element. Temperatures are surface temperatures measured at the spacer bars with corrected material emissivities [86].

The temperatures reflect only the housing temperatures and not the fluid temperatures inside the pipes. It is nevertheless a strong hint that a thermal separation of different plant sections is possible. Here, the left heated section is separated from the colder right section, as the PTFE element (second element from the left) aollows the development of a sufficiently large temperature gradient of approximately 20 °C in this example.

The heat exchanger in Figure 4.42 was supplied with 1 l h^{-1} of water as the heat-transfer medium at an inlet temperature of 60 °C and 2-propanol as the product with a volume flow of 60 ml h^{-1} and an inlet temperature of 27.4 °C. In the heat exchanger it was heated to 46.1 °C.

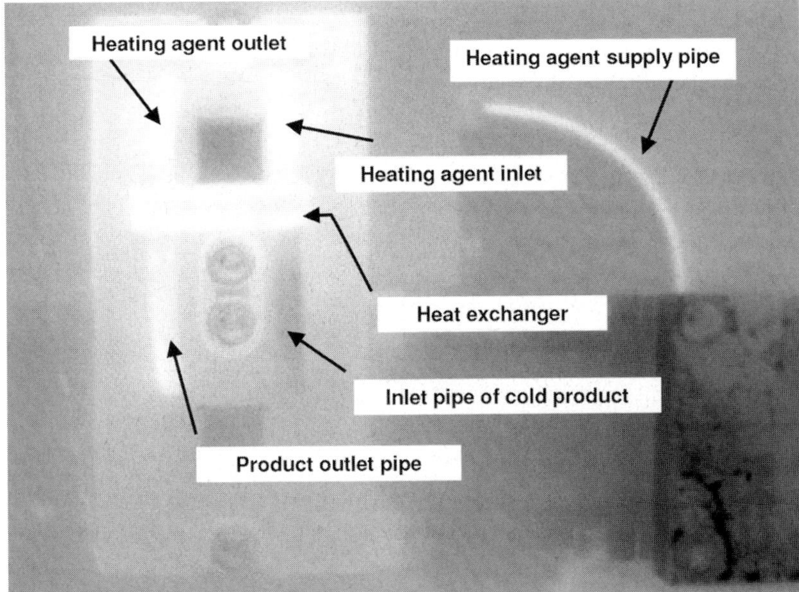

Figure 4.42 Surface temperature variation at heat exchanger 3 (the largest temperature range indicated in the thermographic image varies from approximately 10 to 50 °C) [86].

4.6.10.3 Case Study 2 [C 2]: Chemical Characterization of the Backbone Using the Sulfonation of Toluene with Gaseous SO_3

The highly exothermic sulfonation of toluene with gaseous sulfur trioxide is one reaction which has been investigated. Figure 4.43 shows the process flowsheet of the microstructured reactor plant used at the ACA.

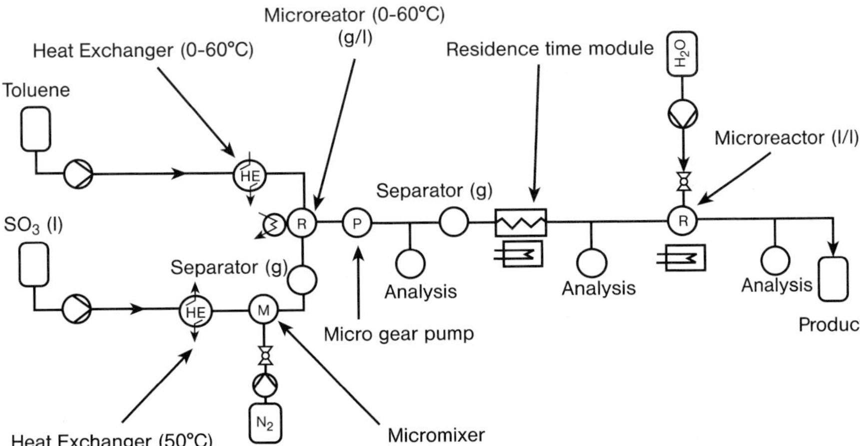

Figure 4.43 Process flow sheet for the sulfonation of toluene with gaseous SO_3 [86].

Experimental protocol

[P 1] [C 2] Toluene is heated to 40 °C using a micro structured heat exchanger while at the same time liquid sulfur trioxide is heated to 60 °C in order to evaporate it. Nitrogen is further added so as to dilute the system and the stream is then passed into a separator with the purpose of removing any traces of liquid. Thus, a gas stream is allowed to flow through to a micro structured reactor where it reacts with the liquid toluene. As shown in Figure 4.44, reaction (1), sulfonic acid is produced here via the desired reaction step. At the same time, however, sulfone [reaction (2)], a mixed anhydride and sulfonic acid anhydride are also formed by side reactions. Sulfone cannot be converted further but the mixed anhydride reacts in the residence time module with toluene and forms the desired product, sulfonic acid, as shown in reaction (3). To convert the sulfonic acid anhydride to sulfonic acid, a hydration step is required [reaction (4)]. To achieve this, water is added to the reaction mixture after the residence time module as shown in Figure 4.43.

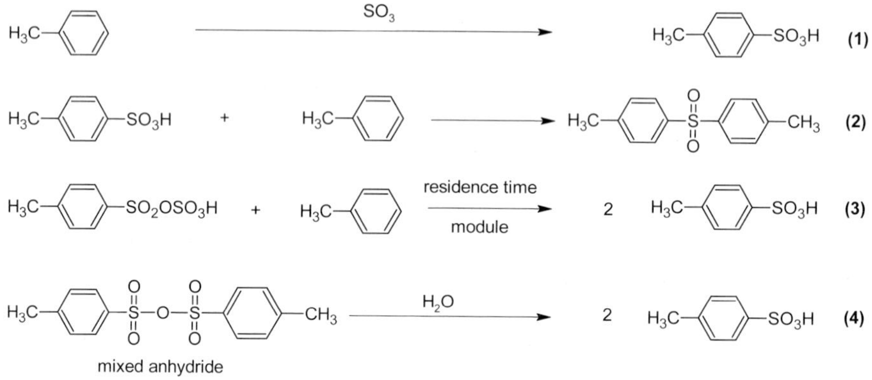

$$H_3C\text{–}C_6H_5 \xrightarrow{SO_3} H_3C\text{–}C_6H_4\text{–}SO_3H \quad (1)$$

$$H_3C\text{–}C_6H_4\text{–}SO_3H + H_3C\text{–}C_6H_5 \longrightarrow H_3C\text{–}C_6H_4\text{–}S(=O)_2\text{–}C_6H_4\text{–}CH_3 \quad (2)$$

$$H_3C\text{–}C_6H_4\text{–}SO_2OSO_3H + H_3C\text{–}C_6H_5 \xrightarrow[\text{module}]{\text{residence time}} 2\ H_3C\text{–}C_6H_4\text{–}SO_3H \quad (3)$$

$$H_3C\text{–}C_6H_4\text{–}S(=O)_2\text{–O–}S(=O)_2\text{–}C_6H_4\text{–}CH_3 \xrightarrow{H_2O} 2\ H_3C\text{–}C_6H_4\text{–}SO_3H \quad (4)$$

mixed anhydride

Figure 4.44 Reaction paths in the sulfonation of toluene with gaseous SO_3 [86].

Experimental set-up

[R 18, modified] [C 2] An overview of the micro structured reactor plant used is given in Figure 4.45. Emerging from the front is a typical micro tooth gear pump, while on the left-hand side a falling-film micro structured reactor can be seen.

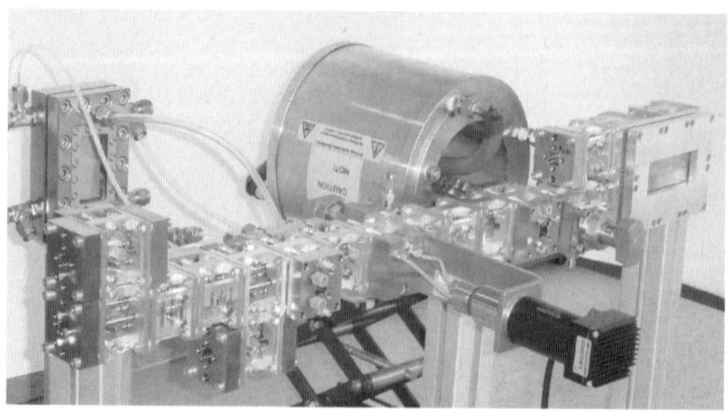

Figure 4.45 Micro structured reactor plant used for the sulfonation of toluene [86].

This is used to carry out the gas/liquid reaction at the start of the process. Further along the backbone, a large cylindrical residence-time module is visible in the background. The macro-scale residence-time module consists of a wound tube heated by a heat-transfer agent. The last reactor seen at the end of the backbone on the right-hand side is the reactor where the final hydration step takes place. The entire set-up has been tested for leak tightness up to 5 bar, the pressure limit for glass reactors.

Typical results

[P 2] [R 18, modified] [C 2] To-date, the reaction has been carried out up until the residence-time module. The final hydration step [Figure 4.44, reaction (4)] has not taken place. Even so, the first results are very encouraging as shown in Figure 4.46. In order to evaluate the reaction conditions, the mole ratio of the two reactants, sulfur trioxide and toluene, was varied and the selectivity of the desired product (sulfonic acid) and of the by-products (sulfone and the anhydride mixture) was determined. Evidently, with increasing SO_3/toluene mole ratio, the selectivity of the undesired by-products decreases whereas the selectivity of sulfonic acid stays nearly constant. At a mole ratio of 13/100, the selectivity of sulfonic acid is approximately 80% whereas that of sulfone decreases to approximately 3% and that of the sulfonic acid anhydride to approximately 1.3%.

The isomer selectivity was also determined as 8.1% for the *ortho*-sulfonic acid, 1.5% for the *meta*-sulfonic acid and 90.4% for the *para*-sulfonic acid.

From the literature, at an SO_3/toluene mole ratio of 13.4, the selectivity of the *ortho*-sulfonic acid was 17.6%, of the *meta*-sulfonic acid 1.2% and of the *para*-sulfonic acid 81.2% [92]. Hence the improvement of the selectivity for the *para*-sulfonic acid can already be seen from these results. Very recently also the last hydration step was executed successfully.

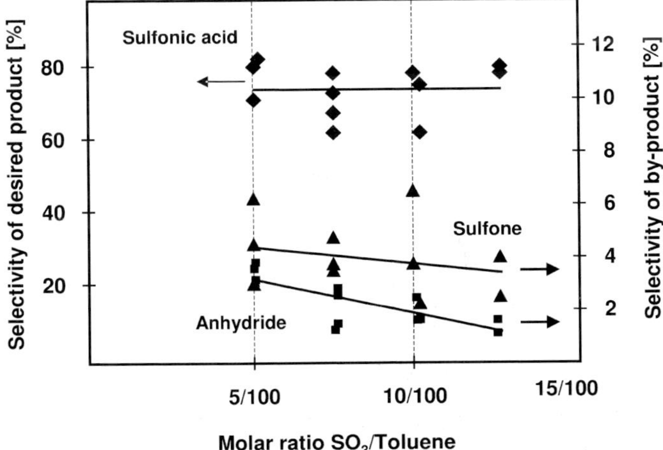

Figure 4.46 Selectivity of products at various SO_3/toluene mole ratios [88] (by courtesy of ACS).

4.7 Hybrid Plants

A plant which involves devices or structures with different geometric scales can be called a hybrid plant. The reason for such an approach often derives from the wish to combine specialized micro structured devices with commercial laboratory-scale products. Reactors could also be designed with different *internal* geometric length scales, for example, using a combination of micro structured heat exchangers with mini-scale tube reactors in a common housing. Sometimes it is advantageous to combine unit operations from different plant concepts or like units from miniplant or microstructured reactor plant technology. In the following, three examples are introduced which cover these three different integration concepts.

The volume flow in a typical miniplant is of the order of $10\,l\,h^{-1}$. The limiting factor is the gravity-driven flow in the separation units, for example, a rectification column. As separation units usually accompany a chemical process, this flow limit dominates the overall capacity of a miniplant. It is surprising that the flow rate is not limited here by the pressure loss.

In a microstructured reactor plant, in contrast, the flow rate will be dominated by the pressure loss. Typical pressure losses in micro devices are of the order of 1 bar at a flow of $1\,l\,h^{-1}$ (water) [50, 93]. If sufficient pump capacity is available, the pressure loss in a micro structured device is limited by the mechanical stability of the reactor housing, which is often made of steel and hence a loss of several bar is certainly acceptable. Even the combination of up to 10 different micro devices only amounts to about 10 bar in this example. The main advantage of a micro structured reactor plant is that the flow rate can be adjusted more freely because the flow is pressure driven and not influenced by a single gravity-driven device as in a miniplant.

This means that miniplants and microstructured reactor plants can be operated in the same flow rate ranges and it should therefore be possible to combine both plant concepts in order to unify the advantages from both technologies. The micro scene would benefit from the integration of separation devices such as rectification columns, a degasser or extraction units in the micro structured reactor plant concept and the miniplant technology could extend their experimental set-ups towards high-pressure applications or replace agitated mixers by static micro mixers.

4.7.1 Reactor 19 [R 19]: Micro Structured Reactor – Miniplant Hybrid Combination

In the above-described μChemTec project, a combination of miniplant and micro structured reactor plant equipment is used for the organoborane synthesis as an example for a future production facility for fine chemicals. This approach differs from the usual application of a miniplant as a mediator between laboratory-scale and pilot plant-scale operation. The microstructured reactor plant shown in Figure 4.47, a combination with a typical miniplant set-up, exhibits greater complexity than the miniplant but it nevertheless consumes a much smaller proportion of the volume of the plant set-up.

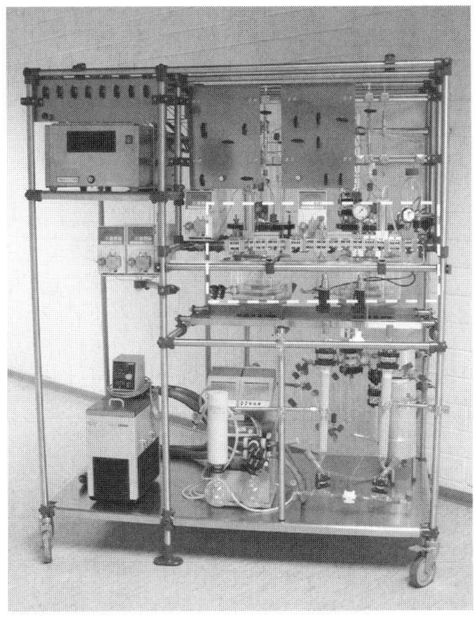

Figure 4.47 Combination of a miniplant with a micro structured reactor plant. The micro structured reactor plant is shown with a dashed box) [86].

4.7.2
Reactor 20 [R 20]: Hybrid Methanol Steam Reformer

As an example of a hybrid plant combining micro structured devices with laboratory components, a methanol steam reformer for low-power fuel cells is described in the following. Most fuel cells require hydrogen as the power source. In mobile applications, hydrogen storage still remains a problem owing to low storage efficiency. These difficulties can be avoided if the hydrogen is produced on-board in a reformer. In the framework of a European project, 'Microreactor Technology for Hydrogen and Electricity' (MiRTH-e), an alternative system to battery packs was developed in the operating range up to 100 W [94].

Experimental set-up

[R 20] The fuel processing system consists of a fuel evaporator, a reformer, a reactor for the preferential oxidation of carbon monoxide and a catalytic burner (Figure 4.48) [95].

The dimensions of the stainless-steel micro structured reformer are 75 mm × 45 mm × 110 mm. It consists of a stack of micro structured steel foils coated with catalyst and tempered by heating cartridges in the reactor housing. Laser-cut graphite foils were used to seal the reformer under operating conditions up to 200 °C at a flow rate up to 900 ml min^{-1} and a residence time of 0.07 s.

Experimental protocol

[P 3] The liquid methanol water mixture leaves the storage tank and enters the evaporator together with nitrogen which is used as carrier gas for the evaporated

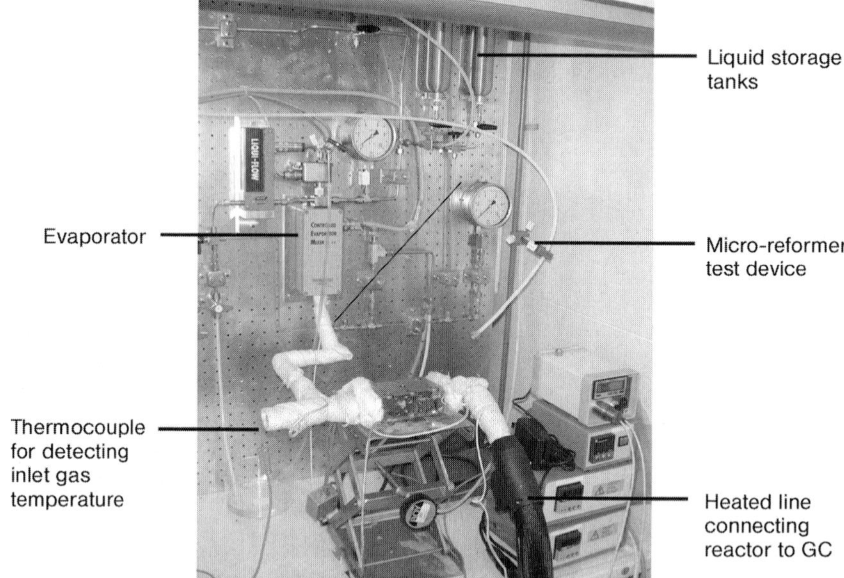

Figure 4.48 Experimental set-up of a hybrid reformer plant (micro structured reformer combined with laboratory equipment) (source IMM).

liquid. The gas is further conveyed via a heated pipe to the micro structured reformer and after the catalytic reaction via a second heated pipe to the GC analysis. The reformer was operated in integral mode and could be equipped with different numbers of coated micro structured plates in order to vary the amount of catalyst. The gas temperature inside the reformer was measured with a 0.5 mm thermocouple to control the reaction process. A 10 μm thin γ-alumina washcoat with a specific surface area of 72 $m^2 g^{-1}$ was impregnated with Cu/ZnO catalysts.

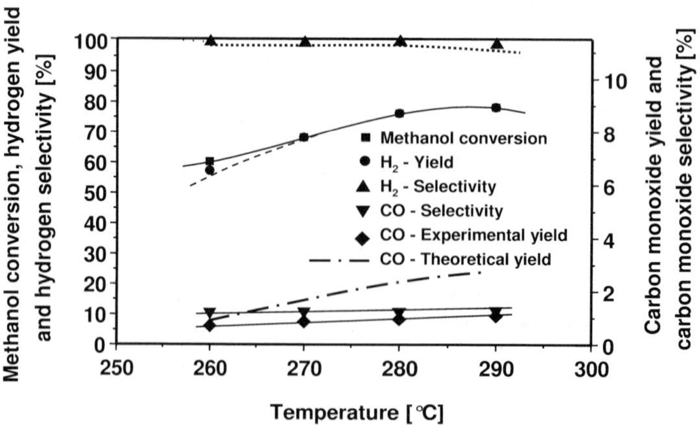

Figure 4.49 Experimental results of the reformer plant (source IMM).

Typical results

[R 20] [P 3] In Figure 4.49 some experimental results are shown. The γ-alumina washcoat impregnated with Cu/ZnO delivers a hydrogen yield of 78% at 98% selectivity.

4.7.3
Reactor 21 [R 21]: Hybrid Set-up of Mini-scaled and Micro Structured Components Inside a Reactor Housing

In the following, a hybrid set-up of micro- and mini-scale components *inside* the reactor housing is presented. Heatric Company [96] developed the so-called printed circuit heat exchanger technology (PCHE), which uses a diffusion-welded stack of etched metal foils to build up large-scale micro structured heat exchangers. They developed a 5 kW stand-alone power generation system for the use of natural gas as the power source for a stationary Proton Exchange Membrane Fuel Cell (PEMFC). The reformer produces hydrogen with a low level of carbon monoxide suitable for the supply of a proton exchange membrane fuel cell.

Heterogeneous gas-phase reactions are usually executed in tubular fixed-bed reactors often under unfavorable adiabatic conditions owing to low heat conductivity in the fixed bed. To improve the temperature course of the reaction, the concept here is based on a separation of the exothermic reforming reaction in a number of adiabatic fixed beds followed each time by a micro structured heat exchanger stage [97]. This means that the adiabatic temperature rise is also followed each time by an isothermal temperature decrease in several steps, resulting in a saw-tooth temperature pattern. A number of these stages are combined in a common diffusion-welded housing.

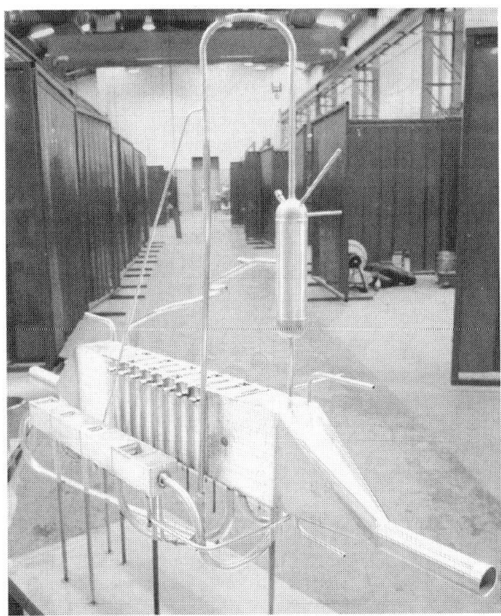

Figure 4.50 Plant set-up of the stand-alone reformer based on PCHE technology [96] (by courtesy of Heatric).

The plant consists of a pre-reforming step which converts C2+ into methane, which also reduces the coking risk downstream (Figure 4.50). The actual reforming step, the water gas shift reaction, follows and also a step for the selective oxidation of carbon monoxide before the hydrogen-rich gas enters the fuel cell. A steam evaporator is also included in the set-up.

4.8 Mobile Plants

One of the most widely used chemical plants is certainly the exhaust pipe section of an automobile. The exhaust pipe system of automobiles shows all features of a chemical plant, as it consists of a catalytic reactor, supply pipes for the gases, heaters and control equipment such as the lambda sensor. The catalyst carriers of an exhaust pipe are not called micro structured reactors but, considering their micro dimensions, they actually belong to this group of reactors.

4.8.1 Reactor 22 [R 22]: Catalytic Automobile Exhaust Gas Converter

Emitec [98] presented a catalytic converter manufactured from stainless-steel sheet metal. The 30 μm foil is wrapped to a cylinder with axial gas throughput (Figure 4.51).

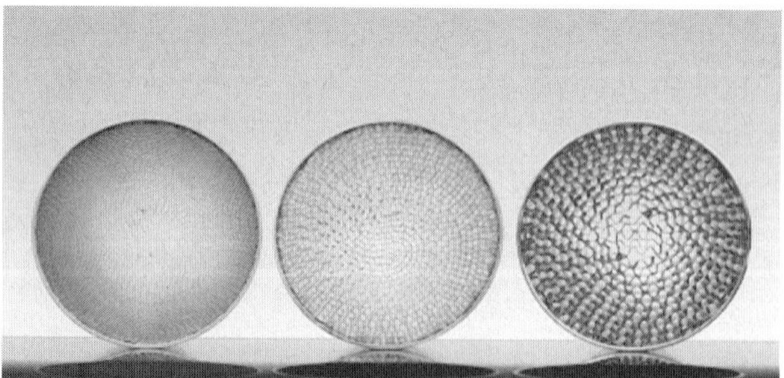

Figure 4.51 Structured metal foil catalytic converter [98].

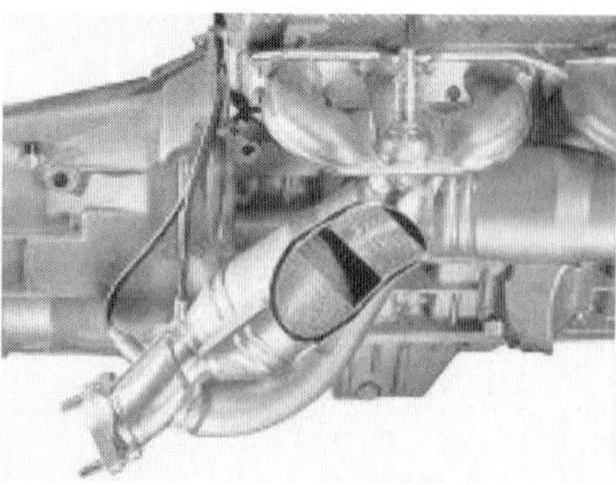

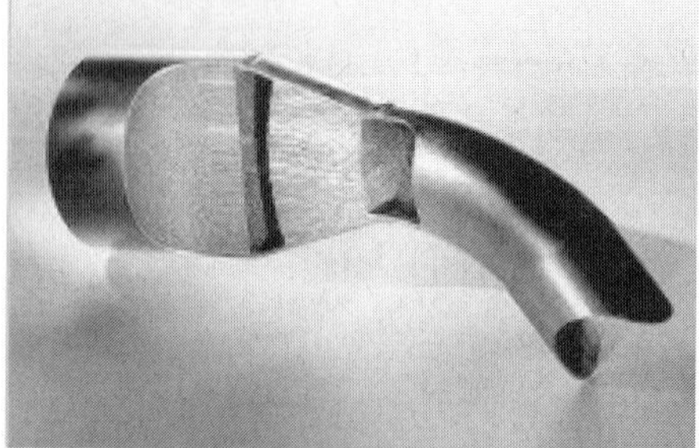

Figure 4.52 Catalytic converter in the exhaust pipe [98] (by courtesy of EMITEC GmbH).

The catalyst is coated as a thin ceramic layer at the walls of the metal foil. These so-called 'metalits' exhibit a steel content of only 40 vol.%, which lowers the counter pressure of the exhaust pipe. At the same time the heating-up period of the catalyst decreases, which reduces emissions during the engine start. The position of the converter in the exhaust pipe section is shown in Figure 4.52.

4.8.1.1 Heating Performance

[R 22] Here, a very interesting solution for the heating of the catalytic converter is presented. The thin metal foil is connected to electrodes and used as a resistance heater. The actual supply circuit is not described but the basic principle of such an internal reactor heating can easily be derived by an approximate calculation to clarify the heating mechanism. If one assumes a well-insulated stainless-steel foil with a length of 50 m, a width of 100 mm, a thickness of 30 μm and a resistance of 0.12 Ω mm^2 m^{-1} connected to a car battery which delivers 12 V, one can directly calculate a heating performance of 72 W. This heat performance is balanced by the amount of heat necessary to heat the catalytic converter defined by the heat capacity:

$$P = \frac{(12\ \mathrm{V})^2}{0.12\,\frac{\Omega\,\mathrm{mm}^2}{\mathrm{m}} \cdot \frac{50\ \mathrm{m}}{0.03\ \mathrm{mm} \cdot 100\ \mathrm{mm}}} = 72$$
$$W = 7.9\,\frac{\mathrm{kg}}{\mathrm{m}^3} \cdot 0.15 \cdot 10^{-3}\ \mathrm{m}^3 \cdot 465\,\frac{\mathrm{J}}{\mathrm{kg\,K}} \cdot \frac{\Delta T}{\Delta t} \tag{4.1}$$

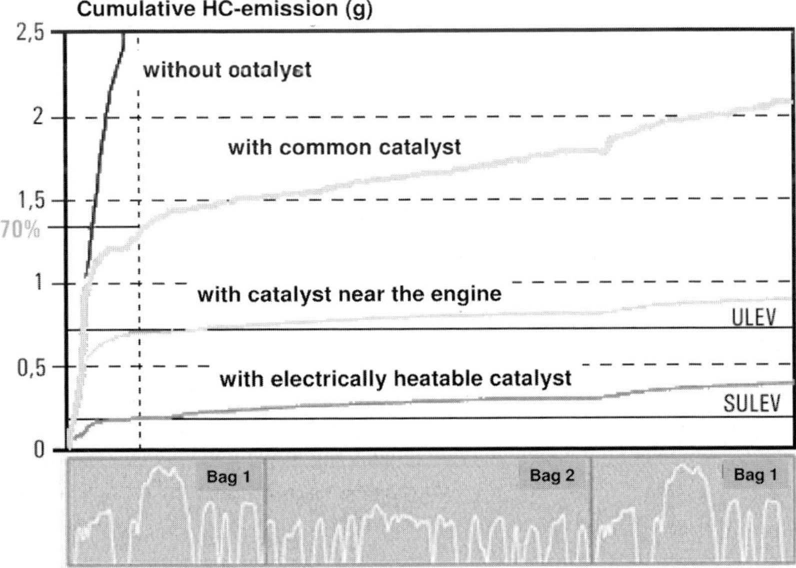

Figure 4.53 Hydrocarbon emission of different catalytic converters [98] (by courtesy of EMITEC GmbH).

From this equation, the adiabatic heating rate $\Delta T/\Delta t$ can be obtained as 130 K s^{-1}. This is only an estimated value neglecting heat losses to the walls and to the gas but it indicates that fast heat during the start of the engine should be possible. This heating mechanism is certainly also useful in immobile applications and it demonstrates clearly that efficient heat management by internal resistance heating is possible in micro structured devices.

The converter was tested and benchmarked against other catalytic converters. A test without catalyst was also executed (Figure 4.53). Especially the performance of the electrically heated converter is up to six times more efficient than that of a standard converter of a monolithic structure.

4.8.1.2 **Benchmarking to Existing Catalytic Converters**

[R 22] This 'microstructured reactor' will in the future be supplemented by a hydrocarbon adsorber in order to reach future zero emission regulations. A hydrocarbon adsorber equipped with a ceramic layer of zeolite will adsorb unburnt hydrocarbons during the engine start-up and desorb the gases when the pipe temperature rises above 150 °C which is obtained by the heated converter right behind the adsorber.

4.9 Production Plants

Micro process technology still remains concentrated on laboratory applications. In the last 3 years, only a few applications have become known of industrial implementations. It is expected that more industrial micro structured reactor plants exist but the details are not published. Published examples exist from Merck [53], UOP, Degussa/Uhde and Clariant, but details were only available at a low level for confidentiality reasons.

4.9.1 Reactor 23 [R 23]: Micro Structured Reactor Plant for Pigment Production

At Clariant, a three-stage micro structured reactor plant for the synthesis of pigments is used for production [99]. They transferred the batch-wise operation of a pigment process into a continuous flow process for the synthesis of yellow and red azo pigments with improved color properties. A primary aromatic amine ($ArNH_2$) was suspended in water and acid (HCl, HBr, HNO_2, H_2SO_4) added. The mixture was cooled to 10 °C in a vessel. Continuous diazotation was executed in an isothermal micro structured reactor and a hydrochloride suspension and sodium nitride solution were added. In a final separation step the pigments were dried and milled. The pigment yield in the plant (shown in Figure 4.54) reached up to 15 l h^{-1} depending on the type of mixer employed. The efficiency of mixing in the micro mixer influenced the pigment quality much more than the temperature balance (adiabatic or isothermal) during the reaction.

Figure 4.54 Multi-scale pigment plant: 1, micro structured reactor; 2, membrane dosing pump; 3, pressure sensor and indicator; 4, flow meter [99] (by courtesy of ACS).

4.9.2
Reactor 24 [R 24]: Micro Structured Reactor Plant for Heterogeneously Catalyzed Gas-phase Reactions

A consortium of two industrial partners (Uhde and Degussa) and four research institutes (Max-Planck-Institut für Kohleforschung, Mühlheim, Technische Universität Darmstadt, Technische Universität Chemnitz and Universität Erlangen-Nürnberg) evaluated the applicability of micro structured reactors for a chemical process with economic relevance. The project, named DEMiS (Demonstration and Evaluation of Microreaction Technology in Industrial Systems), was carried out under the auspices and funding of the German Ministry of Science and Technology (bmbf) [43].

The aim of the project was to investigate different gas-phase reactions. In particular, alternative routes for the synthesis of propylene oxide [101, 102] and the synthesis of phenol from benzene and N_2O [103] should be found. As a first milestone a throughput of approximately 15–20 $mol_{Product}\ kg_{Cat}^{-1}\ h^{-1}$ was targeted.

Experimental protocol

[P 4] For preliminary investigations a small-scale micro structured reactor was used to perform gas-phase epoxidation of propylene to propylene oxide with evaporated H_2O_2. As catalyst titanium silicate (TS-1) was used.

$$\text{propylene} \xrightarrow[H_2O]{H_2O_2\ (vap)} \text{propylene oxide} \qquad \Delta H_R = -220\ \text{kJ/mol}$$

Typical results

[P 4] A maximum catalyst activity of about 80 $mol_{PO}\ kg_{Cat}^{-1}\ h^{-1}$ was determined. This value normally leads in a conventional reactor to hot pots and decomposition of the H_2O_2. By using a micro structured reactor, the formation of hot spots could be avoided. Also, safe working within the explosion envelope could be guaranteed [100].

Figure 4.55 Mounting of modules in the containment [100].

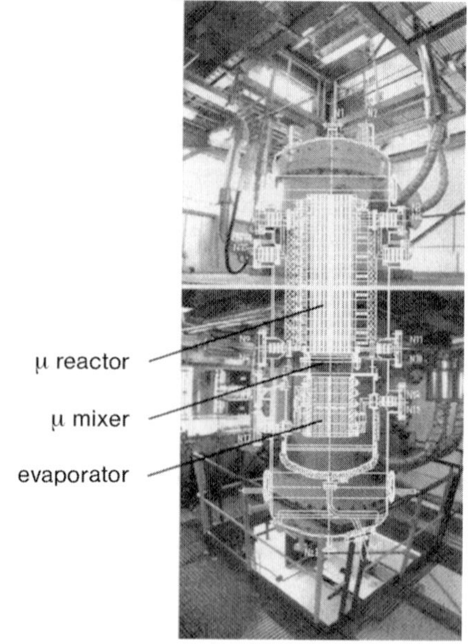

Figure 4.56 Outside view of the micro structured reactor plant with a schematic overlay to show the functionality of the reactor [100].

Microstructured reactor plant design

[R 24] The plant shown in Figure 4.56 consists of different modules, e.g. a module for evaporation of H_2O_2, a micro structured mixer and a micro structured catalytic reactor. Each module has the same dimensions: 1.6 m height and 1.0 m width (Figure 4.55). The whole plant is mounted in a containment of about 1.4 m inner diameter and an overall height comparable to a two-storied house. Because of the modular design, it is possible to increase the throughput by mounting more modules in parallel.

To reduce the decomposition, H_2O_2 is evaporated in a micro structured falling film evaporator at the bottom of the plant. The gaseous H_2O_2 flows upwards to the micro structured mixer where it is mixed with propylene. This mixture is guided through the likewise micro structured catalytic reactor. The mixture of the reaction, mainly propylene oxide, unreacted propylene and water is carried off at the top of the plant [100].

Experimental protocol

[P 5] The plant was on-stream in 6 weeks and with approximately 250 h for producing propylene oxide. The reaction was carried out at temperatures below 160 °C, a pressure of about 1.5 bar with an excess of propylene [100].

Typical results

[R 24] [P 5] Conversion of propylene from 5 to 20% and a selectivity of up to 90% were found.

4.9.3 Reactor 25 [R 25]: Micro Structured Reactor Plant for H_2O_2 Production

A micro structured reactor plant referring to the direct synthesis of H_2O_2 from the elements is under design and construction [104]. H_2O_2 is one of the most commonly employed bleaching agents. Environmental concerns with chlorine-based bleaching spurred the demand for H_2O_2, which decomposes to water and oxygen.

$$H_2 + O_2 \xrightarrow{Pd^{\sigma+}} H_2O_2$$

Traditionally, the anthraquinone process as an indirect route is used for making H_2O_2. Major concerns here are the degradation of the anthraquinone carrier and a complex flow sheet. Some processes for direct H_2O_2 synthesis are claimed in the patent literature, including Eka-Nobel, DuPont and Elf-Atochem process routes. They suffer from either sensitivity to explosions (as operation in the explosive regime is typically required), low selectivity or the need for high-pressure operation. The key to high selectivity is to have a noble-metal catalyst in a partially oxidized state. Otherwise, only water is formed or no reaction is achieved.

Experimental protocol

[P 6] Peroxide testing at IMM used an H_2O_2 selective catalyst placed within a mini-trickle bed reactor equipped with a micro mixer.

Typical results

[R 25] Using UOP process specifications, a space-time yield of 2.0 g H_2O_2 g_{Cat}^{-1} h^{-1} was achieved which exceeds literature values (see Table 4.7).

Table 4.7 Published yield and process parameters compared with results achieved by using a micro structured reactor [104].

Parameter	*Published*	*UOP test*
Pressure (bar)	124	30
Temperature (°C)	35	50
$O_2 : H_2$ ratio	6.8	3
'Space velocity' (g H_2 g_{Cat}^{-1} h^{-1})	2.6	1.8
H_2O_2 concentration (max.) (wt.%)	5.2	1.7 (not optimized)
Yield (g H_2O_2 g_{Cat}^{-1} h^{-1})	1.5	2.0

In addition, operation at only 30 bar, considerably lower than for the published processes, and the use of smaller O_2/H_2 ratios, saving valuable raw materials, is reported. The maximum H_2O_2 weight concentration achieved is 1.7% so far. It could be clearly shown that improved selectivity and conversion are achieved at explosive O_2/H_2 ratios.

Table 4.8 Changing the reaction conditions, e.g. O_2/H_2 ratio and pressure, results in corresponding different yield and selectivity: the lowest costs can be achieved for an O_2/H_2 ratio of 1 and for low-pressure operation (not shown here) [104].

Parameter	*O_2/H_2 ratio*				
	3	*1.50*	*1.0*	*6.79*	*1.89*
Pressure (psi)	300	300	300	1450	1450
Conversion	0.9	0.9	0.9	0.25	0.25
Selectivity	0.85	0.8	0.65	0.8	0.8
Yield	0.765	0.72	0.585	0.2	0.2

Pilot-scale test

UOP then carried out pilot-scale tests at still higher pressures in a fully automated explosion cell to reproduce vendor work and to study conditions and kinetics. Design was based on direct hydrogen peroxide synthesis using a mini-trickle bed reactor with a micro mixer.

A selectivity as high as 85% at 90% conversion has been achieved so far (O_2/H_2 ratio of 3). Several design cases corresponding to various O_2/H_2 ratios and various pressures were sketched (Table 4.8) and their capital costs were analyzed. The lowest costs result for an O_2/H_2 ratio of 1 and for low-pressure operation, the latter parameter having a notable impact.

Micro structured reactor plant design

Based on these considerations, a large-scale plant concept rated at 162 000 t yr^{-1} has been developed and is planned to be realized within the next few years.

4.10
Plant Installations and Supplier-specific Assemblies

A perfect definition transferable to plant modularization is given by the following citation from the American National Standards Institute (ANSI), which characterizes the ISA76 standard, the standard for the MPC system introduced above: "This specification establishes properties and physical dimensions that define the interface for surface mount fluid distribution components with elastomeric sealing devices used within the process analyzer and sample-handling systems. The interface controls the dimensions and location of the sealing surfaces to allow changes of just one element of the system without modification of the entire system. This is what makes the system modular from both a design and a maintenance standpoint." Thus, basically, *modularization* is about the definition of interfaces and pitch dimensions.

Modularization also means creating modules during reactor development. Supplier-specific assemblies reduce costs, increase flexibility and decrease the chance of construction failure, as the same modules are always combined in a new device.

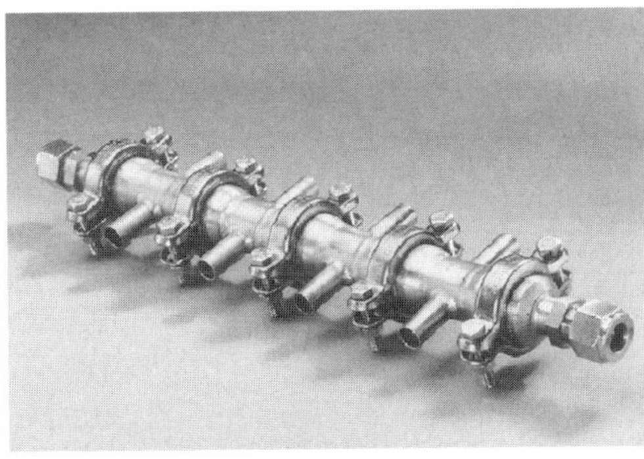

Figure 4.57 Stack of five mixer/reactor modules from the Forschungszentrum Karlsruhe [68] (by courtesy of Forschungszentrum Karlsruhe).

[R 4] The micro structured mixer/reactor of the Forschungszentrum Karlsruhe (Figure 4.57) consists of a serial connection of five identical modules [68]. Here modularity in the sense of using identical units stands more in the foreground than the supplier-specific assembly, which consists of different modules combined with a common base module as in the example above.

The temperature of each module can be adjusted by an integrated cross-flow heat exchanger. A temperature profile can be obtained by the combination of differently tempered modules.

[R 18] A modular set of devices was developed within the μChemTec project introduced above. It consists of a base plate which is identical for all four devices. This base plate acts as the fluidic interface to the piping and is equipped with a micro structured mixer. The base plate can be combined with a heated tube to deliver a mixer-tube reactor. A combination with a porous tube delivers a degasser unit. A combination with a membrane unit (not shown) or a settler results in an extraction device (Figure 4.58).

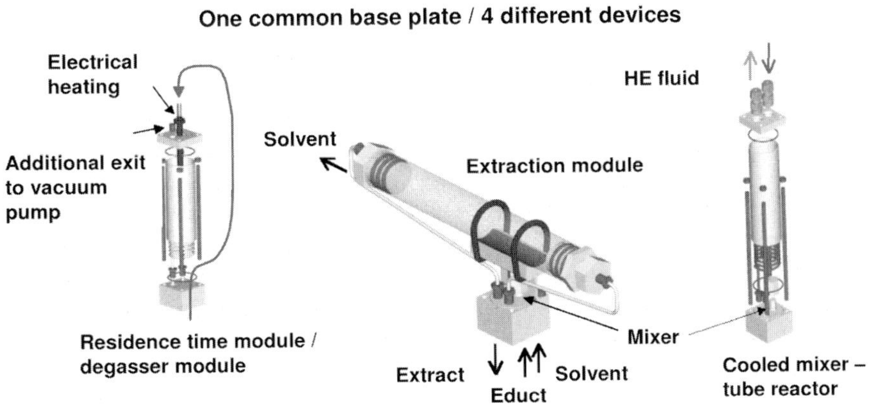

One common base plate / 4 different devices
Electrical heating
HE fluid
Solvent
Additional exit to vacuum pump
Extraction module
Residence time module / degasser module
Mixer
Extract
Solvent
Educt
Cooled mixer – tube reactor

Figure 4.58 Supplier-specific assembly of the μChemTec consortium [24] (by courtesy of microchemtec).

4.11
Process Management

Micro specific differences to macro-scale plants are not encountered in this chapter. This section is added to complete the survey of software tools used in the present work of a process engineer in the industry. It is certainly also important for micro technology as it gives insights into thinking and expectations of the production management. If a micro scale process in future will be used to increase production efficiency, management tools will also be an issue.

Process management as a supplement to a process control system tools allow the product manufacturer to visualize and optimize their production efficiency. A new market space just develops which addresses enterprise operation management. These tools help to integrate existing plant floor and business applications in order to prepare management decisions. The idea is to create a transparent production tool throughout the whole production process in the form of a so-called supply chain management.

Deriving from its historical simulation environment, Aspen presents with AspenTech® an operations manager suite which builds up on software investments already present in a company [105]. The difference from conventional tools is the real-time performance management, which gives the user direct real-time access to production data obtained by a local process control system. Keywords are scheduling, logistics, capacity utilization and sales.

A similar solution but deriving from a business environment is offered by SAP, a leader in business software. The business software mySAP™ permits a transparent and synchronized supply chain with direct access of the production management to yield, personal costs, maintenance, raw material and energy consumption [105].

4.11.1
Process Control and Automation

The management of a chemical plant is unthinkable without an automated process control system (PCS). Also on a laboratory scale such a control system is desired but not very often implemented. The use of a PCS on a laboratory scale is often restricted owing to different demands in laboratory production compared with large-scale production plants. Space is limited in a small plant but nevertheless a large number of sensors have to be supplied to control the product quality. The number of sensors in a small plant often will not be much different from the number of sensors in a large-scale plant but the available space is much smaller. To avoid a confusing cable set-up, some authors recommend the on-site processing of sensor data in the devices [13] and the transport of digital data via a data bus sysem [24]. Actuators and sensors could also be reduced in size by applying MEMS (micro electro mechanical systems) technology, which locates the sensors and actuators directly on the micro structured reactor plates. Actuators such as valves could act as a control mechanism to adjust the volume flow or pressure. However, despite applications outside the reaction zone, there is certainly still a lot of development needed until MEMS can achieve a degree of robustness to enter reaction technology or even a production line. Today, passive devices for flow distribution or sensors which measure integral product values at the reactor exit seem to be a more realistic approach. In this case, process control in a micro structured reactor plant is not much different from process control in a miniplant.

4.11.1.1 Automation 1 [A 1]: Automated Micro Reaction System (AuMµRes)

A first example of an integrated approach to process control and process optimization was executed in research project funded by the German bmbf (Bundesministerium für Bildung und Forschung) with the partners Siemens, Siemens Axiva, Merck and the Fraunhofer ICT [106–108]. The so-called AuMµRes (automated micro reaction system) uses the process control system Simatic PCS 7 from Siemens and is capable of executing automated process parameter changes following a statistical plan for efficient parameter screening.

The micro fluidic devices are integrated into cubic modules of identical shape and size, and can be easily replaced [108]. The modules are equipped with individual temperature control, integrated pressure and/or temperature sensors and, upon demand, additional actors such as non-return valves. This modular concept ensures a flexible interconnection of the micro fluidic devices.

In this way, the AuMµRes process control system can perform systematic variations of process parameters based on predefined statistical DoEs (design of experiments) [108].

Operation modes

[R 18] [A 1] Different modes of process operation such as fully automated, semi-automated or the execution of a cleaning operation are possible. A safety operation mode allows the controlled shut-down of the plant if predefined safety limits are exceeded.

Graphical user interface

[R 18] [A 1] In Figure 4.59 the flow sheet of the process is shown.

Adjustable parameters

[R 18] [A 1] Each module is equipped with a heater (H3–H8) and a fluidic cooling (CO3–CO6). Temperature sensors integrated in the modules deliver the sensor signals for the heater control. Fluidic data such as flow and pressure are measured integrally outside the micro structured devices by laboratory-made flow sensors manufactured by silicon machining. The micro structured pressure sensor can tolerate up to 10 bar at 200 °C with a small dead volume of only 0.5 µl. The micro structured mass flow sensor relies on the Coriolis principle and is positioned behind the pumps in Figure 4.59 (FIC). For more detailed information about the product quality it was recommended to use optical flow cells inline with the chemical process combined with an NIR analytic or a Raman spectrometer.

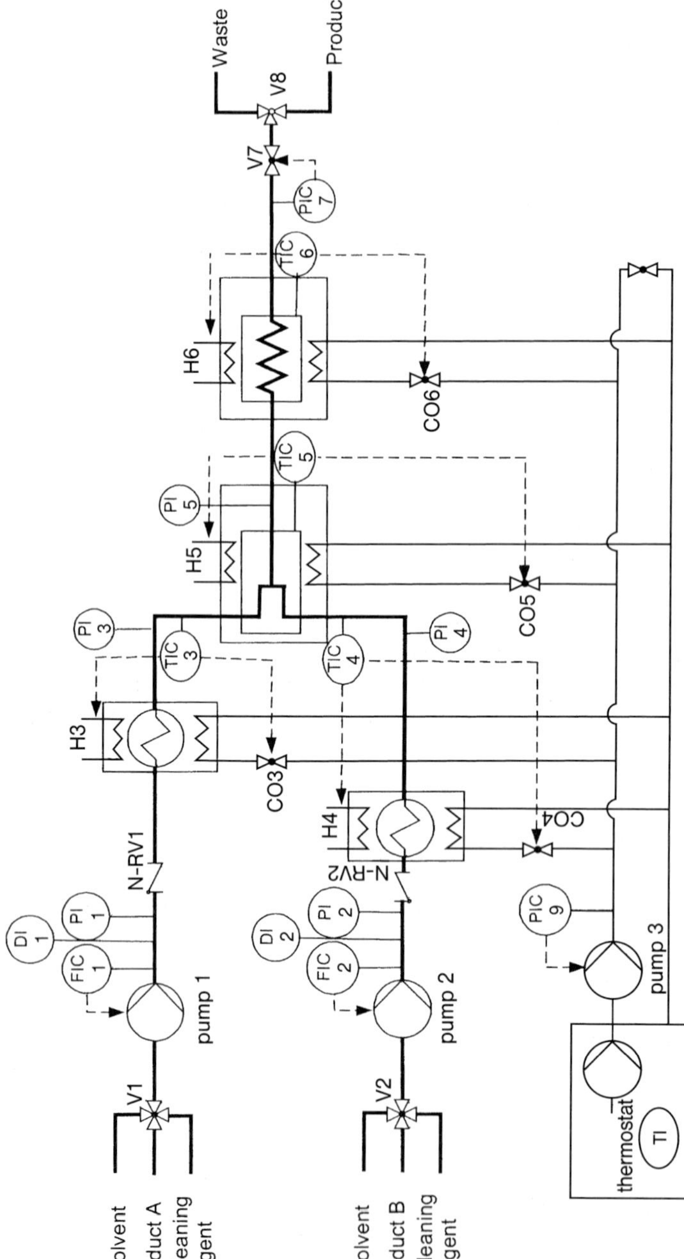

Figure 4.59 Flow sheet of the AuMμRes process [106].

The parameters windows of the automated microreaction system AuMμRes are given in Table 4.9. These specifications allow the operation of most liquid reactions [108].

Table 4.9 Parameter windows of the AuMμRes automated micro reaction system.

Parameter	*Minimum*	*Maximum*	*Tolerance*
Volume flow (ml min^{-1})	0.1	40	±0.25%
Retention time (s)	1	600	–
Reaction temperature (°C)	–20	150	±1
Excess pressure (bar)	0	6	±0.1
Viscosity (mPa s)	0.5	100	–
Heat exchange (kW)	0	1	–
Temperature of educts at the reactor inlet (°C)	–20	100	±1

Sensorics and analytics

A micro structured silicon pressure sensor was designed for flow-through applications in the pressure range 0–10 bar [108]. This sensor can be operated at temperatures up to 200 °C and has a dead volume of < 0.5 l.

Analytical interfaces are integrated into the AuMμRes set-up for at-line analysis by sampling and subsequent chromatography (HPLC) [108]. Moreover, this allows online analysis by infrared or Raman spectroscopy. Real-time monitoring of the chemical processes can be achieved via spectroscopic measurements, for which suitable optical flow-through cells have to be installed at selected positions of the micro reaction system.

Selected results

Pump pulsation may significantly impact on mixing quality and in this way the whole reaction performance [108]. In order to minimize this detrimental effect, commercially available syringe pumps were modified in order to suppress temporary pulsations. By this measure, pulsation spikes could be damped after installation of an additional switching valve in combination with a microcontroller for the individual control of both pistons.

As a test reaction, the regioselective mononitration of 2-(4-chlorobenzoyl)benzoic acid to 2-(4-chloro-3-nitrobenzoyl)benzoic acid was investigated, which is a precursor for the synthesis of a pharmaceutical agent [108].

O COOH
Cl
HNO_3
H_2SO_4
O COOH
Cl
NO_2

This involves dosing HNO_3 to a solution of 2-(4-chlorobenzoyl)benzoic acid in concentrated sulfuric acid (97%). With only one micro mixer conversion of 30–40% is achieved. Connecting 2–3 elements in series gives conversions up to 70%.

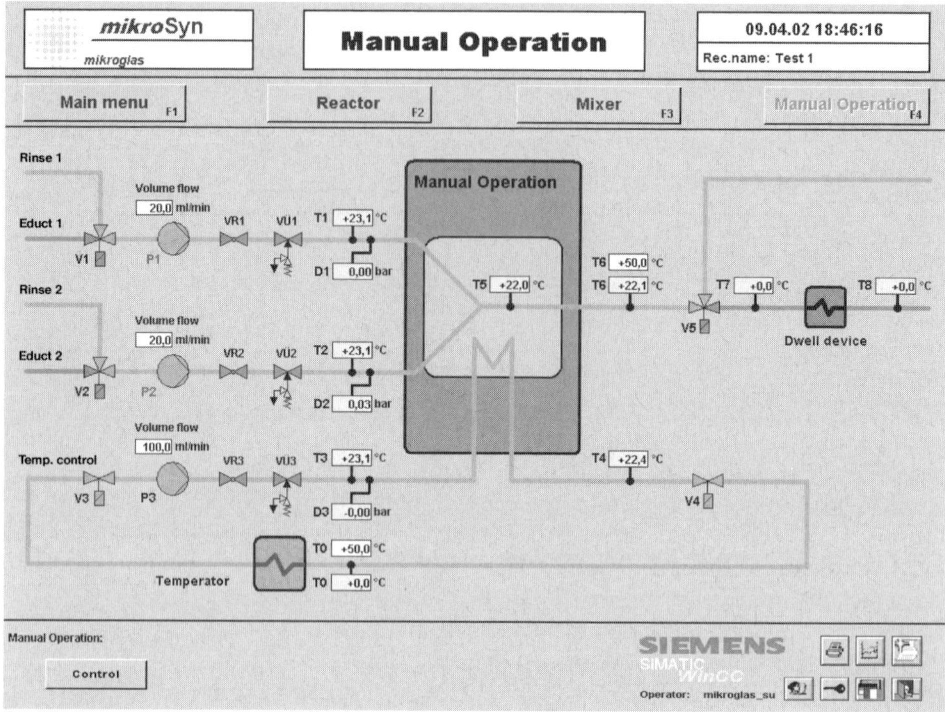

Figure 4.60 Screen layout of the main operation level of a process control system from Mikroglas-chemtec [21] (by courtesy of DECHEMA).

4.11.1.2 Automation 2 [A 2]: MikroSyn Control System

Mikroglas-chemtec GmbH also relies on the Simatic control system from Siemens [21]. As in an industrial process, a flow sheet is indicates all information like temperature, pressure, volume flows and valve settings (see also Figure 4.60).

Graphical user interface

[R 12] [A 2] The screen is subdivided into three areas. The overview area (area 1) indicates the current operation mode (manual/automatic) and the protocol currently processed in the plant. The machine area (area 2) gives an overview of the process parameters which can be modified in this area. The button area (area 3) provides access to functions such as user login, choice of language and the protocol administration.

Operational modes

[R 12] [A 2] In automatic mode, a protocol is executed without the possibility of and need for manual influence. The protocol has to be defined in a window which opens if the button protocol 1 or 2 is activated (Figure 4.61). Now process values for the automatic protocol processing can be defined in the right window and uploaded to the S7 controller after confirmation. A number of formulas can be

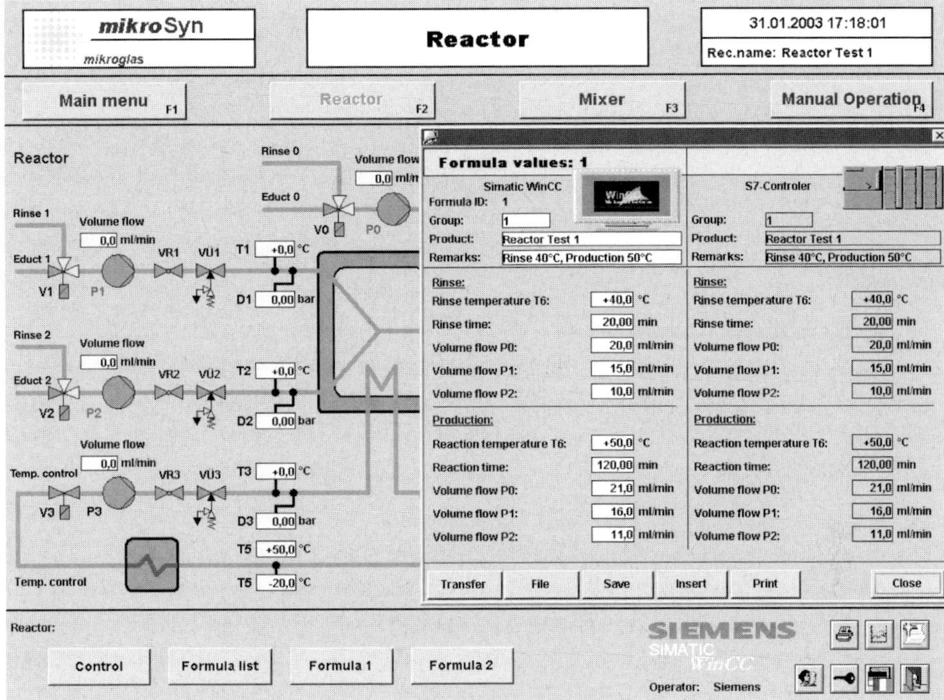

Figure 4.61 Protocol administration with the Mikroglas-chemtec control system [21] (by courtesy of DECHEMA).

stored in the formula list in the right window. Process values are graphically represented in adjustable charts which show the trend presentation. The number of variables and the time frame can be selected. An alarm management system is also supplied which permits the recording of alarm messages and the adjustment of parameter limits. Simatic is a powerful, widespread and well-accepted industrial software package. However, as with most industrial control systems, changes in the plant set-up require new input from the supplier of the software, in this case Siemens.

4.11.1.3 **Automation 3 [A 3]: User-ajustable Process Control System**

Another important aspect during the development in the laboratory is that the plant set-up changes frequently. Hence it is highly desirable to work with a flexible and self-explanatory PCS system which is fully adjustable by the user to any specific process set-up. For this purpose, modular built-up software based on the computer language LabView®, often used for laboratory applications, was developed and combined with also modular constructed hardware [77].

Graphical user interface

[A 3] In contrast to the example above, flow sheets are not shown here. This allows faster adaptation to a new process, but, of course, at the expense of process

transparency. The graphical presentation of the process flow is replaced by tables which indicate process values and set points. To reduce the learning period of the plant operator, the hierarchical window-type structure is replaced by a linear horizontal arrangement of tables accessible by a horizontal bar.

The graphical user interface is created during the software installation using a so-called wizard, which checks the configuration of the plant in cooperation with the user. The wizard permits the adaptation of the graphical user interface to the existing plant set-up and the activation of controller, I/O channels, limit switches, etc.

Modular software construction

[R 18] [A 3] This PCS was especially developed as an integral part of the μChemTec backbone concept [24]. The most obvious feature is – similar to the backbone concept – the modular construction of the software, which permits its application by the user to different plant set-ups.

Data management

[R 18] [A 3] As in every macro-scale chemical plant, all process I/O operations are visualized in the graphical user interface (Figure 4.62): The process data can be recorded as a function of time and also exported in ASCII code (Figure 4.63).

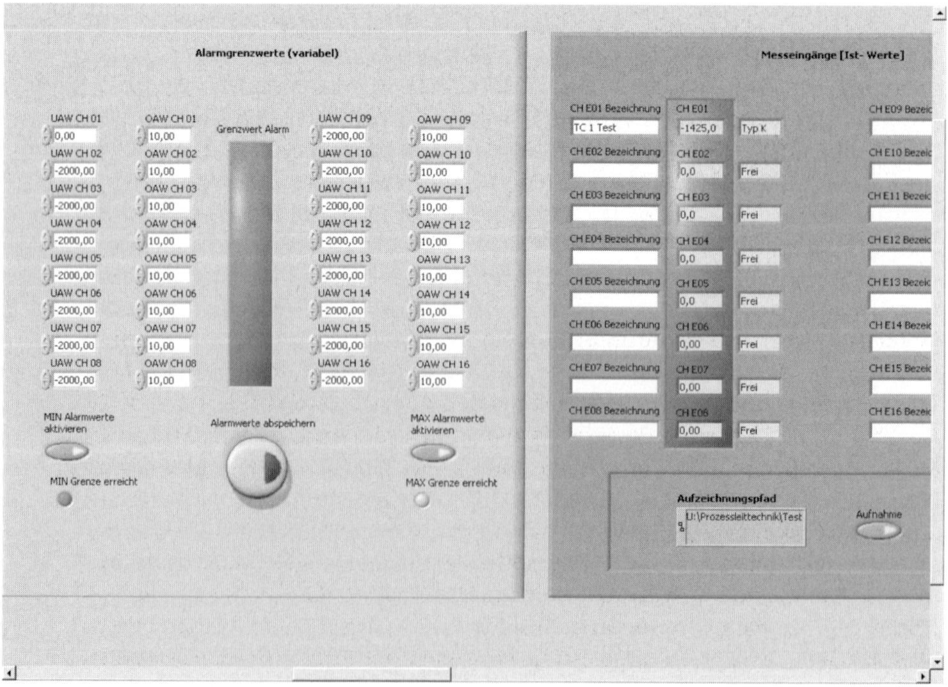

Figure 4.62 Graphical user interface under Windows XP® of the IMM PCS [109].

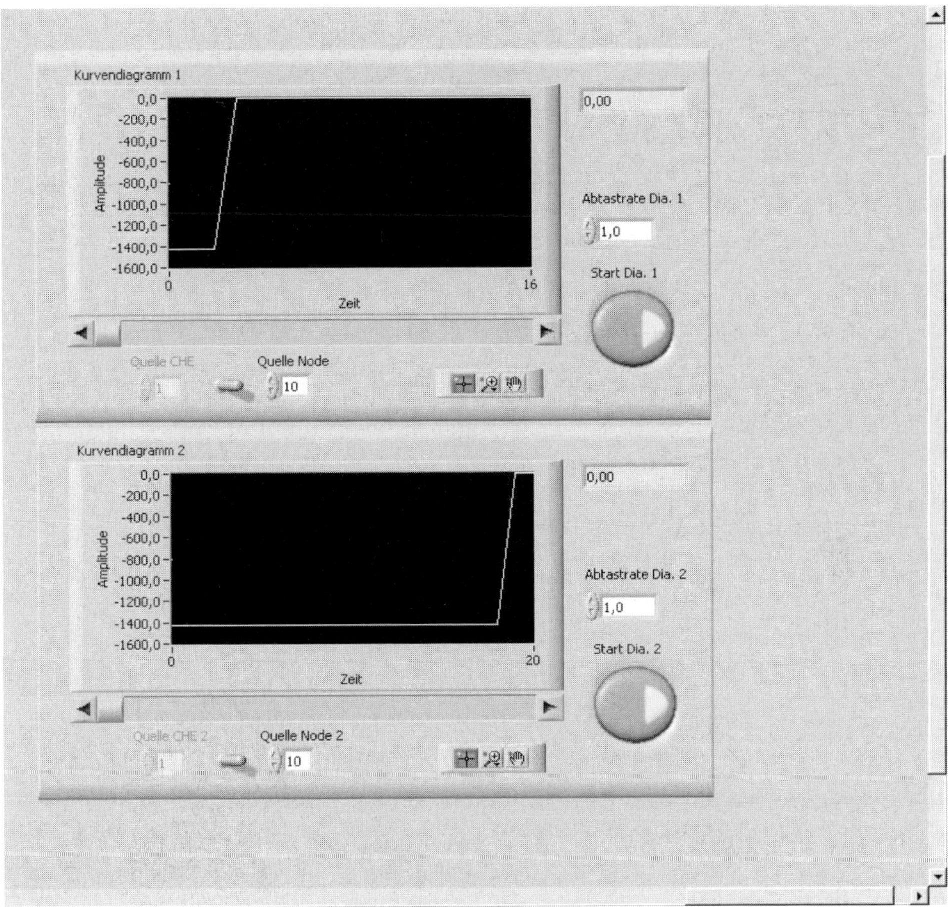

Figure 4.63 Visualization of continuous storage of process data [109].

Operational modes

[R 18] [A 3] For complex processes, a protocol sequence control system is implemented which allows the automated course of a multiple-step process (Figure 4.64). The protocol is also recorded in a text file and can be reloaded directly from the stored text file. The same procedure is attainable with the alarm messages which are also recorded as a function of time and stored in the alarm message text file.

Implementation and installation

[R 18] [A 3] The modular set-up of the software is complemented consistently by a modular build-up of the hardware which comes with the control system. The PCS hardware is supplemented by an embedded PC (see Figure 4.65), but can also be operated without a PC, if an existing PC should be used instead. The central motherboard is equipped with a number of slots which can individually be implemented with small circuit boards (size ca. 40 mm × 20 mm). In such a way

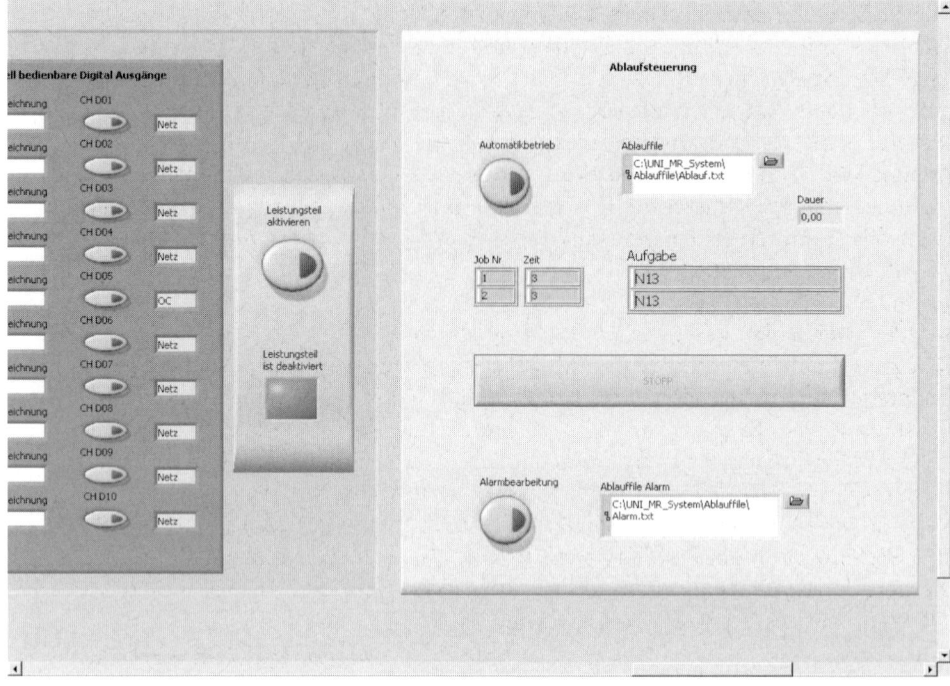

Figure 4.64 Protocol sequence control system and alarm management [109].

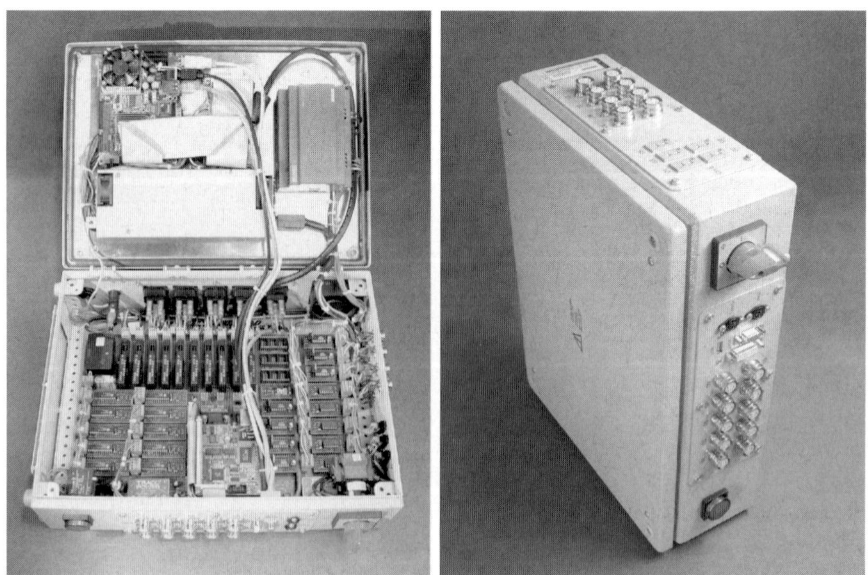

Figure 4.65 Modular PCS hardware installed inside an industry-conformal case. Left, open case; right, closed case [109].

the hardware is configured to fit to any type of plant set-up. The circuit boards are available for the following signals: analogue I/O (thermocouple, PT100, 4–20 mA, 0–10 V), digital output, RS232, 24 V switch output, 230 V output (e.g. heating) and limit switches. For the connection to a LAN an Ethernet connection is implemented. Several of these hardware boxes can be interconnected if the number of I/O connections of one box is not sufficient. However, enough slots are prepared for a standard laboratory-scale plant. In case of computer failure the programmable logic controller (implemented independently of the PC) allows a predefined safe shutdown of the plant.

4.11.1.4 **Automation 4 [A 4]: Sensor Analytical Manager**

The American CPAC initiative NeSSI [23] developed a micro reactor sampling and calibration system intended for analytical applications in the oil industry. Industrial partners such as Swagelok and Parker/Hannifin developed the system originally designed for the gas supply in clean room facilities. This approach is well advanced with respect to valves, gauges, analytical sensors and pipe fittings.

[R 17] [A 4] Currently, the integration of micro structured reactors is under way. The system is equipped with a so-called sensor analytical manager (SAM) which delivers the digital data, transformed in the devices, via a local area network to the PC controller (Figure 4.66).

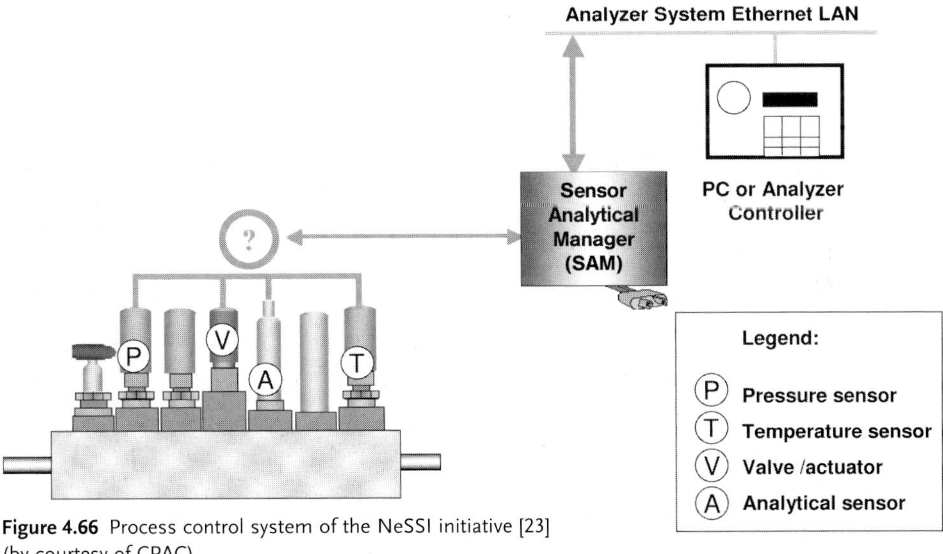

Figure 4.66 Process control system of the NeSSI initiative [23] (by courtesy of CPAC).

4.11.2 Inline Analysis, Actuators and Sensorics

In principle, data can be obtained inside or outside the reactors (or any other device) at the reactor exit. Internal data recording would demand small, robust sensors. The first condition is sometimes fulfilled but the second, the sensor robustness, which would allow continuous operation for months, is normally not fulfilled. Hence the only choice currently available is recording of signals at the reactor exit.

This is not such a disadvantage as it may first seem. The final product delivered to a costumer will have a quality influenced by all small volume elements flowing through a reactor, thus introducing a kind of integral product quality. This does not mean that internal sensors are not extremely useful, for example, in the detection of hot-spots in gas-phase reactors or for the evaluation of mixing quality at different locations inside of a micro mixer. This information is highly desirable but as long as robust sensors are not offered commercially, internal sensors will be useful as expensive prototypes for the optimization of single reactors but will not to the same extent be useful for the optimization of plants.

Integral values can be recorded by any type of flow-through devices connected inline to the product flow, for example, a flow cell. Flow cells are consequently the main feature in this section.

4.11.2.1 Some Analytical Techniques Relevant for Micro-channel Processing

Temperature monitoring is among the most important issues for chemical micro process engineering. Already at a laboratory level, in-line temperature profiles could give valuable hints similar to a calorimetric analysis. However, the use of thermocouples has limits and hardly gives useful data. Thermographic imaging [110], if accessible at all, was used in selected cases, but may require one to reduce thermal insulation and thus can give wrong temperature profiles. Accordingly, the insertion of inline sensors, as initially already made [111, 112], is required.

There is also a need for concentration monitoring. At-line analysis by sampling and subsequent chromatography (HPLC) is briefly mentioned in [108]. First attempts using FTIR imaging have now reached a remarkable level [108, 113–115]. Online near-IR spectroscopy, for example, turned out to be a suitable method for the monitoring of the nitrate and HNO_3 concentrations during the above-described nitration of chlorobenzoylbenzoic acid [108]. Raman spectroscopy is an even more differentiating analytical technique that provides in-depth structural information and thus allows one to distinguish betweenreactants, products and intermediates [108]. Figure 4.67 shows spectra of pure sulfuric acid and mixtures with HNO_3, the starting material, and the actual reaction product.

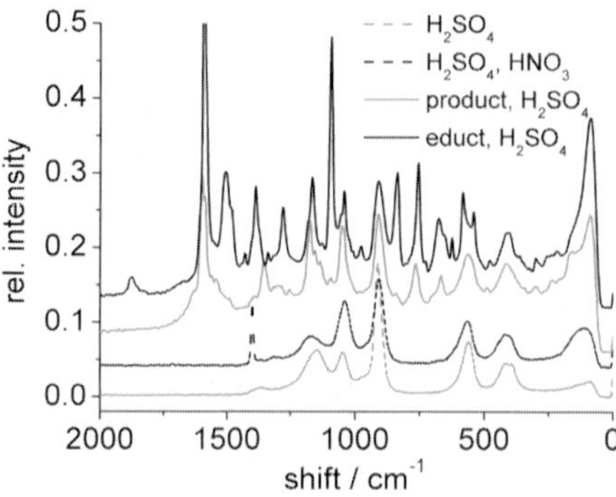

Figure 4.67 Raman spectroscopic monitoring of the nitration of chlorobenzoylbenzoic acid using nitrating acid (H_2SO_4/HNO_3) [108] (by courtesy of ACS).

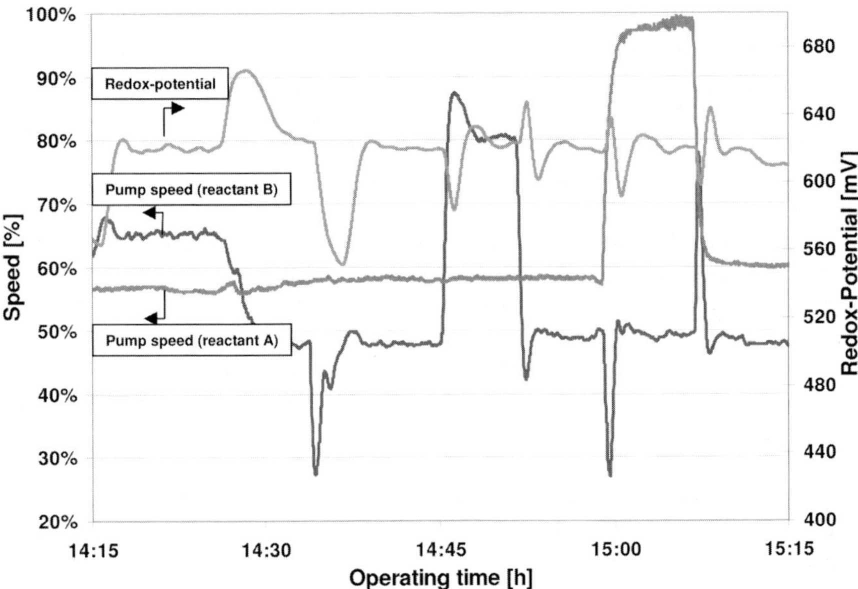

Figure 4.68 Process control for a Clariant pilot process: redox-potential plot of continuous fully automated diazotization [99] (by courtesy of ACS).

Inline UV/Vis absorption measurements to monitor photochemical reactions in microfluidic chips are described in [116, 117].

Analytics for residence time monitoring have also been developed [118].

It is not clear whether analytical techniques for other parameters are of major importance. In addition to data gathering, the same analytical techniques and miniature sensors may be used to monitor and control processes at a production level. Additional monitoring demands arise here such as flow equipartition and pressure monitoring. It has to be questioned if such complex sensing will be feasible and practicable, since it may have to deal with large numbers (of micro channels). It may, moreover, be asked if it is needed at all. There are, e.g., means for flow calculation and some of these issues can be taken for granted. On the other hand, there are regulatory demands during commissioning.

Nonetheless, there are running plants at laboratory and pilot-scale levels at institutes/universities and industry where process control is already exerted. Usually this is done in a rather conventional fashion, e.g. using commercial pressure hold valves and temperature determination at the in- and outlets and process-specific concentration monitoring outside the micro reactor. For example, an analysis of the redox potential was used for process monitoring for continuous azo pigment production at Clariant (see Figure 4.68) [99].

4.11.2.2 **Automation 5 [A 5]: Inline Sensors According to the ISA SP76 Standard**

The NeSSI initiative [23], introduced above, already offers such inline sensor equipment according to the ISA SP76 standard.

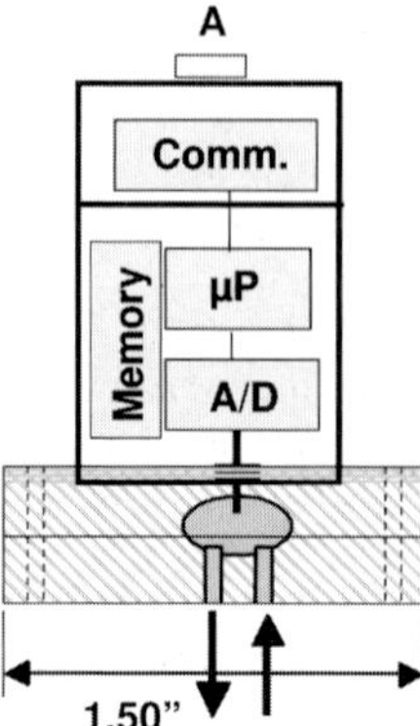

Figure 4.69 Set-up of inline sensors from the NeSSI initiative [23] (by courtesy of CPAC).

Signal processing and data transmission

[R 17] [A 5] The general set-up of these inline sensors always follows a uniform structure (Figure 4.69). A sensor is in direct contact with the fluid in a flow cell in the base plate. From here the analog signal is converted into a digital signal and further processed in a micro processor before it is send to the communication bus.

Sensor implementation

[R 17] [A 5] A number of sensors for oxygen and water detection and also for the measurement of acidity and conductivity are already available, as shown in Figure 4.70.

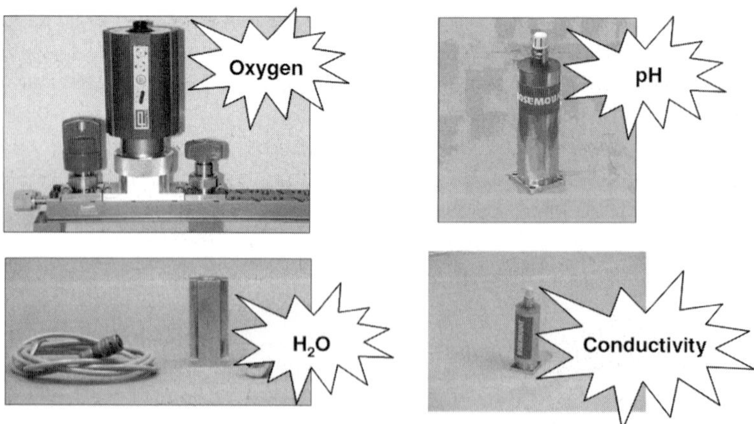

Figure 4.70 Flow cells of the NeSSi initiative for measurement of O_2 and H_2O content and also pH and conductivity [23] (by courtesy of CPAC).

Raman and infrared sensor modules

There is also a prototype of a Raman sensor module scheduled and in future a miniature spectrometer as shown in Figure 4.71 could be integrated [119].

Two other spectrometers not yet adapted for direct integration into a plant but with the potential for future adaptation are described in the following.

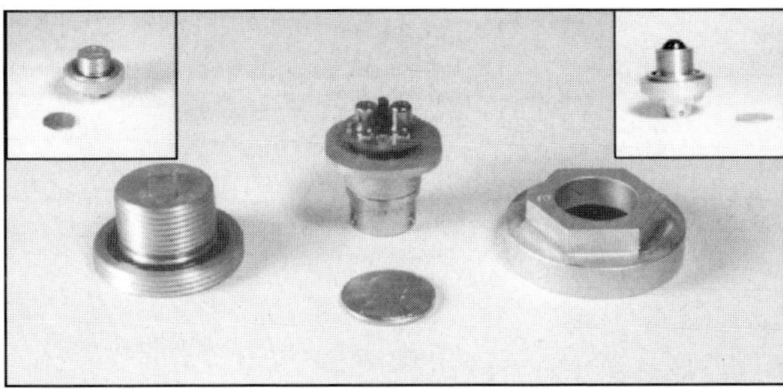

Figure 4.71 Micro spectrometer (2000–14 000 nm) for inline mid-infrared measurements [119] (by courtesy of Applied Analytics Inc.).

4.11.2.3 **Automation 6 [A 6]: Micro Fabricated Near-infrared Fourier Transform Spectrometer**

A miniaturized Fourier transform spectrometer for near-infrared measurements (FTIR, 2500–8330 nm) was developed at the Forschungszentrum Karlsruhe [120]. Near-infrared measurements give information, for example, about the oil, water and protein content of liquids or solids. The dimensions of the detector chip are 11.5 × 9.4 mm, the device is essentially a miniaturized Michelson interferometer and it consists of a micro optical bench with beamsplitter, ball lenses, mirrors and the detector chip. The light beam is coupled in via a glass-fiber and an electromagnetic actuator. The signal is derived from the signal response of the detector by Fourier transformation.

4.11.2.4 **Automation 7 [A 7]: Micro Fabricated Near-infrared and Visible Spectrometer**

The visible to near-infrared spectral region (375–740 or 680–1080 nm, resolution < 10 nm) is the domain of a micro spectrometer presented by the Institut für Mikrotechnik Mainz [121]. This device (overall dimensions 16 × 28 × 38 mm) is commercially available but not yet adapted to direct inline measurements in a plant (Figure 4.72).

[A 7] The fiber-coupled spectrometer is actually designed for applications with hand-held devices. A combination of micro injection molding and an electroforming process allows the production of a high-performance but low-cost device. The optical grating is, for example, replicated in plastic to reduce costs.

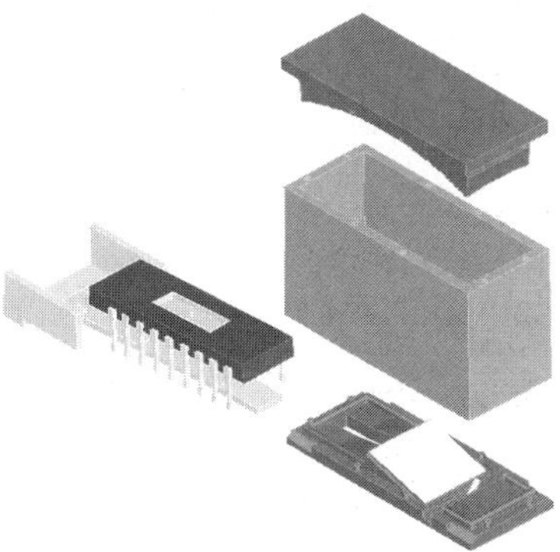

Figure 4.72 Hand-held mini spectrometer for the visible and near-infrared region (source IMM).

4.11.2.5 Automation 8 [A 8]: Micro Gas Chromatograph

A micro gas chromatograph, as small as a check card (dimensions 91 × 59 × 27 mm), was presented by SLS MICRO TECHNOLOGY GmbH [122]. This device is one of the first commercial products of micro system technology and it possesses sufficient robustness to withstand chemical and thermal stress. Hence, it should be ideal for inline measurements in a plant.

Micro injection system

[A 8] The micro injection system relies on an electromagnetic drive and injects a small volume of only 0.5 µl per injection. The valve section is heated. The injected gas pulse is separated in a thin-layer separation column.

Column and separation procedure

[A 8] The separation layer consists of platinum coated on a silicon substrate. Owing to the low heat capacity, quick temperature cycles can be executed and a resolution down to 7 ppm can be achieved with the integrated thermal conductivity detector. The standard length of the capillary is 860 mm with a channel width of 60 µm. A measurement cycle takes less than 60 s.

System integration

[A 8] The column, the injection system, the flow sensor and the processor chips are all integrated on a common integrated circuit board (Figure 4.73, left). A user interface is located on the surface of the GC and a serial RS-232 data interface is also provided. An evaporator shall be available by the end of 2004.

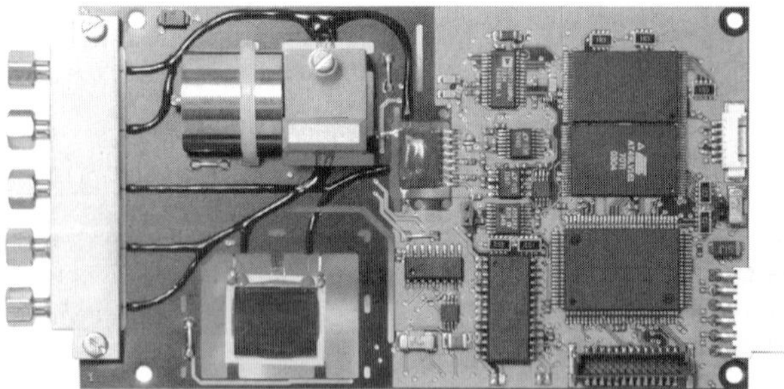

Figure 4.73 GCM 5000 micro gas chromatograph on an integrated circuit board [122] (by courtesy of SLS MICRO TECHNOLOGY GmbH).

There are also a number of commercial flow cells which after slight modifications can be integrated in the process flow of a microstructured reactor plant.

4.11.2.6 Automation 9 [A 9]: Electroanalytical Flow Cell

The ClinLab flow cell Sputnik® [123] was originally developed for HPLC applications but is a separate device which in principle can be combined with every amperometric detector which is designed for three-electrode operation (Figure 4.74). Two PEEK capillaries and three electronic plugs are the only external interfaces.

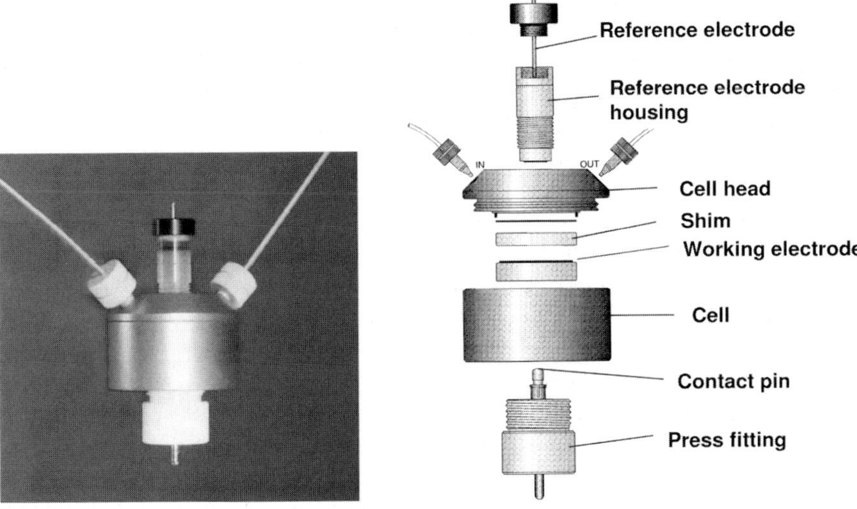

Figure 4.74 Photograph of an electrochemical flow cell (model Sputnik®) [123] (by courtesy of RECIPE CHEMICALS + INSTRUMENTS GmbH).

Electrical circuit and measuring principle

[A 9] The thin-film three-electrode cell possesses a cell volume adjustable from 0.75 to 2.5 μl. The fluid dynamic design of the cell should prevent turbulence. The reference electrode, an Ag/AgCl cell element, can be filled manually with KCl. A transparent housing permits optical control of the liquid level. The operational pressure is limited to 5 bar. The supplier recommends this robust device for routine measurements when high selectivity is desired.

4.11.2.7 Automation 10 [A 10]: High-pressure Flow Cell for Optical Microscopic Observations

A new high-pressure flow cell for optical microscopic observations of living biological cells was presented by the TU Munich [101]. Such a device could be very interesting in chemistry for the detection of mixing problems, for example, in two-phase systems. The fluidic interface consists of a flexible high-pressure tube. The housing dimensions are approximately 60 mm (diameter) × 25 mm (height). Two sapphire windows are inserted in the cover and the bottom of the flow cell. The geometry of the cell housing permits good accessibility of the window with, for example, a LEICA microscope (LEICA DM IRB HC).

Pressure resistance

[A 10] A pressure sensor detects the internal pressure up to the maximum operational pressure of 4000 bar.

Temperature resolution

[A 10] A thermocouple measures the temperature with a resolution of 0.1 K.

Figure 4.75 Flow cell connected to a micro reactor of the FAMOS micro reaction system [124] (by courtesy of FhG Allianz FAMOS).

4.11.2.8 Automation 11 [A 11]: Flow Cell for Optical Inspections

The FAMOS initiative, introduced above, presented a flow cell (Figure 4.75) for optical inspections connected to one of the micro structured reactors in their new micro reaction system [124]. The flow cell measures only a few cubic centimeters and the actual optical interface to a spectrometer consists of a macro-scale table-top device with a clamping mechanism which positions the flow cell in the optical pathway.

4.11.2.9 Automation 12 [A 12]: Golden Gate® Single Reflection Diamond ATR Unit

The micro flow cell Golden Gate® [125] can tolerate up to 66 bar at temperatures up to 200 °C. The cell volume amounts to 28 ml and fluidic interfaces are 1/16 in Swagelok® capillary fittings. While the dimensions of the actual flow cell are only a few centimeters, the actual optical interface to a spectrometer consist of a macro-scale table-top device with a clamping mechanism which positions the flow cell in the optical pathway.

4.11.2.10 Automation 13 [A 13]: Combination of Inline Sensors with Electronic and Fluidic Bus System

A combined approach for the integration of inline sensors with an electronic and fluidic bus system is presented in the following. This development aims at the further intensification of plant compactness using the above-mentioned *backbone* concept of the μChemTec consortium [24]. The goal was the reduction or complete avoidance of cables and wires which connect sensors in a plant to the peripheral data processing system by the supply of an internal wiring scheme, the internal electronic bus [86].

The Match-X concept

[R 18] Instead of developing a new electronic interface, the micro electro mechanical concept of the Match-X initiative, represented in the consortium by the Fraunhofer Institute, mechanical reliability and automation were integrated. The Match-X inter-

Figure 4.76 Selection of geometries of the Match-X interface standard [126] (with courtesy of FhG-IZM).

face defines electronic standards for packaging and interface geometries in order to reduce the production costs of miniaturized sensors and actuators, the so-called micro electro mechanical system (MEMS). By adopting this geometric standard for the electronic part of the backbone concept (Figure 4.76), future integration of all sensors and actuators of the Match-X concept will be available [126, 127].

Principle of the fluidic and electronic backbone concept

[R 18] [A 13] The basic principle of the backbone concept was parallel guidance of fluidic and electronic streams. A backbone consists of single standardized elements each incorporating a number of pipes and housing parts and can also be equipped with internal trace heating for the fluidic channels. Figure 4.77 gives a comparison between a fluidic/mechanical backbone element and its archetype – a human vertebra.

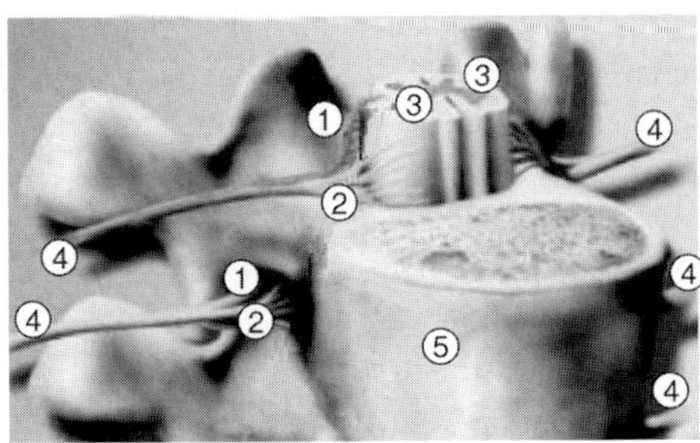

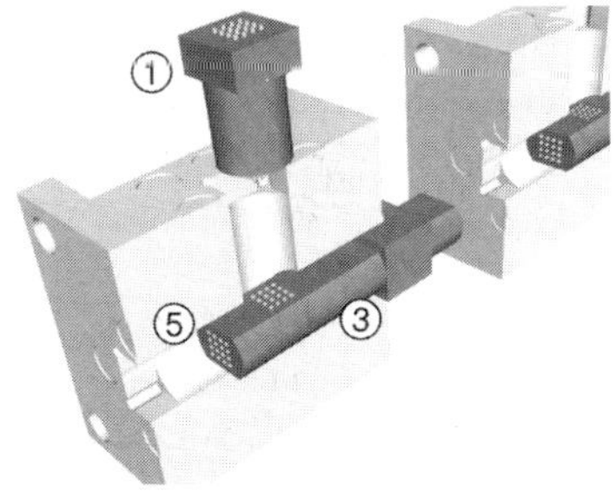

Figure 4.77 Left: archetype human vertebra; 1, sensor nerves; 2, motor nerves; 3, spinal cord; 4, spinal nerves; 5, bone structure. Right: backbone element as electro mechanical equivalent to a spinal cord: 1, sensor signals; 3, signal bus; 5, mechanical support) [86].

Sensor-adapter plate

[R 18] [A 13] A state-of-the-art plant concept must permit controlled process operations. This demands the integration of sensor and control equipment directly into the process. In a first attempt to obtain process parameters actually the integral values at the entry and the exits of devices such as heat exchangers or reactors are

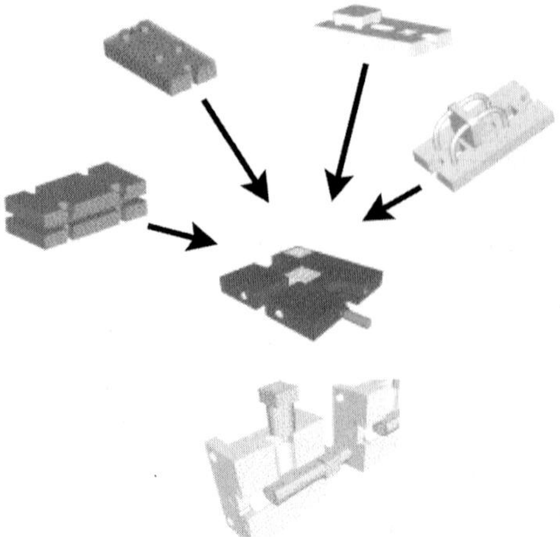

Figure 4.78 Backbone elements, adapter plate incorporating sensors and standardized process technology elements [86].

sufficient information. The first step of sensor integration here was to position a so-called sensor-adapter plate between the backbone and the devices.

Figure 4.78 shows such a combination of **standardized** elements together with an adapter plate and a backbone element. The adapter plate allows the integration of, for example, pressure and temperature sensors.

IC chip integration

[R 18] [A 13] A future extension will be the integration of chips for the conversion of analog sensor signals to digital signals directly neighboring the sensor (Figure 4.79). The electrical data and power bus are routed through the whole backbone system. By using the electrical bus, the control system communicates with the sensors.

Electronic bus system and data transmission

[R 18] [A 13] The data transmission of the signals will be executed by an electronic bus consisting of standardized elements as shown in Figure 4.80. The aperture in the center of the plate is reserved for the electrical connection. This connector system allows horizontal connection from the beginning to the end of the complete

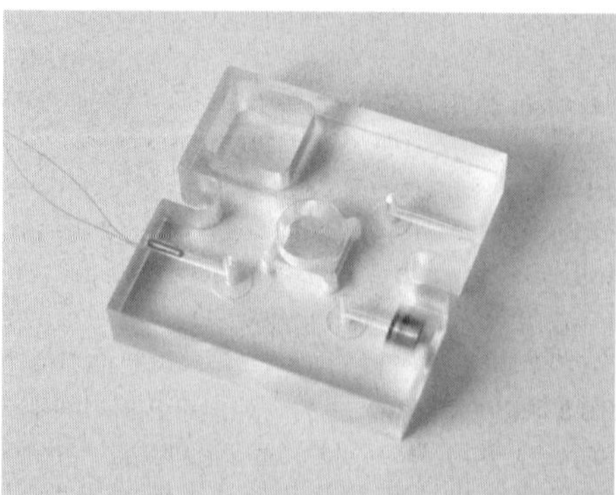

Figure 4.79 Model of sensor adapter plate with temperature (left) and pressure (right) sensor (the cavity in the back of the plate is provided for the future integration of an A/D-converter chip) [86].

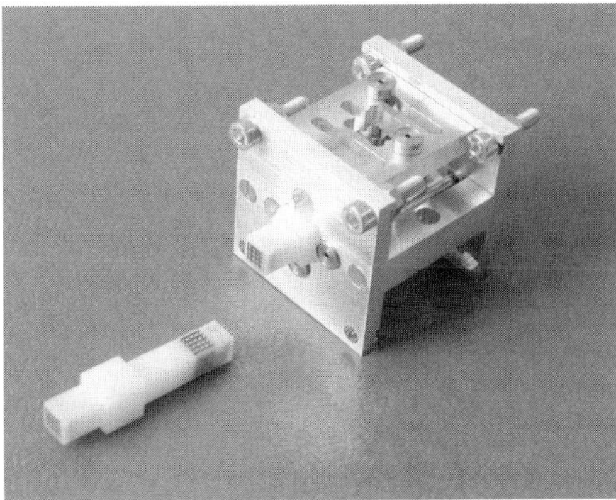

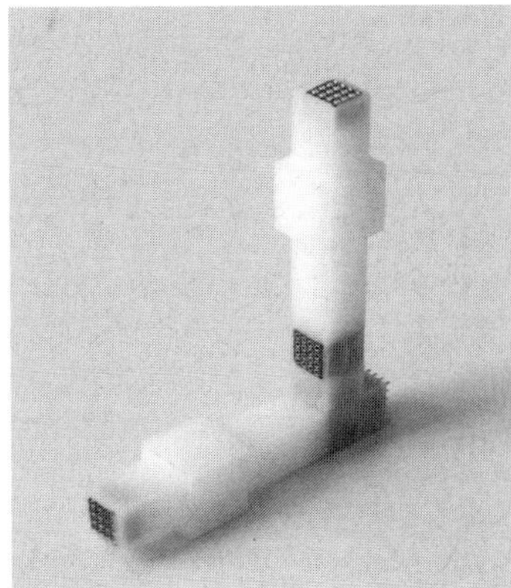

Figure 4.80 Principle of backbone element with models of internal bus connectors (left) and the combination of the same bus connectors for an external bus set-up (right) [86].

backbone, whereas vertical connection is either used for a branch connection or for contacting the sensors and the miniaturized sensor electronics with the bus. These bus connectors can be positioned inside the mechanical backbone but also – in the case of high-temperature applications – outside the backbone as a separate electronic bus. Consequently, cables are no longer installed outside the chemical micro structured reactor plant with the exception of power cables for electrical heaters.

Integration of flow cells and inline sensors

[R 18] [A 13] This bus concept allows the integration of standardized flow cells and inline sensors as described above. Currently, the integration of a commercial flow cell coupled to an inline NIR spectrometer is under development.

4.11.2.11 Automation 14 [A 14]: Booster Pumps

A tool often applied in industrial plants but rarely found in the micro world is the booster pump. Typically in a micro structured reactor plant the pressure is imprinted by dosing pumps at the entry to the plant. The pressure or, more exactly, the kinetic energy of the fluid is then reduced downstream by pressure losses in the pipes and micro channels. This means that the pressure level in the first part of the plant is very high and the devices here are operated under a much higher pressure than is actually necessary for conveyance, a fact which influences, of course, the manufacturing costs. An alternative mode of operation could be to increase the pressure stepwise by supplying a number of small pumps in the process flow instead of a single large pump at the entry to the plant. As the large dosing pump is often not operated at the optimum operational set-point, concerning pump performance (fluctuations), this operation also delivers a large shear rate and possibly heating of the product. This mode of operation can also be a disadvantage if shear-sensitive products are conveyed, such as plastic melts or olefins.

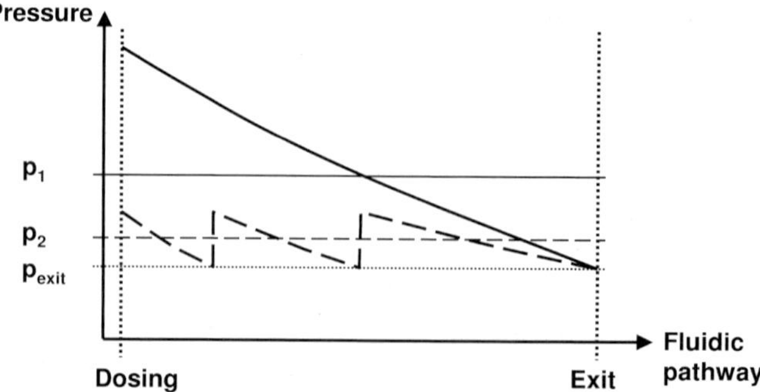

Figure 4.81 Influence of booster pumps on the pressure course of a plant (the average pressure p_2 of the booster pump arrangement (broken lines) is much lower than the average pressure p_1 in the initial pressure arrangement) [128] (by courtesy of Carl Hanser Verlag).

Commercial booster pumps

In polymer technology, therefore, so-called booster or inline pumps (usually tooth gear pumps) are applied [128]. Each time the pressure level decreases below a certain limit, usually the pressure loss of all the downstream devices, an additional booster pump is integrated in the product pathway. This type of pressure management changes the mode of plant operation from an initial imprint of a large over-pressure at the beginning which is then simply decreased downstream into a pressure function which resembles a saw-tooth function with a lower average pressure (Figure 4.81).

Tooth-ring gear pumps

[A 14] Micro fluidic pumps with dimensions small enough to be connected inline with the process flow of a micro structured reactor plant are commercially available [78]. The tooth-ring gear pumps of the mzr-series, for example, convey up to 17 l h^{-1} at pressures up to 80 bar. They can be used either as dosing pumps or as booster pumps inline in a hybrid plant or in a micro structured reactor plant with open architecture. The use of the stepwise pressure increase reduces the average plant pressure as indicated schematically in Figure 4.81.

4.11.3 Process Simulation

Software tools are applied in every step of process development. Tools for individual *reactor* simulations such as computational fluid dynamic simulations are not the topic in this chapter. These tools supply only numerical data for specific defined reactor geometry and defined specific process conditions. A change of parameter would demand a complete recalculation, which is often a very time-consuming process and not applicable to a parameter screening. Methods for reactor optimization by CFD are described in detail in the first volume of this series. Tools for *process* simulation allow the early selection of feasible process routes from a large number of alternative routes. Software for process control increases plant safety and also the economic benefit with respect to product quality and consumption of raw material. Management tools exist for the control of the whole production process from buying via production to the sales department. This so-called supply-chain management makes the production process transparent for the management board. With some restriction to the latter, all of these tools are applied in micro reaction technology and in the following special attention is given to the specific needs of small-scale plants.

Economics introduced the need for the application of process simulation tools. In 1979, BASF and Dow introduced a combined miniplant simulation tool approach [29, 129]. Their miniplant technology was developed as an alternative to expensive pilot plants. The intention was to jump directly from the laboratory to the production plant with a combination of software tools and experimental work executed in a miniplant. The complete process must be automated as much as possible, as the costs for the operation of the miniplant normally exceeds the costs for the erection of the plant. Process data obtained in the experiments are used to check the assumptions of the simulation model, which then will be used for further up-scaling. The whole process design is influenced, as already in the beginning of the design phase the software permits the selection of appropriate process routes from a large number of alternatives. Both companies used their own software programs for this purpose; the type of software was not described. Today this in-house software is more and more being replaced by commercially available software, for example, from Aspen Technologies [61].

Macro-scale tools for process simulation are not readily transferable to micro applications as they normally neglect the dependence of process parameters on the location inside the reactor and on the reactor geometry [13]. Only with neglect of the internal geometry does the principle of always using the same standardized software modules work, which then allows the calculation of a single unit operation without the necessity of considering boundary wall effects and reactor geometries. In macro-scale reactors the volume/wall ratio is much larger than in micro structured reactors. Hence the process in a large reactor is generally not much influenced by the walls of the reactor. This is totally different on a micro scale. One of the main advantages of micro devices is just their ability for fast quenching via the walls, so this effect must be considered. Another example is the influence of the wall material on process conditions, as exemplified in a heat exchanger [13, 130]. The performance of a cross-flow heat exchanger depends considerably on the temperature gradient between the hot and cold streams. The heat transfer efficiency drops drastically if this gradient decreases owing to axial heat conduction inside the wetted wall. In micro devices there is an optimal heat conductivity coefficient for the wall of approximately 0.5 W mK^{-1} [130]. This value is the optimum between undesired axial conductivity and desired vertical conductivity through the wall. In large devices the efficiency is influenced by the vertical heat conduction, in which case the wall conductivity must be as high as possible. In standard process simulation tools axial heat conductivity is ignored to reduce calculation time and convergence problems.

Sequential modular approach

Commercial software tools for steady calculations often use a so-called *sequential modular approach* [16]. The process flow sheet is subdivided into single unit operation modules which are processed consequentially and linked via the exiting and entering streams to neighboring modules. Such a sequence of modules has to be solved iteratively as, for example, recycle streams from a consecutive module which enter a preceding module are at first not defined. This approach gives insights into the individual units but delivers poor convergence in processes with a large number of recycle loops. To account for the poor convergence, recycle loops are cut and start values at this point are first given arbitrarily and then further optimized iteratively.

Equation-oriented tools

So-called *equation-oriented tools* set up a set of equations in a large sparse matrix. The matrix contains a significant number of zero-valued elements deriving from the fact that a cross-link between modules is given by streams to directly neighboring modules. These simultaneous linear equations are solved by Gaussian elimination (direct method) or by iterative procedures (indirect method) [131]. The advantage here is fast and safe convergence behavior on account of a lack of insights in the single units. This in consequence disables individual optimization of unit operations.

4.11.3.1 Simulation 1 [S 1]: Micro Reaction Simulation Toolkit

The Fraunhofer Alliance Modular Microreaction System (FAMOS) is currently working on a micro reaction simulation toolkit (Figure 4.82) with special attention to micro-scale phenomena [124]. The virtual toolkit comes with a physical micro reaction toolkit. The MicroSim software reflects the process by considering reaction conditions and reactor geometries. Of course, this approach on the other hand limits the software to the dimensions and geometries of the reactors supplied with the physical toolkit.

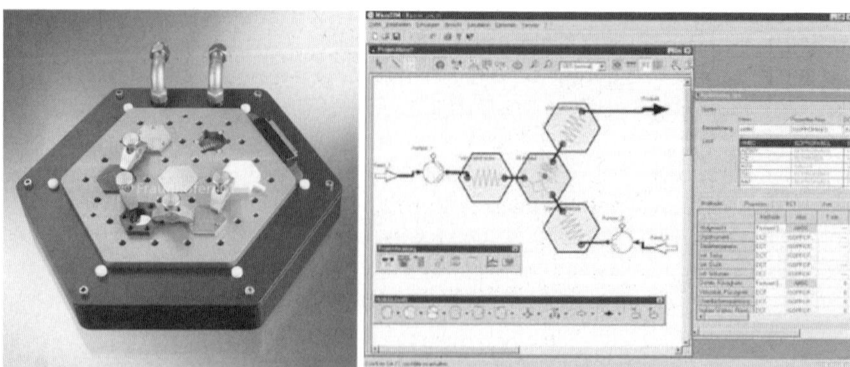

Figure 4.82 Modular micro reaction system of the Fraunhofer Alliance and software process simulation toolkit [124] (by courtesy of FhG-Allianz FAMOS).

CFD simulation

[S 1] As a mediator between CFD calculations and macro-scale process simulations, the reactor geometry is represented by a relatively small number of cells which are assumed to be ideally mixed. The basic equations for mass, impulse and energy balance are calculated for these cells. Mass transport between the cells is considered in a network-of-cells model by coupling equations which account for convection and dispersion. The software is capable of optimizing a process in iterative simulation cycles in a short time on a standard PC, but it also requires experimentally-based data to calibrate the software modules to a specific micro reactor.

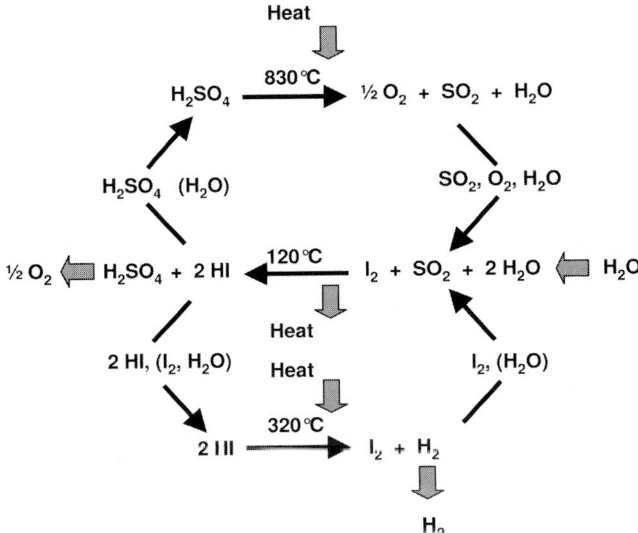

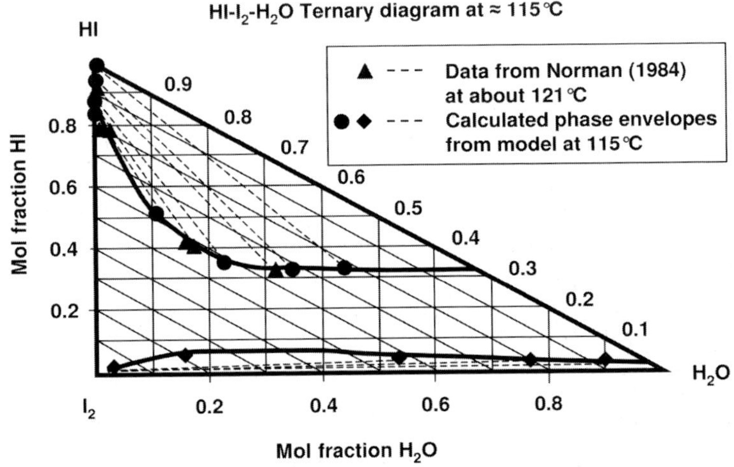

Figure 4.83 Sulfur–iodine cycle and simulated phase behavior of the ternary system $H_2O/HI/I_2$ [132].

4.11.3.2 Simulation 2 [S 2]: Steady-state Process Simulator

Such a separation of a process into single unit operations as just described is not always possible and also not always necessary. Highly cross-linked processes can be simulated using an approach described with the example of a thermochemical reaction for the production of hydrogen [132].

Combination different models

[S 2] With the steady-state process simulator Aspen Plus®, thermodynamic models for the sulfur–iodine cycle given in [132] are combined with chemistry models which describe the dissociation and precipitation reactions.

Simulation results

[S 2] The result was a prediction of phase behavior of the ternary system $H_2O/HI/I_2$ given in Figure 4.83 together with experimental data. The direct application of this tool to micro technology is uncritical as this simulation only considers thermodynamic and chemical model assumptions. It does not make any assumptions such as the neglect of axial heat transfer.

4.11.3.3 Simulation 3 [S 3]: Reactor Modeling for a Homogeneous Catalytic Reaction

As a more critical example concerning the transfer of macroscopic modeling to micro-scale applications, the following example of a simulation of a homogeneous catalytic reaction is described [133]. This example also represents a typical approach in process simulation if a new reactor model or a model for a new unit operation has to be developed.

Kinetic model

[S 3] A process for the oligomerization of ethylene for the production of linear α-olefins had to be developed in stirred tank reactor assuming isothermal conditions and was executed in the kinetic regime. The latter was assured by increasing the rotational speed of the impeller until the rate of reaction did not increase further. The autoclave reactor was heated by an external blanket and supplied with cooling water circulation through an internal coil.

Reactor modeling results

[S 3] The rate constants for the proposed Ziegler–Natta-type kinetics were regressed from experimental data. As a result of the regression, a new model for a unit operation was elaborated and used for the prediction of the behavior of a series of flow reactors. This model can in future also be used in combination with already existing unit operation models to simulate the whole process flow. Isothermal behavior as in this example cannot always be assumed on a micro scale even if the small dimensions seem to imply this behavior (see remarks above). This also holds true for the definition of the mass transfer mechanism. If diffusion is not fast enough, the use of mechanically agitated mixing in order to increase mass transfer rates is normally not applicable on a micro scale. Mixing in non-agitated static mixers improves if the flow velocity increases, but this on

the other hand also influences the flow characteristics throughput and residence time.

These examples underline the fact that macro-scale process simulation tools such as Aspen Plus® will have to be supplemented by micro-type unit operations as introduced by the FAMOS initiative which consider the location of a fluidic cell in the device and does not assume a perfect mixing, piston flow or uniform heat transfer coefficient [13].

4.12 Process Engineering

In the 1980s and 1990s, process operations were largely influenced by process automation. This development can also be observed in process engineering. Software-driven methods for the design of processes replace manual procedures. Especially for the new microstructured reactor plant technology, the chances are that automated engineering tools will establish parallel to the micro structured reactor plant development and application. Computer-aided engineering (CAE) environments already assist engineers in chemical process development and during basic and detailed engineering. Bayer developed a new CAE integrated environment for the complete engineering workflow [134]. A new development, a tool for pipe routing, will be presented in the next section.

4.12.1 Basic Engineering

A conceptual planning of a chemical plant was first undertaken in 1898 by Duisberg [16, p. 317]. Before that a chemical plant usually was adapted to current needs, which seldom resulted in a well-designed and cost-efficient production facility.

The process design step is called basic design or basic engineering in contradistinction to detailed engineering, which means the elaboration of specific planning functions for apparatus design, civil engineering, piping, erection and commissioning. The detailed engineering phase thus occupies a much larger part of process engineering than the basic design.

The methods and the expenditure of basic engineering for a complex process executed in a micro structured reactor plant cannot be distinguished from those for an industrial plant. A flow chart for a small-scale process indicates a similar complexity (Figure 4.84).

The costs of erection and commissioning of even a very complex micro structured reactor plant are, of course, comparably small. This means that the basic design of a micro structured reactor plant is extremely expensive and attention should be given to minimizing costs, for example, by the application of calculation sheets, the standardization of complete process units or the use of process simulation or planning tools such as AspenPlus®. The reduction of this cost factor remains a future task of micro technology.

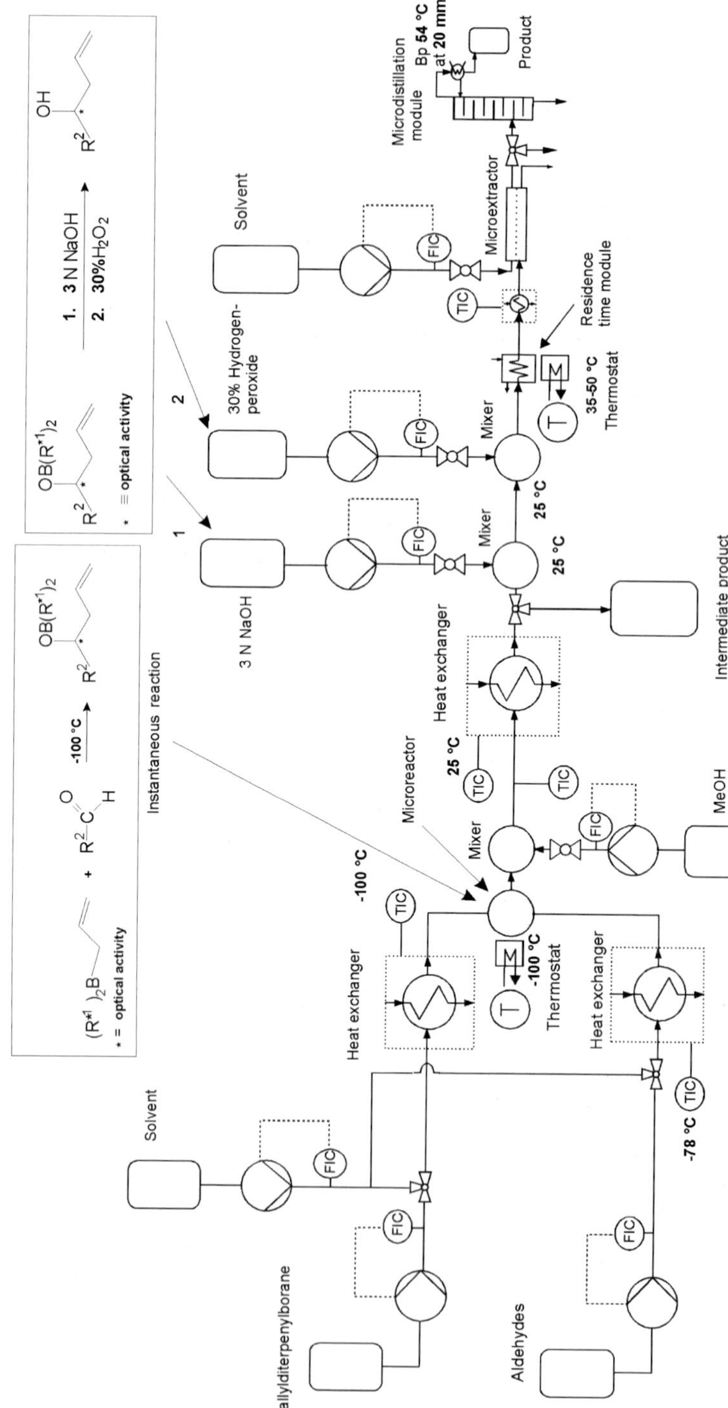

Figure 4.84 Process flow chart for asymmetric organoborane synthesis [87] (by courtesy of DECHEMA).

4.12.2
Detailed Engineering

2-D pipe schemes are still widely used despite the existing 3-D CAD piping tools. The reason is simply that these schemes can be printed and delivered to customers who do not possess 3-D piping visualization facilities. Nevertheless, plastic models formerly also used for the visualization of complex chemical plants certainly belong to the past. Today, a virtual reality plant allows the 'inspection' of a plant in form of a 3-D virtual 'walk'. Obstacles will be visualized; space for inspection and service can be checked. In the past, 3-D planning happened to be a manual CAE-assisted procedure. Since a few years automated 3-D pipe routers have also existed [135].

4.12.2.1 Engineering 1 [E 1]: Computer-aided Plant Design Software

For a long time, these auto-routers have been a standard tool in the electronics industry. They are applied in the design phase of electronic printed boards. An electronic router calculates the most efficient layout of an electronic pathway considering obstacles and predefined constraints.

[E 1] This is exactly were the new software tool CAPD (computer-aided plant design) of the University of Dortmund intervenes, using the obvious analog between fluidic and electronic currents. The tool for plant layout consists of four modules: equipment modeling, layout modeling, pipe routing and analysis (Figure 4.85).

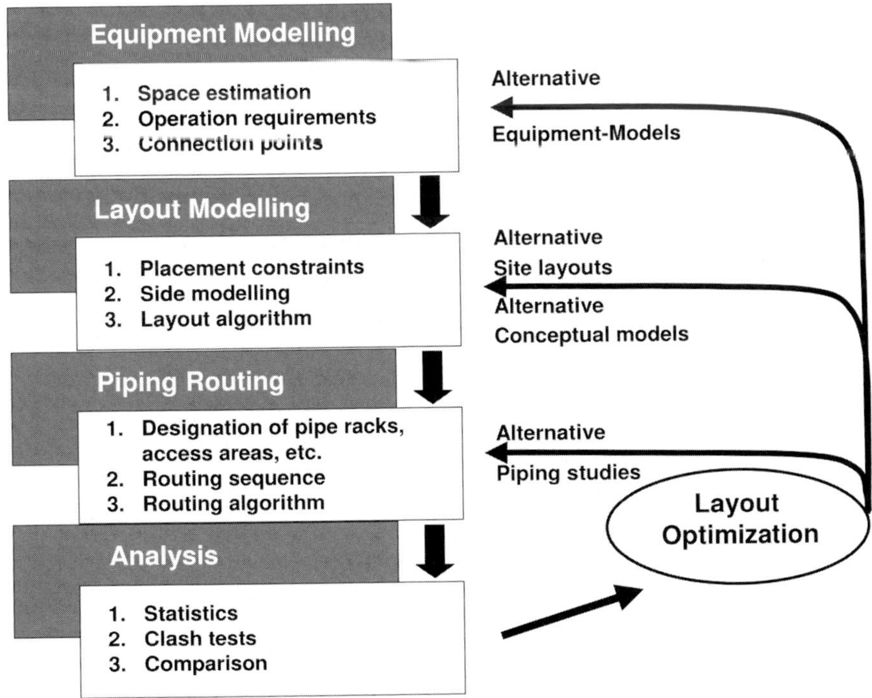

Figure 4.85 Structure of the software tool CAPD (computer-aided plant design) [135].

Step 1: total volume estimation

The procedure starts with an estimate of the total volume of each piece of equipment and also considers enough space for erection and maintenance (step 1).

Step 2: placement constraints

[E 1] Next, values are set for the router to distinguish space where pipes are not allowed. Additional constraints such as necessary NPSH values for pumps or specified gravity flow are defined. Placement constraints are divided in relative constraints (for example, the device must be located next to a heat exchanger) and absolute constraints (for example, the device must be erected in the basement). These constraints result in pulling and pushing 'forces' between the devices [136]. The solution algorithm now tries to minimize the total energy of all these forces (step 2).

Step 3: creating piping studies

[E 1] The third module considers pipe racks, access areas and prohibited areas and creates piping studies. The amount of fittings used and pipes used is indicated in a report (step 4). As final result, a 3-D virtual model of the plant is then obtained (Figure 4.86).

This 3-D layout model is a valuable draft but it has to be revised during detailed engineering. The tool is important for the micro world if in the future micro structured reactor plants become more and more standardized. In this case, standardized autorouter tools could help to automate the time-consuming part of detailed engineering in complex micro structured reactor plants.

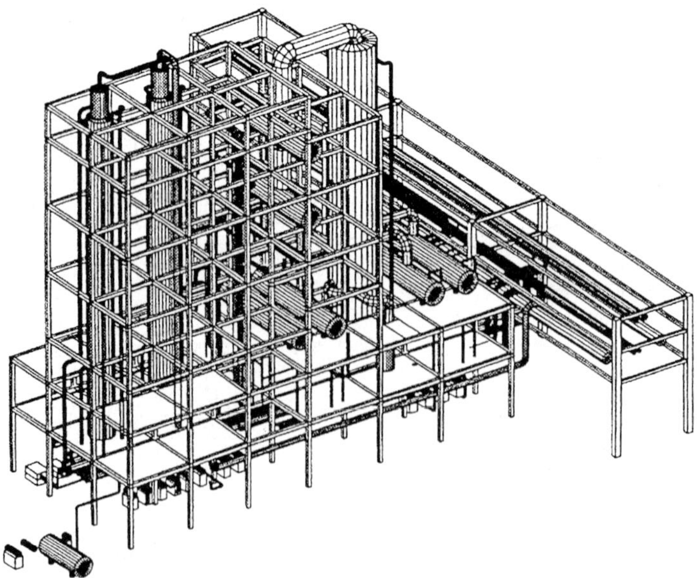

Figure 4.86 Virtual 3-D model of an industrial project [135].

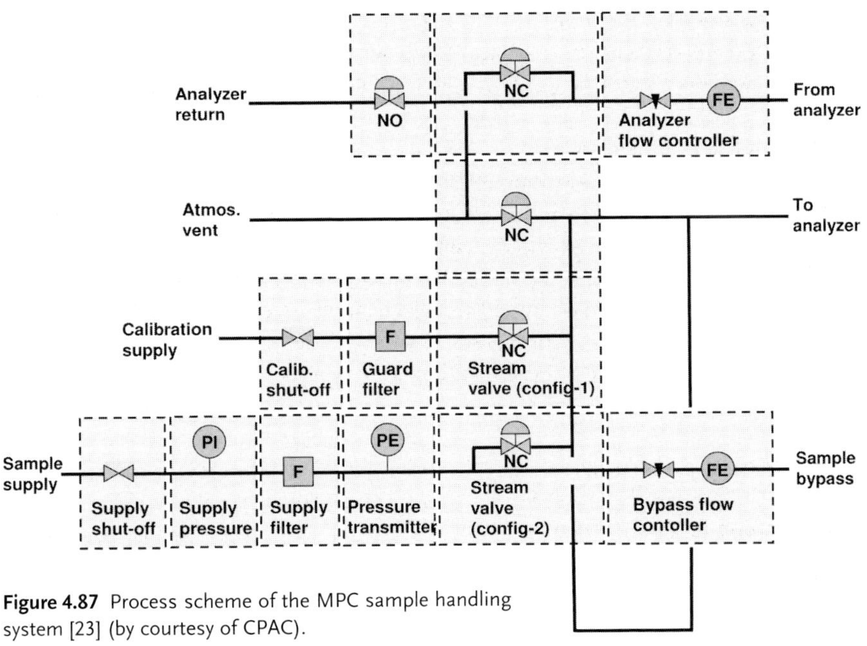

Figure 4.87 Process scheme of the MPC sample handling system [23] (by courtesy of CPAC).

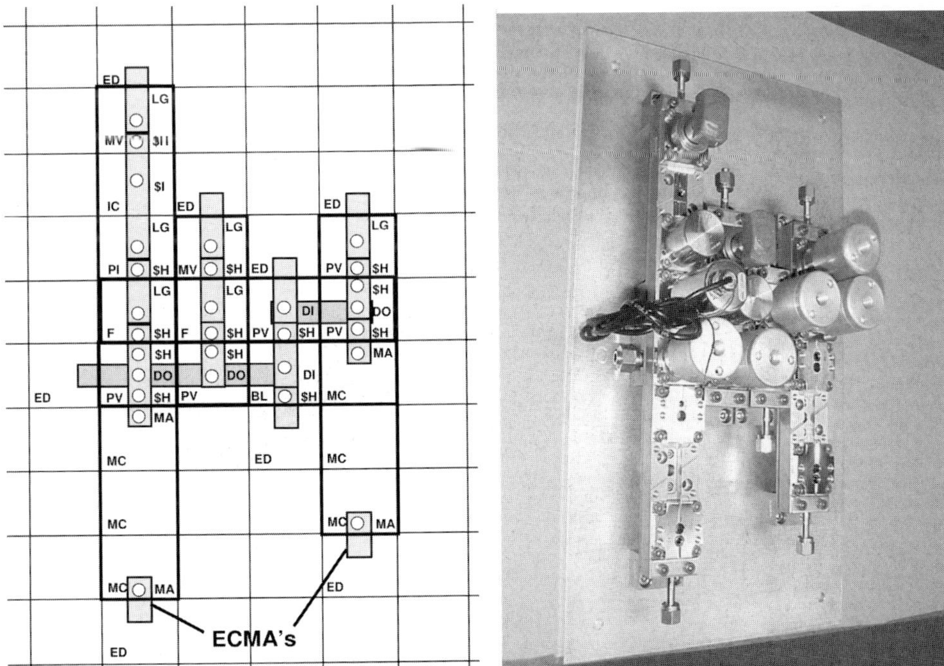

Figure 4.88 Software pipe-installation plan and photograph of the plant set-up of the MPC system [23] (by courtesy of CPAC).

4.12.2.2 **Engineering 2 [E 2]: Process Analyzer and Sample-handling System**

The tool introduced in the following might be applicable as long as 3-D tools are not available for micro technology and also for plants with a lower degree of complexity.

[E 2] [R 17] The process analyzer and sample-handling system MPC introduced by Swagelok also offers the possibility of performing as a process toolkit. Sensors are already supplied (see Section 4.11.2). Micro reactors are still not available from the supplier but a software configurator exists with pipe layout functions [23].

A process scheme (Figure 4.87) consists of standardized surface-mount components and can be arranged on the screen to a customized system. The software also delivers a summary considering necessary seals and pipe fittings which can be used for ordering directly (Figure 4.88).

Process presentation

[E 2] The MPC concept, as originally developed for sampling and analysis of gases and not for chemical processing, consists of a single channel suspended and sealed in a mechanical support. It is arranged in a plane configuration which permits the presentation of the process in a two-dimensional pipe scheme.

4.12.2.3 **Engineering 3 [E 3]: The μChemTech Piping Concept**

The concept of the μChemTec consortium [24] was developed for chemical processes and is equipped with multiple channels in a mechanical frame construction. This concept allows three-dimensional branching similar to industrial plants. A two-dimensional representation of the pipe scheme thus demands a more abstract visualization method.

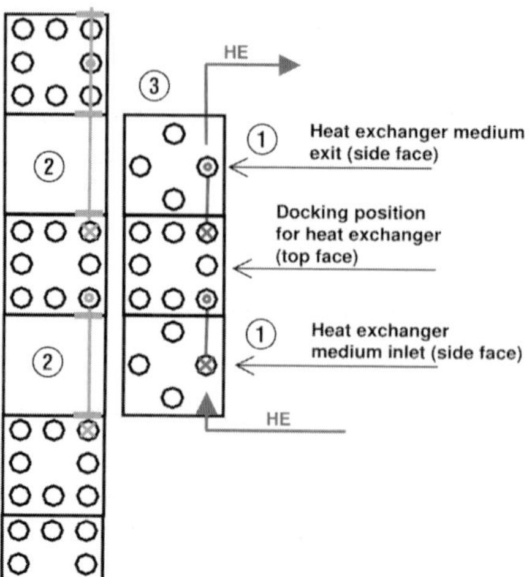

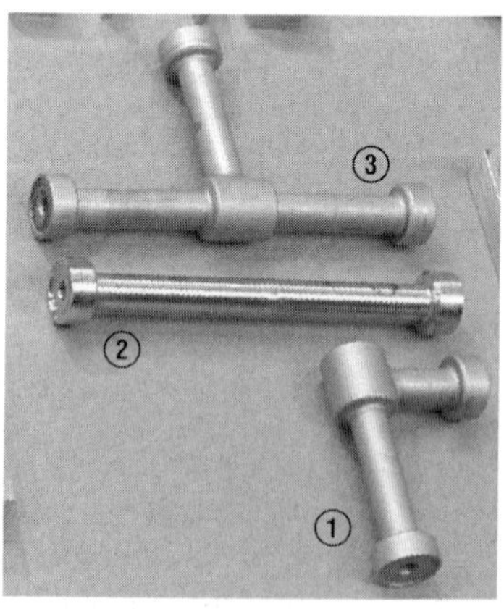

Figure 4.89 One layer of a 3-D pipe scheme of the μChemTec system [24] (by courtesy of microchemtec).

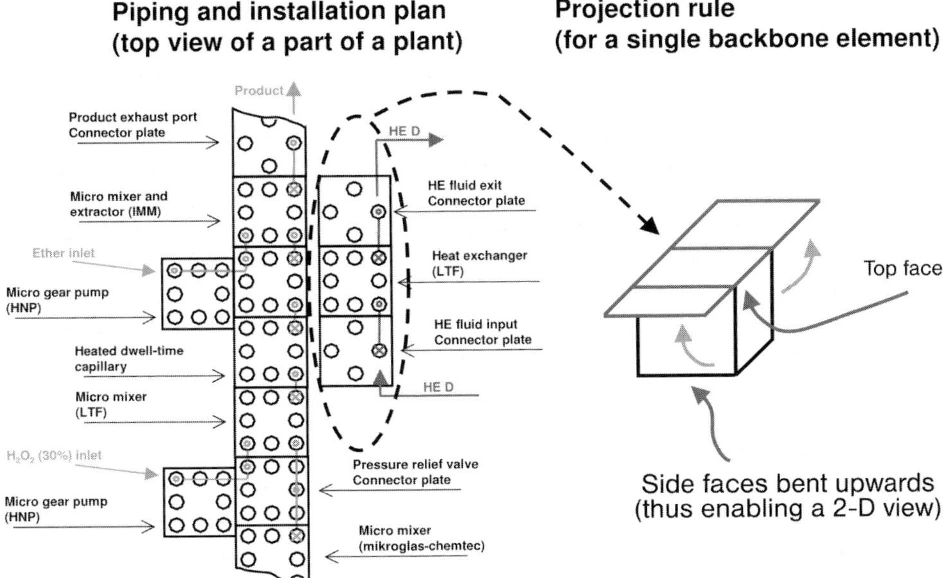

Figure 4.90 Projection of a 3-D cube in a 2-D flow scheme of the μChemTec system [24] (by courtesy of microchemtec).

Layered flow presentation

[F 3] First, the 3-D plant is divided into horizontal layers as indicated in Figure 4.89. The standardized pipe inlays (elbow, straight pipe and T-piecc) arc denoted ①, ② and ③, respectively. The plane-leaving vertical streams are marked with × or o and empty positions not occupied by pipe inlays with O. The fixed pitch dimensions divide the plane into a square grid where every square represents a single cubic element which can be equipped with a number of pipe inlays.

External streams enter the plant at the side faces of the cubic elements. These side faces are projected into the drawing plane (see Figure 4.90) to indicate the position of the respective pipe inlay which transports the entering streams, for example the stream HE indicated in Figure 4.89. All other side faces are not shown for simplicity. These simple rules allow the conversion of the 3-D pipe scheme into a 2-D layered flow representation.

4.12.3
Scale-up, Flow Distribution and Interface to the Macroscopic World

Micro Structured Reactor Scale-up

Before scale-up in micro structured reactor plants is discussed, a short introduction to problems which occur during scale-up from a miniplant to a production plant will be given. Experimental correlations found in a miniplant will have to be transferred to the geometric scale of a production plant. Some scale-up factors which have been tested successfully for a number of devices are given in Table 4.10 [16, 137–139].

Table 4.10 Factors for safe scale-up of the mass flow from a miniplant to a production plant [137].

Distillation and rectification	1000–50 000
Tubular bundle reactors, homogeneous tubular reactors, stirred tank reactors	> 10 000
Bubble column	< 1000
Extraction	500–1000
Fluidized bed reactors	50–100
Drying	20–50
Crystallization	20–30

It is obvious from Table 4.10 that scale-up is more and more critical if either heterogeneous phase systems or solid phases are incorporated. Scale-up is also strongly influenced by heat and mass transfer if the construction of the devices has a dominant effect on the transfer [137].

Internal vs. external numbering-up

Numbering-up can be performed in two ways, either externally or internally. *External numbering-up* refers to the connection of many devices in a parallel fashion [9, 53, 140, 141] (see also Figure 1.4 in [9, p. 7]). A device in the sense used here is defined as a functional element, e.g. a micro-mixing flow configuration such as an interdigital feed array with mixing chamber attached, which is encased and has outer connections, for fluid supply and withdrawal. Connection of devices is therefore achieved via their outer connections, which most often follow commercial standards. Accordingly, conventional tubing with standard process-control equipment may be used here, but seems to suffer from fluid equipartition problems as mentioned above. For this reason, the development and use of professional specially-splitting tools were proposed. The construction of such a six-fold liquid splitting unit is given in Figure 4.91.

External numbering-up is numbering-up in the truest sense. Virtually, the complete fluid path is repeated. This resembles also the real meaning of scaling-out.

Internal numbering-up means the parallel connection of the functional elements only, rather than of the complete devices. These elements are grouped in a new way, usually as a stack, and are encompassed in a new housing. This housing typically contains one flow manifold and one collection zone, most often having a simple design like a header of a diffuser. Although often overlooked, internal numbering-up actually is state of the art. To name only a few realized examples, a mixer array with 10 parallel interdigital units [93], a gas/liquid contactor array with 10 packed beds [142] and micro heat exchangers with hundreds of parallel platelets [143] were realized. The internal numbering-up of the last device type, for instance, permits a throughput of up to 7 t h^{-1} water flow [143]. Hence one device, of size from a shoebox up to a computer, may be sufficient for a complete production.

To give an example of an internally numbered-up device and to illustrate the concept, a new IMM development is shown in Figure 4.92. A novel micro-flow heat exchanger has 6685 parallel micro channels of 250 μm depth, 2 mm width and 240 mm length, giving a gas throughput in the range of about 1 m^3 min^{-1} at a pressure drop of about 100 mbar.

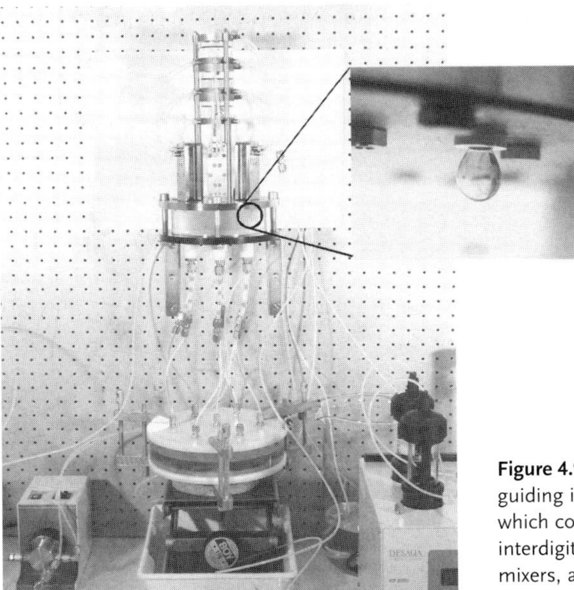

Figure 4.91 Demonstration of flow guiding in a liquid flow-splitting unit which consists of three tanks, six interdigital separation-layer micro mixers, a multiple channel segmenter and six tubular sections [140].

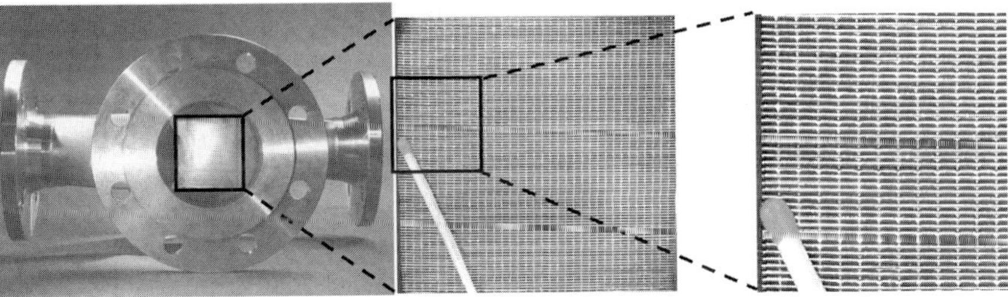

Figure 4.92 Photograph and detailed view of 6685 parallel micro channels which are numbered up internally to give a micro-flow heat exchanger of large capacity (source IMM).

External numbering-up benefits from true repetition of the fluidic path, and hence preserves all the transport properties and hydrodynamics, determined in advance for a single-device operation. The disadvantage is the need for a sophisticated monitoring and control system, in particular to achieve fluid distribution. Another drawback stems from the unfavorable ratio of housing material to active internal volume. Furthermore, the costs of the micro reactor fabrication are considerably raised owing to the increasing material demand and, more notably, owing to the need to structure the housings also multiple times.

In contrast, internal numbering-up provides, as the existing examples demonstrate, compact reactor architecture at reasonably high throughput. The fluid distribution is manageable, in particular using modern simulation. The disadvantage is that possibly a slight redesign of the fluidic path, sometimes even of the functional element, has to be made.

Obviously, internal numbering-up has more arguments in favor than external numbering-up. So why propose a liquid-splitting tool for external numbering-up? There are still good reasons to do so. Potentially, not all micro reactor processes may be amenable to internal numbering-up, e.g. for fluid dynamic reasons. Complex fluidic operations such as droplet formation may require very special fluidic architectures, which require single feeds. Also, external numbering-up may be chosen for simple, practical reasons, if the degree of parallelism involves only low numbers, e.g. not exceeding 10 devices. The expenditure for a new development of an internally numbered-up device may be prohibitively large, so that the drawbacks of an external solution are simply accepted.

Heat and mass transfer considerations

Heat transfer and its counterpart diffusion mass transfer are in principle not correlated with a scale or a dimension. On a molecular level, long-range dimensional effects are not effective and will not affect the molecular 'carriers' of heat. One could say that physical processes are dimensionless. This is essentially the background of the so-called Buckingham theorem, also known as the Π-theorem. This theorem states that a product of dimensionless numbers can be used to describe a process. The dimensionless numbers can be derived from the dimensional numbers which describe the process (for example, viscosity, density, diameter, rotational speed). The amount of dimensionless numbers is equal to the number of dimensional numbers minus their basic dimensions (mass, length, time and temperature). This procedure is the background for the development of Nusselt correlations in heat transfer problems. It is important to note that in fluid dynamics especially laminar flow and turbulent flow cannot be described by the same set of dimensionless correlations because in laminar flow the density can be neglected whereas in turbulent flow the viscosity has a minor influence [144]. This is the most severe problem for the scale-up of laminar micro results to turbulent macro results.

Wall influence

The Π-theorem helps to reduce the parameter space which has to be examined during the experimental evaluation. Scale-up of a tubular reactor, for example, from a micro channel with 1 g h^{-1} to a production scale with 10 kg h^{-1} (factor 10 000) should be simple, according to Table 4.10. This is certainly true if there were no wall influence. However, the influence of the wall cannot be ignored, as micro structures, owing to their large surface to volume ratio, are just used for efficient cooling and quenching by heat transfer through the wall. Here, this feature disables direct up-scaling. The dimensionless parameter space for heat transfer of a macro-scale production tube and a micro channel is different. The parameter space of the latter would have to be supplemented by the Nusselt number (laminar flow assumed).

With increasing tube size, the flow appears more and more adiabatic. There is actually no direct solution found for an adjustment of this scale-up problem. The walls in a micro structure cannot be made adiabatic in order to justify the neglect of vertical heat transfer through the wall and also inside the wall in an axial direction. An approximate solution could be the use of ceramic materials with low conductivity. However, even then, heat transfer through the wall is not completely negligible. For a strong exothermic reaction this resembles as closely as possible the situation in a large production tube in which problems exist with sufficient heat transfer through the walls, which in turn produces large temperature gradients between the wall and the core flow.

Scaling-up vs. numbering-up

The question arises of whether up-scaling could be replaced by up-numbering. As for a larger throughput in a flow reactor also a larger cross-sectional area is required, this area could also be obtained by combining a large number of channels in one reactor. This approach is very efficient because it bypasses up-scaling and the flow characteristics remain the same.

Linear scaling-up

An example from heterogeneous catalysis [145] demonstrates that the method of *linear up-scaling* can be even more precise than the conventional method of up-scaling in chemical reactor design. A micro-machined silicon micro reactor was etched with KOH to produce parallel straight-walled channels with channel widths from 5 to 20 μm. The platinum catalyst was sputter-coated to the silicon substrate. After coating, the reactor was covered either with a glass layer by anodic bonding to the silicon substrate or with a silicon cover by diffusion bonding. Such a unit was tested individually and then in combination with a multitude of identical units, which proved that the results from a single unit are transferable to a stack of units. It was stated, that this result stems from the fact that the internal reactor temperature is constant and thus the reaction rates measured are reliable. Also the contact between catalyst and reactants is not influenced by channeling or wake effects.

Parallelization of reactors

It should also be mentioned that the parallelization of reactors raises other problems such as the maldistribution of the reactants to the single units. While here usually passive devices such as flow restrictors are effectively applied, more severe problems are encountered if the process temperature should be controlled actively. One solution is to heat the whole micro reactor (assuming isothermal conditions due to large heat transfer coefficients) at a constant temperature controlled by the temperature of the flow at the reactor exit [13].

Embedding of sensors and actuators

MEMS technology also allows embedding of actuators and sensors in single reactor channels. Despite problems with temperature robustness, a solution must be found to transport the signals from the micro channels to the central process control system. To avoid a confusing cable set-up ('spaghetti conditions'), it is desirable to process the sensor data on-site, for example in an A/D converter, and to feed the digital data in a common bus system [13].

Geometric factors
Another reason for the preference of up-numbering is actually due to a manufacturing problem. Tolerances in silicon etching are absolute tolerances which means that with decreasing size of the channels the geometric errors increase. This influence is balanced if a large number of channels are combined.

Macro-scale flow distribution
In macro-scale distribution problems, a sparger or a perforated pipe distributor is often applied. Generally the effect of a sparger is simply that pressure over the length of the pipe distributor is small compared with the pressure drop across the small exiting pipes. A good compromise between an undesired large pressure drop and good distribution is given for turbulent flow [Eq. 4.2] [146, pp. 6–32]. The following rule delivers a maldistribution of smaller than 1.5%:

$$\frac{V_0}{V_i} = \sqrt{10}\, C_0 \qquad C_0 = 0.62 \tag{4.2}$$

The orifice discharge coefficient C_0 is equal to 0.62 only for large ratios between inlet/outlet velocities and inlet/outlet pipe diameters.

Slot-type flow distribution
More convenient in micro structured devices due to the planar manufacturing methods is certainly the slot distributor introduced in [146, pp. 248–381].

4.12.4 Calculation of Fluid Dynamics in Rectangular Channels

One reason to use micro structured reaction chambers is certainly the possibility of describing the fluid dynamic behavior in these structures due to the laminar flow regime. With the following calculations the reactive gas flow in a square micro structure with coated catalytically active walls will be studied in detail. The task was to find a channel arrangement and to calculate the residence time distribution of this arrangement numerically (Figure 4.93).

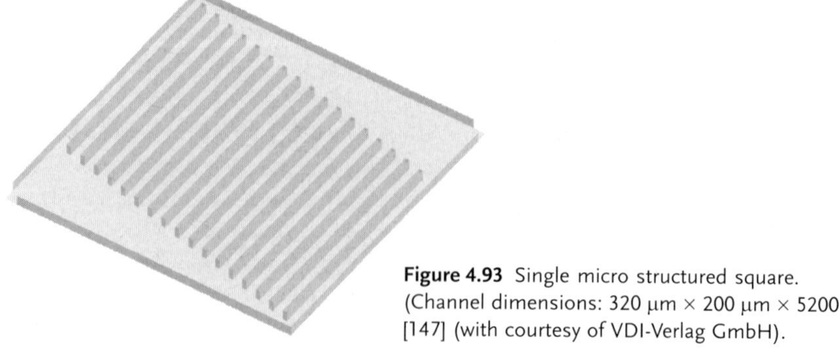

Figure 4.93 Single micro structured square. (Channel dimensions: 320 μm × 200 μm × 5200 μm) [147] (with courtesy of VDI-Verlag GmbH).

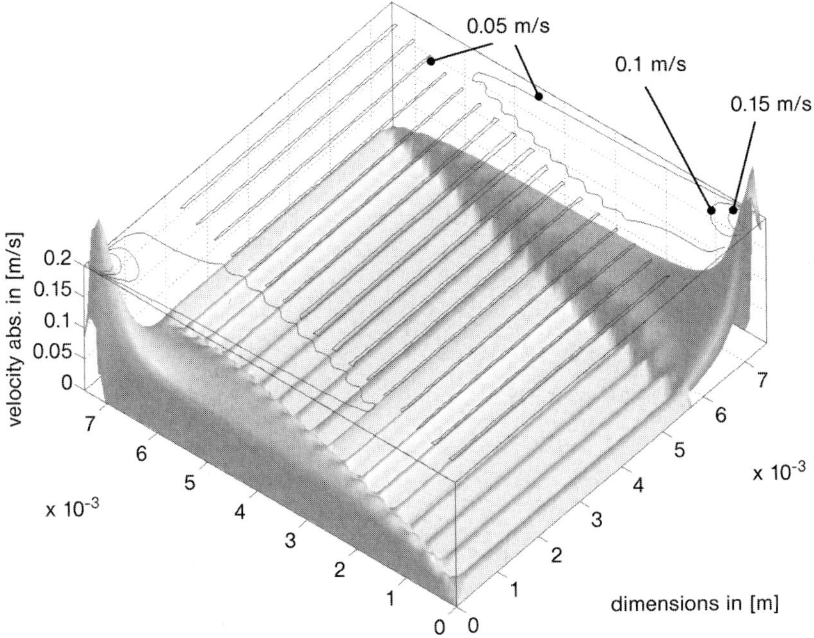

Figure 4.94 Numerically simulated surface of the absolute velocity in a structured square (the lines of constant velocity are marked above the profile) [147] (by courtesy of VDI-Verlag GmbH).

4.12.4.1 **Simulation of a Gas-phase Reaction**

The feed gas was a stoichiometric mixture of methane and oxygen which reacted irreversibly at the walls, releasing the reaction heat. To account for this reaction heat in the simulation, appropriate heat sources were added at the walls. This mixture enters the square at its left corner (Figure 4.94) from where it is delivered to the individual channels in the distribution area with a considerable velocity decrease. As soon as the gas reaches the channels, the velocity increases owing to the local narrowing. The velocity profile inside the channels is apparently parabolic. Behind the channel, the gas enters the collection area. Here the velocity increases again towards the velocity at the exit of the square which was the corner on the right side. Here it is already obvious that the walls of the channels narrow the gas flow strongly, thus causing an effective mechanism to distribute the flow evenly to the single channels. In order to obtain similar residence times for the reactants, it seems reasonable to coat only those portions of the square with a catalyst in which every fluidic element spends a similar amount of time. This would exclude the coating of the distribution and the collection area because here the residence time of the trajectories is not equal.

4.12.4.2 **Residence Time Distribution for Guided Flow in Channels**

The influence of channels, i.e. flow-guiding internal structures, also accounts for the overall residence time distribution in the square. This will be demonstrated by the observation of particles emitted at the structure inlet. The path of such an

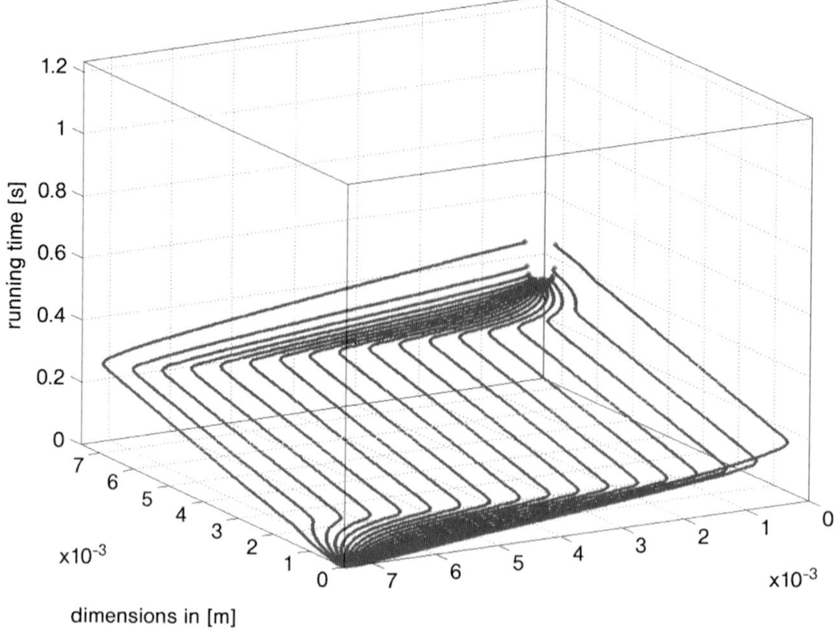

Figure 4.95 Trajectories in a structured well (visible are the particle paths through the respective channels) [147] (by courtesy of VDI-Verlag GmbH).

imaginary particle during its whole residence in the structure will characterize the fluid dynamics in the structure. These trajectories represent a convenient method to study the time-dependent movement of particles in a steady laminar flow. In Figure 4.95 the particles were emitted at the front corner of the diagram. The particles leave the structure at the corner in the back and here the different residence times of the particles become visible. The quasi 'discretisized' parabolic residence time distribution is similar to that of a tubular reactor but with a much closer distribution.

Typical results

The maximum relative deviation is 36%. The parabolic velocity profile in the micro channels will not influence the residence time distribution, otherwise the deviation would be much larger. The reason for this is the fast balancing of the concentration in a small channel by dispersion.

4.12.4.3 Residence Time Distribution for Non-guided Flow

The same calculation was executed for a structure with the same outer dimensions but without flow-guiding internal structures, i.e. channels (Figure 4.96). The particles were emitted at the same positions as before but they did not follow the same routes as before because no guiding elements were present. In fact some parts of the square were not even reached by these particles, showing that the residence time in these areas is theoretically infinite.

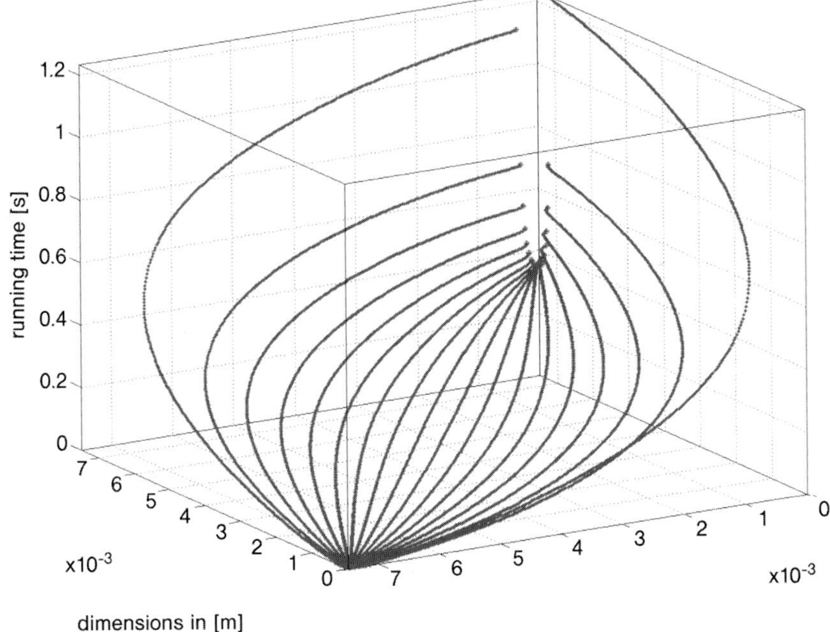

Figure 4.96 Trajectories in an unstructured well (the trajectories do not reach the corners) [147] (by courtesy of VDI-Verlag GmbH).

Typical results

The residence time of such a well is again best visible at the exit. The parabolic profile this time is much wider than for the structured case. The maximum relative deviation amounts to 233%, which is 6.5 times larger than for the structured well. This is important because it demonstrates that micro structures are indeed a means to obtain a narrow overall residence time distribution. The error introduced by manufacturing tolerances (estimated 5 μm absolute tolerance in a 320 μm wide channel) is 1.6% in width, a value which does not influence this evaluation.

4.12.4.4 **Calculation of Cumulative Residence Time Distribution**

The derivation of the residence time behavior of the single streamlines now allows the formulation of the cumulative residence-time distribution function $F(t)$ according to the following formula:

$$F(t) = \int_0^t E(t)\,\mathrm{d}t \tag{4.3}$$

The exit-age distribution function $E(t)$ is approximately obtained by following the experimentally often used method of 'backward differencing':

$$E(t) = \frac{\mathrm{d}F(t)}{\mathrm{d}t} = \frac{F_i - F_{i-1}}{t_i - t_{i-1}} \tag{4.4}$$

The division of the flow area into individual channels forces a discretization of the distribution function. The integration will be replaced by the sum of the contributions of every channel i:

$$F(t) = \sum_{i=1}^{N+1} E(t_i)\,\Delta t \tag{4.5}$$

The amount of the exit flow F_i with the respective residence time t_i can be calculated if the cross-section of the respective streamline and their mean velocity v_i is known. This is the case inside the channels (values given above):

$$F_i = \frac{\dot{V}_i}{\dot{V}} = \frac{v_i\,A_i}{\dot{V}} \tag{4.6}$$

4.12.4.5 **Calculations for Laminar- and Plug-flow Reactors**

A graphical representation of the cumulative residence time distribution function is given in Figure 4.97 for a structured well, a laminar flow reactor and an ideal plug flow reactor assuming the same average residence time and mean velocity in each reactor.

Obviously the characteristic distribution of the structured square, as expected, is much closer to the ideal plug flow reactor than to the laminar flow reactor. This desired behavior is a result of the channel walls, which are flow-guiding elements and pressure resistors to the flow at the same time. Two of the streamlines are projecting with a residence time of more than 0.4 s. These are the streamlines passing the area close to the wall of the distribution area, which introduces a larger resistance to these particles due to wall friction. This could, for example, be accounted for by a different channel width between the near wall channels and the central channels.

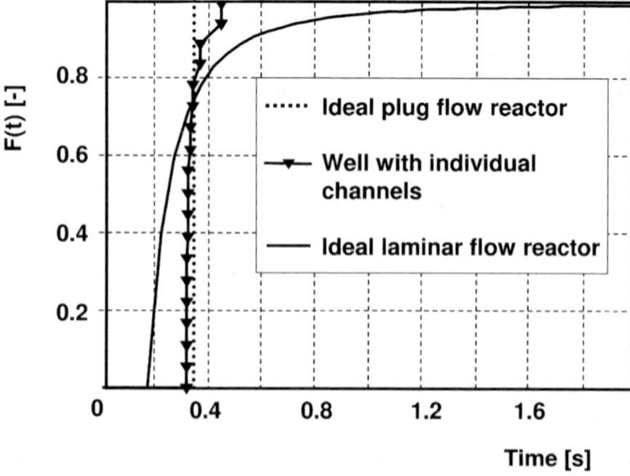

Figure 4.97 Calculated cumulative residence time distribution function for a multi-channel well, a laminar flow reactor and a plug flow reactor [147] (by courtesy of VDI-Verlag GmbH).

4.12.4.6 External Numbering-up and Flow Distribution

Most of the above-mentioned methods are used inside a reactor for distribution to individual channels. One of a few examples in which distribution between several devices is reported is described in the following [140, 141, 148–150]. This application is also a good example of the processing of fouling-sensitive reactions in micro devices. A process for the production of powder with a narrow particle size distribution had to be up-scaled. The devices developed for this purpose were two different types of micro mixers.

Separation layer mixer

The separation layer mixer used a barrier liquid to delay the mixing until the exit of the mixer is reached by the fluid. Here the actual mixing is processed outside the reactor inside the developing bubble (for more detailed information, see Section 1.3.13, *Droplet Separation-Layer Mixing*).

Impinging jet mixer

The second type of mixer uses two impinging jets exiting from two borings to contact the reactants (for more detailed information, see Section 1.3.33, *Jet Collision Turbulent or Swirling-flow Mixing*).

Both devices allow mixing without clogging the micro structures.

4.12.4.7 External Flow Distribution

The simplest approach of splitting one stream into six streams is the use of a fluidic element that directly connects one inlet with six symmetrically arranged outlets [140, 141, 148]. Commercial fluid distributors, however, suffer from flow maldistribution, as the smallest deviations in their manufacture have a large impact. Hence specially devices are required for flow splitting to micro devices. A liquid-flow distribution module was one first such device to be realized.

One solution to the problem mentioned above is not to rely solely on flow symmetry, but to achieve equidistribution by pressure-loss adjustment by means of flow resistors (see Figure 4.98) [140, 141, 148]. In the flow sequence consisting of a main stream tube, fluid inlet, damping tank, distribution tubes and micro device such as a micro mixer, the main pressure drop is nearly always on the last side. The separation layer mixer acts here as a pressure restrictor similar to the sparger mentioned above. This requires accurate control over structural precision of micro fabrication.

Tolerances in diameter will, according to the Hagen–Poiseuille law, influence the flow rate to the power of 4.

Since oscillation originating from the piston pumps is another cause of flow maldistribution, the flow distribution module comprises a damping element between the main stream and the six split sub-streams [140, 141, 148]. This element is a tank of cylindrical shape (Figure 4.98). Depending on the type of liquid processing, one, two or three tanks have to be used. Micro heat exchangers, micro mixers and delay-type micro mixers are examples of these three types. In the case of two or more liquids, stacking of the multiple tanks on each other is applied (see Figure 4.99).

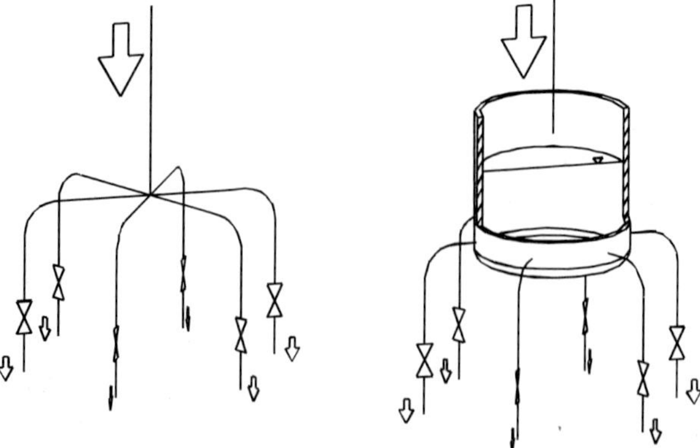

Figure 4.98 Generic presentation of two flow splitting concepts: both concepts are based on symmetry and use of flow resistors by choice. The left one is without a damping element; the right one includes a damping element [141].

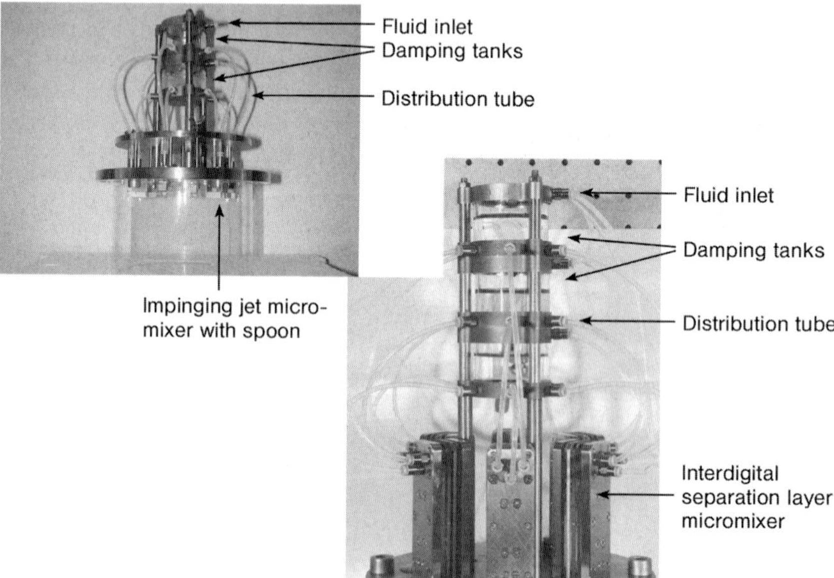

Figure 4.99 Stack of three liquid-flow splitting units comprising three damping tanks and three inlet and 18 outlet tubes. Some of the tubes are not visible since they are on the rear of the tank [141].

Typical results

Figure 4.100 shows the simulation of the individual volume flows through the six outlets of a damping tank for different pressure drops and a special injection mode into the tank, angular injection [141]. Clearly, the imbalance of the volume flows is strongly reduced for an increased pressure drop. For quantification of the imbalance in the flow distribution, the difference between two individual flow rates $\Delta j = j1 - j4$ divided by the average flow rate is chosen, where $j1$ and $j4$ are the volume flows through the corresponding outlets. The quantity $\Delta j/j_{\text{average}}$ as a function of the average pressure drop in the connected tubes is displayed in Figure 4.100. The inlet figure shows the inverse $(\Delta j/j_{\text{average}}) - 1$. The almost perfect linear behavior can be explained on the basis of the simplified model discussed above. According to the equation $\Delta j = j1 - j4$, $(\Delta j/j_{\text{average}}) - 1$ depends linearly on the flow resistance in the tubes, which in turn is proportional to the average pressure drop for a given flow rate.

As can be read off from the graphs, a maximum flow imbalance of 4% is achieved for a relatively low pressure drop of 0.22 mbar [141]. Furthermore, the linear behavior in the inset in Figure 4.100 allows an easy extrapolation to arbitrary small deviations from equal distributions. For example, a relatively high pressure drop of 120 mbar results in a flow imbalance below 0.01%.

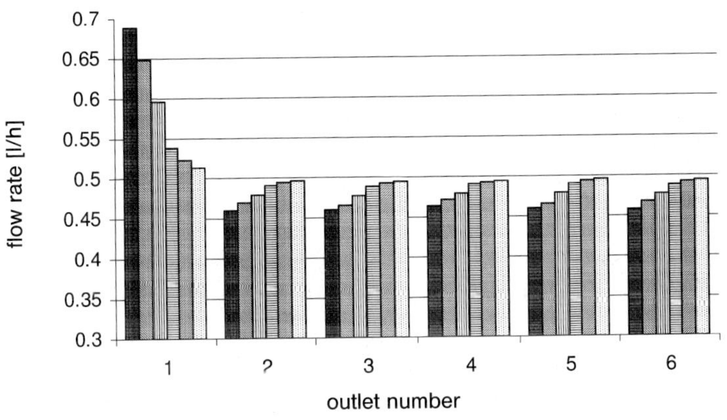

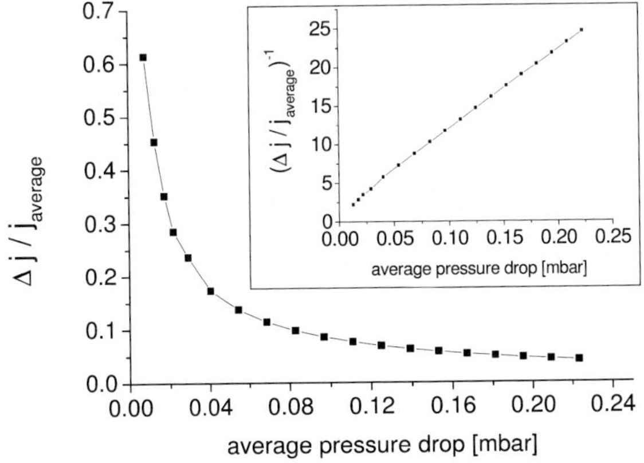

Figure 4.100 Volume flow rates through the six outlets (left). Relative imbalance of volume flows of outlets 1 and 4 as function of the pressure drop in the outlet tubes (right) [141].

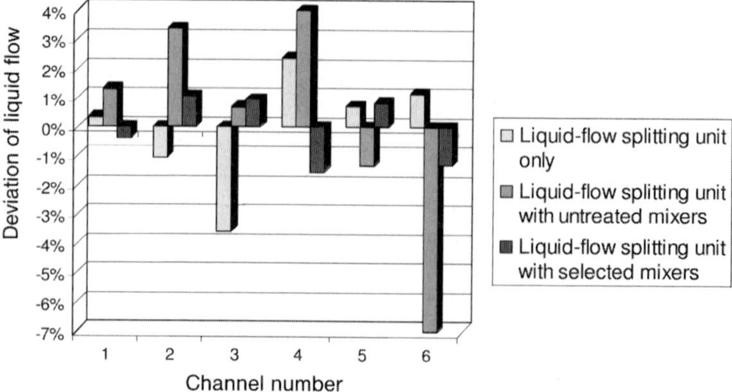

Figure 4.101 Experimental determination of the deviation of the total volume flow for the liquid-flow splitting unit alone and with six impinging-jet micro mixers attached. Concerning the latter, two cases were studied. First, the six best micro mixers were selected from 18 according to the minimal pressure drop deviation taking into account both outlet holes. Second, the same selection was done but relying on one hole only [141].

The predictions of the simulations were corroborated by experimental findings. For the liquid-flow splitting unit without an additional micro device as flow resistor, a minimum/maximum flow deviation of about 6% (see Figure 4.101) was found at a low pressure drop below 10 mbar [141]. The maximum flow deviation for the liquid flow-spitting unit equipped with these six selected impinging-jet mixers amounted for initial tests to 11% (4% standard deviation) at 64–74 mbar pressure drop (see Figure 4.101). By optimization of material pretreatment and micro fabrication, a minimum/maximum deviation of the water distribution below 5% and a standard deviation below 2% were finally obtained.

The functioning of the liquid-flow guiding was demonstrated with the formation of iron rhodanide as a test reaction in a device built of three liquid-flow splitting units and six interdigital separation layer micro mixers (see Figure 4.91) [141].

4.13
New Processes for Cost-efficient Reactor Manufacturing

For the fabrication of micro reactors, advanced fabrication procedures were used in the past. These procedures are well described elsewhere, e.g. [151–157]. Currently the situation is changing but sometimes high manufacturing costs of micro structured reactors are not counterbalanced by a large throughput or high production efficiency. Industrial processes facing such a situation usually will tend to automate their production, usually by continuous processing routes. The bottleneck is obvious and therefore some manufacturers and working groups have already started to find new production methods applying, for example, well-established reel-to-reel manufacturing methods for low-cost products.

Catalytic processes are an essential operation of large-scale processes in the chemical industry. Appropriate micro structured reactors offer an increase in efficiency compared with conventional reactors if a method can be found to integrate low-cost devices in industrial practise. Automated procedures will enable professional manufacturing of reactors with an acceptable cost/performance ratio.

Also an increase in volume flow in these reactors is inevitable. Micro structured reactors for high-throughput applications can consist of hundreds of plates; all of them have to be of a uniform size with a uniform coating and, of course, should be inexpensive [158], especially as the micro structured plates sometimes will have to be exchanged in case of fouling or deactivation of the coated catalyst.

4.13.1 Ceramic Foil Manufacturing

Ceramic foils are produced continuously by tape-casting methods. These ceramic tapes consist of an organic binder and oxide powder material, for example, zirconia, titania or alumina. If it is possible to integrate the production process of a micro structured reactor into such a continuous process, production costs would decrease strongly. An early approach used unstructured ceramic foils to build up a micro reactor [159]. The reactor consisted essentially of two functional layers, a reaction layer and a heating layer (Figure 4.102).

The heater for the reaction chamber and the conductors were printed on the lowest layer and electrically contacted to the pads on the cover layer by vertical metal-filled vias in every intermediate layer. Still some manual work was necessary to shape the reaction channel from the pre-fired tapes. Also the soldering of clamp-type fittings to the cover layer is certainly difficult to automate. However, a manufacturing process for ceramic foil reactors seems to be within reach.

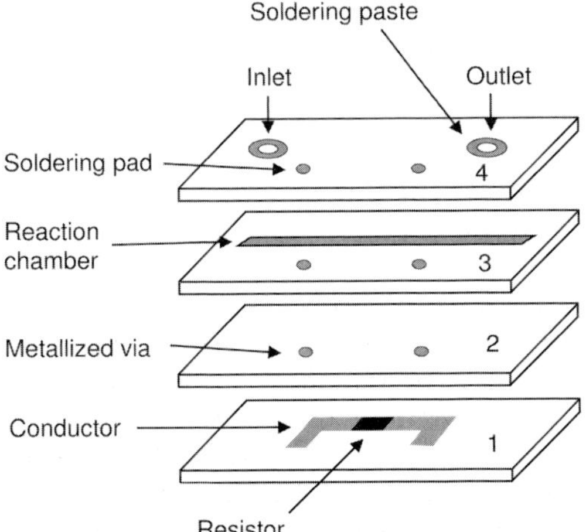

Figure 4.102 Micro structured reactor manufactured from a ceramic foil [159] (by courtesy of Kluwer Academic Publishers).

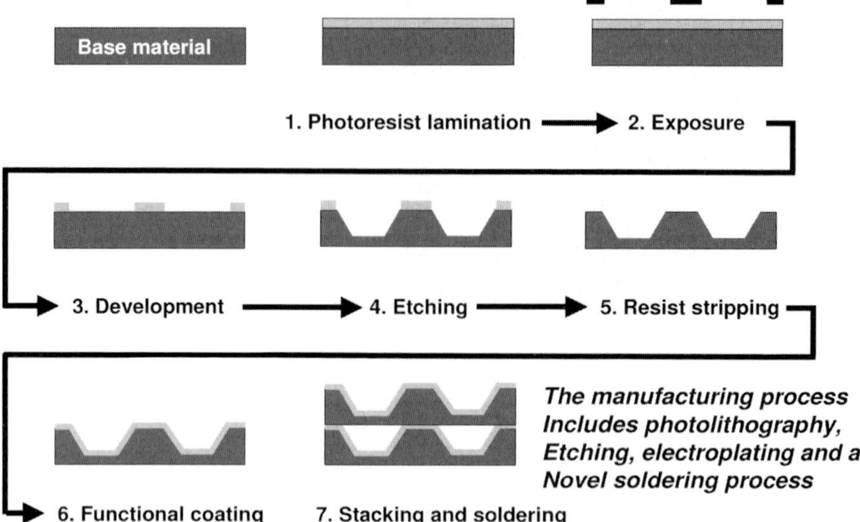

Figure 4.103 Continuous production process for a heat exchanger derived from printed circuit board production [161] (by courtesy of ATDTECH).

4.13.2 Solder-based Interconnection Techniques

Atotech [160, 161] already offers liquid coolers for heat removal from electronic components equipped with micro structured plates. The concept is intended to improve the cooling efficiency of CPUs or any other chips, hard discs and also high-powered electronic devices which in the near future will reach their performance limits if cooling performance is not increased. Atotech integrate the micro structure technology into their production facilities for general metal finishing and printed circuit board production. The production process as described in Figure 4.103 [161] includes also a step for the functional coating of the shaped substrate with a soldering material for reactor bonding. Such a coat could also be, for example, a catalyst sputtered on the substrate to permit the manufacture of a catalytic reactor.

4.13.2.1 Channel Manufacturing by Copper Etching

The micro structure is etched from a copper plate with channel dimensions ranging from 25 to 1000 µm in depth and from 100 µm to 2000 µm in width. The thickness of the base plate varies from 200 µm to 2000 µm.

4.13.2.2 Typical Application – Micro CPU Cooler

First prototypes were pressure-tested in a burst test of more than 60 bar. The pressure drop amounted to 110 mbar at a flow rate 3 l min^{-1}. As an example, an eight AMD 2000+ CPU with 70 W thermal power is cooled with eight plates of the above-given

Figure 4.104 Micro structured liquid coolers in a workstation [161] (by courtesy of ATOTECH).

dimensions at a flow rate of 100 ml min^{-1} using a glycol/water mixture. Figure 4.104 demonstrates the positioning of the liquid coolers in a workstation.

4.13.3 Printed Circuit Heat Exchanger Technology

In the early 1980s, a research group at the University of Sydney developed the so-called printed-circuit heat exchanger technology. In 1985, the company Heatric was founded in Australia. The printed circuit board technology was adapted by Heatric to create an etching technology for stainless-steel heat exchangers [97, 162]. The so-called printed circuit heat exchangers (PCHE) are structured by photo-chemical machining and formed to large heat exchanger monoliths by diffusion welding. The chemical etching process is similar to the manufacturing process used for electronic printed circuit boards. In the successive diffusion bonding process, the plates are assembled to a single block (Figure 4.105) [97, 162].

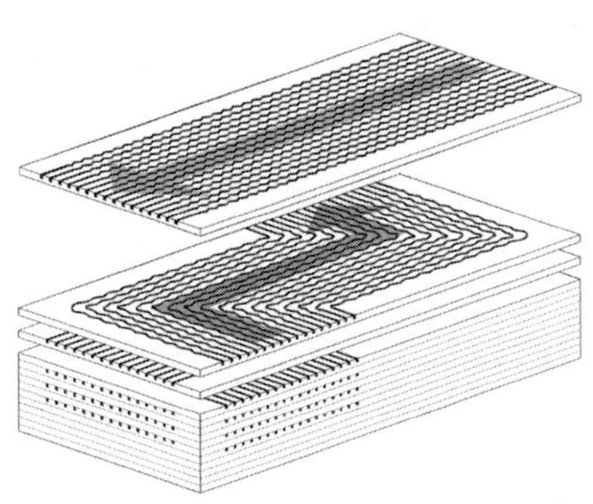

Figure 4.105 Printed circuit board heat exchanger [96, 162] (by courtesy of HEATRIC).

4.13.3.1 Stainless-steel Diffusion Bonding

The bonding takes place in a large oven equipped with an internal press mechanism. A huge press force is needed during the heating to allow intensive contact between the plate surfaces. During diffusion bonding, the metal grains grow across the metal boundaries, which permits a solid-state joining procedure. The strength of the parent material is reached and gas-tight sealing is achieved. A number of these bonded blocks can be welded together and equipped with headers and nozzles for the distribution of gases and liquids. Materials used are austenitic steel suitable for operational temperatures up to 800 °C and pressures up to 500 bar. The PCHE technology was also extended to chemical processing then called printed circuit reactor (PCR) technology, which is essentially a technology free of gaskets. The so-called In Passage (IP) reactor is coated inside the semi-circular channels with a thin layer of catalyst.

4.13.3.2 Catalyst Carrier Coating Inside Bonded Reactors

Heatric claims to have developed a robust and renewable technology to apply catalyst coats to the passages within a PCR [163]. This coating has to be executed after the diffusion bonding but the coating procedure is not described. Such a development would certainly be a milestone in micro reaction technology as the standard wash-coat methods are usually applied to free accessible surfaces and not to closed channels.

4.13.4 Online Reactor Manufacturing

To demonstrate the huge problems behind online coating and in consequence also online reactor manufacturing, a study executed at the IMM [164] will be presented in more detail in the following.

4.13.4.1 Continuous Coating Processes in the Polymer Industry

The study aims at the development of a production line for readily manufactured reactors with integrated catalysts as marketable products. The archetype for such an online technology was found in continuous coating and mass-produce processes in the polymer industry for the production of non-woven textiles and foils. A typical scheme for the melt-coating of a polymer foil is shown in Figure 4.106 [165].

4.13.4.2 Adaptation of Industrial Online Processes to Micro Structured Reactor Manufacturing

The central point was the preparation of a concept for the adaptation of industrial online processes for continuous goods to micro reaction technology. In a first approach, the mechanism of a non-woven production line was chosen as a typical example and predator for a cheap continuous good, which during its production path also undergoes steps such as coating, consolidation and folding into other geometric shapes. For this purpose, existing online coating and drying technologies had to be adapted to the specific needs of micro structures.

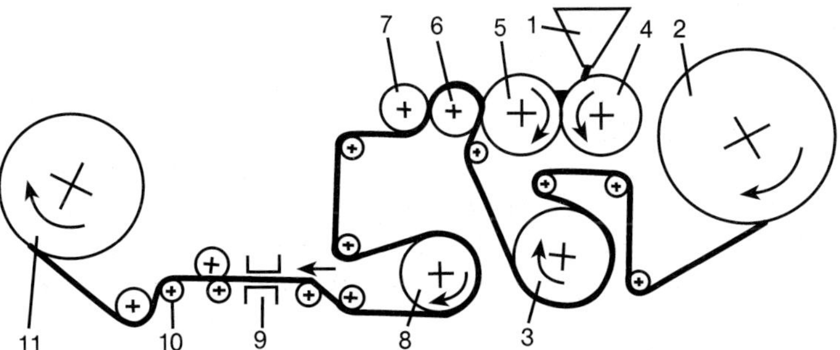

Figure 4.106 Scheme of a continuous melt-coating process of polymer foils: 1, powder for coating; 2, substrate unwinding; 3, preheating; 4 + 5, heated roller; 6, driven roller; 7, embossing roller; 8, cooling roller; 9, thickness measurement; 10, edge cutting; 11, up-winding [165] (by courtesy of Carl Hanser Verlag).

Sealing of high-temperature microstructured reactors

The large potential of micro structured reactors is best described by an example from heterogeneous catalysis, which is also a typical application in reaction technology. In reaction technology, temperatures exceeding 600 °C are often demanded, for example, for reforming reactions or gas-phase partial oxidation reactions. The standard seal material which can tolerate these temperatures is usually a structured graphite foil. However, the limit of graphite foils is reached at approximately 500 °C and is also dependent on the exposure of the graphite to an oxidizing or inert atmosphere. To exceed this temperature barrier, it is desirable to find a manufacturing process – similar to those for large-scale reactors – based on welded seals but applied to micro structured reactors in an automated continuous process.

Continuous catalyst carrier coating procedure

The same is valid for catalyst carriers used in heterogeneous catalysis, which are micro structured foils coated by a wash-coating procedure. The catalyst powder is distributed in a suspension and applied evenly on the channel surface. Usually this procedure is executed manually by a blade touching the channel rims and moving the slurry along the channels. This procedure leads to filling of the channels completely with slurry. During the consecutive drying and sintering steps, the coat shrinks by up to 90%, which opens a passage for the fluid again.

If one is confronted with expensive manual work with low complexity, a typical industrial approach would be to automate such a standard manual procedure. This will not only reduce costs of staff but will also increase the reproducibility of the coating procedure, which is an extremely important fact in catalyst development.

It is readily imaginable that this process can be transferred into a continuous partially automated procedure similar to a non-woven line. The new production process will also permit the broad availability of catalysts on micro structures. A wide application of these catalysts was up to now hindered by the costs per unit and the non-existent technology for industrial production. The process is also applicable to heat exchangers, which can just be considered as uncoated reactors.

Reel-to-reel etching technology

The whole process consists of a sequence of consecutive steps which allows separation into production modules. Some of them already exist and will just need slight modifications. Others have to be developed. At the very beginning, one makes use of the newly developed reel-to-reel etching technology for micro structured stainless-steel foils similar to the process described in [166]. The structured foils are then coated with catalyst, structured again with a laser tool and finally folded by sheet metal forming to a reactor monolith. The readily mass-produced reactors will be sealed by laser welding.

4.13.4.3 Production Modules

The complete manufacturing is divided into a number of production modules, e.g. for coating, laser structuring and reactor mass-producing. This facilitates the identification of possible defects in the production line and also permits parallel development.

Coating module

The first of three modules is the coating module, which is followed by the laser structuring and reactor folding modules.

Continuous coating is a well-established process in the non-woven industry. Non-woven textiles reach their final product quality tensile strength, resistance against humidity or surface structure only after the application of appropriate additives to the surface or interior of the textile. In principle, a number of methods are convenient, as shown in Figure 4.107 [167].

In order to transfer the manual coating procedure of applying the slurry to the micro structured surface of a foil, the third method described in Figure 4.107 should be the appropriate method. During the coating, the characteristics of the metal substrate and the coat are combined. The foil essentially serves as the mechanical support and of course also enables the transport to the next module. The coat, in contrast to a non-woven product, only influences the surface characteristics of the final product, the catalytic activity. The physical characteristics of the final substrate are only influenced by the foil substrate. This strongly facilitates the development of a production line. Another advantage is the mechanical stability of the foil. The rigidity of the foil, on the other hand, also introduces new problems. The band guiding characteristic, for example, is much more influenced by oblique-mounted rollers and unbalanced leveling out of the support frames, as the metal foil exhibits no flexibility compared with a textile.

Modes of coating operation

With the coating module in Figure 4.108, two modes of operation are possible: the coating of a continuous strip in a reel-to-reel fashion and the discontinuous coating of short foil strips (minimum length: 2 m). In the left picture, a 72 m long micro structured foil with a width of 150 mm is shown, which is sufficient to produce several hundred conventional micro structured reactors. The foil thickness for the reel-to-reel process is limited to 300 μm. A larger thickness would result in increasing

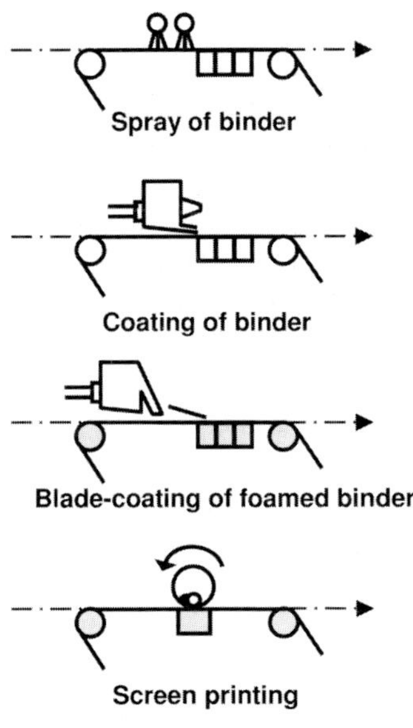

Figure 4.107 Methods for the application of binder to a non-woven web [167] (by courtesy of Georg Thieme Verlag.

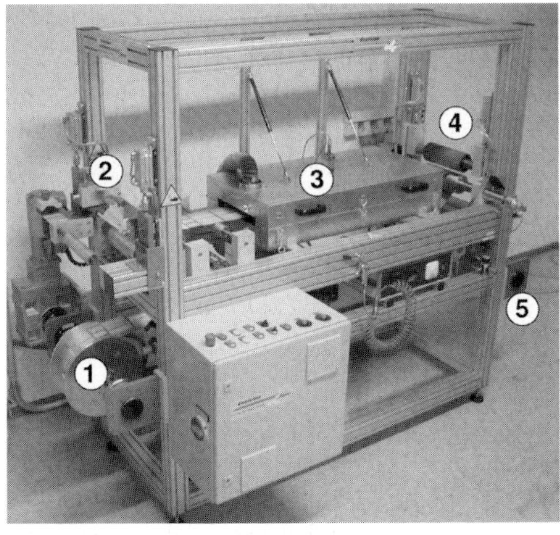

Figure 4.108 Coating module (processing from left to right: 1, unwinding; 2, coating; 3, drying compartment with temperature sensor; 4, additional driven roller pair; 5, up-winding) and continuous etched foil (length 72 m, width 150 mm, thickness 300 μm, structure dimensions: breadth × height × length = 300 × 200 × 10000 μm and 500 × 300 × 5000 μm) [109].

bending forces on the rollers and introduce permanent deformation of the metal strip. However, the foil thickness is not limited if only short strips up to several meters are processed. For this purpose, the module is equipped with an additional driven roller pair which is activated if the upper rubber-coated roller is lowered by two compressed air pistons (position 4 in Figure 4.108).

Discharge excess of catalyst carrier slurry

The discharge of the slurry is influenced by the slurry viscosity. The alumina slurry used for the first tests was prepared using the same standard procedure as normally used for manual coating. An adjustment of the viscosity was not necessary. The material for the foil substrate was a typical steel often used for steel etching (1.4571) without further surface treatments. The slurry was poured into the broad jet distributor and brushed against the foil, filling the channels completely. The layer thickness is in this process step adjusted by the height of the rims between the channels. The latter were oriented in the processing direction. Excess slurry was wiped off with a hardened stainless-steel knife which touched the surface of the foil.

The horizontal axes of knife and roller were not aligned in the vertical direction; instead, the knife touched the roller some millimeters behind the top surface of the roller in order to permit a spring-loaded contact between knife and foil. The contact force is then adjusted by the foil tension. The horizontal, vertical and angular position is adjusted by six micrometer screws. In Figure 4.109, coating has just started and the distinction between the coated and uncoated foil areas is still visible (left). The reservoir with the slurry is shown on the right.

Figure 4.109 Slurry coating of a continuous micro structured foil [109].

Drying of raw coated catalyst carrier

The coated foil is then processed to the next step, the drying compartment. The drying compartment essentially consists of a convection heater with two small slits for the traversing of the foil. The surface temperature of the foil is measured by an infrared detector. Temperature control is crucial in this process step, as the aqueous slurry must not be allowed to boil, which would deteriorate the coat evenness. During drying, the coat thickness shrinks (depending on the water content of the slurry) by up to 90% of its original height, giving way to gas passage in the channels again. After the drying section, the heated foil is cooled by a water-cooled roller (position 4 in Figure 4.109, roller pair open) and then either up-wound or further processed to the laser structuring module.

In the experiment the process stopped here and the coated foil was up-wound on the coil. These tests already proved that an even coating with a coating knife is possible. The drying compartment permits quick drying of the coat, which allows further processing without any adhesion between consecutive convolutions on the coil. The most important result is that the coat itself possesses enough elasticity and adhesive force to prevent the coat from cracking off during up-winding.

Future work will be aimed at the improvement of the band transport characteristics. It is known that especially thin metal foil tends to oblique running during up-winding. To ensure a smooth run, the tension of the foil has to be controlled carefully. This is controlled by a tension roller which measures the foil tension and adjusts the differential speed between the unwinding and up-winding drives. However, a band guiding system will also have to be added. The latter consists of a contactless distance sensor which controls the position of the foil edges and is combined with active roller adjustment, which adjusts the tension of the foil on either the left or the right edge, in order to improve the running behavior on the winder.

Laser cutting of catalyst carrier coated foils

Laser cutting tools are commercially available, so there is no need for individual development. The device should be equipped with a two axes driven laser head and possess the ability to adjust the laser power in order to permit also surface treatment such as laser ablation (Figure 4.110). For a continuous process, also enough space must be supplied for the continuous inlet and outlet of the foil. Further, a means to position the laser head with respect to the foil is necessary. This part of the process has only been tested using individual short foils and was still not integrated in the continuous process. However, continuous laser cutting of foils is a known commercial process and it is expected that the amount of future development in this part of the process will be limited.

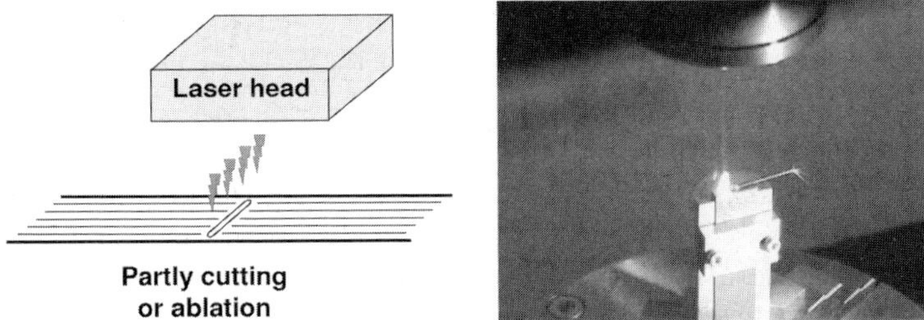

Figure 4.110 Principle of laser structuring and photograph of the laser module showing the two axes-driven laser head [109].

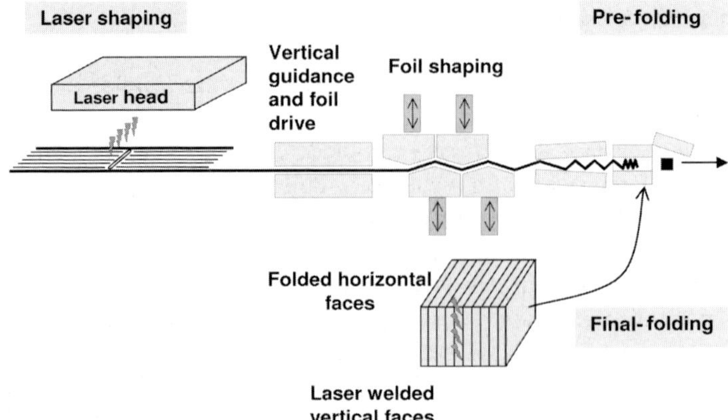

Figure 4.111 Principle of continuously working laser module and folding module [109].

Positioning problems during laser cutting procedure

Some aspects of continuous laser cutting will be discussed. Direct positioning with respect to the laser device is not possible, as the foil will move laterally depending on slight differences in the foil tension between the left and the right sides. The laser head will have to be positioned on the foil itself. For this purpose, laser marks were etched on the foil surface (already supplied by the commercial etching supplier). The laser head is moved to this mark to reach the zero position for the structuring process. Structuring means, for example, cutting of the contours of the micro structured foils to produce single non-coherent coated foil substrates (this part was already tested). In this case, the process is already finished and the produced foils are stacked to a reactor manually and sealed by graphite foils. Structuring can also mean a local thinning of lateral 'channels', which will be used in the following folding step as crimping joints (see laser module in Figure 4.111).

Mass-produce of a reactor

The mass-produce of the foils to a reactor is certainly the most challenging task in this whole process. Up to now only conceptional work was done here. A mechanism similar to a crimper in staple-fiber processing line [168] is proposed. In a textile crimper unit, which is used to give the textiles a more bulky appearance, the crimp of the fiber band is – in a geometric sense – not very well defined. This is certainly acceptable for the processing of textiles but, if a metal foil has to be used to build up a reactor, the 'crimp' or better the positions at which the foil is folded (the film 'joints') must be defined much more accurately. In Figure 4.111, a mechanism is proposed for such a foil-crimping procedure. A foil drive pushes the foil through horizontal guiding elements which prevent vertical movement to a pair of pistons. By vertical movement, the pistons imprint a saw-tooth profile on the foil in one or more steps. The pre-crimped foil then enters the crimping unit in which the foil is further compressed to a metal monolith. As soon as the desired stack size is reached, the process is interrupted, the foil cut mechanically and the stack laser welded at the vertical faces and then taken out.

4.13.4.4 Monolithic Heat Exchanger Manufacturing

As a prototype for such a reactor, the monolithic heat exchanger in Figure 4.112, was manufactured from a stack of micro structured foils and laser welded at the front and side faces [169]. The flanges were welded manually.

As the foil shaping by pistons demands the repeated stopping of the conveyance for the period of the shaping, the laser structuring, foil shaping and the coating will also have to be controlled and aligned to assure a common frequency for these steps.

For the period of the foil shaping, the coating knife must be positioned on a section of the foil which is not structured. This prevents the slurry from delivering an undefined coat thickness and helps to seal the jet exit. This is one of the reasons for the existence of the laser structuring module as this foil section will be thinned by the laser after the coating step to create a predefined joint for the subsequent folding process.

The monolith is now ready for insertion in a sintering oven in order to adjust the final characteristics of the catalyst. This means, for example, that the catalyst will have to be sintered at a temperature above the desired reaction temperature in order to prevent sintering during the reaction, which usually results in a decrease in surface area and deactivation.

A double-sided U-shaped foil, as shown in Figure 4.113, permits the execution of an exothermic reaction on the coated laser-structured upper side of the foil and

Figure 4.112 Monolithic counter-current heat exchanger manufactured from a stack of micro structured plates and sealed by laser welding (source IMM).

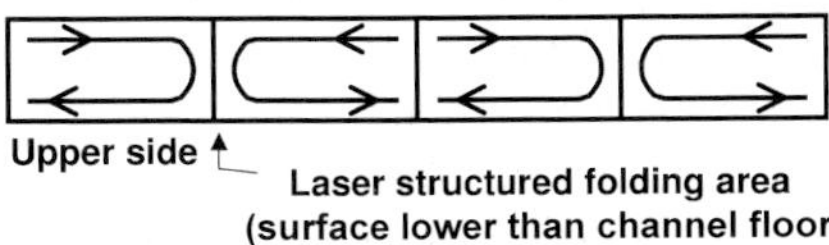

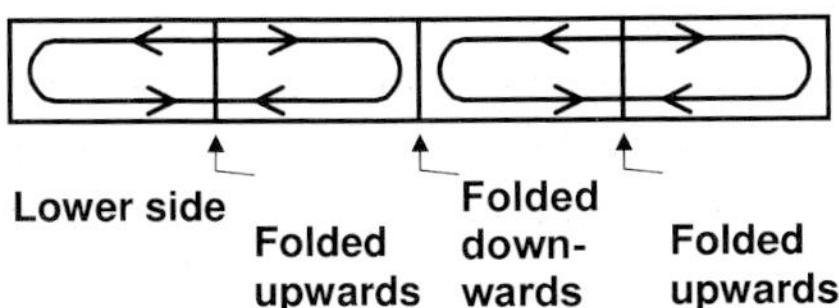

Figure 4.113 Scheme of a double-sided continuous etched foil (four sections shown; the arrows indicate the flow direction chosen in order to realize a counter-current flow on both sides of the foil; only one channel is indicated) [109].

Figure 4.114 Micro (mini) structured heat exchanger for oil platforms [96].

the processing of a cooling agent (flowing in counter-current mode) on the lower side of the foil. A continuous etching process for the production of double-sided foils has recently become available.

The band width in this study was limited to 150 mm but could exceed 1000 mm, thus opening a route to large-scale micro structured plate-like reactors comparable to the large-scale micro (or mini) structured heat exchangers from Heatric [96]. These are manufactured by diffusion welding of stacked plates (Figure 4.114).

References

1 Benson, R. S., Ponton, J. W., Process miniaturization – a route to total environmental acceptability? *Trans. Ind. Chem. Eng.* **1993**, *71*, 160–168.

2 Lerou, J. J., Harold, M. P., Ryley, J., Ashmead, J., O'Brien, T. C., Johnson, M., Perrotto, J., Blaisdell, C. T., Rensi, T. A., Nyquist, J., Microfabricated minichemical systems: technical feasibility, in Ehrfeld, W. (Ed.), *Microsystem Technology for Chemical and Biological Microreactors, DECHEMA Monographs*, Vol. 132, Verlag Chemie, Weinheim, **1996**, 51–69.

3 Wegeng, R. W., Call, C. J., Drost, M. K., Chemical system miniaturization, in *Proceedings of the AIChE Spring National Meeting* (25–29 Feb. 1996), New Orleans, **1996**, 1–13.

4 Ehrfeld, W., Hessel, V., Möbius, H., Richter, T., Russow, K., Potentials and realization of micro reactors, in Ehrfeld, W. (Ed.), *Microsystem Technology for Chemical and Biological Microreactors, DECHEMA Monographs*, Vol. 132, Verlag Chemie, Weinheim, **1996**, 1–28.

5 Ehrfeld, W., Hessel, V., Haverkamp, V., Microreactors, *Ullmann's Encyclopedia of Industrial Chemistry*, Wiley-VCH, Weinheim, **1999**.

6 Ehrfeld, W., Hessel, V., Löwe, H., *Microreactors*, Wiley-VCH, Weinheim, **2000**.

7 Koch, T. A., Krause, K. R., Mehdizadeh, M., Improved safety through distributed manufacturing of hazardous chemicals, in *Proceedings of the 5th World Congress on Chemical Engineering*, **1996**, San Diego, CA.

8 Löwe, H., Hessel, V., Müller, A., Microreactors – prospects already achieved and possible misuse, *Pure Appl. Chem.* **2002**, *74*, 2271–2276.

9 Hessel, V., Hardt, S., Löwe, H., *Chemical Micro Process Engineering – Fundamentals, Modeling and Reactions*, Wiley-VCH, Weinheim **2004**.

10 Hessel, V., Löwe, H., Micro chemical engineering: components – plant concepts – user acceptance: Part I, *Chem. Eng. Technol.* **2003**, *26*, 13–24.

11 Hessel, V., Löwe, H., Micro chemical engineering: components – plant concepts – user acceptance: Part II, *Chem. Eng. Technol.* **2003**, *26*, 391–408.

12 Hessel, V., Löwe, H., Micro chemical engineering: components – plant concepts – user acceptance: Part III, *Chem. Eng. Technol.* **2003**, *26*, 531–544.

13 Hasebe, S., Design and control of micro chemical plants, in *Proceedings of the International Workshop on Micro Chemical Plant* (3 Feb. 2003), Kyoto, **2003**, 33–41.

14 Hessel, V., Hardt, S., Hofmann, C., Pennemann, H., Schönfeld, F., Micro mixing principles and applications – a review on specific multi-lamination, split-recombine and chaotic approaches, in *Proceedings of the 226th ACS National Meeting* (7–11 Nov. 2003), New York, **2004**, *5*, 535–554.

15 Hessel, V., Löwe, H., Organische Synthese mit mikrostrukturierten Reaktoren – ein Vergleich zu bekannten Labor- und Industrieprozessierungen, in Hessel, V. (Ed.), *Chem. Ing. Tech., Sonderheft Mikroverfahrenstechnik* **2004**, *76*, *5*, 535–554.

16 Vogel, G. H., *Verfahrensentwicklung – von der ersten Idee zur chemischen Produktionsanlage*, Wiley-VCH, Weinheim **2002**.

17 FAMOS – *Fraunhofer Alliance Modular Microreaction System – Virtual Toolkit, http://www.microreaction-technology.info*, 14 April **2004**.

18 Loebbecke, S., Baukastenkonzepte für die Mikroverfahrenstechnik: Das FAMOS-System, *Chem. Ing. Tech.* **2004**, *76*, 581–583.

19 *Modulares Mikroreaktorsystem, DE 20201753*, Ehrfeld Mikrotechnik GmbH, **2002**.

20 Taghavi-Moghadam, S., Golbig, K., Microreactors: application of CYTOS^{TM}-technology from laboratory to production scale, *MST News* **2002**, *3*, 36–38.

21 Dietrich, T., Freitag, A., Scholz, R., Microreactors and microreaction systems for development and production, in *Proceedings of the 27th International Exhibition-Congress on Chemical Engineering, Environmental Protection and Biotechnology, ACHEMA* (19–24 May 2003), DECHEMA, Frankfurt, **2003**, 19.

22 Dietrich, T. R., Freitag, A., Scholz, R., Microreactors and microreaction systems for development and production, *MST News* **2002**, *3*, 12–14.

23 *CPAC initiative, NeSSI, www.cpac.washington.edu*.

24 *www.microchemtec.de*, 15 May **2004**.

25 Behr, A., Ebbers, W., Wiese, N., Miniplants- Ein Beitrag zur inhärenten Sicherheit? *Chem. Ing. Tech.* **2000**, *72*, 1157–1166.

26 Wiessmeier, G., *Untersuchungen zur heterogen katalysierten Oxidation von Propen in einem Mikrostrukturreaktor*, Diplomarbeit, Karlsruhe, **1992**.

27 *SAP Deutschland, www.sap.de*, 20 May **2004**.

28 Robbins, L. A., Pilot plant operations – the miniplant concept, *Chem. Eng. Progr.* **1979**, 45–48.

29 Gress, D., Hartmann, H., Kaibel, G., Seid, B., Einsatz von mathematischer Simulation und Miniplant-Technik in der Verfahrensentwicklung, *Chem. Ing. Tech.* **1979**, *51*, 601–611.

30 Rinard, I. H., Miniplant design methodology, in Ehrfeld, W., Rinard, I. H., Wegeng, R. S. (Eds.), *Process Miniaturization: 2nd International Conference on Microreaction Technology, IMRET 2, Topical Conf. Preprints*, AIChE, New Orleans, **1998**, 299–312.

31 Bier, W., Linder, G., Schubert, K., Mechanische Mikrotechnik, *KfK – Nachrichten* **1991**, *23*, 165–173.

32 Miyake, R., Lammerink, T. S. J., Elwenspoek, M., Fluitman, J. H. J., Micromixer with fast diffusion, in *Proceedings of IEEE-MEMS '93* (Feb. 1993), Fort Lauderdale, FL, **1993**, 248–253.

33 Wörz, O., Process development via miniplant, *Chem. Eng. Proc.* **1995**, *34*, 261–268.

34 Autze, V., Golbig, K., Kleemann, A., Oberbeck, S., Mikroreaktoren für die chemische Synthese, *Nachr. Chem.* **2000**, *48*, 683–685.

35 Oberbeck, S., Schwalbe, T., *Mikroreaktor, EP 1031375*, CPC Cellular Process Chemistry, **1999**.

36 Wille, C., Autze, V., Kim, H., Nickel, U., Oberbeck, S., Schwalbe, T., Unverdorben, L., Progress in transferring microreactors from lab into production – an example in the field of pigments technology, in *Proceedings of the 6th International Conference on Micro-*

reaction Technology, IMRET 6 (11–14 March 2002), AIChE Pub. No. 164, New Orleans, **2002**, 7–17.

37 Schwalbe, T., Autze, V., Wille, G., Chemical synthesis in microreactors, *Chimia* **2002**, *56*, 636–646.

38 Dietrich, T., Freitag, A., Microreactors made of glass, in *Proceedings of the MicroChemtec – Statusseminar Modulare Mikroverfahrenstechnik* (28 Feb. 2002), DECHEMA, Frankfurt/Main, **2002**, 22 (Poster).

39 Freitag, A., Dietrich, T. R., Glass as a material for microreaction technology, in *Proceedings of the 4th International Conference on Microreaction Technology, IMRET 4* (5–9 March 2000), AIChE Topical Conf. Proc., Atlanta, GA, **2000**, 48–54.

40 Ehrfeld, W., Microreaction technology for process development and production, in *Proceedings of the 27th International Exhibition-Congress on Chemical Engineering, Environmental Protection and Biotechnology, ACHEMA* (19–24 May 2003), DECHEMA, Frankfurt, **2003**, 4.

41 Kroschel, M., Bieber, T., Ehrfeld, W., Development of a microreaction platform for evaluation and implementation of novel process routes, in *Proceedings of the 27th International Exhibition-Congress on Chemical Engineering, Environmental Protection and Biotechnology, ACHEMA* (19–24 May 2003), DECHEMA, Frankfurt, **2003**, 26.

42 Rinke, G., Kraut, M., Grosser, V., Hillmann, V., Müller, A., Horn, B., Jähnisch, K., Bazzanella, A., Modular micro process engineering and applications, in *Proceedings of the 7th International Conference on Microreaction Technology, IMRET 7* (7–10 Sept. 2003), Lausanne, **2003**, 339–341.

43 Markowz, G., Neuer Schub für die Mikroreaktionstechnik – Projekt DEMIS™ geht in die entscheidende Phase, *Degussa Sci. Newsl.* **2003**, *4*, 4–6.

44 Schirrmeister, S., Markowz, G., Bulk chemicals in microreactors: development and scale-up of a new synthesis route into the pilot plant scale, in *Proceedings of the 27th International Exhibition-Congress on Chemical Engineering, Environmental Protection and Biotechnology, ACHEMA* (19–24 May 2003), DECHEMA, Frankfurt, **2003**, 3.

45 Schütte, R., Balduf, T., Becker, C., Hemme, I., Bertsch-Frank, B., Wildner, W., Rollmann, J., Markowz, G., *Verfahren und Vorrichtungen zum Durchführen von Reaktionen in einem Reaktor mit spaltförmigen Reaktionsräumen, DE 10042746*, Degussa, Düsseldorf, **2000**.

46 Tonomura, O., Tanaka, S., Noda, M., Kano, M., Hasebe, S., Hashimoto, I., CFD-based analysis of heat transfer and flow pattern in plate-fin micro heat exchangers, in *Proceedings of the PSEAsia* (4–6 Dec. 2002), Taipei, **2002**, 109–114.

47 Stief, T., Langer, O.-U., Schubert, K., Numerical investigations of opimal heat conductivity in micro heat exchangers, *Chem. Eng. Technol.* **1999**, *21*, 297–302.

48 Stief, T., Langer, O.-U., Schubert, K., Numerische Untersuchungen zur optimalen Wärmeleitfähigkeit in Mikrowärmeubertragerstrukturen, *Chem. Ing. Tech.* **1998**, *70*, 1539–1544.

49 Stief, T., Langer, O.-U., Schubert, K., Numerical investigations of opimal heat conductivity in micro heat exchangers, in *Proceedings of the 4th International Conference on Microreaction Technology, IMRET 4* (5–9 March 2000), AIChE Topical Conf. Proc., Atlanta, GA, **2000**, 314–321.

50 Werner, B., Hessel, V., Löb, P., Mischer mit mikrostrukturierten Folien für chemische Produktionsaufgaben, *Chem. Ing. Tech.* **2004**, *76*, 567–574.

51 Stange, T., Kiesewalter, S., Russow, K., PAMIR: *Potential and applications of microreaction technology – a market survey*, IMM Institut für Mikrotechnik, Mainz and Yole Developpement, Lyon, **2002**.

52 Appelhaus, P., Meßtechnik in Miniplants – Erfolgsbestimmender Faktor für das Gesamtproject, *Chem. Tech.* **1998**, *27*, 26–30.

53 Krummradt, H., Kopp, U., Stoldt, J., Experiences with the use of microreactors in organic synthesis, in Ehrfeld, W. (Ed.), *Microreaction Technology: 3rd International Conference on Microreaction Technology, Proc. of IMRET 3*, Springer-Verlag, Berlin, **2000**, 181–186.

54 Bayer, T., Pysall, D., Wachsen, O., Micro mixing effects in continuous radical polymerization, in Ehrfeld, W. (Ed.), *Microreaction Technology: 3rd International Conference on Microreaction Technology, Proc. of IMRET 3*, Springer-Verlag, Berlin, **2000**, 165–170.

55 Schwesinger, K., Frank, T., A modular microfluidic system with an integrated micromixer, in *Proceedings of the MME '95*, Copenhagen, **1995**, 144–147.

56 Schwesinger, N., Dresser, L., Frank, T., Wurmus, H., Integrated microreactor for chemical and biochemical applications, in *Proceedings of the International Conference on Integrated Micro/Nanotechnology for Space Applications* (30 Nov. – 3 Oct. 1995), Springer-Verlag, Houston, **1996**, 353–363.

57 Schwesinger, N., Frank, T., Wurmus, H., A modular microfluidic system with an integrated micromixer, *J. Micromech. Microeng.* **1996**, *6*, 99–102.

58 Ponton, J. W., The disposable batch plant, in *Proceedings of the ChE National Spring Meeting*, **1996**, *107d*, New Orleans.

59 Ponton, J. W., Some thoughts on the batch plant of the future, in *Proceedings of the 5th World Congress on Chemical Engineering*, **1996**, *52e*, San Diego.

60 Ponton, J. W., Observations on hypothetical miniaturized disposable chemical plant, in Ehrfeld, W. (Ed.), *Microreaction Technology – Proc. of the 1st International Conference on Microreaction Technology, IMRET 1*, Springer-Verlag, Berlin, **1997**, 10–19.

61 Buschulte, T. K., Heimann, F., Verfahrensentwicklung durch Kombination von Prozeßsimulation und Miniplant-Technik, *Chem. Ing. Tech.* **1995**, *67*, 718–724.

62 Kletz, T., *Plant Design for Safety*, Hemisphere, New York **1991**.

63 Morari, M., Effect of design on the controllability of continuous plants, in Perkins, J. D. (Ed.), *IFAC Workshop: Interactions Between Process Design and Control*, Pergamon Press, Oxford, **1992**, 3–16.

64 Rijnsdorp, J. E., Bekkers, P., Early integration of process and control design, effect of design on the controllability of continuous plants, in Perkins, J. D. (Ed.), *IFAC Workshop: Interaction Between Process Design and Process Control*, Pergamon Press, Oxford, **1992**, 17–22.

65 Klenk, H., Cyano compounds, inorganic, in Gerhart, W. (Ed.), *Ullmann's Encyclopedia of Industrial Chemistry*, Vol. A8, VCH, New York, **1987**, 159–163.

66 Wan, J. K. S., Koch, T. A., Application of microwave radiation for the synthesis of hydrogen cyanide, in Burka, M. (Ed.), *Proceedings of the Microwave Induced Reactions Workshop*, Vol. A-3-1, EPRI, Palo Alto, CA, **1992**.

67 Fink, H., Hampe, M. J., Designing and constructing microplants, in Ehrfeld, W. (Ed.), *Microreaction Technology: 3rd International Conference on Microreaction Technology, Proc. of IMRET 3*, Springer-Verlag, Berlin, **2000**, 664–673.

68 Forschungszentrum Karlsruhe, *www.fzk.de*, 17 July **2004**.

69 Rinke, G., Schubert, K., Bauteile für die Mikroverfahrenstechnik, in *Proceedings of the MicroChemtec – Statusseminar Modulare Mikroverfahrenstechnik* (28 Feb. 2002), DECHEMA, Frankfurt/Main, **2002**, 38–39 (Poster).

70 Nittis, V., Fortt, R., Legge, C. H., de Mello, A. J., A high-pressure interconnect for chemical microsystem applications, *Lab Chip* **2001**, *1*, 148–152.

71 Ahn, C. H., Puniambekar, A., Lee, S. M., Cho, H. J., Hong, C.-C., Structurally programmable micofluidic systems, in van den Berg, A., Olthuis, W., Bergveld, P. (Eds.), *Micro Total Analysis Systems*, Kluwer, Dordrecht, **2000**, 205–208.

72 Fujii, T., Sando, Y., Higashino, K., Fujii, Y., A plug and play microfluidic device, *Lab Chip* **2003**, *3*, 193–197.

73 Chemnitz, S., Eberhardt, D., Böhm, M., Microfluidic on ASIC – a novel concept for an integrated lab on a chip, in *Proceedings of the 6th International Conference on Microreaction Technology, IMRET 6* (11–14 March 2002), AIChE Pub. No. 164, New Orleans, **2002**, 326–328.

74 CPC – Cellular Process Chemistry GmbH, *info@cpc-net.com*, Mainz, **2004**.

75 Golbig, K., Taghavi-Moghadam, S., Born, P., CYTOS^{TM}-technology – microreaction technology in practical sense, in *Proceedings of the 6th Inter-*

national Conference on Microreaction Technology, IMRET 6 (11–14 March 2002), AIChE Pub. No. 164, New Orleans, **2002**, 131–134.

76 Golbig, K., Hohmann, M., Kursawe, A., Schwalbe, T., Verweilzeitverhalten in Mikrokanälen als Voraussetzung für den Bau sequenziell arbeitender Syntheseautomaten, *Chem. Ing. Tech.* **2004**, *76*, 598–603.

77 IMM-Catalogue, *The Catalogue – Process Technology of Tomorrow*, Vol. III, **2004**, IMM Institut für Mikrotechnik Mainz GmbH, Mainz.

78 *www.hnp-mikrosysteme.de*, **2004**.

79 Weiner, M., Chemiebaukasten für Profis, *Fraunhofer Magazin*, **2004**, 44–45.

80 Ehrfeld, W., Electrochemistry and microsystems, *Electrochim. Acta* **2003**, *48*, 2857–2868.

81 *Mikroreaktorsystem für die Erkennung und Kontrolle von Mikroreaktormodulen, DE 20202807*, Ehrfeld Mikrotechnik, **2002**.

82 *Vorrichtung zur modularen Integration von Sensoren in mikroreaktions- und mikroverfahrenstechnischen Anlagen, DE 20218400*, Ehrfeld Mikrotechnik, **2002**.

83 Ehrfeld Mikrotechnik AG, *www.ehrfeld-mikrotechnik.de*, 14 January **2004**.

84 Bard, A. J., *Integrated chemical synthesizers, US 5580523*, **1994**.

85 Zanzucchi, P. J., Cherukuri, S. C., McBride, S. E., *Etching to form cross-over, non-intersecting channel networks for use in partitioned microelectronic and fluidic device arrays for clinical diagnostics and chemical synthesis, US 5681484*, David Sarnoff Research Center, Princeton, NJ, **1995**.

86 Müller, A., Cominos, V., Hessel, V., Horn, B., Schürer, J., Ziogas, A., Jähnisch, K., Hillmann, V., Grosser, V., Jam, K. A., Bazanella, A., Rinke, G., Kraut, M., Fluidisches Bussystem für die chemische Verfahrenstechnik und für die Produktion von Feinchemikalien, *Chem. Ing. Tech.* **2004**, *76*, 641–651.

87 Müller, A., Cominos, V., Drese, K., Hessel, V., Horn, B., Löwe, H., Ziogas, A., Groser, V., Hillmann, V., Jam, K. A., Fluidic bus system for chemical micro process engineering, in *Proceedings of the 27th International Exhibition-Congress on Chemical Engineering, Environmental Protection and Biotechnology, ACHEMA 2003* (19–24 May 2003), DECHEMA, Frankfurt, **2003**, 25.

88 Müller, A., Cominos, V., Horn, B., Ziogas, A., Jähnisch, K., Grosser, V., Hillmann, V., Jam, K. A., Bazzanella, A., Rinke, G., Kraut, M., Fluidic bus system for chemical micro process engineering in the laboratory and for small-scale production, *Chem. Eng.* **2004**, in press.

89 Bazanella, A., Die Industrieplattform Mikroverfahrenstechnik – Eine deutsche Initiative zur Förderung der Mikrotechnik in der Chemie, Pharma und Life Sciences, *Chem. Ing. Tech.* **2004**, *76*, 511–513.

90 Kestenbaum, H., Lange de Olivera, A., Schmidt, W., Schüth, F., Ehrfeld, W., Gebauer, K., Löwe, H., Richter, T., Silver-catalyzed oxidation of ethylene to ethylene oxide in a microreaction system, *Ind. Eng. Chem. Res.* **2000**, *41*, 710–719.

91 Hagendorf, U., Janicke, M., Schüth, F., Schubert, K., Fichtner, M., A Pt/Al_2O_3 coated microstructured reactor/heat exchanger for the controlled H_2/O_2-reaction in the explosion regime, in Ehrfeld, W., Rinard, I. H., Wegeng, R. S. (Eds.), *Process Miniaturization: 2nd International Conference on Microreaction Technology, IMRET 2, Topical Conf. Preprints*, AIChE, New Orleans, **1998**, 81–87.

92 Wu, J. C., Zang, D. L., Fong, G. F., Yuan, J. T., Lin, B. F., *J. Chem. Technol. Biotechnol.* **2001**, *76*, 619.

93 Ehrfeld, W., Golbig, K., Hessel, V., Löwe, H., Richter, T., Characterization of mixing in micromixers by a test reaction: single mixing units and mixer arrays, *Ind. Eng. Chem. Res.* **1999**, *38*, 1075–1082.

94 Delsman, E. R., de Croon, M. H. J. M., Kramer, G. J., Cobden, P. D., Hofmann, C., Cominos, V., Schouten, J. C., Experiments and modeling of an integrated preferential oxidation-heat exchanger microdevice, *Chem. Eng.* **2004**, *101*, 123–131.

95 Cominos, V., Hardt, S., Hessel, V., Kolb, G., Löwe, H., Wichert, M., Zapf, R., A methanol steam micro-reformer for

low power fuel cell applications, *Chem. Eng. Commun.* **2005**, in press.

96 Heatric, *www.heatric.com*, Poole, UK, 30 May **2004**.

97 Johnston, A. M., Haynes, B. S., Heatric steam reforming technology, in *Proceedings of the AIChE Spring National Meeting – 2nd Topical Conference on Natural Gas Utilization* (10–12 March 2001), AIChE, New Orleans, **2001**, 112.

98 EMITEC, *www.emitec.com*, **2004**.

99 Wille, C., Gabski, H.-P., Haller, T., Kim, H., Unverdorben, L., Winter, R., Synthesis of pigments in a three-stage microreactorpilot plant – an experimental technical report, *Chem. Eng.* **2004**, *101*, 179–185.

100 Markowz, G., Schirrmeister, S., Albrecht, J., Becker, F., Schütte, R., Caspary, K. J., Klemm, E., Mikrostrukturierte Reaktoren für heterogen katalysierte Gasphasenreaktionen im industriellen Maßstab, *Chem. Ing. Tech.* **2004**, *76*, 620–625.

101 Hartmann, M., Pfeifer, F., Dornheim, G., Sommer, K., HPDS-Hochdruckzelle zur Beobachtung mikroskopischer Phänomene unter Hochdruck, *Chem. Ing. Tech.* **2003**, *75*, 1763–1767.

102 Schüssler, M., Lamla, O., Stefanovski, T., zur Megede, D., Vorstellung eines kaltstartfähigen Reaktors zur autothermen Reformierung von Methanol, *Chem. Ing. Tech.* **2000**, *72*, 982.

103 Gjerwan, T., Venvik, H., Aartun, I., Holmen, A., Görke, O., Pfeifer, P., Schubert, K., in *Proceedings of the 225th National Meeting of the American Chemical Society* (23–27 March 2003), New Orleans, **2003**.

104 Hessel, V., Löb, P., Löwe, H., Meudt, A., Pennemann, H., Scherer, S., Werner, B., Chemical micro process engineering in the field of organic synthesis, in *Proceedings of the ISPE–API Seminar* (7–10 April 2004), Brussels, **2004**, submitted.

105 Aspen Technologies, *www.aspen.com*, 15 July **2004**.

106 Loebbecke, S., Ferstl, W., Lohf, A., Steckenborn, A., Hassel, J., Häberl, M., Schmalz, D., Muntermann, H., Bayer, T., Kinzl, M., Leipprand, I., Automatisierung in der Mikroreaktionstechnik: das AuMµRes-System, *Chem. Ing. Tech.* **2004**, *76*, 637–640.

107 Antes, J., Ferstl, W., Krause, H., Loebbecke, S., Grund, M., Haeberl, M., Muntermann, H., Schmalz, D., Hassel, S., Lohf, A., Steckenborn, A., Bayer, T., Leipprand, I., AuMµRes – a new automated microreaction system for chemical engineering and production, in *Proceedings of the 27th International Exhibition-Congress on Chemical Engineering, Environmental Protection and Biotechnology, ACHEMA* (19–24 May 2003), DECHEMA, Frankfurt, **2003**, 20.

108 Ferstl, W., Loebbecke, S., Antes, J., Krause, H., Haeberl, M., Schmalz, D., Muntermann, H., Grund, M., Steckenborn, A., Lohf, A., Hassel, S., Bayer, T., Leipprand, I., Development of an automated microreaction system with integrated sensorics for process screening and production, *Chem. Eng.* **2004**, *101*, 431–438.

109 IMM Institut für Mikrotechnik Mainz, unpublished.

110 Wille, C., Ehrfeld, W., Herweck, T., Haverkamp, V., Hessel, V., Löwe, H., Lutz, N., Möllmann, K.-P., Pinno, F., Dynamic monitoring of fluid Equi partition and heat release in a falling film microreactor using real-time thermography, in *Proceedings of the VDE World Microtechnologies Congress, MICRO.tec 2000* (25–27 Sept. 2000), VDE Verlag, Berlin, EXPO Hannover, **2000**, 349–354.

111 Srinivasan, R., Hsing, I.-M., Berger, P. E., Jensen, K. F., Firebaugh, S. L., Schmidt, M. A., Harold, M. P., Lerou, J. J., Ryley, J. F., Micromachined reactors for catalytic partial oxidation reactions, *AIChE J.* **1997**, *43*, 3059–3069.

112 Jensen, K. F., Hsing, I.-M., Srinivasan, R., Schmidt, M. A., Harold, M. P., Lerou, J. J., Ryley, J. F., Reaction engineering for microreactor systems, in Ehrfeld, W. (Ed.), *Microreaction Technology – Proc. of the 1st International Conference on Microreaction Technology, IMRET 1*, Springer-Verlag, Berlin, **1997**, 2–9.

113 Loebbecke, S., Antes, J., Tuercke, T., Boskovich, D., Schweikert, W., Marioth, E., Schnuerer, F., Krause, H. H., Black box microreactor? Possibilities for a better control and understanding of microreaction processes by applying suitable analytics, in *Proceedings of the 6th International Conference on Microreaction Technology, IMRET 6* (11–14 March 2002), AIChE Pub. No. 164, New Orleans, **2002**, 37–38.

114 Antes, J., Türcke, T., Marioth, E., Lechner, F., Scholz, M., Schnürer, F., Krause, H. H., Loebbecke, S., Investigation, analysis and optimization of exothermic nitrations in microreactor processes, in Matlosz, M., Ehrfeld, W., Baselt, J. P. (Eds.), *Microreaction Technology – IMRET 5: Proc. of the 5th International Conference on Microreaction Technology*, Springer-Verlag, Berlin, **2001**, 446–454.

115 Türcke, T., Schweikert, W., Lechner, F., Antes, J., Krause, H. H., Loebbecke, S., Monitoring of chemical processes in microreactors by adapting FTIR microscopy and HPLC, in Matlosz, M., Ehrfeld, W., Baselt, J. P. (Eds.), *Microreaction Technology – IMRET 5: Proc. of the 5th International Conference on Microreaction Technology*, Springer-Verlag, Berlin, **2001**, 479–488.

116 Jackman, R. J., Floyd, T. M., Ghodssi, R., Schmidt, M. A., Jensen, K. F., Microfluidic systems with on-line UV detection fabricated in photodefinable epoxy, *J. Micromech. Microeng.* **2001**, *11*, 263–269.

117 Lu, H., Schmidt, M. A., Jensen, K. F., Photochemical reactions and on-line UV detection in microfabricated reactors, *Lab Chip* **2001**, *1*, 22–28.

118 Günther, M., Schneider, S., Wagner, J., Georges, R., Henkel, T., Kielpinski, M., Albert, J., Bierbaum, R., Köhler, J. M., Characterization of residence time and residence time distribution in chip reactors with modular arrangements by integrated optical detection, *Chem. Eng.* **2004**, *101*, 373–378.

119 Applied Analytics, *www.a-a-inc.com*, 15 July **2004**.

120 Solf, C., *Entwicklung von miniaturisierten Fourier-Transformations-Spektrometern und ihre Herstellung mit dem LiGA Verfahren*, Forschungszentrum, Karlsruhe, **2004**.

121 Gindele, F., Novotny, C., Koch, A., Kunz, S., Kufner, S., Polymeric optical components for sensors based on molding techniques, in *Proceedings of the IEEE/LEOS International Conference on Optical MEMS* (20–23 Aug. 2002), IEEE, Lugano, **2002**, 101–102.

122 *www.slsmt.com*, **2004**.

123 RECIPE CHEMICALS + INSTRUMENTS, *www.recipe.de*, **2004**.

124 *www.microreaction-technology.info*, **2004**.

125 *Golden Gate – single reflection diamond ATR unit, http://www.lot-oriel.de*, 14 July **2004**.

126 Grosser, V., Microfabrication of modular MEMS – assembly and reliability, in *Proceedings of the Conference on Design, Test, Integration and Packaging of MEMS/MOEMS, DTPI 2002* (5–8 May 2002), Cannes Mandelieu, **2002**.

127 Amiri Jam, K., Hillmann, V., Grosser, V., Michel, B., Miniaturized motion converter and positioning sensors, in *Proceedings of the VDE World Microtechnologies Congress, MICRO.tec 2003*, VDE Verlag, Berlin, EXPO Hannover, **2003**.

128 Fourné, F., *Synthetische Fasern, Herstellung, Maschinen und Apparate, Eigenschaften, Handbuch für Anlagenplanung, Maschinenkonstruktion und Betrieb*, Carl Hanser Verlag, Munich, **1995**.

129 Wiessmeier, G., Hönicke, D., Strategy for the development of micro channel reactors for heterogeneously catalyzed reactions, in Ehrfeld, W., Rinard, I. H., Wegeng, R. S. (Eds.), *Process Miniaturization: 2nd International Conference on Microreaction Technology, IMRET 2, Topical Conf. Preprints*, AIChE, New Orleans, **1998**, 24–32.

130 Hardt, S., Ehrfeld, W., Hessel, V., vanden Bussche, K. M., Strategies for size reduction of microreactors by heat transfer enhancement effects, *Chem. Eng. Commun.* **2003**, *190*, 540–559.

131 The language of technical computing, www.mathworks.com, *Matlab*, **2002**, 6.

132 Mathias, P. M., Brown, L. C., Thermodynamics of the sulfur-iodine cycle for thermochemical hydrogen production, in *Proceedings of the 68th Annual Meeting Of the Society of Chemical Engineers* (23 March 2003), Japan, **2003**.

133 Aldrett, S., Worstell, J. H., Improved ethylene oligomerization modeling using Aspentech's polymers plus, in *Proceedings of the AIChE Annual Meeting* (16–21 Nov. 2003), AIChE, San Francisco, **2003**.

134 LOHMANN, B., KÄMPFER, J., Integrated CAE environments as a means for the rationalization of engineering workflows, in *Proceedings of the 27th International Exhibition-Congress on Chemical Engineering, Environmental Protection and Biotechnology, ACHEMA* 2003 (19–24 May 2003), DECHEMA, Frankfurt, **2003**, 247.

135 SCHMIDT-TRAUB, H., HOLTKÖTTER, T., LEDERHOSE, M., LEUDERS, P., An approach to plant layout optimization, *Chem. Eng. Technol.* **1999**, *22*, 105–109.

136 BURDORF, A., KAMPCZYK, B., SCHMIDT-TRAUB, H., CAPD – Computer-aided plant design, in *Proceedings of the 27th International Exhibition–Congress on Chemical Engineering, Environmental Protection and Biotechnology, ACHEMA* 2003 (19–24 May 2003), DECHEMA, Frankfurt/Main, **2003**, 213.

137 HOFEN, W., KÖRFER, M., ZETZMANN, K., Scale-up Probleme bei der experimentellen Verfahrensentwicklung, *Chem. Ing. Tech.* **1990**, *62*, 805–812.

138 KREKEL, J., SIEKMANN, G., Die Rolle des Experiments in der Verfahrensentwicklung, *Chem. Ing. Tech.* **1985**, *57*, 511–519.

139 KREKEL, J., POLKE, R., Qualitätssicherung – Unternehmensziel integriert ın Verfahren, *Jahrbuch Chemiewirtschaft 1992*, VCH, Weinheim, **1992**.

140 SCHENK, R., HESSEL, V., HOFMANN, C., KISS, J., LÖWE, H., SCHÖNFELD, F., ZIOGAS, A., Numbering-up von Mikroreaktoren: Ein neues Flüssigkeitsverteilsystem, *Chem. Ing. Tech.* **2004**, *76*, 584–597.

141 SCHENK, R., HESSEL, V., HOFMANN, C., LÖWE, H., SCHÖNFELD, F., Novel liquid-flow splitting unit specifically made for numbering-up of liquid/liquid chemical microprocessing, *Chem. Eng. Technol.* **2003**, *26*, 1271–1280.

142 LOSEY, M. W., SCHMIDT, M. A., JENSEN, K. F., Microfabricated multiphase packed-bed reactors: characterization of mass transfer and reactions, *Ind. Chem. Res.* **2001**, *40*, 2555–2562.

143 SCHUBERT, K., BRANDNER, J., FICHTNER, M., LINDER, G., SCHYGULLA, U., WENKA, A., Microstructure devices for applications in thermal and chemical process engineering, *Microscale Therm. Eng.* **2001**, *5*, 17–39.

144 ZLOKARNIK, M., Scale-up und Miniplants, *Chem. Ing. Tech.* **2003**, *75*, 370–375.

145 BESSER, R., PREVOT, M., Linear scale-up of micro reaction systems, in *Proceedings of the 4th International Conference on Microreaction Technology, IMRET 4* (5–9 March 2000), AIChE Topical Conf. Proc., Atlanta, GA, **2000**, 278–283.

146 PERRY, R. H., GREEN, D. W., *Perry's Chemical Engineers' Handbook*, 7th ed., McGraw-Hill, New York, **1997**.

147 MÜLLER, A., *A Modular Approach to Heterogeneous Catalyst Screening in the Laminar Flow Regime*, VDI Verlag, Düsseldorf, **2004**.

148 SCHENK, R., HESSEL, V., HOFMANN, C., KISS, J., LÖWE, H., ZIOGAS, A., Numbering-up of micro devices: a first liquid-flow splitting unit, *Chem. Eng.* **2004**, *101*, 421–429.

149 SCHENK, R., HESSEL, V., HOFMANN, C., KISS, J., LÖWE, H., SCHÖNFELD, F., Numbering-up of micro devices: a first liquid-flow splitting unit, *Chem. Eng. Technol.* **2003**, *26*, 1271–1280.

150 SCHENK, R., HESSEL, V., HOFMANN, C., KISS, J., LÖWE, H., SCHÖNFELD, F., Numbering-up of micro devices: a first liquid-flow splitting unit, in *Proceedings of the 7th International Conference on Microreaction Technology, IMRET 7* (7–10 Sept. 2003), Lausanne, **2003**, 71–73.

151 EHRFELD, W., GOLBIG, K., HESSEL, V., KONRAD, R., LÖWE, H., RICHTER, T., Fabrication of components and systems for chemical and biological microreactors, in EHRFELD, W. (Ed.), *Microreaction Technology – Proc. of the 1st International Conference on Microreaction Technology, IMRET 1*, Springer-Verlag, Berlin, **1997**, 72–90.

152 BREUER, N., MEYER, H., *Chemical microreactors and method for producing same, DE 19 708 472*, ATOTECH Deutschland, Berlin, **1997**.

153 PROVIN, C., MONNERET, S., LE GALL, H., RIGNEAULT, H., LENNE, P.-F., GIOVANNINI, H., New process for manufacturing ceramic microfluidic devices for microreactor and bioanalytical applications, in MATLOSZ, M., EHRFELD, W., BASELT, J. P. (Eds.), *Microreaction Technology – IMRET 5: Proc. of the 5th International Conference on Microreaction Technology*, Springer-Verlag, Berlin, **2001**, 103–112.

154 PAUL, B. K., HASAN, H., DEWEY, T., AHMAN, D., WILSON, R. D., Development of aluminide microchannel arrays for high-temperature microreactors and micro-scale heat exchangers, in *Proceedings of the 6th International Conference on Microreaction Technology, IMRET 6* (11–14 March 2002), AIChE Pub. No. 164, New Orleans, **2002**, 202–211.

155 RANGELOV, I. W., KASSING, R., Silicon microreactors made by reactive ion etching, in MATLOSZ, M., EHRFELD, W., BASELT, J. P. (Eds.), *Microreaction Technology – IMRET 5: Proc. of the 5th International Conference on Microreaction Technology*, Springer-Verlag, Berlin, **2001**, 169–174.

156 RICHTER, T., EHRFELD, W., WOLF, A., GRUBER, H. P., WÖRZ, O., Fabrication of microreactor components by electro discharge machining, in EHRFELD, W. (Ed.), *Microreaction Technology – Proc. of the 1st International Conference on Microreaction Technology, IMRET 1*, Springer-Verlag, Berlin, **1997**, 158–168.

157 KNITTER, R., LIAUW, M. A., Ceramic microreactors – fabrication and application, in *Proceedings of the 27th International Exhibition-Congress on Chemical Engineering, Environmental Protection and Biotechnology, ACHEMA* (19–24 May 2003), DECHEMA, Frankfurt, **2003**, 15.

158 ZAPF, R., BECKER-WILLINGER, C., BERRESHEIM, K., HOLZ, H., GNASER, H., HESSEL, V., KOLB, G., LÖB, P., PANNWITT, A.-K., ZIOGAS, A., Detailed characterization of various porous alumina based catalyst coatings within microchannels and their testing for methanol steam reforming, *Trans IChemE* **2003**, *A81*, 721–729.

159 WANG, X., ZHU, J., BAU, H., GORTE, R. J., Fabrication of micro-reactors using tape-casting methods, *Catal. Lett.* **2001**, *77*, 173–177.

160 CRÄMER, K., FÜRSTENAU, F., MEYER, H., Mass production of microreactors on the basis of printed circuit board technology and electrochemistry, in *Proceedings of the VDE World Microtechnologies Congress, MICRO.tec 2000* (25–27 Sept. 2000), VDE Verlag, Berlin, EXPO Hannover, **2000**, 775–777.

161 *www.atotech.com*, **2004**.

162 JOHNSTON, A. M., LEVY, W., RUMBOLD, S. O., Application of printed circuit heat exchanger technology within heterogeneous catalytic reactors, in *Proceedings of the AIChE Annual Meeting*, **2001**.

163 PUA, L. M., RUMBOLD, S. O., Industrial microchannel devices – where are we today? in *Proceedings of the 1st International Conference on Microchannels and Minichannels, ICMM 2003* (24–25 April 2003), Rochester, NY, **2003**.

164 MÜLLER, A., unpublished results.

165 SAECHTLING, P., *Kunststoff Taschenbuch*, 24th ed., Carl Hanser Verlag, Munich, **1989**.

166 STRIEDIECK, W., Die Herstellung von präzisen Bauteilen durch Fotoätztechnik, *Galvanotechnik* **1991**, *82*, 820–827.

167 LÜNENSCHLOSS, J., ALBRECHT, W., *Vliesstoffe*, Georg Thieme Verlag, Stuttgart, **1982**.

168 VON FALKAI, B., *Synthesefasern, Grundlagen, Technologie, Verarbeitung und Anwendung*, Verlag Chemie, Weinheim, **1981**.

169 LÖWE, H., KOLB, G., COMINOS, V., HESSEL, V., Microstructured Reformer for fuel-cell applications, in *Proceedings of the 3rd International Workshop on Micro and Nanotechnology for Power Generation and Energy Conversion, PowerMEMS* (4–5 Dec. 2003), Global Emerging Technology Institute, New York, Makuhari, Japan, **2003**.

Subject Index

Chemical Micro Process Engineering. V. Hessel, H. Löwe, A. Müller, G. Kolb.

ISBN: 3-527-30998-5

f

g